# aris.mhhe.com



## ARIS, Your Home Page for Studying Environmental Science!

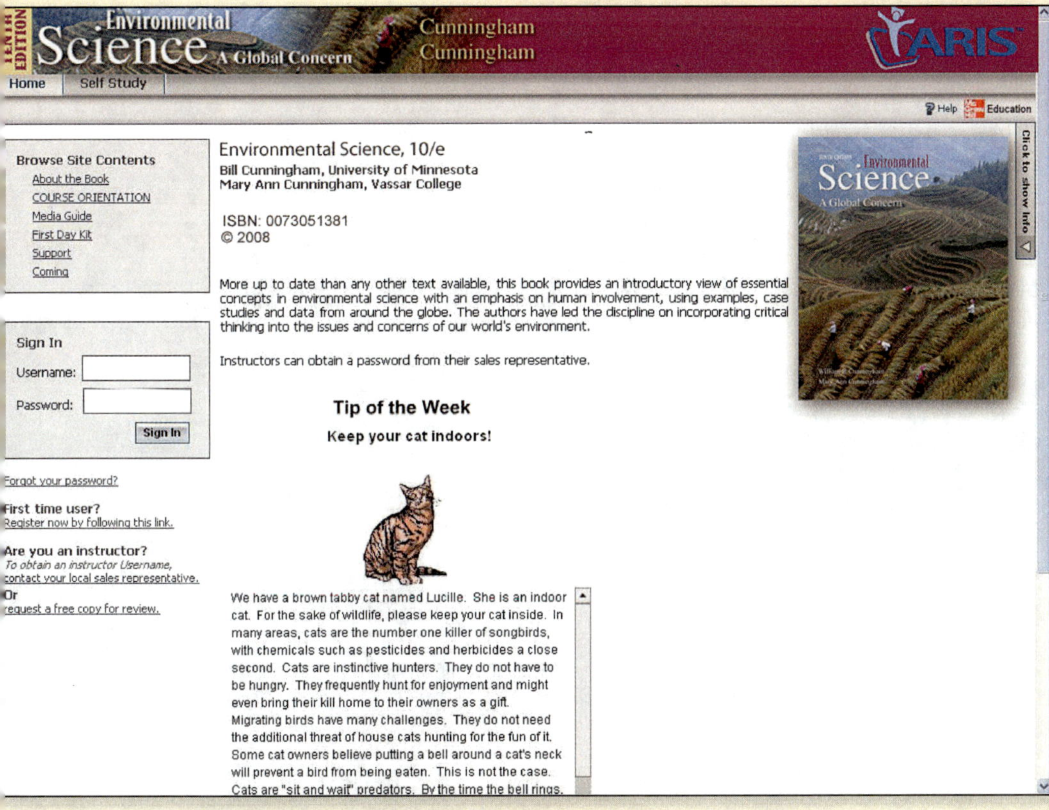


ARIS, the online resource for *Environmental Science: A Global Concern* can be found directly at www.mhhe.com/cunningham10e. The site offers an extensive array of quizzing and learning tools that will help you master the topics covered in the textbook.

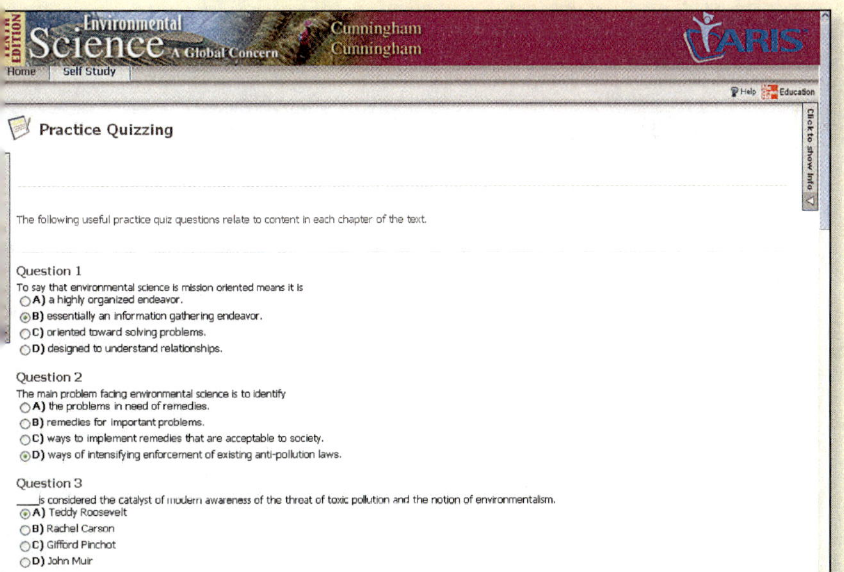


## Test Yourself

Take a chapter quiz on the *Environmental Science: A Global Concern* ARIS site. Each multiple choice quiz is specially constructed to test your comprehension of a wide range of key concepts.

tista

## Interactive Activities

**Google Earth™** interactive satellite imagery gives you a geographic context for global places and topics discussed in the text. Globe icons indicate when to visit the *ARIS* site, where you will find links to locations mentioned in the text and corresponding exercises that will help you understand environmental topics.

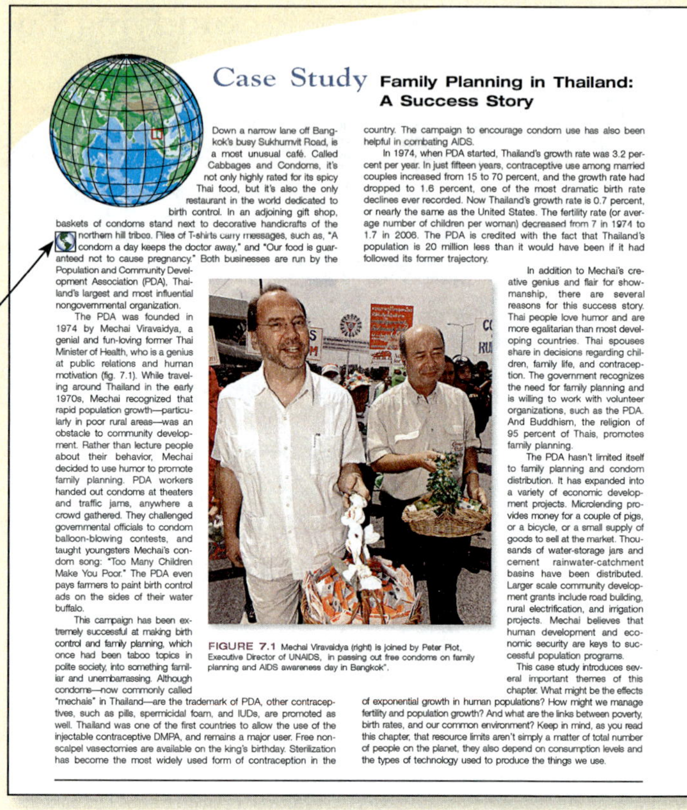

Case Study **Family Planning in Thailand: A Success Story**

Down a narrow lane off Bangkok's busy Sukhumvit Road, is a most unusual café. Called Cabbages and Condoms, it's not only highly rated for its spicy Thai food, but it's also the only restaurant in the world dedicated to birth control. In an adjoining gift shop, baskets of condoms stand next to decorative handicrafts of the northern hill tribes. Piles of T-shirts carry messages, such as, "A condom a day keeps the doctor away," and "Our food is guaranteed not to cause pregnancy." Both businesses are run by the Population and Community Development Association (PDA), Thailand's largest and most influential nongovernmental organization.

The PDA was founded in 1974 by Mechai Viravaidya, a genial and fun-loving former Thai Minister of Health, who is a genius at public relations and human motivation (fig. 7.1). While traveling around Thailand in the early 1970s, Mechai recognized that rapid population growth—particularly in poor rural areas—was an obstacle to community development. Rather than lecture people about their behavior, Mechai decided to use humor to promote family planning. PDA workers handed out condoms at theaters and traffic jams, anywhere a crowd gathered. They challenged governmental officials to condom balloon-blowing contests, and taught youngsters Mechai's condom song: "Too Many Children Make You Poor." The PDA even pays farmers to paint birth control ads on the sides of their water buffalo.

This campaign has been extremely successful at making birth control and family planning, which once had been taboo topics in polite society, into something familiar and unembarrassing. Although condoms—now commonly called "mechais" in Thailand—are the trademark of PDA, other contraceptives, such as pills, spermicidal foam, and IUDs, are promoted as well. Thailand was one of the first countries to allow the use of the injectable contraceptive DMPA, and remains a major user. Free non-scalpel vasectomies are available on the king's birthday. Sterilization has become the most widely used form of contraception in the country. The campaign to encourage condom use has also been helpful in combating AIDS.

In 1974, when PDA started, Thailand's growth rate was 3.2 percent per year. In just fifteen years, contraceptive use among married couples increased from 15 to 70 percent, and the growth rate had dropped to 1.6 percent, one of the most dramatic birth rate declines ever recorded. Now Thailand's growth rate is 0.7 percent, or nearly the same as the United States. The fertility rate (or average number of children per woman) decreased from 7 in 1974 to 1.7 in 2006. The PDA is credited with the fact that Thailand's population is 20 million less than it would have been if it had followed its former trajectory.

FIGURE 7.1 Mechai Viravaidya (right) is joined by Peter Piot, Executive Director of UNAIDS, in passing out free condoms on family planning and AIDS awareness day in Bangkok".

In addition to Mechai's creative genius and flair for showmanship, there are several reasons for this success story. Thai people love humor and are more egalitarian than most developing countries. Thai spouses share in decisions regarding children, family life, and contraception. The government recognizes the need for family planning and is willing to work with volunteer organizations, such as the PDA. And Buddhism, the religion of 95 percent of Thais, promotes family planning.

The PDA hasn't limited itself to family planning and condom distribution. It has expanded into a variety of economic development projects. Microlending provides money for a couple of pigs, or a bicycle, or a small supply of goods to sell at the market. Thousands of water-storage jars and cement rainwater-catchment basins have been distributed. Larger scale community development grants include road building, rural electrification, and irrigation projects. Mechai believes that human development and economic security are keys to successful population programs.

This case study introduces several important themes of this chapter. What might be the effects of exponential growth in human populations? How might we manage fertility and population growth? And what are the links between poverty, birth rates, and our common environment? Keep in mind, as you read this chapter, that resource limits aren't simply a matter of total number of people on the planet, they also depend on consumption levels and the types of technology used to produce the things we use.

Case studies, flash cards, animation-based quizzes, and other activities are organized by chapter and designed to reinforce the concepts in the text.

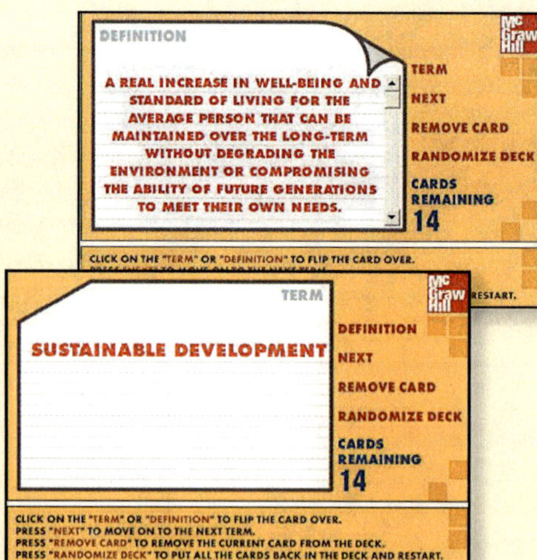


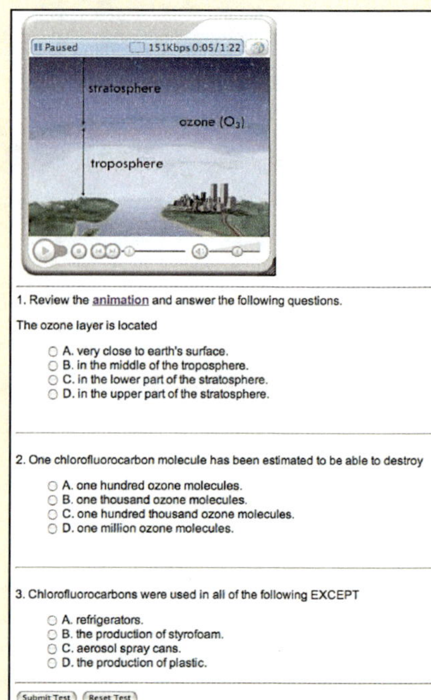

# Periodic Table of the Elements

- Metals (main-group)
- Metals (transition)
- Metals (inner transition)
- Metalloids
- Nonmetals

MAIN–GROUP ELEMENTS | TRANSITION ELEMENTS | MAIN–GROUP ELEMENTS

| Period | 1A (1) | 2A (2) | 3B (3) | 4B (4) | 5B (5) | 6B (6) | 7B (7) | 8B (8) | (9) | (10) | 1B (11) | 2B (12) | 3A (13) | 4A (14) | 5A (15) | 6A (16) | 7A (17) | 8A (18) |
|---|---|---|---|---|---|---|---|---|---|---|---|---|---|---|---|---|---|---|
| 1 | 1 **H** 1.008 | | | | | | | | | | | | | | | | | 2 **He** 4.003 |
| 2 | 3 **Li** 6.941 | 4 **Be** 9.012 | | | | | | | | | | | 5 **B** 10.81 | 6 **C** 12.01 | 7 **N** 14.01 | 8 **O** 16.00 | 9 **F** 19.00 | 10 **Ne** 20.18 |
| 3 | 11 **Na** 22.99 | 12 **Mg** 24.31 | | | | | | | | | | | 13 **Al** 26.98 | 14 **Si** 28.09 | 15 **P** 30.97 | 16 **S** 32.07 | 17 **Cl** 35.45 | 18 **Ar** 39.95 |
| 4 | 19 **K** 39.10 | 20 **Ca** 40.08 | 21 **Sc** 44.96 | 22 **Ti** 47.88 | 23 **V** 50.94 | 24 **Cr** 52.00 | 25 **Mn** 54.94 | 26 **Fe** 55.85 | 27 **Co** 58.93 | 28 **Ni** 58.69 | 29 **Cu** 63.55 | 30 **Zn** 65.41 | 31 **Ga** 69.72 | 32 **Ge** 72.61 | 33 **As** 74.92 | 34 **Se** 78.96 | 35 **Br** 79.90 | 36 **Kr** 83.80 |
| 5 | 37 **Rb** 85.47 | 38 **Sr** 87.62 | 39 **Y** 88.91 | 40 **Zr** 91.22 | 41 **Nb** 92.91 | 42 **Mo** 95.94 | 43 **Tc** (98) | 44 **Ru** 101.1 | 45 **Rh** 102.9 | 46 **Pd** 106.4 | 47 **Ag** 107.9 | 48 **Cd** 112.4 | 49 **In** 114.8 | 50 **Sn** 118.7 | 51 **Sb** 121.8 | 52 **Te** 127.6 | 53 **I** 126.9 | 54 **Xe** 131.3 |
| 6 | 55 **Cs** 132.9 | 56 **Ba** 137.3 | 57 **La** 138.9 | 72 **Hf** 178.5 | 73 **Ta** 180.9 | 74 **W** 183.9 | 75 **Re** 186.2 | 76 **Os** 190.2 | 77 **Ir** 192.2 | 78 **Pt** 195.1 | 79 **Au** 197.0 | 80 **Hg** 200.6 | 81 **Tl** 204.4 | 82 **Pb** 207.2 | 83 **Bi** 209.0 | 84 **Po** (209) | 85 **At** (210) | 86 **Rn** (222) |
| 7 | 87 **Fr** (223) | 88 **Ra** (226) | 89 **Ac** (227) | 104 **Rf** (263) | 105 **Db** (262) | 106 **Sg** (266) | 107 **Bh** (267) | 108 **Hs** (277) | 109 **Mt** (268) | 110 **Ds** (281) | 111 **Rg** (272) | 112 (285) | | 114 (289) | | 116 (292) | | |

As of late 2005, elements 112, 114, and 116 have not been named.

INNER TRANSITION ELEMENTS

| Period | | | | | | | | | | | | | | | |
|---|---|---|---|---|---|---|---|---|---|---|---|---|---|---|---|
| 6 | Lanthanides | 58 **Ce** 140.1 | 59 **Pr** 140.9 | 60 **Nd** 144.2 | 61 **Pm** (145) | 62 **Sm** 150.4 | 63 **Eu** 152.0 | 64 **Gd** 157.3 | 65 **Tb** 158.9 | 66 **Dy** 162.5 | 67 **Ho** 164.9 | 68 **Er** 167.3 | 69 **Tm** 168.9 | 70 **Yb** 173.0 | 71 **Lu** 175.0 |
| 7 | Actinides | 90 **Th** 232.0 | 91 **Pa** (231) | 92 **U** 238.0 | 93 **Np** (237) | 94 **Pu** (242) | 95 **Am** (243) | 96 **Cm** (247) | 97 **Bk** (247) | 98 **Cf** (251) | 99 **Es** (252) | 100 **Fm** (257) | 101 **Md** (258) | 102 **No** (259) | 103 **Lr** (260) |

# UNITS OF MEASUREMENT METRIC/ENGLISH CONVERSIONS

## Length

1 meter = 39.4 inches = 3.28 feet = 1.09 yard
1 foot = 0.305 meters = 12 inches = 0.33 yard
1 inch = 2.54 centimeters
1 centimeter = 10 millimeters = 0.394 inch
1 millimeter = 0.001 meter = 0.01 centimeter = 0.039 inch
1 fathom = 6 feet = 1.83 meters
1 rod = 16.5 feet = 5 meters
1 chain = 4 rods = 66 feet = 20 meters
1 furlong = 10 chains = 40 rods = 660 feet = 200 meters
1 kilometer = 1,000 meters = 0.621 miles = 0.54 nautical miles
1 mile = 5,280 feet = 8 furlongs = 1.61 kilometers
1 nautical mile = 1.15 mile

## Area

1 square centimeter = 0.155 square inch
1 square foot = 144 square inches = 929 square centimeters
1 square yard = 9 square feet = 0.836 square meters
1 square meter = 10.76 square feet = 1.196 square yards = 1 million square millimeters
1 hectare = 10,000 square meters = 0.01 square kilometers = 2.47 acres
1 acre = 43,560 square feet = 0.405 hectares
1 square kilometer = 100 hectares = 1 million square meters = 0.386 square miles = 247 acres
1 square mile = 640 acres = 2.59 square kilometers

## Volume

1 cubic centimeter = 1 milliliter = 0.001 liter
1 cubic meter = 1 million cubic centimeters = 1,000 liters
1 cubic meter = 35.3 cubic feet = 1.307 cubic yards = 264 US gallons
1 cubic yard = 27 cubic feet = 0.765 cubic meters = 202 US gallons
1 cubic kilometer = 1 million cubic meters = 0.24 cubic mile = 264 billion gallons
1 cubic mile = 4.166 cubic kilometers
1 liter = 1,000 milliliters = 1.06 quarts = 0.265 US gallons = 0.035 cubic feet
1 US gallon = 4 quarts = 3.79 liters = 231 cubic inches = 0.83 imperial (British) gallons
1 quart = 2 pints = 4 cups = 0.94 liters
1 acre foot = 325,851 US gallons = 1,234,975 liters = 1,234 cubic meters
1 barrel (of oil) = 42 US gallons = 159 liters

## Mass

1 microgram = 0.001 milligram = 0.000001 gram
1 gram = 1,000 milligrams = 0.035 ounce
1 kilogram = 1,000 grams = 2.205 pounds
1 pound = 16 ounces = 454 grams
1 short ton = 2,000 pounds = 909 kilograms
1 metric ton = 1,000 kilograms = 2,200 pounds

## Temperature

Celsius to Fahrenheit $^\circ F = (^\circ C \times 1.8) + 32$
Fahrenheit to Celsius $^\circ C = (^\circ F - 32) \div 1.8$

## Energy and Power

1 erg = 1 dyne per square centimeter
1 joule = 10 million ergs
1 calorie = 4.184 joules
1 kilojoule = 1,000 joules = 0.949 British Thermal Units (BTU)
1 megajoule = MJ = 1,000,000 joules
1 kilocalorie = 1,000 calories = 3.97 BTU = 0.00116 kilowatt-hour
1 BTU = 0.293 watt-hour
1 kilowatt-hour = 1,000 watt-hours = 860 kilocalories = 3,400 BTU
1 horsepower = 640 kilocalories
1 quad = 1 quadrillion kilojoules = 2.93 trillion kilowatt-hours

## Quantitative Prefixes

| Large Numbers | Description | Small Numbers |
|---|---|---|
| exa $10^{18}$ | quintillion | alto $10^{-18}$ |
| peta $10^{15}$ | quadrillion | femto $10^{-15}$ |
| tera $10^{12}$ | trillion | pico $10^{-12}$ |
| giga $10^{9}$ | billion | nano $10^{-9}$ |
| mega $10^{6}$ | million | micro $10^{-6}$ |
| kilo $10^{3}$ | thousand | milli $10^{-3}$ |

(e.g., a kilogram = 1,000 gm; a milligram = one-thousandth of a gram)

SELECTED CHAPTERS FROM

# ENVIRONMENTAL SCIENCE A GLOBAL CONCERN

Tenth Edition

William P. Cunningham
*University of Minnesota*

Mary Ann Cunningham
*Vassar College*

with additional material from

## Taking Sides: Clashing Views on Environmental Issues

Thirteenth Edition, Expanded

Selected, Edited, and with Introductions by

Thomas Easton
*Thomas College*

MICHIGAN STATE UNIVERSITY EDITION

McGraw Hill Learning Solutions

Boston Burr Ridge, IL Dubuque, IA New York San Francisco St. Louis
Bangkok Bogotá Caracas Lisbon London Madrid
Mexico City Milan New Delhi Seoul Singapore Sydney Taipei Toronto

The McGraw-Hill Companies

Selected Material from
Environmental Science: A Global Concern, Tenth Edition
With Additional Material
Michigan State University Edition



This book is a McGraw-Hill Learning Solutions textbook and contains select material from the following sources:
*Environmental Science: A Global Concern*, Tenth Edition by William P. Cunningham and Mary Ann Cunningham. Copyright © 2008, 2007, 2005, 2003, 2001, 1999, 1997 by The McGraw-Hill Companies, Inc.
*Taking Sides: Environmental Issues*, Thirteenth Edition by Thomas Easton. Copyright © 2010, 2009, 2008, 2007 by The McGraw-Hill Companies, Inc.
Both are reprinted with permission of the publisher. Many custom published texts are modified versions or adaptations of our best-selling textbooks. Some adaptations are printed in black and white to keep prices at a minimum, while others are in color.

2 3 4 5 6 7 8 9 0 KNG KNG 0 9

ISBN-13: 978-0-07-736380-2
ISBN-10: 0-07-736380-9

*Learning Solutions Manager: Melani Theis*
*Production Editor: Jennifer Pickel*
*Cover Design: Lou Civitello*
*Cover Photos: Praying Mantis: © 2009 JupiterImages Corporatiton, Pesticide: © 2009 Digital Vision/PunchStock, Crops: © 2009 Pixtal/age fotostock, Children: © 2009 Royalty Free/Corbis, Bear: © 2009 Royalty Free/Corbis*
*Title Page Photos: Praying Mantis: © 2009 JupiterImages Corporation*
*Printer/Binder: King Printing*

# Brief Contents

from Cunningham: *Environmental Science*, 10e



from Easton: *Taking Sides: Clashing Views on Environmental Issues*, 13e

# Contents

# PART TWO PEOPLE IN THE ENVIRONMENT

from Easton: *Taking Sides: Clashing Views on Environmental Issues*, 13e

# Preface

## Environmental Science Has Never Been More Important

We seem to have reached a turning point in environmental attitudes. A decade ago, few people took climate change seriously. Today, climate is a topic of headline news and political campaigns. Hundreds of colleges, communities, and local governments are working locally to reduce carbon emissions and improve efficiency. More than 400 bills have been passed in 40 states to require renewable energy. "Green" buildings are transforming architecture, and green business models are beginning to transform industry. The 2008 Summer Olympics in Beijing aims to be the greenest games ever with radical new strategies to conserve energy, water, and air quality. Most importantly, it's not just environmentalists who are involved in efforts to protect and improve our environment; it is business leaders finding ways to reduce costs by reducing waste, insurance companies concerned about rising sea levels, and inner-city communities who are trying to lower asthma rates in children. Environmental science is increasingly understood to be a pragmatic field that helps us understand issues that affect our lives.

In the twenty-five years since we began work on this book, the United States and the world have undergone several cycles of concern and neglect for our global environment. We have seen growth, declines, and a recent resurgence in public support for energy conservation, farmland protection, and environmental health. Global biodiversity, once a special interest of ecologists, is now an economic concern of drug companies and the global fishing industry. After years of purchasing ever-larger vehicles, Americans have recently found a renewed interest in automobile efficiency as we become increasingly aware of the costs of climate change—and of the political and economic costs of fighting to preserve our fossil fuel supplies. Major events like the Earth Summit in Rio de Janeiro in 1992 and the Kyoto conference on climate change in 1997 have raised awareness of how globally interconnected all of us are in terms of our environment, our economies, and our well-being. All these changes present new possibilities for building coalitions and finding new approaches to living sustainably. We hope this book will inform and inspire students as they consider their role in protecting our shared environment.

## What Sets This Book Apart?

### A positive viewpoint

We wrote this book because we think it's important for students to realize the difference they can make in their community. We believe a book focused on gloom and decay provides little inspiration to students, and in this time of exciting change, we think such a gloomy view is inaccurate. Many environmental problems remain severe, but there have been many improvements over past decades including cleaner water and cleaner air for most Americans. The Kyoto Protocol, despite its imperfections, is now pushing nations to reduce their climate impacts. The earth's population exceeds 6 billion people, but birth rates have plummeted as education and health care for women have improved. This book highlights these developments and presents positive steps that individuals can take, while acknowledging the many challenges we face. Case studies that show successful projects, a new chapter on Restoration Ecology, and "What Can You Do?" boxes are some of the features written to give students an applicable sense of direction. A number of other features also set this book apart.

### An integrated, global perspective

Globalization spotlights the interconnectedness of environmental concerns, as well as economies. To remain competitive in a global economy, it is critical that we understand conditions in other countries and cultures. This book provides case studies and topics from regions around the world, as well as maps and data showing global issues. These examples also show the integration between environmental, social, and economic conditions at home and abroad.

### A balanced presentation that encourages critical thinking

Environmental science often involves special interests, contradictory data, and conflicting interpretations of data. Throughout the text, one of the most important skills a student can learn is to think analytically and clearly about evidence, weigh the data, consider uncertainty, and skeptically evaluate the sources of information. We give students opportunities to practice critical thinking in brief "Think About It" boxes and in "What Do You Think?" readings. We present balanced evidence, while not suggesting that any opinion is on par with ideas accepted by the community of informed scientists, and we provide the tools for students to discuss and form their own opinions.

### Emphasis on science

Science is critical for understanding environmental change. We emphasize principles and methods of science through coverage on uncertainty and probability, new graphing exercises, data analysis exercises, and "Exploring Science" readings that show how scientists observe the world and gather data.

## Overview of Changes to *Environmental Science*, Tenth Edition

### What's new to this edition?

The tenth edition includes marked changes in approach as well as a thorough update of topics and data. Among the important changes is an emphasis on positive lessons presented through recent events in environmental science. Also important are several new pedagogical features including data analysis exercises and web-based exercises.

### New restoration ecology chapter

Chapter 13 is a brand new chapter on the important topic of ecological restoration. It explains how forests, grasslands, savannas, wetlands, and other ecosystems are being repaired and restored to ecological health, and provides many positive examples of restoration efforts.

### New case studies and readings

Seventeen of the 25 chapter opening case studies are new, as are 10 of the boxed readings within chapters. A majority of the case studies and boxed readings in this edition are focused on current events and success stories that display the global progress being made in environmental protection.

### Google Earth™ placemarks

An exciting new feature in this edition are the Google Earth™ placemarks in every chapter. Google Earth™ is an online program that provides interactive satellite imagery of the earth and will help students understand the geographic context of places and topics in the text. Wherever students see this icon in the text, they can go to our website (http://www.mhhe.com/cunningham10e) to find a Google Earth™ placemark that will take them to the specific place being discussed. Students can zoom in to see amazing detail and zoom out to gain regional perspective.

### Data Analysis exercises

A Data Analysis box has been added to the end of every chapter. These exercises ask students to graph and evaluate data, to practice looking at numbers and graphs, and to critically analyze what they see.

### Learning Outcomes and other pedagogical tools

Chapter material now includes Learning Outcomes presented at the beginning of each chapter and a Reviewing Learning Outcomes section at the end of each chapter. These Learning Outcomes are connected to the major headings of each numbered section within the chapter to help students better organize the content and their study. Each chapter also includes a conclusion, which summarizes major points, a Practice Quiz to aid in understanding key concepts from the chapter, and Critical Thinking and Discussion Questions that challenge students to apply what they have learned.

### Specific changes by chapter

- Chapter 1 includes a new case study on the Green Olympics 2008 that illustrates China's new concern for environmental quality and its importance in our global environment. The environmental history section has been expanded to include contemporary, diverse leaders. Chapter 1 also has a major revision of current environmental conditions using 2006 data. The Data Analysis box introduces graphing.
- Chapter 2 has a new case study on the ethics of climate change. It also has important new material on statistics, probability and uncertainty. The Data Analysis box discusses bar graphs, pie charts, and scatter plots.
- Chapter 3 adds a Data Analysis exercise on extracting data from a graph.
- Chapter 4 benefits from a strengthened section on evolution and revised material on competition, predation, symbiosis, keystone groups, and succession. The Data Analysis box presents Gause's historic experiments on interspecific competition with population growth graphs.
- Chapter 5 opens with a positive case study on saving the reefs of Apo Island (Philippines). It also has new material on coastal zones, coral reefs, estuaries, and shorelines. The Data Analysis box explains how to read climate graphs.
- Chapter 6 combines an extensive revision of population growth dynamics together with a new section on changes in human life expectancies. The Data Analysis box demonstrates graphing exponential and logistic growth.
- Chapter 7 opens with a case study on successful family planning in Thailand. Chapter 7 also presents a revised and strengthened discussion of factors influencing population, and different approaches to family planning in India are revealed in a What Do You Think box. The Data Analysis

box displays how graphs can be used to explain, persuade, and inform.

- Chapter 8 starts with an encouraging case study on Guinea worm eradication. The discussion and data on infectious and emergent diseases is extensively revised and updated, and the Data Analysis box presents graphing multiple variables.
- Chapter 9 updates information on world food supply, farm policy, and soil conservation. A new section has been added on consumer choices and local food supplies. The Data Analysis box shows how to graph relative values on an index scale.
- Chapter 10 begins with a new case study on forgotten pollinators. The Data Analysis box assesses pesticide use in schools.
- Chapter 11 includes a significantly revised and updated opening case study on biodiversity and ecological resilience. The discussion of endangered and threatened species has been extended and strengthened. The Data Analysis box addresses confidence limits in data.
- Chapter 12 has undergone a major reorganization. It opens with a new case study on saving the Great Bear Rainforest and it has a new What Do You Think? box on forest thinning and salvage logging. A major new section on parks and preserves includes successes in preserving landscapes and relative percentages of different biomes protected, together with ecotourism and the role of indigenous people in biodiversity protection. The Data Analysis box is a practical exercise in detecting edge effects.
- Chapter 13 is a brand new chapter on the important topic of ecological restoration. It explains how forests, grasslands, savannas, wetlands, and other ecosystems are being repaired and restored to ecological health, and provides many positive examples of restoration efforts. Data Analysis box clarifies concept maps.
- Chapter 14 opens with a new case study on cyanide heap-leach gold mining. It continues with a rewritten section on earth structure and tectonic processes. A table of the world's worst polluted places shows the environmental effects of mining and smelting. The Data Analysis box explores mapping volcanoes and earthquakes.
- Chapter 15 contains a thoroughly updated discussion of critical issues of climate change, informed by the 2007 reports from the IPCC. It opens with an inspiring case study on California's new law to regulate greenhouse gases. It also includes a new Exploring Science box on free-air carbon enrichment studies and a Data Analysis box on graphing methane emissions.
- Chapter 16 opens with a new positive case study on controlling mercury pollution and market mechanisms for reducing greenhouse gases. It adds an encouraging case study on reducing air pollution in New Delhi, India. The Data Analysis box graphs air pollution abatement in Europe.
- Chapter 17 presents a new case study about China's gargantuan South-to-North Water Diversion project. It adds material on drying of the Aral Sea and Lake Chad, and it illustrates freshwater shortages, as well as the problems with dams and diversions. The Data Analysis box examines a water scarcity graph.
- Chapter 18 explains how low-cost natural systems can be used for wastewater treatment. The section on inorganic water pollutants has been extensively rewritten. The Data Analysis box elucidates water pollution graphs.
- Chapter 19 starts with a new positive case study on integrated gasification combined cycle (IGCC) "clean" coal power plants. It continues with information on carbon sequestration. The discussion on nuclear energy has been updated with the report of the first new reactor approval in 30 years. The Data Analysis box compares energy use by different countries.
- Chapter 20 presents an encouraging case study of Danish islands that depend entirely on renewable energy. It has a new discussion of hybrid gasoline-electric vehicles and adds an extensive new section on biofuels including the net energy balance of new energy sources. The Data Analysis box illustrates energy calculations.
- Chapter 21 has a new positive case study on waste recycling in New York City. The chapter includes new information on ocean pollution, waste export to developing countries, and electronic waste, and the Data Analysis box invites personal waste calculations.
- Chapter 22 revises and extends the heartening opening case study on Curitiba, Brazil. It retains the positive example of sustainable housing in London. The Data Analysis box elucidates graphing with a logarithmic scale.
- Chapter 23 begins with a new case study on Grameen bank microlending. It adds a new section on the Environmental Performance Index, and explores market mechanisms for pollution reduction. The Data Analysis box illustrates graphing the human development index.
- Chapter 24 retains its opening story on the snail darter and the Endangered Species Act. The Data Analysis box discusses scatter plots and regression analysis.
- Chapter 25 opens with a new case study on saving the gray whale nursery in Laguna San Ignacio. The chapter includes new sections on environmental leadership, campus greening, and the millennium development goals. The Data Analysis box shows how students can do a campus environmental audit.

## Acknowledgements

We owe a great debt to the hardworking, professional team that has made this the best environmental science text possible. We express special thanks for editorial support to Marge Kemp, Janice Roerig-Blong, Joan Weber, Rose Koos, and Ashley Zellmer. We are grateful to the excellent production team led by April Southwood and marketing leadership by Tami Petsche. We also thank Kandis Elliot for her outstanding artwork, LouAnn Wilson for photo research, and Cathy Conroy for copyediting. We also thank Dr. Kim Chapman for essays that contributed to the text. Finally, we thank the many contributions of careful reviewers who shared their ideas with us during revisions.

## Tenth Edition Reviewers

Professor Dwight W. Allen, Eminent Scholar of Educational Reform
*Old Dominion University*

Daphne Hall Babcock
*Collin County Community College*

David C. Belt
*Johnson County Community College, Overland Park, Kansas*

Donna H. Bivans
*Pitt Community College*

J. Christopher Brown
*University of Kansas*

Kelly S. Cartwright
*College of Lake County*

Robert S. Dill
*Bergen Community College*

Dr. Iver W. Duedall, Professor Emeritus
*Florida Institute of Technology, Melbourne*

Daniel Habib
*Queens College of the City University of New York*

Barbara Hollar
*University of Detroit Mercy*

Timothy V. Horger
*Illinois Valley Community College*

Charles Ide, Director, Environmental Institute
*Western Michigan University*

Richard R. Jurin
*University of Northern Colorado*

Dawn G. Keller
*Hawkeye Community College*

David Knowles
*East Carolina University*

Tim Marbach
*California State University, Sacramento*

Chris Migliaccio
*Miami Dade College*

Jay C. Odaffer
*Manatee Community College*

Julie Phillips
*De Anza College*

Michael Phillips
*Illinois Valley Community College*

Sarah Quast
*Middlesex Community College*

Dawn Ranish
*Broward Community College*

Michelle L. Stevens
*Imperial Valley College*

Julie Stoughton
*University of Nevada, Reno*

S. Kant Vajpayee
*University of Southern Mississippi*

Kelly Wessell
*Grand Valley State University*

Van Wheat
*South Texas College*

John J. Wielichowski
*Milwaukee Area Technical College*

Lorne Wolfe
*Georgia Southern University*

# Guided Tour

## *A Global Perspective Is Vital to Learning about Environmental Science*

### Case Studies

In the front of each chapter, case studies utilize stories to portray real-life global issues that affect our food, our quality of life, and our future. Seventeen new case studies have been added to further focus on current events and the success stories of environmental protection progress.

Case Study **Family Planning in Thailand: A Success Story**

Down a narrow lane off Bangkok's busy Sukhumvit Road, is a most unusual café. Called Cabbages and Condoms, it's not only highly rated for its spicy Thai food, but it's also the only restaurant in the world dedicated to birth control. In an adjoining gift shop, baskets of condoms stand next to decorative handicrafts of the northern hill tribes. Piles of T-shirts carry messages, such as, "A condom a day keeps the doctor away," and "Our food is guaranteed not to cause pregnancy." Both businesses are run by the Population and Community Development Association (PDA), Thailand's largest and most influential nongovernmental organization.

The PDA was founded in 1974 by Mechai Viravaidya, a genial and fun-loving former Thai Minister of Health, who is a genius at public relations and human motivation (fig. 7.1). While traveling around Thailand in the early 1970s, Mechai recognized that rapid population growth—particularly in poor rural areas—was an obstacle to community development. Rather than lecture people about their behavior, Mechai decided to use humor to promote family planning. PDA workers handed out condoms at theaters and traffic jams, anywhere a crowd gathered. They challenged governmental officials to condom balloon-blowing contests, and taught youngsters Mechai's condom song: "Too Many Children Make You Poor." The PDA even pays farmers to paint birth control ads on the sides of their water buffalo.

This campaign has been extremely successful at making birth control and family planning, which once had been taboo topics in polite society, into something familiar and unembarrassing. Although condoms—now commonly called "mechais" in Thailand—are the trademark of PDA, other contraceptives, such as pills, spermicidal foam, and IUDs, are promoted as well. Thailand was one of the first countries to allow the use of the injectable contraceptive DMPA, and remains a major user. Free non-scalpel vasectomies are available on the king's birthday. Sterilization has become the most widely used form of contraception in the country. The campaign to encourage condom use has also been helpful in combating AIDS.

In 1974, when PDA started, Thailand's growth rate was 3.2 percent per year. In just fifteen years, contraceptive use among married couples increased from 15 to 70 percent, and the growth rate had dropped to 1.6 percent, one of the most dramatic birth rate declines ever recorded. Now Thailand's growth rate is 0.7 percent, or nearly the same as the United States. The fertility rate (or average number of children per woman) decreased from 7 in 1974 to 1.7 in 2006. The PDA is credited with the fact that Thailand's population is 20 million less than it would have been if it had followed its former trajectory.

**FIGURE 7.1** Mechai Viravaidya (right) is joined by Peter Piot, Executive Director of UNAIDS, in passing out free condoms on family planning and AIDS awareness day in Bangkok".

In addition to Mechai's creative genius and flair for showmanship, there are several reasons for this success story. Thai people love humor and are more egalitarian than most developing countries. Thai spouses share in decisions regarding children, family life, and contraception. The government recognizes the need for family planning and is willing to work with volunteer organizations, such as the PDA. And Buddhism, the religion of 95 percent of Thais, promotes family planning.

The PDA hasn't limited itself to family planning and condom distribution. It has expanded into a variety of economic development projects. Microlending provides money for a couple of pigs, or a bicycle, or a small supply of goods to sell at the market. Thousands of water-storage jars and cement rainwater-catchment basins have been distributed. Larger scale community development grants include road building, rural electrification, and irrigation projects. Mechai believes that human development and economic security are keys to successful population programs.

This case study introduces several important themes of this chapter. What might be the effects of exponential growth in human populations? How might we manage fertility and population growth? And what are the links between poverty, birth rates, and our common environment? Keep in mind, as you read this chapter, that resource limits aren't simply a matter of total number of people on the planet, they also depend on consumption levels and the types of technology used to produce the things we use.

### Google Earth™ Placemarks

This feature provides interactive satellite imagery of the earth to give students a geographic context of places and topics in the text. Students can zoom in for detail or they can zoom out for a more global perspective. Placemark links can be found on the website http://www.mhhe.com/cunningham10e.

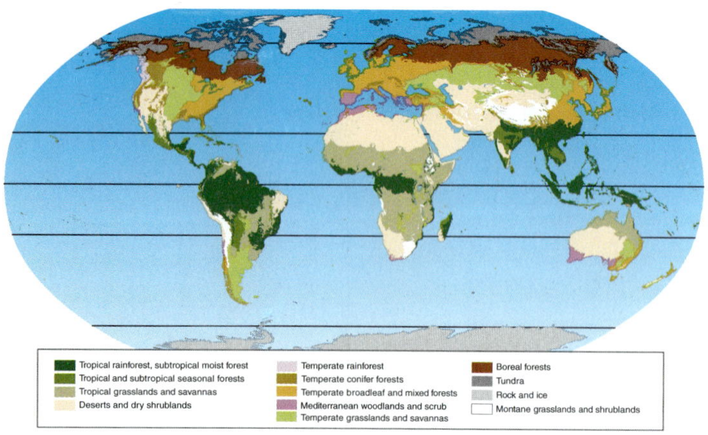


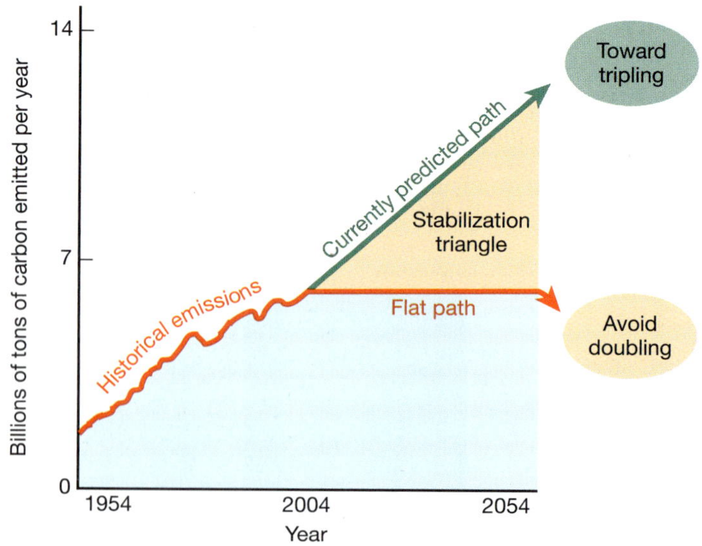


### The Latest Global Data

Easy to follow graphs, charts, and maps display numerous examples from many regions of the world. Students are exposed to the fact that environmental issues cross borders and oceans.

# Critical Thinking Skills Support Understanding of Environmental Change

## Exploring Science

Real-life environmental issues drive these readings as students learn about the principles of scientific observation and proper data-gathering techniques.

### *Exploring* SCIENCE — The Cichlids of Lake Victoria

If you visit your local pet store, chances are you'll see some cichlids (*Haplochromis* sp.). These small colorful, prolific fish come in a wide variety of colors and shapes from many parts of the world. The greatest cichlid diversity on earth—and probably the greatest vertebrate diversity anywhere—is found in the three great African rift lakes: Victoria, Malawi, and Tanganyika. Together, these lakes once had about 1,000 types of cichlids—more than all the fish species in Europe and North America combined. All these cichlids apparently evolved from a few ancestral varieties in the 15,000 years or so since the lakes were formed by splitting of the continental crust. This is one of the fastest and most extensive examples of vertebrate speciation known.

We believe that one of the factors that allowed cichlids to evolve so quickly is that they found few competitors or predators and a multitude of ecological roles to play in these

Cichlid fishes of Lake Victoria. More than 300 species have evolved from an original common ancestor to take advantage of different food sources and habitats.

been particularly hard hit. Cichlids once made up 80 percent of the animal biomass in the

large power boats and nylon nets to harvest great schools of perch, which are filleted, frozen, and shipped to markets in Europe and the Middle East. Because the perch are oily, local fishers can't sun dry them as they once did the cichlids. Instead, they are cooked or smoked over wood fires for local consumption. Forests are being denuded for firewood, and protein malnutrition is common in a region that exports 200,000 tons of fish each year.

Perhaps worst of all, Lake Victoria, which covers an area the size of Switzerland, is dying. Algae blooms clot the surface, oxygen levels have fallen alarmingly, and thick layers of soft silt are filling-in shallow bays. Untreated sewage, chemical pollution, and farm runoff are the immediate causes of these deleterious changes, but destabilization of the natural community plays a role as well. The swarms of cichlids that once ate algae and rotting detritus were

## Data Analysis

At the end of every chapter, these exercises ask students to graph and evaluate data while critically analyzing what they observe.

### DATA analysis — Reading Climate Graphs

As you've learned in this chapter, temperature and moisture are critical factors in determining the distribution and health of ecosystems. But how do you read the climate and precipitation graphs that accompany the description of each biome? To begin, examine the three climate graphs in this box. These graphs show annual trends in temperature and precipitation (rainfall and snow). They also indicate the relationship between potential evaporation, which depends on temperature and precipitation. When evaporation exceeds precipitation, dry conditions result (yellow areas). Extremely wet months are shaded dark blue on the graphs. Moist climates may vary in precipitation rates, but evaporation rarely exceeds precipitation. Months above freezing temperature (shaded brown on the X-axis) have most evaporation. Comparing these climate graphs helps us understand the different seasonal conditions that control plant and animal lives in different biomes.

1. What are the maximum and minimum temperatures in each of the three locations shown?
2. What do these temperatures correspond to in Fahrenheit? (*Hint:* look at the table in the back of your book).
3. Which area has the wettest climate; which is driest?
4. How do the maximum and minimum monthly rainfalls in San Diego and Belém compare?
5. Describe these three climates.
6. What kinds of biomes would you expect to find in these areas?

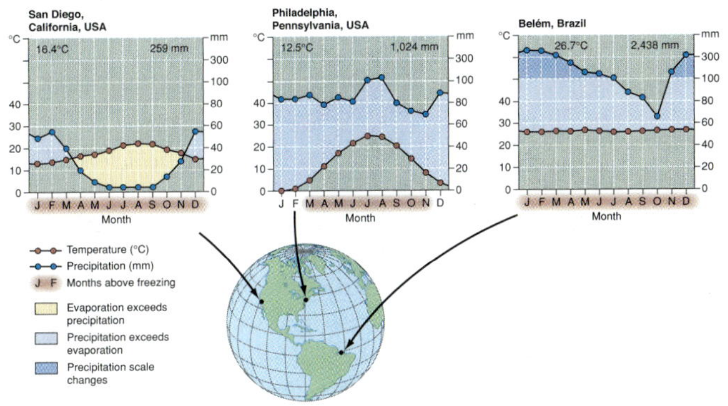


Moisture availability depends on temperature as well as precipitation. The horizontal axis on these climate diagrams shows months of the year; vertical axes show temperature (*left side*) and precipitation (*right*). The number of dry months (*shaded yellow*) and wetter months (*blue*) varies with geographic location. Mean annual temperature (°C) and precipitation (mm) are shown at the top of each graph.

## What Do You Think?

This feature provides challenging environmental studies that offer an opportunity for students to consider contradictory data, special interests, and conflicting interpretations within a real scenario.

### *What Do You Think?*

#### Too Many Deer?

A century ago, few Americans had ever seen a wild deer. Uncontrolled hunting and habitat destruction had reduced the deer population to about 500,000 animals nationwide. Some states had no deer at all. To protect the remaining deer, laws were passed in the 1920s and 1930s to restrict hunting, and the main deer predators—wolves and mountain lions—were exterminated throughout most of their former range.

As Americans have moved from rural areas to urban centers, forests have regrown, and deer populations have undergone explosive growth. Maturing at age two, a female deer can give birth to twin fawns every year for a decade or more. Increasing more than 20 percent annually, a deer population can double in just three years, an excellent example of irruptive, exponential growth.

Wildlife biologists estimate that the contiguous 48 states now have a population of more than 30 million white-tailed deer (*Odocoileus virginianus*), probably triple the number present in pre-Columbian times. Some areas have as many as 200 deer per square mile ($77/km^2$). At this density, woodland plant diversity is generally reduced to a few species that deer won't eat. Most deer, in such conditions, suffer from malnourishment, and many die every year of disease and starvation. Other species are diminished as well. Many small mammals and ground-dwelling birds begin to disappear when deer populations reach 25 animals per square mile. At 50 deer per square mile, most ecosystems are seriously impoverished.

A white-tailed deer (*Odocoileus virginianus*)

The social costs of large deer populations are high. In Pennsylvania alone, where deer numbers are now about 500 times greater than a century ago, deer destroy about $70 million worth of crops and $75 million worth of trees annually. Every year some 40,000 collisions with motor vehicles cause $80 million in property damage. Deer help spread Lyme disease, and, in many states, chronic wasting disease is found in wild deer herds. Some of the most heated criticisms of current deer management policies are in the suburbs. Deer love to browse on the flowers, young trees, and ornamental bushes in suburban yards. Heated disputes often arise between those who love to watch deer and their neighbors who want to exterminate them all.

In remote forest areas, many states have extended hunting seasons, increased the bag limit to four or more animals, and encouraged hunters to shoot does (females) as well as bucks (males). Some hunters criticize these changes because they believe that fewer deer will make it harder to hunt successfully and less likely that they'll find a trophy buck. Others, however, argue that a healthier herd and a more diverse ecosystem is better for all concerned.

In urban areas, increased sport hunting usually isn't acceptable. Wildlife biologists argue that the only practical way to reduce deer herds is culling by professional sharpshooters. Animal rights activists protest lethal control methods as cruel and inhumane. They call instead for fertility controls, reintroduction of predators, such as wolves and mountain lions, or trap and transfer programs. Birth control works in captive populations but is expensive and impractical with wild animals. Trapping, also, is expensive, and there's rarely anyplace willing to take surplus animals. Predators may kill domestic animals or even humans.

This case study shows that carrying capacity can be more complex than the maximum number of organisms an ecosystem can support. While it may be possible for 200 deer to survive in a square mile, there's an ecological carrying capacity lower than that if we consider the other species dependent on that same habitat. There's also an ethical carrying capacity if we don't want to see animals suffer from malnutrition and starve to death every winter. And there's a cultural carrying capacity if we consider the tolerable rate of depredation on crops and lawns or an acceptable number of motor vehicle collisions.

If you were a wildlife biologist charged with managing the deer herd in your state, how would you reconcile the different interests in this issue? How would you define the optimum deer population, and what methods would you suggest to reach this level? What social or ecological indicators would you look for to gauge whether deer populations are excessive or have reached an appropriate level?

# Sound Pedagogy Encourages Science Inquiry and Application

## Learning Outcomes

Found at the beginning of each chapter, and organized by major headings, these outcomes give students an overview of the key concepts they will need to understand.

> **LEARNING OUTCOMES**
>
> After studying this chapter, you should be able to:
>
> **12.1** Discuss the types and uses of world forests.
>
> **12.2** Describe the location and state of grazing lands around the world.
>
> **12.3** Summarize the types and locations of nature preserves.

## Conclusion

This section summarizes the chapter by highlighting key ideas and relating them to one another.

> **CONCLUSION**
>
> Forests and grasslands cover nearly 60 percent of global land area. The vast majority of humans live in these biomes, and we obtain many valuable materials from them. And yet, these biomes also are the source of much of the world's biodiversity on which we depend for life-supporting ecological services. How we can live sustainably on our natural resources while also preserving enough nature so those resources can be replenished represents one of the most important questions in environmental science.
>
> There is some good news in our search for a balance between … and land … particular… are more … ago. Prot…
>
> Australia's great barrier reef shows that we can choose to protect some biodiverse areas in spite of forces that want to exploit them. Overall, nearly 12 percent of the earth's land area is now in some sort of protected status. While the level of protection in these preserves varies, the rapid recent increase in number and area in protected status exceeds the goals of the United Nations Millennium Project.
>
> While we haven't settled the debate between focusing on individual endangered species versus setting aside representative samples of habitat, pursuing both strategies seems to be working.

## Reviewing Learning Outcomes

Connected to the Learning Outcomes at the beginning of each chapter, this review clearly restates the important concepts associated with each outcome.

> **REVIEWING LEARNING OUTCOMES**
>
> By now you should be able to explain the following points:
>
> **12.1** Discuss the types and uses of world forests.
> - Boreal and tropical forests are most abundant.
> - Forests provide many valuable products.
> - Tropical forests are being cleared rapidly.
> - Temperate forests have competing uses.
>
> **12.2** Describe th… world.
> - Grazing c…
> - Overgrazi…
> - Ranchers …
>
> **12.3** Summarize the types and locations of nature preserves.
> - Many countries have created nature preserves.
> - Not all preserves are preserved.
> - Marine ecosystems need greater protection.
> - Conservation and economic development can work together.
> - Native people can play important roles in nature protection.
> - Species survival can depend on preserve size and shape.

## Critical Thinking and Discussion Questions

Brief scenarios of everyday occurrences or ideas challenge students to apply what they have learned to their lives.

> **CRITICAL THINKING AND DISCUSSION QUESTIONS**
>
> 1. Paper and pulp are the fastest growing sector of the wood products market, as emerging economies of China and India catch up with the growing consumption rates of North America, Europe, and Japan. What should be done to reduce paper use?
> 2. Conservationists argue that watershed protection and other ecological functions of forests are more economically valuable than timber. Timber companies argue that continued production supports stable jobs and local economies. If you were a judge attempting to decide which group was right, what evidence would you need on both sides? How would you gather this evidence?
> 3. Divide your class into a ranching group, a conservation grou… prot… natu… land… to m…
> 4. Calculating forest area and forest losses is complicated by the difficulty of defining exactly what constitutes a forest. Outline a definition for what counts as forest in your area, in terms of size, density, height, or other characteristics. Compare your definition to those of your colleagues. Is it easy to agree? Would your definition change if you lived in a different region?
> 5. Why do you suppose dry topical forest and tundra are well represented in protected areas, while grasslands and wetlands are protected relatively rarely? Consider social, cultural, geographic, and economic reasons in your answer.
> 6. Oil and gas companies want to drill in several parks, monu…

## Practice Quiz

Short-answer questions allow students to check their knowledge of chapter concepts.

> **PRACTICE QUIZ**
>
> 1. What do we mean by *closed-canopy* forest and *old-growth* forest?
> 2. Which commodity is used most heavily in industrial economies: steel, plastic, or wood? What portion of the world's population depends on wood or charcoal as the main energy supply?
> 3. What is a *debt-for-nature swap*?
> 4. Why is fire suppression a controversial strategy? Why are forest thinning and salvage logging controversial?
> 5. Are pastures and rangelands always damaged by grazing animals? What are some results of overgrazing?
> 6. What is *rotational grazing*, and how does it mimic natural processes?
> 7. What was the first national park in the world, and when was it established? How have the purposes of this park and others changed?
> 8. How do the size and design of nature preserves influence their effectiveness? What do landscape ecologists mean by *interior habitat* and *edge effects*?
> 9. What is *ecotourism*, and why is it important?
> 10. What is a *biosphere reserve*, and how does it differ from a wilderness area or wildlife preserve?

## What Can You Do?

This feature gives students realistic steps for applying their knowledge to make a positive difference in our environment.

> **What Can You Do?**
>
> 
>
> Reducing Waste
>
> 1. Buy foods that come with less packaging; shop at farmers' markets or co-ops, using your own containers.
> 2. Take your own washable refillable beverage container to meetings or convenience stores.
> 3. When you have a choice at the grocery store between plastic, glass, or metal containers for the same food, buy the reusable or easier-to-recycle glass or metal.
>
> **Source:** Minnesota Pollution Control Agency.

## Think About It

These boxes provide several opportunities in each chapter for students to review material, practice critical thinking, and apply scientific principles.

> ***Think About It***
>
> Why might you and your mother rank some risks differently? List some activities on which the two of you might disagree.

## Teaching and Learning Supplements

**McGraw-Hill offers various tools and technology products to support *Environmental Science: A Global Concern.* Students can order supplemental study materials by contacting their local bookstore or by calling 800-262-4729. Instructors can obtain teaching aids by calling the Customer Service Department at 800-338-3987, visiting the McGraw-Hill website at www.mhhe.com, or by contacting their local McGraw-Hill sales representative.**

### Teaching supplements for instructors

**Presentation Center**

ARIS Presentation Center is an online digital library containing assets such as photos, artwork, PowerPoints, animations, and other media types that can be used to create customized lectures, visually enhanced tests and quizzes, compelling course websites, and attractive printed support materials. The following digital assets are grouped by chapter:

- **Color Art** Full-color digital files of illustrations in the text can be readily incorporated into lecture presentations, exams, or custom-made classroom materials. These include all of the 3-D realistic art found in this edition, representing some of the most important concepts in environmental science.
- **Photos** Digital files of photographs from the text can be reproduced for multiple classroom uses.
- **Tables** Every table that appears in the text is provided in electronic format.
- **Videos** This special collection of 69 underwater video clips displays interesting habitats and behaviors for many animals in the ocean.
- **Animations** One hundred full-color animations that illustrate many different concepts covered in the study of environmental science are available for use in creating classroom lectures, testing materials, or online course communication. The visual impact of motion will enhance classroom presentations and increase comprehension.
- **Test Bank** A computerized test bank that uses testing software to quickly create customized exams is available on for this text. The user-friendly program allows instructors to search for questions by topic or format, edit existing questions or add new ones; and scramble questions for multiple versions of the same test. Word files of the test bank questions are provided for those instructors who prefer to work outside the test-generator software.
- **Global Base Maps** Eighty-eight base maps for all world regions and major subregions are offered in four versions: black-and-white and full-color, both with labels and without labels. These choices allow instructors the flexibility to plan class activities, quizzing opportunities, study tools, and PowerPoint enhancements.
- **PowerPoint Lecture Outlines** Ready-made presentations that combine art and photos and lecture notes are provided for each of the 25 chapters of the text. These outlines can be used as they are or tailored to reflect your preferred lecture topics and sequences.
- **PowerPoint Slides** For instructors who prefer to create their lectures from scratch, all illustrations, photos, and tables are preinserted by chapter into blank PowerPoint slides for convenience.

ARIS™

McGraw-Hill's **ARIS—Assessment, Review, and Instruction System** (aris.mhhe.com) for *Environmental Science: A Global Concern* is a complete, online tutorial, electronic homework, and course management system, designed for greater ease of use than any other system available. Free upon adoption of *Environmental Science: A Global Concern,* instructors can create and share course materials and assignments with colleagues with a few clicks of the mouse. All PowerPoint lectures, assignments, quizzes, tutorials, and interactives are directly tied to text-specific materials in *Environmental Science,* but instructors can also edit questions, import their own content, and create announcements and due dates for assignments. ARIS has automatic grading and reporting of easy-to-assign homework, quizzing, and testing. All student activity within McGraw-Hill's ARIS website is automatically recorded and available to the instructor through a fully integrated grade book that can be downloaded to Excel.

**eInstruction**

This classroom performance system (CPS) utilizes wireless technology to bring interactivity into the classroom or lecture hall. Instructors and students receive immediate feedback through wireless response pads that are easy to use and engage students. eInstruction can assist instructors by:

- taking attendance
- administering quizzes and tests
- creating a lecture with intermittent questions
- using the CPS grade book to manage lectures and student comprehension
- integrating interactivity into PowerPoint presentations

Contact your local McGraw-Hill sales representative for more information.

**Course Delivery Systems**

With help from WebCT and Blackboard, professors can take complete control of their course content. Course cartridges containing website content, online testing, and powerful student tracking features are readily available for use within these platforms.

Go to aris.mhhe.com to learn more and register!

**Earth and Environmental Science DVD by Discovery Channel Education (ISBN-13: 978-0-07-352541-9; ISBN-10: 0-07-352541-3)**

Begin your class with a quick peek at science in action. The exciting NEW DVD by Discovery Channel Education offers 50 short (3–5 minute) videos on topics ranging from conservation to volcanoes. Search by topic and download into your PowerPoint lecture. Available to colleges and universities. See your McGraw-Hill sales representative for a detailed listing.

**McGraw-Hill's Biology Digitized Videos (ISBN-13: 978-0-07-312155-0; ISBN-10: 0-07-312155-X)**

Licensed from some of the highest quality life-science video producers in the world, these brief video clips on DVD range in length from 15 seconds to 2 minutes and cover all areas of general biology, from cells to ecosystems. Engaging and informative, McGraw-Hill's digitized biology videos will help capture students' interest while illustrating key biological concepts, applications, and processes.

## Learning supplements for students

**ARIS (aris.mhhe.com)**

This site includes quizzes for each chapter, additional case studies, interactive base maps, and much more. Learn more about the exciting features provided for students through the enhanced *Environmental Science: A Global Concern* website.

***Exploring Environmental Science with GIS* by Stewart, Cunningham, Schneiderman, and Gold (ISBN-13: 978-0-07-297564-2; ISBN-10: 0-07-297564-4)**

This short book provides exercises for students and instructors who are new to GIS, but are familiar with the Windows operating system. The exercises focus on improving analytical skills, understanding spatial relationships, and understanding the nature and structure of environmental data. Because the software used is distributed free of charge, this text is appropriate for courses and schools that are not yet ready to commit to the expense and time involved in acquiring other GIS packages.

***Annual Editions: Environment* 07/08 by Allen (ISBN-13: 978-0-07-351544-1; ISBN-10: 0-07-351544-2)**

This twenty-sixth edition is a compilation of current articles from the best of the public press. The selections explore the global environment, the world's population, energy, the biosphere, natural resources, and pollutions.

***Taking Sides: Clashing Views on Controversial Environmental Issues,* Twelfth Edition by Easton (ISBN-13: 978-0-07-351443-7; ISBN-10: 0-07-351443-8)**

This book represents the arguments of leading environmentalists, scientists, and policymakers. The issues reflect a variety of viewpoints and are staged as "pro" and "con" debates. Issues are organized around four core areas: general philosophical and political issues, the environment and technology, disposing of wastes, and the environment and the future.

***Classic Edition Sources: Environmental Studies,* Third Edition by Easton (ISBN-13: 978-0-07-352758-1; ISBN-10: 0-07-352758-0)**

This volume brings together primary source selections of enduring intellectual value—classic articles, book excerpts, and research studies—that have shaped environmental studies and our contemporary understanding of it. The book includes carefully edited selections from the works of the most distinguished environmental observers, past and present. Selections are organized topically around the following major areas of study: energy, environmental degradation, population issues and the environment, human health and the environment, and environment and society.

***Student Atlas of Environmental Issues,* by Allen (ISBN-13: 978-0-69-736520-0; ISBN-10: 0-69-736520-4)**

This atlas is an invaluable pedagogical tool for exploring the human impact on the air, waters, biosphere, and land in every major world region. This informative resource provides a unique combination of maps and data that help students understand the dimensions of the world's environmental problems and the geographical basis of these problems.

Environmental science students learn about wetland biology.

# Learning to Learn

*What kind of world do you want to live in? Demand that your teachers teach you what you need to know to build it.*

*—Peter Kropotkin—*

## LEARNING OUTCOMES

After studying this introduction, you should be able to:

**L.1** Form a plan to organize your efforts and become a more effective and efficient student.

**L.2** Make an honest assessment of the strengths and weaknesses of your current study skills.

**L.3** Assess what you need to do to get the grade you want in this class.

**L.4** Set goals, schedule your time, and evaluate your study space.

**L.5** Use this textbook effectively, practice active reading, and prepare for exams.

**L.6** Be prepared to apply critical and reflective thinking in environmental science.

**L.7** Understand the advantages of concept mapping and use it in your studying.

# Case Study Why Study Environmental Science?

Welcome to environmental science. We hope you'll enjoy learning about the material presented in this book, and that you'll find it both engaging and useful. There should be something here for just about everyone, whether your interests are in basic ecology, natural resources, or the broader human condition. You'll see, as you go through the book, that it covers a wide range of topics. It defines our environment, not only the natural world, but also the built world of technology, cities, and machines, as well as human social or cultural institutions. All of these interrelated aspects of our life affect us, and, in turn, are affected by what we do.

You'll find that many issues discussed here are part of current news stories on television or in newspapers. Becoming an educated environmental citizen will give you a toolkit of skills and attitudes that will help you understand current events and be a more interesting person. Because this book contains information from so many different disciplines, you will find connections here with many of your other classes. Seeing material in an environmental context may assist you in mastering subject matter in many courses, as well as in life after you leave school.

One of the most useful skills you can learn in any of your classes is critical thinking—a principal topic of this chapter. Much of the most important information in environmental science is highly contested. Facts vary depending on when and by whom they were gathered. For every opinion there is an equal and opposite opinion. How can you make sense out of this welter of ever-changing information? The answer is that you need to develop a capacity to think independently, systematically, and skillfully to form your own opinions (fig. L.1). These qualities and abilities can help you in many aspects of life. Throughout this book you will find "What Do You Think?" boxes that invite you to practice your critical and reflective thinking skills.

**FIGURE L.1** What does it all mean? Studying environmental science gives you an opportunity to develop creative, reflective, and critical thinking skills.

There is much to be worried about in our global environment. Evidence is growing relentlessly that we are degrading our environment and consuming resources at unsustainable rates. Biodiversity is disappearing at a pace unequaled since the end of the age of dinosaurs 65 million years ago. Irreplaceable topsoil erodes from farm fields, threatening global food supplies. Ancient forests are being destroyed to make newsprint and toilet paper. Rivers and lakes are polluted with untreated sewage, while soot and smoke obscure our skies. Even our global climate seems to be changing to a new regime that could have catastrophic consequences.

At the same time, we have better tools and knowledge than any previous generation to do something about these crises. Worldwide public awareness of—and support for—environmental protection is at an all-time high. Over the past 50 years, human ingenuity and enterprise have brought about a breathtaking pace of technological innovations and scientific breakthroughs. We have learned to produce more goods and services with less material. The breathtaking spread of communication technology makes it possible to share information worldwide nearly instantaneously. Since World War II, the average real income in developing countries has doubled; malnutrition has declined by almost one-third; child death rates have been halved; average life expectancy has increased by 30 percent; and the percentage of rural families with access to safe drinking water has risen from less than 10 percent to almost 75 percent.

The world's gross domestic product has increased more than tenfold over the past five decades, but the gap between the rich and poor has grown ever wider. More than a billion people now live in abject poverty without access to adequate food, shelter, medical care, education, and other resources required for a healthy, secure life. The challenge for us is to spread the benefits of our technological and economic progress more equably and to find ways to live sustainably over the long run without diminishing the natural resources and vital ecological services on which all life depends. We've tried to strike a balance in this book between enough doom and gloom to give you a realistic view of our problems, and enough positive examples to give hope that we can discover workable solutions.

What would it mean to become a responsible environmental citizen? What rights and privileges do you enjoy as a member of the global community? What duties and responsibilities go with that citizenship? In many chapters of this book you will find practical advice on things you can do to conserve resources and decrease adverse environmental impacts. Ethical perspectives are an important part of our relationship to the environment and the other people with whom we share it. The discussion of ethical principles and worldviews in chapter 2 is a key section of this book. We hope you'll think about the ethics of how we treat our common environment.

Clearly, to become responsible and productive environmental citizens each of us needs a basis in scientific principles, as well as some insights into the social, political, and economic systems that impact our global environment. We hope this book and the class you're taking will give you the information you need to reach those goals. As the noted Senegalese conservationist and educator, Baba Dioum, once said, "in the end, we will conserve only what we love, we will love only what we understand, and we will understand only what we are taught."

## L.1 How Can I Get an A in This Class?

"What have I gotten myself into?" you are probably wondering as you began to read this book. "Will environmental science be worth my while? Do I have a chance to get a good grade?" The answers to these questions depend, to a large extent, on you and how you decide to apply yourself. Expecting to be interested and to do either well or poorly in your classes often turns out to be a self-fulfilling prophecy. As Henry Ford once said, "If you think you can do a thing, or think you can't do a thing, you're right." Cultivating good study skills can help you to reach your goals and make your experience in environmental science a satisfying and rewarding one. The purpose of this introduction is to give you some tips to help you get off to a good start in studying. You'll find that many of these techniques are also useful in other courses and after you graduate, as well.

Environmental science, as you can see by skimming through the table of contents of this book, is a complex, transdisciplinary field that draws from many academic specialties. It is loaded with facts, ideas, theories, and confusing data. It is also a dynamic, highly contested subject. Topics such as environmental contributions to cancer rates, potential dangers of pesticides, or when and how much global warming may be caused by human activities are widely disputed. Often you will find distinguished and persuasive experts who take completely opposite positions on any particular question. It will take an active, organized approach on your part to make sense of the vast amount of information you'll encounter here. And it will take critical, thoughtful reasoning to formulate your own position on the many controversial theories and ideas in environmental science. Learning to learn will help you keep up-to-date on important issues after you leave this course. Becoming educated voters and consumers is essential for a sustainable future.

### Develop good study habits

Many students find themselves unprepared for studying in college. In a survey released in 2003 by the Higher Education Research Institute, more than two-thirds of high school seniors nationwide reported studying outside of class less than one hour per day. Nevertheless, because of grade inflation, nearly half those students claim to have an A average. It comes as a rude shock to many to discover that the study habits they developed in high school won't allow them to do as well—or perhaps even to pass their classes—in college. Many will have to triple or even quadruple their study time. In addition, they need urgently to learn to study more efficiently and effectively.

What are your current study skills and habits? Making a frank and honest assessment of your strengths and weaknesses will help you set goals and make plans for achieving them during this class. Answer the questions in table L.1 as a way of assessing where you are as you begin to study environmental science and where you need to work to improve your study habits.

One of the first requirements for success is to set clear, honest, attainable goals for yourself. Are you willing to commit the time and effort necessary to do well in this class? Make goals for yourself in terms that you can measure and in time frames

**TABLE L.1**

**Assess Your Study Skills**

Rate yourself on each of the following study skills and habits on a scale of 1 (excellent) to 5 (needs improvement). If you rate yourself below 3 on any item, think about an action plan to improve that competence or behavior.

____ How strong is your commitment to be successful in this class?

____ How well do you manage your time (e.g., do you always run late or do you complete assignments on time)?

____ Do you have a regular study environment that reduces distraction and encourages concentration?

____ How effective are you at reading and note-taking (e.g., do you remember what you've read; can you decipher your notes after you've made them)?

____ Do you attend class regularly and listen for instructions and important ideas? Do you participate actively in class discussions and ask meaningful questions?

____ Do you generally read assigned chapters in the textbook before attending class or do you wait until the night before the exam?

____ Are you usually prepared before class with questions about material that needs clarification or that expresses your own interest in the subject matter?

____ How do you handle test anxiety (e.g., do you usually feel prepared for exams and quizzes or are you terrified of them? Do you have techniques to reduce anxiety or turn it into positive energy)?

____ Do you actively evaluate how you are doing in a course based on feedback from your instructor and then make corrections to improve your effectiveness?

____ Do you seek out advice and assistance outside of class from your instructors or teaching assistants?

within which you can see progress and adjust your approach if it isn't taking you where you want to go. Be positive but realistic. It's more effective to try to accomplish a positive action than to avoid a negative one. When you set your goals, use proactive language that states what you want rather than negative language about what you're trying to avoid. It's good to be optimistic, but setting impossibly high standards will only lead to disappointment. Be objective about the obstacles you face and be willing to modify your goals if necessary. As you gain more experience and information, you may need to adjust your expectations either up or down. Take stock from time to time to see whether you are on track to accomplish what you expect from your studies. In environmental planning, this is called adaptive management.

One of the most common mistakes many of us make is to procrastinate and waste time. Be honest, are you habitually late for meetings or in getting assignments done? Do you routinely leave your studying until the last minute and then frantically cram the night before your exams? If so, you need to organize your schedule so that you can get your work done and still have a life. Make a study schedule for yourself and stick to it. Allow enough time for sleep, regular meals, exercise, and recreation so that you will be rested, healthy, and efficient when you do study. Schedule regular study times between your classes and work. Plan some study times during the day when you are fresh; don't leave all your work until late night hours when you don't get much done. Divide your work into reasonable sized segments that you can accomplish on a daily basis. Plan to have all your reading and assignments completed several days before your exams so you will have adequate time to review and process information. Carry a calendar so you will remember appointments and assignments.

Establish a regular study space in which you can be effective and productive. It might be a desk in your room, a carrel in the library, or some other quiet, private environment. Find a place that works for you and be disciplined about sticking to what you need to do. If you get in the habit of studying in a particular place and time, you will find it easier to get started and to stick to your tasks. Many students make the mistake of thinking that they can study while talking to their friends or watching TV. They may put in many hours but not really accomplish much. On the other hand, some people think most clearly in the anonymity of a crowd. The famous philosopher, Immanuel Kant, found that he could think best while wandering through the noisy, crowded streets of Königsberg, his home town.

How you behave in class and interact with your instructor can have a big impact on how much you learn and what grade you get. Make an effort to get to know your instructor. She or he is probably not nearly as formidable as you might think. Sit near the front of the room where you can see and be seen. Pay attention and ask questions that show your interest in the subject matter. Practice the skills of good note-taking (table L.2). Attend every class and arrive on time. Don't fold up your papers and prepare to leave until after the class period is over. Arriving late and leaving early says to your instructor that you don't care much about either the class or your grade. If you think of yourself as a good student and act like one, you may well get the benefit of the doubt when your grade is assigned.

**TABLE L.2**

**Learning Skills—Taking Notes**

1. Identify the important points in a lecture and organize your notes in an outline form to show main topics and secondary or supporting points. This will help you follow the sense of the lecture.
2. Write down all you can. If you miss something, having part of the notes will help your instructor identify what you've missed.
3. Leave a wide margin in your notes in which you can generate questions to which your notes are the answers. If you can't write a question about the material, you probably don't understand it.
4. Study for your test under test conditions by answering your own questions without looking at your notes. Cover your notes with a sheet of paper on which you write your answers, then slide it to the side to check your accuracy.
5. Go all the way through your notes once in this test mode, then go back to review those questions you missed.
6. Compare your notes and the questions you generated with those of a study buddy. Did you get the same main points from the lecture? Can you answer the questions someone else has written?
7. Review your notes again just before test time, paying special attention to major topics and questions you missed during study time.

**Source:** Dr. Melvin Northrup, Grand Valley State University.

Practice active, purposeful learning. It isn't enough to passively absorb knowledge provided by your instructor and this textbook. You need to actively engage the material in order to really understand it. The more you invest yourself in the material, the easier it will be to comprehend and remember. It is very helpful to have a study buddy with whom you can compare notes and try out ideas (fig. L.2). You will get a valuable perspective

**FIGURE L.2** Cooperative learning, in which you take turns explaining ideas and approaches with a friend, can be one of the best ways to comprehend material.

on whether you're getting the main points and understanding an adequate amount by comparing. It's an old adage that the best way to learn something is to teach it to someone else. Take turns with your study buddy explaining the material you're studying. You may think you've mastered a topic by quickly skimming the text but you're likely to find that you have to struggle to give a clear description in your own words. Anticipating possible exam questions and taking turns quizzing each other can be a very good way to prepare for tests.

## Recognize and hone your learning styles

Each of us has ways that we learn most effectively. Discovering techniques that work for you and fit the material you need to learn is an important step in reaching your goals. Do any of the following fit your preferred ways of learning?

- **Visual Learner:** understands and remembers best by reading, looking at photographs, figures, and diagrams. Good with maps and picture puzzles. Visualizes image or spatial location for recall. Uses flash cards for memorization.
- **Verbal Learner:** understands and remembers best by listening to lectures, reading out loud, and talking things through with a study partner. May like poetry and word games. Memorizes by repeating item verbally.
- **Logical Learner:** understands and remembers best by thinking through a subject and finding reasons that make sense. Good at logical puzzles and mysteries. May prefer to find patterns and logical connections between items rather than memorize.
- **Active Learner:** understands and remembers best those ideas and skills linked to physical activity. Takes notes, makes lists, uses cognitive maps. Good at working with hands and learning by doing. Remembers best by writing, drawing, or physically manipulating items.

The list above represents only a few of the learning styles identified by educational psychologists. How can you determine which approaches are right for you? Think about the one thing in life that you most enjoy and in which you have the greatest skills. What hobbies or special interests do you have? How do you learn new material in that area? Do you read about a procedure in a book and then do it, or do you throw away the manual and use trial and error to figure out how things work? Do you need to see a diagram or a picture before things make sense, or are spoken directions most memorable and meaningful for you? Some people like to learn by themselves in a quiet place where there are no distractions, while others need to discuss ideas with another person to feel really comfortable about what they're learning.

Sometimes you have to adjust your preferred learning style to the specific material you're trying to master. You may be primarily an oral learner, but if what you need to remember for a particular subject is spatial or structural, you may need to try some visual learning techniques. Memorizing vocabulary items might be best accomplished by oral repetition, while developing your ability to work quantitative problems should be approached by practicing analytical or logical skills.

## Use this textbook effectively

An important part of productive learning is to read assigned material in a purposeful, deliberate manner. Ask yourself questions as you read. What is the main point being made here? Does the evidence presented adequately support the assertions being made? What personal experience have you had or what prior knowledge can you bring to bear on this question? Can you suggest alternative explanations for the phenomena being discussed? What additional information would you need in order to make an informed judgment about this subject and how might you go about obtaining that information or making that judgment?

A study technique developed by Frances Robinson and called the SQ3R method (table L.3) can be a valuable aid in improving your reading comprehension. Start your study session with a *survey* of the entire chapter or section you are about to read so you'll have an idea of how the whole thing fits together. What are the major headings and subdivisions? Notice that there is usually a hierarchical organization that gives you clues about the relationship between the various parts. This survey will help you plan your strategy for approaching the material. Next, *question* what the main points are likely to be in each of the sections. Which parts look most important or interesting? Ask yourself where you should invest the most time and effort. Is one section or topic likely to be more relevant to your particular class? Has your instructor emphasized any of the topics you see? Being alert for important material can help you plan the most efficient way to study.

After developing a general plan, begin *active reading* of the text. Read in small segments and stop frequently for reflection and to make notes. Don't fall into a trance in which the words swim by without leaving any impression. Highlight or underline

**TABLE L.3**

**The SQ3R Method for Studying Texts**

| Step | Description |
|---|---|
| **Survey** | Preview the information to be studied before reading. |
| **Question** | Ask yourself critical questions about the content of what you are reading. |
| **Read** | Conduct the actual reading in small segments. |
| **Recite** | Stop periodically to recite to yourself what you have just read. |
| **Review** | Once you have completed the section, review the main points to make sure you remember them clearly. |

**FIGURE L.3** Cooperative learning is an important part of mastering environmental science. You never grasp material as clearly as when you explain it to someone else.

the main points but be careful that you don't just paint the whole page yellow. If you highlight too much, nothing will stand out. Try to distinguish what is truly central to the argument being presented. Make brief notes in the margins that identify main points. This can be very helpful in finding important sections or ideas when you are reviewing. Check your comprehension at the end of each major section. Ask yourself: Did I understand what I just read? What are the main points being made here? Does this relate to my own personal experiences or previous knowledge? Are there details or ideas that need clarification or elaboration?

As you read, stop periodically to *recite* the information you've just acquired. Summarize the information in your own words to be sure that you really understand and are not just depending on rote memory. This is a good time to have a study group (fig. L.3). Taking turns to summarize and explain material really helps you internalize it. If you don't have a study group and you feel awkward talking to yourself, you can try writing your summary. Finally, *review* the section. Did you miss any important points? Do you understand things differently the second time through? This is a chance to think critically about the material. Do you agree with the conclusions suggested by the authors? Can you think of alternative explanations for the same evidence? As you review each section, think about how this may be covered on the test. Put yourself in the position of the instructor. What would be some good questions based on this material? Don't try to memorize everything but try to anticipate what might be the most important points.

After class, compare your lecture notes with your study notes. Do they agree? If not, where are the discrepancies? Is it possible that you misunderstood what was said in class, or does your instructor differ with what's printed in the textbook? Are there things that your instructor emphasized in lecture that you missed in your preclass reading? This is a good time to go back over the readings to reinforce your understanding and memory of the material.

## Will this be on the test?

Students often complain that test results don't adequately reflect what they know and how much they've learned in studying. It may well be that test questions won't cover what you think is important or use a style that appeals to you, but you'll probably be more successful if you adapt yourself to the realities of your instructor's test methods rather than trying to force your instructor to accommodate to your preferences. One of your first priorities in studying, therefore, should be to learn your instructor's test style. Are you likely to have short-answer objective questions (multiple choice, true or false, fill in the blank) or does your instructor prefer essay questions? If you have an essay test, will the questions be broad and general or more analytical? You should develop a very different study strategy depending on whether you are expected to remember and choose between a multitude of facts and details, or whether you will be asked to write a paragraph summarizing some broad topic.

Organize the ideas you're reading and hearing in lecture. This course will probably include a great deal of information. Unless you have a photographic memory, you won't be able to remember every detail. What's most important? What's the big picture? If you see how pieces of the course fit together, it will all make more sense and be easier to remember. As you read and review, ask yourself what might be some possible test questions in each section. If you're likely to have factual questions, what are the most significant facts in the material you've read? Memorize some benchmark figures. Just a few will help a lot. Pay special attention to tables, graphs, and diagrams. They were chosen because they illustrate important points.

You probably won't be expected to remember all the specific numbers in this book but you probably should know orders of magnitude. The world population is about 6.5 *billion* people, not thousands, millions, or trillions. Highlight facts and figures in your lecture notes about which your instructor seemed especially interested. There is a good chance you'll see those topics again on a test. It often helps to remember facts and figures if you can relate them to some other familiar example. The United States, for instance, has about 295 million residents. The European Union is slightly larger, India is about three times and China is more than four times as large. Be sure you're familiar with the bold-face key terms in the textbook. Vocabulary terms make good objective questions. The Practice Quiz at the end of each chapter generally covers objective material that makes good short-answer questions.

A number of strategies can help you be successful in test-taking. Look over the whole test at the beginning and answer the questions you know well first, then tackle the harder ones. On multiple choice tests, find out whether there is a penalty for guessing. Use the process of elimination to narrow down the possible choices and improve the odds for guessing. Often you can get hints from the context of the question or from other similar questions. Notice that the longest or most specific answer often is right while those that are vague or general are more likely wrong. Be alert for absolutes (such as always, never, all) which could indicate wrong choices. Qualifiers (such as sometimes, may, or could) on the other hand, often point to correct answers. Exactly opposite answers may indicate that one of them is correct.

If you anticipate essay questions, practice writing one- or two-paragraph summaries of major points in each chapter. Develop your ability to generalize and to make connections between important facts and ideas. Notice that the Critical Thinking and Discussion Questions at the end of each chapter are open-ended topics that can work well either for discussion groups or as questions for an essay test. You'll have a big advantage on a test if you have some carefully thought out arguments for and against the major ideas presented in each chapter. If you don't have any idea what a particular essay question means, you often can make a transition to something you do understand. Look for a handle that links the question to a topic you are prepared to answer. Even if you have no idea what the question means, make an educated guess. You might get some credit. Anything is better than a zero. Sometimes if you explain your answer, you'll get at least some points. "If the question means such and such, then the answer would be ________" may get you partial credit.

Does your instructor like thought questions? Does she/he expect you to be able to interpret graphs or to draw inferences from a data table? Might you be asked to read a paragraph and describe what it means or relate it to other cases you've covered in the class? If so, you should practice these skills. Making up and sharing these types of questions with your study group can greatly increase your understanding of the material as well as improve your performance on exams. Writing a paragraph answer for each of the Critical Thinking and Discussion Questions could be a very good way to study for an essay test.

Concentrate on positive attitudes and building confidence before your tests. If you have fears and test anxiety, practice relaxation techniques and visualize success. Be sure you are rested and well prepared. You certainly won't do well if you're sleep-deprived and a bundle of nerves. Often the worst thing you can do is to stay up all night to cram your brain with a jumble of data. Being able to think clearly and express yourself well may count much more than knowing a pile of unrelated facts. Review your test when it is returned to learn what you did well and where you need to improve. Ask your instructor for pointers on how you might have answered the questions better. Carefully add your score to be sure you got all the points you deserve. Sometimes graders make simple mathematical errors in adding up points.

## L.2 Thinking About Thinking

Perhaps the most valuable skill you can learn in any of your classes is the ability to think clearly, creatively, and purposefully. In a rapidly moving field such as environmental science, facts and explanations change constantly. It's often said that in six years approximately half the information you learn from this class will be obsolete. During your lifetime you will probably change careers four to six times. Unfortunately, we don't know which of the ideas we now hold will be outdated or what qualifications you will need for those future jobs. Developing the ability to learn new skills, examine new facts, evaluate new theories, and formulate your own interpretations is essential to keep up in a changing world. In other words, you need to learn how to learn on your own.

Even in our everyday lives most of us are inundated by a flood of information and misinformation. Competing claims and contradictory ideas battle for our attention. The rapidly growing complexity of our world and our lives intensifies the difficulties in knowing what to believe or how to act. Consider how the communications revolution has brought us computers, e-mail, cell phones, mobile faxes, pagers, the World Wide Web, hundreds of channels of satellite TV, and direct mail or electronic marketing that overwhelm us with conflicting information. We have more choices than we can possibly manage, and know more about the world around us than ever before but, perhaps, understand less. How can we deal with the barrage of often contradictory news and advice that inundates us?

To complicate our difficulty in knowing what to believe, distinguished authorities disagree vehemently about many important topics. A law of environmental science might be that for any expert there is always an equal and opposite expert. How can you decide what is true and meaningful in such a welter of confusing information? Is it simply a matter of what feels good at the moment or supports our preconceived notions? Or are there ways to use logical, orderly, creative thinking procedures to reach decisions?

By now, most of us know not to believe everything we read or hear (fig. L.4). "Tastes great . . . Low, low sale price . . . Vote for me . . . Lose 30 pounds in 3 weeks . . . You may already be a winner . . . Causes no environmental harm . . . I'll never lie to you . . . Two out of three doctors recommend . . ." More and more of the information we use to buy, elect, advise, judge, or heal has been created not to expand our knowledge but to sell a product or advance a cause. It would be unfortunate if we become cynical and apathetic due to information overload. It does make a difference what we think and how we act.

**FIGURE L.4** "There is absolutely no cause for alarm at the nuclear plant!"

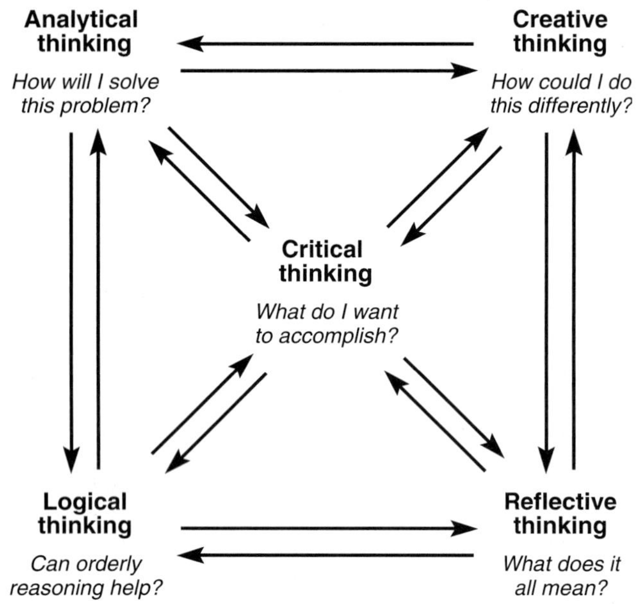


**FIGURE L.5** Different approaches to thinking are used to solve different kinds of problems or to study alternate aspects of a single issue.

## Approaches to truth and knowledge

A number of skills, attitudes, and approaches can help us evaluate information and make decisions. **Analytical thinking** asks, "How can I break this problem down into its constituent parts?" **Creative thinking** asks, "How might I approach this problem in new and inventive ways?" **Logical thinking** asks, "How can orderly, deductive reasoning help me think clearly?" **Critical thinking** asks, "What am I trying to accomplish here and how will I know when I've succeeded?" **Reflective thinking** asks, "What does it all mean?" In this section, we'll look more closely at critical and reflective thinking as a foundation for your study of environmental science. We hope you will apply these ideas consistently as you read this book.

As figure L.5 suggests, critical thinking is central in the constellation of thinking skills. It challenges us to examine theories, facts, and options in a systematic, purposeful, and responsible manner. It shares many methods and approaches with other methods of reasoning but adds some important contextual skills, attitudes, and dispositions. Furthermore, it challenges us to plan methodically and to assess the process of thinking as well as the implications of our decisions. Thinking critically can help us discover hidden ideas and means, develop strategies for evaluating reasons and conclusions in arguments, recognize the differences between facts and values, and avoid jumping to conclusions. Professor Karen J. Warren of Macalester College identifies ten steps in critical thinking (table L.4).

Notice that many critical thinking processes are self-reflective and self-correcting. This form of thinking is sometimes called "thinking about thinking." It is an attempt to plan rationally how to analyze a problem, to monitor your progress while you are doing it, and to evaluate how your strategy worked and what you have learned when you are finished. It is not critical in the sense of finding fault, but it makes a conscious, active, disciplined effort to be aware of hidden motives and assumptions, to uncover bias, and to recognize the reliability or unreliability of sources (What Do You Think? p. 9).

**TABLE L.4**

**Steps in Critical Thinking**

1. What is the *purpose* of my thinking?
2. What precise *question* am I trying to answer?
3. Within what *point of view* am I thinking?
4. What *information* am I using?
5. How am I *interpreting* that information?
6. What *concepts* or ideas are central to my thinking?
7. What *conclusions* am I aiming toward?
8. What am I taking for granted; what *assumptions* am I making?
9. If I accept the conclusions, what are the *implications*?
10. What would the *consequences* be, if I put my thoughts into action?

**Source:** Courtesy of Karen Warren, Philosophy Department, Macalester College, St. Paul, MN.

## What do I need to think critically?

Certain attitudes, tendencies, and dispositions are essential for well-reasoned analysis. Professor Karen Warren suggests the following list:

- *Skepticism and independence.* Question authority. Don't believe everything you hear or read—including this book; even experts sometimes are wrong.
- *Open-mindedness and flexibility.* Be willing to consider differing points of view and to entertain alternative explanations. Try arguing from a viewpoint different from your own. It will help you identify weaknesses and limitations in your own position.
- *Accuracy and orderliness.* Strive for as much precision as the subject permits or warrants. Deal systematically with parts of a complex whole. Be disciplined in the standards you apply.
- *Persistence and relevance.* Stick to the main point. Don't allow diversions or personal biases to lead you astray. Information may be interesting or even true, but is it relevant?
- *Contextual sensitivity and empathy.* Consider the total situation, relevant context, feelings, level of knowledge, and sophistication of others as you evaluate information. Imagine being in someone else's place to try to understand how they feel.

## What Do You Think?

### Don't Believe Everything You See or Hear on the News

For most of us, access to news is becoming ever more abundant and ubiquitous. Internet web logs comment on events even as they're happening. Cable television news is available around the clock. Live images are projected to our homes from all over the world. We watch video coverage of distant wars and disasters as if they are occurring in our living rooms, but how much do we really know about what's going on? At the same time that media is becoming more technically sophisticated, news providers are also becoming more adept at manipulating images and content to convey particular messages.

Many people watch TV news programs and read newspapers or web logs today not so much to be educated or to get new ideas as to reinforce their existing beliefs. A State of the Media study by the Center for Journalistic Excellence at Columbia University concluded that the news is becoming increasingly partisan and ideological.[1] The line between news and entertainment has become blurred in most media. Disputes and disasters are overdramatized, while too little attention is paid to complex issues. News reports are increasingly shallow and one-sided, with little editing or fact checking. On live media, such as television and radio, attack journalism is becoming ever more common. Participants try to ridicule and demean their opponents rather than listening respectfully and comparing facts and sources. Many shows simply become people shouting at each other. Print media also is moving toward tabloid journalism, featuring many photographs and sensationalist coverage of events.

According to the State of the Media report, most television stations have all but abandoned the traditional written and edited news story. Instead, more than two-third of all news segments now consist of on-site "stand-up" reports or live interviews in which a single viewpoint is presented as news without any background or perspective. Visual images seem more immediate and are regarded as more believable by most people: after all, pictures don't lie, but they can give a misleading impression of what's really important. Many topics, such as policy issues, don't make good visuals, and therefore never make it into TV coverage. Crime, accidents, disasters, lifestyle stories, sports, and weather make up more than 90 percent of the coverage on a typical television news program. If you watched cable TV news for an entire day, for instance, you'd see, on average, only 1 minute each about the environment and health care, 2 minutes each on science and education, and 4 minutes on art and culture. More than 70 percent of the segments are less than 1 minute long, meaning that they convey more emotion than substance. People who get their news primarily from TV are significantly more fearful and pessimistic than those who get news from print media.

Partisan journalism has become much more prevalent since the deregulation of public media. From the birth of the broadcasting industry, the airwaves were regarded and regulated as a public trust. Broadcasters, as a condition of their licenses, were required to operate in the "public interest" by covering important policy issues and providing equal time to both sides of contested issues. In 1988, however, the Federal Communications Commission ruled that the proliferation of mass media gives the public adequate access to diverse sources of information. Media outlets no longer are obliged to provide fair and balanced coverage of issues. Presenting a single perspective or even a deceptive version of events is no longer regarded as a betrayal of public trust.

A practice that further erodes the honesty and truthfulness of media coverage is the use of video news releases that masquerade as news stories. In these videos, actors, hired by public relations firms, pose as reporters or experts to promote a special interest. Businesses have long used this tactic to sell products, but a recent disturbing development is placement of news videos by governmental agencies. For example, in 2004, the federal Department of Health and Human Services sent video stories to TV stations promoting the benefits of the recently passed but controversial Medicare drug law. The actors in these videos appeared to be simply reporting news, but, in fact, were presenting a highly partisan viewpoint. Critics complained that these "stealth ads" undermine the credibility of both journalists and public officials. Kevin W. Keane, a Health Department spokesman, dismissed the criticism, saying this is "a common, routine practice in government and the private sector." In 2004, the federal government paid $88 million to public relations firms and news commentators to represent administration positions on policy issues.

How can you detect bias in a news report? Ask yourself the following questions:

1. What political positions are represented in the story?
2. What special interests might be involved here? Who stands to gain presenting a particular viewpoint? Who is paying for the message?
3. What sources are used as evidence in this story? How credible are they?
4. Are statistics cited in the presentation? Are citations provided so you can check the source?
5. Is the story one-sided, or are alternate viewpoints presented? Are both sides represented by credible spokespersons, or is one simply a patsy set up to make the other side look good?
6. Are the arguments presented based on facts and logic, or are they purely emotional appeals?

We need to practice critical thinking to detect bias and make sense out of what we see and hear. Although the immediacy and visual impact of television or the Internet may seem convincing, we have to use caution and judgment to interpret the information they present. Don't depend on a single source for news. Compare what different media outlets say about an issue before making up your mind.

[1]The State of the News Media 2004 available at http://www.journalism.org.

- *Decisiveness and courage.* Draw conclusions and take a stand when the evidence warrants doing so. Although we often wish for more definitive information, sometimes a well-reasoned but conditional position has to be the basis for action.
- *Humility.* Realize that you may be wrong and that reconsideration may be called for in the future. Be careful about making absolute declarations; you may need to change your mind someday.

While critical thinking shares many of the orderly, systematic approaches of formal logic, it also invokes traits like empathy, sensitivity, courage, and humility. Formulating intelligent opinions about some of the complex issues you'll encounter in environmental science requires more than simple logic. Developing these attitudes and skills is not easy or simple. It takes practice. You have to develop your mental faculties just as you need to train for a sport. Traits such as intellectual integrity, modesty, fairness, compassion, and fortitude are not things you can use only occasionally. They must be cultivated until they become your normal way of thinking.

## Applying critical thinking

We all use critical or reflective thinking at times. Suppose a television commercial tells you that a new breakfast cereal is tasty and good for you. You may be suspicious and ask yourself a few questions. What do they mean by good? Good for whom or what? Does "tasty" simply mean more sugar and salt? Might the sources of this information have other motives in mind besides your health and happiness? Although you may not have been aware of it, you already have been using some of the techniques of critical analysis. Working to expand these skills helps you recognize the ways information and analysis can be distorted, misleading, prejudiced, superficial, unfair, or otherwise defective.

Here are some steps in critical thinking:

1. *Identify and evaluate premises and conclusions in an argument.* What is the basis for the claims made here? What evidence is presented to support these claims and what conclusions are drawn from this evidence? If the premises and evidence are correct, does it follow that the conclusions are necessarily true?
2. *Acknowledge and clarify uncertainties, vagueness, equivocation, and contradictions.* Do the terms used have more than one meaning? If so, are all participants in the argument using the same meanings? Are ambiguity or equivocation deliberate? Can all the claims be true simultaneously?
3. *Distinguish between facts and values.* Are claims made that can be tested? (If so, these are statements of fact and should be able to be verified by gathering evidence.) Are claims made about the worth or lack of worth of something? (If so, these are value statements or opinions and probably cannot be verified objectively.) For example, claims of what we *ought* to do to be moral or righteous or to respect nature are generally value statements.
4. *Recognize and assess assumptions.* Given the backgrounds and views of the protagonists in this argument, what underlying reasons might there be for the premises, evidence, or conclusions presented? Does anyone have an "axe to grind" or a personal agenda in this issue? What do they think you know, need, want, or believe? Is there a subtext based on race, gender, ethnicity, economics, or some belief system that distorts this discussion?
5. *Distinguish the reliability or unreliability of a source.* What makes the experts qualified in this issue? What special knowledge or information do they have? What evidence do they present? How can we determine whether the information offered is accurate, true, or even plausible?
6. *Recognize and understand conceptual frameworks.* What are the basic beliefs, attitudes, and values that this person, group, or society holds? What dominating philosophy or ethics control their outlook and actions? How do these beliefs and values affect the way people view themselves and the world around them? If there are conflicting or contradictory beliefs and values, how can these differences be resolved?

## Some clues for unpacking an argument

In logic, an argument is made up of one or more introductory statements (called **premises**), and a **conclusion** that supposedly follows logically from the premises. Often in ordinary conversation, different kinds of statements are mixed together, so it is difficult to distinguish between them or to decipher hidden or implied meanings. Social theorists call the process of separating and analyzing textual components *unpacking.* Applying this type of analysis to an argument can be useful.

An argument's premises are usually claimed to be based on facts; conclusions are usually opinions and values drawn from, or used to interpret, those facts. Words that often introduce a premise include: *as, because, assume that, given that, since, whereas,* and *we all know that* . . . Words that generally indicate a conclusion or statement of opinion or values include: *and, so, thus, therefore, it follows that, consequently, the evidence shows,* and *we can conclude that.*

For instance, in the example we used earlier, the television ad might have said: "*Since* we all need vitamins, and *since* this cereal contains vitamins, *consequently* the cereal must be good for you." Which are the premises and which is the conclusion? Does one necessarily follow from the other? Remember that even if the facts in a premise are correct, the conclusions drawn from them may not be. Information may be withheld from the argument such as the fact that the cereal is also loaded with unhealthy amounts of sugar.

## Avoiding logical errors and fallacies

Formal logic catalogs a large number of fallacies and errors that invalidate arguments. Although we don't have room here to include all of these fallacies and errors, it may be helpful to review a few of the more common ones.

- *Red herring:* Introducing extraneous information to divert attention from the important point.
- *Ad hominem attacks:* Criticizing the opponent rather than the logic of the argument.
- *Hasty generalization:* Drawing conclusions about all members of a group based on evidence that pertains only to a selected sample.

- *False cause:* Drawing a link between premises and conclusions that depends on some imagined causal connection that does not, in fact, exist.
- *Appeal to ignorance:* Because some facts are in doubt, therefore a conclusion is impossible.
- *Appeal to authority:* It's true because _______ says so.
- *Begging the question:* Using some trick to make a premise seem true when it is not.
- *Equivocation:* Using words with double meanings to mislead the listener.
- *Slippery slope:* A claim that some event or action will cause some subsequent action.
- *False dichotomy:* Giving either/or alternatives as if they are the only choices.

Avoiding these fallacies yourself or being aware of them in another's argument can help you be more logical and have more logical and reasonable discussions.

### Using critical thinking in environmental science

As you go through this book, you will have many opportunities to practice critical thinking skills. Every chapter includes many facts, figures, opinions, and theories. Are all of them true? No, probably not. They were the best information available when this text was written, but much in environmental science is in a state of flux. Data change constantly as does our interpretation of them. Do the ideas presented here give a complete picture of the state of our environment? Unfortunately, they probably don't. No matter how comprehensive our discussion is of this complex, diverse subject, it can never capture everything worth knowing, nor can it reveal all possible points of view.

When reading this text, try to distinguish between statements of fact and opinion. Ask yourself if the premises support the conclusions drawn from them. Although we have tried to be fair and even-handed in presenting controversies, we, like everyone, have biases and values—some of which we may not even recognize—that affect how we see issues and present arguments. Watch for cases in which you need to think for yourself and utilize your critical and reflective thinking skills to find the truth.

## L.3 Concept Maps

Concept mapping is a learning strategy that many students find useful in understanding complex ideas and clarifying ambiguous relationships. Creating a graphic representation of a topic often can help you visualize key concepts and organize your knowledge more clearly than will other methods of study. You may find that the physical process of drawing a map of a topic engages a different part of your brain than does ordinary reading or taking notes. Taking time to think carefully about what is most important about a particular topic will help you remember it better at a later date. It can also point out weaknesses in your understanding as well as areas in which you need more study. Practicing this technique as you think about environmental science can help you comprehend difficult material and prepare for exams.

From ancient times, people have made maps to help them understand and remember important aspects of the world around them. Maps integrate and summarize knowledge. They suggest linkages that we may not have seen before, and they suggest routes for further exploration. But maps can never record every possible detail of the world or the ideas they represent. Only the most useful and meaningful information is put onto the map so that important points can easily be seen and remembered. No map is ever complete and finished. As we learn more, we revise our maps, correcting errors, adding new information, removing unnecessary features, and refining the presentation. The act of drawing a map exposes doubtful knowledge and shows us where we need more data and a clearer understanding.

If you ask different people to construct a map of the same area of a city, you would probably get very different results. A young child, for example, might draw in only the locations of her school, home, and playground. A commuter from the suburbs, on the other hand, might show the locations of the major office buildings, filling stations, freeway ramps, and the cheapest parking lots in that same city. Neither of these representations is wrong, they just emphasize the aspects of greatest importance to each person.

When we think about maps, we usually visualize physical features such as mountains, lakes, roads, buildings, and so forth. But we can also create maps of biodiversity, magnetic fields, economic flows, cultures, language families, or anything else that can be presented in graphical form. A **concept map** is a two-dimensional representation of the relationship between key ideas. It shows us how we think and suggests affinities and associations that might not otherwise be obvious. At first glance, a concept map looks like a flow chart in which key terms are placed in boxes connected by directional arrows. Based on educational psychology theories of how we organize information, concept maps are hierarchical, with broader, more general items at the top and more specific topics arranged in a cascade below them. They are metacognitive tools that empower the learner to take charge of his/her own learning in a highly organized and meaningful manner.

### How do I create a concept map?

To create a concept map, start with what you already know. Build from what's familiar. What are the key components or ideas in the topic you're trying to understand? Place each concept in its own individual circle, box, or other geometrical shape. You might want to use different shapes to indicate relative levels or types of ideas. Connect concept boxes with directional arrows to show relationships. Label each arrow with descriptive terms so that your diagram can be read as a statement or proposition by following interconnections from the top down. In figure L.6, for example, you can read the proposition that "concept maps are used to develop study strategies that lead to learning that determines your grade" as one set of associations.

As you can see in this example, branches to one side or the other of the key concepts show related ideas. Where appropriate, cross links or bridges can connect branches of your map. Linking arrows can be bidirectional to indicate mutual interactions, but be careful not to make everything connected to everything else. Focus on the most important concepts and the most significant relationships. View this as an exercise in discrimination. Don't try to make your map perfect. Sometimes working briskly will help you cut away the superfluous while fostering creativity and synthesis. The point is not to create a work of art but to organize, discover, and understand central meanings. The map helps you learn how to learn.

A concept map can show just a small part or a subset of a broader field of knowledge. The top or key concept in one map may be subsumed into a lower position in a map with a different focus. A small branch in a general map could be expanded into a much more specific map of its own.

Remember that concept maps are works in progress. There are no right or wrong maps. Each one represents one possible way of understanding a particular set of ideas by an individual at the time the map was made. Expect to do several iterations, right from the outset. There are probably as many ways of representing a collection of concepts as there are concepts in the collection, but some—typically those discovered by a process of trial and error—are more elegant and easier to understand than others.

The benefits of mapping are mainly to the individual making the map. The process of simplifying concepts and arranging them on a page forces you to think about what's most important.

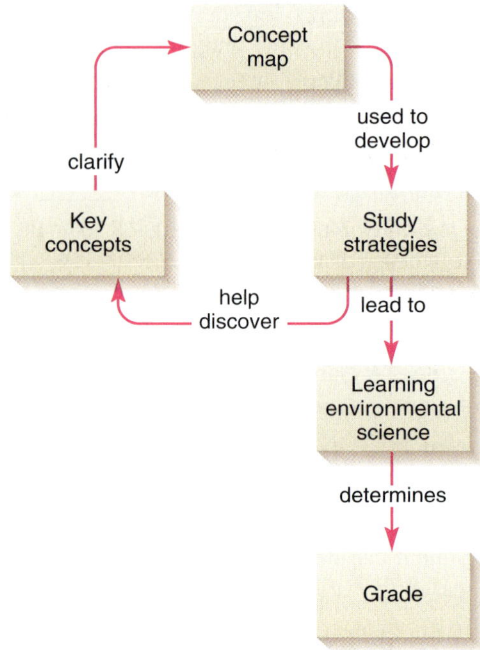


**FIGURE L.6** A concept map can help you organize your thinking.

It helps you clarify your thoughts and understandings and makes learning more meaningful. A concept map can be a heuristic device; that is, a process in which you make discoveries and uncover meanings through trial and error. Although you benefit most directly from making your own map, it can be instructive to compare your map to that of a fellow student to see how her or his take on a topic compares to yours.

## Conclusion

Whether you find environmental science interesting and useful depends largely on your own attitudes and efforts. Developing good study habits, setting realistic goals for yourself, taking the initiative to look for interesting topics, finding an appropriate study space, and working with a study partner can both make your study time more efficient and improve your final grade. Each of us has his or her own learning style. You may understand and remember things best if you see them in writing, hear them spoken by someone else, reason them out for yourself, or learn by doing. By determining your preferred style, you can study in the way that is most comfortable and effective for you.

## Practice Quiz

1. Which study skills in table L.1 do you need to improve?
2. Describe some ways you can avoid procrastination and keep on schedule.
3. List four learning styles. Which fits you best?
4. Describe the SQ3R study techniques.
5. What are ten steps in critical thinking?
6. Name (and describe) seven attributes or dispositions essential for critical thinking.
7. List some adverbs or adverbial phrases that introduce premises and conclusions.
8. Distinguish between an *ad hominem* attack and an appeal to ignorance.
9. Describe three questions you'd ask to evaluate the reliability of Internet information.

## Critical Thinking and Discussion Questions

1. What is critical thinking? Why is it important?
2. Suppose your preferred learning style doesn't match the teaching style of your instructor. How might you find a common ground?
3. You may find that you fit two or more of the learning profiles described in this chapter. How would you decide which is most effective or appropriate for you?
4. Why is critical thinking important in environmental science?
5. What is the difference between critical and reflective thinking?
6. Suppose you see a claim on television or the Internet, how would you evaluate its reliability?
7. Why are empathy and contextual sensitivity important in critical thinking?
8. If some facts in this book are vague or doubtful, why mention them at all? How would you decide which "facts" to use and which to ignore?
9. Why are concept maps usually hierarchical and linked by active verbs?

As part of it's green Olympics effort, China is planting trees to stabilize soil, reduce erosion and flooding, and hold back the dust storms that plague northern China. In 2006, some 500 million Chinese planted more than 2 billion trees.

CHAPTER

# 1 Understanding Our Environment

*Today we are faced with a challenge that calls for a shift in our thinking, so that humanity stops threatening its life-support system.*

*—Wangari Maathai—*
*Winner of 2004 Nobel Peace Prize*

## LEARNING OUTCOMES

After studying this chapter, you should be able to:

**1.1** Define environmental science and identify some important environmental concerns we face today.

**1.2** Discuss the history of conservation and the different attitudes toward nature at various times in our past.

**1.3** Think critically about the major environmental dilemmas and issues that shape our current environmental agenda.

**1.4** Appreciate the human dimensions of environmental science, including the connection between poverty and environmental degradation.

**1.5** Explain sustainable development and evaluate some of its requirements.

# Case Study **A Green Olympics?**

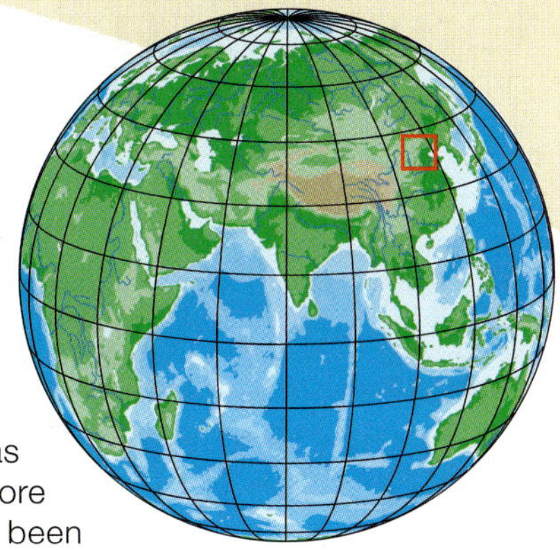

The People's Republic of China has ambitious plans to make the 2008 summer Olympics the "greenest" ever. Plans for sustainable resource use and minimal environmental effects include buildings constructed with environmentally friendly materials and the latest energy-saving technology. Solar water heaters will provide hot showers for athletes, while windmills and photovoltaic cells will contribute 20 percent of the electricity used in the Olympic Village. Rain will be collected and used to water playing fields. Toilets will use recycled water. New wastewater treatment plants will reduce sewage effluents, and two new recycling plants will reduce solid waste disposal. Reforestation projects and fuel-switching are expected to offset carbon dioxide emissions and make the games climate neutral.

Although critics doubt how effective these measures will be, the fact that they're being mentioned at all is a hopeful sign. A generation ago, environmental conditions weren't even mentioned by Chinese officials. Mere survival dominated the political agenda. Providing enough food, jobs, and housing for a rapidly growing population and stabilizing a chaotic political system preoccupied public attention. Now, at least environmental quality and resource use is something that leaders feel they need to address.

**FIGURE 1.1** Sixteen of the 20 smoggiest cities in the world are in China. Sulfur dioxide emissions are highest in the world, and the number of Chinese automobiles increases by about 5 million per year.

China has many serious environmental challenges. Seventy percent of all Chinese cities don't meet national air quality standards. According to the United Nations Environment Programme, 16 of the 20 smoggiest cities in the world are in China, and one-third of the country is affected by acid precipitation (fig. 1.1). It's estimated that 400,000 people die each year from the effects of air pollution. China is now the world's largest producer of sulfur dioxide, chlorofluorocarbons, and carbon dioxide. Seventy percent of Chinese rivers and lakes fail to meet government water quality standards. About half the surface water is so contaminated that it isn't useful even for agriculture. Two-thirds of Chinese cities don't have enough water to meet demands, and at least 300 million people live in areas with severe water shortages. One-third of China's land has been degraded by unsustainable farming, grazing, and logging, and 400 million people are threatened by expanding deserts.

But there also is positive news among this litany of woes. In addition to planning for a green Olympics, Chinese leaders have pledged to take steps to improve environmental quality for the whole country. They plan to spend at least (U.S.)$125 billion over the next five years to reduce water pollution and bring clean drinking water to everyone in the country by 2015. Officials also promise that 10 percent of China's energy will come from renewable sources within a decade. Logging on steep hillsides has been officially banned, and more than 50 billion trees have been planted on 500,000 km$^2$ of marginal land to hold back deserts and reduce blowing dust. Construction has already begun on new cities designed to be largely self-sufficient in energy, water, and food and to be climate neutral with respect to greenhouse gases. Although there are doubts about whether the government can actually deliver on all these promises, it's encouraging that they're moving toward sustainability.

Some of the new-found governmental concern about the Chinese environment comes from public demand. In 2005 there were at least 85,000 public protests in China, many of which were about pollution, environmental health, land degradation, and similar issues. A campaign led by artists, students, and writers forced the government to cancel plans for a series of 13 large dams on the Nu River (the Chinese portion of the Salween) in an area of high biological and cultural diversity. There are now more than 2,000 nongovernmental organizations (NGOs) in China working on social and environmental issues, and for the first time, they have officially recognized status.

But why should you be concerned about what happens half a world away? Part of the answer is that we all share a single planet. What happens in one place affects all of us. You may have noticed that rising Chinese demand for commodities, such as oil, copper, steel, and seafood, are driving up the prices worldwide. Less well known is the fact that on some days, three-fourths of the air pollution in some North American west coast cities can be traced to China. And if China meets its rapidly growing power demand by burning more coal, the effects on our global climate could be disastrous.

Although China currently has the world's largest population, and, therefore, has a huge impact on our global resources and environment, its problems aren't unique. Many of the countries in which the poorer four-fifths of all humans live face similar environmental

# Case Study *continued*

challenges. Finding ways that all of us can live sustainably within the limits of our resource base and without damaging nature's life-support systems is the preeminent challenge of environmental science. And, as is the case with China, while the world faces many serious environmental problems, there are also signs of progress that give us hope for the future.

**For more information, see**

Liu, Jianguo, and Jared Diamond. 2005. China's environment in a globalizing world: How China and the rest of the world affect each other. *Nature* 435:1179–86.

Min, Shao, et al. 2006. City clusters in China: Air and surface water pollution. *Frontiers in Ecological Environment* 4 (7):353–61.

## 1.1 What Is Environmental Science?

Humans have always inhabited two worlds. One is the natural world of plants, animals, soils, air, and water that preceded us by billions of years and of which we are a part. The other is the world of social institutions and artifacts that we create for ourselves using science, technology, and political organization. Both worlds are essential to our lives, but integrating them successfully causes enduring tensions.

Where earlier people had limited ability to alter their surroundings, we now have power to extract and consume resources, produce wastes, and modify our world in ways that threaten both our continued existence and that of many organisms with which we share the planet. To ensure a sustainable future for ourselves and future generations, we need to understand something about how our world works, what we are doing to it, and what we can do to protect and improve it.

**Environment** (from the French *environner:* to encircle or surround) can be defined as (1) the circumstances or conditions that surround an organism or group of organisms, or (2) the complex of social or cultural conditions that affect an individual or community. Since humans inhabit the natural world as well as the "built" or technological, social, and cultural world, all constitute important parts of our environment (fig. 1.2).

**Environmental science,** then, is the systematic study of our environment and our proper place in it. A relatively new field, environmental science is highly interdisciplinary, integrating natural sciences, social sciences, and humanities in a broad, holistic study of the world around us. In contrast to more theoretical disciplines, environmental science is mission-oriented. That is, it seeks new, valid, contextual knowledge about the natural world and our impacts on it, but obtaining this information creates a responsibility to get involved in trying to do something about the problems we have created.

As distinguished economist Barbara Ward pointed out, for an increasing number of environmental issues, the difficulty is not to identify remedies. Remedies are now well understood. The problem is to make them socially, economically, and politically acceptable. Foresters know how to plant trees, but not how to establish conditions under which villagers in developing countries can manage plantations for themselves. Engineers know how to control pollution, but not how to persuade factories to install the necessary equipment. City planners know how to build housing and design safe drinking water systems, but not how to make them affordable for the poorest members of society. The solutions to these problems increasingly involve human social systems as well as natural science.

Criteria for environmental literacy suggested by the National Environmental Education Advancement Project in Wisconsin include: awareness and appreciation of the natural and built environment; knowledge of natural systems and ecological concepts; understanding of current environmental issues; and the ability to use critical-thinking and problem-solving skills on environmental issues. These are good overall goals to keep in mind as you study this book.

Chapter 2 looks more closely at science as a way of knowing, environmental ethics, and other tools that help us analyze and understand the world around us. For the remainder of this chapter, we'll complete our overview with a short history of environmental thought and a survey of some important current issues that face us.

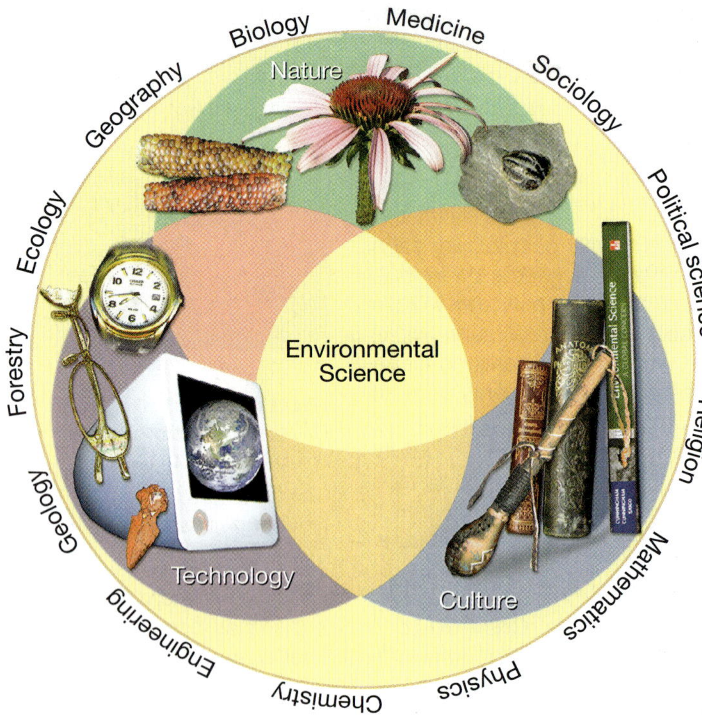


**FIGURE 1.2** The intersections of the natural, cultural, and technological worlds outline the province of environmental science. Many disciplines contribute to our understanding and management of our environment.

## 1.2 A Brief History of Conservation and Environmentalism

Although many early societies had negative impacts on their surroundings, others lived in relative harmony with nature. In modern times, however, growing human populations and the power of our technology have increased our impacts on our environment. We can divide conservation history and environmental activism into at least four distinct stages: (1) pragmatic resource conservation, (2) moral and aesthetic nature preservation, (3) a growing concern about health and ecological damage caused by pollution, and (4) global environmental citizenship. Each era focused on different problems and each suggested a distinctive set of solutions. These stages are not necessarily mutually exclusive, however; parts of each persist today in the environmental movement and one person may embrace them all simultaneously.

### Nature protection has historic roots

Recognizing human misuse of nature is not unique to modern times. Plato complained in the fourth century B.C. that Greece once was blessed with fertile soil and clothed with abundant forests of fine trees. After the trees were cut to build houses and ships, however, heavy rains washed the soil into the sea, leaving only a rocky "skeleton of a body wasted by disease." Springs and rivers dried up while farming became all but impossible. Many classical authors regarded Earth as a living being, vulnerable to aging, illness, and even mortality. Periodic threats about the impending death of nature as a result of human misuse have persisted into our own time. Many of these dire warnings have proven to be premature or greatly exaggerated, but others remain relevant to our own times. As Mostafa K. Tolba, former Executive Director of the United Nations Environment Programme has said, "The problems that overwhelm us today are precisely those we failed to solve decades ago."

Some of the earliest scientific studies of environmental damage were carried out in the eighteenth century by French and British colonial administrators who often were trained scientists and who considered responsible environmental stewardship as an aesthetic and moral priority, as well as an economic necessity. These early conservationists observed and understood the connection between deforestation, soil erosion, and local climate change. The pioneering British plant physiologist, Stephen Hales, for instance, suggested that conserving green plants preserved rainfall. His ideas were put into practice in 1764 on the Caribbean island of Tobago, where about 20 percent of the land was marked as "reserved in wood for rains."

Pierre Poivre, an early French governor of Mauritius, an island in the Indian Ocean, was appalled at the environmental and social devastation caused by destruction of wildlife (such as the flightless dodo) and the felling of ebony forests on the island by early European settlers. In 1769, Poivre ordered that one-quarter of the island was to be preserved in forests, particularly on steep mountain slopes and along waterways. Mauritius remains a model for balancing nature and human needs. Its forest reserves shelter a larger percentage of its original flora and fauna than most other human-occupied islands.

### Resource waste inspired pragmatic, utilitarian conservation

Many historians consider the publication of *Man and Nature* in 1864 by geographer George Perkins Marsh as the wellspring of environmental protection in North America. Marsh, who also was a lawyer, politician, and diplomat, traveled widely around the Mediterranean as part of his diplomatic duties in Turkey and Italy. He read widely in the classics (including Plato) and personally observed the damage caused by the excessive grazing by goats and sheep and by the deforesting of steep hillsides. Alarmed by the wanton destruction and profligate waste of resources still occurring on the American frontier in his lifetime, he warned of its ecological consequences. Largely as a result of his book, national forest reserves were established in the United States in 1873 to protect dwindling timber supplies and endangered watersheds.

Among those influenced by Marsh's warnings were President Theodore Roosevelt (fig. 1.3*a*) and his chief conservation advisor, Gifford Pinchot (fig. 1.3*b*). In 1905, Roosevelt, who was the leader of the populist, progressive movement, moved

(a) President Teddy Roosevelt (b) Gifford Pinchot

(c) John Muir (d) Aldo Leopold

**FIGURE 1.3** Some early pioneers of the American conservation movement. President Teddy Roosevelt (a) and his main advisor Gifford Pinchot (b) emphasized pragmatic resource conservation, while John Muir (c) and Aldo Leopold (d) focused on ethical and aesthetic relationships.

the Forest Service out of the corruption-filled Interior Department into the Department of Agriculture. Pinchot, who was the first native-born professional forester in North America, became the founding head of this new agency. He put resource management on an honest, rational, and scientific basis for the first time in our history. Together with naturalists and activists such as John Muir, William Brewster, and George Bird Grinnell, Roosevelt and Pinchot established the framework of our national forest, park, and wildlife refuge systems, passed game protection laws, and tried to stop some of the most flagrant abuses of the public domain. In 1908, Pinchot organized and chaired the White House Conference on Natural Resources, perhaps the most prestigious and influential environmental meeting ever held in the United States.

The basis of Roosevelt's and Pinchot's policies was pragmatic **utilitarian conservation.** They argued that the forests should be saved "not because they are beautiful or because they shelter wild creatures of the wilderness, but only to provide homes and jobs for people." Resources should be used "for the greatest good, for the greatest number for the longest time." "There has been a fundamental misconception," Pinchot said, "that conservation means nothing but husbanding of resources for future generations. Nothing could be further from the truth. The first principle of conservation is development and use of the natural resources now existing on this continent for the benefit of the people who live here now. There may be just as much waste in neglecting the development and use of certain natural resources as there is in their destruction." This pragmatic approach still can be seen in the multiple use policies of the Forest Service.

## Ethical and aesthetic concerns inspired the preservation movement

John Muir (fig. 1.3*c*), geologist, author, and first president of the Sierra Club, strenuously opposed Pinchot's influence and policies. Muir argued that nature deserves to exist for its own sake, regardless of its usefulness to us. Aesthetic and spiritual values formed the core of his philosophy of nature protection. This outlook has been called **biocentric preservation** because it emphasizes the fundamental right of other organisms to exist and to pursue their own interests. Muir wrote: "The world, we are told, was made for man. A presumption that is totally unsupported by the facts. . . . Nature's object in making animals and plants might possibly be first of all the happiness of each one of them. . . . Why ought man to value himself as more than an infinitely small unit of the one great unit of creation?"

Muir, who was an early explorer and interpreter of the Sierra Nevada Mountains in California, fought long and hard for establishment of Yosemite and Kings Canyon National Parks. The National Park Service, established in 1916, was first headed by Muir's disciple, Stephen Mather, and has always been oriented toward preservation of nature in its purest state. It has often been at odds with Pinchot's utilitarian Forest Service. Environmental ethics is discussed further in chapter 2.

**FIGURE 1.4** Aldo Leopold's Wisconsin shack, the main location for his *Sand County Almanac,* in which he wrote, "A thing is right when it tends to preserve the integrity, stability, and beauty of the biotic community. It is wrong when it tends otherwise." How might you apply this to your life?

In 1935, pioneering wildlife ecologist Aldo Leopold (fig. 1.3*d*) bought a small, worn-out farm in central Wisconsin. A dilapidated chicken shack, the only remaining building, was remodeled into a rustic cabin (fig. 1.4). Working together with his children, Leopold planted thousands of trees in a practical experiment in restoring the health and beauty of the land. "Conservation," he wrote, "is the positive exercise of skill and insight, not merely a negative exercise of abstinence or caution." The shack became a writing refuge and became the main focus of *A Sand County Almanac,* a much beloved collection of essays about our relation with nature. In it, Leopold wrote, "We abuse land because we regard it as a commodity belonging to us. When we see land as a community to which we belong, we may begin to use it with love and respect." Together with Bob Marshall and two others, Leopold was a founder of the Wilderness Society.

## Rising pollution levels led to the modern environmental movement

The undesirable effects of pollution probably have been recognized at least as long as those of forest destruction. In 1273, King Edward I of England threatened to hang anyone burning coal in London because of the acrid smoke it produced. In 1661, the English diarist John Evelyn complained about the noxious air pollution caused by coal fires and factories and suggested that sweet-smelling trees be planted to purify city air. Increasingly dangerous smog attacks in Britain led, in 1880, to formation of a national Fog and Smoke Committee to combat this problem.

The tremendous industrial expansion during and after the Second World War added a new set of concerns to the environmental agenda. *Silent Spring,* written by Rachel Carson (fig. 1.5*a*) and published in 1962, awakened the public to the threats of pollution and toxic chemicals to humans as well as other species.

(a) Rachel Carson (b) David Brower

(c) Barry Commoner (d) Wangari Maathai

**FIGURE 1.5** Among many distinguished environmental leaders in modern times, Rachel Carson (a), David Brower (b), Barry Commoner (c), and Wangari Maathai (d) stand out for their dedication, innovation, and bravery.

The movement she engendered might be called **environmentalism** because its concerns are extended to include both environmental resources and pollution. Among the pioneers of this movement were activist David Brower (fig. 1.5*b*) and scientist Barry Commoner (fig. 1.5*c*). Brower, while executive director of the Sierra Club, Friends of the Earth, and the Earth Island Institute, introduced many of the techniques of modern environmentalism, including litigation, intervention in regulatory hearings, book and calendar publishing, and using mass media for publicity campaigns. Commoner, who was trained as a molecular biologist, has been a leader in analyzing the links between science, technology, and society. Both activism and research remain hallmarks of the modern environmental movement.

In 1977, Professor Wangari Maathai (fig. 1.5*d*) founded the Green Belt Movement in her native Kenya as a way of both organizing poor rural women and restoring their environment. Beginning at a small, local scale, this organization has grown to more than 600 grassroots networks across Kenya. They have planted more than 30 million trees while mobilizing communities for self-determination, justice, equity, poverty reduction, and environmental conservation. Dr. Maathai was elected to the Kenyan Parliament and served as Assistant Minister for Environment and Natural Resources. Her leadership has helped bring democracy and good government to her country. In 2004, she received the Nobel Peace Prize for her work, the first time a Nobel has been awarded for environmental action. In her acceptance speech, she said, "Working together, we have proven that sustainable development is possible; that reforestation of degraded land is possible; and that exemplary governance is possible when ordinary citizens are informed, sensitized, mobilized and involved in direct action for their environment."

Under the leadership of a number of other brilliant and dedicated activists and scientists, the environmental agenda was expanded in the 1960s and 1970s to include issues such as human population growth, atomic weapons testing and atomic power, fossil fuel extraction and use, recycling, air and water pollution, wilderness protection, and a host of other pressing problems that are addressed in this textbook. Environmentalism has become well established on the public agenda since the first national Earth Day in 1970. A majority of Americans now consider themselves environmentalists, although there is considerable variation in what that term means.

***Think About It***

Suppose a beautiful grove of trees near your house is scheduled to be cut down for a civic project such as a swimming pool. Would you support this? Why or why not? Which of the philosophies described in this chapter best describes your attitude?

## Global interconnections have expanded environmentalism

Increased opportunities to travel, as well as greatly expanded international communications, now enable us to know about daily events in places unknown to our parents or grandparents. We have become, as Marshal McLuhan announced in the 1960s, a global village. As in a village, we are all interconnected in various ways. Events that occur on the other side of the globe have profound and immediate effects on our lives.

Photographs of the earth from space (fig. 1.6) provide a powerful icon for the fourth wave of ecological concern that might be called **global environmentalism.** These photos remind us how small, fragile, beautiful, and rare our home planet is. We all share a common environment at this global scale. As our attention shifts from questions of preserving particular landscapes or preventing pollution of a specific watershed or airshed, we begin to worry about the life-support systems of the whole planet.

We now understand that we are changing planetary weather systems and atmospheric chemistry, reducing the natural variety of organisms, and degrading ecosystems in ways that could have devastating effects, both on humans and on all other life-forms. Protecting our environment has become an international cause and it will take international cooperation to bring about many necessary changes.

Some Chinese leaders are part of this global environmental movement. In 2006, Yu Xiaogang was awarded the Goldman Prize, the world's top honor for environmental protection. Yu was recognized for his work on Yunan's Lashi Lake where he brought

**FIGURE 1.6** The life-sustaining ecosystems on which we all depend are unique in the universe, as far as we know.

**FIGURE 1.7** Perhaps the most amazing feature of our planet is its rich diversity of life.

together residents, government officials, and entrepreneurs to protect wetlands, restore fisheries, and improve water quality. He also worked on sustainable development programs, such as women's schools and microcredit loans. His leadership was instrumental in stopping plans for dams on the Nu River, mentioned in the opening case study for this chapter. Another Goldman Prize winner is Dai Qing, who was jailed for her book that revealed the social and environmental costs of the Three Gorges Dam on the Yangtze River.

Other global environmental leaders include Professor Muhammad Yunus of Bangladesh, who won the Nobel Peace Prize in 2006 for his microcredit loan program at the Grameen Bank, and former Norwegian Prime Minister Gro Harlem Brundtland, who chaired the World Commission on Environment and Development, which coined the most widely accepted definition of sustainability.

## 1.3 Current Conditions

As you probably already know, many environmental problems now face us. Before surveying them in the following section, we should pause for a moment to consider the extraordinary natural world that we inherited and that we hope to pass on to future generations in as good—perhaps even better—a condition than when we arrived.

### We live on a marvelous planet

Imagine that you are an astronaut returning to Earth after a long trip to the moon or Mars. What a relief it would be to come back to this beautiful, bountiful planet after experiencing the hostile, desolate environment of outer space. Although there are dangers and difficulties here, we live in a remarkably prolific and hospitable world that is, as far as we know, unique in the universe. Compared to the conditions on other planets in our solar system, temperatures on the earth are mild and relatively constant. Plentiful supplies of clean air, fresh water, and fertile soil are regenerated endlessly and spontaneously by geological and biological cycles (discussed in chapters 3 and 4).

Perhaps the most amazing feature of our planet is the rich diversity of life that exists here. Millions of beautiful and intriguing species populate the earth and help sustain a habitable environment (fig. 1.7). This vast multitude of life creates complex, interrelated communities where towering trees and huge animals live together with, and depend upon, tiny life-forms such as viruses, bacteria, and fungi. Together all these organisms make up delightfully diverse, self-sustaining communities, including dense, moist forests, vast sunny savannas, and richly colorful coral reefs. From time to time, we should pause to remember that, in spite of the challenges and complications of life on earth, we are incredibly lucky to be here. We should ask ourselves: what is our proper place in nature? What *ought* we do and what *can* we do to protect the irreplaceable habitat that produced and supports us?

But we also need to get outdoors and appreciate nature. As author Ed Abbey said, "It is not enough to fight for the land; it is even more important to enjoy it. While you can. While it is still there. So get out there and mess around with your friends, ramble out yonder and explore the forests, encounter the grizz, climb the mountains. Run the rivers, breathe deep of that yet sweet and lucid air, sit quietly for a while and contemplate the precious stillness, that lovely, mysterious and awesome space. Enjoy yourselves, keep your brain in your head and your head firmly attached to your body, the body active and alive."

## We face many serious environmental problems

It's important for you to be aware of current environmental conditions. We'll cover all these issues in subsequent chapters of this book, but here's an overview to get you started. With more than 6.5 billion humans currently, we're adding about 75 million more to the world every year. While demographers report a transition to slower growth rates in most countries, present trends project a population between 8 and 10 billion by 2050. The impacts of that many people on our natural resources and ecological systems is a serious concern.

Water may well be the most critical resource in the twenty-first century. Already at least 1.1 billion people lack an adequate supply of safe drinking water, and more than twice that many don't have modern sanitation. Polluted water and lack of sanitation are estimated to contribute to the ill health of more than 1.2 billion people annually, including the death of 15 million children per year. About 40 percent of the world population lives in countries where water demands now exceed supplies, and by 2025 the UN projects that as many as three-fourths of us could live under similar conditions. Water wars may well become the major source of international conflict in coming decades.

Over the past century, global food production has more than kept pace with human population growth, but there are worries about whether we will be able to maintain this pace. Soil scientists report that about two-thirds of all agricultural lands show signs of degradation. Biotechnology and intensive farming techniques responsible for much of our recent production gains often are too expensive for poor farmers. Can we find ways to produce the food we need without further environmental degradation? And will that food be distributed equitably? In a world of food surpluses, the United Nations estimates that some 850 million people are now chronically undernourished, and at least 60 million face acute food shortages due to natural disasters or conflicts.

How we obtain and use energy is likely to play a crucial role in our environmental future. Fossil fuels (oil, coal, and natural gas) presently provide around 80 percent of the energy used in industrialized countries (fig. 1.8). Supplies of these fuels are diminishing, however, and problems associated with their acquisition and use—air and water pollution, mining damage, shipping accidents, and geopolitics—may limit what we do with remaining reserves. Cleaner renewable energy resources—solar power, wind, geothermal, and biomass—together with conservation, could give us cleaner, less destructive options if we invest in appropriate technology.

Burning fossil fuels, making cement, cultivating rice paddies, clearing forests, and other human activities release carbon dioxide and other so-called "greenhouse gases" that trap heat in the atmosphere. Over the past 200 years, atmospheric $CO_2$ concentrations have increased about 35 percent. By 2100, if current trends continue, climatologists warn that mean global temperatures will probably warm 1.5° to 6°C (2.7°–11°F). Although it's controversial whether specific recent storms were influenced by global warming, climate changes caused by greenhouse gases are very likely to cause increasingly severe weather events including droughts in some areas and floods in others. Melting alpine glaciers and snowfields could threaten water supplies on which millions of people depend.

**FIGURE 1.8** Fossil fuels supply about 80 percent of world commercial energy. They also produce a large percentage of all air pollutants and greenhouse gases, and contribute to economic and political instability.

Already, we are seeing dramatic climate changes in the Antarctic and Arctic where seasons are changing, sea ice is disappearing, and permafrost is melting (fig. 1.9). Rising sea levels

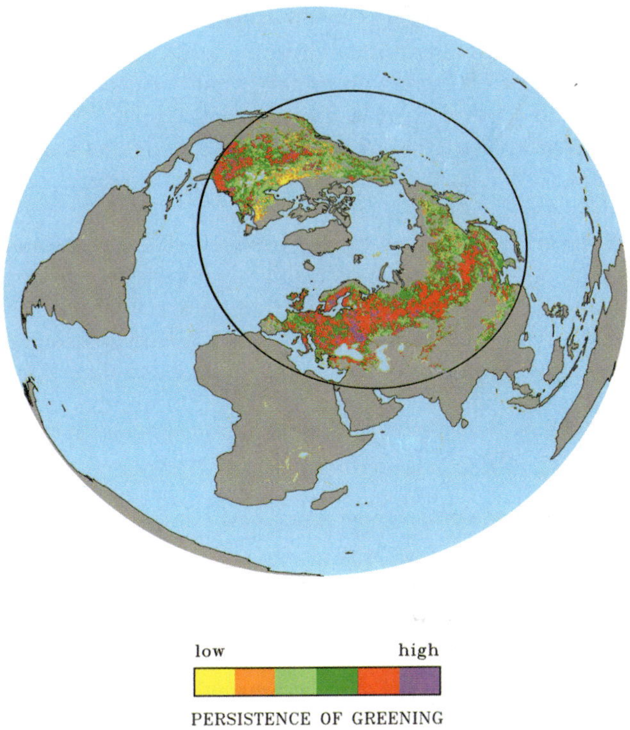


**FIGURE 1.9** Satellite images and surface temperature data show that polar regions, especially in Eurasia, are becoming green earlier and staying green longer than ever in recorded history. This appears to be evidence of a changing global climate.
**Source:** NASA, 2002.

**FIGURE 1.10** At least half of all primates are considered threatened or endangered. Hunting and habitat destruction are the biggest problems.

are flooding low-lying islands and coastal regions, while habitat losses and climatic changes are affecting many biological species. Canadian Environment Minister David Anderson has said that global climate change is a greater threat than terrorism because it could threaten the homes and livelihood of billions of people and trigger worldwide social and economic catastrophe.

Air quality has worsened dramatically in many areas. Over southern Asia, for example, satellite images recently revealed a 3-km (2-mile)-thick toxic haze of ash, acids, aerosols, dust, and photochemical products regularly covers the entire Indian subcontinent for much of the year. Nobel laureate Paul Crutzen estimates that at least 3 million people die each year from diseases triggered by air pollution. Worldwide, the United Nations estimates that more than 2 billion metric tons of air pollutants (not including carbon dioxide or wind-blown soil) are emitted each year. Air pollution no longer is merely a local problem. Mercury, polychlorinated biphenyls (PCB), DDT, and other long-lasting pollutants accumulate in arctic ecosystems and native people after being transported by air currents from industrial regions thousands of kilometers to the south. And during certain days, as much as 75 percent of the smog and particulate pollution recorded on the west coast of North America can be traced to Asia.

Biologists report that habitat destruction, overexploitation, pollution, and introduction of exotic organisms are eliminating species at a rate comparable to the great extinction that marked the end of the age of dinosaurs. The UN Environment Programme reports that over the past century, more than 800 species have disappeared and at least 10,000 species are now considered threatened. This includes about half of all primates and freshwater fish together with around 10 percent of all plant species (fig. 1.10). More than three-quarters of all global fisheries are overfished or harvested at their biological limit. At least half of the forests existing before the introduction of agriculture have been cleared, and much of the diverse "old growth" on which many species depend for habitat, is rapidly being cut and replaced by secondary growth or monoculture. All these biodiversity losses could threaten the ecological life-support systems on which we all depend.

Finding solutions to these problems requires good science as well as individual and collective actions. Becoming educated about our global environment is the first step in understanding how to control our impacts on it. We hope this book will help you in that quest.

***Think About It***

With your classmates or friends, list five important environmental issues in your area. What kinds of actions might you take to improve your local situation?

## There are many signs of hope

Is there hope that we can find solutions to these dilemmas? We think so. As the opening case study for this chapter shows, even countries, such as China, are making progress on social and environmental problems. China now has more than 200,000 wind generators and 10 million biogas generators (most in the world). Solar collectors on 35 million buildings furnish hot water. China could easily get all its energy from renewable sources, and it may be better able to provide advice and technology to other developing countries than can rich nations.

Many cities in Europe and North America are cleaner and much more livable now than they were a century ago. Population has stabilized in most industrialized countries and even in some very poor countries where social security and democracy have been established. Over the last 20 years, the average number of children born per woman worldwide has decreased from 6.1 to 2.7. By 2050, the UN Population Division predicts that all developed countries and 75 percent of the developing world will experience a below-replacement fertility rate of 2.1 children per woman. This prediction suggests that the world population will stabilize at about 8.9 billion rather than 9.3 billion, as previously estimated.

The incidence of life-threatening infectious diseases has been reduced sharply in most countries during the past century, while life expectancies have nearly doubled on average. Smallpox has been completely eradicated and polio has been vanquished except in a few countries. Since 1990, more than 800 million people have gained access to improved water supplies and modern sanitation. In spite of population growth that added nearly a billion people to the world during the 1990s, the number facing food insecurity and chronic hunger during this period actually declined by about 40 million.

Deforestation has slowed in Asia, from more than 8 percent during the 1980s to less than 1 percent in the 1990s. Nature preserves and protected areas have increased nearly fivefold over the past 20 years, from about 2.6 million $km^2$ to about 12.2 million $km^2$. This represents only 8.2 percent of all land area—less than the 12 percent thought necessary to protect a viable sample of the world's biodiversity—but is a dramatic expansion nonetheless.

Dramatic progress is being made in a transition to renewable energy sources. The European Union has pledged to get 20 percent of its energy from renewable sources (30 percent if other countries participate) by 2020. Former British Prime Minister Tony Blair laid out even more ambitious plans to fight global warming by cutting carbon dioxide emissions in his country by 60 percent through energy conservation and a switch to renewables. If nonpolluting, sustainable energy technology is made available to

## What Do You Think?

### Calculating Your Ecological Footprint

Can the earth sustain our current lifestyles? Will there be adequate natural resources for future generations? These questions are among the most important in environmental science today. We depend on nature for food, water, energy, oxygen, waste disposal, and other life-support systems. Sustainability implies that we cannot turn our resources into waste faster than nature can recycle that waste and replenish the supplies on which we depend. It also recognizes that degrading ecological systems ultimately threatens everyone's well-being. Although we may be able to overspend nature's budget temporarily, future generations will have to pay the debts we leave them. Living sustainably means meeting our own vital needs without compromising the ability of future generations to meet their own needs.

How can we evaluate and illustrate our ecological impacts? Redefining Progress, a nongovernmental environmental organization, has developed a measure called the **ecological footprint** to compute the demands placed on nature by individuals and nations. A simple questionnaire of 16 items gives a rough estimate of your personal footprint. A more complex assessment of 60 categories including primary commodities (such as milk, wood, or metal ores), as well as the manufactured products derived from them, gives a measure of national consumption patterns.

According to Redefining Progress, the average world citizen has an ecological footprint equivalent to 2.3 hectares (5.6 acres), while the biologically productive land available is only 1.9 hectares (ha) per person. How can this be? The answer is that we're using nonrenewable resources (such as fossil fuels) to support a lifestyle beyond the productive capacity of our environment. It's like living by borrowing on your credit cards. You can do it for a while, but eventually you have to pay off the deficit. The unbalance is far more pronounced in some of the richer countries. The average resident of the United States, for example, lives at a consumption level that requires 9.7 ha of bioproductive land. A dramatic comparison of consumption levels versus population size is shown in figure 1. If everyone in the world were to adopt a North American lifestyle, we'd need about four more planets to support us all. You can check out your own ecological footprint by going to www.redefiningprogress.org/.

Like any model, an ecological footprint gives a useful description of a system. Also like any model, it is built on a number of assumptions: (1) Various measures of resource consumption and waste flows can be converted into the biologically productive area required to maintain them; (2) different kinds of resource use and dissimilar types of productive land can be standardized into roughly equivalent areas; (3) because these areas stand for mutually exclusive uses, they can be added up to a total—a total representing humanity's demand—that can be compared to the total world area of bioproductive land. The model also implies that our world has a fixed supply of resources that can't be expanded. Part of the power of this metaphor is that we all can visualize a specific area of land and imagine it being divided into smaller and smaller parcels as our demands increase. But this perspective doesn't take into account technological progress. For example, since 1950, world food production has increased about fourfold. Some of this growth has come from expansion of croplands, but most has come from technological advances such as irrigation, fertilizer use, and higher-yielding crop varieties. Whether this level of production is sustainable is another question, but this progress shows that land area isn't always an absolute limit. Similarly, switching to renewable energy sources such as wind and solar power would make a huge impact on estimates of our ecological footprint. Notice that in figure 2 energy consumption makes up about half of the calculated footprint.

What do you think? Does analyzing our ecological footprint inspire you to correct our mistakes, or does it make sustainability seem an impossible goal? If we in the richer nations have the technology and political power to exploit a larger share of resources, do we have a right to do so, or do we have an ethical responsibility to restrain our consumption? And what about future generations? Do we have an obligation to leave resources for them, or can we assume they'll make technological discoveries to solve their own problems if resources become scarce? You'll find that many of the environmental issues we discuss in this book aren't simply a matter of needing more scientific data. Ethical considerations and intergenerational justice often are just as important as having more facts.

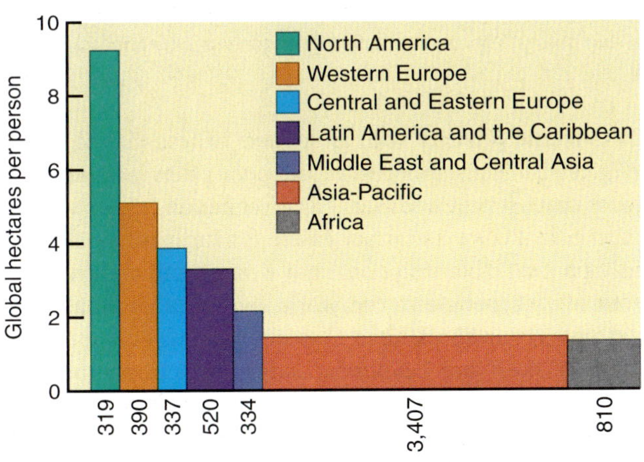


**FIGURE 1** Ecological footprint by region in 2001. The height of each bar is proportional to each region's average footprint per person, the width of the bar is proportional to its population, and the area of the bar is proportional to the region's total ecological footprint.

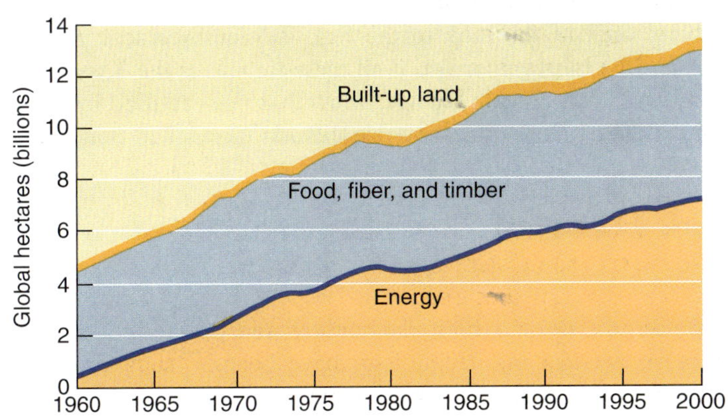


**FIGURE 2** Humanity's ecological footprint grew by about 160 percent from 1961 to 2001, somewhat faster than population, which doubled over the same period.
WWF, 2004.

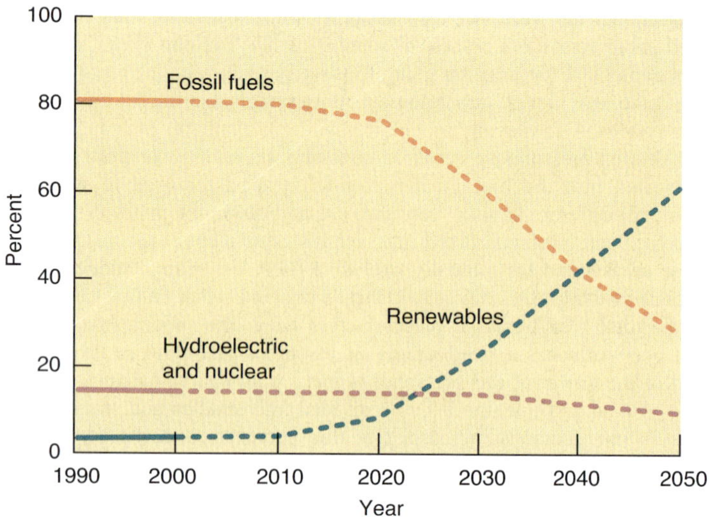


**FIGURE 1.11** A possible energy future. Global warming and other environmental problems may require that we switch from our current dependence on fossil fuels to renewable sources such as wind and solar energy. **Source:** World Bank, 2000.

the world's poorer countries, it may be possible to promote human development while simultaneously reducing environmental damage (fig. 1.11).

Over the past two decades, the world has made dramatic progress in opening up political systems and expanding political freedoms. During this time, some 81 countries took significant steps toward democracy. Currently, nearly three-quarters of the world's 200 countries now hold multiparty elections. At least 60 developing countries claim to be transferring decision-making authority to local units of government. Of course, decentralization doesn't always guarantee better environmental stewardship, but it puts people with direct knowledge of local conditions in a position of power rather than distant elites or bureaucrats.

Currently, more than 500 international environmental protection agreements are now in force. Some, such as the Montreal Protocol on Stratospheric Ozone, have been highly successful. Others, such as the Law of the Sea, lack enforcement powers. Perhaps the most important of all these treaties is the Kyoto Protocol on global climate change, which has been ratified by every industrialized nation except Australia and the United States.

## 1.4 Human Dimensions of Environmental Science

Because we live in both the natural and social worlds, and because we and our technology have become such dominant forces on the planet, environmental science must take human institutions and the human condition into account. We live in a world of haves and have-nots; a few of us live in increasing luxury, while many others lack the basic necessities for a decent, healthy, productive life. The World Bank estimates that more than 1.4 billion people—almost one-fifth of the world's population—

**FIGURE 1.12** Three-quarters of the world's poorest nations are in Africa. Millions of people lack adequate food, housing, medical care, clean water, and safety. The human suffering engendered by this poverty is tragic.

live in **extreme poverty** with an income of less than (U.S.)$1 per day (fig. 1.12). These poorest of the poor often lack access to an adequate diet, decent housing, basic sanitation, clean water, education, medical care, and other essentials for a humane existence. Seventy percent of those people are women and children. In fact, four out of five people in the world live in what would be considered poverty in the United States or Canada.

Policymakers are becoming aware that eliminating poverty and protecting our common environment are inextricably interlinked because the world's poorest people are both the victims and the agents of environmental degradation. The poorest people are often forced to meet short-term survival needs at the cost of long-term sustainability. Desperate for croplands to feed themselves and their families, many move into virgin forests or cultivate steep, erosion-prone hillsides, where soil nutrients are exhausted after only a few years. Others migrate to the grimy, crowded slums and ramshackle shantytowns that now surround most major cities in the developing world. With no way to dispose of wastes, the residents often foul their environment further and contaminate the air they breathe and the water on which they depend for washing and drinking.

The cycle of poverty, illness, and limited opportunities can become a self-sustaining process that passes from one generation to another. People who are malnourished and ill can't work productively to obtain food, shelter, or medicine for themselves or their children, who also are malnourished and ill. About 250 million children—mostly in Asia and Africa and some as young as 4 years old—are forced to work under appalling conditions weaving carpets, making ceramics and jewelry, or working in the sex trade. Growing up in these conditions leads to educational, psychological, and developmental deficits that condemn these children to perpetuate this cycle.

Faced with immediate survival needs and few options, these unfortunate people often have no choice but to overharvest resources; in doing so, however, they diminish not only their own options but also those of future generations. And in an increasingly interconnected world, the environments and resource bases damaged by poverty and ignorance are directly linked to those on which we depend.

The Worldwatch Institute warns that "poverty, disease and environmental decline are the true axis of evil." Terrorist attacks—and the responses they provoke—are the symptoms of the underlying sources of global instability, including the dangerous interplay among poverty, hunger, disease, environmental degradation, and rising resource competition. Failure to deal with these sources of insecurity could plunge the world into a dangerous downward spiral in which instability and radicalization grows. Unless the world takes action to promote sustainability and equity, Worldwatch suggests we will face an uphill battle to deal with the consequences of wars, terrorism, and natural disasters.

## We live in an inequitable world

About one-fifth of the world's population lives in the 20 richest countries, where the average per capita income is above (U.S.) $25,000 per year. Most of these countries are in North America or Western Europe, but Japan, Singapore, and Australia also fall into this group. Almost every country, however, even the richest, such as the United States and Canada, has poor people. No doubt everyone reading this book knows about homeless people or other individuals who lack resources for a safe, productive life. According to the U.S. Census Bureau, 37 million Americans—one-third of them children—live in households below the poverty line.

The other four-fifths of the world's population lives in middle- or low-income countries, where nearly everyone is poor by North American standards. Nearly 3 billion people live in the poorest nations, where the average per capita income is below (U.S.)$620 per year. China and India are the largest of these countries, with a combined population of about 2.3 billion people. Among the 41 other nations in this category, 33 are in sub-Saharan Africa. All the other lowest-income nations, except Haiti, are in

**TABLE 1.1**

**Quality of Life Indicators**

| | Least-Developed Countries | Most-Developed Countries |
|---|---|---|
| GDP/Person[1] | (U.S.)$329 | (U.S.)$30,589 |
| Poverty Index[2] | 78.1% | ~0 |
| Life Expectancy | 43.6 years | 76.5 years |
| Adult Literacy | 58% | 99% |
| Female Secondary Education | 11% | 95% |
| Total Fertility[3] | 5.0 | 1.7 |
| Infant Mortality[4] | 97 | 5 |
| Improved Sanitation | 23% | 100% |
| Improved Water | 61% | 100% |
| $CO^2$/capita[5] | 0.2 tons | 13 tons |

[1]Annual gross domestic product
[2]Percent living on less than (U.S.)$2/day
[3]Average births/woman
[4]Per 1,000 live births
[5]Metric tons/yr/person
**Source:** UNDP Human Development Index, 2006.

Asia. Although poverty levels in countries such as China and Indonesia have fallen in recent years, most countries in sub-Saharan Africa and much of Latin America have made little progress. The destabilizing and impoverishing effects of earlier colonialism continue to play important roles in the ongoing problems of these unfortunate countries. Meanwhile, the relative gap between rich and poor has increased dramatically.

As table 1.1 shows, the gulf between the richest and poorest nations affects many quality-of-life indicators. The average individual in the highest-income countries has an annual income nearly 100 times that of those in the lowest-income nations. Infant mortality in the least-developed countries is nearly 20 times as high as in the most-developed countries. Only 23 percent of residents in poorer countries have access to modern sanitation, while this ammenity is essentially universal in richer countries. Carbon dioxide emissions (a measure of both energy use and contributions to global warming) are 65 times greater in rich countries.

The gulf between rich and poor is even greater at the individual level. The richest 200 people in the world have a combined wealth of $1 trillion. This is more than the total owned by the 3 billion people who make up the poorest half of the world's population.

## Is there enough for everyone?

Those of us in the richer nations now enjoy a level of affluence and comfort unprecedented in human history. But we consume an inordinate share of the world's resources, and produce an unsustainable

**FIGURE 1.13** "And may we continue to be worthy of consuming a disproportionate share of this planet's resources."

**FIGURE 1.14** A rapidly growing economy has brought increasing affluence to China that has improved standards of living for many Chinese people, but it also brings environmental and social problems associated with western life styles.

amount of pollution to support our lifestyle. What if everyone in the world tried to live at that same level of consumption? The United States, for example, with about 4.6 percent of the world's population, consumes about 25 percent of all oil while producing about 25 percent of all carbon dioxide and 50 percent of all toxic wastes in the world (fig. 1.13). What will the environmental effects be if other nations try to emulate our prosperity?

Take the example of China that we discussed in the opening case study for this chapter. In the early 1960s, it's estimated that 300 million Chinese suffered from chronic hunger, and at least 30 million starved to death in the worst famine in world history. Since then, however, China has experienced amazing economic growth. The national GDP has been growing at about 10 percent per year. If current trends continue, the Chinese economy will surpass the United States and become the world's largest by 2020. This rapid growth has brought many benefits. Hundreds of millions of people have been lifted out of extreme poverty. Chronic hunger has decreased from about 30 percent of the population 40 years ago to less than 10 percent today. Average life expectancy has increased from 42 to 72.5 years. And infant mortality dropped from 150 per 1,000 live births in 1960 to 24.5 today, while the annual per capita GDP has grown from less than (U.S.)$200 per year to more than $4,500.

Still, most Chinese live at a low level of material consumption by European or American standards. One way of measuring material consumption is by environmental footprint. It now takes about 9.7 global hectares to support the average American. By contrast, the average Chinese citizen has a footprint of only 1.6 global hectares. If all the 1.3 billion residents of China were to try to match the American level of consumption it would take about four extra planets using the same technology we now employ.

Many of the environmental problems mentioned in the opening case study for this chapter arise from poverty. China couldn't afford to worry (at least so they thought) about pollution and land degradation in the past. Today, however, the greatest environmental worries are about the effects of rising affluence (fig. 1.14). In 1985, there were essentially no private automobiles in China. Bicycles and public transportation were how nearly everyone got around. Now, there are about 30 million automobiles in China, and by 2015, if current trends continue, there could be 150 million (fig. 1.15). Already, Chinese auto efficiency standards are higher than in the United States, but is there enough petroleum in the world to support all these vehicles? China is now the second largest source of $CO_2$

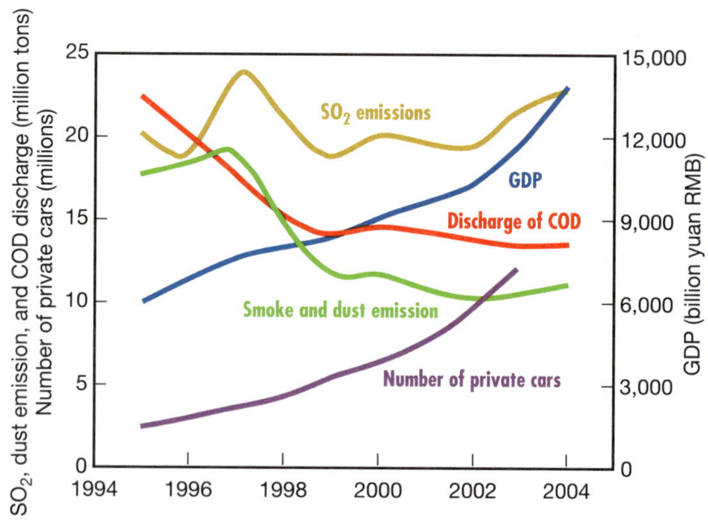


**FIGURE 1.15** Between 1994 and 2004, Chinese GDP more than doubled, while the number of private automobiles grew about fourfold. Chemical oxygen demand (COD) fell by more than half as citizens demanded effluent controls on industry, but sulfur dioxide ($SO_2$) emissions increased as more coal was burned.
**Source:** Shao, M., et al., 2006.

(the United States is first) in the world. Both China and the United States depend on coal for about 75 percent of their electricity. Both have very large supplies of coal. There are many benefits of expanding China's electrical supply, but if they reach the same level of power consumption—which is now about one-tenth the amount per person as in the United States—by burning coal, the effects on our global climate will be disastrous.

On the other hand, China is now building a series of new cities that are expected to be self-sustaining in food production, water and energy supplies, and climate-neutral with respect to climate-changing gases. If more countries in both the developed and developing countries adopt these environmentally friendly technologies, the world could easily have enough resources for everyone.

### Recent progress is encouraging

Over the past 50 years, human ingenuity and enterprise have brought about a breathtaking pace of technological innovations and scientific breakthroughs. The world's gross domestic product increased more than tenfold during that period, from $2 trillion to $22 trillion per year. While not all that increased wealth was applied to human development, there has been significant progress in increasing general standard of living nearly everywhere. In 1960, for instance, nearly three-quarters of the world's population lived in abject poverty. Now, less than one-third are still at this low level of development.

Since World War II, average real income in developing countries has doubled; malnutrition declined by almost one-third; child death rates have been reduced by two-thirds; average life expectancy increased by 30 percent. Overall, poverty rates have decreased more in the last 50 years than in the previous 500. Nonetheless, while general welfare has increased, so has the gap between rich and poor worldwide. In 1960, the income ratio between the richest 20 percent of the world and the poorest 20 percent was 30 to 1. In 2000, this ratio was 100 to 1. Because perceptions of poverty are relative, people may feel worse off compared to their rich neighbors than development indices suggest they are.

## 1.5 Sustainable Development

Can we improve the lives of the world's poor without destroying our shared environment? A possible solution to this dilemma is **sustainable development,** a term popularized by *Our Common Future,* the 1987 report of the World Commission on Environment and Development, chaired by Norwegian Prime Minister Gro Harlem Brundtland (and consequently called the Brundtland Commission). In the words of this report, sustainable development means "meeting the needs of the present without compromising the ability of future generations to meet their own needs."

Another way of saying this is that we are dependent on nature for food, water, energy, fiber, waste disposal, and other life-support services. We can't deplete resources or create wastes faster than nature can recycle them if we hope to be here for the long term. Development means improving people's lives. Sustainable development, then, means progress in human well-being that can be extended or prolonged over many generations rather than just a few years. To be truly enduring, the benefits of sustainable development must be available to all humans rather than to just the members of a privileged group.

To many economists, it seems obvious that economic growth is the only way to bring about a long-range transformation to more advanced and productive societies and to provide resources to improve the lot of all people. As former President John F. Kennedy said, "A rising tide lifts all boats." But economic growth is not sufficient in itself to meet all essential needs. As the Brundtland Commission pointed out, political stability, democracy, and equitable economic distribution are needed to ensure that the poor will get a fair share of the benefits of greater wealth in a society. A study released in 2006 by researchers at Yale and Columbia Universities reported a significant correlation between environmental sustainability, open political systems, and good government. Of the 133 countries in this study, New Zealand, Sweden, Finland, Czech Republic, and the United Kingdom held the top five places (in that order). The United States ranked 28th, behind countries such as Japan, and most of Western Europe.

### Can development be truly sustainable?

Many ecologists regard "sustainable" growth of any sort as impossible in the long run because of the limits imposed by nonrenewable resources and the capacity of the biosphere to absorb our wastes. Using ever-increasing amounts of goods and services to make human life more comfortable, pleasant, or agreeable must inevitably interfere with the survival of other species and, eventually, of humans themselves in a world of fixed resources. But, supporters of sustainable development assure us, both technology and social organization can be managed in ways that meet essential needs and provide long-term—but not infinite—growth within natural limits, if we use ecological knowledge in our planning.

While economic growth makes possible a more comfortable lifestyle, it doesn't automatically result in a cleaner environment. As figure 1.16 shows, people will purchase clean water and sanitation if they can afford to do so. For low-income people, however, more money tends to result in higher air pollution because they can afford to burn more fuel for transportation and heating. Given enough money, people will be able to afford both convenience *and* clean air. Some environmental problems, such as waste generation and carbon dioxide emissions, continue to rise sharply with increasing wealth because their effects are diffuse and delayed. If we are able to sustain economic growth, we will need to develop personal restraint or social institutions to deal with these problems.

**Think About It**

Examine figure 1.16. Describe in your own words how increasing wealth affects the three kinds of pollution shown. Why do the trends differ?

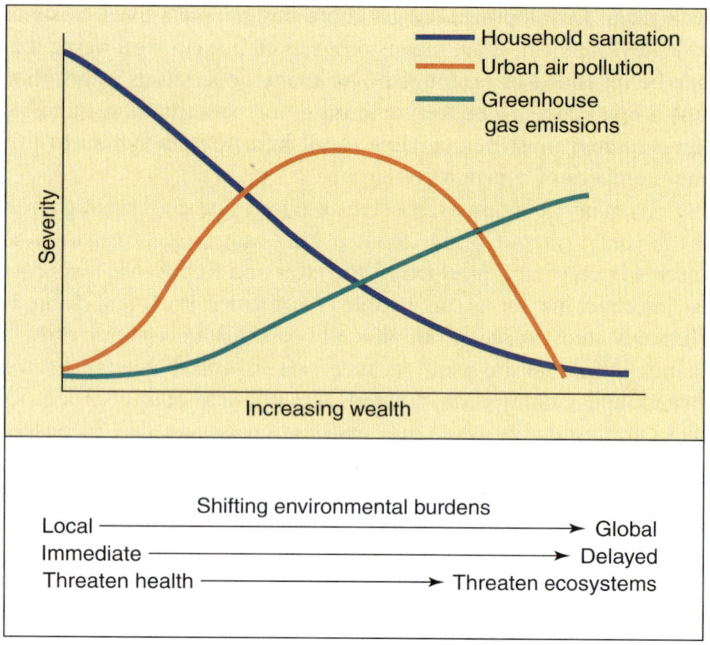


**FIGURE 1.16** Environmental indicators show different patterns as incomes rise. Sanitation problems decrease when people can afford septic systems and clean water. Local air pollution, on the other hand, increases as more fuel is burned; eventually, however, development reaches a point at which people can afford both clean air and the benefits of technology. Delayed, distant problems, such as greenhouse gas emissions that lead to global climate change, tend to rise steadily with income because people make decisions based on immediate needs and wants rather than long-term consequences. Thus, we tend to shift environmental burdens from local and immediate to distant and delayed if we can afford to do so. Graph from World Energy Assessment, UNDP 2000, Figure 3.10, p. 95.

Some projects intended to foster development have been environmental, economic, and social disasters. Large-scale hydropower projects, like that in the James Bay region of Quebec or the Brazilian Amazon that were intended to generate valuable electrical power, also displaced indigenous people, destroyed wildlife, and poisoned local ecosystems with acids from decaying vegetation and heavy metals leached out of flooded soils. Similarly, introduction of "miracle" crop varieties in Asia and huge grazing projects in Africa financed by international lending agencies crowded out wildlife, diminished the diversity of traditional crops, and destroyed markets for small-scale farmers.

Other development projects, however, work more closely with both nature and local social systems. Socially conscious businesses and environmental, nongovernmental organizations sponsor ventures that allow people in developing countries to grow or make high-value products—often using traditional techniques and designs—that can be sold on world markets for good prices (fig. 1.17). Pueblo to People, for example, is a nonprofit organization that buys textiles and crafts directly from producers in Latin America. It sells goods in America, with the profits going to community development projects in Guatemala, El Salvador, and Peru. It also informs customers in wealthy countries about the conditions in the developing world.

**FIGURE 1.17** A Mayan woman from Guatemala weaves on a backstrap loom. A member of a women's weaving cooperative, she sells her work to nonprofit organizations in the United States at much higher prices than she would get at the local market.

As the economist John Stuart Mill wrote in 1857, "It is scarcely necessary to remark that a stationary condition of capital and population implies no stationary state of human improvement. There would be just as much scope as ever for all kinds of mental culture and moral and social progress; as much room for improving the art of living and much more likelihood of its being improved when minds cease to be engrossed by the art of getting on." Somehow, in our rush to exploit nature and consume resources, we have forgotten this sage advice.

## What's the role of international aid?

Could we eliminate the most acute poverty and ensure basic human needs for everyone in the world? Many experts say this goal is eminently achievable. Economist Jeffery Sachs, director of the UN Millennium Development Project, says we could end extreme poverty worldwide by 2025 if the richer countries would donate just 0.7 percent of their national income for development aid in the poorest nations. These funds could be used for universal childhood vaccination against common infectious diseases, access to primary schools for everyone, family planning services

**FIGURE 1.18** Every year, military spending equals the total income of half the world's people. The cost of a single large aircraft carrier equals ten years of human development aid given by all the world's industrialized countries.

for those who wish them, safe drinking water and sanitation for all, food supplements for the hungry, and strategic microcredit loans for self-employment.

How much would this cost? A rough estimate provided by the United Nations Development Agency is that it would take about (U.S.)$135 billion per year to abolish extreme poverty and the worst infectious diseases over the next 20 years. That's a lot of money—much more than we currently give—but it's not an impossible goal. Annual global military spending is now over $1 trillion (fig. 1.18). If we were to shift one-tenth of that to development aid, we'd not only reduce incalculable suffering but also be safer in the long run, according to many experts.

In 2005 the G-8, made up of the eight wealthiest nations, pledged (U.S.)$50 billion per year by 2010 to combat poverty. In addition, they promised to cancel at least $40 billion in debt of the 18 poorest nations. Meanwhile the United States, while the world's largest total donor, sets aside only 0.16 percent of its gross domestic product for development aid. Put another way, the United States currently donates about 18 cents per citizen per day for both private and government aid to foreign nations.

What do you think? Would you be willing to donate an extra dollar per day to reduce suffering and increase political stability? As former Canadian Prime Minister Jean Chrétien says, "Aid to developing countries isn't charity; it's an investment. It will make us safer, and when standards of living increase in those countries, they'll become customers who will buy tons of stuff from us."

## Indigenous people are important guardians of nature

Often at the absolute bottom of the social strata, whether in rich or poor countries, are the indigenous or native peoples who are generally the least powerful, most neglected groups in the world. Typically descendants of the original inhabitants of an area taken over by more powerful outsiders, they often are distinct from their country's dominant language, culture, religion, and racial communities. Of the world's nearly 6,000 recognized cultures, 5,000 are indigenous ones that account for only about 10 percent of the total world population. In many countries, these indigenous people are repressed by traditional caste systems, discriminatory laws, economics, or prejudice. Unique cultures are disappearing, along with biological diversity, as natural habitats are destroyed to satisfy industrialized world appetites for resources. Traditional ways of life are disrupted further by dominant Western culture sweeping around the globe.

At least half of the world's 6,000 distinct languages are dying because they are no longer taught to children. When the last few elders who still speak the language die, so will the culture that was its origin. Lost with those cultures will be a rich repertoire of knowledge about nature and a keen understanding about a particular environment and a way of life (fig. 1.19).

Nonetheless, in many places, the 500 million indigenous people who remain in traditional homelands still possess valuable ecological wisdom and remain the guardians of little-disturbed habitats that are the refuge for rare and endangered species and relatively undamaged ecosystems. Author Alan Durning estimates that indigenous homelands harbor more biodiversity than all the

**FIGURE 1.19** Do indigenous people have unique knowledge about nature and inalienable rights to traditional territories?

world's nature reserves and that greater understanding of nature is encoded in the languages, customs, and practices of native people than is stored in all the libraries of modern science. Interestingly, just 12 countries account for 60 percent of all human languages (fig. 1.20). Seven of those are also among the "megadiversity" countries that contain more than half of all unique plant and animal species. Conditions that support evolution of many unique species seem to favor development of equally diverse human cultures as well.

Recognizing native land rights and promoting political pluralism is often one of the best ways to safeguard ecological processes and endangered species. As the Kuna Indians of Panama say, "Where there are forests, there are native people, and where there are native people, there are forests." A few countries, such as Papua New Guinea, Fiji, Ecuador, Canada, and Australia acknowledge indigenous title to extensive land areas.

In other countries, unfortunately, the rights of native people are ignored. Indonesia, for instance, claims ownership of nearly three-quarters of its forest lands and all waters and offshore fishing rights, ignoring the interests of indigenous people who have lived in these areas for millennia. Similarly, the Philippine government claims possession of all uncultivated land in its territory, while Cameroon and Tanzania recognize no rights at all for forest-dwelling pygmies who represent one of the world's oldest cultures.

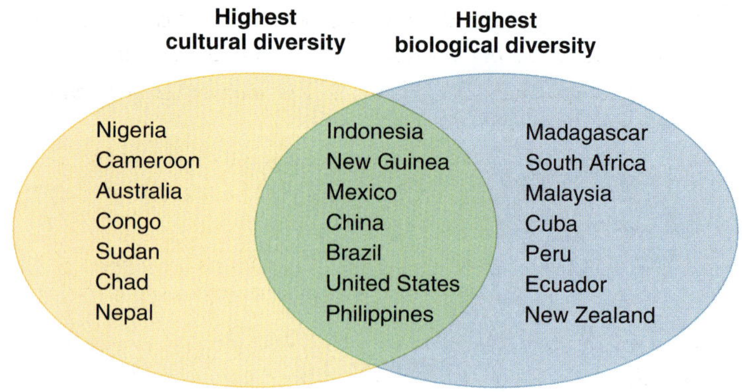


**FIGURE 1.20** Cultural diversity and biodiversity often go hand in hand. Seven of the countries with the highest cultural diversity in the world are also on the list of "megadiversity" countries with the highest number of unique biological organisms. (Listed in decreasing order of importance.) **Source:** Norman Myers, Conservation International and Cultural Survival Inc., 2002.

## Conclusion

We face many environmental dilemmas. As the case of China shows, air and water pollution, chronic hunger, water shortages, land degradation, and other environmental problems exact a terrible toll. China's problems aren't unique. About 3 billion people (nearly half the world's population) live on less than (U.S.)$2 per day, and face environmental hardships similar to those of the poorest Chinese. Finding ways that all of us can live sustainably within the limits of our resource base and without damaging nature's life-support systems is the preeminent challenge of environmental science.

Nature protection has deep roots reaching back into ancient history. Pragmatic resource conservation and moral or aesthetic concerns motivated early efforts at environmental defense. More recently, the health risks from pollution and the ecological dangers of habitat destruction and biodiversity losses have entered the dialog. Global environmentalism raises questions about sustainability. There are many signs of hope. Population growth is slowing nearly everywhere, many terrible diseases have been conquered, progress is being made in a transition to renewable energy and pollution control. Some countries have made encouraging advances in reducing greenhouse gas emissions.

Understanding the links between poverty and environmental degradation, and recognizing the rights of indigenous people are essential if we are to protect natural resources and improve environmental quality. Former UN Ambassador Adlai Stevenson once said, "We travel together passengers on a little spaceship, dependent upon its vulnerable reserve of air and soil; all committed for our safety to its security and peace; preserved from annihilation only by the care, the work, and I will say, the love we give to our fragile craft. We cannot maintain it half fortunate, half miserable; half confident, half despairing; half slave to the ancient enemies of man; half free in a liberation of resources. No craft, no crew can travel safely with such vast contradictions. On their resolution, then, depends the survival of us all."

## Reviewing Learning Outcomes

By now you should be able to explain the following points:

**1.1** Define environmental science and identify some important environmental concerns we face today.

- Environmental science is the systematic study of our environment and our proper place in it.
- China is a good case study of environmental concerns including population growth, poverty, food supplies, air and water pollution, energy choices, and the threats of global climate change.

**1.2** Discuss the history of conservation and the different attitudes toward nature at various times in our past.

- Nature protection has historic roots.
- Resource waste inspired pragmatic, utilitarian conservation.
- Ethical and aesthetic concerns inspired the preservation movement.
- Rising pollution levels led to the modern environmental movement.
- Global interconnections have expanded environmentalism.

**1.3** Think critically about the major environmental dilemmas and issues that shape our current environmental agenda.

- We live on a marvelous planet of rich biodiversity and complex ecological systems.
- We face many serious environmental problems including water supplies, safe drinking water, hunger, land degradation, energy, air quality, and biodiversity losses.
- There are many signs of hope in terms of social progress, environmental protection, energy choices, and the spread of democracy.

**1.4** Appreciate the human dimensions of environmental science, including the connection between poverty and environmental degradation.

- We live in an inequitable world.
- Recent progress is encouraging.
- Is there enough for everyone?

**1.5** Summarize sustainable development and evaluate some of its requirements.

- Can development be truly sustainable?
- What's the role of international aid?
- Indigenous people are important guardians of nature.

## Practice Quiz

1. Define *environment* and *environmental science*.
2. Describe four stages in conservation history and identify one leader associated with each stage.
3. List six environmental dilemmas that we now face and summarize how each concerns us.
4. Identify some signs of hope for solving environmental problems.
5. What is extreme poverty, and why should we care?
6. How much difference is there in per capita income, infant mortality, and $CO_2$ production between the poorest and richest countries?
7. Why should we be worried about economic growth in China?
8. Define *sustainable development*.
9. How much would it cost to eliminate acute poverty and ensure basic human needs for everyone?
10. Why are indigenous people important as guardians of nature?

## Critical Thinking and Discussion Questions

1. Should environmental science include human dimensions? Explain.
2. Overall, do environmental and social conditions in China give you hope or fear about the future?
3. What are the underlying assumptions and values of utilitarian conservation and altruistic preservation? Which do you favor?
4. What resource uses are most strongly represented in the ecological footprint? What are the advantages and disadvantages of using this assessment?
5. Are there enough resources in the world for 8 or 10 billion people to live decent, secure, happy lives? What do these terms mean to you? Try to imagine what they mean to residents of other countries.
6. What would it take for human development to be truly sustainable?
7. Are you optimistic or pessimistic about our chances of achieving sustainability? Why?

## DATA analysis Working with Graphs

Graphs are one of the most common and important ways that scientists communicate their results. To be a scientifically literate citizen, it's important for you to be familiar with graphing techniques.

Graphs are visual presentations of data that help us identify trends and understand relationships. A table of numbers is more precise, but most of us have difficulty visualizing patterns in a field of numbers. We can look at the patterns in a graphic representation and make comparisons much more easily than we could with raw data. The ability to read graphs and plot data are essential skills for students of environmental science. You'll see graphs throughout this book. In this exercise, we'll explore simple line plots.

A line plot represents a data set that involves a sequence of some sort, or a change over time. Table 1.2 shows growth of a population of a single-cell protozoan called *Tetrahymena* in a laboratory culture. To start the experiment, 10 cells were placed in a growth medium (this is time 0 in the table). Each hour for the next 12 hours, the cell number was counted and recorded. How can you convert data such as these into a graph? First, you lay out the axes. By convention, the horizontal (X) axis shows

**TABLE 1.2 Growth of *Tetrahymena***

| Hours | Cells |
|---|---|
| 0 | 10 |
| 1 | 15 |
| 2 | 20 |
| 3 | 40 |
| 4 | 80 |
| 5 | 160 |
| 6 | 320 |
| 7 | 450 |
| 8 | 550 |
| 9 | 600 |
| 10 | 620 |
| 11 | 630 |
| 12 | 640 |

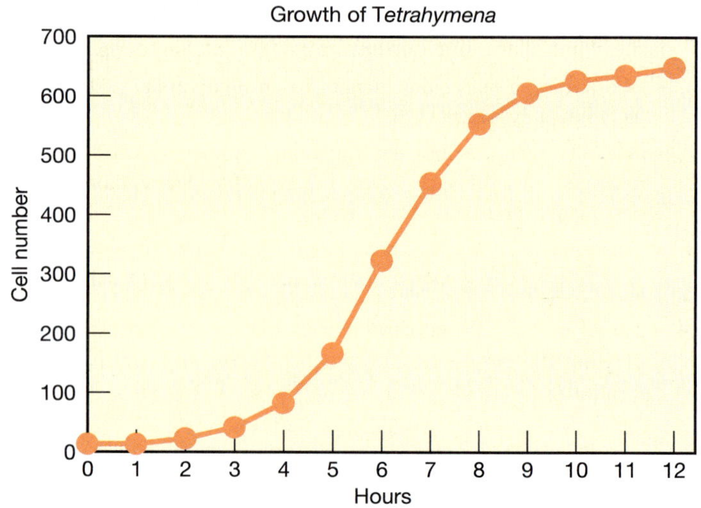


the **independent variable** (one whose values presumably don't depend on the other factor). The vertical axis (Y) records the **dependent variable** (which responds to changes in the other factor). Choose a range for each axis that approximately fits the range of values: this way your data will fill most of the graph, and the patterns should be big enough to see. Now plot the data: for hour 0, find the number of cells (10). Draw a dot *above the 0 on the X-axis* and *level with the 10 on the Y-axis.* Repeat these steps for hour 1, hour 2, and so on. After you've added all your dots, you can connect them with a line. The line makes it easier to see trends in the data. Your graph should look something like the one in this box.

Can you extract data from a line plot? Approximately. The process is simply the reverse of the way you just learned to create the graph. You draw a line from any point on the curve to the spot on the axes directly under and parallel to that point. For instance, the point on the curve for hour 6 corresponds to slightly more than 300 cells on the Y-axis.

To make comparisons between two different sets of dependent variables, it's often useful to plot them on the same graph. Look, for example, at figures 1.15 and 1.16 on pp. 26 and 27. You can see they have as many as five categories of independent variables in a single graph. To make the graphs simpler, the individual data points have been removed and each curve is a simple continuous line. You could still find the data for a specific point on either of these graphs with the process outlined above.

**For Additional Help in Studying This Chapter,** please visit our website at www.mhhe.com/cunningham10e. You will find additional practice quizzes and case studies, flashcards, regional examples, place markers for Google Earth™ mapping, and an extensive reading list, all of which will help you learn environmental science.

The Pacific Island nation of Tuvalu is being abandoned as global warming raises sea level. Do we have a moral obligation to poor nations or future generations for a livable environment?

CHAPTER 2

# Frameworks for Understanding Science, Systems, and Ethics

*The ultimate test of a moral society is the kind of world that it leaves to its children.*

*—Dietrich Bonhoeffer—*

## LEARNING OUTCOMES

After studying this chapter, you should be able to:

**2.1** Describe the scientific method and explain how it works.

**2.2** Evaluate the role of scientific consensus and conflict.

**2.3** Explain systems and how they're useful in science.

**2.4** Discuss environmental ethics and worldviews.

**2.5** Identify the roles of religious and cultural perspectives in conservation and environmental justice.

# Case Study Is Climate Change a Moral Issue?

Across America, thousands of religious groups are meeting in churches, synagogues, mosques, and meeting houses to discuss the science and ethics of global warming. Congregations are watching movies about global climate change, sponsoring discussions about the science behind the issue, distributing educational kits, encouraging members to become ecologically-concerned citizens, and installing energy conservation measures. But why do all this in a religious context?

"At its core, global climate change is not about economic theory or political platforms," according to the United States Conference of Catholic Bishops. "It is about our human stewardship of God's creation and our responsibility to those who come after us." Ethical questions lie at the heart of most important environmental challenges, many religious leaders insist. "We share a deep conviction that global climate change presents an unprecedented threat to the integrity of life on Earth and a challenge to the universal values that bind us as human beings," says a statement written by a group of religious leaders from many faiths. "Global warming is harming God's creation: first the poor of the world, and eventually, all of us, and all life," says the Reverend Sally Bingham of the Grace Cathedral in San Francisco.

Although they have tended in the past to be politically, socially, and economically conservative, many evangelical Christians also have joined the crusade against global warming. In a widely circulated document entitled "An Evangelical Call to Action," the presidents of 39 evangelical colleges, leaders of religious social service agencies, and pastors of some of the largest evangelical congregations in the country wrote that (1) Human-induced climate change is real. (2) The consequences of climate change will be significant and will hit the poor the hardest. (3) Christian moral convictions demand our response to the climate change problem and (4) The time to act now is urgent. Governments, businesses, churches, and individuals all have a role to play in addressing climate change—starting now." Calling for "Creation Care," these leaders call for federal legislation that will require reductions in carbon dioxide emissions through "cost-effective, market-based mechanisms."

Of course, not all religious believers agree with their brethren. A group called the Interfaith Stewardship Alliance, for instance, which includes a number of high-profile evangelical leaders, argues that those who demand environmental stewardship are practicing a kind of neo-heathenism, placing nature, including the "creepy, crawly things," ahead of humans and the Creator. They claim that the science isn't settled about whether global warming is actually a problem or that human beings are causing it. Moreover, they assert, the solutions proposed by global warming opponents will harm business and cause energy costs to rise. In this view, free-market solutions to environmental problems are much better than government mandates, and individuals and private organizations should be trusted to care for their own property without outside intervention.

While there are debates about the timing and detailed effects of global warming, a vast majority of the world's climate scientists now agree that climate change is already occurring and that humans are the main cause. In its 2007 report on the consequences of global warming, the Intergovernmental Panel on Climate Change (IPCC) concluded that poorer countries (especially in the tropics, such as Tuvalu), which have contributed least to climate change and can least afford to accommodate to it, will likely suffer most while richer countries, which have produced the vast majority of greenhouse gases, will have much less trouble adapting. The burdens of environmental damage will also fall mostly on unborn generations, especially in developing countries. Is it fair for those of us who have benefited from burning fossil fuels leave the burden of dealing with this problem to our descendents?

This case study introduces two important aspects of environmental science. The first of these is ethics and values. What obligations do we have as world citizens? What rights and values do other people, other species, and future generations have? Should we be stewards of nature or have dominion over it? How you answer this question depends on your worldview, values, and religious beliefs. In this chapter, we'll examine some moral and ethical perspectives that shape different understandings of important environmental questions. An equally important topic is the nature of science. To be an informed citizen, you need to understand how science works and what questions it can—and can't—answer. We'll start this chapter by examining how science helps us understand our world.

## 2.1 What Is Science?

**Science** is a process for producing knowledge methodically and logically. Derived from *scire,* "to know" in Latin, science depends on making precise observations of natural phenomena. We develop or test theories (proposed explanations of how a process works) using these observations. "Science" also refers to the cumulative body of knowledge produced by many scientists. Science is valuable because it helps us understand the world and meet practical needs, such as new medicines, new energy sources, or new foods. In this section, we'll investigate how and why science follows standard methods.

Science rests on the assumption that the world is knowable and that we can learn about the world by careful observation (table 2.1). For early philosophers of science, this assumption was a radical departure from religious and philosophical approaches. In the Middle Ages, the ultimate sources of knowledge about matters, such as how crops grow, how diseases spread, or how the stars move, were religious authorities or cultural

**TABLE 2.1**

**Basic Principles of Science**

1. *Empiricism:* We can learn about the world by careful observation of empirical (real, observable) phenomena; we can expect to understand fundamental processes and natural laws by observation.
2. *Uniformitarianism:* Basic patterns and processes are uniform across time and space; the forces at work today are the same as those that shaped the world in the past, and they will continue to do so in the future.
3. *Parsimony:* When two plausible explanations are reasonable, the simpler (more parsimonious) one is preferable. This rule is also known as Ockham's razor, after the English philosopher who proposed it.
4. *Uncertainty:* Knowledge changes as new evidence appears, and explanations (theories) change with new evidence. Theories based on current evidence should be tested on additional evidence, with the understanding that new data may disprove the best theories.
5. *Repeatability:* Tests and experiments should be repeatable; if the same results cannot be reproduced, then the conclusions are probably incorrect.
6. *Proof is elusive:* We rarely expect science to provide absolute proof that a theory is correct, because new evidence may always undermine our current understanding.
7. *Testable questions:* To find out whether a theory is correct, it must be tested; we formulate testable statements (hypotheses) to test theories.

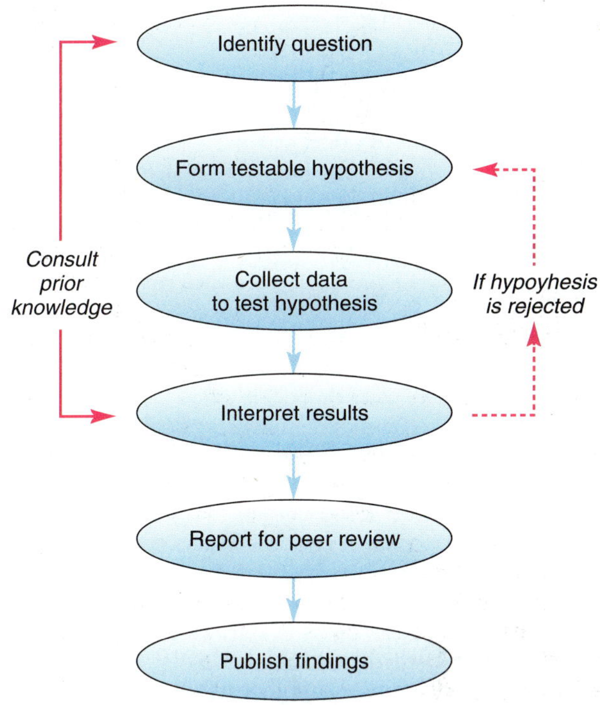


**FIGURE 2.1** Ideally, scientific investigation follows a series of logical, orderly steps to formulate and test hypotheses.

traditions. While these sources provided many useful insights, there was no way to test their explanations independently and objectively. The benefit of scientific thinking is that it searches for testable evidence: If you suspect a disease spreads through contaminated water, you can close off access to the water source and see if the disease stops spreading.

## Science depends on skepticism and accuracy

Ideally, scientists are skeptical. They are cautious about accepting proposed explanations until there is substantial evidence to support them. Even then, as we saw in the case study about global warming that opened this chapter, explanations are considered only provisionally true, because there is always a possibility that some additional evidence may appear to disprove them. Scientists also aim to be methodical and unbiased. Because bias and methodical errors are hard to avoid, scientific tests are subject to review by informed peers, who can evaluate results and conclusions (fig. 2.1). The peer review process is an essential part of ensuring that scientists maintain good standards in study design, data collection, and interpretation of results.

Scientists demand **reproducibility** because they are cautious about accepting conclusions. Making an observation or obtaining a result just once doesn't count for much. You have to produce the same result consistently to be sure that your first outcome wasn't a fluke. Even more important, you must be able to describe the conditions of your study so that someone else can reproduce your findings. Repeating studies or tests is known as **replication**.

Science also relies on accuracy and precision. Accuracy is correctness of measurements. Inaccurate data can produce sloppy and misleading conclusions (fig. 2.2). Precision means repeatability of results and level of detail. The classic analogy for repeatability is throwing darts at a dart board. You might throw ten darts and miss the center every time, but if all the darts hit nearly the same spot, they were very precise. Another way to think of precision is levels of detail. Suppose you want to measure how much snow fell last night, so you take out your ruler, which is marked in centimeters, and you find that the snow is just over 6 cm deep. You cannot tell if it is 6.3 cm or 6.4 cm because the ruler doesn't report that level of detail. If you average several measurements, you might find an average depth of 6.4333 cm. If you report all four decimal places, it will imply that you know more than you really do about the snow depth. If you had a ruler marked in millimeters (one-tenth of a centimeter), you could find a depth of 6.4 cm. Here, the one decimal place would be a **significant number,** or a level of detail you actually knew. Reporting 6.4333 cm would still involve three insignificant digits.

## Deductive and inductive reasoning are both useful

Ideally, scientists deduce conclusions from general laws that they know to be true. For example, if we know that massive objects attract each other (because of gravity), then it follows

**FIGURE 2.2** Making careful, accurate measurements and keeping good records are essential in scientific research.

that an apple will fall to the ground when it releases from the tree. This logical reasoning from general to specific is known as **deductive reasoning.** Often, however, we do not know general laws that guide natural systems. We observe, for example, that birds appear and disappear as a year goes by. Through many repeated observations in different places, we can infer that the birds move from place to place. We can develop a general rule that birds migrate seasonally. Reasoning from many observations to produce a general rule is **inductive reasoning.** Although deductive reasoning is more logically sound than inductive reasoning, it only works when our general laws are correct. We often rely on inductive reasoning to understand the world because we have few immutable laws.

Sometimes it is insight, as much as reasoning, that leads us to an answer. Many people fail to recognize the role that insight, creativity, aesthetics, and luck play in research. Some of our most important discoveries were made not because of superior scientific method and objective detachment, but because the investigators were passionately interested in their topics and pursued hunches that appeared unreasonable to fellow scientists. A good example is Barbara McClintock, the geneticist who discovered that genes in corn can move and recombine spontaneously. Where other corn geneticists saw random patterns of color and kernel size, McClintock's years of experience in corn breeding and an uncanny ability to recognize patterns, led her to guess that genes could recombine in ways that no one had yet imagined. Her intuitive understanding led to a theory that took other investigators years to accept.

## Testable hypotheses and theories are essential tools

You may already be using the scientific method without being aware of it. Suppose you have a flashlight that doesn't work. The flashlight has several components (switch, bulb, batteries) that could be faulty. If you change all the components at once, your flashlight might work, but a more methodical series of tests will tell you more about what was wrong with the system—knowledge that may be useful next time you have a faulty flashlight. So you decide to follow the standard scientific steps:

1. *Observe* that your flashlight doesn't light; also, there are three main components of the lighting system (batteries, bulb, and switch).
2. Propose a ***hypothesis,*** a testable explanation: "The flashlight doesn't work because the batteries are dead."
3. Develop a *test* of the hypothesis and *predict* the result that would indicate your hypothesis was correct: "I will replace the batteries; the light should then turn on."
4. Gather *data* from your test: After you replaced the batteries, did the light turn on?
5. *Interpret* your results: If the light works now, then your hypothesis was right; if not, then you should formulate a new hypothesis, perhaps that the bulb is faulty, and develop a new test for that hypothesis.

In systems more complex than a flashlight, it is almost always easier to prove a hypothesis wrong than to prove it unquestionably true. This is because we usually test our hypotheses with observations, but there is no way to make every possible observation. The philosopher Ludwig Wittgenstein illustrated this problem as follows: Suppose you saw hundreds of swans, and all were white. These observations might lead you to hypothesize that all swans were white. You could test your hypothesis by viewing thousands of swans, and each observation might support your hypothesis, but you could never be entirely sure that it was correct. On the other hand, if you saw just one black swan, you would know with certainty that your hypothesis was wrong.

As you'll read in later chapters, the elusiveness of absolute proof is a persistent problem in environmental policy and law. You can never absolutely prove that the toxic waste dump up the street is making you sick. The elusiveness of proof often decides environmental liability lawsuits.

When an explanation has been supported by a large number of tests, and when a majority of experts have reached a general consensus that it is a reliable description or explanation, we call it a **scientific theory.** Note that scientists' use of this term is very different from the way the public uses it. To many people, a theory is speculative and unsupported by facts. To a scientist, it means just the opposite: While all explanations are tentative and open to revision and correction, an explanation that counts as a scientific theory is supported by an overwhelming body of data and experience, and it is generally accepted by the scientific community, at least for the present (fig. 2.3).

## Understanding probability helps reduce uncertainty

One strategy to improve confidence in the face of uncertainty is to focus on probability. Probability is a measure of how likely something is to occur. Usually, probability estimates are based

**FIGURE 2.3** Data collection and repeatable tests support scientific theories. Here students use telemetry to monitor radio-tagged fish.

on a set of previous observations or on standard statistical measures. Probability does not tell you what *will* happen, but it tells you what *is likely* to happen. If you hear on the news that you have a 20 percent chance of catching a cold this winter, that means that 20 of every 100 people are likely to catch a cold. This doesn't mean that you will catch one. In fact, it's more likely that you won't catch a cold than that you will. If you hear that 80 out of every 100 people will catch a cold, you still don't know whether you'll get sick, but there's a much higher chance that you will.

Science often involves probability, so it is important to be familiar with the idea. Sometimes probability has to do with random chance: If you flip a coin, you have a random chance of getting heads or tails. Every time you flip, you have the same 50 percent probability of getting heads. The chance of getting ten heads in a row is small (in fact, the chance is 1 in $2^{10}$, or 1 in 1,024), but on any individual flip, you have exactly the same 50 percent chance, since this is a random test. Sometimes probability is weighted by circumstances: Suppose that about 10 percent of the students in this class earn an A each semester. Your likelihood of being in that 10 percent depends a great deal on how much time you spend studying, how many questions you ask in class, and other factors. Sometimes there is a combination of chance and circumstances: The probability that you will catch a cold this winter depends partly on whether you encounter someone who is sick (largely random chance) and whether you take steps to stay healthy (get enough rest, wash your hands frequently, eat a healthy diet, and so on).

Scientists often increase their confidence in a study by comparing results to a random sample or a larger group. Suppose that 40 percent of the students in your class caught a cold last winter. This *seems* like a lot of colds, but is it? One way to decide is to compare to the cold rate in a larger group. You call your state epidemiologist, who took a random sample of the state population last year: She collected 200 names from the telephone book and called each to find out if each got a cold last year. A larger sample, say 2,000 people, would have been more likely to represent the actual statewide cold rate. But a sample of 200 is much better than a sample of 50 or 100. The epidemiologist tells you that in your state as a whole, only 20 percent of people caught a cold.

Now you know that the rate in your class was quite high, and you can investigate possible causes for the difference. Perhaps people in your class got sick because they were short on sleep, because they tended to stay up late studying. Could you test whether studying late was a contributing factor? One way to test the relationship is to separate the class into two groups: those who study long and late, and those who don't. Then compare the rate of colds in these groups. Suppose it turns out that among the 40 late-night studiers, 30 got colds (a rate of 75 percent). Among the 60 casual studiers, only 10 got colds (17 percent). This difference would give you a good deal of confidence that staying up late contributes to getting sick. (Note, however, that all 40 of the studying group got good grades!)

## Statistics can calculate the probability that your results were random

Statistics can help in experimental design as well as in interpreting data (see Exploring Science, p. 38). Many statistical tests focus on calculating the probability that observed results could have occurred by chance. Often, the degree of confidence we can assign to results depends on sample size as well as the amount of variability between groups.

Ecological tests are often considered significant if there is less than 5 percent probability that the results were achieved by random chance. A probability of less than 1 percent gives still greater confidence in the results.

As you read this book, you will encounter many statistics, including many measures of probability. When you see these numbers, stop and think: Is the probability high enough to worry about? How high is it compared to other risks or chances you've read about? What are the conditions that make probability higher or lower? Science involves many other aspects of statistics.

## Experimental design can reduce bias

The study of colds and sleep deprivation is an example of an observational experiment, one in which you observe natural events and interpret a causal relationship between the variables. This kind of study is also called a **natural experiment,** one that involves observation of events that have already happened. Many scientists depend on natural experiments: A geologist, for instance, might want to study mountain building, or an ecologist might want to learn about how species coevolve, but neither scientist can spend millions of years watching the process happen. Similarly, a toxicologist cannot give people a disease just to see how lethal it is.

Other scientists can use **manipulative experiments,** in which conditions are deliberately altered, and all other variables are held constant (fig. 2.4). In one famous manipulative study, ecologists Edward O. Wilson and Robert MacArthur were interested in how quickly species colonize small islands, depending on distance to the mainland. They fumigated several tiny islands in the Florida Keys, killing all resident insects, spiders, and other invertebrates. They then monitored the islands to learn how quickly ants and spiders recolonized them from the mainland or other islands.

## *Exploring* SCIENCE — What Are Statistics, and Why Are They Important?

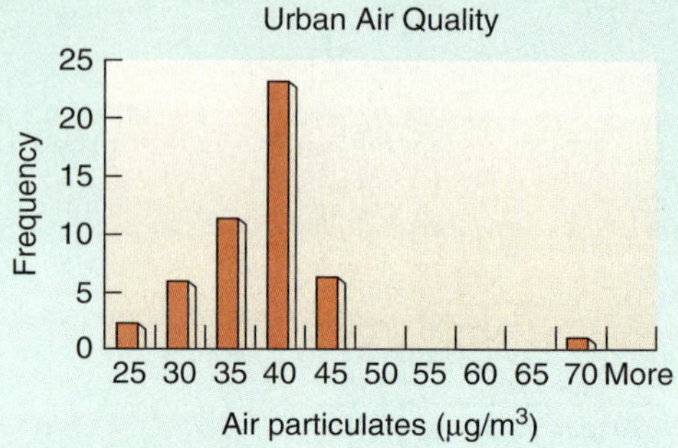


**FIGURE 1** Average annual airborne dust levels for 50 cities in 2001.

**Source:** Data from U.S. Environmental Protection Agency.

**Statistics** are numbers that let you evaluate and compare things. "Statistics" is also a field of study that has developed meaningful methods of comparing those numbers. By both definitions, statistics are widely used in environmental sciences, partly because they can give us a useful way to assess patterns in a large population, and partly because the numbers can give us a measure of confidence in our research or observations. Understanding the details of statistical tests can take years of study, but a few basic ideas will give you a good start toward interpreting statistics.

1. *Descriptive statistics help you assess the general state of a group.* In many towns and cities, the air contains a dust, or particulate matter, as well as other pollutants. From personal experience you might know your air isn't as clean as you'd like, but you may not know how clean or dirty it is. You could start by collecting daily particulate measurements to find average levels. An averaged value is more useful than a single day's values, because daily values may vary a great deal, but general, long-term conditions affect your general health. Collect a sample every day for a year; then divide the sum by the number of days, to get a **mean** (average) dust level. Suppose you found a mean particulate level of 30 micrograms per cubic meter ($\mu g/m^3$) of air. Is this level high or low? In 1997 the EPA set a standard of 50 $\mu g/m^3$ as a limit for allowable levels of coarse particulates (2.5–10 micrometers in diameter). Higher levels tend to be associated with elevated rates of asthma and other respiratory diseases. Now you know that your town, with an annual average of 30 $\mu g/m^3$, has relatively clean air, after all.
2. *Statistical samples.* Although your town is clean by EPA standards, how does it compare with the rest of the cities in the country? Testing the air in *every* city is probably not possible. You could compare your town's air quality with a **sample,** or subset of cities, however. A large, random sample of cities should represent the general "population" of cities reasonably well. Taking a large sample reduces the effects of outliers (unusually high or low values) that might be included. A random sample minimizes the chance that you're getting only the worst sites, or only a collection of sites that are close together, which might all have similar conditions. Suppose you getaverage annual particulate levels from a sample of 50 randomly selected cities. You can draw a frequency distribution, or histogram, to display your results (fig. 1). The mean value of this group is 36.8 $\mu g/m^3$, so by comparison your town (at 30 $\mu g/m^3$) is relatively clean.

   Many statistical tests assume that the sample has a normal, or Gaussian, frequency distribution, often described as a bell-shaped curve (fig. 2). In this distribution, the mean is near the center of the range of values, and most values are fairly close to the mean. Large and random samples are more likely to fit this shape than are small and nonrandom samples.
3. *Confidence.* How do you know that the 50 cities you sampled really represent all the cities in the country? You can't ever be completely certain, but you can use estimates, such as confidence limits, to express the reliability of your mean statistic. Depending on the size of your sample (not 10, not 100, but 50) and the amount of variability in the sample data, you can calculate a confidence interval that the mean represents the whole population (all cities). Confidence levels, or confidence intervals,

**FIGURE 2.4** Manipulative experiments attempt to control all variables except the tested variables. Here students control the number of species in a biodiversity experiment.

Most manipulative experiments are done in the laboratory, where conditions can be carefully controlled. Suppose you were interested in studying whether lawn chemicals contributed to deformities in tadpoles. You might keep two groups of tadpoles in fish tanks, and expose one to chemicals. In the lab, you could ensure that both tanks had identical temperatures, light, food, and oxygen. By comparing a treatment (exposed) group and a control (unexposed) group, you have also made this a **controlled study.**

Often, there is a risk of experimenter bias. Suppose the researcher sees a tadpole with a small nub that looks like it might become an extra leg. Whether she calls this nub a deformity

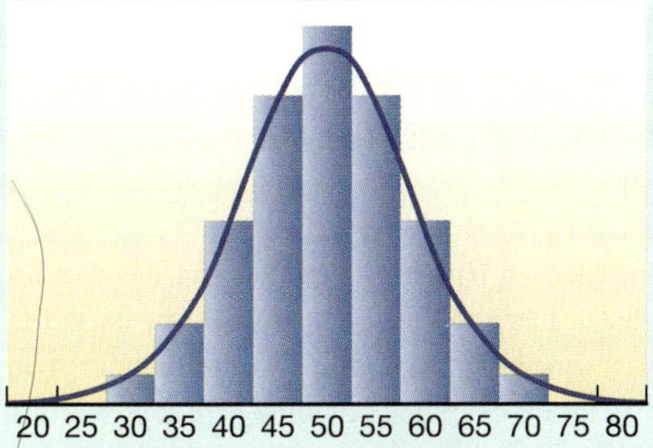


**FIGURE 2** A normal distribution.

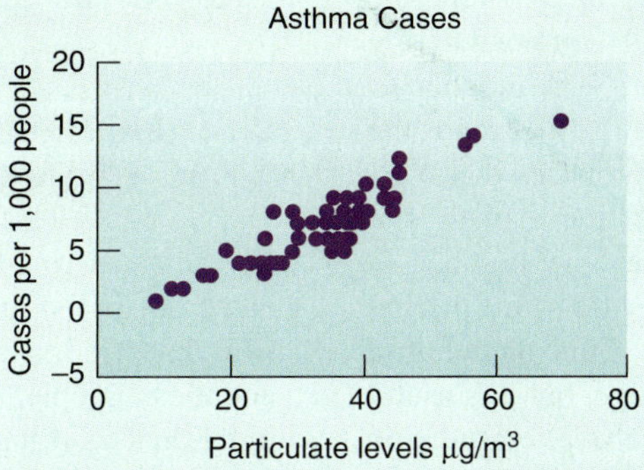


**FIGURE 3** A plot showing relationships between variables.

represent the likelihood that your statistics represent the entire population correctly. For the mean of your sample, a confidence interval tells you the probability that your sample is similar to other random samples of the population. A common convention is to compare values with a 95 percent confidence level, or a probability of 5 percent or less that your conclusions are misleading. Using statistical software, we can calculate that, for our 50 cities, the mean is 36.8 µg/m$^3$, and the confidence interval is 35.0 to 38.6. This suggests that, if you take 1,000 samples from the entire population of cities, 95 percent of those samples ought to be within 2 µg/m$^3$ of your mean. This indicates that your mean is reliable and representative.

4. *Is your group unusual?* Once you have described your group of cities, you can compare it with other groups. For example, you might believe that Canadian cities have cleaner air than U.S. cities. You can compare mean air quality levels for the two groups. Then you can calculate confidence intervals for the difference between the means, to see if the difference is meaningful.
5. *Evaluating relationships between variables.* Are respiratory diseases correlated with air pollution? For each city in your sample, you could graph pollution and asthma rates (fig. 3). If the graph looks like a loose cloud of dots, there is no clear relationship. A tight, linear pattern of dots trending upward to the right indicates a strong and positive relationship. You can also use a statistical package to calculate an equation to describe the relationship and, again, confidence intervals for the equation. This is known as a regression equation.
6. *Lies, damned lies, and statistics.* Can you trust a number to represent a complex or large phenomenon? One of the devilish details of representing the world with numbers is that those numbers can be tabulated in many ways. If we want to assess the greatest change in air quality statistics, do we report rates of change or the total amount of change? Do we look at change over five years? Twenty-five years? Do we accept numbers selected by the EPA, by the cities themselves, by industries, or by environmental groups? Do we trust that all the data were collected with a level of accuracy and precision that we would accept if we knew the hidden details in the data-gathering process? Like all information, statistics need to be interpreted in terms of who produced them, when, and why. Awareness of some of the standard assumptions behind statistics, such as sampling, confidence, and probability, will help you interpret statistics that you see and hear.

For more information about creating and interpreting graphs, see the Data Analysis exercise at the end of this chapter.

might depend on whether she knows that the tadpole is in the treatment group or the control group. To avoid this bias, **blind experiments** are often used, in which the researcher doesn't know which group is treated until after the data have been analyzed. In health studies, such as tests of new drugs, **double-blind experiments** are used, in which neither the subject (who receives a drug or a placebo) nor the researcher knows who is in the treatment group and who is in the control group.

In each of these studies there is one **dependent variable** and one, or perhaps more, **independent variables.** The dependent variable, also known as a response variable, is affected by the independent variables. In a graph, the dependent variable is on the vertical (Y) axis, by convention. Independent variables are rarely really independent (they are affected by the same environmental conditions as the dependent variable, for example). Many people prefer to call them explanatory variables, because we hope they will explain differences in the dependent variable.

## Models are an important experimental strategy

Another way to gather information about environmental systems is to use **models.** A model is a simple representation of something. Perhaps you have built a model airplane. The model doesn't have all the elements of a real airplane, but it has the most important ones for your needs. A simple wood or plastic airplane has the proper shape, enough to allow a child to imagine it is flying (fig. 2.5). A more complicated model airplane might have a small gas engine, just enough to let a teenager fly it around for short distances.

Similarly, scientific models vary greatly in complexity, depending on their purposes. Some models are physical models: Engineers test new cars and airplanes in wind tunnels to see how they perform, and biologists often test theories about evolution and genetics using "model organisms" such as fruit flies or rats as a surrogate for humans.

**FIGURE 2.5** A model uses just the essential elements to represent a complex system.

Most models are numeric, though. A model could be a mathematical equation, such as a simple population growth model ($N_t = rN_{(t-1)}$). Here the essential components are number (N) of individuals at time *t* ($N_t$), and the model proposes that $N_t$ is equal to the growth rate (r) times the number in the previous time period ($N_{(t-1)}$). This model is a very simplistic representation of population change, but it is useful because it precisely describes a relationship between population size and growth rate. Also, by converting the symbols to numbers, we can predict populations over time. For example, if last year's rabbit population was 100, and the growth rate is 1.6 per year, then this year's population will be 160. Next year's population will be $160 \times 1.6$, or 460. This is a simple model, then, but it can be useful. A more complicated model might account for deaths, immigration, emigration, and other factors.

More complicated mathematical models can be used to describe and calculate more complex processes, such as climate change or economic growth (fig. 2.6). These models are also useful because they allow the researcher to manipulate variables without actually destroying anything. An economist can experiment with different interest rates to see how they affect economic growth. A climatologist can raise $CO_2$ levels and see how quickly temperatures respond. These models are often called simulation models, because they simulate a complex system. Of course, the results depend on the assumptions built into the models. One model might show temperature rising quickly in response to $CO_2$; another might show temperature rising more slowly, depending on how evaporation, cloud cover, and other variables are taken into account. Consequently, simulations can produce powerful but controversial results. If multiple models generally agree, though, as in the cases of climate models that agree on generally upward temperature trends, we can have confidence that the overall predictions are reliable. These models are also very useful in laying out and testing our ideas about how a system works.

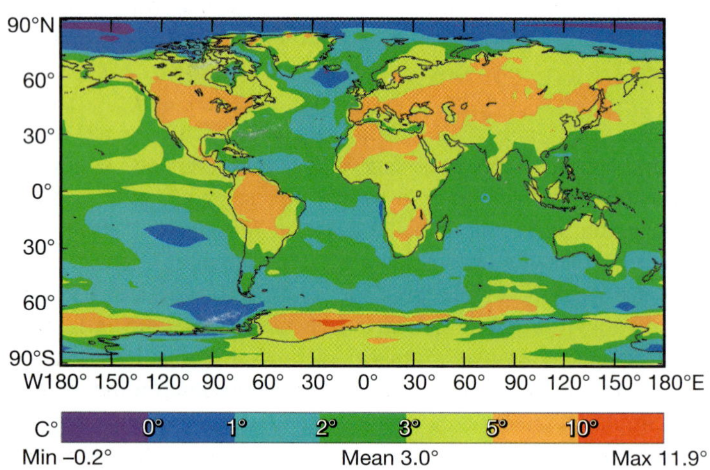


**FIGURE 2.6** Numerical models, calculated from observed data, project scenarios such as global climate change. The Hadley model of summer temperature change for 70 to 100 years from now is shown.
**Source:** © Crown. Copyright 2005. Published by the Met Office.

## 2.2 Scientific Consensus and Conflict

The scientific method outlined in figure 2.1 is the process used to carry out individual studies. Larger-scale accumulation of scientific knowledge involves cooperation and contributions from countless people. Good science is rarely carried out by a single individual working in isolation. Instead, a community of scientists collaborates in a cumulative, self-correcting process. You often hear about big breakthroughs and dramatic discoveries that change our understanding overnight, but in reality these changes are usually the culmination of the labor of many people, each working on different aspects of a common problem, each adding small insights to solve a problem. Ideas and information are exchanged, debated, tested, and re-tested to arrive at **scientific consensus,** or general agreement among informed scholars.

The idea of consensus is important. For those not deeply involved in a subject, the multitude of contradictory results can be bewildering: Are shark populations disappearing, and does it matter? Is climate changing, and how much? Among those who have performed and read many studies, there tends to emerge a general agreement about the state of a problem. Scientific consensus now holds that many shark populations are in danger, though opinions vary on how severe the problem is. Consensus is that global climates are changing, though models differ somewhat on how rapidly they will change under different policy scenarios.

Sometimes new ideas emerge that cause major shifts in scientific consensus. Two centuries ago, geologists explained many earth features in terms of Noah's flood. The best scientists held that the flood created beaches well above modern sea level, scattered boulders erratically across the landscape, and gouged enormous valleys where there is no water now (fig. 2.7). Then the Swiss glaciologist Louis Agassiz and others suggested that the earth had once been much colder and that glaciers had covered large areas. Periodic ice ages better explained changing sea levels, boulders transported far from their source rock, and the great, gouged valleys. This new idea completely altered the way geologists explained their subject. Similarly, the idea of tectonic plate movement, in which continents shift slowly around the earth's surface, revolutionized the ways geologists, biogeographers, ecologists, and others explained the development of the earth and its life-forms.

These great changes in explanatory frameworks were termed **paradigm shifts** by Thomas Kuhn (1967), who studied revolutions in scientific thought. According to Kuhn, paradigm shifts occur when a majority of scientists accept that the old explanation no longer explains new observations very well. The shift is often contentious and political, because whole careers and worldviews, based on one

**FIGURE 2.7** Paradigm shifts change the ways we explain our world. Geologists now attribute Yosemite's valleys to glaciers, where once they believed Noah's flood carved its walls.

sort of research and explanation, can be undermined by a new model. Sometimes a revolution happens rather quickly. Quantum mechanics and Einstein's theory of relativity, for example, overturned classical physics in only about 30 years. Sometimes a whole generation of scholars has to retire before new paradigms can be accepted.

As you study this book, try to identify some of the paradigms that guide our investigations, explanations, and actions today. This is one of the skills involved in critical thinking, discussed in the introductory chapter of this book.

## Detecting pseudoscience relies on independent, critical thinking

Ideally, science should serve the needs of society. Deciding what those needs are, however, is often a matter of politics and economics. Should water be taken from a river for irrigation or left in the river for wildlife habitat? Should we force coal-burning power plants to reduce air pollution in order to lower health costs and respiratory illnesses, or are society and our economy better served by having cheap but dirty energy? These thorny questions are decided by a combination of scientific evidence, economic priorities, political positions, and ethical viewpoints. Among these factors, science is most widely regarded as a source of truth: Scientific conclusions are based on observations, so science must be objective and rational.

On the other hand, in every political debate, lawyers and lobbyists can find scientists who will back either side. Politicians hold up favorable studies, proclaiming them "sound science," while they dismiss others as "junk science." Opposing sides dispute the scientific authority of the study they dislike. What is "sound" science, anyway? If science is often embroiled in politics, does this mean that science is always a political process?

The opening case study for this chapter illustrates some of the connections between science, politics, ethics, and values. If you judge only from reports in newspapers or on television about this issue, you'd probably conclude that scientific opinion is about equally divided on whether global warming is a threat or not. In fact, the vast majority of scientists working on this issue agree with the proposition that the earth's climate is being affected by human activities. Only a handful of maverick scientists disagree. In a study of 928 papers published in refereed scientific journals between 1993 and 2003, not one disagreed with the broad scientific consensus on global warming.

Why, then, is there so much confusion among the public about this issue? Why do politicians continue to assert that the dangers of climate change are uncertain at best, or "the greatest hoax ever perpetrated on the American people," as James Inhofe, former chair of the Senate Committee on Environment and Public Works, claims. A part of confusion lies in the fact that media often present the debate as if it's evenly balanced. The fact that an overwhelming majority of working scientists are mostly in agreement on this issue doesn't make good drama, so the media give equal time to minority viewpoints just to make an interesting fight.

Perhaps a more important source of misinformation comes from corporate funding for articles and reports denying climate change. The ExxonMobil corporation, for example, has donated at least $20 million over the past decade to more than 100 think tanks, media outlets, and consumer, religious, and civil rights groups that promote skepticism about global warming. Some of these organizations sound like legitimate science or grassroots groups but are really only public relations operations. Others are run by individuals who find it rewarding to offer contrarian views. This tactic of spreading doubt and disbelief through innocuous-sounding organizations or seemingly authentic experts isn't limited to the climate change debate. It was pioneered by the tobacco industry to mislead the public about the dangers of smoking. Interestingly, some of the same individuals, groups, and lobbying firms employed by tobacco companies are now working to spread confusion about climate change.

Given this highly sophisticated battle of "experts," how do you interpret these disputes, and how do you decide whom to trust? The most important strategy is to apply critical thinking as you watch or read the news. What is the position of the person making the report? What is the source of their expertise? What economic or political interests do they serve? Do they appeal to your reason or to your emotions? Do they use inflammatory words (such as "junk"), or do they claim that scientific uncertainty makes their opponents' study meaningless? If they use statistics, what is the context for their numbers?

It helps to seek further information as you answer some of these questions. When you watch or read the news, you can look for places where reporting looks incomplete, you can consider sources and ask yourself what unspoken interests might lie behind the story.

Another strategy for deciphering the rhetoric is to remember that there are established standards of scientific work, and to investigate whether an "expert" follows these standards: Is the report peer-reviewed? Do a majority of scholars agree? Are the methods used to produce statistics well documented?

**TABLE 2.2**

**Questions for Baloney Detection**

1. How reliable are the sources of this claim? Is there reason to believe that they might have an agenda to pursue in this case?
2. Have the claims been verified by other sources? What data are presented in support of this opinion?
3. What position does the majority of the scientific community hold in this issue?
4. How does this claim fit with what we know about how the world works? Is this a reasonable assertion or does it contradict established theories?
5. Are the arguments balanced and logical? Have proponents of a particular position considered alternate points of view or only selected supportive evidence for their particular beliefs?
6. What do you know about the sources of funding for a particular position? Are they financed by groups with partisan goals?
7. Where was evidence for competing theories published? Has it undergone impartial peer review or it is only in proprietary publication?

Harvard's Edward O. Wilson writes, "We will always have contrarians whose sallies are characterized by willful ignorance, selective quotations, disregard for communications with genuine experts, and destructive campaigns to attract the attention of the media rather than scientists. They are the parasite load on scholars who earn success through the slow process of peer review and approval." How can we identify misinformation and questionable claims? The astronomer Carl Sagan proposed a "Baloney Detection Kit" containing the questions in table 2.2.

Most scientists have an interest in providing knowledge that is useful, and our ideas of what is useful and important depend partly on our worldviews and priorities. Science is not necessarily political, but it is often used for political aims. The main task of educated citizens is to discern where it is being misused or disregarded for purposes that undermine public interests.

### What's the relation between environmental science and environmentalism?

As you've learned already, environmental science uses scientific methods to study processes and systems in our environment. Environmentalism, on the other hand, works to influence attitudes and policies that affect our environment. Obviously, the former should be objective, while the latter has an subjective agenda. And yet, how can you gather information about what's happening to our environment without wanting to share that knowledge with policymakers? As the great conservationist Aldo Leopold wrote, "one of the penalties of an ecological education is that one lives alone in a world of wounds." He took the position that scientists have a duty not only to study the world, but to try to make it a better place.

This creates a dilemma for scientists. How can you gather information about habitat destruction, extinction of species, health risks of pollution, and other environmental threats without speaking out about the evidence you've collected? If you don't take a position, aren't you an accomplice in the damage done to our environment? Perhaps the best solution for this dilemma is to be very clear when you are acting as a scientist and when you are speaking as an advocate. When you're carrying out scientific studies, you need to be objective and impartial. When you're promoting solutions to the problems you've discovered, you need to make it clear if you're taking a partisan role. We'll try to separate these two functions as we discuss issues in this book, but you, also, need to be careful to distinguish between objective facts and opinions and interpretations as you read the text.

## 2.3 Systems

Environmental science seeks to understand **systems,** or networks of interactions among many interdependent factors. An ecosystem, for example, consists of living organisms and the soil, water, and air on which, or in which, these organisms live. Our global climate is an extremely complex system involving feedbacks between solar energy, atmosphere, oceans, fresh water, and living organisms. It's further complicated by our actions. You will encounter many systems in this book, such as ecological systems that include both living and nonliving components; geologic systems, which involve endless erosion and recycling of rocks in and on the earth's surface; and economic systems, which draw on natural resources and cultural information to circulate resources through societies and from one place to another.

A focus on systems is useful because it encourages us to inspect the relationships among components. Marine biologists worry about the disappearance of sharks because, as top predators, they affect population dynamics of many prey species. Loss of a top predator means more than just the loss of a species: it means possible disruption of the ecosystem as a whole. Biologists worry about sharks because we do not know how far cascading collapse could spread through marine ecosystems if sharks populations continue to fall.

### Systems are composed of processes

We can think of systems as consisting of flows and storage compartments. A familiar example of a system is a fish tank (fig. 2.8*a*) in which plants grow (using sunlight). As the plants grow, they store solar energy and carbon (from carbon dioxide dissolved in the water) in their leaves. A fish grazes on the plant, taking energy and nutrients from the leaves. Waste from the fish fertilizes algae growing on the sides of the tank. A snail consumes the algae, and waste from the snail (and fish) fertilizes the plants, which feed the fish . . . In this balanced fish tank, nutrients and energy flow between the plants and animals. Plants and animals store nutrients temporarily. The fish tank stays healthy as long as the size of compartments and the rates of the flows remain in balance.

If the plants grow faster than the fish can eat them, and if there is plenty of sunlight, then the plants grow new stems, which grow new leaves, which provide energy to store more stems, which grow more leaves . . . until the tank becomes clogged with vegetation. This situation, when a flow leads to compartment

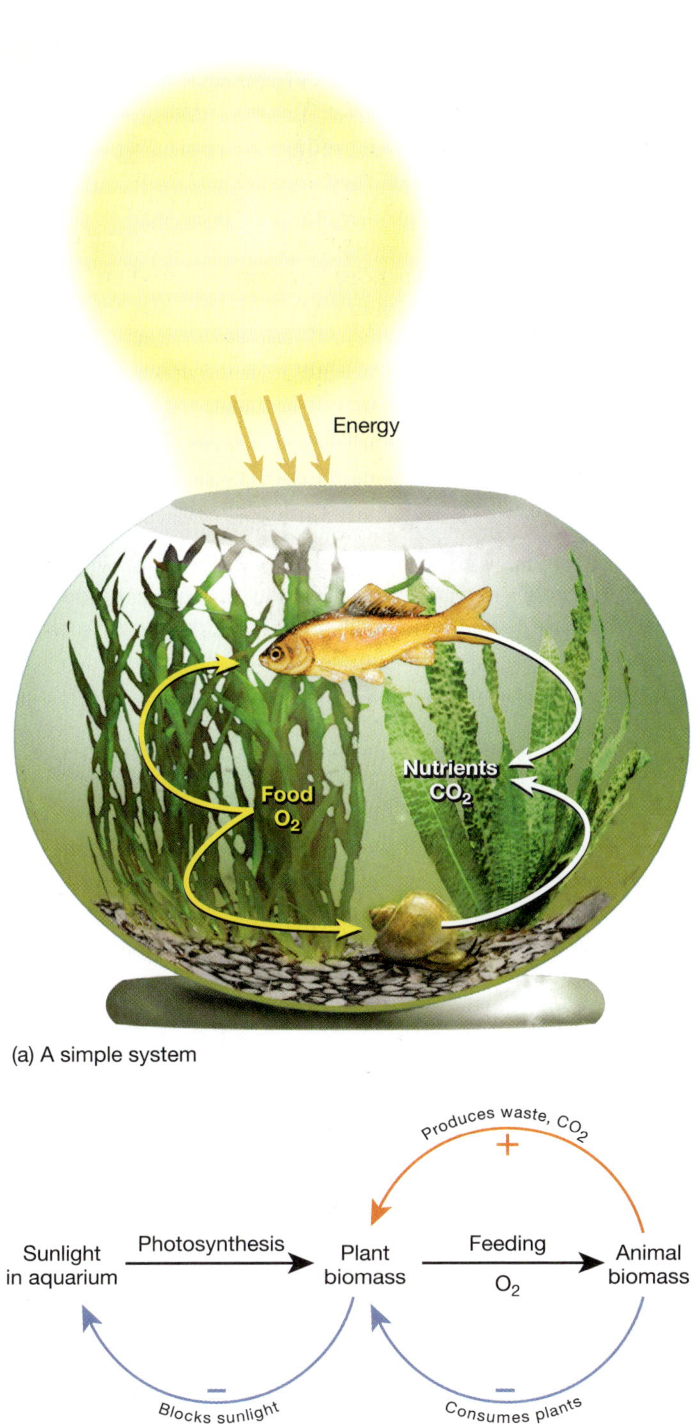


(a) A simple system



(b) Feedback loops

**FIGURE 2.8** Systems consist of compartments (also known as state variables) such as fish and plants, and flows of resources, such as nutrients or $O_2$ (a). Feedback loops (b) enhance or inhibit flows and the growth of compartments.

changes that further enhance the flow, we call a **positive feedback loop** (fig. 2.8*b*). If unchecked, positive feedback loops can lead a system to become increasingly unbalanced. Systems also have **negative feedback loops,** which dampen flows. In the fish tank system, suppose fish reproduce after feeding on the plants. Reproduction leads to more fish, which graze more heavily. Once the fish begin to overgraze the plants, there will be less food to go around, and reproduction will fall. Thus there is a negative feedback loop: reproduction → overgrazing → less reproduction.

When negative feedbacks prevent instability from positive feedback loops, we say the system is in homeostasis: It changes little from its stable condition. Your body is a system that is generally homeostatic. If your temperature changes more than slightly, you put on a sweater or take a cool drink; you tend to maintain fairly stable storage by eating when you are hungry and stopping when you are full.

If the system changes little over time, in the rate of flows or the relative size of storage compartments, we say the system is in **equilibrium.** Ecologists long considered that most forests normally existed in equilibrium: The number of trees, deer, birds, and other organisms should change little over the centuries if humans did not interfere. Increasingly, we see many ecosystems in terms of dynamic equilibrium: They undergo disturbance, then return to something like their previous state, then undergo disturbance again. Western lodgepole pine forests, for example, naturally experience periodic fires, which destroy expansive areas of pines. After the burn, sun-loving ground plants flourish for a while, and pine seeds released in the fire gradually grow to replace the dead pines. A century or so later, the mature pines begin to die, producing fuel that supports another stand-replacing fire.

## Disturbances and emergent properties are important characteristics of many systems

**Disturbances,** periodic, destructive events such as fire or floods, are a natural part of many systems. Floodplains need periodic floods to replenish soil nutrients and maintain a diversity of understory vegetation; prairies depend on fires to recycle nutrients, slow tree growth, and reduce built-up, undecayed plant litter; ponderosa pine forests need periodic fires to clear understory vegetation below the mature pines. When systems recover quickly from disturbance, as in the case of a floodplain, prairie, or lodgepole pine ecosystem, we say the system has **resilience.**

We can think of systems as open or closed. A closed system, in theory, is entirely self-contained. It receives no inputs of energy or material from outside. There are few examples of closed systems, however. Your fish tank receives energy and oxygen from outside. Ecosystems receive energy, nutrients, and materials from their surroundings. Systems that take inputs from elsewhere are **open systems.**

**Emergent properties** are characteristics of a whole, functioning system that are quantitatively or qualitatively greater than the sum of the system's parts. For example, you are a system, consisting of flows (energy, nutrients, water, and information) and compartments (your body stores water, nutrients, and energy). But you are also much more than the matter of which you are made. You can sing, dance, talk, produce ideas and art, and share feelings with those around you. All of these are properties that emerge because you function well as a system. Similarly, a tree is more than a mass of stored carbon. It provides structure to a forest, habitat for other organisms, shades and cools the ground, and

**FIGURE 2.9** Emergent properties of systems, including beautiful sights and sounds, make them exciting to study.

holds soil on a hillside with its roots. Often, we examine systems in terms of flows, compartments, and feedbacks, but it is the emergent properties that make systems exciting to study (fig. 2.9).

As you read this book, you will encounter many types of systems. Think about how flows, feedbacks, disturbance, and emergent properties occur in these systems.

## 2.4 Environmental Ethics and Worldviews

As we study our environment, science informs many of our actions, but values and ethics can be just as important when we decide what to consume and how to vote. **Ethics** is a branch of philosophy concerned with what actions are right and wrong. The study of ethics helps us identify principles that guide what we should and should not do. Because most people, at some level, want to do the right thing, ethics are an important consideration in our dealings with environment, as well as with other people.

Environmental ethics has to do with our moral obligations to the world around us. Do we have duties, obligations, or responsibilities toward other species or to nature in general? Are there ethical principles that constrain how we use resources or modify our environment? How do we weigh obligations to nature against obligations to human interests? The way we answer these questions depends on our view of the world.

### Worldviews express our deepest values

**Worldviews** are sets of basic beliefs, images, and understandings that shape how we see the world around us. They are very closely related to Thomas Kuhn's paradigms. Most of us learn these explanatory frameworks early in life, and we cannot change them easily. When we encounter evidence that doesn't fit our worldview, we often reject the evidence and cling to prior beliefs.

Our basic beliefs impact not only the way we think about ourselves and our place in the world; they also determine what questions are valid to ask. Slave owners in ancient Greece could not question whether slaves had rights; modern Americans have difficulty asking if wealth is the best way to determine who has access to resources; many devout religious people find it unimaginable that the basic tenets of their faith might be wrong.

Part of the reason worldviews are hard to change is that they are hard to recognize. We often have a hard time verbalizing the underlying reasons for our actions and our ideas. When you find yourself saying, "I can't explain this, it's just the way things are," you are probably describing part of your worldview.

Each of us has deeply held core values that we learn, often without being aware of them, from life experiences. Try to identify two or three of your core values. How do these values influence your actions with regard to: the car you drive? relationships with your family? relationships with friends? the way you vote? where you donate, or spend, money? what you do for recreation? How do your core values or worldviews differ from, or overlap with, those of other students in your class? What can you learn from comparing lists of core values and worldviews?

### Who (or what) has moral value?

Extending moral value to others is known as **moral extensionism.** Ancient Greeks granted moral value (the idea of worth, as well as responsibility) to adult male citizens. Slaves, women, and children had few rights and were essentially treated as property. Gradually, we have come to believe that all humans have certain inalienable rights, such as life, liberty, and the pursuit of happiness. No one can ethically treat another human as a mere object for their own pleasure or profit.

Do animals have rights? The philosopher René Descartes (1596–1650) argued that animals are mere automata, or machines, that can neither reason nor feel pain, so they also lack moral rights. In this view, other species are simply objects, and it is pointless to worry about their feelings. This debate is not just historical. In 2004, a report published in *Science* found that fish feel pain. While this may not seem surprising, it caused public outrage among recreational anglers, many of whom had, until then, managed to suppress worries about inflicting pain on fish.

For many people, moral extensionism now extends to granting some degree of moral value to animals, and even plants or inanimate objects (fig. 2.10).

### Living things can have intrinsic or instrumental value

Rather than couch ethics strictly in terms of rights, some philosophers prefer to consider values. Value is a measure of the worth of something. But value can be either inherent or conferred. All humans, we believe, have **inherent value**—an intrinsic or innate worth—simply because they are human. They deserve moral consideration no matter who they are or what they do. Tools, on the other hand, have conferred, or **instrumental value.** They are worth something only because they are valued by someone who matters. If I hurt you without good reason, I owe you an apology. If I borrow your car and smash it into a tree, however, I don't

**FIGURE 2.10** Moral extensionism describes an increasing consideration of moral value in other living things—or even nonliving things.

owe the car an apology. I owe *you* the apology—or reimbursement—for ruining your car. The car is valuable only because you want to use it. It doesn't have inherent values or rights of itself.

How does this apply to nonhumans? Domestic animals clearly have an instrumental value because they are useful to their owners. But some philosophers would say they also have inherent values and interests. By living, breathing, struggling to stay alive, the animal carries on its own life independent of its usefulness to someone else.

Some people believe that even nonliving things also have inherent worth. Rocks, rivers, mountains, landscapes, and certainly the earth itself, have value. These things were in existence before we came along and we couldn't re-create them if they are altered or destroyed. In a landmark 1969 court case, the Sierra Club sued the Disney Corporation on behalf of the trees, rocks, and wildlife of Mineral King Valley in the Sierra Nevada Mountains (fig. 2.11) where Disney wanted to build a ski resort. The Sierra Club argued that it represented the interests of beings that could not speak for themselves in court.

A legal brief entitled *Should Trees Have Standing?,* written for this case by Christopher D. Stone, proposed that organisms as well as ecological systems and processes should have standing (or rights) in court. After all, corporations—such as Disney—are treated as persons and given legal rights even though they are really only figments of our imagination. Why shouldn't nature have similar standing? The case went all the way to the Supreme Court but was overturned on a technicality. In the meantime, Disney lost interest in the project and the ski resort was never built. What do you think? Where would you draw the line of what deserves moral considerability? Are there ethical limits on what we can do to nature?

**FIGURE 2.11** Mineral King Valley at the southern border of Sequoia National Park was the focus of an important environmental law case in 1969. The Disney Corporation wanted to build a ski resort here, but the Sierra Club sued to protect the valley on behalf of the trees, rocks, and native wildlife.

### Is discrimination against other people related to our attitudes toward nature?

Many feminist philosophers argue that thoughtless mistreatment of nature is related to how we treat minorities, women, children, and others who lack power and prestige. **Ecofeminism,** the main framework for this perspective, holds that patriarchal attitudes in society are based on domination and ideas of superiority. If you believe that nature has only instrumental value, then you probably also believe you can do anything you please to it. In societies where some people are granted greater value than others, discrimination and exploitation of those with lower status also are likely to be regarded as acceptable. Ecofeminists argue that we need less domination and more cooperation with both nature and other people to achieve a peaceful, sustainable society.

## 2.5 Faith-Based Conservation and Environmental Justice

As the opening case study for this chapter shows, religious and ethical values can play important roles in environmental debates. Often people's most deeply held worldviews and values are expressed in their religious beliefs. Some of our most pressing environmental problems don't need technological or scientific solutions; they're not so much a question of what we're able to do, but what we're willing to do. Are we willing to take the steps necessary to stop global climate change? Do our values and ethics require us to do so? In this section, we'll look at some religious perspectives and how they influence our attitudes toward nature.

Environmental scientists have long been concerned about religious perspectives. In 1967, historian Lynn White, Jr., published a widely influential paper, "The Historic Roots of Our Ecological Crisis." He argued that Christian societies have often exploited natural resources carelessly because the Bible says that God commanded Adam and Eve to dominate nature: "Be fruitful, and multiply, and replenish the earth and subdue it: and have dominion over the fish of the sea, and over the fowl of the air, and over every living thing that moveth upon the earth" (Genesis 1:28). Since then, many religious scholars have pointed out that God also commanded Adam and Eve to care for the garden they were given, "to till it and keep it" (Genesis 2:15). Furthermore, Noah was commanded to preserve individuals of all living species, so that they would not perish in the great Flood. Passages such as these inspire many Christians to insist that it is our responsibility to act as stewards of nature, and to care for God's creations.

Both calls for environmental stewardship and anthropocentric domination over nature can be found in the writings of most major faiths. The Koran teaches that "each being exists by virtue of the truth and is also owed its due according to nature," a view that extends moral rights and value to all other creatures. Hinduism and Buddhism teach *ahimsa,* or the practice of not harming other living creatures, because all living beings are divinely connected (fig. 2.12).

**FIGURE 2.12** Many religions emphasize the divine relationships among humans and the natural world. The Tibetan Buddhist goddess Tara represents compassion for all beings.

## Many faiths support environmental conservation

The idea of **stewardship,** or taking care of the resources we are given, inspires many religious leaders to promote conservation. "Creation care" is a term that has become prominent among evangelical Christians in the United States. In 1995, representatives of nine major religions met in Ohito, Japan, to discuss views of environmental stewardship in their various traditions. The resulting document, the Ohito Declaration, outlined common beliefs and responsibilities of these different faiths toward protecting the earth and its life (table 2.3). In recent years, religious organizations have played important roles in nature protection. A coalition of evangelical Christians has been instrumental in opposing Congressional attacks on the U.S. Endangered Species Act. Similarly, the Central Conference of American Rabbis has declared that desecration of the Headwaters Redwood Forest in California breaks our covenant with the Creator.

Religious concern extends beyond our treatment of plants and animals. Pope John Paul II and Orthodox Patriarch Bartholomew called on countries bordering the Black Sea to stop pollution, saying that "to commit a crime against nature is a sin." In addition to its campaign to combat global warming described at the beginning of this chapter, the Creation Care Network has also launched initiatives against energy inefficiency, mercury pollution, mountaintop removal mining, and endangered species destruction. For many people, religious beliefs provide the best justification for environmental protection.

## Environmental justice combines civil rights and environmental protection

People of color in the United States and around the world are subjected to a disproportionately high level of environmental health risks in their neighborhoods and on their jobs. Minorities, who tend to be poorer and more disadvantaged than other residents, work in the dirtiest jobs where they are exposed to toxic chemicals and other hazards. More often than not they also live in urban ghettos, barrios, reservations, and rural poverty pockets that have shockingly high pollution levels and are increasingly the site of unpopular industrial facilities, such as toxic waste dumps, landfills, smelters, refineries, and incinerators. **Environmental justice** combines civil rights with environmental protection to demand a safe, healthy, life-giving environment for everyone.

**TABLE 2.3**

**Principles and Actions in the Ohito Declaration**

**Spiritual Principles**

1. Religious beliefs and traditions call us to care for the earth.
2. For people of faith, maintaining and sustaining environmental life systems is a religious responsibility.
3. Environmental understanding is enhanced when people learn from the example of prophets and of nature itself.
4. People of faith should give more emphasis to a higher quality of life, in preference to a higher standard of living, recognizing that greed and avarice are root causes of environmental degradation and human debasement.
5. People of faith should be involved in the conservation and development process.

**Recommended Courses of Action**

The Ohito Declaration calls upon religious leaders and communities to

1. emphasize environmental issues within religious teaching: faith should be taught and practiced as if nature mattered.
2. commit themselves to sustainable practices and encourage community use of their land.
3. promote environmental education, especially among youth and children.
4. pursue peacemaking as an essential component of conservation action.
5. take up the challenge of instituting fair trading practices devoid of financial, economic, and political exploitation.

Among the evidence of environmental injustice is the fact that three out of five African-Americans and Hispanics, and nearly half of all Native Americans, Asians, and Pacific Islanders live in communities with one or more uncontrolled toxic waste sites, incinerators, or major landfills, while fewer than 10 percent of all whites live in these areas. Using zip codes or census tracts as a unit of measurement, researchers found that minorities make up twice as large a population share in communities with these locally unwanted land uses (**LULUs**) as in communities without them. A 2006 study using "distance-based" methods found an even greater correlation between race and location of hazardous waste facilities.

Although it is difficult to distinguish between race, class, historical locations of ethnic groups, economic disparities, and other social factors in these disputes, racial origins often seem to play a role in exposure to environmental hazards. Simple correlation doesn't prove causation; still, while poor people in general are more likely to live in polluted neighborhoods than rich people, the discrepancy between the pollution exposure of middle class blacks and middle class whites is even greater than the difference between poorer whites and blacks. Where upper class whites can "vote with their feet" and move out of polluted and dangerous neighborhoods, blacks and other minorities are restricted by color barriers and prejudice (overt or covert) to the less desirable locations (fig. 2.13).

**FIGURE 2.13** Poor people and people of color often live in the most dangerous and least desirable places. Here children play next to a chemical refinery in Texas City, Texas.

## Environmental racism distributes hazards inequitably

Racial prejudice is a belief that someone is inferior merely because of their race. Racism is prejudice with power. **Environmental racism** is inequitable distribution of environmental hazards based on race. Evidence of environmental racism can be seen in lead poisoning in children. The Federal Agency for Toxic Substances and Disease Registry considers lead poisoning to be the number one environmental health problem for children in the United States. Some 4 million children—many of whom are African American, Latino, Native American, or Asian, and most of whom live in inner-city areas—have dangerously high lead levels in their bodies. This lead is absorbed from old lead-based house paint, contaminated drinking water from lead pipes or lead solder, and soil polluted by industrial effluents and automobile exhaust. The evidence of racism is that at every income level, whether rich or poor, black children are two to three times more likely than whites to suffer from lead poisoning.

Because of their quasi-independent status, most Native-American reservations are considered sovereign nations that are not covered by state environmental regulations. Court decisions holding that reservations are specifically exempt from hazardous waste storage and disposal regulations have resulted in a land rush of seductive offers from waste disposal companies to Native-American reservations for onsite waste dumps, incinerators, and landfills. The short-term economic incentives can be overwhelming for communities in which adult unemployment runs between 60 and 80 percent. Uneducated, powerless people often can be tricked or intimidated into signing environmentally and socially disastrous contracts. Nearly every tribe in America has been approached with proposals for some dangerous industry or waste facility.

The practice of targeting poor communities of color in the developing nations for waste disposal and/or experimentation with risky technologies has been described as **toxic colonialism.** Internationally, the trade in toxic waste has mushroomed in recent years as wealthy countries have become aware of the risks of industrial refuse. Poor, minority communities at home and abroad are being increasingly targeted as places to dump unwanted wastes. Although a treaty regulating international shipping of toxics was signed by 105 nations in 1989, millions of tons of toxic and hazardous materials continue to move—legally or illegally—from the richer countries to the poorer ones every year. This issue is discussed further in chapter 23.

Another form of toxic colonialism is the flight of polluting industries from developed nations and states where control requirements are stringent to less developed areas where regulations are lax and local politicians are easily co-opted. For example, more than 2,000 *maquiladoras,* or assembly plants operated by American, Japanese, or other foreign companies, are now located along the United States/Mexico border to take advantage of favorable import quotas, low wages, and weak pollution-control laws.

Mexican laborers work under appalling conditions in some of these factories, assembling imported components into consumer goods to be exported to the United States. Although the jobs are demeaning, dangerous, and pay far less than the minimum U.S. wage, large populations have been attracted to squalid shantytowns along the border where industrial effluents poison the air and water (fig. 2.14).

**FIGURE 2.14** Living conditions in the *colonias* or unplanned settlements along the U.S./Mexican border often are substandard. Is this evidence of racism or simply a result of poverty and poor planning?

The U.S. Environmental Justice Act was established in 1992 to identify areas threatened by the highest levels of toxic chemicals, assess health effects caused by emissions of those chemicals, and ensure that groups or individuals residing within those areas have opportunities and resources to participate in public discussions concerning siting and cleanup of industrial facilities. Perhaps we need something similar worldwide.

## Conclusion

Many of the most difficult environmental problems we face are moral or ethical dilemmas rather than technological crises. Are we willing to do what we need to leave a habitable world for our children and grandchildren? Do we have a moral obligation to the poor people of the world whose lives are harmed by the way we've chosen to use (or abuse) resources? Scientific uncertainties about what may happen in the future make many of these decisions even more difficult. However, those who oppose policies that may be expensive or personally inconvenient often claim more doubt or ambiguity than is scientifically justified.

Rigorous, objective, empirical science based on unbiased, systematic research and open, honest communication among a community of scientists is one of the most important features of the modern age. It's how we discover information about the world around us. It has given rise to most of the comforts and conveniences we enjoy today. Not everyone is able to—or wants to—be a practicing scientist, but because science is such a powerful force in modern society, all citizens should understand the basics about how science works, and what questions it is able, and not able, to answer.

"No man is an island," wrote John Donne; nor does any organism live in complete isolation. All of us are part of interlocking systems that involve interactions between many interdependent processes and factors. Often those systems have properties that are greater than or different from the individual components that produce them. Understanding how systems work is another essential element in an environmental education. Some systems exist in equilibrium, changing little over time. Others may be highly variable, responding to disturbances. Resilience, or the ability to recover from disturbance is an important quality of some systems.

As environmental citizens, science informs many of our decisions, but values and ethics can be just as important when we choose what to do or not do. Environmental ethics has to do with our moral obligations to the world around us. Do we have duties, obligations, or responsibilities toward nature? Do other species—or even inanimate objects or features in nature—have rights and values? How we answer these questions depends on our worldview, which generally expresses our most deeply held beliefs. While some people use religious beliefs to justify domination and exploitation of nature, others believe that God calls us to be stewards and caretakers of nature. Different religious interpretations have become critical aspects of debates about how to manage resources.

Environmental justice combines civil rights with environmental protection to demand a safe, healthy, life-giving environment for everyone. Too often, minorities and poor people are exposed unfairly to toxins, hazards, and degraded environments.

Together the fish, bears, birds, and other organisms transport and distribute nutrients in and around the McNeil River, Alaska.

CHAPTER 3

# Matter, Energy, and Life

*When one tugs at a single thing in nature, he finds it attached to the rest of the world.*

*—John Muir—*

## LEARNING OUTCOMES

After studying this introduction, you should be able to:

**3.1** Describe matter, atoms, and molecules and give simple examples of the role of four major kinds of organic compounds in living cells.

**3.2** Define energy and explain how thermodynamics regulates ecosystems.

**3.3** Understand how living organisms capture energy and create organic compounds.

**3.4** Define species, populations, communities, and ecosystems, and summarize the ecological significance of trophic levels.

**3.5** Compare the ways that water, carbon, nitrogen, sulfur, and phosphorus cycle within ecosystems.

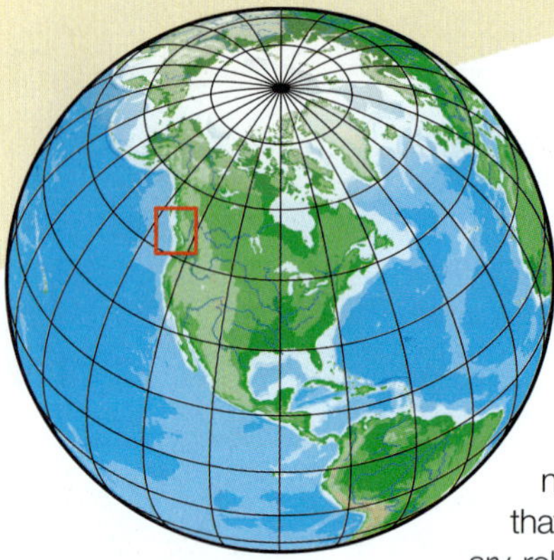

# Case Study Why Trees Need Salmon

Ecologists have long known that salmon need clean, fast-moving streams to breed, and that clear streams need healthy forests. Surprising new evidence now indicates that some forests themselves need salmon to remain healthy, and that bears play an important intermediary role in this dynamic relationship.

The yearly return of salmon from the open Pacific Ocean to coastal waters of western North America is one of nature's grand displays. Salmon (*Onchorhyncus* sp.) are anadromous: They hatch in freshwater lakes and streams, spend most of their lives at sea, then return to the stream where they were born, to breed and die. To reproduce successfully, these fish require clear, cold, shaded streams and clean gravel riverbeds. If forests are stripped from riverbanks and surrounding hillsides, sediment washes down into streams, clogging gravel beds and suffocating eggs. Open to the sunlight, the water warms, lowering its oxygen levels, and reducing survival rates of eggs and young fish.

Every year, as millions of fish return to spawn and die in rivers of the Pacific Northwest, they provide a bonanza for bears, eagles, and other species (see photograph previous page). Ecologist Tom Reimchen estimates that each bear fishing in British Columbia's rivers catches about 700 fish during the 45-day spawn, and that 70 percent of the bear's annual protein comes from salmon. After a quick bite on the head to kill the fish, the bears drag their prey back into the forest, where they can feed undisturbed. Some bears have been observed carrying fish as much as 800 m (0.5 mi) from the river before feeding on them.

Bears don't eat everything they catch. They leave about half of each carcass to be scavenged by eagles, martens, crows, ravens, and gulls. A diversity of insects, including flies and beetles, also feed on the leftovers. Within a week, all the soft tissue is consumed, leaving only a bony skeleton. Reimchen calculates that between the nutrients leeching directly from decomposing carcasses and the excreta from bears and other scavengers, the fish provide about 120 kg of nitrogen per hectare of forest along salmon-spawning rivers. This is comparable to the rate of fertilizer applied by industry to commercial forest plantations. Altogether, British Columbia's 80,000 to 120,000 brown and black bears could be transferring 60 million kg of salmon tissue into the rainforest every year.

How do ecologists know that trees absorb nitrogen from salmon? Analyzing different kinds of nitrogen atoms, researchers can distinguish between marine-derived nitrogen (MDN) and that from terrestrial sources. Marine phytoplankton (tiny floating plant cells) have more of a rare, heavy form of nitrogen called $^{15}N$ compared to most terrestrial vegetation, in which $^{14}N$, the more common, lighter form, predominates. Using a machine called a mass spectrometer, researchers can separate and measure the kinds and amounts of nitrogen in different tissues. We'll discuss different forms of atoms (called isotopes) later in this chapter. Because salmon spend most of their lives feeding on dense clouds of plankton far out to sea, they have higher ratio of $^{15}N/^{14}N$ in their bodies than do most freshwater or terrestrial organisms. When the fish die and decompose, they contribute their nitrogen to the ecosystem. Bears and other scavengers distribute this nitrogen throughout the forest where they drop fish carcasses or defecate in the woods.

Robert Naiman and James Helfield from the University of Washington found that foliage of spruce trees growing in bear-impacted areas is significantly enriched with MDN relative to similar trees growing at comparable distances from streams with and without spawning salmon (fig. 3.1). These results suggest that in feeding on salmon, bears play an important role in transferring MDN from the stream to the riparian (streamside) forest. Nitrogen is often a limiting nutrient for rainforest vegetation. Tree ring studies show that when salmon are abundant, trees grow up to three times as fast as when salmon are scarce. For some streamside trees, researchers estimate that between one-quarter to one-half of all their nitrogen is derived from salmon. Not only do salmon replenish the forest, but they also vitalize the streams and lakes with carbon, nitrogen, phosphorous, and micronutrients. Nearly 50 percent of the nutrients that juvenile salmon consume comes from dead parents.

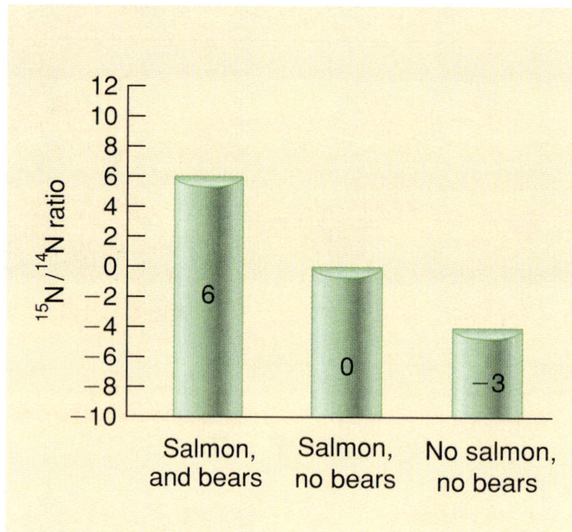


**FIGURE 3.1** Nitrogen in trees near rivers with both salmon and bears have a significantly higher $^{15}N/^{14}N$ ratio in their foliage than do trees with salmon and no bears, or those without either salmon or bears. These findings suggest an important role for both fish and bears in distributing marine-derived nitrogen in riparian forests.

This research is important because salmon stocks are dwindling throughout the Pacific Northwest. In Washington, Oregon, and California, most salmon populations have fallen by 90 percent from their historic numbers, and some stocks are now extinct. Because of the close relationship of salmon and the trees, biologists argue, forest, wildlife, and fish management need to be integrated. Each population—rainforest trees, bears, hatchlings, and ocean-going fish—affects the stability of the others. Salmon need healthy forests and streams to reproduce successfully, and forests and bears need abundant salmon. Stream ecosystems need standing trees to retain soil and provide shade. So healthy streams depend on fish, just as the fish depend on the streams. As this case shows, the flow of nutrients and energy between organisms can be intricate and complex. Relationships between apparently separate environments, such as rivers and forests, can be equally complex and important. In this chapter we'll explore some of these relationships among organisms and between organisms and their environment.

# Case Study *continued*


**For more information, see**

Helfield, J. M., and R. J. Naiman. 2001. Effects of salmon-derived nitrogen on riparian forest growth and implications for stream productivity. *Ecology* 82(9):2403–9.

Reimchen, T., et al. 2003. Isotopic evidence for enrichment of salmon-derived nutrients in vegetation, soil and insects in riparian zones in coastal British Columbia. *American Fisheries Society Symposium* 34:59–69.


## 3.1 Elements of Life

How are nutrients being exchanged between salmon, bears, and trees in the opening case study of this chapter? And what energy source keeps this whole system running? These questions are at the core of **ecology,** the scientific study of relationships between organisms and their environment. In this chapter we'll introduce a number of concepts that are essential to understanding how living things function in their environment.

As an introduction to principles of ecology, this chapter first reviews the nature of matter and energy, then explores the ways organisms acquire and use energy and chemical elements. Then we'll investigate feeding relationships among organisms—the ways that energy and nutrients are passed from one living thing to another—forming the basis of ecosystems. Finally, we'll review some of the key substances that cycle through organisms, ecosytems, and our environment.

In chapters 4 and 5 we'll continue investigating concepts of ecology: general organization of ecosystems by environmental types and landscapes, and principles of population growth.

In a sense, every organism is a chemical factory that captures matter and energy from its environment and transforms them into structures and processes that make life possible. To understand how these processes work, we will begin with some of the fundamental properties of matter and energy.

### Matter is made of atoms, molecules, and compounds

Everything that takes up space and has mass is **matter.** Matter exists in three distinct states—solid, liquid, and gas—due to differences in the arrangement of its constitutive particles. Water, for example, can exist as ice (solid), as liquid water, or as water vapor (gas).

Under ordinary circumstances, matter is neither created nor destroyed but rather is recycled over and over again. Some of the molecules that make up your body probably contain atoms that once made up the body of a dinosaur and most certainly were part of many smaller prehistoric organisms, as chemical elements are used and reused by living organisms. Matter is transformed and combined in different ways, but it doesn't disappear; everything goes somewhere. These statements paraphrase the physical principle of **conservation of matter.**

How does this principle apply to human relationships with the biosphere? Particularly in affluent societies, we use natural resources to produce an incredible amount of "disposable" consumer goods. If everything goes somewhere, where do the things we dispose of go after the garbage truck leaves? As the sheer amount of "disposed-of stuff" increases, we are having greater problems finding places to put it. Ultimately, there is no "away" where we can throw things we don't want any more.

Matter consists of **elements,** which are substances that cannot be broken down into simpler forms by ordinary chemical reactions. Each of the 118 known elements (92 natural, plus 26 created under special conditions) has distinct chemical characteristics. Just four elements—oxygen, carbon, hydrogen, and nitrogen—are responsible for more than 96 percent of the mass of most living organisms.

All elements are composed of **atoms,** which are the smallest particles that exhibit the characteristics of the element. Atoms are composed of positively charged protons, negatively charged electrons, and electrically neutral neutrons. Protons and neutrons, which have approximately the same mass, are clustered in the nucleus in the center of the atom (fig. 3.2). Electrons, which are tiny in comparison to the other particles, orbit the nucleus at the speed of light.

Each element has a characteristic number of protons per atom, called its **atomic number.** The number of neutrons in different atoms of the same element can vary slightly. Thus, the atomic mass, which is the sum of the protons and neutrons in each nucleus, also can vary. We call forms of a single element that differ in atomic mass **isotopes.** For example, hydrogen, the lightest element, normally has only one proton (and no neutrons) in its nucleus. A small percentage of hydrogen atoms have one proton and one neutron. We call this isotope deuterium ($^{2}$H). An even smaller percentage of natural hydrogen called tritium ($^{3}$H) has one proton plus two neutrons. The heavy form of nitrogen ($^{15}$N) mentioned in the opening story of this

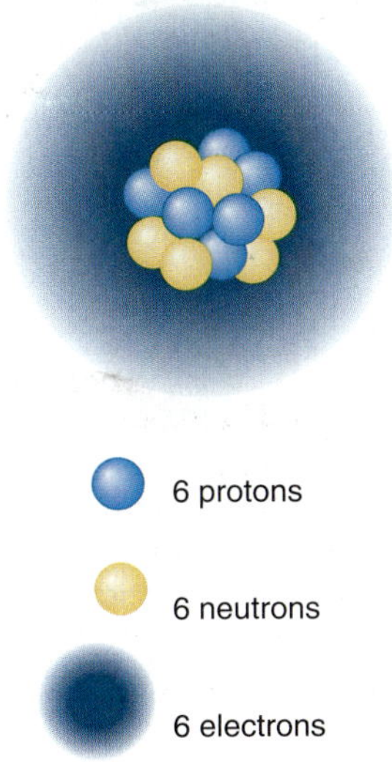


**FIGURE 3.2** As difficult as it may be to imagine when you look at a solid object, all matter is composed of tiny, moving particles, separated by space and held together by energy. It is hard to capture these dynamic relationships in a drawing. This model represents carbon-12, with a nucleus of six protons and six neutrons; the six electrons are represented as a fuzzy cloud of potential locations rather than as individual particles.

chapter has one more neutron in its nucleus than does the more common $^{14}N$. Both these nitrogen isotopes are stable but some isotopes are unstable—that is, they may spontaneously emit electromagnetic energy, or subatomic particles, or both. Radioactive waste, and nuclear energy, result from unstable isotopes of elements such as uranium and plutonium.

## Chemical bonds hold molecules together

Atoms often join to form **compounds,** or substances composed of different kinds of atoms (fig. 3.3). A pair or group of atoms that can exist as a single unit is known as a **molecule.** Some elements commonly occur as molecules, such as molecular oxygen ($O_2$) or molecular nitrogen ($N_2$), and some compounds can exist as molecules, such as glucose ($C_6H_{12}O_6$). In contrast to these molecules, sodium chloride (NaCl, table salt) is a compound that cannot exist as a single pair of atoms. Instead it occurs in a large mass of Na and Cl atoms or as two ions, $Na^+$ and $Cl^-$, in solution. Most molecules consist of only a few atoms. Others, such as proteins and nucleic acids, can include millions or even billions of atoms.

When ions with opposite charges form a compound, the electrical attraction holding them together is an *ionic bond.* Sometimes atoms form bonds by *sharing* electrons. For example, two hydrogen atoms can bond by sharing a pair of electrons—they orbit the two hydrogen nuclei equally and hold the atoms together. Such electron-sharing bonds are known as *covalent bonds.* Carbon (C) can form covalent bonds simultaneously with four other atoms, so carbon can create complex structures such as sugars and proteins. Atoms in covalent bonds do not always share electrons evenly. An important example in environmental science is the covalent bonds in water ($H_2O$). The oxygen atom attracts the shared electrons more strongly than do the two hydrogen atoms. Consequently, the hydrogen portion of the molecule has a slight positive charge, while the oxygen has a slight negative charge. These charges create a mild attraction between water molecules, so that water tends to be somewhat cohesive. This fact helps explain some of the remarkable properties of water (Exploring Science, p. 55).

When an atom gives up one or more electrons, we say it is *oxidized* (because it is very often oxygen that takes the electron, as bonds are formed with this very common and highly reactive element). When an atom gains electrons, we say it is *reduced.* Chemical reactions necessary for life involve oxidation and reduction: Oxidation of sugar and starch molecules, for example, is an important part of how you gain energy from food.

Forming bonds usually releases energy. Breaking bonds generally requires energy. For example, burning wood is exothermic (releases heat) because the energy released in the formation of carbon dioxide and water is greater than the energy required to break the bonds in cellulose (and other organic compounds in wood). Generally, some energy input (activation energy) is needed to initiate these reactions. In your fireplace, a match might provide the needed activation energy. In your car, a spark from the battery provides activation energy to initiate the oxidation (burning) of gasoline.

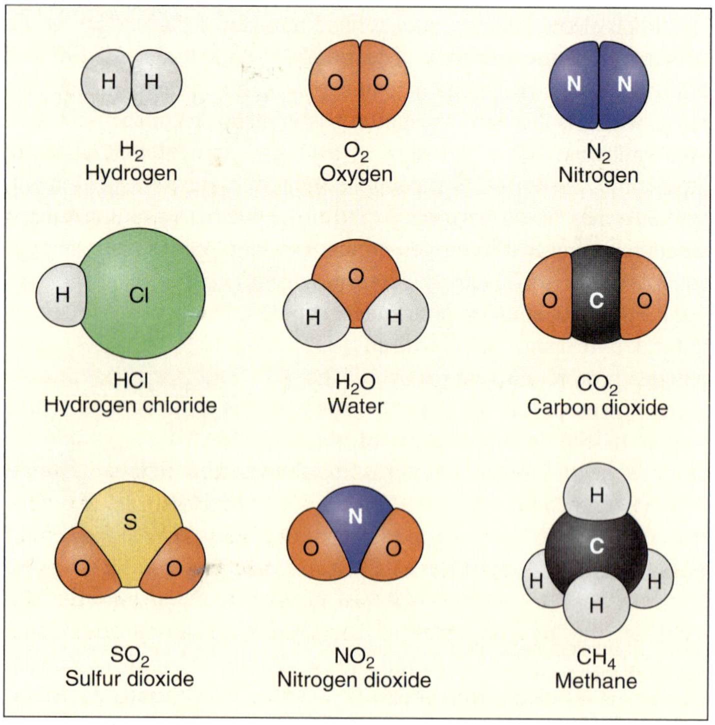


**FIGURE 3.3** These common molecules, with atoms held together by covalent bonds, are important components of the atmosphere or important pollutants.

## Electrical charge is an important chemical characteristic

Atoms frequently gain or lose electrons, acquiring a negative or positive electrical charge. Charged atoms (or combinations of atoms) are called **ions.** Negatively charged ions (with one or more extra electrons) are *anions.* Positively charged ions are *cations.* A hydrogen (H) atom, for example, can give up its sole electron to become a hydrogen ion ($H^+$). Chlorine (Cl) readily gains electrons, forming chlorine ions ($Cl^-$).

Substances that readily give up hydrogen ions in water are known as **acids.** Hydrochloric acid, for example, dissociates in water to form $H^+$ and $Cl^-$ ions. In later chapters, you may read about acid rain (which has an abundance of $H^+$ ions), acid mine drainage, and many other environmental problems involving acids. In general, acids cause environmental damage because the $H^+$ ions react readily with living tissues (such as your skin or tissues of fish larvae) and with nonliving substances (such as the limestone on buildings, which erodes under acid rain).

Substances that readily bond with $H^+$ ions are called **bases** or alkaline substances. Sodium hydroxide (NaOH), for example, releases hydroxide ions ($OH^-$) that bond with $H^+$ ions in water. Bases can be highly reactive, so they also cause significant

## *Exploring* SCIENCE — A "Water Planet"

If travelers from another solar system were to visit our lovely, cool, blue planet, they might call it Aqua rather than Terra because of its outstanding feature: the abundance of streams, rivers, lakes, and oceans of liquid water. Our planet is the only place we know where water exists as a liquid in any appreciable quantity. Water covers nearly three-fourths of the earth's surface and moves around constantly via the hydrologic cycle (discussed in chapter 15) that distributes nutrients, replenishes freshwater supplies, and shapes the land. Water makes up 60 to 70 percent of the weight of most living organisms. It fills cells, giving form and support to tissues. Among water's unique, almost magical qualities, are the following:

1. Water molecules are polar, that is, they have a slight positive charge on one side and a slight negative charge on the other side. Therefore, water readily dissolves polar or ionic substances, including sugars and nutrients, and carries materials to and from cells.
2. Water is the only inorganic liquid that exists under normal conditions at temperatures suitable for life. Most substances exist as either a solid or a gas, with only a very narrow liquid temperature range. Organisms synthesize organic compounds such as oils and alcohols that remain liquid at ambient temperatures and are therefore extremely valuable to life, but the original and predominant liquid in nature is water.
3. Water molecules are cohesive, tending to stick together tenaciously. You have experienced this property if you have ever done a belly flop off a diving board. Water has the highest surface tension of any common, natural liquid. Water also adheres to surfaces. As a result, water is subject to *capillary action:* it can be drawn into small channels. Without capillary action, movement of water and nutrients into groundwater reservoirs and through living organisms might not be possible.
4. Water is unique in that it expands when it crystallizes. Most substances shrink as they change from liquid to solid. Ice floats because it is less dense than liquid water. When temperatures fall below freezing, the surface layers of lakes, rivers, and oceans cool faster and freeze before deeper water. Floating ice then insulates underlying layers, keeping most water bodies liquid (and aquatic organisms alive) throughout the winter in most places. Without this feature, many aquatic systems would freeze solid in winter.
5. Water has a high heat of vaporization, using a great deal of heat to convert from liquid to vapor. Consequently, evaporating water is an effective way for organisms to shed excess heat. Many animals pant or sweat to moisten evaporative cooling surfaces. Why do you feel less comfortable on a hot, humid day than on a hot, dry day? Because the water vapor–laden air inhibits the rate of evaporation from your skin, thereby impairing your ability to shed heat.
6. Water also has a high specific heat; that is, a great deal of heat is absorbed before it changes temperature. The slow response of water to temperature change helps moderate global temperatures, keeping the environment warm in winter and cool in summer. This effect is especially noticeable near the ocean, but it is important globally.

Surface tension is demonstrated by the resistance of a water surface to penetration, as when it is walked upon by a water strider.

All these properties make water a unique and vitally important component of the ecological cycles that move materials and energy and make life on earth possible.

environmental problems. Acids and bases can also be essential to living things: The acids in your stomach dissolve food, for example, and acids in soil help make nutrients available to growing plants.

We describe the strength of an acid and base by its **pH,** the negative logarithm of its concentration of $H^+$ ions (fig. 3.4). Acids have a pH below 7; bases have a pH greater than 7. A solution of exactly pH 7 is "neutral." Because the pH scale is logarithmic, pH 6 represents *ten times* more hydrogen ions in solution than pH 7.

A solution can be neutralized by adding buffers, or substances that accept or release hydrogen ions. In the environment, for example, alkaline rock can buffer acidic precipitation, decreasing its acidity. Lakes with acidic bedrock, such as granite, are especially vulnerable to acid rain because they have little buffering capacity.

## Organic compounds have a carbon backbone

Organisms use some elements in abundance, others in trace amounts, and others not at all. Certain vital substances are concentrated within cells, while others are actively excluded. Carbon is a particularly important element because chains and rings of carbon atoms form the skeletons of **organic compounds,** the material of which biomolecules, and therefore living organisms, are made.

The four major categories of organic compounds in living things ("bio-organic compounds") are lipids, carbohydrates, proteins, and nucleic acids. Lipids (including fats and oils) store energy for cells, and they provide the core of cell membranes and other structures. Lipids do not readily dissolve in water, and their

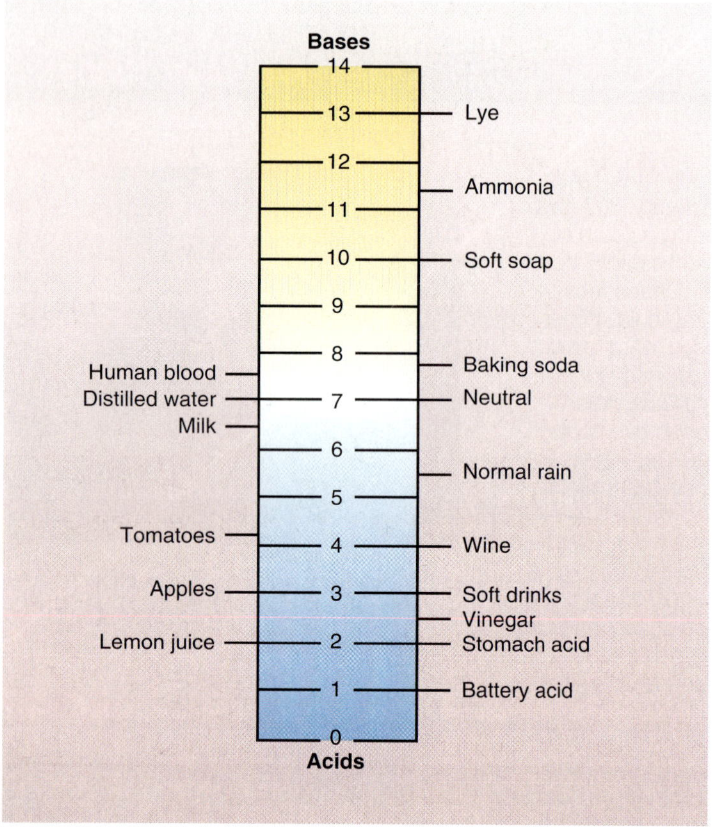


**FIGURE 3.4** The pH scale. The numbers represent the negative logarithm of the hydrogen ion concentration in water.

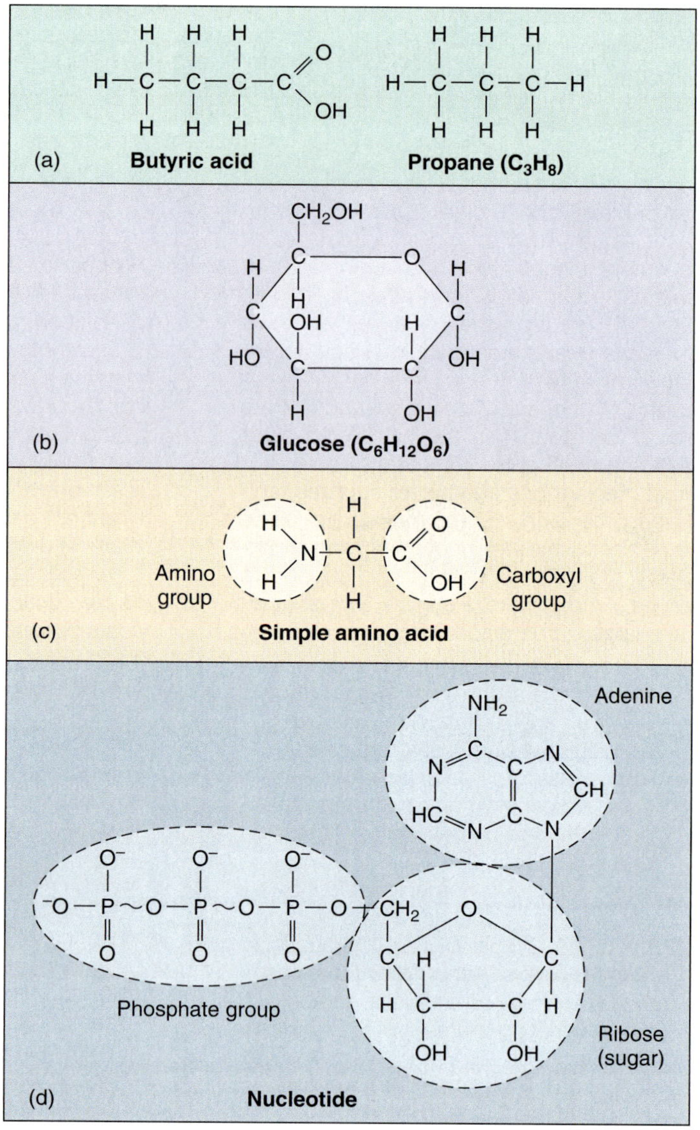


**FIGURE 3.5** The four major groups of organic molecules are based on repeating subunits of these carbon-based structures. Basic structures are shown for (a) butyric acid (a building block of lipids) and a hydrocarbon, (b) a simple carbohydrate, (c) a protein, and (d) a nucleic acid.

basic structure is a chain of carbon atoms with attached hydrogen atoms. This structure makes them part of the family of hydrocarbons (fig. 3.5*a*). Carbohydrates (including sugars, starches, and cellulose) also store energy and provide structure to cells. Like lipids, carbohydrates have a basic structure of carbon atoms, but hydroxyl (OH) groups replace half the hydrogen atoms in their basic structure, and they usually consist of long chains of sugars. Glucose (fig. 3.5*b*) is an example of a very simple sugar.

Proteins are composed of chains of subunits called amino acids (fig. 3.5*c*). Folded into complex three-dimensional shapes, proteins provide structure to cells and are used for countless cell functions. Most enzymes, such as those that release energy from lipids and carbohydrates, are proteins. Proteins also help identify disease-causing microbes, make muscles move, transport oxygen to cells, and regulate cell activity.

Nucleotides are complex molecules made of a five-carbon sugar (ribose or deoxyribose), one or more phosphate groups, and an organic nitrogen-containing base called either a purine or pyrimidine (fig. 3.5*d*). Nucleotides are extremely important as signaling molecules (they carry information between cells, tissues, and organs) and as sources of intracellular energy. They also form long chains called *r*ibo*n*ucleic *a*cid (RNA) or ***d****eoxyribo**n**ucleic **a**cid* **(DNA)** that are essential for storing and expressing genetic information. Only four kinds of nucleotides (adenine, guanine, cytosine, and thyamine) occur in DNA, but there can be billions of these molecules lined up in a very specific sequence. Groups of three nucleotides (called codons) act as the letters in messages that code for the amino acid sequences in proteins. Long chains of DNA bind together to form a stable double helix (fig. 3.6). These chains separate for replication in preparation for cell division or to express their genetic information during protein synthesis. Molecular biologists have developed techniques for extracting DNA from cells and reading its nucleotide sequence. These techniques are proving to be tremendously powerful in medical genetics and agriculture. They also are extremely

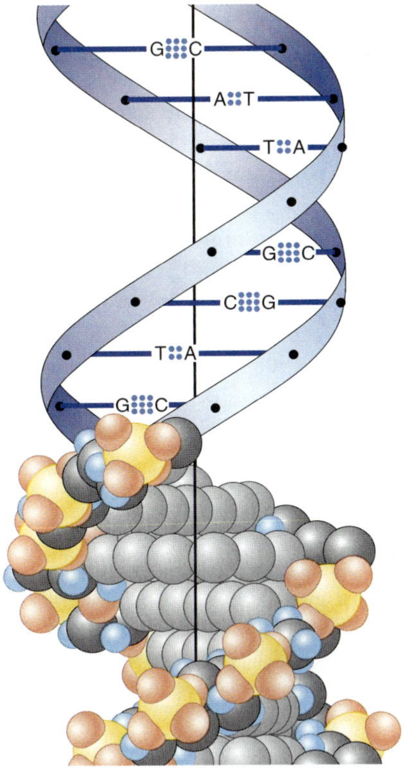


**FIGURE 3.6** A composite molecular model of DNA. The lower part shows individual atoms, while the upper part has been simplified to show the strands of the double helix held together by hydrogen bonds (small dots) between matching nucleotides (A, T, G, and C). A complete DNA molecule contains millions of nucleotides and carries genetic information for many specific, inheritable traits.

useful in fields such as forensics and taxonomy. Because every individual has a unique set of DNA molecules, sequencing their nucleotide content can provide a distinctive individual identification. We'll discuss this technology further in chapters 9 and 11.

## Cells are the fundamental units of life

All living organisms are composed of **cells,** minute compartments within which the processes of life are carried out (fig. 3.7). Microscopic organisms such as bacteria, some algae, and protozoa are composed of single cells. Most higher organisms are multicellular, usually with many different cell varieties. Your body, for instance, is composed of several trillion cells of about two hundred distinct types. Every cell is surrounded by a thin but dynamic membrane of lipid and protein that receives information about the exterior world and regulates the flow of materials between the cell and its environment. Inside, cells are subdivided into tiny organelles and subcellular particles that provide the machinery for life. Some of these organelles store and release energy. Others manage and distribute information. Still others create the internal structure that gives the cell its shape and allows it to fulfill its role.

All of the chemical reactions required to create these various structures, provide them with energy and materials to carry out their functions, dispose of wastes, and perform other functions of life at the cellular level are carried out by a special class of proteins called **enzymes.** Enzymes are molecular catalysts, that is, they regulate chemical reactions without being used up or inactivated in the process. Think of them as tools: Like hammers or wrenches, they do their jobs without being consumed or damaged as they work. There are generally thousands of different kinds of enzymes in every cell, all necessary to carry out the many processes on which life depends. Altogether, the multitude of enzymatic reactions performed by an organism is called its **metabolism.**

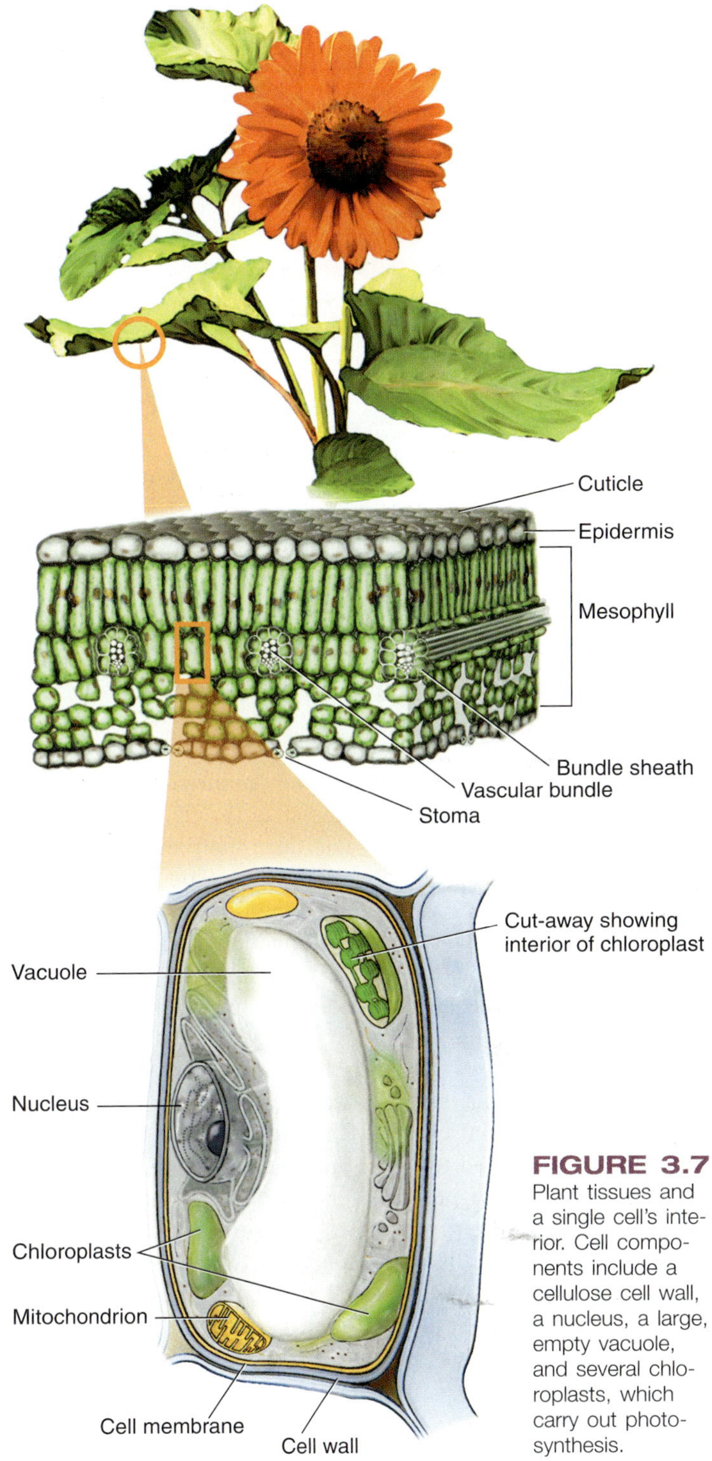


**FIGURE 3.7** Plant tissues and a single cell's interior. Cell components include a cellulose cell wall, a nucleus, a large, empty vacuole, and several chloroplasts, which carry out photosynthesis.

## 3.2 Energy

If matter is the material of which things are made, energy provides the force to hold structures together, tear them apart, and move them from one place to another. In this section we will look at some fundamental characteristics of these components of our world.

**FIGURE 3.8** Water stored behind this dam represents potential energy. Water flowing over the dam has kinetic energy, some of which is converted to heat.

## Energy occurs in different types and qualities

**Energy** is the ability to do work such as moving matter over a distance or causing a heat transfer between two objects at different temperatures. Energy can take many different forms. Heat, light, electricity, and chemical energy are examples that we all experience. The energy contained in moving objects is called **kinetic energy.** A rock rolling down a hill, the wind blowing through the trees, water flowing over a dam (fig. 3.8), or electrons speeding around the nucleus of an atom are all examples of kinetic energy. **Potential energy** is stored energy that is latent but available for use. A rock poised at the top of a hill and water stored behind a dam are examples of potential energy. **Chemical energy** stored in the food that you eat and the gasoline that you put into your car are also examples of potential energy that can be released to do useful work. Energy is often measured in units of heat (calories) or work (joules). One joule (J) is the work done when one kilogram is accelerated at one meter per second per second. One calorie is the amount of energy needed to heat one gram of pure water one degree Celsius. A calorie can also be measured as 4.184 J.

**Heat** describes the energy that can be transferred between objects of different temperature. When a substance absorbs heat, the kinetic energy of its molecules increases, or it may change state: A solid may become a liquid, or a liquid may become a gas. We sense change in heat content as change in temperature (unless the substance changes state).

An object can have a high heat content but a low temperature, such as a lake that freezes slowly in the fall. Other objects, like a burning match, have a high temperature but little heat content. Heat storage in lakes and oceans is essential to moderating climates and maintaining biological communities. Heat absorbed in changing states is also critical. As you will read in chapter 15, evaporation and condensation of water in the atmosphere helps distribute heat around the globe.

Energy that is diffused, dispersed, and low in temperature is considered low-quality energy because it is difficult to gather and use for productive purposes. The heat stored in the oceans, for instance, is immense but hard to capture and use, so it is low quality. Conversely, energy that is intense, concentrated, and high in temperature is high-quality energy because of its usefulness in carrying out work. The intense flames of a very hot fire or high-voltage electrical energy are examples of high-quality forms that are valuable to humans. Many of our alternative energy sources (such as wind) are diffuse compared to the higher-quality, more concentrated chemical energy in oil, coal, or gas.

***Think About It***

Can you describe one or two practical examples of the laws of physics and thermodynamics in your own life? Do they help explain why you can recycle cans and bottles but not energy? Which law is responsible for the fact that you get hot and sweaty when you exercise?

## Thermodynamics regulates energy transfers

Atoms and molecules cycle endlessly through organisms and their environment, but energy flows in a one-way path. A constant supply of energy—nearly all of it from the sun—is needed to keep biological processes running. Energy can be used repeatedly as it flows through the system, and it can be stored temporarily in the chemical bonds of organic molecules, but eventually it is released and dissipated.

The study of thermodynamics deals with how energy is transferred in natural processes. More specifically, it deals with the rates of flow and the transformation of energy from one form or quality to another. Thermodynamics is a complex, quantitative discipline, but you don't need a great deal of math to understand some of the broad principles that shape our world and our lives.

The **first law of thermodynamics** states that energy is *conserved;* that is, it is neither created nor destroyed under normal conditions. Energy may be transformed, for example, from the energy in a chemical bond to heat energy, but the total amount does not change.

The **second law of thermodynamics** states that, with each successive energy transfer or transformation in a system, less energy is available to do work. That is, energy is degraded to lower-quality forms, or it dissipates and is lost, as it is used. When you drive a car, for example, the chemical energy of the gas is degraded to kinetic energy and heat, which dissipates, eventually, to space. The second law recognizes that disorder, or **entropy,** tends to increase in all natural systems. Consequently, there is always less *useful* energy available when you finish a process than there was before you started. Because of this loss, everything in the universe tends to fall apart, slow down, and get more disorganized.

How does the second law of thermodynamics apply to organisms and biological systems? Organisms are highly organized, both structurally and metabolically. Constant care and maintenance is required to keep up this organization, and a constant supply of energy is required to maintain these processes. Every time some energy is used by a cell to do work, some of that energy is dissipated or lost as heat. If cellular energy supplies are interrupted or depleted, the result—sooner or later—is death.

## 3.3 Energy for Life

Where does the energy needed by living organisms come from? How is it captured and used to do work? For nearly all plants and animals living on the earth's surface, the sun is the ultimate energy source, but for organisms living deep in the earth's crust or at the bottom of the oceans, where sunlight is unavailable, chemicals derived from rocks provide alternate energy sources. We'll consider this alternate energy pathway first because it seems to be more ancient. Before green plants existed, we believe that ancient bacteria-like cells probably lived by processing chemicals in hot springs.

### Extremophiles live in severe conditions

Until recently, the deep ocean floor was believed to be a biological desert. Cold, dark, subject to crushing pressures, and without any known energy supply, it was a place where scientists thought nothing could survive. Undersea explorations in the 1970s, however, revealed dense colonies of animals—blind shrimp, giant tubeworms, strange crabs, and bizarre clams—clustered around vents called black chimneys, where boiling hot, mineral-laden water bubbles up through cracks in the earth's crust. But how were these organisms getting energy? The answer is **chemosynthesis,** the process in which inorganic chemicals, such as hydrogen sulfide ($H_2S$) or hydrogen gas ($H_2$), provide energy for synthesis of organic molecules.

Discovering organisms living under the severe conditions of deep-sea hydrothermal vents led to an interest in other sites that seem exceptionally harsh to us. A variety of interesting organisms have been discovered in hot springs in thermal areas such as Yellowstone National Park, in intensely salty lakes, and even in deep rock formations (up to 1,500 m, or nearly a mile deep) in Columbia River basalts, for example. Some species are amazingly hardy. The recently described *Pyrolobus fumarii* can withstand temperatures up to 113°C (235°F). Most of these extremophiles are archaea, single-celled organisms intermediate between bacteria or eukaryotic organisms (those with their genetic material enclosed in a nucleus (see fig. 3.7). Archaea are thought to be the most primitive of all living organisms, and the conditions under which they live are thought to be similar to those in which life first evolved.

Deep-sea explorations of areas without thermal vents also have found abundant life (fig. 3.9).We now know that archaea live in oceanic sediments in astonishing numbers. The deepest of these species (they can be 800 m or more below the ocean floor) make methane from gaseous hydrogen ($H_2$) and carbon dioxide

**FIGURE 3.9** A colony of tube worms and mussels clusters over a cool, deep-sea methane seep in the Gulf of Mexico.

($CO_2$), derived from rocks. Other species oxidize methane using sulfur to create hydrogen sulfide ($H_2S$), which is consumed by bacteria that serve as a food source for more complex organisms such as tubeworms. But why should we care about this exotic community? It's estimated that the total mass of microbes (microscopic organisms) living beneath the seafloor represents nearly one-third of all the biomass (organic material) on the planet. Furthermore, the vast supply of methane generated by this community could be either a great resource or a terrible threat to us.

The total amount of methane made by these microbes is probably greater than all the known reserves of coal, gas, and oil. If we could safely extract the huge supplies of methane hydrate in ocean sediments, it could supply our energy needs for hundreds of years. Of greater immediate importance is that if methane-eating microbes weren't intercepting the methane produced by their neighbors, more than 300 million tons per year of this potent greenhouse gas would probably be bubbling to the surface, and we'd have run-away global warming. Some geologists believe that sudden "burps" of methane from melting hydrate deposits may have been responsible for catastrophic landslides and tsunamis, sudden climatic shifts, and mass extinctions in the past. We'll look further at issues of global climate change in chapter 15.

### Green plants get energy from the sun

Our sun is a star, a fiery ball of exploding hydrogen gas. Its thermonuclear reactions emit powerful forms of radiation, including potentially deadly ultraviolet and nuclear radiation (fig. 3.10), yet life here is nurtured by, and dependent upon, this searing, energy source. Solar energy is essential to life for two main reasons.

First, the sun provides warmth. Most organisms survive within a relatively narrow temperature range. In fact, each species has its own range of temperatures within which it can function normally. At high temperatures (above 40°C), biomolecules

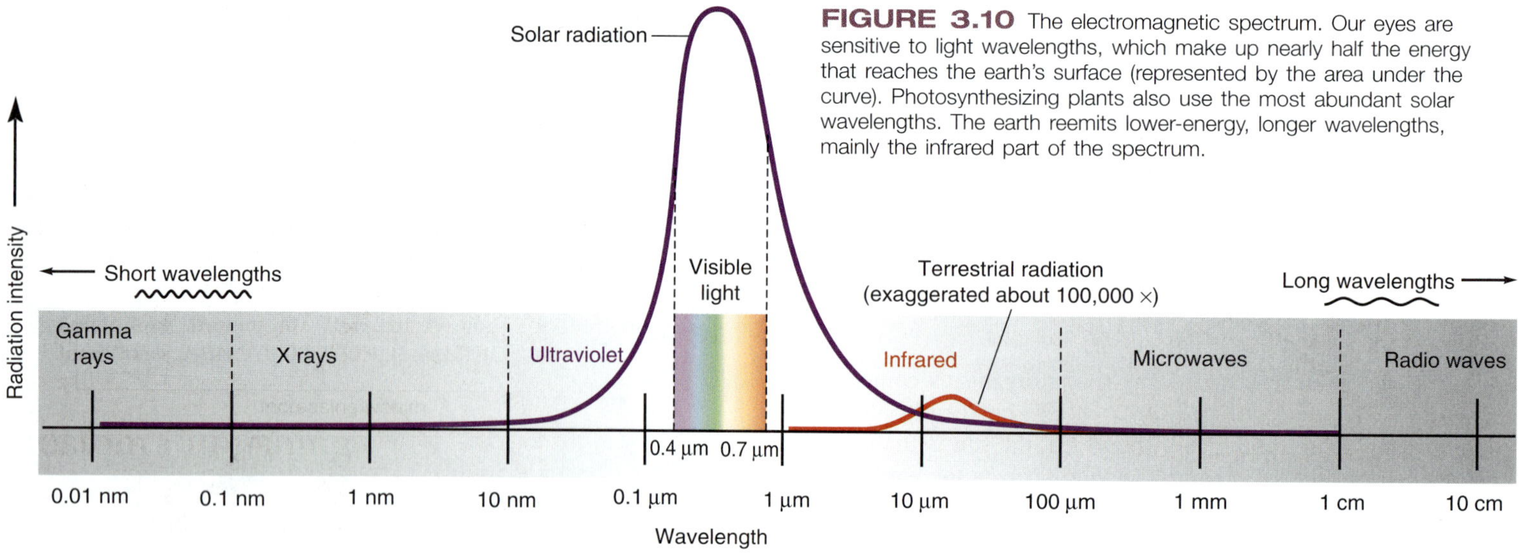


**FIGURE 3.10** The electromagnetic spectrum. Our eyes are sensitive to light wavelengths, which make up nearly half the energy that reaches the earth's surface (represented by the area under the curve). Photosynthesizing plants also use the most abundant solar wavelengths. The earth reemits lower-energy, longer wavelengths, mainly the infrared part of the spectrum.

begin to break down or become distorted and nonfunctional. At low temperatures (near 0°C), some chemical reactions of metabolism occur too slowly to enable organisms to grow and reproduce. Other planets in our solar system are either too hot or too cold to support life as we know it. The earth's water and atmosphere help to moderate, maintain, and distribute the sun's heat.

Second, nearly all organisms on the earth's surface organisms depend on solar radiation for life-sustaining energy, which is captured by green plants, algae, and some bacteria in a process called **photosynthesis.** Photosynthesis converts radiant energy into useful, high-quality chemical energy in the bonds that hold together organic molecules.

How much of the available solar energy is actually used by organisms? The amount of incoming solar radiation is enormous, about 1,372 watts/$m^2$ at the top of the atmosphere (1 watt = 1 J per second). However, more than half of the incoming sunlight may be reflected or absorbed by atmospheric clouds, dust, and gases. In particular, harmful, short wavelengths are filtered out by gases (such as ozone) in the upper atmosphere; thus, the atmosphere is a valuable shield, protecting life-forms from harmful doses of ultraviolet and other forms of radiation. Even with these energy reductions, however, the sun provides much more energy than biological systems can harness, and more than enough for all our energy needs if technology could enable us to tap it efficiently.

Of the solar radiation that does reach the earth's surface, about 10 percent is ultraviolet, 45 percent is visible, and 45 percent is infrared. Most of that energy is absorbed by land or water or is reflected into space by water, snow, and land surfaces. (Seen from outer space, Earth shines about as brightly as Venus.)

Of the energy that reaches the earth's surface, photosynthesis uses only certain wavelengths, mainly red and blue light. (Most plants reflect green wavelengths, so that is the color they appear to us.) Half of the energy plants absorb is used in evaporating water. In the end, only 1 to 2 percent of the sunlight falling on plants is available for photosynthesis. This small percentage represents the energy base for virtually all life in the biosphere!

## Photosynthesis captures energy while respiration releases that energy

Photosynthesis occurs in tiny membranous organelles called chloroplasts that reside within plant cells (see fig. 3.7). The most important key to this process is chlorophyll, a unique green molecule that can absorb light energy and use it to create high-energy chemical bonds in compounds that serve as the fuel for all subsequent cellular metabolism. Chlorophyll doesn't do this important job all alone, however. It is assisted by a large group of other lipid, sugar, protein, and nucleotide molecules. Together these components carry out two interconnected cyclic sets of reactions (fig. 3.11).

Photosynthesis begins with a series of steps called light-dependent reactions: These occur only while the chloroplast is receiving light. Enzymes split water molecules and release molecular oxygen ($O_2$). This is the source of all the oxygen in the atmosphere on which all animals, including you, depend for life. The light-dependent reactions also create mobile, high-energy molecules (adenosine triphosphate, or ATP, and nicotinamide adenine dinucleotide phosphate, or NADPH), which provide energy for the next set of processes, the light-independent reactions. As their name implies, these reactions do not use light directly. Here, enzymes extract energy from ATP and NADPH to add carbon atoms (from carbon dioxide) to simple sugar molecules, such as glucose. These molecules provide the building blocks for larger, more complex organic molecules.

In most temperate-zone plants, photosynthesis can be summarized in the following equation:

$$6H_2O + 6CO_2 + \text{solar energy} \xrightarrow[\text{chlorophyll}]{} C_6H_{12}O_6 \text{ (sugar)} + 6O_2$$

We read this equation as "water plus carbon dioxide plus energy produces sugar plus oxygen." The reason the equation uses six water and six carbon dioxide molecules is that it takes six carbon atoms to make the sugar product. If you look closely,

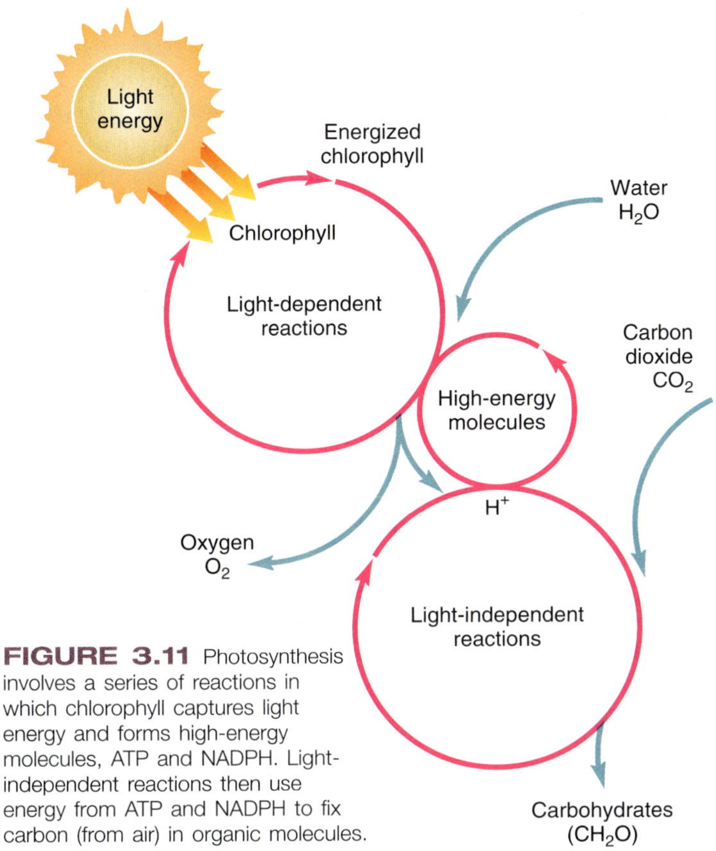


**FIGURE 3.11** Photosynthesis involves a series of reactions in which chlorophyll captures light energy and forms high-energy molecules, ATP and NADPH. Light-independent reactions then use energy from ATP and NADPH to fix carbon (from air) in organic molecules.

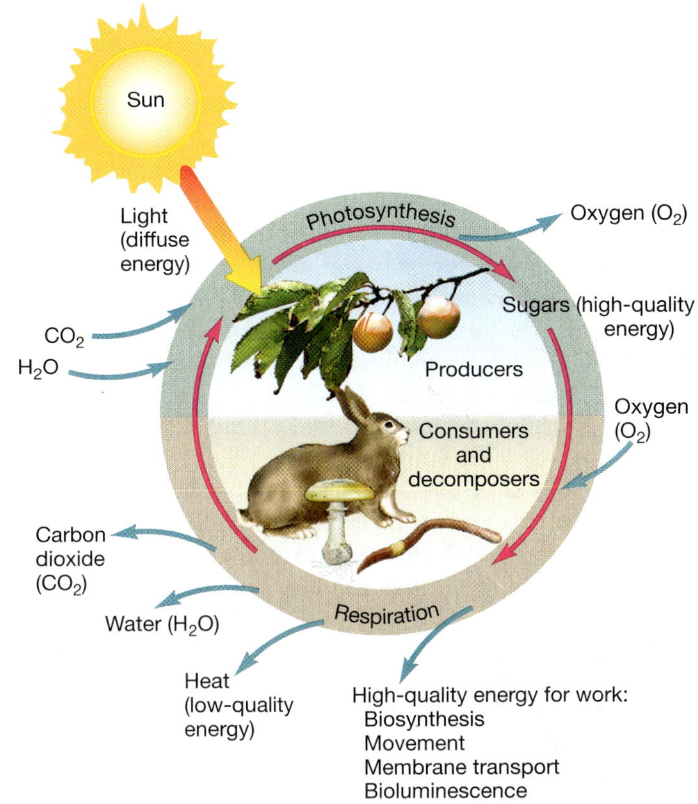


**FIGURE 3.12** Energy exchange in ecosystems. Plants use sunlight, water, and carbon dioxide to produce sugars and other organic molecules. Consumers use oxygen and break down sugars during cellular respiration. Plants also carry out respiration, but during the day, if light, water, and $CO_2$ are available, they have a net production of $O_2$ and carbohydrates.

you will see that all the atoms in the reactants balance with those in the products. This is an example of conservation of matter.

You might wonder how making a simple sugar benefits the plant. The answer is that glucose is an energy-rich compound that serves as the central, primary fuel for all metabolic processes of cells. The energy in its chemical bonds—the ones created by photosynthesis—can be released by other enzymes and used to make other molecules (lipids, proteins, nucleic acids, or other carbohydrates), or it can drive kinetic processes such as movement of ions across membranes, transmission of messages, changes in cellular shape or structure, or movement of the cell itself in some cases. This process of releasing chemical energy, called **cellular respiration,** involves splitting carbon and hydrogen atoms from the sugar molecule and recombining them with oxygen to re-create carbon dioxide and water. The net chemical reaction, then, is the reverse of photosynthesis:

$$C_6H_{12}O_6 + 6O_2 \longrightarrow 6H_2O + 6CO_2 + \text{released energy}$$

Note that in photosynthesis, energy is *captured,* while in respiration, energy is *released.* Similarly, photosynthesis *consumes* water and carbon dioxide to *produce* sugar and oxygen, while respiration does just the opposite. In both sets of reactions, energy is stored temporarily in chemical bonds, which constitute a kind of energy currency for the cell. Plants carry out both photosynthesis and respiration, but during the day, if light, water, and $CO_2$ are available, they have a net production of $O_2$ and carbohydrates.

We animals don't have chlorophyll and can't carry out photosynthetic food production. We do have the components for cellular respiration, however. In fact, this is how we get all our energy for life. We eat plants—or other animals that have eaten plants—and break down the organic molecules in our food through cellular respiration to obtain energy (fig. 3.12). In the process, we also consume oxygen and release carbon dioxide, thus completing the cycle of photosynthesis and respiration. Later in this chapter we will see how these feeding relationships work.

## 3.4 From Species to Ecosystems

While cellular and molecular biologists study life processes at the microscopic level, ecologists study interactions at the species, population, biotic community, or ecosystem level. In Latin, *species* literally means *kind.* In biology, **species** refers to all organisms of the same kind that are genetically similar enough to breed in nature and produce live, fertile offspring. There are several qualifications and some important exceptions to this definition of species (especially among bacteria and plants), but for our purposes this is a useful working definition.

## Organisms occur in populations, communities, and ecosystems

A **population** consists of all the members of a species living in a given area at the same time. Chapter 6 deals further with population growth and dynamics. All of the populations of organisms living and interacting in a particular area make up a **biological community.** What populations make up the biological community of which you are a part? The population sign marking your city limits announces only the number of humans who live there, disregarding the other populations of animals, plants, fungi, and microorganisms that are part of the biological community within the city's boundaries. Characteristics of biological communities are discussed in more detail in chapter 4.

An ecological system, or **ecosystem,** is composed of a biological community and its physical environment. The environment includes abiotic factors (nonliving components), such as climate, water, minerals, and sunlight, as well as biotic factors, such as organisms, their products (secretions, wastes, and remains), and effects in a given area. As you learned in chapter 2, systems often have properties above and beyond those of the individual components and processes that produce them. It is useful to think about the biological community and its environment together, because energy and matter flow through both. Understanding how those flows work is a major theme in ecology.

For simplicity's sake, we think of ecosystems as fixed ecological units with distinct boundaries. If you look at a patch of woods surrounded by farm fields, for instance, a relatively sharp line separates the two areas, and conditions such as light levels, wind, moisture, and shelter are quite different in the woods than in the fields around them. Because of these variations, distinct populations of plants and animals live in each place. By studying each of these areas, we can make important and interesting discoveries about who lives where and why and about how conditions are established and maintained there.

The division between the fields and woods is not always clear, however. Air, of course, moves freely from one to another, and the runoff after a rainfall may carry soil, leaf litter, and even live organisms between the areas. Birds may feed in the field during the day but roost in the woods at night, giving them roles in both places. Are they members of the woodland community or the field community? Is the edge of the woodland ecosystem where the last tree grows, or does it extend to every place that has an influence on the woods?

As you can see, it may be difficult to draw clear boundaries around communities and ecosystems. To some extent we define these units by what we want to study and how much information we can handle. Thus, an ecosystem might be as large as a whole watershed or as small as a pond or even the surface of your skin. Even though our choices may be somewhat arbitrary, we still can make useful discoveries about how organisms interact with each other and with their environment within these units. The woods are, after all, significantly different from the fields around them.

Like the woodland we just considered, most ecosystems are open, in the sense that they exchange materials and organisms with other ecosystems. A stream ecosystem is an extreme example. Water, nutrients, and organisms enter from upstream and are lost downstream. The species and numbers of organisms present may be relatively constant, but they are made up of continually changing individuals. Other ecosystems are relatively closed, in the sense that very little enters or leaves them. A balanced aquarium is a good example of a closed ecosystem. Aquatic plants, animals, and decomposers can balance material cycles in the aquarium if care is taken to balance their populations. Because of the second law of thermodynamics, however, every ecosystem must have a constant inflow of energy and a way to dispose of heat. Thus, at least with regard to energy flow, every ecosystem is open.

Many ecosystems have mechanisms that maintain composition and functions within certain limits. A forest tends to remain a forest, for the most part, and to have forestlike conditions if it isn't disturbed by outside forces. Some ecologists suggest that ecosystems—or perhaps all life on the earth—may function as superorganisms because they maintain stable conditions and can be resilient to change.

## Food chains, food webs, and trophic levels link species

Photosynthesis (and rarely chemosynthesis) is the base of all ecosystems. Organisms that photosynthesize, mainly green plants and algae, are therefore known as **producers.** One of the major properties of an ecosystem is its **productivity,** the amount of **biomass** (biological material) produced in a given area during a given period of time. Photosynthesis is described as *primary productivity* because it is the basis for almost all other growth in an ecosystem. Manufacture of biomass by organisms that eat plants is termed *secondary productivity.* A given ecosystem may have very high total productivity, but if decomposers decompose organic material as rapidly as it is formed, the *net primary productivity* will be low.

Think about what you have eaten today and trace it back to its photosynthetic source. If you have eaten an egg, you can trace it back to a chicken, which ate corn. This is an example of a **food chain,** a linked feeding series. Now think about a more complex food chain involving you, a chicken, a corn plant, and a grasshopper. The chicken could eat grasshoppers that had eaten leaves of the corn plant. You also could eat the grasshopper directly—some humans do. Or you could eat corn yourself, making the shortest possible food chain. Humans have several options of where we fit into food chains.

In ecosystems, some consumers feed on a single species, but most consumers have multiple food sources. Similarly, some species are prey to a single kind of predator, but many species in an ecosystem are beset by several types of predators and parasites. In this way, individual food chains become interconnected to form a **food web.** Figure 3.13 shows feeding relationships among some of the larger organisms in a woodland and lake community. If we were to add all the insects, worms, and microscopic organisms that belong in this picture, however, we would have overwhelming

**FIGURE 3.13** Each time an organism feeds, it becomes a link in a food chain. In an ecosystem, food chains become interconnected when predators feed on more than one kind of prey, thus forming a food web. The arrows in this diagram and in figure 3.14 indicate the direction in which matter and energy are transferred through feeding relationships. Only a few representative relationships are shown here. What others might you add?

complexity. Perhaps you can imagine the challenge ecologists face in trying to quantify and interpret the precise matter and energy transfers that occur in a natural ecosystem!

An organism's feeding status in an ecosystem can be expressed as its **trophic level** (from the Greek *trophe,* food). In our first example, the corn plant is at the producer level; it transforms solar energy into chemical energy, producing food molecules. Other organisms in the ecosystem are **consumers** of the chemical energy harnessed by the producers. An organism that eats producers is a primary consumer. An organism that eats primary consumers is a secondary consumer, which may, in turn, be eaten by a tertiary consumer, and so on. Most terrestrial food chains are relatively short (seeds → mouse → owl), but aquatic food chains may be quite long (microscopic algae → copepod → minnow → crayfish → bass → osprey). The length of a food chain also may reflect the physical characteristics of a particular ecosystem. A harsh arctic landscape has a much shorter food chain than a temperate or tropical one (fig. 3.14).

Organisms can be identified both by the trophic level at which they feed and by the *kinds* of food they eat (fig. 3.15). **Herbivores** are plant eaters, **carnivores** are flesh eaters, and **omnivores** eat both plant and animal matter. What are humans? We are natural omnivores, by history and by habit. Tooth structure is an important clue to understanding animal food preferences, and humans are no exception. Our teeth are suited for an omnivorous diet, with a combination of cutting and crushing surfaces that are not highly adapted for one specific kind of food, as are the teeth of a wolf (carnivore) or a horse (herbivore).

### *Think About It*

What would have been the leading primary producers and top consumers in the native ecosystem where you now live? What are they now? Are fewer trophic levels now represented in your ecosystem than in the past?

One of the most important trophic levels is occupied by the many kinds of organisms that remove and recycle the dead bodies and waste products of others. **Scavengers** such as crows, jackals, and vultures clean up dead carcasses of larger animals. **Detritivores** such as ants and beetles consume litter, debris, and dung, while **decomposer** organisms such as fungi and bacteria complete the final breakdown and recycling of organic materials. It could be argued that these microorganisms are second in importance only to producers, because without their activity nutrients would remain locked up in the organic compounds of dead organisms and discarded body wastes, rather than being made available to successive generations of organisms.

**FIGURE 3.14** Harsh environments tend to have shorter food chains than environments with more favorable physical conditions. Compare the arctic food chains depicted here with the longer food chains in the food web in figure 3.13.

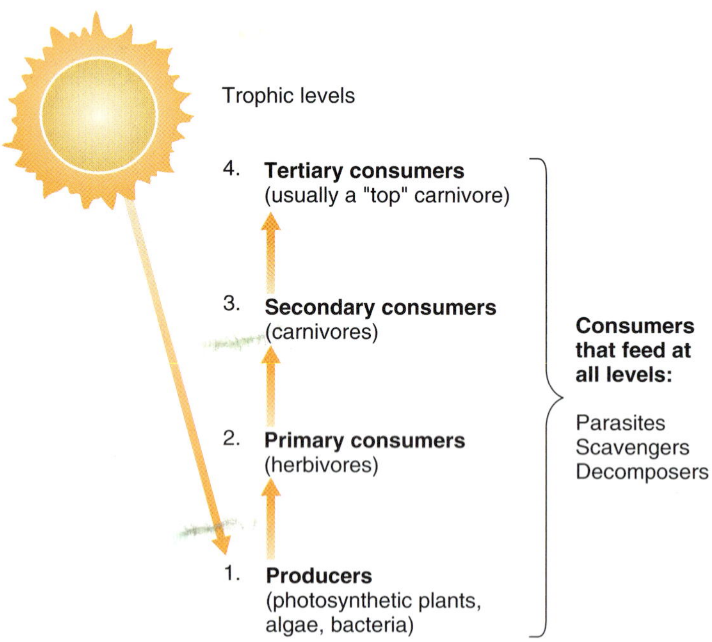


**FIGURE 3.15** Organisms in an ecosystem may be identified by how they obtain food for their life processes (producer, herbivore, carnivore, omnivore, scavenger, decomposer, reducer) or by consumer level (producer; primary, secondary, or tertiary consumer) or by trophic level (1st, 2nd, 3rd, 4th).

## Ecological pyramids describe trophic levels

If we arrange the organisms in a food chain according to trophic levels, they often form a pyramid with a broad base representing primary producers and only a few individuals in the highest trophic levels. This pyramid arrangement is especially true if we look at the energy content of an ecosystem (fig. 3.16). True to the second principle of thermodynamics, less food

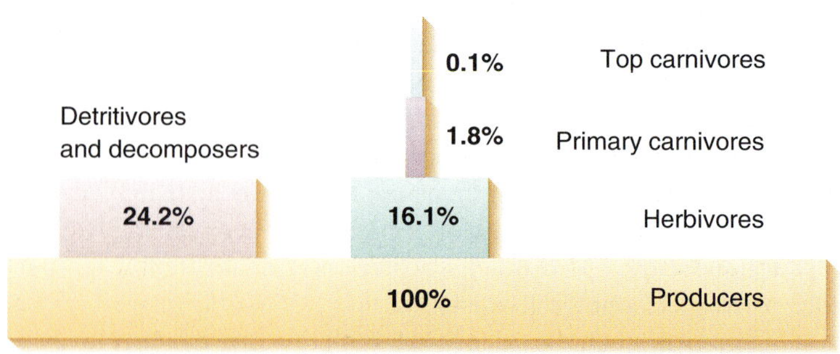


**FIGURE 3.16** A classic example of an energy pyramid from Silver Springs, Florida. The numbers in each bar show the percentage of the energy captured in the primary producer level that is incorporated into the biomass of each succeeding level. Detritivores and decomposers feed at every level but are shown attached to the producer bar because this level provides most of their energy.

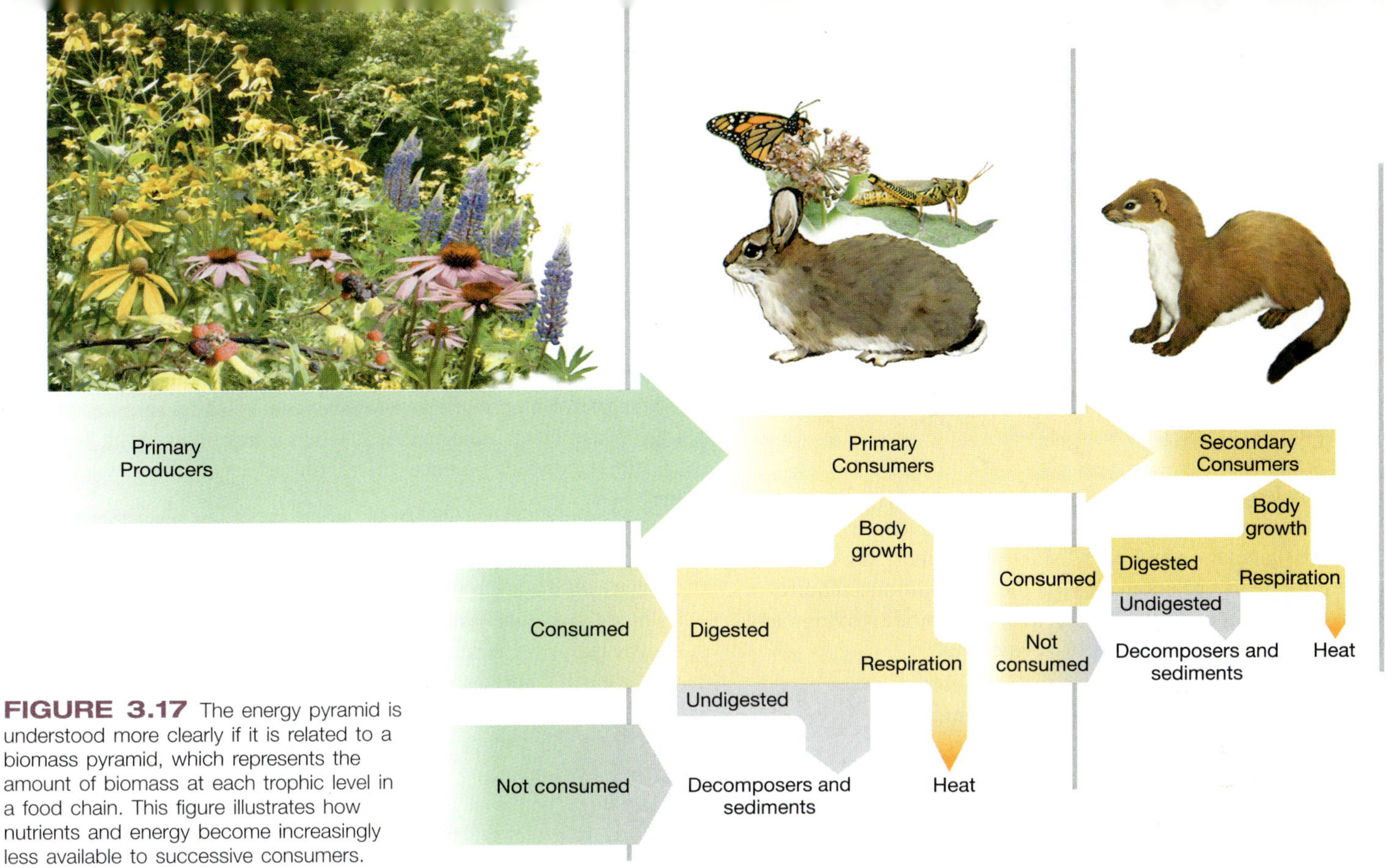


**FIGURE 3.17** The energy pyramid is understood more clearly if it is related to a biomass pyramid, which represents the amount of biomass at each trophic level in a food chain. This figure illustrates how nutrients and energy become increasingly less available to successive consumers.

energy is available to the top trophic level than is available to preceding levels. For example, it takes a huge number of plants to support a modest colony of grazers such as prairie dogs. Several colonies of prairie dogs, in turn, might be required to feed a single coyote. And a very large top carnivore like a tiger may need a home range of hundreds of square kilometers to survive.

Why is there so much less energy in each successive level in figure 3.16? In the first place, some of the food that organisms eat is undigested and doesn't provide usable energy. Much of the energy that is absorbed is used in the daily processes of living or lost as heat when it is transformed from one form to another and thus isn't stored as biomass that can be eaten.

Furthermore, predators don't operate at 100 percent efficiency. If there were enough foxes to catch all the rabbits available in the summer when the supply is abundant, there would be too many foxes in the middle of the winter when rabbits are scarce. A general rule of thumb is that only about 10 percent of the energy in one consumer level is represented in the next higher level (fig. 3.17). The amount of energy available is often expressed in biomass. For example, it generally takes about 100 kg of clover to make 10 kg of rabbit and 10 kg of rabbit to make 1 kg of fox.

The total number of organisms and the total amount of biomass in each successive trophic level of an ecosystem also may form pyramids (fig. 3.18) similar to those describing energy content. The relationship between biomass and numbers is not as dependable as energy, however. The biomass pyramid, for

**FIGURE 3.18** Usually, smaller organisms are eaten by larger organisms and it takes numerous small organisms to feed one large organism. The classic study represented in this pyramid shows numbers of individuals at each trophic level per 1,000 $m^2$ of grassland, and reads like this: to support one individual at the top carnivore level, there were 90,000 primary carnivores feeding upon 200,000 herbivores that in turn fed upon 1,500,000 producers.

instance, can be inverted by periodic fluctuations in producer populations (for example, low plant and algal biomass present during winter in temperate aquatic ecosystems). The numbers pyramid also can be inverted. One coyote can support numerous tapeworms, for example. Numbers inversion also occurs at the lower trophic levels (for example, one large tree can support thousands of caterpillars).

## 3.5 Material Cycles and Life Processes

To our knowledge, Earth is the only planet in our solar system that provides a suitable environment for life as we know it. Even our nearest planetary neighbors, Mars and Venus, do not meet these requirements. Maintenance of these conditions requires a constant recycling of materials between the biotic (living) and abiotic (nonliving) components of ecosystems.

### The hydrologic cycle moves water around the earth

The path of water through our environment, known as the **hydrologic cycle,** is perhaps the most familiar material cycle, and it is discussed in greater detail in chapter 17. Most of the earth's water is stored in the oceans, but solar energy continually evaporates this water, and winds distribute water vapor around the globe. Water that condenses over land surfaces, in the form of rain, snow, or fog, supports all terrestrial (land-based) ecosystems (fig. 3.19). Living organisms emit the moisture they have consumed through respiration and perspiration. Eventually this moisture reenters the atmosphere or enters lakes and streams, from which it ultimately returns to the ocean again.

As it moves through living things and through the atmosphere, water is responsible for metabolic processes within cells, for maintaining the flows of key nutrients through ecosystems, and for global-scale distribution of heat and energy (chapter 15). Water performs countless services because of its unusual properties. Water is so important that when astronomers look for signs of life on distant planets, traces of water are the key evidence they seek.

Everything about global hydrological processes is awesome in scale. Each year, the sun evaporates approximately 496,000 $km^3$ of water from the earth's surface. More water evaporates in the tropics than at higher latitudes, and more water evaporates over the oceans than over land. Although the oceans cover about 70 percent of the earth's surface, they account for 86 percent of total evaporation. Ninety percent of the water evaporated from the ocean falls back on the ocean as rain. The remaining 10 percent is carried by prevailing winds over the continents where it combines with water evaporated from soil, plant surfaces, lakes, streams, and wetlands to provide a total continental precipitation of about 111,000 $km^3$.

What happens to the surplus water on land—the difference between what falls as precipitation and what evaporates? Some of it is incorporated by plants and animals into biological tissues. A large share of what falls on land seeps into the ground to be

**FIGURE 3.19** The hydrologic cycle. Most exchange occurs with evaporation from oceans and precipitation back to oceans. About one-tenth of water evaporated from oceans falls over land, is recycled through terrestrial systems, and eventually drains back to oceans in rivers.

stored for a while (from a few days to many thousands of years) as soil moisture or groundwater. Eventually, all the water makes its way back downhill to the oceans. The 40,000 $km^3$ carried back to the ocean each year by surface runoff or underground flow represents the renewable supply available for human uses and sustaining freshwater-dependent ecosystems.

## Carbon moves through the carbon cycle

Carbon serves a dual purpose for organisms: (1) it is a structural component of organic molecules, and (2) the energy-holding chemical bonds it forms represent energy "storage." The **carbon cycle** begins with the intake of carbon dioxide ($CO_2$) by photosynthetic organisms (fig. 3.20). Carbon (and hydrogen and oxygen) atoms are incorporated into sugar molecules during photosynthesis. Carbon dioxide is eventually released during respiration, closing the cycle. The carbon cycle is of special interest because biological accumulation and release of carbon is a major factor in climate regulation (Exploring Science, p. 68).

The path followed by an individual carbon atom in this cycle may be quite direct and rapid, depending on how it is used in an organism's body. Imagine for a moment what happens to a simple sugar molecule you swallow in a glass of fruit juice. The sugar molecule is absorbed into your bloodstream where it is made available to your cells for cellular respiration or for making more complex biomolecules. If it is used in respiration, you may exhale the same carbon atom as $CO_2$ the same day.

Can you think of examples where carbon may not be recycled for even longer periods of time, if ever? Coal and oil are the compressed, chemically altered remains of plants or microorganisms that lived millions of years ago. Their carbon atoms (and hydrogen, oxygen, nitrogen, sulfur, etc.) are not released until the coal and oil are burned. Enormous amounts of carbon also are locked up as calcium carbonate ($CaCO_3$), used to build shells and skeletons of marine organisms from tiny protozoans to corals. Most of these deposits are at the bottom of the oceans. The world's extensive surface limestone deposits are biologically formed calcium carbonate from ancient oceans, exposed by geological events. The carbon in limestone has been locked away for millennia, which is probably the fate of carbon currently being deposited in ocean sediments. Eventually, even the deep ocean deposits are recycled as they are drawn into deep molten layers and released via volcanic activity. Geologists estimate that every carbon atom on the earth has made about thirty such round trips over the last 4 billion years.

How does tying up so much carbon in the bodies and byproducts of organisms affect the biosphere? Favorably. It helps balance $CO_2$ generation and utilization. Carbon dioxide is one of the so-called greenhouse gases because it absorbs heat radiated from the earth's surface, retaining it instead in the atmosphere. This phenomenon is discussed in more detail in chapter 15. Photosynthesis and deposition of $CaCO_3$ remove atmospheric carbon dioxide; therefore, vegetation (especially large forested areas such as the boreal forests) and the oceans are very important **carbon sinks** (storage deposits). Cellular respiration and combustion both release $CO_2$, so they are referred to as carbon sources of the cycle.

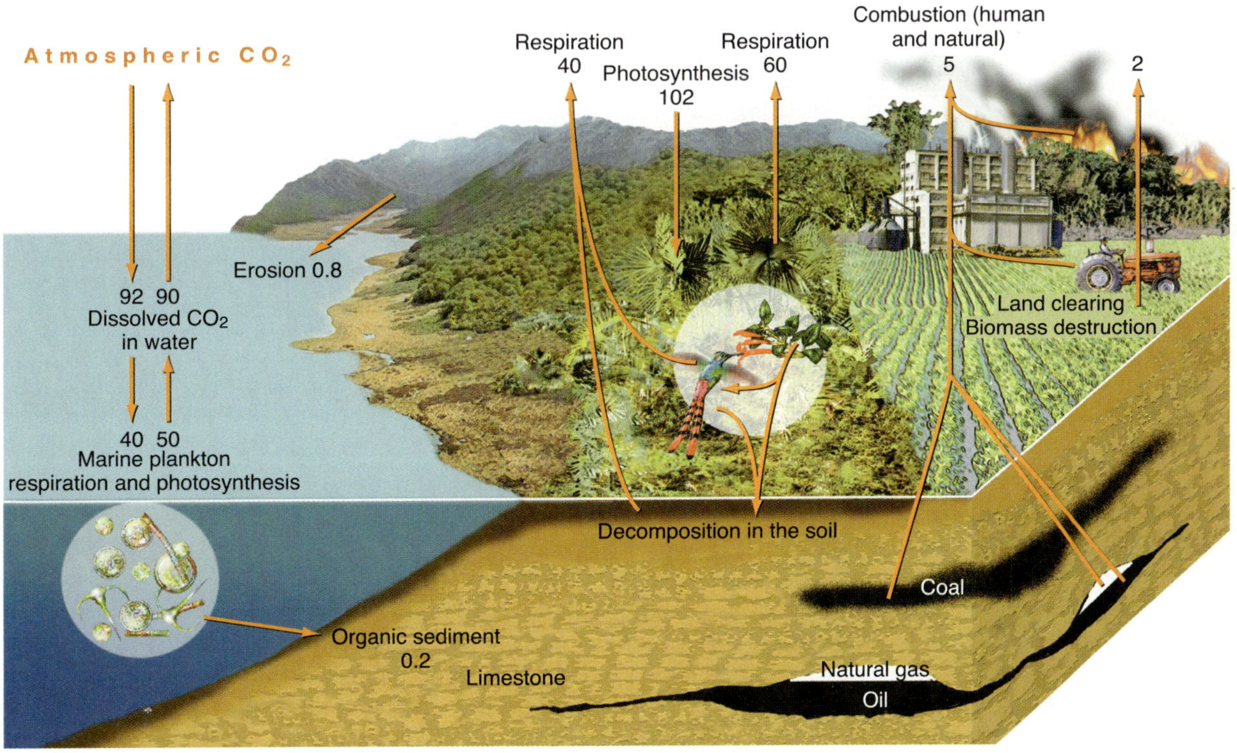


**FIGURE 3.20** The carbon cycle. Numbers indicate approximate exchange of carbon in gigatons (Gt) per year. Natural exchanges are balanced, but human sources produce a net increase of $CO_2$ in the atmosphere.

# *Exploring* SCIENCE

## Remote Sensing, Photosynthesis, and Material Cycles

Measuring primary productivity is important for understanding individual plants and local environments. Understanding rates of primary productivity is also key to understanding global processes, material cycling, and biological activity:

- In global carbon cycles, how much carbon is stored by plants, how quickly is it stored, and how does carbon storage compare in contrasting environments, such as the Arctic and the tropics?
- How does this carbon storage affect global climates (chapter 15)?
- In global nutrient cycles, how much nitrogen and phosphorus wash offshore, and where?

How can environmental scientists measure primary production (photosynthesis) at a global scale? In small-scale systems, you can simply collect all the biomass and weigh it. But that method is impossible for large ecosystems, especially for oceans, which cover 70 percent of the earth's surface. One of the newest methods of quantifying biological productivity involves remote sensing, or data collected from satellite sensors that observe the energy reflected from the earth's surface.

As you have read in this chapter, chlorophyll in green plants *absorbs* red and blue wavelengths of light and *reflects* green wavelengths. Your eye receives, or senses, these green wavelengths. A white-sand beach, on the other hand, reflects approximately equal amounts of all light wavelengths (see fig. 3.10) that reach it from the sun, so it looks white (and bright) to your eye. In a similar way, different surfaces of the earth reflect characteristic wavelengths. Snow-covered surfaces reflect light wavelengths; dark-green forests with abundant chlorophyll-rich leaves—and ocean surfaces rich in photosynthetic algae and plants—reflect greens and near-infrared wavelengths. Dry, brown forests with little active chlorophyll reflect more red and less infrared energy than do dark-green forests (fig. 1).

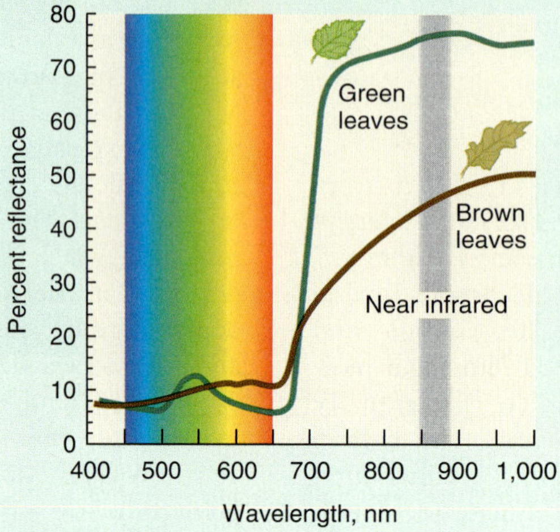


**FIGURE 1** Energy wavelengths reflected by green and brown leaves.

To detect land-cover patterns on the earth's surface, we can put a sensor on a satellite that orbits the earth. As the satellite travels, the sensor receives and transmits to earth a series of "snapshots." One of the best known earth- imaging satellites, *Landsat 7,* produces images that cover an area 185 km (115 mi) wide, and each pixel represents an area of just 30 × 30 m on the ground. *Landsat* orbits approximately from pole to pole, so as the earth spins below the satellite, it captures images of the entire surface every 16 days. Another satellite, *SeaWIFS,* was designed mainly for monitoring biological activity in oceans (fig. 2). *SeaWIFS* follows a path similar to *Landsat*'s but it revisits each point on the earth every day and produces images with a pixel resolution of just over 1 km.

Since satellites detect a much greater range of wavelengths than our eyes can, they are able to monitor and map chlorophyll abundance. In oceans, this is a useful measure of ecosystem health, as well as carbon dioxide uptake. By quantifying and mapping primary production in oceans, climatologists are working to estimate the role of ocean ecosystems in moderating climate change: for example, they can estimate the extent of biomass production in the cold, oxygen-rich waters of the North Atlantic (fig. 2). Oceanographers can also detect near-shore areas where nutrients washing off the land surface fertilize marine ecosystems and stimulate high productivity, such as near the mouth of the Amazon or Mississippi Rivers. Monitoring and mapping these patterns helps us estimate human impacts on nutrient flows (figs. 3.21, 3.23) from land to sea.

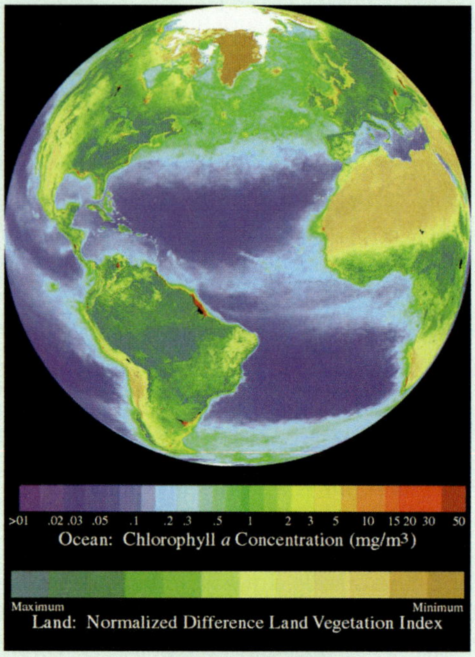


**FIGURE 2** *SeaWIFS* image showing chlorophyll abundance in oceans and plant growth on land (normalized difference vegetation index).

Presently, natural fires and human-created combustion of organic fuels (mainly wood, coal, and petroleum products) release huge quantities of $CO_2$ at rates that seem to be surpassing the pace of $CO_2$ removal. Scientific concerns over the linked problems of increased atmospheric $CO_2$ concentrations, massive deforestation, and reduced productivity of the oceans due to pollution are discussed in chapters 15 and 16.

## Nitrogen moves via the nitrogen cycle

As the opening case study of this chapter shows, nitrogen often is one of the most important limiting factors in ecosystems. The complex interrelationships through which organisms exchange this vital element help shape these biological communities. Organisms cannot exist without amino acids, peptides, nucleic

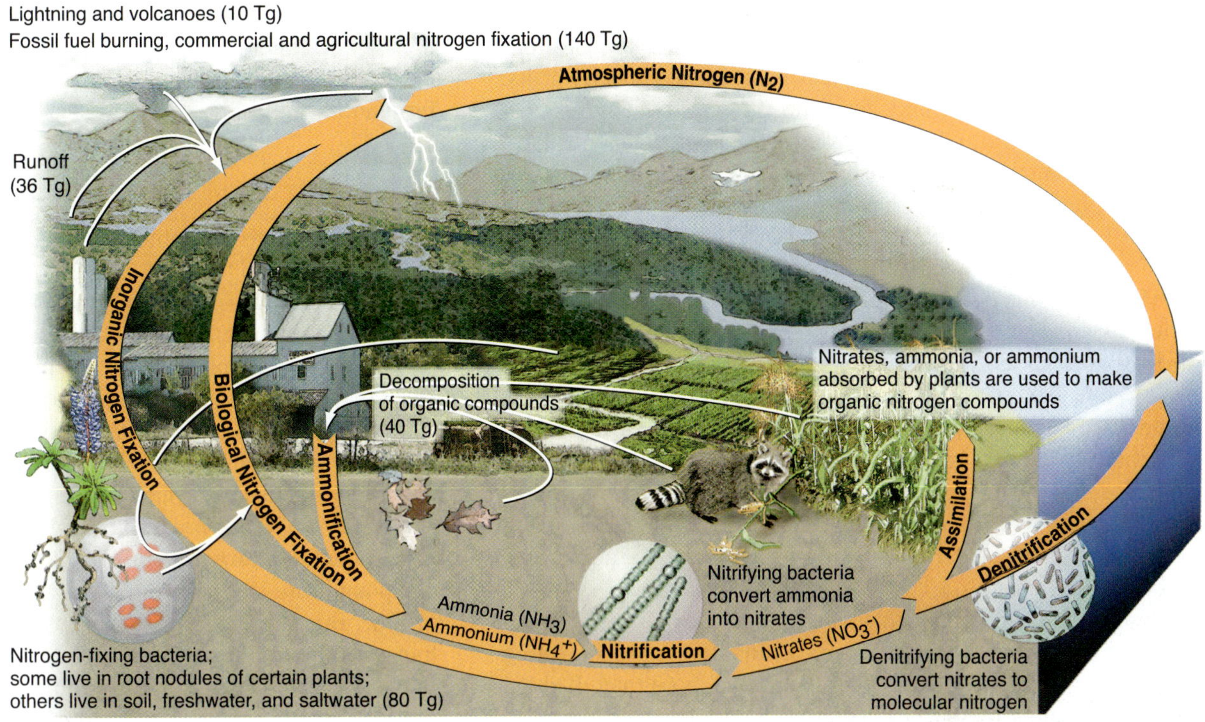


**FIGURE 3.21** The nitrogen cycle. Human sources of nitrogen fixation (conversion of molecular nitrogen to ammonia or ammonium) are now about 50 percent greater than natural sources. Bacteria convert ammonia to nitrates, which plants use to create organic nitrogen. Eventually, nitrogen is stored in sediments or converted back to molecular nitrogen (1 Tg = $10^{12}$ g).

acids, and proteins, all of which are organic molecules containing nitrogen. The nitrogen atoms that form these important molecules are provided by producer organisms. Plants assimilate (take up) inorganic nitrogen from the environment and use it to build their own protein molecules, which are eaten by consumer organisms, digested, and used to build their bodies. Even though nitrogen is the most abundant gas (about 78 percent of the atmosphere), however, plants cannot use $N_2$, the stable diatomic (2-atom) molecule in the air.

Where and how, then, do green plants get *their* nitrogen? The answer lies in the most complex of the gaseous cycles, the **nitrogen cycle.** Figure 3.21 summarizes the nitrogen cycle. The key natural processes that make nitrogen available are carried out by nitrogen-fixing bacteria (including some blue-green algae or cyanobacteria). These organisms have a highly specialized ability to "fix" nitrogen, meaning they change it to less mobile, more useful forms by combining it with hydrogen to make ammonia ($NH_3$).

Nitrite-forming bacteria combine the ammonia with oxygen, forming nitrites, which have the ionic form $NO_2^-$. Another group of bacteria then convert nitrites to nitrates, which have the ionic form $NO_3^-$, that can be absorbed and used by green plants. After nitrates have been absorbed into plant cells, they are reduced to ammonium ($NH_4^+$), which is used to build amino acids that become the building blocks for peptides and proteins.

Members of the bean family (legumes) and a few other kinds of plants are especially useful in agriculture because they have nitrogen-fixing bacteria actually living *in* their root tissues (fig. 3.22). Legumes and their associated bacteria enrich the soil, so interplanting and rotating legumes with crops such as corn that use but cannot replace soil nitrates are beneficial farming practices that take practical advantage of this relationship.

Nitrogen reenters the environment in several ways. The most obvious path is through the death of organisms. Their bodies are decomposed by fungi and bacteria, releasing ammonia and

**FIGURE 3.22** The roots of this adzuki bean plant are covered with bumps called nodules. Each nodule is a mass of root tissue containing many bacteria that help to convert nitrogen in the soil to a form the bean plants can assimilate and use to manufacture amino acids.

ammonium ions, which then are available for nitrate formation. Organisms don't have to die to donate proteins to the environment, however. Plants shed their leaves, needles, flowers, fruits, and cones; animals shed hair, feathers, skin, exoskeletons, pupal cases, and silk. Animals also produce excrement and urinary wastes that contain nitrogenous compounds. Urinary wastes are especially high in nitrogen because they contain the detoxified wastes of protein metabolism. All of these by-products of living organisms decompose, replenishing soil fertility.

How does nitrogen reenter the atmosphere, completing the cycle? Denitrifying bacteria break down nitrates into $N_2$ and nitrous oxide ($N_2O$), gases that return to the atmosphere; thus, denitrifying bacteria compete with plant roots for available nitrates. However, denitrification occurs mainly in waterlogged soils that have low oxygen availability and a high amount of decomposable organic matter. These are suitable growing conditions for many wild plant species in swamps and marshes, but not for most cultivated crop species, except for rice, a domesticated wetland grass.

In recent years, humans have profoundly altered the nitrogen cycle. By using synthetic fertilizers, cultivating nitrogen-fixing crops, and burning fossil fuels, we have more than doubled the amount of nitrogen cycled through our global environment. This excess nitrogen input is causing serious loss of soil nutrients such as calcium and potassium, acidification of rivers and lakes, and rising atmospheric concentrations of the greenhouse gas, nitrous oxide. It also encourages the spread of weeds into areas such as prairies occupied by native plants adapted to nitrogen-poor environments. In coastal areas, blooms of toxic algae and dinoflagellates result from excess nitrogen carried by rivers from farmlands and cities.

## Phosphorus is an essential nutrient

Minerals become available to organisms after they are released from rocks. Two mineral cycles of particular significance to organisms are phosphorus and sulfur. Why do you suppose phosphorus is a primary ingredient in fertilizers? At the cellular level, energy-rich, phosphorus-containing compounds are primary participants in energy-transfer reactions, as we have discussed. The amount of available phosphorus in an environment can, therefore, have a dramatic effect on productivity. Abundant phosphorus stimulates lush plant and algal growth, making it a major contributor to water pollution.

The **phosphorus cycle** (fig. 3.23) begins when phosphorus compounds are leached from rocks and minerals over long periods of time. Because phosphorus has no atmospheric form, it is usually transported in aqueous form. Inorganic phosphorus is taken in by producer organisms, incorporated into organic molecules, and then passed on to consumers. It is returned to the environment by decomposition. An important aspect of the phosphorus cycle is the very long time it takes for phosphorus atoms to pass through it. Deep sediments of the oceans are significant phosphorus sinks of extreme longevity. Phosphate ores that now are mined to make detergents

**FIGURE 3.23** The phosphorus cycle. Natural movement of phosphorus is slight, involving recycling within ecosytems and some erosion and sedimentation of phosphorus-bearing rock. Use of phosphate ($PO_4^{-3}$) fertilizers and cleaning agents increases phosphorus in aquatic systems, causing eutrophication. Units are teragrams (Tg) phosphorus per year.

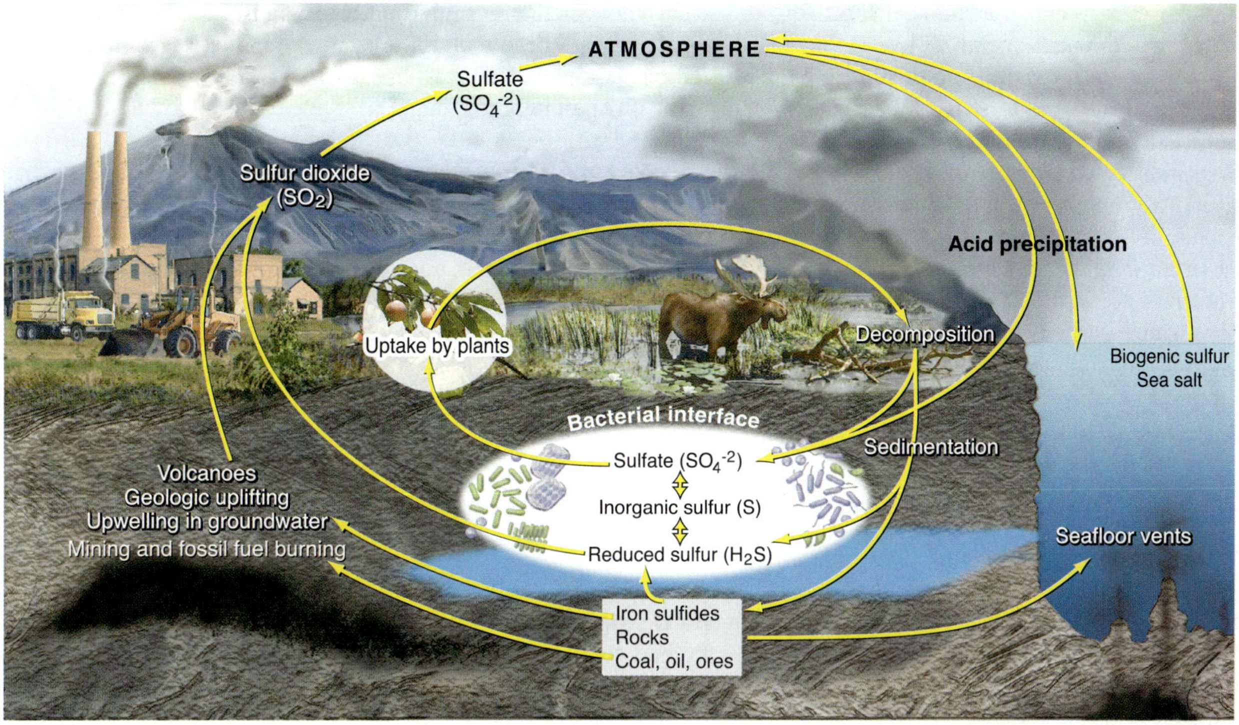


**FIGURE 3.24** The sulfur cycle. Sulfur is present mainly in rocks, soil, and water. It cycles through ecosystems when it is taken in by organisms. Combustion of fossil fuels causes increased levels of atmospheric sulfur compounds, which create problems related to acid precipitation.

and inorganic fertilizers represent exposed ocean sediments that are millennia old. You could think of our present use of phosphates, which are washed out into the river systems and eventually the oceans, as an accelerated mobilization of phosphorus from source to sink. Aquatic ecosystems often are dramatically affected in the process because excess phosphates can stimulate explosive growth of algae and photosynthetic bacteria populations, upsetting ecosystem stability. Notice also that in this cycle, as in the others, the role of organisms is only one part of a larger picture.

## Sulfur also cycles

Sulfur plays a vital role in organisms, especially as a minor but essential component of proteins. Sulfur compounds are important determinants of the acidity of rainfall, surface water, and soil. In addition, sulfur in particles and tiny air-borne droplets may act as critical regulators of global climate. Most of the earth's sulfur is tied up underground in rocks and minerals such as iron disulfide (pyrite) or calcium sulfate (gypsum). This inorganic sulfur is released into air and water by weathering, emissions from deep seafloor vents, and by volcanic eruptions (fig. 3.24).

The **sulfur cycle** is complicated by the large number of oxidation states the element can assume, including hydrogen sulfide ($H_2S$), sulfur dioxide ($SO_2$), sulfate ion ($SO_4^{-2}$), and sulfur, among others. Inorganic processes are responsible for many of these transformations, but living organisms, especially bacteria, also sequester sulfur in biogenic deposits or release it into the environment. Which of the several kinds of sulfur bacteria prevail in any given situation depends on oxygen concentrations, pH, and light levels.

Human activities also release large quantities of sulfur, primarily through burning fossil fuels. Total yearly anthropogenic sulfur emissions rival those of natural processes, and acid rain caused by sulfuric acid produced as a result of fossil fuel use is a serious problem in many areas (see chapter 16). Sulfur dioxide and sulfate aerosols cause human health problems, damage buildings and vegetation, and reduce visibility. They also absorb UV radiation and create cloud cover that cools cities and may be offsetting greenhouse effects of rising $CO_2$ concentrations.

# Conclusion

Matter is conserved as it cycles over and over through ecosystems, but energy is always degraded or dissipated as it is transformed or transferred from one place to another. These laws of physics and thermodynamics mean that elements are continuously recycled, but that living systems need a constant supply of external energy to replace that lost to entropy. Some extremophiles, living in harsh conditions, such as hot springs or the bottom of the ocean, capture energy from chemical reactions. For most organisms, however, the ultimate source of energy is the sun. Plants capture sunlight through the process of photosynthesis, and use the captured energy for metabolic processes and to build biomass (organic material). Herbivores eat plants to obtain energy and nutrients, carnivores eat herbivores or each other, and decomposers eat the waste products of this food web.

This dependence on solar energy is a fundamental limit for most life on earth. It's estimated that humans now dominate roughly 40 percent of the potential terrestrial net productivity. We directly eat only about 10 percent of that total (mainly because of the thermodynamic limits on energy transfers in food webs), but the crops and livestock that feed, clothe, and house us represent the rest of that photosynthetic output. By dominating nature, as we do, we exclude other species.

While energy flows in a complex, but ultimately one-way path through nature, materials are endlessly recycled. Five of the major material cycles (water, carbon, nitrogen, phosphorus, and sulfur) are summarized in this chapter. Each of these materials is critically important to living organisms. As humans interfere with these material cycles, we make it easier for some organisms to survive and more difficult for others. Often, we're intent on manipulating material cycles for our own short-term gain, but we don't think about the consequences for other species or even for ourselves in the long-term. An example of that is the carbon cycle. Our lives are made easier and more comfortable by burning fossil fuels, but in doing so we release carbon dioxide into the atmosphere, causing global warming that could have disastrous results. Clearly, it's important to understand these environmental systems and to take them into account in our public policy.

## Reviewing Learning Outcomes

By now you should be able to explain the following points:

**3.1** Describe matter, atoms, and molecules and give simple examples of the role of four major kinds of organic compounds in living cells.

- Matter is made of atoms, molecules, and compounds.
- Chemical bonds hold molecules together.
- Electrical charge is an important characteristic.
- Organic compounds have a carbon backbone.
- Cells are the fundamental units of life.

**3.2** Define energy and explain how thermodynamics regulates ecosystems.

- Energy occurs in different types and qualities.
- Thermodynamics regulates energy transfers.

**3.3** Understand how living organisms capture energy and create organic compounds.

- Extremophiles live in severe conditions.
- Green plants get energy from the sun.
- Photosynthesis captures energy while respiration releases that energy.

**3.4** Define species, populations, communities, and ecosystems, and summarize the ecological significance of trophic levels.

- Organisms occur in populations, communities, and ecosystems.
- Food chains, food webs, and trophic levels link species.
- Ecological pyramids describe trophic levels.

**3.5** Compare the ways that water, carbon, nitrogen, sulfur, and phosphorus cycle within ecosystems.

- The hydrologic cycle moves water around the earth.
- Carbon moves through the carbon cycle.
- Nitrogen moves via the nitrogen cycle.
- Phosphorus is an essential nutrient.
- Remote sensing allows us to evaluate photosynthesis and material cycles.
- Sulfur also cycles.

## Practice Quiz

1. Define *atom* and *element*. Are these terms interchangeable?
2. Your body contains vast numbers of carbon atoms. How is it possible that some of these carbon atoms may have been part of the body of a prehistoric creature?
3. What are six characteristics of water that make it so valuable for living organisms and their environment?
4. In the biosphere, matter follows a circular pathway while energy follows a linear pathway. Explain.
5. The oceans store a vast amount of heat, but (except for climate moderation) this huge reservoir of energy is of little use to humans. Explain the difference between high-quality and low-quality energy.
6. Ecosystems require energy to function. Where does this energy come from? Where does it go? How does the flow of energy conform to the laws of thermodynamics?
7. Heat is released during metabolism. How is this heat useful to a cell and to a multicellular organism? How might it be detrimental, especially in a large, complex organism?
8. Photosynthesis and cellular respiration are complementary processes. Explain how they exemplify the laws of conservation of matter and thermodynamics.
9. What do we mean by carbon-fixation or nitrogen-fixation? Why is it important to humans that carbon and nitrogen be "fixed"?
10. The population density of large carnivores is always very small compared to the population density of herbivores occupying the same ecosystem. Explain this in relation to the concept of an ecological pyramid.
11. A species is a specific kind of organism. What general characteristics do individuals of a particular species share? Why is it important for ecologists to differentiate among the various species in a biological community?

## Critical Thinking and Discussion Questions

1. A few years ago, laundry detergent makers were forced to reduce or eliminate phosphorus. Other cleaning agents (such as dishwasher detergents) still contain substantial amounts of phosphorus. What information would make you change your use of nitrogen, phosphorus, or other useful pollutants?
2. The first law of thermodynamics is sometimes summarized as "you can't get something for nothing." The second law is summarized as "you can't even break even." Explain what these phrases mean. Is it dangerous to oversimplify these important concepts?
3. The ecosystem concept revolutionized ecology by introducing holistic systems thinking as opposed to individualistic life history studies. Why was this a conceptual breakthrough?
4. If ecosystems are so difficult to delimit, why is this such a persistent concept? Can you imagine any other ways to define or delimit environmental investigation?
5. The properties of water are so unique and so essential for life as we know it that some people believe it proves that our planet was intentionally designed for our existence. What would an environmental scientist say about this belief?
6. Choose one of the material cycles (carbon, nitrogen, phosphorus, or sulfur) and identify the components of the cycle in which you participate. For which of these components would it be easiest to reduce your impacts?

## DATA analysis Extracting Data from a Graph

1. How can you extract data from a line plot? The process is simply the reverse of the way you just learned to create the graph. You draw lines from the axes to where they intersect on the graph at the point whose value you want to know. Let's look at figure 1, the line plot we examined in chapter 1.

- How many cells were there at the half point of the growth curve?
- How does that result compare to the cell population three hours earlier?
- When did the growth of this population start to slow?

A curve, such as this, that increases slowly at first but then rapidly accelerates is called a logistic curve. We'll discuss the mathematics that create this pattern in chapter 6.

2. How do you extract data from a bar graph? Look again at the bar graph in figure 2. Using the same procedure described for a line plot, draw a horizontal line from the top of any bar straight to the Y-axis. Read the value for the frequency of that category on the Y-axis.

- Why don't you need to draw a vertical line with this type of graph?
- How many cities have 30 ug/m$^3$?
- Why aren't there any bars between 50 and 65 ug/m$^3$?
- Why are there so many more cities at 40 ug/m$^3$?
- Why is one city at 70 ug/m$^3$?
- How many more cities have 40 ug/m$^3$ than 30 ug/m$^3$?

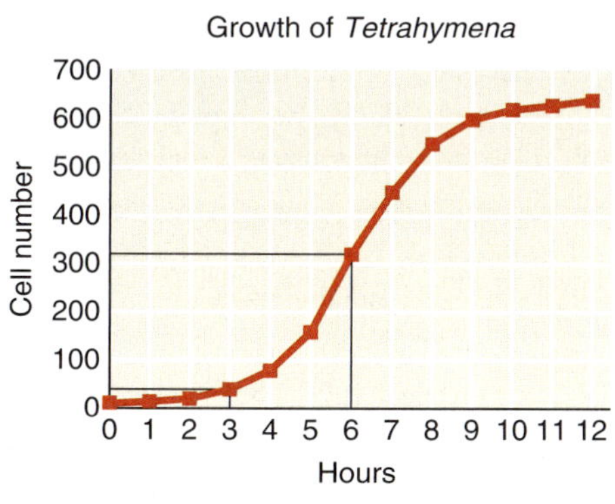


**FIGURE 1** Growth of *Tetrahymena.*

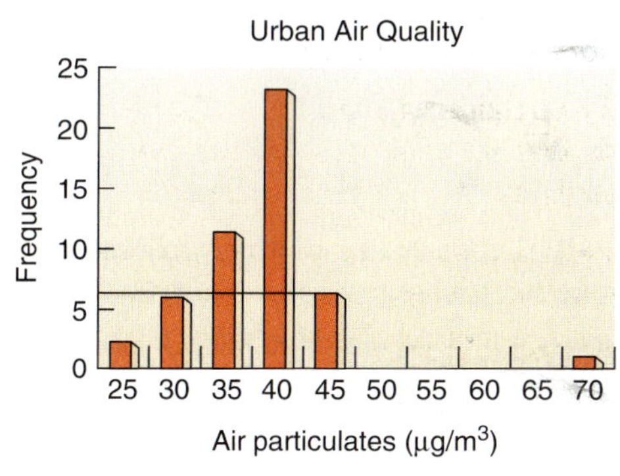


**FIGURE 2** Urban air quality.

**For Additional Help in Studying This Chapter,** please visit our website at www.mhhe.com/cunningham10e. You will find additional practice quizzes and case studies, flashcards, regional examples, place markers for Google Earth™ mapping, and an extensive reading list, all of which will help you learn environmental science.

The relatively young and barren volcanic islands of the Galápagos, isolated from South America by strong, cold currents and high winds, have developed a remarkable community of unique plants and animals.

CHAPTER 4

# Evolution, Biological Communities, and Species Interactions

*Any species of bug is an irreplaceable marvel, equal to the works of art which we religiously preserve in our museums.*

*—Claude Levi-Strauss—*

## LEARNING OUTCOMES

After studying this chapter, you should be able to:

**4.1** Describe how evolution produces species diversity.

**4.2** Discuss how species interactions shape biological communities.

**4.3** Summarize how community properties affect species and populations.

**4.4** Explain why communities are dynamic and change over time.

# Case Study Darwin's Voyage of Discovery

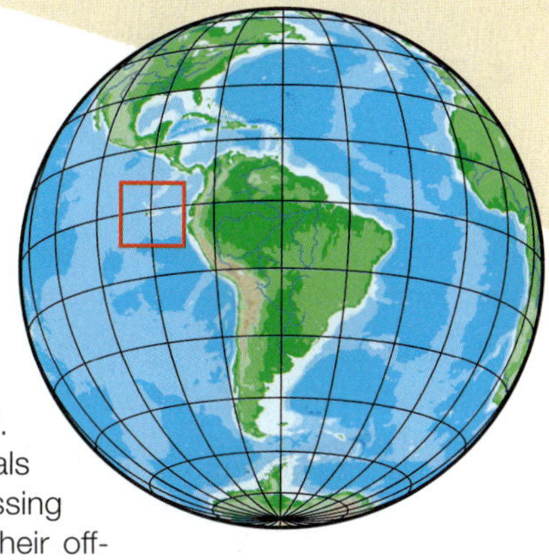

Charles Darwin was only 22 years old when he set out in 1831 on his epic five-year, around-the-world voyage aboard the H.M.S. *Beagle* (fig. 4.1). It was to be the adventure of a lifetime, and would lead to insights that would revolutionize the field of biology. Initially an indifferent student, Darwin had found inspiring professors in his last years of college. One of them helped him get a position as an unpaid naturalist on board the *Beagle.* Darwin turned out to be a perceptive observer, an avid collector of specimens, and an extraordinary scientist.

As the *Beagle* sailed slowly along the coast of South America, mapping coastlines and navigational routes, Darwin had time to go ashore on long field trips to explore natural history. He was amazed by the tropical forests of Brazil and the fossils of huge, extinct mammals in Patagonia. He puzzled over the fact that many fossils looked similar, but not quite identical, to contemporary animals. Could species change over time? In Darwin's day, most people believed that everything in the world was exactly as it had been created by God only a few thousand years earlier. But Darwin had read the work of Charles Lyell (1797–1875), who suggested that the world was much older than previously thought, and capable of undergoing gradual, but profound, change over time.

**FIGURE 4.1** Charles Darwin, in a portrait painted shortly after the voyage on the *Beagle.*

After four years of exploring and mapping, Darwin and the *Beagle* reached the Galápagos Islands, 900 km (540 mi) off the coast of Ecuador. The harsh, volcanic landscape of these remote islands (see opposite page) held an extraordinary assemblage of unique plants and animals. Giant land tortoises fed on tree-size cacti. Sea-going iguanas scraped algae off underwater shoals. Sea birds were so unafraid of humans that Darwin could pick them off their nests. The many finches were especially interesting: Every island had its own species, marked by distinct bill shapes, which graded from large and parrot-like to small and warbler-like. Each bird's anatomy and behavior was suited to exploit specific food sources available in its habitat. It seemed obvious that these birds were related, but somehow had been modified to survive under different conditions.

Darwin didn't immediately understand the significance of these observations. Upon returning to England, he began the long process of cataloging and describing the specimens he had collected. Over the next 40 years, he wrote important books on a variety of topics including the formation of oceanic islands from coral reefs, the geology of South America, and the classification and natural history of barnacles. Throughout this time, he puzzled about how organisms might adapt to specific environmental situations.

A key in his understanding was Thomas Malthus's *Essay on the Principle of Population* (1798). From Malthus, Darwin saw that most organisms have the potential to produce far more offspring than can actually survive. Those individuals with superior attributes are more likely to live and reproduce than those less well-endowed. Because the more fit individuals are especially successful in passing along their favorable traits to their offspring, the whole population will gradually change to be better suited for its particular environment. Darwin called this process *natural selection* to distinguish it from the artificial selection that plant and animal breeders used to produce the wide variety of domesticated crops and livestock.

Darwin completed a manuscript outlining his theory of **evolution** (gradual change in species) through natural selection in 1842, but he didn't publish it for another 16 years, perhaps because he was worried about the controversy he knew it would provoke. When his masterpiece, *On the Origin of Species,* was finally made public in 1859, it was both strongly criticized and highly praised. Although Darwin was careful not to question the existence of a Divine Creator, many people interpreted his theory of gradual change in nature as a challenge to their faith. Others took his theory of survival of the fittest much further than Darwin intended, applying it to human societies, economics, and politics.

One of the greatest difficulties for the theory of evolution was that little was known in Darwin's day of the mechanisms of heredity. No one could explain how genetic variation could arise in a natural population, or how inheritable traits could be sorted and recombined in offspring. It took nearly another century before biologists could use their understanding of molecular genetics to put together a modern synthesis of evolution that clarifies these details.

An overwhelming majority of biologists now consider the theory of evolution through natural selection to be the cornerstone of their science. The theory explains how the characteristics of organisms have arisen from individual molecules, to cellular structures, to tissues and organs, to complex behaviors and population traits. In this chapter, we'll look at the evidence for evolution and how it shapes species and biological communities. We'll examine the ways in which interactions between species and between organisms and their environment allow species to adapt to particular conditions as well as to modify both their habitat and their competitors.

**For more information, see**


Darwin, Charles. *The Voyage of the* Beagle (1837) and *On the Origin of Species* (1859).

Quammen, David. *1996. The Song of the Dodo: Island Biogeography in an Age of Extinctions.* Scribners.

## 4.1 Evolution Produces Species Diversity

Why do some species live in one place but not another? A more important question to environmental scientists is, what are the mechanisms that promote the great variety of species on earth and that determine which species will survive in one environment but not another? In this section you will come to understand (1) concepts behind the theory of speciation by means of natural selection and adaptation (evolution); (2) the characteristics of species that make some of them weedy and others endangered; and (3) the limitations species face in their environments and implications for their survival. First we'll start with the basics: How do species arise?

### Natural selection leads to evolution

How does a polar bear stand the long, sunless, super-cold arctic winter? How does the saguaro cactus survive blistering temperatures and extreme dryness of the desert? We commonly say that each species is *adapted* to the environment where it lives, but what does that mean? **Adaptation,** the acquisition of traits that allow a species to survive in its environment, is one of the most important concepts in biology.

We use the term *adapt* in two ways. An individual organism can respond immediately to a changing environment in a process called acclimation. If you keep a houseplant indoors all winter and then put it out in full sunlight in the spring, the leaves become damaged. If the damage isn't severe, your plant may grow new leaves with thicker cuticles and denser pigments that block the sun's rays. However, the change isn't permanent. After another winter inside, it will still get sun-scald in the following spring. The leaf changes are not permanent and cannot be passed on to offspring, or even carried over from the previous year. Although the capacity to acclimate is inherited, houseplants in each generation must develop their own protective leaf epidermis.

Another type of adaptation affects populations consisting of many individuals. Genetic traits are passed from generation to generation and allow a species to live more successfully in its environment. As the opening case study for this chapter shows, this process of adaptation to environment is explained by the theory of evolution. The basic idea of evolution is that species change over generations because individuals compete for scarce resources. Better competitors in a population survive—they have greater reproductive potential or fitness—and their offspring inherit the beneficial traits. Over generations, those traits become common in a population (fig. 4.2). The process of better-selected individuals passing their traits to the next generation is called **natural selection.** The traits are encoded in a species' DNA, but from where does the original DNA coding come, which then gives some individuals greater fitness? Every organism has a dizzying array of genetic diversity in its DNA. It has been demonstrated in experiments and by observing natural populations that changes to the DNA coding sequence of individuals occurs, and that the changed sequences are inherited by offspring. Exposure to ionizing radiation and toxic materials, and random recombination and mistakes in replication of DNA strands during reproduction are the main causes of genetic mutations. Sometimes a single mutation has a large effect, but evolutionary change is mostly brought about by many mutations accumulating over time. Only mutations in reproductive cells (gametes) matter; body cell changes—cancers, for example—are not inherited. Most mutations have no effect on fitness, and many actually have a negative effect. During the course of a species' life span—a million or more years—some mutations are thought to have given those individuals an advantage under the **selection pressures** of their environment at that time. The result is a species population that differs from those of numerous preceding generations.

**FIGURE 4.2** Giraffes don't have long necks because they stretch to reach tree-top leaves, but those giraffes that happened to have longer necks got more food and had more offspring, so the trait became fixed in the population.

### All species live within limits

Environmental factors exert selection pressure and influence the fitness of individuals and their offspring. For this reason, species are limited in where they can live. Limitations include the following: (1) physiological stress due to inappropriate levels of some critical environmental factor, such as moisture, light, temperature, pH, or specific nutrients; (2) competition with other species; (3) predation, including parasitism and disease; and (4) luck. In some cases, the individuals of a population that survive environmental catastrophes or find their way to a new habitat, where they start a new population, may simply be lucky rather than more fit than their contemporaries.

An organism's physiology and behavior allow it to survive only in certain environments. Temperature, moisture level, nutrient supply, soil and water chemistry, living space, and other environmental factors must be at appropriate levels for organisms to persist. In 1840, the chemist Justus von Liebig proposed that the single factor in shortest supply relative to demand is the **critical factor** determining where a species lives. The giant saguaro cactus (*Carnegiea gigantea*), which grows in the dry, hot Sonoran desert of southern Arizona and northern Mexico, offers an example (fig. 4.3). Saguaros are extremely sensitive to freezing temperatures. A single winter night with temperatures below freezing for 12 or more hours kills growing tips on the branches,

**FIGURE 4.3** Saguaro cacti, symbolic of the Sonoran desert, are an excellent example of distribution controlled by a critical environmental factor. Extremely sensitive to low temperatures, saguaros are found only where minimum temperatures never dip below freezing for more than a few hours at a time.

preventing further development. Thus the northern edge of the saguaro's range corresponds to a zone where freezing temperatures last less than half a day at any time.

Ecologist Victor Shelford (1877–1968) later expanded Liebig's principle by stating that each environmental factor has both minimum and maximum levels, called **tolerance limits,** beyond which a particular species cannot survive or is unable to reproduce (fig. 4.4). The single factor closest to these survival limits, Shelford postulated, is the critical factor that limits where a particular organism can live. At one time, ecologists tried to identify unique factors limiting the growth of every plant and animal population. We now know that several factors working together, even in a clear-cut case like the saguaro, usually determine a species' distribution. If you have ever explored the rocky coasts of New England or the Pacific Northwest, you have probably noticed that mussels and barnacles grow thickly in the intertidal zone, the place between high and low tides. No one factor decides this pattern. Instead, the distribution of these animals is determined by a combination of temperature extremes, drying time between tides, salt concentrations, competitors, and food availability.

In some species, tolerance limits affect the distribution of young differently than adults. The desert pupfish, for instance, lives in small, isolated populations in warm springs in the northern Sonoran desert. Adult pupfish can survive temperatures between 0° and 42°C (a remarkably high temperature for a fish) and tolerate an equally wide range of salt concentrations. Eggs and juvenile fish, however, can survive only between 20° and 36°C and are killed by high salt levels. Reproduction, therefore, is limited to a small part of the range of the adult fish.

Sometimes the requirements and tolerances of species are useful **indicators** of specific environmental characteristics. The presence or absence of such species indicates something about the community and the ecosystem as a whole. Lichens and eastern white pine, for example, are indicators of air pollution because they are extremely sensitive to sulfur dioxide and ozone, respectively. Bull thistle and many other plant weeds grow on disturbed soil but are not eaten by cattle; therefore, a vigorous population of bull thistle or certain other plants in a pasture indicates it is being overgrazed. Similarly, anglers know that trout species require cool, clean, well-oxygenated water; the presence or absence of trout is used as an indicator of good water quality.

## The ecological niche is a species' role and environment

**Habitat** describes the place or set of environmental conditions in which a particular organism lives. A more functional term, **ecological niche,** describes either the role played by a species in a biological community or the total set of environmental factors that determine a species distribution. The concept of niche was

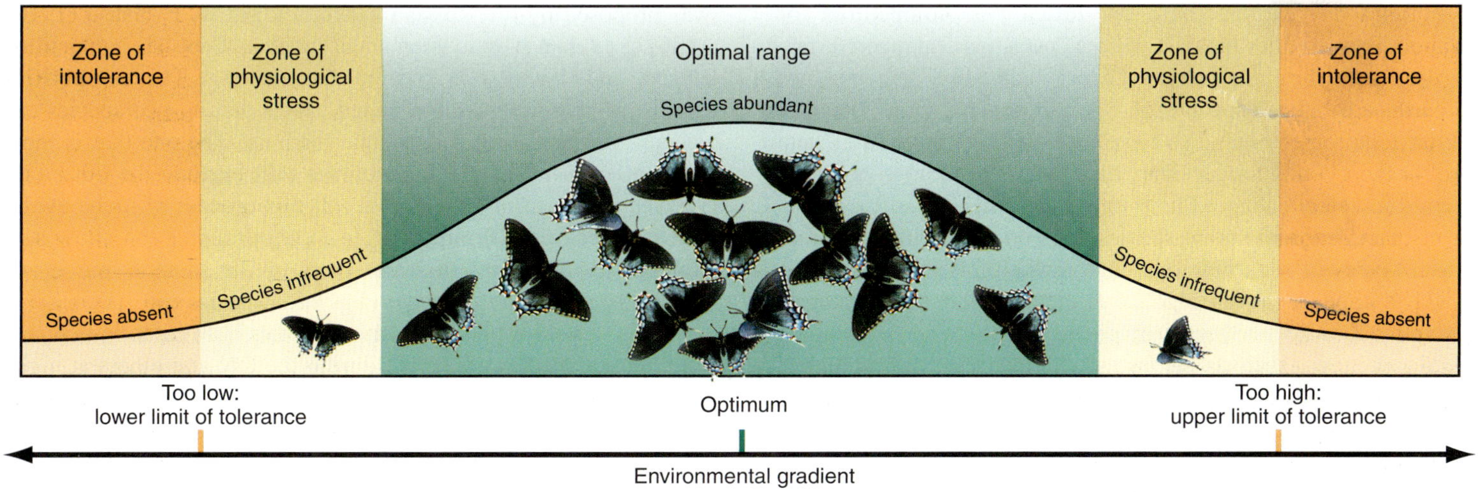


**FIGURE 4.4** The principle of tolerance limits states that for every environmental factor, an organism has both maximum and minimum levels beyond which it cannot survive. The greatest abundance of any species along an environmental gradient is around the optimum level of the critical factor most important for that species. Near the tolerance limits, abundance decreases because fewer individuals are able to survive the stresses imposed by limiting factors.

**FIGURE 4.5** Each of the animal and plant species living in this African savanna occupies an ecological niche. Some niches are broad and general, some are highly specialized.

first defined in 1927 by the British ecologist Charles Elton (1900–1991). To Elton, each species had a role in a community of species, and the niche defined its way of obtaining food, the relationships it had with other species, and the services it provided to its community. Thirty years later, the American limnologist G. E. Hutchinson (1903–1991) proposed a more biophysical definition of niche. Every species, he pointed out, exists within a range of physical and chemical conditions (temperature, light levels, acidity, humidity, salinity, etc.) and also biological interactions (predators and prey present, defenses, nutritional resources available, etc.). The niche is more complex than the idea of a critical factor (fig. 4.5). A graph of a species niche would be multidimensional, with many factors being simultaneously displayed, almost like an electron cloud.

For a generalist, like the brown rat, the ecological niche is broad. In other words, a generalist has a wide range of tolerance for many environmental factors. For others, such as the giant panda (*Ailuropoda melanoleuca*), only a narrow ecological niche exists (fig. 4.6). Bamboo is low in nutrients, but provides 95 percent of a panda's diet, requiring it to spend as much as 16 hours a day eating. There are virtually no competitors for bamboo, except other pandas, yet the species is endangered, primarily due to shrinking habitat. Giant pandas, like many species on earth, are habitat specialists. Specialists have more exacting habitat requirements, tend to have lower reproductive rates, and care for their young longer. They may be less resilient in response to environmental change. Plants can also be habitat specialists—for instance, **endemic** plant species exist on serpentine and other unusual types of rock outcrops, and nowhere else.

Over time, niches change as species develop new strategies to exploit resources. Species of greater intelligence or complex social structures, such as elephants, chimpanzees, and dolphins, learn from their social group how to behave and can invent new ways of doing things when presented with novel opportunities or challenges. In effect, they alter their ecological niche by passing on cultural behavior from one generation to the next. Most organisms, however, are restricted to their niche by their genetically determined bodies and instinctive behaviors. When two such species compete for limited resources, one eventually gains the larger share, while the other finds different habitat, dies out, or experiences a change in its behavior or physiology so that competition is minimized. The idea that "complete competitors cannot coexist" was proposed by the Russian microbiologist G. F. Gause (1910–1986) to explain why mathematical models of species competition always ended with one species disappearing. The **competitive exclusion principle,** as it is called, states that no two species can occupy the same ecological niche for long. The one that is more efficient in using available resources will exclude the other (see Species Competition p. 97). We call this process of niche evolution **resource partitioning** (fig. 4.7). Partitioning can allow several species to utilize different parts of the same resource and coexist within a single habitat (fig. 4.8). Species can specialize in time, too. Swallows and insectivorous bats both catch insects, but some insect species are active during the day and others at night, providing noncompetitive feeding opportunities for day-active swallows and night-active bats. The competitive exclusion principle does not explain all situations, however. For example, many similar plant species coexist in some habitats. Do they avoid competition in ways we cannot observe, or are resources so plentiful that no competition need occur?

**FIGURE 4.6** The giant panda feeds exclusively on bamboo. Although its teeth and digestive system are those of a carnivore, it is not a good hunter, and has adapted to a vegetarian diet. In the 1970s, huge acreages of bamboo flowered and died, and many pandas starved.

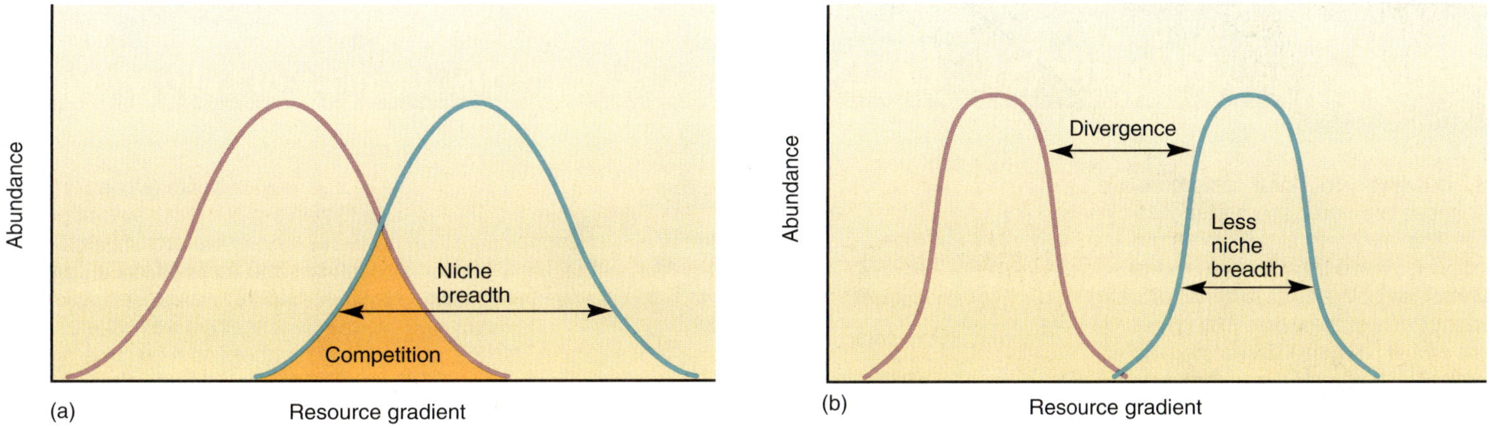


**FIGURE 4.7** Competition causes resources partitioning and niche specialization. (a) Where niches of two species overlap along a resource gradient, competition occurs (*shaded area*). Individuals in this part of the niche have less success producing young. (b) Over time the traits of the populations diverge, leading to specialization, narrower niche breadth, and less competition between species.

## Speciation maintains species diversity

As an interbreeding species population becomes better adapted to its ecological niche, its genetic heritage (including mutations passed from parents to offspring) gives it the potential to change further as circumstances, dictate. In the case of Galápagos finches studied a century and a half ago by Charles Darwin, evidence from body shape, behavior, and genetics leads to the idea that modern Galapagos finches look, behave, and bear DNA related to an original seed-eating finch species (fig. 4.9). It is proposed to have blown to the islands from the mainland where a similar species still exists. Today there are 13 distinct species on the islands that differ markedly in appearance, food preferences, and habitat. Fruit eaters have thick, parrot-like bills; seed eaters have heavy, crushing bills; insect eaters have thin, probing beaks to catch their prey. One of the most unusual species is the woodpecker finch, which pecks at tree bark for hidden insects. Lacking the woodpecker's long tongue, the finch uses a cactus spine as a tool to extract bugs.

The development of a new species is called **speciation.** No one has observed a new species springing into being, but in some organisms, especially plants, it is inferred to occur somewhat frequently. Nevertheless, given the evidence, speciation is a reasonable proposal for how species arise. Speciation may be relatively rapid on millennial timescales (punctuated equilibrium). For example, after a long period of stability, a new species may arise from parents following the appearance of a new food source, predator, or competitor, or a change in climate. One mechanism of speciation is **geographic isolation.** This is termed **allopatric speciation**—species arise in non-overlapping geographic locations. The original Galapagos finches were separated from the rest of the population on the mainland, could no longer share genetic material, and became reproductively isolated.

The barriers that divide subpopulations are not always physical. For example, two virtually identical tree frogs (*Hyla versicolor, H. chrysoscelis*) live in similar habitats of eastern North America but have different mating calls. This is an example of behavioral isolation. It also happens that one species has twice the chromosomes of

**FIGURE 4.8** Several species of insect-eating wood warblers occupy the same forests in eastern North America. The competitive exclusion principle predicts that the warblers should partition the resource—insect food—in order to reduce competition. And in fact, the warblers feed in different parts of the forest.

**Source:** Original observations by R. H. MacArthur (1958).

## *Exploring* SCIENCE — The Cichlids of Lake Victoria

Cichlid fishes of Lake Victoria. More than 300 species have evolved from an original common ancestor to take advantage of different food sources and habitats.

If you visit your local pet store, chances are you'll see some cichlids (*Haplochromis* sp.). These small colorful, prolific fish come in a wide variety of colors and shapes from many parts of the world. The greatest cichlid diversity on earth—and probably the greatest vertebrate diversity anywhere—is found in the three great African rift lakes: Victoria, Malawi, and Tanganyika. Together, these lakes once had about 1,000 types of cichlids—more than all the fish species in Europe and North America combined. All these cichlids apparently evolved from a few ancestral varieties in the 15,000 years or so since the lakes were formed by splitting of the continental crust. This is one of the fastest and most extensive examples of vertebrate speciation known.

We believe that one of the factors that allowed cichlids to evolve so quickly is that they found few competitors or predators and a multitude of ecological roles to play in these new lakes. There are mud biters, algae scrapers, leaf chewers, snail crushers, zooplankton eaters, prawn predators, and fish feeders. Because they live in different habitats in the lakes, are active at different times of the day, and have developed different body sizes and shapes to feed on specialized prey, the cichlids have been ecologically isolated for long enough to evolve into an amazing variety of species. Cichlids are a good example of radiative evolution.

Unfortunately, a well-meaning but disastrous fish-stocking experiment has wiped out at least half the cichlid species in these lakes in just a few decades and set off a series of changes that are upsetting important ecological relationships. Lake Victoria, which lies between Kenya, Tanzania, and Uganda, has been particularly hard hit. Cichlids once made up 80 percent of the animal biomass in the lake and were the base for a thriving local fishery, supplying much-needed protein for local people. Management agencies regarded the bony little cichlids as "trash fish," and decided in the 1960s to introduce Nile perch (*Lates niloticus*), a voracious, exotic predator that can weigh up to 100 kg (220 lbs) and grow up to 2 m long. The perch, they believed, would support a lucrative commercial export trade.

The perch gobbled up the cichlids so quickly that, by 1980, two-thirds of the haplochromine species in the lake were extinct. Although there are still lots of fish in the lake, 80 percent of the biomass is now made up of perch, which are too large and powerful for the small boats, papyrus nets, and woven baskets traditionally used to harvest cichlids. International fishing companies now use large power boats and nylon nets to harvest great schools of perch, which are filleted, frozen, and shipped to markets in Europe and the Middle East. Because the perch are oily, local fishers can't sun dry them as they once did the cichlids. Instead, they are cooked or smoked over wood fires for local consumption. Forests are being denuded for firewood, and protein malnutrition is common in a region that exports 200,000 tons of fish each year.

Perhaps worst of all, Lake Victoria, which covers an area the size of Switzerland, is dying. Algae blooms clot the surface, oxygen levels have fallen alarmingly, and thick layers of soft silt are filling-in shallow bays. Untreated sewage, chemical pollution, and farm runoff are the immediate causes of these deleterious changes, but destabilization of the natural community plays a role as well. The swarms of cichlids that once ate algae and rotting detritus were the lake's self-cleaning system. Eliminating them threatens the long-term ability of the lake to support any useful aquatic life.

As this example shows, species and ecological diversity are important. Misguided management and development schemes that destroyed native species in Lake Victoria have resulted in an ecosystem that no longer supports the natural community or the local people dependent on it. It's difficult to see how we could replace the variety of species and the ecological roles they played, which evolution provided for free.

**For more information, see**


Stiassny, M. L. J., and A. Meyer. 1999. Cichlids of the rift lakes. *Scientific American* 280(2):64–69.


the other. This example of **sympatric speciation** takes place in the same location as the ancestor species (Exploring Science above). Fern species and other plants seem prone to sympatric speciation by doubling or quadrupling the chromosome number of their ancestors.

Once isolation is imposed, the two populations begin to diverge in genetics and physical characteristics. Genetic drift ensures that DNA of two formerly joined populations eventually diverges; in several generations, traits are lost from a population during the natural course of reproduction. Under more extreme circumstances, a die-off of most members of an isolated population strips much of the variation in traits from the survivors. The cheetah experienced a genetic bottleneck about 10,000 years ago and exists today as virtually identical individuals.

In isolation, selection pressures shape physical, behavioral, and genetic characteristics of individuals, causing population

**FIGURE 4.9** Each of the 13 species of Galápagos finches, although originally derived from a common ancestor, has evolved distinctive anatomies and behaviors to exploit different food sources. The woodpecker finch (d) uses cactus thorns to probe for insects under tree bark.

traits to shift over time (fig. 4.10). From an original range of characteristics, the shift can be toward an extreme of the trait (directional selection), it can narrow the range of a trait (stabilizing selection), or it can cause traits to diverge to the extremes (disruptive selection). Directional selection is implied by increased pesticide resistance in German cockroaches (*Blattella germanica*). Apparently some individuals can make an enzyme that detoxifies pesticides. Individual cockroaches that lack this characteristic are dying out, and as a result, populations of cockroaches with pesticide resistance are developing.

A small population in a new location—island, mountaintop, unique habitat—encounters new environmental conditions that favor some individuals over others (fig. 4.11). The physical and behavioral traits these individuals have are passed to the next generation, and the frequency of the trait shifts in the population. Where a species may have existed but has died out, others arise and contribute to the incredible variety of life-forms seen in nature. The fossil record is one of ever-increasing species diversity, despite several catastrophes, which were recorded in different geological strata and which wiped out a large proportion of the earth's species each time.

## Evolution is still at work

You may think that evolution only occurred in the distant past, but it's an ongoing process. Ample evidence from both laboratory experiments and from nature shows evolution at work. Geneticists have modified many fruit fly properties—including body size, eye color, growth rate, life span, and feeding behavior—using artificial selection. In one experiment, researchers selected for flies with many bristles (stiff, hairlike structures) on their abdomen. In each generation, the flies with the most bristles were allowed to mate. After 86 generations, the number of bristles had quadrupled. In a similar experiment with corn, agronomists chose seeds with the highest oil content to plant and mate. After 90 generations, the average oil content had increased 450 percent.

Evolutionary change is also occurring in nature. A classic example is seen in some of the finches on the Galapágos Island of Daphne. Twenty years ago, a large-billed species (*Geospiza magnirostris*) settled on the island, which previously had only a medium-billed species (*Geospiza fortis*). The *G. magnirosris* were better at eating larger seeds and pushed *G. fortis* to depend more

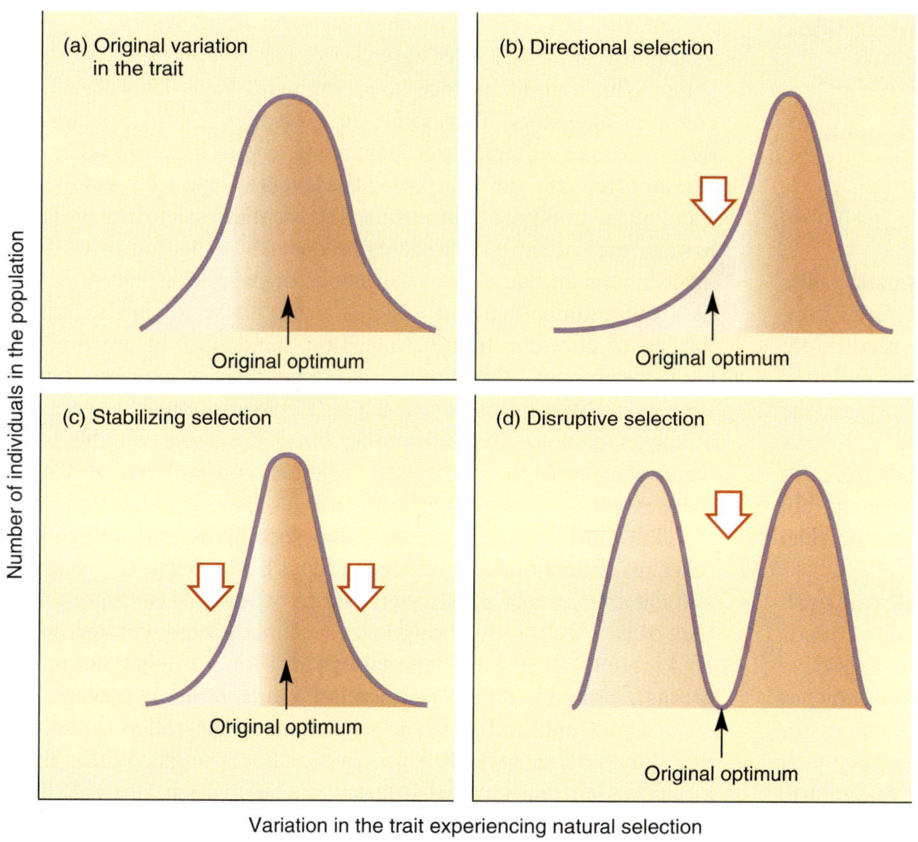


**FIGURE 4.10** A species trait, such as beak shape, changes in response to selection pressure. (a) The original variation is acted on by selection pressure (*arrows*) that (b) shifts the characteristics of that trait in one direction, (c) or to an intermediate condition. (d) Disruptive selection moves characteristics to the extremes of the trait. Which selection type plausibly resulted in two distinct beak shapes among Galápagos finches—narrow in tree finches versus stout in ground finches?

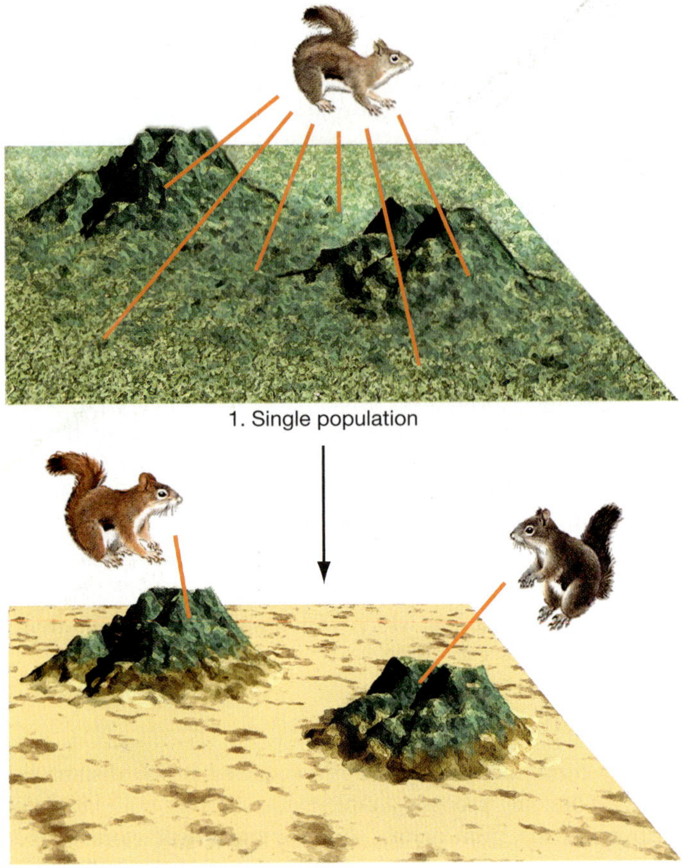


**FIGURE 4.11** Geographic barriers can result in allopatric speciation. During cool, moist glacial periods, what is now Arizona was forest-covered and squirrels could travel and interbreed freely. As the climate warmed and dried, desert replaced forest on the plains. Squirrels were confined to cooler mountaintops, which acted as island refugia, where new, reproductively isolated species gradually evolved.

and more on smaller seeds. Gradually, birds with smaller bills suited to small seeds became more common in the *G. fortis* population. During a severe drought in 2003–2004, large seeds were scarce, and most birds with large beaks disappeared. This included almost all of the recently arrived *G. magnirosris* as well as the larger-beaked *G. fortis*. In just two generations, the *G. fortis* population changed to entirely small-beaked individuals. At first, this example of rapid evolution was thought to be a rarity, but subsequent research suggests that it may be more common than previously thought.

Similarly, the widespread application of pesticides in agriculture and urban settings has led to the rapid evolution of resistance in more than 500 insect species. Similarly, the extensive use of antibiotics in human medicine and livestock operations has led to antibiotic resistance in many microbes. The Centers for Disease Control estimates that 90,000 Americans die every year from hospital-acquired infections, most of which are resistant to one or more antibiotics. We're engaged in a kind of an arms race with germs. As quickly as new drugs are invented, microbes become impervious to them. Currently, vancomycin is the drug of last resort. When resistance to it becomes widespread, we may have no protection from infections.

***Think About It***

Try to understand the position of someone who holds an opposite view from your own about evolution. Why would they argue for or against this theory? If you were that person, what evidence would you want to see before you'd change your beliefs?

On the other hand, evolution sometimes works in our favor. We've spread a number of persistent organic pollutants (called POPs), such as pesticides and industrial solvents, throughout our environment. One of the best ways to get rid of them is with microbes that can destroy or convert them to a nontoxic form. It turns out that the best place to look for these species is in the most contaminated sites. The presence of a new food source has stimulated evolution of organisms that can metabolize it. A little artificial selection and genetic modification in the laboratory can turn these species into very useful bioremediation tools.

## Taxonomy describes relationships among species

Taxonomy is the study of types of organisms and their relationships. With it you can trace how organisms have descended from common ancestors. Taxonomic relationships among species are displayed like a family tree. Botanists, ecologists, and other scientists often use the most specific levels of the tree, genus and species, to compose **binomials.** Also called scientific or Latin names, they identify and describe species using Latin, or Latinized nouns and adjectives, or names of people or places. Scientists communicate about species using these scientific names instead of common names (e.g., lion, dandelion, or ant lion), to avoid confusion. A common name can refer to any number of species in different places, and a single species might have many common names. The bionomial *Pinus resinosa,* on the other hand, always is the same tree, whether you call it a red pine, Norway pine, or just pine.

Taxonomy also helps organize specimens and subjects in museum collections and research. You are *Homo sapiens* (human) and eat chips made of *Zea mays* (corn or maize). Both are members of two well-known kingdoms. Scientists, however, recognize six kingdoms (fig. 4.12): animals, plants, fungi (molds and mushrooms), protists (algae, protozoans, slime molds), bacteria (or eubacteria), and archaebacteria (ancient, single-celled organisms that live in harsh environments, such as hot springs). Within these kingdoms are millions of different species, which you will learn more about in chapters 5 and 11.

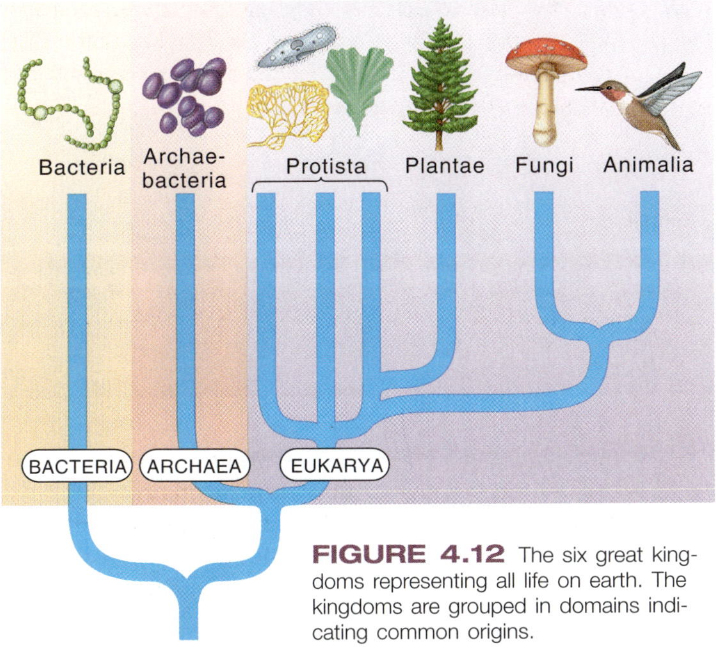


**FIGURE 4.12** The six great kingdoms representing all life on earth. The kingdoms are grouped in domains indicating common origins.

**FIGURE 4.13** In this tangled Indonesian rainforest, space and light are at a premium. Plants growing beneath the forest canopy have adaptations that help them secure these limited resources. The ferns and bromeliads seen here are epiphytes; they find space and get closer to the sun by perching on limbs and tree trunks. Strangler figs start out as epiphytes, but send roots down to the forest floor and, once contact is made, put on a growth spurt that kills the supporting tree. These are just some of the adaptations to life in the dark jungle.

## 4.2 Species Interactions Shape Biological Communities

We have learned that adaptation to one's environment, determination of ecological niche, and even speciation is affected not just by bodily limits and behavior, but also by competition and predation. Don't despair. Not all biological interactions are antagonistic, and many, in fact, involve cooperation or at least benign interactions and tolerance. In some cases, different organisms depend on each other to acquire resources. Now we will look at the interactions within and between species that affect their success and shape biological communities.

### Competition leads to resource allocation

Competition is a type of antagonistic relationship within a biological community. Organisms compete for resources that are in limited supply: energy and matter in usable forms, living space, and specific sites to carry out life's, activities. Plants compete for growing space to develop root and shoot systems so that they can absorb and process sunlight, water, and nutrients (fig. 4.13). Animals compete for living, nesting, and feeding sites, and also for mates. Competition among members of the same species is called **intraspecific competition,** whereas competition between members of different species is called **interspecific competition.** Recall the competitive exclusion principle as it applies to interspecific competition. Competition shapes a species population and biological community by causing individuals and species to shift their focus from one segment of a resource type to another. Thus, warblers all competing with each other for insect food in New England tend to specialize on different areas of the forest's trees, reducing or avoiding competition. Since the 1950s there have been hundreds of interspecific competition studies in natural populations. In general, scientists assume it does occur, but not always, and in some groups—carnivores and plants—it has little effect.

In intraspecific competition, members of the same species compete directly with each other for resource. Several avenues exist to reduce competition in a species population. First, the young of the year disperse. Even plants practice dispersal; seeds are carried by wind, water, and passing animals to less crowded conditions away from the parent plants. Second, by exhibiting strong territoriality, many animals force their offspring or trespassing adults out of their vicinity. In this way territorial species, which include bears, songbirds, ungulates, and even fish, minimize competition between individuals and generations. A third way to reduce intraspecific competition is resource partitioning between generations. The adults and juveniles of these species occupy different ecological niches. For instance, monarch caterpillars munch on milkweed leaves, while metamorphosed butterflies lap nectar. Crabs begin as floating larvae and do not compete with bottom-dwelling adult crabs.

We think of competition among animals as a battle for resources—"nature red in tooth and claw" is the phrase. In fact, many animals avoid fighting if possible, or confront one another with noise and predictable movements. Bighorn sheep and many other ungulates, for example, engage in ritualized combat, with the weaker animal knowing instinctively when to back off. It's worse to be injured than to lose. Instead, competition often is

**FIGURE 4.14** Insect herbivores are predators as much as are lions and tigers. In fact, insects consume the vast majority of biomass in the world. Complex patterns of predation and defense have often evolved between insect predators and their plant prey.

**FIGURE 4.15** Microscopic plants and animals form the basic levels of many aquatic food chains and account for a large percentage of total world biomass. Many oceanic plankton are larval forms that have habitats and feeding relationships very different from their adult forms.

simply about getting to food or habitat first, or being able to use it more efficiently. As we discussed, each species has tolerance limits for nonbiological (abiotic) factors. Studies often show that, when two species compete, the one living in the center of its tolerance limits for a range of resources has an advantage and, more often than not, prevails in competition with another species living outside its optimal environmental conditions.

## Predation affects species relationships

All organisms need food to live. Producers make their own food, while consumers eat organic matter created by other organisms. As we saw in chapter 3, photosynthetic plants and algae are the producers in most communities. Consumers include herbivores, carnivores, omnivores, scavengers, detritivores, and decomposers. You may think only carnivores are predators, but ecologically a predator is any organism that feeds directly on another living organism, whether or not this kills the prey (fig. 4.14). Herbivores, carnivores, and omnivores, which feed on live prey, are predators, but scavengers, detritivores, and decomposers, which feed on dead things, are not. In this sense, parasites (organisms that feed on a host organism or steal resources from it without necessarily killing it) and even pathogens (disease-causing organisms) can be considered predator organisms. Herbivory is the type of predation practiced by grazing and browsing animals on plants.

Predation is a powerful but complex influence on species populations in communities. It affects (1) all stages in the life cycles of predator and prey species; (2) many specialized food-obtaining mechanisms; and (3) the evolutionary adjustments in behavior and body characteristics that help prey escape being eaten, and predators more efficiently catch their prey. Predation also interacts with competition. In **predator-mediated competition,** a superior competitor in a habitat builds up a larger population than its competing species; predators take note and increase their hunting pressure on the superior species, reducing its abundance and allowing the weaker competitor to increase its numbers. To test this idea, scientists remove predators from communities of competing species. Often the superior competitors eliminated other species from the habitat. In a classic example, the ochre starfish (*Pisaster ochraceus*) was removed from Pacific tidal zones and its main prey, the common mussel (*Mytilus californicus*), exploded in numbers and crowded out other intertidal species.

Knowing how predators affect prey populations has direct application to human needs, such as pest control in cropland. The cyclamen mite (*Phytonemus pallidus*), for example, is a pest of California strawberry crops. Its damage to strawberry leaves is reduced by predatory mites (*Typhlodromus* and *Neoseiulus*). which arrive naturally or are introduced into fields. Pesticide spraying to control the cyclamen mite can actually increase the infestation because it also kills the beneficial predatory mites.

Predatory relationships may change as the life stage of an organism changes. In marine ecosystems, crustaceans, mollusks, and worms release eggs directly into the water where they and hatchling larvae join the floating plankton community (fig. 4.15). Planktonic animals eat each other and are food for larger carnivores, including fish. As prey species mature, their predators change. Barnacle larvae are planktonic and are eaten by small fish, but as adults their hard shells protect them from fish, but not starfish and predatory snails. Predators often switch prey in the course of their lives. Carnivorous adult frogs usually begin their lives as herbivorous tadpoles. Predators also switch prey when it becomes rare, or something else becomes abundant. Many predators have morphologies and behaviors that make them highly adaptable to a changing prey base, but some, like the polar bear are highly specialized in their prey preferences.

## Some adaptations help avoid predation

Predator-prey relationships exert selection pressures that favor evolutionary adaptation. In this world, predators become more efficient at searching and feeding, and prey become more effective at escape and avoidance. Toxic chemicals, body armor, extraordinary speed, and the ability to hide are a few strategies organisms use to protect themselves. Plants have thick bark,

**FIGURE 4.16** Poison arrow frogs of the family Dendrobatidae use brilliant colors to warn potential predators of the extremely toxic secretions from their skin. Native people in Latin America use the toxin on blowgun darts.

spines, thorns, or distasteful and even harmful chemicals in tissues—poison ivy and stinging nettle are examples. Arthropods, amphibians, snakes, and some mammals produce noxious odors or poisonous secretions that cause other species to leave them alone. Animal prey are adept at hiding, fleeing, or fighting back. On the Serengeti Plain of East Africa, the swift Thomson's gazelle and even swifter cheetah are engaged in an arms race of speed, endurance, and quick reactions. The gazelle escapes often because the cheetah lacks stamina, but the cheetah accelerates from 0 to 72 kph in 2 seconds, giving it the edge in a surprise attack. The response of predator to prey and vice versa, over tens of thousands of years, produces physical and behavioral changes in a process known as **coevolution.** Coevolution can be mutually beneficial: many plants and pollinators have forms and behaviors that benefit each other. A classic case is that of fruit bats, which pollinate and disperse seeds of fruit-bearing tropical plants.

Often species with chemical defenses display distinct coloration and patterns to warn away enemies (fig. 4.16). In a neat evolutionary twist, certain species that are harmless resemble poisonous or distasteful ones, gaining protection against predators who remember a bad experience with the actual toxic organism. This is called **Batesian mimicry,** after the English naturalist H. W. Bates (1825–1892), a traveling companion of Alfred Wallace. Many wasps, for example, have bold patterns of black and yellow stripes to warm off potential predators (fig. 4.17*a*). The much rarer longhorn beetle has no stinger but looks and acts much like a wasp, tricking predators into avoiding it (fig. 4.17*b*). The distasteful monarch and benign viceroy butterflies are a classic case of Batesian mimicry. Another form of mimicry, **Müllerian mimicry** (after the biologist Fritz Müller) involves two unpalatable or dangerous species who look alike. When predators learn to avoid either species, both benefit. Species also display forms, colors, and patterns that help avoid being discovered. Insects that look like dead leaves or twigs are among the most remarkable examples (fig. 4.18). Unfortunately for prey, predators also often use camouflage to conceal themselves as they lie in wait for their next meal.

(a)

(b)

**FIGURE 4.17** An example of Batesian mimicry. The dangerous wasp (a) has bold yellow and black bands to warn predators away. The much rarer longhorn beetle (b) has no poisonous stinger, but looks and acts like a wasp and thus avoids predators as well.

## Symbiosis involves intimate relations among species

In contrast to predation and competition, some interactions between organisms can be nonantagonistic, even beneficial. In such relationships, called **symbiosis,** two or more species live intimately together, with their fates linked. Symbiotic relationships often enhance the survival of one or both partners. In lichens, a fungus and a photosynthetic partner (either an alga or a cyanobacterium) combine tissues to mutual benefit (fig. 4.19*a*). This association is called **mutualism.** Some ecologists believe that cooperative, mutualistic relationships may be more important in evolution than commonly thought (fig. 4.19*b*). Survival of the fittest may also mean survival of organisms that can live together.

Symbiotic relationships often entail some degree of coevolution of the partners, shaping—at least in part—their structural and behavioral characteristics. This mutualistic coadaptation is evident

**FIGURE 4.18** This walking stick is highly camouflaged to blend in with the forest floor. Natural selection and evolution have created this remarkable shape and color.

between swollen thorn acacias (*Acacia collinsii*) and the ants (*Pseudomyrmex ferruginea*) that tend them in Central and South America. Acacia ant colonies live inside the swollen thorns on the acacia tree branches. Ants feed on nectar that is produced in glands at the leaf bases and also eat special protein-rich structures that are produced on leaflet tips. The acacias thus provide shelter and food for the ants. Although they spend energy to provide these services, the trees are not harmed by the ants. What do the acacias get in return? Ants aggressively defend their territories, driving away herbivorous insects that would feed on the acacias. Ants also trim away vegetation that grows around the tree, reducing competition by other plants for water and nutrients. You can see how mutualism is structuring the biological community in the vicinity of acacias harboring ants, just as competition or predation shapes communities.

Mutualistic relationships can develop quickly. In 2005 the Harvard entomologist E. O. Wilson pieced together evidence to explain a 500-year-old agricultural mystery in the oldest Spanish settlement in the New World, Hispaniola. Using historical accounts and modern research, Dr. Wilson reasoned that mutualism developed between the tropical fire ant (*Solenopsis geminata*), native to the Americas, and a sap-sucking insect that was probably introduced from the Canary Islands in 1516 on a shipment of plantains. The plantains were planted, the sap-suckers were distributed across Hispaniola, and in 1518 a great die-off of crops occurred. Apparently the native fire ants discovered the foreign sap-sucking insects, consumed their excretions of sugar and protein, and protected them from predators, thus allowing the introduced insect population to explode. The Spanish assumed the fire ants caused the agricultural blight, but a little ecological knowledge would have led them to the real culprit.

**Commensalism** is a type of symbiosis in which one member clearly benefits and the other apparently is neither benefited nor harmed. Many mosses, bromeliads, and other plants growing on trees in the moist tropics are considered commensals (fig. 4.19*c*). These epiphytes are watered by rain and obtain nutrients from leaf litter and falling dust, and often they neither help nor hurt the trees on which they grow. Robins and sparrows that inhabit suburban yards are commensals with humans. **Parasitism,** a form of predation, may also be considered symbiosis because of the dependency of the parasite on its host.

## Keystone species have disproportionate influence

A **keystone species** plays a critical role in a biological community that is out of proportion to its abundance. Originally, keystone species were thought to be top predators—lions, wolves, tigers—which limited herbivore abundance and reduced the herbivory of plants. Scientists now recognize that less-conspicuous species also play keystone roles. Tropical figs, for example, bear fruit year-round at a low but steady rate. If rigs are removed from a forest, many fruit-eating animals (frugivores) would starve in the dry season when fruit of other species is scarce. In turn, the disappearance of frugivores would affect plants that

(a) Lichen on a rock (b) Oxpecker and impala (c) Bromeliad

**FIGURE 4.19** Symbiotic relationships. (a) Lichens represent an obligatory mutualism between a fungus and alga or cyanobacterium. (b) Mutualism between a parasite-eating red-billed oxpecker and parasite-infested impala. (c) Commensalism between a tropical tree and free-loading bromeliad.

**FIGURE 4.20** Sea otters protect kelp forests in the northern Pacific Ocean by eating sea urchins that would otherwise destroy the kelp. But the otters are being eaten by killer whales. Which is the keystone in this community—or is there a keystone set of organisms?

depend on them for pollination and seed dispersal. It is clear that the effect of a keystone species on communities often ripples across trophic levels.

Keystone functions have been documented for vegetation-clearing elephants, the predatory ochre sea star, and frog-eating salamanders in coastal North Carolina. Even microorganisms can play keystone roles. In many temperate forest ecosystems, groups of fungi that are associated with tree roots (mycorrhizae) facilitate the uptake of essential minerals. When fungi are absent, trees grow poorly or not at all. Overall, keystone species seem to be more common in aquatic habitats than in terrestrial ones.

The role of keystone species can be difficult to untangle from other species interactions. Off the northern Pacific coast, a giant brown alga (*Macrocystis pyrifera*) forms dense "kelp forests," which shelter fish and shellfish species from predators, allowing them to become established in the community. It turns out, however, that sea otters eat sea urchins living in the kelp forests (fig. 4.20); when sea otters are absent, the urchins graze on and eliminate kelp forests. To complicate things, around 1990, killer whales began preying on otters because of the dwindling stocks of seals and sea lions, thereby creating a cascade of effects. Is the kelp, otter, or orca the keystone here? Whatever the case, keystone species exert their influence by changing competitive relationships. In some communities, perhaps we should call it a "keystone set" of organisms.

## 4.3 Community Properties Affect Species and Populations

The processes and principles that we have studied thus far in this chapter—tolerance limits, species interactions, resource partitioning, evolution, and adaptation—play important roles in determining the characteristics of populations and species. In this section we will look at some fundamental properties of biological communities and ecosystems—productivity, diversity, complexity, resilience, stability, and structure—to learn how they are affected by these factors.

### Productivity is a measure of biological activity

A community's **primary productivity** is the rate of biomass production, an indication of the rate of solar energy conversion to chemical energy. The energy left after respiration is net primary production. Photosynthetic rates are regulated by light levels, temperature, moisture, and nutrient availability. Figure 4.21 shows approximate productivity levels for some major ecosystems. As you can see, tropical forests, coral reefs, and estuaries (bays or inundated river valleys where rivers meet the ocean) have high levels of productivity because they have abundant supplies of all these resources. In deserts, lack of water limits photosynthesis. On the arctic tundra or in high mountains, low temperatures inhibit plant growth. In the open ocean, a lack of nutrients reduces the ability of algae to make use of plentiful sunshine and water.

Some agricultural crops such as corn (maize) and sugar cane grown under ideal conditions in the tropics approach the productivity levels of tropical forests. Because shallow water ecosystems such as coral reefs, salt marshes, tidal mud flats, and other highly productive aquatic communities are relatively rare compared to the vast extent of open oceans—which are effectively biological deserts—marine ecosystems are much less productive on average than terrestrial ecosystems.

Even in the most photosynthetically active ecosystems, only a small percentage of the available sunlight is captured and used to make energy-rich compounds. Between one-quarter and three-quarters of the light reaching plants is reflected by leaf surfaces. Most of the light absorbed by leaves is converted to heat that is either radiated away or dissipated by evaporation of water. Only 0.1 to 0.2 percent of the absorbed energy is used by chloroplasts to synthesize carbohydrates.

In a temperate-climate oak forest, only about half the incident light available on a midsummer day is absorbed by the leaves. Ninety-nine percent of this energy is used to evaporate water. A large oak tree can transpire (evaporate) several thousand liters of water on a warm, dry, sunny day while it makes only a few kilograms of sugars and other energy-rich organic compounds.

### Abundance and diversity measure the number and variety of organisms

**Abundance** is an expression of the total number of organisms in a biological community, while **diversity** is a measure of the number of different species, ecological niches, or genetic variation present. The abundance of a particular species often is inversely related to the total diversity of the community. That is, communities with a very large number of species often have only a few members of any given species in a particular area. As a general rule, diversity decreases but abundance within species increases as we go from the equator toward the poles. The Arctic has vast

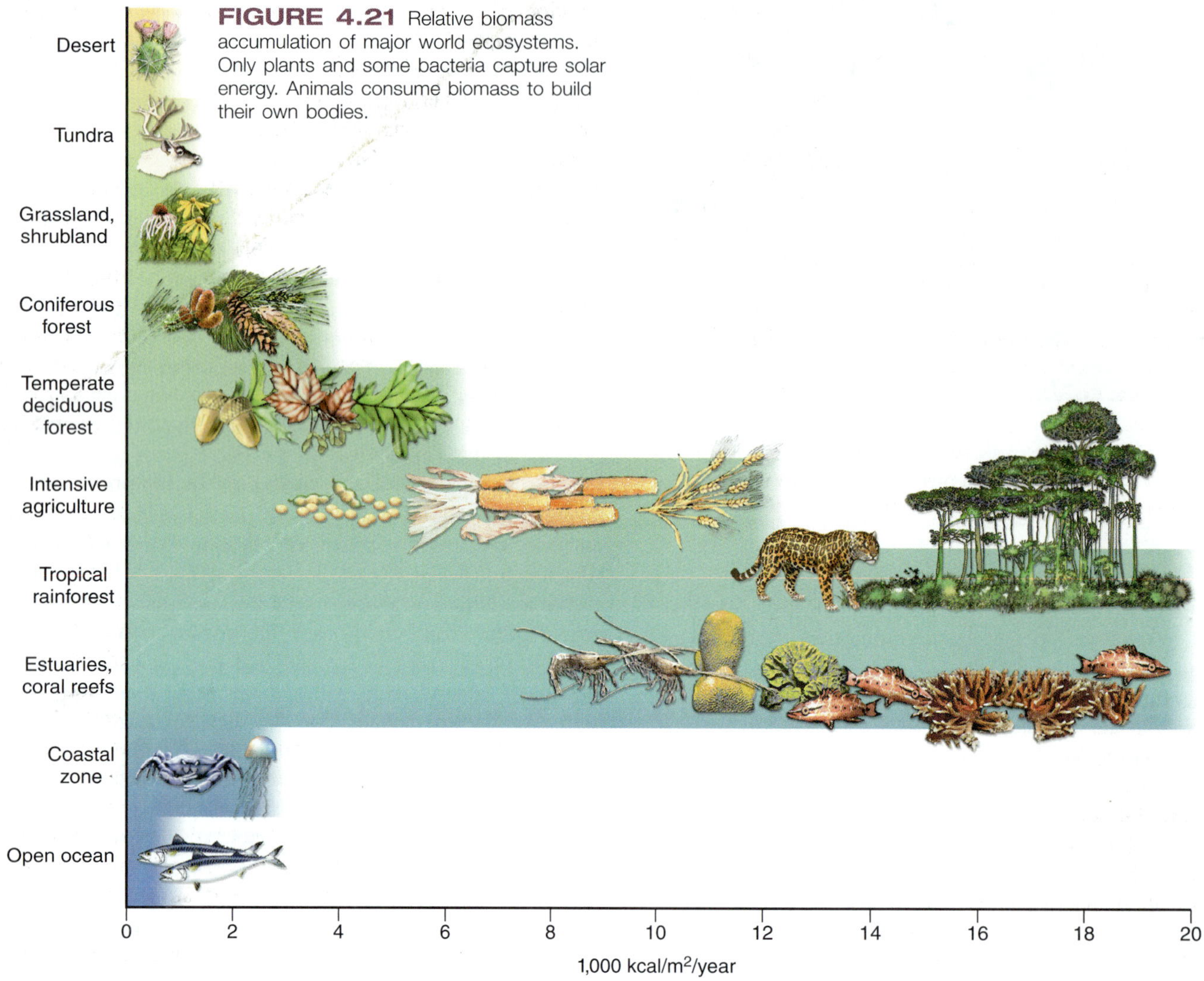


**FIGURE 4.21** Relative biomass accumulation of major world ecosystems. Only plants and some bacteria capture solar energy. Animals consume biomass to build their own bodies.

numbers of insects such as mosquitoes, for example, but only a few species. The tropics, on the other hand, have vast numbers of species—some of which have incredibly bizarre forms and habits—but often only a few individuals of any particular species in a given area.

Consider bird populations. Greenland is home to 56 species of breeding birds, while Colombia, which is only one-fifth the size of Greenland, has 1,395. Why are there so many species in Colombia and so few in Greenland?

Climate and history are important factors. Greenland has such a harsh climate that the need to survive through the winter or escape to milder climates becomes the single most important critical factor that overwhelms all other considerations and severely limits the ability of species to specialize or differentiate into new forms. Furthermore, because Greenland was covered by glaciers until about 10,000 years ago, there has been little time for new species to develop.

Many areas in the tropics, by contrast, have relatively abundant rainfall and warm temperatures year-round so that ecosystems there are highly productive. The year-round dependability of food, moisture, and warmth supports a great exuberance of life and allows a high degree of specialization in physical shape and behavior. Coral reefs are similarly stable, productive, and conducive to proliferation of diverse and amazing life-forms. The enormous abundance of brightly colored and fantastically shaped fish, corals, sponges, and arthropods in the reef community is one of the best examples we have of community diversity.

Productivity is related to abundance and diversity, both of which are dependent on the total resource availability in an ecosystem as well as the reliability of resources, the adaptations of the member species, and the interactions between species. You shouldn't assume that all communities are perfectly adapted to their environment. A relatively new community that hasn't had time for niche specialization, or a disturbed one where roles such as top predators are missing, may not achieve maximum efficiency of resource use or reach its maximum level of either abundance or diversity.

## What Can You Do?

### Working Locally for Ecological Diversity

You might think that diversity and complexity of ecological systems are too large or too abstract for you to have any influence. But you can contribute to a complex, resilient, and interesting ecosystem, whether you live in the inner city, a suburb, or a rural area.

- Keep your cat indoors. Our lovable domestic cats are also very successful predators. Migratory birds, especially those nesting on the ground, have not evolved defenses against these predators (Exploring Science p. 91).
- Plant a butterfly garden. Use native plants that support a diverse insect population. Native trees with berries or fruit also support birds. (Be sure to avoid non-native invasive species: see chapter 11.) Allow structural diversity (open areas, shrubs, and trees) to support a range of species.
- Join a local environmental organization. Often, the best way to be effective is to concentrate your efforts close to home. City parks and neighborhoods support ecological communities, as do farming and rural areas. Join an organization working to maintain ecosystem health; start by looking for environmental clubs at your school, park organizations, a local Audubon chapter, or a local Nature Conservancy branch.
- Take walks. The best way to learn about ecological systems in your area is to take walks and practice observing your environment. Go with friends and try to identify some of the species and trophic relationships in your area.
- Live in town. Suburban sprawl consumes wildlife habitat and reduces ecosystem complexity by removing many specialized plants and animals. Replacing forests and grasslands with lawns and streets is the surest way to simplify, or eliminate, ecosystems.

## Community structure describes spatial distribution of organisms

*Ecological structure* refers to patterns of spatial distribution of individuals and populations within a community, as well as the relation of a particular community to its surroundings. At the local level, even in a relatively homogeneous environment, individuals in a single population can be distributed randomly, clumped together, or in highly regular patterns. In randomly arranged populations, individuals live wherever resources are available (fig. 4.22*a*). Ordered patterns may be determined by the physical environment but are more often the result of biological competition. For example, competition for nesting space in seabird colonies on the Falkland Islands is often fierce. Each nest tends to be just out of reach of the neighbors sitting on their own nests. Constant squabbling produces a highly regular pattern (fig. 4.22*b*). Similarly, sagebrush releases toxins from roots and fallen leaves, which inhibit the growth of competitors and create a circle of bare ground around each bush. As neighbors fill in empty spaces up to the limit of this chemical barrier, a regular spacing results.

Some other species cluster together for protection, mutual assistance, reproduction, or access to a particular environmental resource (fig. 4.22*c*). Dense schools of fish, for instance, cluster closely together in the ocean, increasing their chances of detecting and escaping predators. Similarly, predators, whether sharks, wolves, or humans, often hunt in packs to catch their prey. A flock of blackbirds descending on a cornfield or a troop of baboons traveling across the African savanna band together both to avoid predators and to find food more efficiently.

Plants can cluster for protection, as well. A grove of wind-sheared evergreen trees is often found packed tightly together at the crest of a high mountain or along the seashore. They offer mutual protection from the wind not only to each other but also to other creatures that find shelter in or under their branches.

Most environments are patchy at some scale. Organisms cluster or disperse according to patchy availability of water, nutrients, or other resources. Distribution in a community can be vertical as well as horizontal. The tropical forest, for instance, has many layers, each with different environmental conditions and combinations of species. Distinct communities of smaller plants, animals, and microbes live at different levels. Similarly, aquatic communities are often stratified into layers based on light penetration in the water, temperature, salinity, pressure, or other factors.

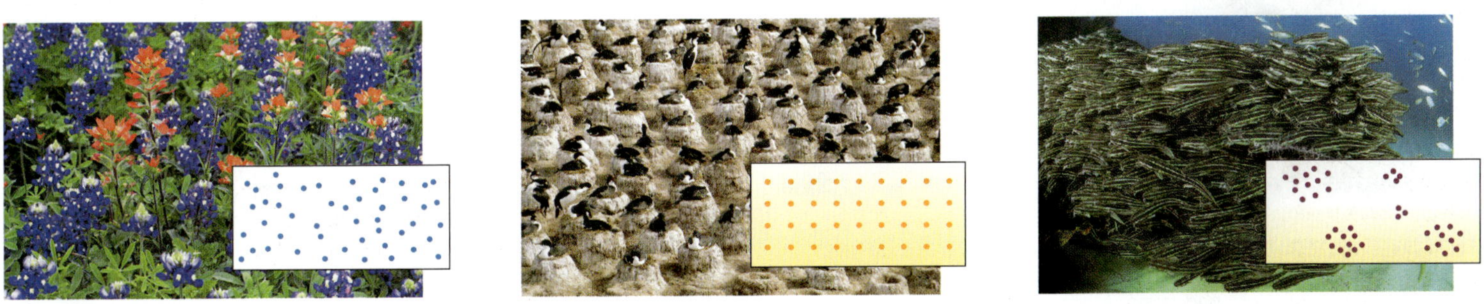

(a) Random (b) Uniform (c) Clustered

**FIGURE 4.22** Distribution of members of a population in a given space can be (a) random, (b) uniform, or (c) clustered. The physical environment and biological interactions determine these patterns. The patterns may produce a graininess or patchiness in community structure.

## Complexity and connectedness are important ecological indicators

Community complexity and connectedness generally are related to diversity and are important because they help us visualize and understand community functions. **Complexity** in ecological terms refers to the number of species at each trophic level and the number of trophic levels in a community. A diverse community may not be very complex if all its species are clustered in only a few trophic levels and form a relatively simple food chain.

By contrast, a complex, highly interconnected community (fig. 4.23) might have many trophic levels, some of which can be compartmentalized into subdivisions. In tropical rainforests, for instance, the herbivores can be grouped into "guilds" based on the specialized ways they feed on plants. There may be fruit eaters, leaf nibblers, root borers, seed gnawers, and sap suckers, each composed of species of very different size, shape, and even biological kingdom, but that feed in related ways. A highly interconnected community such as this can form a very elaborate food web.

**FIGURE 4.23** Tropical rainforests are complex structurally and ecologically. Trees form layers, each with a different amount of light and a unique combination of flora and fauna. Many insects, arthropods, birds, and mammals spend their entire life in the canopy. In Brazil's Atlantic Rainforest, a single hectare had 450 tree species and many times that many insects. With so many species, the ecological relationships are complex and highly interconnected.

## Resilience and stability make communities resistant to disturbance

Many biological communities tend to remain relatively stable and constant over time. An oak forest tends to remain an oak forest, for example, because the species that make it up have self-perpetuating mechanisms. We can identify three kinds of stability or resiliency in ecosystems: *constancy* (lack of fluctuations in composition or functions), *inertia* (resistance to perturbations), and *renewal* (ability to repair damage after disturbance).

In 1955, Robert MacArthur, who was then a graduate student at Yale, proposed that the more complex and interconnected a community is, the more stable and resilient it will be in the face of disturbance. If many different species occupy each trophic level, some can fill in if others are stressed or eliminated by external forces, making the whole community resistant to perturbations and able to recover relatively easily from disruptions. This theory has been controversial, however. Some studies support it, while others do not. For example, Minnesota ecologist David Tilman, in studies of native prairie and recovering farm fields, found that plots with high diversity were better able to withstand and recover from drought than those with only a few species.

On the other hand, in a diverse and highly specialized ecosystem, removal of a few keystone members can eliminate many other associated species. Eliminating a major tree species from a tropical forest, for example, may destroy pollinators and fruit distributors as well. We might replant the trees, but could we replace the whole web of relationships on which they depend? In this case, diversity has made the forest less resilient rather than more.

Diversity is widely considered important and has received a great deal of attention. In particular, human impacts on diversity are a primary concern of many ecologists (Exploring Science p. 91).

## Edges and boundaries are the interfaces between adjacent communities

An important aspect of community structure is the boundary between one habitat and its neighbors. We call these relationships **edge effects.** Sometimes, the edge of a patch of habitat is relatively sharp and distinct. In moving from a woodland patch into a grassland or cultivated field, you sense a dramatic change from the cool, dark, quiet forest interior to the windy, sunny, warmer,

# *Exploring* SCIENCE

## Where Have All the Songbirds Gone?

Every June, some 2,200 amateur ornithologists and bird watchers across the United States and Canada join in an annual bird count called the Breeding Bird Survey. Organized in 1966 by the U.S. Fish and Wildlife Service to follow bird population changes, this survey has discovered some shocking trends. While birds such as robins, starlings, and blackbirds that prosper around humans have increased their number and distribution over the past 30 years, many of our most colorful forest birds have declined severely. The greatest decreases have been among the true songbirds such as thrushes, orioles, tanagers, catbirds, vireos, buntings, and warblers. These long-distance migrants nest in northern forests but spend the winters in South or Central America or in the Caribbean Islands. Scientists call them neotropical migrants.

In many areas of the eastern United States and Canada, three-quarters or more of the neotropical migrants have declined significantly since the survey was started. Some that once were common have become locally extinct. Rock Creek Park in Washington, D.C., for instance, lost 75 percent of its songbird population and 90 percent of its long-distance migrant species in just 20 years. Nationwide, cerulean warblers, American redstarts, and ovenbirds declined about 50 percent in the single decade of the 1970s. Studies of radar images from National Weather Service stations in Texas and Louisiana suggest that only about half as many birds fly across the Gulf of Mexico each spring now compared to the 1960s. This could mean a loss of about half a billion birds in total.

What causes these devastating losses? Destruction of critical winter habitat is clearly a major issue. Birds often are much more densely crowded in the limited areas available to them during the winter than they are on their summer range. Unfortunately, forests throughout Latin America are being felled at an appalling rate. Central America, for instance, is losing about 1.4 million hectares (2 percent of its forests or an area about the size of Yellowstone National Park) each year. If this trend continues, there will be essentially no intact forest left in much of the region in 50 years.

But loss of tropical forests is not the only threat. Recent studies show that fragmentation of breeding habitat and nesting failures in the United States and Canada may be just as big a problem for woodland songbirds. Many of the most threatened species are adapted to deep woods and need an area of 10 hectares (25 acres) or more per pair to breed and raise their young. As our woodlands are broken up by roads, housing developments, and shopping centers, it becomes more and more difficult for these highly specialized birds to find enough contiguous woods to nest successfully.

This thrush has been equipped with a lightweight radio transmitter and antenna so that its movements can be followed by researchers

Predation and nest parasitism also present a growing threat to many bird species. In human-dominated landscapes, raccoons, opossums, crows, bluejays, squirrels, and house cats thrive. They are protected from larger predators like wolves or owls and find abundant supplies of food and places to hide. Cats are a particular problem. By some estimates, there are 100 million feral cats in the United States, and 73 million pet cats. A comparison of predation rates in the Great Smoky Mountain National Park and in small rural and suburban woodlands shows how devastating predators can be. In a 1,000-hectare study area of mature, unbroken forest in the national park, only one songbird nest in fifty was raided by predators. By contrast, in plots of 10 hectares or less near cities, up to 90 percent of the nests were raided.

Nest parasitism by brown-headed cowbirds is one of the worst threats for woodland songbirds. Rather than raise their young themselves, cowbirds lay their eggs in the nests of other species. The larger and more aggressive cowbird young either kick their foster siblings out of the nest, or claim so much food that the others starve. Well adapted to live around humans, there are now about 150 million cowbirds in the United States.

A study in southern Wisconsin found that 80 percent of the nests of woodland species were raided by predators and that three-quarters of those that survived were invaded by cowbirds. Another study in the Shawnee National Forest in southern Illinois found that 80 percent of the scarlet tanager nests contained cowbird eggs and that 90 percent of the wood thrush nests were taken over by these parasites. The sobering conclusion of this latter study is that there probably is no longer any place in Illinois where scarlet tanagers and wood thrushes can breed successfully.

What can we do about this situation? Elsewhere in this book, we discuss sustainable forestry and economic development projects that could preserve forests at home and abroad. Preserving corridors that tie together important areas also will help. In areas where people already live, clustering of houses protects remaining woods. Discouraging the clearing of underbrush and trees from yards and parks leaves shelter for the birds.

Could we reduce the number of predators or limit their access to critical breeding areas? Would you accept fencing or trapping of small predators in wildlife preserves? How would you feel about a campaign to keep house cats inside during the breeding season?

**FIGURE 4.24** Ecological edges are known as ecotones. Temperature, wind, and humidity differ at the edges in a landscape. Edge conditions do extend into patches of habitat. Small or linear fragments may be mostly edge.

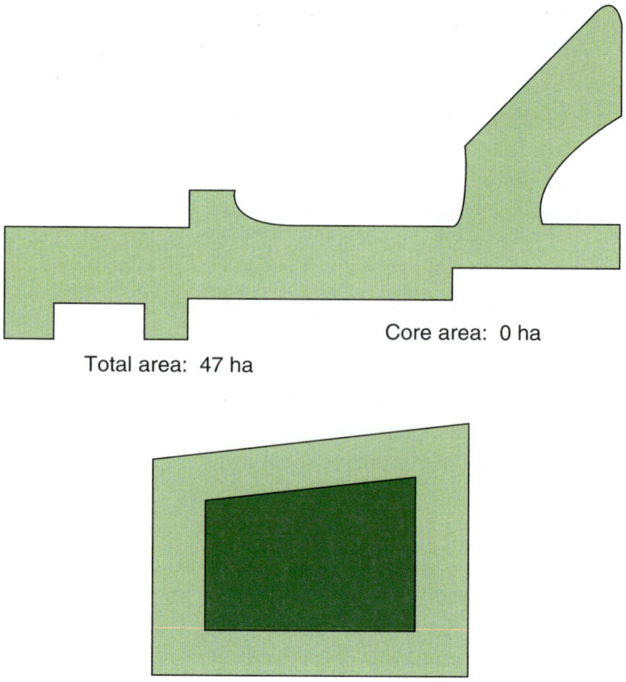


**FIGURE 4.25** Shape can be as important as size in small preserves. While these areas are similar in size, no place in the top figure is far enough from the edge to have characteristics of core habitat, while the bottom patch has a significant core.

open space of the field or pasture (fig. 4.24). In other cases, one habitat type intergrades very gradually into another, so there is no distinct border.

Ecologists call the boundaries between adjacent communities **ecotones.** A community that is sharply divided from its neighbors is called a closed community. In contrast, communities with gradual or indistinct boundaries over which many species cross are called open communities. Often this distinction is a matter of degree or perception. As we saw earlier in this chapter, birds might feed in fields or grasslands but nest in the forest. As they fly back and forth, the birds interconnect the ecosystems by moving energy and material from one to the other, making both systems relatively open. Furthermore, the forest edge, while clearly different from the open field, may be sunnier and warmer than the forest interior, and may have a different combination of plant and animal species than either field or forest "core."

Depending on how far edge effects extend from the boundary, differently shaped habitat patches may have very dissimilar amounts of interior area (fig. 4.25). In Douglas fir forests of the Pacific Northwest, for example, increased rates of blowdown, decreased humidity, absence of shade-requiring ground cover, and other edge effects can extend as much as 200 m into a forest. A 40-acre block (about 400 $m^2$) surrounded by clear-cut would have essentially no true core habitat at all.

Many popular game animals, such as white-tailed deer and pheasants that are adapted to human disturbance, often are most plentiful in boundary zones between different types of habitat. Game managers once were urged to develop as much edge as possible to promote large game populations. Today, however, most wildlife conservationists recognize that the edge effects associated with habitat fragmentation are generally detrimental to biodiversity. Preserving large habitat blocks and linking smaller blocks with migration corridors may be the best ways to protect rare and endangered species (chapter 12).

## 4.4 Communities Are Dynamic and Change Over Time

If fire sweeps through a biological community, it's destroyed, right? Not so fast. Fire may be good for that community. Up until now, we've focused on the day-to-day interactions of organisms with their environments, set in a context of adaptation and selection. In this section, we'll step back and look at more dynamic aspects of communities and how they change over time.

### The nature of communities is debated

For several decades starting in the early 1900s, ecologists in North America and Europe argued about the basic nature of communities. It doesn't make interesting party conversation, but those discussions affected how we study and understand communities, view the changes taking place within them, and ultimately use them. Both J. E. B. Warming (1841–1924) in Denmark and Henry Chandler Cowles (1869–1939) in the United States came up with the idea that communities develop in a sequence of stages, starting either from bare rock or after a severe disturbance. They worked in sand dunes and watched the changes as plants first took root in bare sand and, with further development, created forest. This example represents constant change, not stability. In sand dunes, the community that developed last and lasted the longest was called the **climax community.**

The idea of climax community was first championed by the biogeographer F. E. Clements (1874–1945). He viewed the process as a relay—species replace each other in predictable groups and in a fixed, regular order. He argued that every landscape has a characteristic climax community, determined mainly by climate. If left undisturbed, this community would mature to a characteristic set of organisms, each performing its optimal functions. A climax community to Clements represented the maximum complexity and stability that was possible. He and others made the analogy that the development of a climax community resembled the maturation of an organism. Both communities and organisms, they argued, began simply and primitively, maturing until a highly integrated, complex community developed.

This organismal theory of community was opposed by Clements' contemporary, H. A. Gleason (1882–1975), who saw community history as an unpredictable process. He argued that species are individualistic, each establishing in an environment according to its ability to colonize, tolerate the environmental conditions, and reproduce there. This idea allows for myriad temporary associations of plants and animals to form, fall apart, and reconstitute in slightly different forms, depending on environmental conditions and the species in the neighborhood. Imagine a time-lapse movie of a busy airport terminal. Passengers come and go; groups form and dissipate. Patterns and assemblages that seem significant may not mean much a year later. Gleason suggested that we think ecosystems are uniform and stable only because our lifetimes are too short and our geographic scope too limited to understand their actual dynamic nature.

## Ecological succession describes a history of community development

In any landscape, you can read the history of biological communities. That history is revealed by the process of ecological succession. During succession, organisms occupy a site and change the environmental conditions. In **primary succession** land that is bare of soil—a sandbar, mudslide, rock face, volcanic flow—is colonized by living organisms where none lived before (fig. 4.26). When an existing community is disturbed, a new one develops from the biological legacy of the old in a process called **secondary succession.** In both kinds of succession, organisms change the environment by modifying soil, light levels, food supplies, and microclimate. This change permits new species to colonize and eventually replace the previous species, a process known as ecological development or facilitation.

In primary succession on land, the first colonists are hardy **pioneer species,** often microbes, mosses, and lichens that can withstand a harsh environment with few resources. When they die, the bodies of pioneer species create patches of organic matter. Organics and other debris accumulate in pockets and crevices, creating soil where seeds lodge and grow. As succession proceeds, the community becomes more diverse and interspecies competition arises. Pioneers disappear as the environment favors new colonizers that have competitive abilities more suited to the new environment.

You can see secondary succession all around you, in abandoned farm fields, in clear-cut forests, and in disturbed suburbs and lots. Soil and possibly plant roots and seeds are present. Because soil lacks vegetation, plants that live one or two years (annuals and biennials) do well. Their light seeds travel far on the wind, and their seedlings tolerate full sun and extreme heat. When they die, they lay down organic material that improves the soil's fertility and shelters other seedlings. Soon long-lived and deep-rooted perennial grasses, herbs, shrubs, and trees take hold, building up the soil's organic matter and increasing its ability to store moisture. Forest species that cannot survive bare, dry, sunny ground eventually find ample food, a diverse community structure, and shelter from drying winds and low humidity.

Generalists figure prominently in early succession. Over thousands of years, however, competition should decrease as niches proliferate and specialists arise. In theory, long periods of community development lead to greater community complexity, high nutrient conservation and recycling, stable productivity, and great resistance to disturbance—an ideal state to be in when the slings and arrows of misfortune arrive.

## Appropriate disturbances can benefit communities

Disturbances are plentiful on earth: landslides, mudslides, hailstorms, earthquakes, hurricanes, tornadoes, tidal waves, wildfires, and volcanoes, to name just the obvious. A **disturbance** is any force that disrupts the established patterns of species diversity and abundance, community structure, or community properties. Animals can cause disturbance. African elephants rip out small

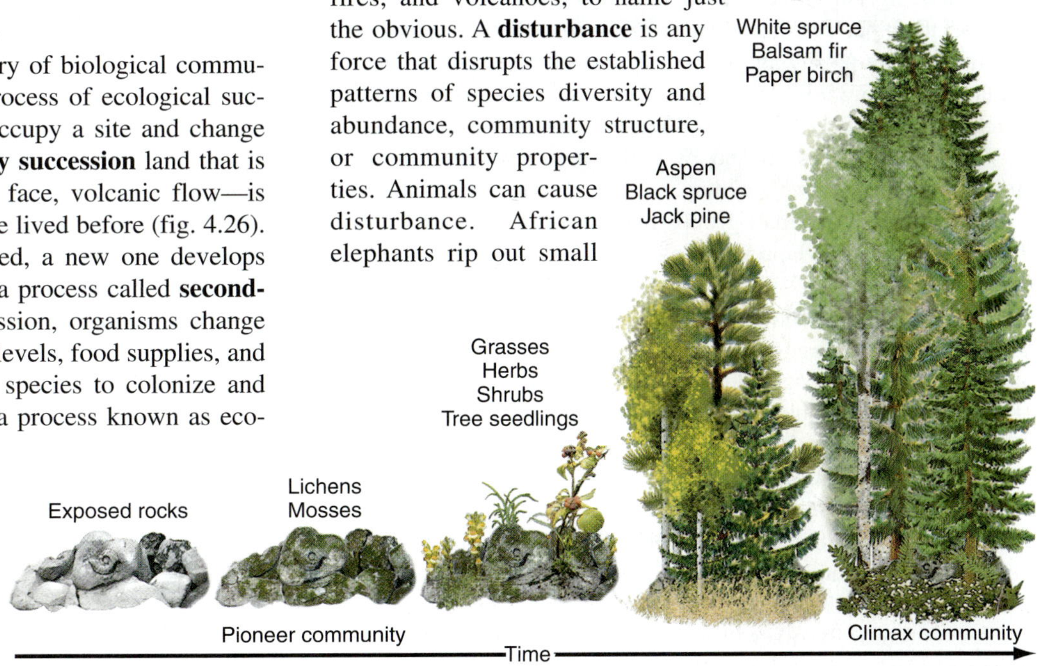


**FIGURE 4.26** One example of primary succession, shown in five stages (*left to right*). Here, bare rocks are colonized by lichens and mosses, which trap moisture and build soil for grasses, shrubs, and eventually trees.

**FIGURE 4.27** These "stump barrens" in Michigan's Upper Peninsula were created over a century ago when clear-cutting of dense white pine forest was followed by repeated burning. The stumps are left from the original forest, which has not grown back in more than 100 years.

trees, trample shrubs, and tear down tree limbs as they forage and move about, opening up forest communities and creating savannas. People also cause disturbances with agriculture, forestry, new roads and cities, and construction projects for dams and pipelines. It is customary in ecology to distinguish between natural disturbances and human-caused (or anthropogenic) disturbances, but a subtle point of clarification is needed.

Aboriginal populations have disturbed and continue to disturb communities around the world, setting fire to grasslands and savannas, practicing slash-and-burn agriculture in forests, and so on. Because their populations often are or were relatively small, the disturbances are patchy and limited in scale in forests, or restricted to quickly passing wildfires in grasslands, savannas, or woodland, which are comprised of species already adapted to fire.

The disturbances caused by technologically advanced and numerous people however, may be very different from the disturbances caused by small groups of aborigines. In the Kingston Plains of Michigan's Upper Peninsula, clear-cut logging followed by repeated human-set fires from 1880 to 1900 caused a change in basic ecological conditions such that the white pine forest has never regenerated (fig. 4.27). Given the right combination of disturbances by modern people, or by nature, it may take hundreds of years for a community to return to its predisturbance state.

Ecologists generally find that disturbance benefits most species, much as predation does, because it sets back supreme competitors and allows less-competitive species to persist. In northern temperate forests, maples (especially sugar maple) are more prolific seeders and more shade tolerant at different stages of growth than nearly any other tree species. Given decades of succession, maples out compete other trees for a place in the forest canopy. Most species of oak, hickory, and other light-requiring trees diminish in abundance, as do species of forest herbs. The dense shade of maples basically starves other species for light. When windstorms, tornadoes, wildfires, or ice storms hit a maple forest, trees are toppled, branches broken, and light again reaches the forest floor and stimulates seedlings of oaks and hickories, as well as forest herbs. Breaking the grip of a supercompetitor is the helpful role disturbances often play.

**FIGURE 4.28** This lodgepole pine forest in Yellowstone National Park was once thought to be a climax forest, but we now know that this forest must be constantly renewed by periodic fire. It is an example of an equilibrium, or disclimax, community.

Some landscapes never reach a stable climax in a traditional sense because they are characterized by periodic disturbance and are made up of **disturbance-adapted species** that survive fires underground, or resist the flames, and then reseed quickly after fires. Grasslands, the chaparral scrubland of California and the Mediterranean region, savannas, and some kinds of coniferous forests are shaped and maintained by periodic fires that have long been a part of their history (fig. 4.28). In fact, many of the dominant plant species in these communities need fire to suppress competitors, to prepare the ground for seeds to germinate, or to pop open cones or split thick seed coats and release seeds. Without fire, community structure would be quite different.

People taking an organismal view of such communities believe that disturbance is harmful. In the early 1900s this view merged with the desire to protect timber supplies from ubiquitous wildfires, and to store water behind dams while also controlling floods. Fire suppression and flood control became the central policies in American natural resource management (along with predator control) for most of the twentieth century. Recently, new concepts about natural disturbances are entering land management discussions and bringing change to land management policies. Grasslands and some forests are now considered "fire-adapted" and fires are allowed to burn in them if weather conditions are appropriate. Floods also are seen as crucial for maintaining floodplain and river health. Policymakers and managers increasingly consider ecological information when deciding on new dams and levee construction projects.

From another view, disturbance resets the successional clock that always operates in every community. Even though all seems chaotic after a disturbance, it may be that preserving species diversity by allowing in natural disturbances (or judiciously applied human disturbances) actually ensures stability over the long run, just as diverse prairies managed with fire recover after drought. In time, community structure and productivity get back to normal, species diversity is preserved, and nature seems to reach its dynamic balance.

**FIGURE 4.29** Mongooses were released in Hawaii in an effort to control rats. The mongooses are active during the day, however, while the rats are night creatures, so they ignored each other. Instead, the mongooses attacked defenseless native birds and became as great a problem as the rats.

## Introduced species can cause profound community change

Succession requires the continual introduction of new community members and the disappearance of previously existing species. New species move in as conditions become suitable; others die or move out as the community changes. New species also can be introduced after a stable community already has become established. Some cannot compete with existing species and fail to become established. Others are able to fit into and become part of the community, defining new ecological niches. If, however, an introduced species preys upon or competes more successfully with one or more populations that are native to the community, the entire nature of the community can be altered.

Human introductions of Eurasian plants and animals to non-Eurasian communities often have been disastrous to native species because of competition or overpredation. Oceanic islands offer classic examples of devastation caused by rats, goats, cats, and pigs liberated from sailing ships. All these animals are prolific, quickly developing large populations. Goats are efficient, nonspecific herbivores; they eat nearly everything vegetational, from grasses and herbs to seedlings and shrubs. In addition, their sharp hooves are hard on plants rooted in thin island soils. Rats and pigs are opportunistic omnivores, eating the eggs and nestlings of seabirds that tend to nest in large, densely packed colonies, and digging up sea turtle eggs. Cats prey upon nestlings of both ground- and tree-nesting birds. Native island species are particularly vulnerable because they have not evolved under circumstances that required them to have defensive adaptations to these predators.

Sometimes we introduce new species in an attempt to solve problems created by previous introductions but end up making the situation worse. In Hawaii and on several Caribbean Islands, for instance, mongooses were imported to help control rats that had escaped from ships and were destroying indigenous birds and devastating plantations (fig. 4.29). Since the mongooses were diurnal (active in the day), however, and rats are nocturnal, they tended to ignore each other. Instead, the mongooses also killed native birds and further threatened endangered species. Our lessons from this and similar introductions have a new technological twist. Some of the ethical questions currently surrounding the release of genetically engineered organisms are based on concerns that they are novel organisms, and we might not be able to predict how they will interact with other species in natural ecosystems—let alone how they might respond to natural selective forces. It is argued that we can't predict either their behavior or their evolution.

# CONCLUSION

Evolution is one of the key organizing principles of biology. It explains how species diversity originates, and how organisms are able to live in highly specialized ecological niches. Natural selection, in which beneficial traits are passed from survivors in one generation to their progeny, is the mechanism by which evolution occurs. Species interactions—competition, predation, symbiosis, and coevolution—are important factors in natural selection. The unique set of organisms and environmental conditions in an ecological community give rise to important properties, such as productivity, abundance, diversity, structure, complexity, connectedness, resilience, and succession. Human introduction of new species as well as removal of existing ones can cause profound changes in biological communities and can compromise the life-supporting ecological services on which we all depend. Understanding these community ecology principles is a vital step in becoming an educated environmental citizen.

## Reviewing Learning Outcomes

By now you should be able to explain the following points:

**4.1** Describe how evolution produces species diversity.

- Natural selection leads to evolution.
- All species live within limits.
- The ecological niche is a species' role and environment.
- Speciation maintains species diversity.
- Evolution is still at work.
- Taxonomy describes relationships among species.

**4.2** Discuss how species interactions shape biological communities.

- Competition leads to resource allocation.
- Predation affects species relationships.
- Some adaptations help avoid predation.
- Symbiosis involves intimate relations among species.
- Keystone species have disproportionate influence.

**4.3** Summarize how community properties affect species and populations.

- Productivity is a measure of biological activity.
- Abundance and diversity measure the number and variety of organisms.
- Community structure describes spatial distribution of organisms.
- Complexity and connectedness are important ecological indicators.
- Resilience and stability make communities resistant to disturbance.
- Edges and boundaries are the interfaces between adjacent communities.

**4.4** Explain why communities are dynamic and change over time.

- The nature of communities is debated.
- Ecological succession describes a history of community development.
- Appropriate disturbances can benefit communities.
- Introduced species can cause profound community change.

## Practice Quiz

1. Explain how tolerance limits to environmental factors determine distribution of a highly specialized species such as the saguaro cactus.
2. Productivity, diversity, complexity, resilience, and structure are exhibited to some extent by all communities and ecosystems. Describe how these characteristics apply to the ecosystem in which you live.
3. Define *selective pressure* and describe one example that has affected species where you live.
4. Define *keystone species* and explain their importance in community structure and function.
5. The most intense interactions often occur between individuals of the same species. What concept discussed in this chapter can be used to explain this phenomenon?
6. Explain how predators affect the adaptations of their prey.
7. Competition for a limited quantity of resources occurs in all ecosystems. This competition can be interspecific or intraspecific. Explain some of the ways an organism might deal with these different types of competition.
8. Describe the process of succession that occurs after a forest fire destroys an existing biological community. Why may periodic fire be beneficial to a community?
9. Which world ecosystems are most productive in terms of biomass (fig. 4.21)? Which are least productive? What units are used in this figure to quantify biomass accumulation?
10. Discuss the dangers posed to existing community members when new species are introduced into ecosystems.

## Critical Thinking and Discussion Questions

1. The concepts of natural selection and evolution are central to how most biologists understand and interpret the world, and yet the theory of evolution is contrary to the beliefs of many religious groups. Why do you think this theory is so important to science and so strongly opposed by others? What evidence would be required to convince opponents of evolution?
2. What is the difference between saying that a duck has webbed feet because it needs them to swim and saying that a duck is able to swim because it has webbed feet?
3. The concept of keystone species is controversial among ecologists because most organisms are highly interdependent. If each of the trophic levels is dependent on all the

others, how can we say one is most important? Choose an ecosystem with which you are familiar and decide whether it has a keystone species or keystone set.

4. Some scientists look at the boundary between two biological communities and see a sharp dividing line. Others looking at the same boundary see a gradual transition with much intermixing of species and many interactions between communities. Why are there such different interpretations of the same landscape?
5. The absence of certain lichens is used as an indicator of air pollution in remote areas such as national parks. How can we be sure that air pollution is really responsible? What evidence would be convincing?
6. We tend to regard generalists or "weedy" species as less interesting and less valuable than rare and highly specialized endemic species. What values or assumptions underlie this attitude?

## DATA analysis Species Competition

In a classic experiment on competition between species for a common food source, the Russian microbiologist G. F. Gause grew populations of different species of ciliated protozoans separately and together in an artificial culture medium. He counted the number of cells of each species and plotted the total volume of each population. The organisms were *Paramecium caudatum* and its close relative, *Paramecium aurelia.* He plotted the aggregate volume of cells rather than the total number in each population because *P. caudatum* is much larger than *P. aurelia* (this size difference allowed him to distinguish between them in a mixed culture). The graphs in this box show the experimental results. As we mentioned earlier in the text, this was one of the first experimental demonstrations of the principle of competitive exclusion. After studying these graphs, answer the following questions.

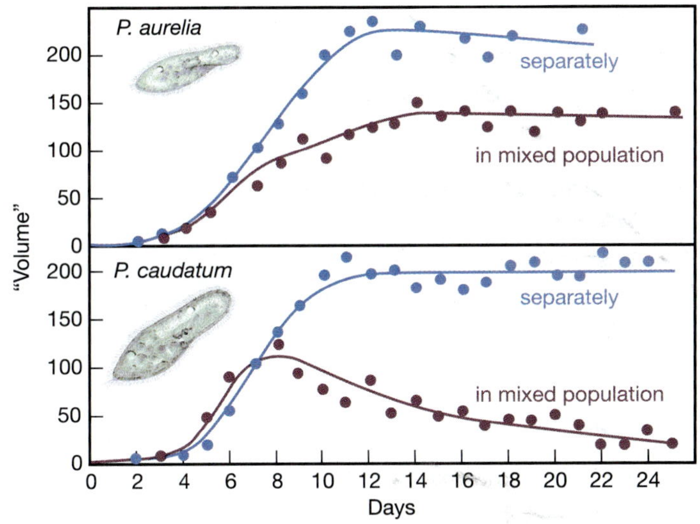


Growth of two *paramecium* species separately and in combination.
**Source:** Gause, Georgyi Frantsevitch. 1934 *The Struggle for Existence.* Dover Publications 1971 reprint of original text.

1. How do you read these graphs? What is shown in the top and bottom panels?
2. How did the total volume of the two species compare after 14 days of separate growth?
3. If *P. caudatum* is roughly twice as large as *P. aurelia,* how did the total number of cells compare after 14 days of separate growth?
4. How did the total volume of the two species compare after 24 days of growth in a mixed population?
5. Which of the two species is the more successful competitor in this experiment?
6. Does the larger species always win in competition for food? Why not?

**For Additional Help in Studying This Chapter,** please visit our website at www.mhhe.com/cunningham10e. You will find additional practice quizzes and case studies, flashcards, regional examples, place markers for Google Earth™ mapping, and an extensive reading list, all of which will help you learn environmental science.

Traditional outrigger canoes and hand lines are still used by villagers on many islands in the southwestern Pacific, but these low-impact fishing methods are being threatened by trawlers, dynamite fishing, and other destructive techniques.

# CHAPTER 5

# Biomes

## Global Patterns of Life

*What is the use of a house if you haven't got a tolerable planet to put it on?*

*—Henry David Thoreau—*

## LEARNING OUTCOMES

After studying this chapter, you should be able to:

**5.1** Recognize the characteristics of some major terrestrial biomes as well as the factors that determine their distribution.

**5.2** Understand how and why marine environments vary with depth and distance from shore.

**5.3** Compare the characteristics and biological importance of major freshwater ecosystems.

**5.4** Summarize the overall patterns of human disturbance of world biomes.

# Case Study Saving the Reefs of Apo Island

As their outrigger canoes glide gracefully onto Apo Island's beach after an early morning fishing expedition, villagers call to each other to ask how fishing was. "Tunay mabuti!" (very good!) is the cheerful reply. Nearly every canoe has a basketful of fish; enough to feed a family for several days with a surplus to send to the market. Life hasn't always been so good on the island. Thirty years ago, this island, like many others in the Philippines, suffered a catastrophic decline in the seafood that was the mainstay of their diet and livelihood. Rapid population growth coupled with destructive fishing methods such as dynamite or cyanide fishing, small mesh gill nets, deep-sea trawling, and *Muro-ami* (a technique in which fish are chased into nets by pounding on coral with weighted lines) had damaged the reef habitat and exhausted fish stocks.

In 1979, scientists from Silliman University on nearby Negros Island visited Apo to explain how establishing a marine sanctuary could help reverse this decline. The coral reef fringing the island acts as a food source and nursery for many of the marine species sought by fishermen. Protecting that breeding ground, they explained, is the key to preserving a healthy fishery. The scientists took villagers from Apo to the uninhabited Sumilon Island, where a no-take reserve was teeming with fish.

After much discussion, several families decided to establish a marine sanctuary along a short section of Apo Island shoreline. Initially, the area had high-quality coral but few fish. The participating families took turns watching to make sure that no one trespassed in the no-fishing zone. Within a few years, fish numbers and sizes in the sanctuary increased dramatically, and "spillover" of surplus fish led to higher catches in surrounding areas. In 1985, Apo villagers voted to establish a 500 m (0.3 mi) wide marine sanctuary around the entire island.

**FIGURE 5.1** Coral reefs are among the most beautiful, species-rich, and productive biological communities on the planet. They serve as the nurseries for many open-water species. At least half the world's reefs are threatened by pollution, global climate change, destructive fishing methods, and other human activities, but they can be protected and restored if we care for them.

Fishing is now allowed in this reserve, but only by low-impact methods such as hand-held lines, bamboo traps, large mesh nets, spearfishing without SCUBA gear, and hand netting. Coral-destroying techniques, such as dynamite, cyanide, trawling, and *Muro-ami* fishing are prohibited. By protecting the reef, villagers are guarding the nursery that forms the base for their entire marine ecosystem. Young fish growing up in the shelter of the coral move out as adults to populate the neighboring waters and yield abundant harvests. Fishermen report that they spend much less time traveling to distant fishing areas now that fish around the island are so much more abundant.

Apo Island's sanctuary is so successful that it has become the inspiration for more than 400 marine preserves throughout the Philippines and many others around the world. Not all are functioning as well as they might, but many, like Apo, have made dramatic progress in restoring abundant fish populations in nearby waters.

The rich marine life and beautiful coral formations in Apo Island's crystal clear water now attract international tourists (fig. 5.1). Two small hotels and a dive shop provide jobs for island residents. Other villagers take in tourists as boarders or sell food and T-shirts to visitors. The island government collects a diving/snorkeling fee, which has been used to build schools, improve island water supplies, and provide electricity to most of the island's 145 households. Almost all the island men still fish as their main occupation, but the fact that they don't have to go so far or work so hard for the fish they need means they have time for other activities, such as guiding diving tours or helping with household chores.

Higher family incomes now allow most island children to attend high school on Negros. Many continue their education with college or technical programs. Some find jobs elsewhere in the Philippines, and the money they send back home is a big economic boost for Apo families. Others return to their home island as teachers or to start businesses such as restaurants or dive shops. Seeing that they can do something positive to improve their environment and living conditions has empowered villagers to take on self-improvement projects that they may not otherwise have attempted.

Finding ways to live sustainably within the limits of the resource base available to us and without damaging the life-support systems provided by our ecosystem is a preeminent challenge of environmental science. Our answers to these challenges must be ecologically sound, economically sustainable, and socially acceptable if they are to succeed in the long term. Sometimes, as this case study shows, actions based on ecological knowledge and local action can spread to have positive effects on a global scale. Economics, policy, planning, and social organization all play vital roles in finding answers to human/environment problems. We'll discuss those disciplines later in this book, but first we'll look in this chapter, at some of the major terrestrial and aquatic biological communities as well as ways that humans are degrading them.

## 5.1 Terrestrial Biomes

Although all local environments are unique, it is helpful to understand them in terms of a few general groups with similar climate conditions, growth patterns, and vegetation types. We call these broad types of biological communities **biomes.** Understanding the global distribution of biomes, and knowing the differences in what grows where and why, is essential to the study of global environmental science. Biological productivity—and ecosystem resilience—varies greatly from one biome to another. Human use of biomes depends largely on those levels of productivity. Our ability to restore ecosystems and nature's ability to restore itself, depend largely on biome conditions. Clear-cut forests regrow relatively quickly in New England, but very slowly in Siberia, where current logging is expanding. Some grasslands rejuvenate quickly after grazing, and some are slower to recover. Why these differences? The sections that follow seek to answer this question.

Temperature and precipitation are among the most important determinants in biome distribution on land (fig. 5.2). If we know the general temperature range and precipitation level, we can predict what kind of biological community is likely to occur there, in the absence of human disturbance. Landforms, especially mountains, and prevailing winds also exert important influences on biological communities.

Because the earth is cooler at high latitudes (away from the equator), many temperature-controlled biomes occur in latitudinal bands. For example, a band of boreal (northern) forests crosses Canada and Siberia, tropical forests occur near the equator, and expansive grasslands lie near—or just beyond—the tropics (fig. 5.3). Many biomes are even named for their latitudes Tropical rainforests occur between the Tropic of Cancer (23° north) and the Tropic of Capricorn (23° south); arctic tundra lies near or above the Arctic Circle (66.6° north).

Temperature and precipitation change with elevation as well as with latitude. In mountainous regions, temperatures are cooler and precipitation is usually greater at high elevations. Communities can transition quickly from warm and dry to cold and wet as you go up a mountain. **Vertical zonation** is a term applied to vegetation zones defined by altitude. A 100 km transect from California's Central Valley up to Mt. Whitney, for example, crosses as many vegetation zones as you would find on a journey from southern California to northern Canada (fig. 5.4).

In this chapter, we'll examine the major terrestrial biomes, then we'll investigate ocean and freshwater communities and environments. Ocean environments are important because they cover two-thirds of the earth's surface, provide food for much of humanity, and help regulate our climate through photosynthesis. Wetlands are often small, but they have great influence on environmental health, biodiversity, and water quality. In chapter 12, we'll look at how we use these communities; and in chapter 13, we'll see how we preserve, manage, and restore them when they're degraded.

> ***Think About It***
>
> As you look at the map in figure 5.3, which biomes do you think are most heavily populated by humans? Why? Which biomes are most altered by humans? (Check your answers by looking at figure 5.22.)

### Tropical moist forests are warm and wet year-round

The humid tropical regions of the world support one of the most complex and biologically rich biome types in the world (fig. 5.5). Although there are several kinds of moist tropical forests, they share common attributes of ample rainfall and uniform temperatures. Cool **cloud forests** are found high in the mountains where fog and mist keep vegetation wet all the time. **Tropical rainforests** occur where rainfall is abundant—more than 200 cm (80 in.) per year—and temperatures are warm to hot year-round. For aid in reading the climate graphs in these figures, see the Data Analysis box at the end of this chapter.

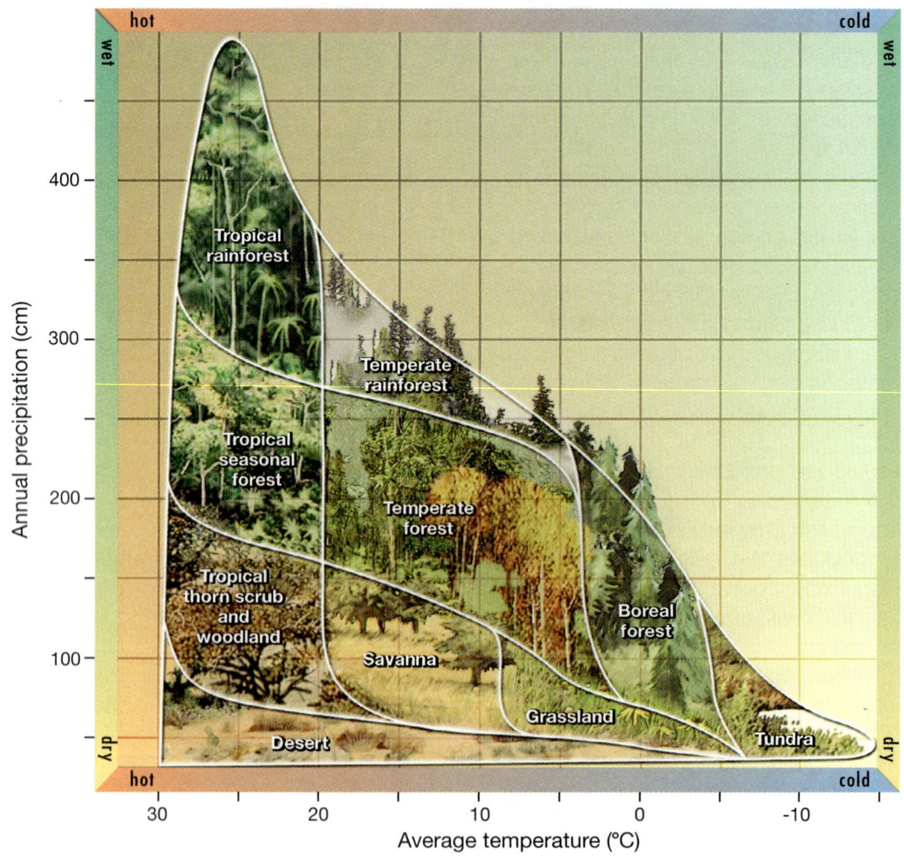


**FIGURE 5.2** Biomes most likely to occur in the absence of human disturbance or other disruptions, according to average annual temperature and precipitation. *Note:* this diagram does not consider soil type, topography, wind speed, or other important environmental factors. Still, it is a useful general guideline for biome location.

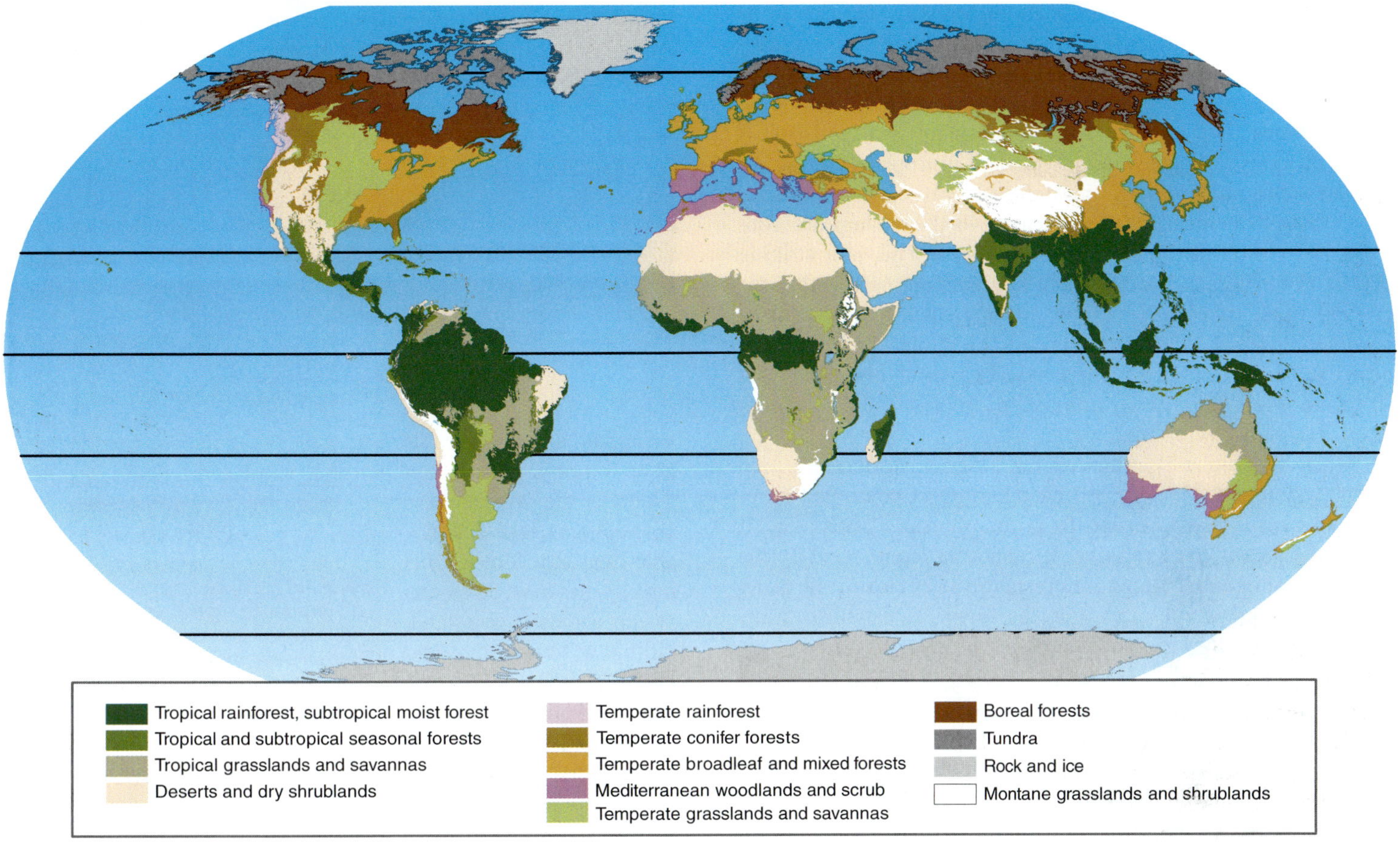


**FIGURE 5.3** Major world biomes. Compare this map to figure 5.2 for generalized temperature and moisture conditions that control biome distribution. Also compare it to the satellite image of biological productivity (fig. 5.13).
**Source:** WWF Ecoregions.

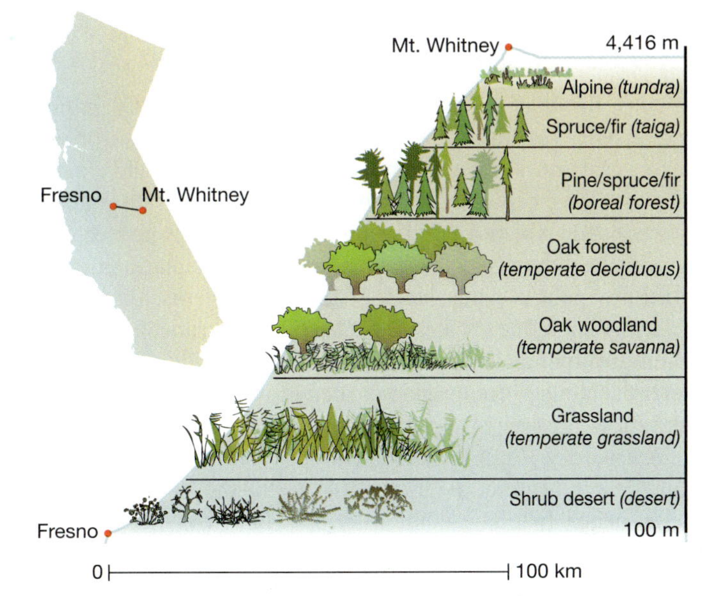


**FIGURE 5.4** Vegetation changes with elevation because temperatures are lower and precipitation is greater high on a mountain side. A 100 km transect from Fresno, California, to Mt. Whitney (California's highest point) crosses vegetation zones similar to about seven different biome types.

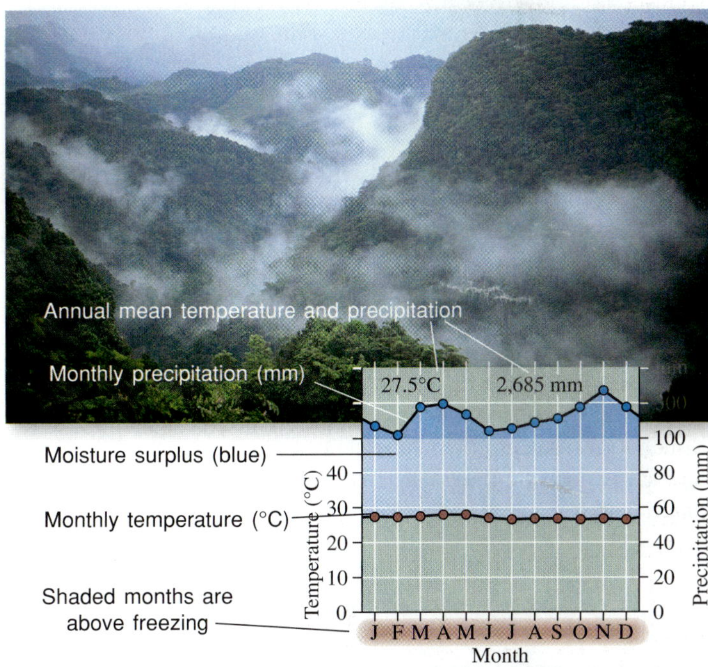


**FIGURE 5.5** Tropical rainforests have luxuriant and diverse plant growth. Heavy rainfall in most months, shown in the climate graph, supports this growth.

The soil of both these tropical moist forest types tends to be old, thin, acidic, and nutrient-poor, yet the number of species present can be mind-boggling. For example, the number of insect species in the canopy of tropical rainforests has been estimated to be in the millions! It is estimated that one-half to two-thirds of all species of terrestrial plants and insects live in tropical forests.

The nutrient cycles of these forests also are distinctive. Almost all (90 percent) of the nutrients in the system are contained in the bodies of the living organisms. This is a striking contrast to temperate forests, where nutrients are held within the soil and made available for new plant growth. The luxuriant growth in tropical rainforests depends on rapid decomposition and recycling of dead organic material. Leaves and branches that fall to the forest floor decay and are incorporated almost immediately back into living biomass.

When the forest is removed for logging, agriculture, and mineral extraction, the thin soil cannot support continued cropping and cannot resist erosion from the abundant rains. And if the cleared area is too extensive, it cannot be repopulated by the rainforest community. Rapid deforestation is occurring in many tropical areas as people move into the forests to establish farms and ranches, but the land soon loses its fertility.

## Tropical seasonal forests have annual dry seasons

Many tropical regions are characterized by distinct wet and dry seasons, although temperatures remain hot year-round. These areas support **tropical seasonal forests:** drought-tolerant forests that look brown and dormant in the dry season but burst into vivid green during rainy months. These forests are often called dry tropical forests because they are dry much of the year; however, there must be some periodic rain to support tree growth. Many of the trees and shrubs in a seasonal forest are drought-deciduous: They lose their leaves and cease growing when no water is available. Seasonal forests are often open woodlands that grade into savannas.

Tropical dry forests have typically been more attractive than wet forests for human habitation and have suffered greater degradation. Clearing a dry forest with fire is relatively easy during the dry season. Soils of dry forests often have higher nutrient levels and are more agriculturally productive than those of a rainforest. Finally, having fewer insects, parasites, and fungal diseases than a wet forest makes a dry or seasonal forest a healthier place for humans to live. Consequently, these forests are highly endangered in many places. Less than 1 percent of the dry tropical forests of the Pacific coast of Central America or the Atlantic coast of South America, for instance, remain in an undisturbed state.

## Tropical savannas and grasslands are dry most of the year

Where there is too little rainfall to support forests, we find open **grasslands** or grasslands with sparse tree cover, which we call **savannas** (fig. 5.6). Like tropical seasonal forests, most tropical savannas and grasslands have a rainy season, but generally the rains are less abundant or less dependable than in a forest. During dry seasons, fires can sweep across a grassland, killing off young trees and keeping the landscape open. Savanna and grassland plants have many adaptations to survive drought, heat, and fires. Many have deep, long-lived roots that seek groundwater and that persist when leaves and stems above the ground die back. After a fire, or after a drought, fresh green shoots grow quickly from the roots. Migratory grazers, such as wildebeest, antelope, or bison thrive on this new growth. Grazing pressure from domestic livestock is an important threat to both the plants and animals of tropical grasslands and savannas.

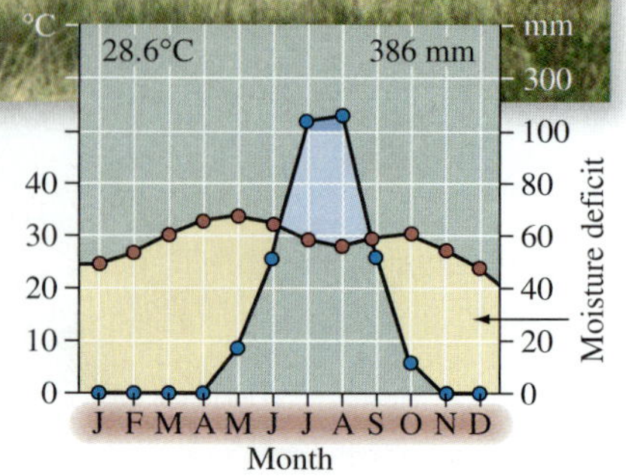


**FIGURE 5.6** Tropical savannas and grasslands experience annual drought and rainy seasons and year-round warm temperatures. Thorny acacias and abundant grazers thrive in this savanna. Yellow areas show moisture deficit.

## Deserts are hot or cold, but always dry

You may think of deserts as barren and biologically impoverished. Their vegetation is sparse, but it can be surprisingly diverse, and most desert plants and animals are highly adapted to survive long droughts, extreme heat, and often extreme cold. **Deserts** occur where precipitation is rare and unpredictable, usually with less than 30 cm of rain per year. Adaptations to these conditions include water-storing leaves and stems, thick epidermal layers to reduce water loss, and salt tolerance. As in other dry environments, many plants are drought-deciduous. Most desert plants also bloom and set seed quickly when a spring rain does fall.

Warm, dry, high-pressure climate conditions (chapter 15) create desert regions at about 30° north and south. Extensive deserts occur in continental interiors (far from oceans, which evaporate the moisture for most precipitation) of North America, Central Asia, Africa, and Australia (fig. 5.7). The rain shadow of the Andes produces the world's driest desert in coastal Chile. Deserts can also be cold. Antarctica is a desert. Some inland valleys apparently get almost no precipitation at all.

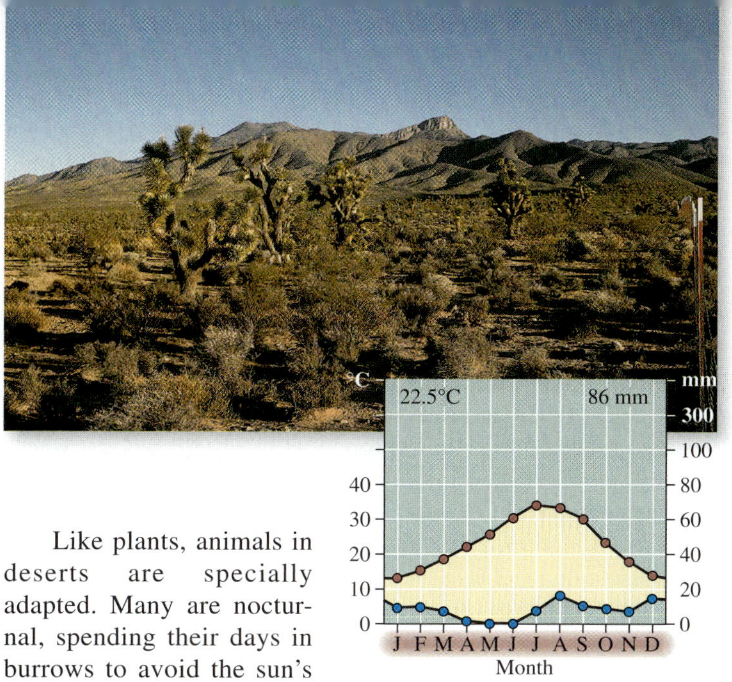


**FIGURE 5.7** Deserts generally receive less than 300 mm (30 cm) of precipitation per year. Hot deserts, as in the American Southwest, endure year-round drought and extreme heat in summer.

Like plants, animals in deserts are specially adapted. Many are nocturnal, spending their days in burrows to avoid the sun's heat and desiccation. Pocket mice, kangaroo rats, and gerbils can get most of their moisture from seeds and plants. Desert rodents also have highly concentrated urine and nearly dry feces that allow them to eliminate body waste without losing precious moisture.

Deserts are more vulnerable than you might imagine. Sparse, slow-growing vegetation is quickly damaged by off-road vehicles. Desert soils recover slowly. Tracks left by army tanks practicing in California deserts during World War II can still be seen today.

Deserts are also vulnerable to overgrazing. In Africa's vast Sahel (the southern edge of the Sahara Desert), livestock are destroying much of the plant cover. Bare, dry soil becomes drifting sand, and restabilization is extremely difficult. Without plant roots and organic matter, the soil loses its ability to retain what rain does fall, and the land becomes progressively drier and more bare. Similar depletion of dryland vegetation is happening in many desert areas, including Central Asia, India, and the American Southwest and Plains states.

## Temperate grasslands have rich soils

As in tropical latitudes, temperate (midlatitude) grasslands occur where there is enough rain to support abundant grass but not enough for forests (fig. 5.8). Usually grasslands are a complex, diverse mix of grasses and flowering herbaceous plants, generally known as forbs. Myriad flowering forbs make a grassland colorful and lovely in summer. In dry grasslands, vegetation may be less than a meter tall. In more humid areas, grasses can exceed 2 m. Where scattered trees occur in a grassland, we call it a savanna.

Deep roots help plants in temperate grasslands and savannas survive drought, fire, and extreme heat and cold. These roots, together with an annual winter accumulation of dead leaves on the surface, produce thick, organic-rich soils in temperate grasslands. Because of this rich soil, many grasslands have been converted to farmland. The legendary tallgrass prairies of the central United States and Canada are almost completely replaced by corn, soybeans, wheat, and other crops. Most remaining grasslands in this region are too dry to support agriculture, and their greatest threat is overgrazing. Excessive grazing eventually kills even deep-rooted plants. As ground cover dies off, soil erosion results, and unpalatable weeds, such as cheatgrass or leafy spurge, spread.

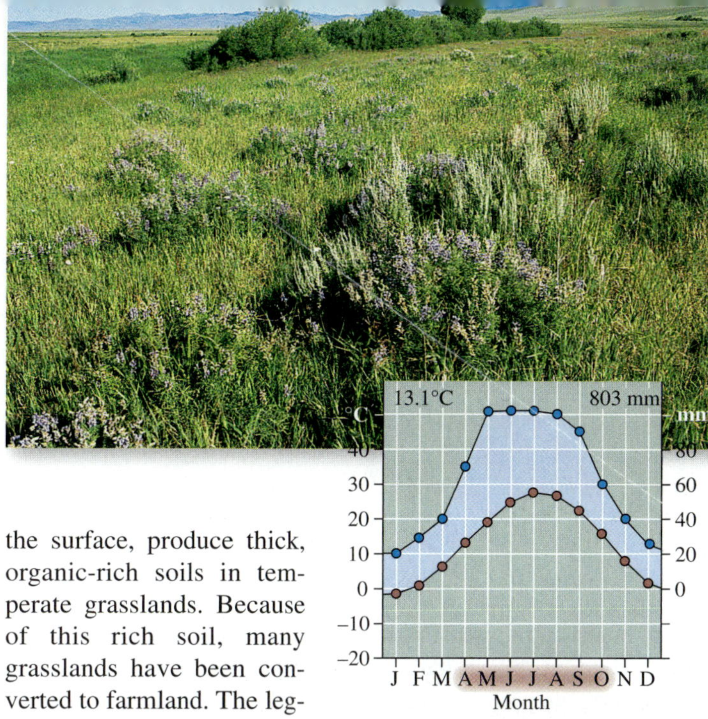


**FIGURE 5.8** Grasslands occur at midlatitudes on all continents. Kept open by extreme temperatures, dry conditions, and periodic fires, grasslands can have surprisingly high plant and animal diversity.

## Temperate shrublands have summer drought

Often, dry environments support drought-adapted shrubs and trees, as well as grass. These mixed environments can be highly variable. They can also be very rich biologically. Such conditions are often described as Mediterranean (where the hot season coincides with the dry season producing hot, dry summers and cool, moist winters). Evergreen shrubs with small, leathery, sclerophyllous (hard, waxy) leaves form dense thickets. Scrub oaks, drought-resistant pines, or other small trees often cluster in sheltered valleys. Periodic fires burn fiercely in this fuel-rich plant assemblage and are a major factor in plant succession. Annual spring flowers often bloom profusely, especially after fires. In California, this landscape is called **chaparral,** Spanish for thicket. Resident animals are drought tolerant such as jackrabbits, kangaroo rats, mule deer, chipmunks, lizards, and many bird species. Very similar landscapes are found along the Mediterranean coast as well as southwestern Australia, central Chile, and South Africa. Although this biome doesn't cover a very large total area, it contains a high number of unique species and is often considered a "hot spot" for biodiversity. It also is highly desired for human habitation, often leading to conflicts with rare and endangered plant and animal species.

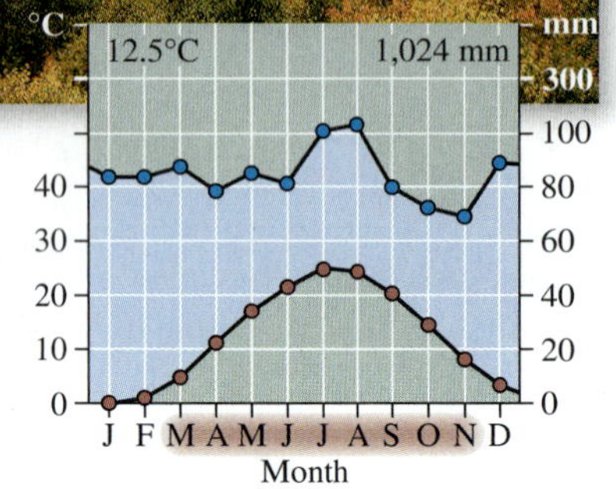


FIGURE 5.9 Temperate deciduous forests have year-round precipitation and winters near or below freezing.

Areas that are drier year-round, such as the African Sahel (edge of the Sahara Desert), northern Mexico, or the American Intermountain West (or Great Basin) tend to have a more sparse, open shrubland, characterized by sagebrush (*Artemisia* sp.), chamiso (*Adenostoma* sp.), or saltbush (*Atriplex* sp.). Some typical animals of this biome in America are a wide variety of snakes and lizards, rodents, birds, antelope, and mountain sheep.

## Temperate forests can be evergreen or deciduous

Temperate, or midlatitudes, forests occupy a wide range of precipitation conditions but occur mainly between about 30 and 55 degrees latitude (see fig. 5.3). In general we can group these forests by tree type, which can be broad-leaf **deciduous** (losing leaves seasonally) or evergreen **coniferous** (cone-bearing).

### *Deciduous Forests*

Broad-leaf forests occur throughout the world where rainfall is plentiful. In midlatitudes, these forests are deciduous and lose their leaves in winter. The loss of green chlorophyll pigments can produce brilliant colors in these forests in autumn (fig. 5.9). At lower latitudes, broad-leaf forest may be evergreen or drought-deciduous. Southern live oaks, for example, are broad-leaf evergreen trees.

Although these forests have a dense canopy in summer, they have a diverse understory that blooms in spring, before the trees leaf out. Spring ephemeral (short-lived) plants produce lovely flowers, and vernal (springtime) pools support amphibians and insects. These forests also shelter a great diversity of songbirds.

North American deciduous forests once covered most of what is now the eastern half of the United States and southern Canada. Most of western Europe was once deciduous forest but was cleared a thousand years ago. When European settlers first came to North America, they quickly settled and cut most of the eastern deciduous forests for firewood, lumber, and industrial uses, as well as to clear farmland. Many of those regions have now returned to deciduous forest, though the dominant species have changed.

Deciduous forests can regrow quickly because they occupy moist, moderate climates. But most of these forests have been occupied so long that human impacts are extensive, and most native species are at least somewhat threatened. The greatest threat to broad-leaf deciduous forests is in eastern Siberia, where deforestation is proceeding rapidly. Siberia may have the highest deforestation rate in the world. As forests disappear, so do Siberian tigers, bears, cranes, and a host of other endangered species.

### *Coniferous Forests*

Coniferous forests grow in a wide range of temperature and moisture conditions. Often they occur where moisture is limited: In cold climates, moisture is unavailable (frozen) in winter; hot climates may have seasonal drought; sandy soils hold little moisture, and they are often occupied by conifers. Thin, waxy leaves (needles) help these trees reduce moisture loss. Coniferous forests provide most wood products in North America. Dominant wood production regions include the southern Atlantic and Gulf coast states, the mountain West, and the Pacific Northwest (northern California to Alaska), but coniferous forests support forestry in many regions.

The coniferous forests of the Pacific coast grow in extremely wet conditions. The wettest coastal forests are known as **temperate rainforest,** a cool, rainy forest often enshrouded in fog (fig. 5.10). Condensation in the canopy (leaf drip) is a major form of precipitation in the understory. Mild year-round temperatures and abundant rainfall, up to 250 cm (100 in.) per year, result in luxuriant plant growth and giant trees such as the California redwoods, the largest trees in the world and the largest aboveground organism ever known to have existed. Redwoods once grew along the Pacific coast from California to Oregon, but logging has reduced them to a few small fragments.

Remaining fragments of ancient temperate rainforests are important areas of biodiversity. Recent battles over old-growth conservation (chapter 12) focus mainly on these areas. As with deciduous forests, Siberian forests are especially vulnerable to old-growth logging. The rate of this clearing, and its environmental effects, remain largely unknown.

## Boreal forests occur at high latitudes

Because conifers can survive winter cold, they tend to dominate the **boreal forest,** or northern forests, that lie between about 50° and 60° north (fig. 5.11). Mountainous areas at lower latitudes may also have many characteristics and species of the boreal forest. Dominant trees are pines, hemlocks, spruce, cedar, and fir. Some deciduous trees are also present, such as maples, birch, aspen, and alder. These forests are slow-growing because of the cold temperatures and short frost-free growing

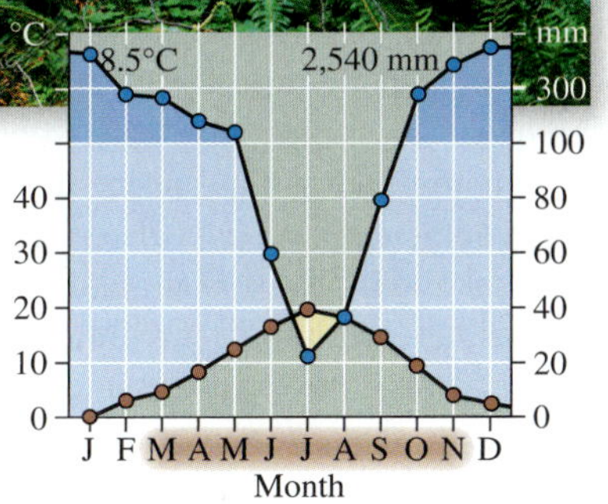


season, but they are still an expansive resource. In Siberia, Canada, and the western United States, large regional economies depend on boreal forests.

The extreme, ragged edge of the boreal forest, where forest gradually gives way to open tundra, is known by its Russian name, **taiga.** Here extreme cold and short summer limits the growth rate of trees. A 10 cm diameter tree may be over 200 years old in the far north.

**FIGURE 5.10** Temperate rainforests have abundant but often seasonal precipitation that supports magnificent trees and luxuriant understory vegetation. Often these forests experience dry summers.

## Tundra can freeze in any month

Where temperatures are below freezing most of the year, only small, hardy vegetation can survive. **Tundra,** a treeless landscape that occurs at high latitudes or on mountaintops, has a growing season of only two to three months, and it may have frost any month of the year. Some people consider tundra a variant of grasslands because it has no trees; others consider it a very cold desert because water is unavailable (frozen) most of the year.

*Arctic* tundra is an expansive biome that has low productivity because it has a short growing season (fig. 5.12). During midsummer, however, 24-hour sunshine supports a burst of plant growth and an explosion of insect life. Tens of millions of waterfowl, shorebirds, terns, and songbirds migrate to the Arctic every year to feast on the abundant invertebrate and plant life and to raise their young on the brief bounty. These birds then migrate to wintering grounds, where they may be eaten by local predators—effectively they carry energy and protein from high latitudes to low latitudes. Arctic tundra is essential for global biodiversity, especially for birds.

*Alpine* tundra, occurring on or near mountaintops, has environmental conditions and vegetation similar to arctic tundra. These areas have a short, intense growing season. Often one sees a splendid profusion of flowers in alpine tundra; everything must flower at once in order to produce seeds in a few

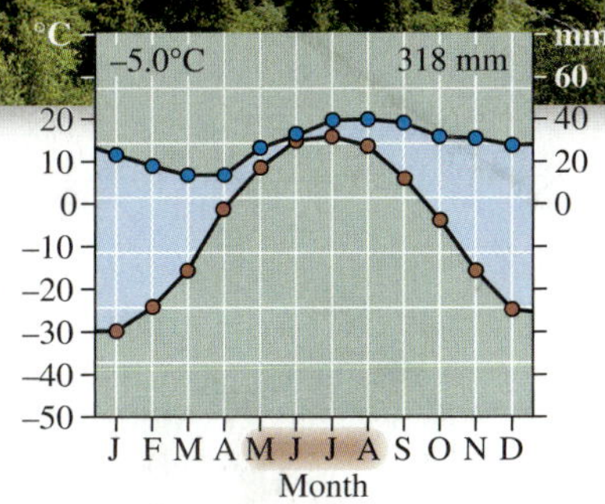


**FIGURE 5.11** Boreal forests have moderate precipitation but are often moist because temperatures are cold most of the year. Cold-tolerant and drought-tolerant conifers dominate boreal forests and taiga, the forest fringe.

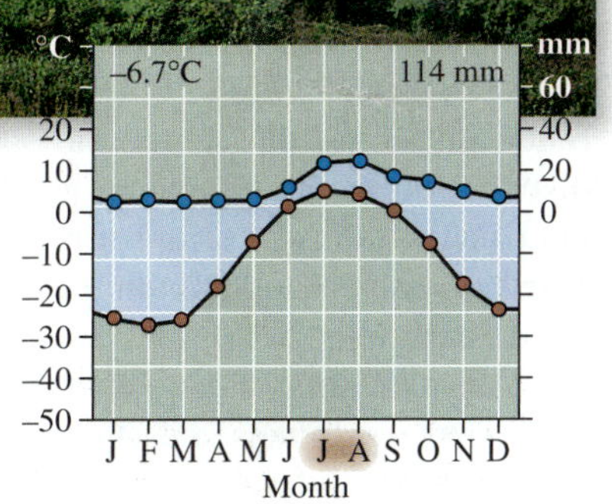


**FIGURE 5.12** This landscape in Canada's Northwest Territories has both alpine and arctic tundra. Plant diversity is relatively low, and frost can occur even in summer.

weeks before the snow comes again. Many alpine tundra plants also have deep pigmentation and leathery leaves to protect against the strong ultraviolet light in the thin mountain atmosphere.

Compared to other biomes, tundra has relatively low diversity. Dwarf shrubs, such as willows, sedges, grasses, mosses, and lichens tend to dominate the vegetation. Migratory musk-ox, caribou, or alpine mountain sheep and mountain goats can live on the vegetation because they move frequently to new pastures.

Because these environments are too cold for most human activities, they are not as badly threatened as other biomes. There are important problems, however. Global climate change may be altering the balance of some tundra ecosystems, and air pollution from distant cities tends to accumulate at high latitudes (chapter 15). In eastern Canada, coastal tundra is being badly depleted by overabundant populations of snow geese, whose numbers have exploded due to winter grazing on the rice fields of Arkansas and Louisiana. Oil and gas drilling—and associated truck traffic—threatens tundra in Alaska and Siberia. Clearly, this remote biome is not independent of human activities at lower latitudes.

## 5.2 Marine Ecosystems

The biological communities in oceans and seas are poorly understood, but they are probably as diverse and as complex as terrestrial biomes. In this section, we will explore a few facets of these fascinating environments. Oceans cover nearly three-fourths of the earth's surface, and they contribute in important, although often unrecognized, ways to terrestrial ecosystems. Like land-based systems, most marine communities depend on photosynthetic organisms. Often it is algae or tiny, free-floating photosynthetic plants **(phytoplankton)** that support a marine food web, rather than the trees and grasses we see on land. In oceans, photosynthetic activity tends to be greatest near coastlines, where nitrogen, phosphorus, and other nutrients wash offshore and fertilize primary producers. Ocean currents also contribute to the distribution of biological productivity, as they transport nutrients and phytoplankton far from shore (fig. 5.13).

As plankton, algae, fish, and other organisms die, they sink toward the ocean floor. Deep-ocean ecosystems, consisting of crabs, filter-feeding organisms, strange phosphorescent fish, and many other life-forms, often rely on this "marine snow" as a primary nutrient source. Surface communities also depend on this material. Upwelling currents circulate nutrients from the ocean floor back to the surface. Along the coasts of South America, Africa, and Europe, these currents support rich fisheries.

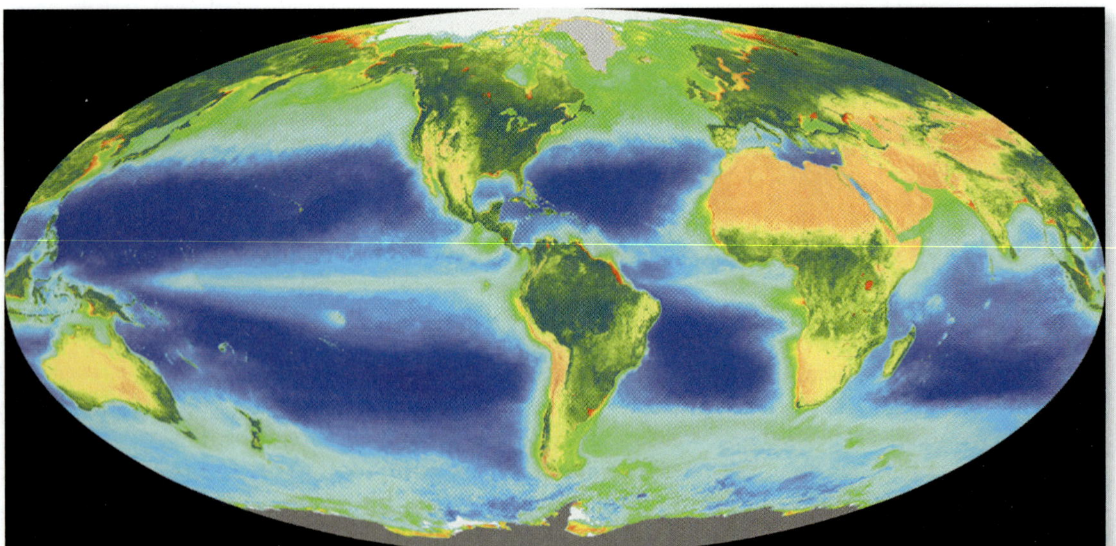

**FIGURE 5.13** Satellite measurements of chlorophyll levels in the oceans and on land. Dark green to blue land areas have high biological productivity. Dark blue oceans have little chlorophyll and are biologically impoverished. Light green to yellow ocean zones are biologically rich. Courtesy Seawifs/NASA.

Vertical stratification is a key feature of aquatic ecosystems. Light decreases rapidly with depth, and communities below the photic zone (light zone, often reaching about 20 m deep) must rely on energy sources other than photosynthesis to persist. Temperature also decreases with depth. Deep-ocean species often grow slowly in part because metabolism is reduced in cold conditions. In contrast, warm, bright, near-surface communities such as coral reefs and estuaries are among the world's most biologically productive environments. Temperature also affects the amount of oxygen and other elements that can be absorbed in water. Cold water holds abundant oxygen, so productivity is often high in cold oceans, as in the North Atlantic, North Pacific, and Antarctic.

Ocean systems can be described by depth and proximity to shore (fig. 5.14). In general, **benthic** communities occur on the bottom, and **pelagic** (from "sea" in Greek) zones are the water column. The epipelagic zone (*epi* = on top) has photosynthetic organisms. Below this are the mesopelagic (*meso* = medium), and bathypelagic (*bathos* = deep) zones. The deepest layers are the abyssal zone (to 4,000 m) and hadal zone (deeper than 6,000 m). Shorelines are known as littoral zones, and the area exposed by low tides is known as the intertidal zone. Often there is a broad, relatively shallow region along a continent's coast, which may reach a few kilometers or hundreds of kilometers from shore. This undersea area is the continental shelf.

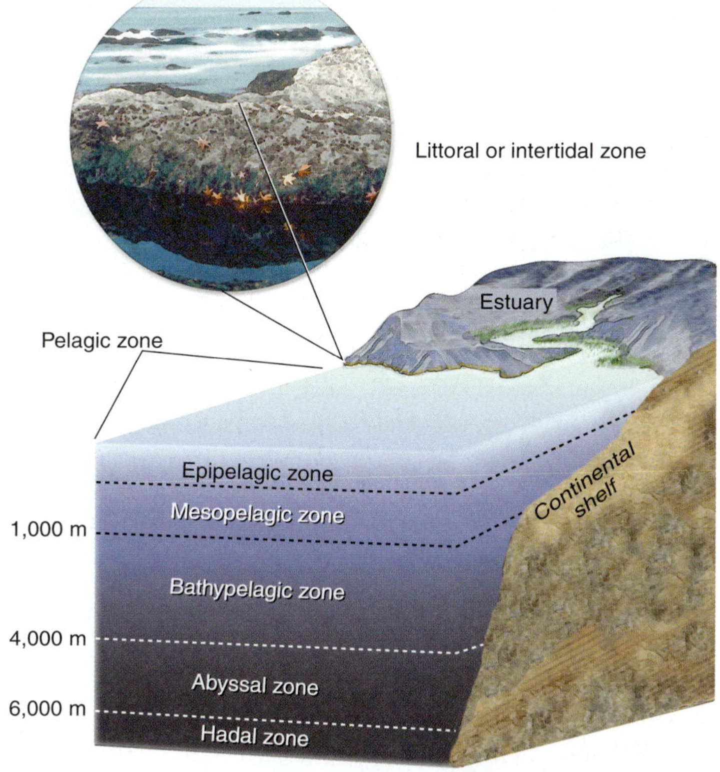


**FIGURE 5.14** Light penetrates only the top 10–20 m of the ocean. Below this level, temperatures drop and pressure increases. Nearshore environments include the intertidal zone and estuaries.

**FIGURE 5.15** Deep-ocean thermal vent communities were discovered only recently. They have great diversity and are unusual because they rely on chemosynthesis, not photosynthesis, for energy.

## Open-ocean communities vary from surface to hadal zones

The open ocean is often referred to as a biological desert because it has relatively low productivity. But like terrestrial deserts, the open ocean has areas of rich productivity and diversity. Fish and plankton abound in regions such as the equatorial Pacific and Antarctic oceans, where nutrients are distributed by currents. Another notable exception, the Sargasso Sea in the western Atlantic, is known for its free-floating mats of brown algae. These algae mats support a phenomenal diversity of animals, including sea turtles, fish, and even eels that hatch amid the algae, then eventually migrate up rivers along the Atlantic coasts of North America and Europe.

Deep-sea thermal vent communities are another remarkable type of marine system (fig. 5.15) that was completely unknown until 1977 explorations with the deep-sea submarine *Alvin.* These communities are based on microbes that capture chemical energy, mainly from sulfur compounds released from thermal vents—jets of hot water and minerals on the ocean floor. Magma below the ocean crust heats these vents. Tube worms, mussels, and microbes on these vents are adapted to survive both extreme temperatures, often above 350°C (700°F), and the intense water pressure at depths of 7,000 m (20,000 ft) or more. Oceanographers have discovered thousands of different types of organisms, most of them microscopic, in these communities (chapter 3).

## Coastal zones support rich, diverse biological communities

As in the open ocean, shoreline communities vary with depth, light, nutrient concentrations, and temperature. Some shoreline communities, such as estuaries, have high biological productivity and diversity because they are enriched by nutrients washing from the land. But nutrient loading can be excessive. Around the world, more than 200 "dead zones" occur in coastal zones where excess nutrients stimulate bacterial growth that consumes almost all oxygen in the water and excludes most other life. We'll discuss this problem further in chapter 18.

**Corals reefs** are among the best-known marine ecosystems because of their extraordinary biological productivity and their diverse and beautiful organisms (see fig. 5.1). Reefs are aggregations of minute colonial animals (coral polyps) that live

(a) Coral reefs

(c) Estuary and salt marsh

(b) Mangroves

(d) Tide pool

**FIGURE 5.16** Coastal environments support incredible diversity and help stabilize shorelines. Coral reefs (a), mangroves (b), and estuaries (c) also provide critical nurseries for marine ecosystems. Tide pools (d) also shelter highly specialized organisms.

symbiotically with photosynthetic algae. Calcium-rich coral skeletons build up to make reefs, atolls, and islands (fig. 5.16*a*). Reefs protect shorelines and shelter countless species of fish, worms, crustaceans, and other life-forms. Reef-building corals live where water is shallow and clear enough for sunlight to reach the photosynthetic algae. They need warm (but not too warm) water, and can't survive where high nutrient concentrations or runoff from the land create dense layers of algae, fungi, or sediment. Coral reefs also are among the most endangered biomes in the world. As the opening case study for this chapter shows, destructive fishing practices can damage or destroy coral communities. In addition, polluted urban runoff, trash, sewage and industrial effluent, sediment from agriculture, and unsustainable forestry are smothering coral reefs along coastlines that have high human populations. Introduced pathogens and predators also threaten many reefs. Perhaps the greatest threat to reefs is global warming. Elevated water temperatures cause **coral bleaching,** in which corals expel their algal partner and then die. The third UNESCO Conference on Oceans, Coasts, and Islands in 2006 reported that one-third of all coral reefs have been destroyed, and that 60 percent are now degraded and probably will be dead by 2030.

The value of an intact reef in a tourist economy, like that of Apo Island, can be upwards of (U.S.)$1 million per km$^2$. The costs of conserving these same reefs in a marine-protected area would be just (U.S.)$775 per km$^2$ per year, the UN Environment Programme estimates. Of the estimated 30 million small-scale fishers in the developing world, most are dependent to a greater or lesser extent on coral reefs. In the Philippines, the UN estimates that more than 1 million fishers depend directly on coral reefs for their livelihoods. We'll discuss reef restoration efforts further in chapter 13.

Sea-grass beds, or eel-grass beds, often occupy shallow, warm, sandy areas near coral reefs. Like reefs, these communities support a rich diversity of grazers, from snails and worms to turtles and manatees. Also like reefs, these environments are easily smothered by sediment originating from onshore agriculture and development.

**Mangroves** are trees that grow in salt water. They occur along calm, shallow, tropical coastlines around the world (fig. 5.16*b*). Mangrove forests or swamps help stabilize shorelines, and they are also critical nurseries for fish, shrimp, and other commercial species. Like coral reefs, mangroves line tropical and subtropical coastlines, where they are vulnerable to development, sedimentation, and overuse. Unlike reefs, mangroves provide commercial timber, and they can be clear-cut to make room for aquaculture (fish farming) and other activities. Ironically, mangroves provide the protected spawning beds for most of the fish and shrimp farmed in these ponds. As mangroves become increasingly threatened in tropical countries, villages relying on fishing for income and sustenance are seeing reduced catches and falling income.

**Estuaries** are bays where rivers empty into the sea, mixing fresh water with salt water. **Salt marshes,** shallow wetlands flooded regularly or occasionally with seawater, occur on shallow coastlines, including estuaries (fig. 5.16*c*). Usually calm, warm, and nutrient-rich, estuaries and salt marshes are biologically diverse and productive. Rivers provide nutrients and sediments, and a muddy bottom supports emergent plants (whose leaves emerge above the water surface), as well as the young forms of crustaceans, such as crabs and shrimp, and mollusks, such as clams and oysters. Nearly two-thirds of all marine fish and shellfish rely on estuaries and saline wetlands for spawning and juvenile development.

Estuaries near major American cities once supported an enormous wealth of seafood. Oyster beds and clam banks in the waters adjacent to New York, Boston, and Baltimore provided free and easy food to early residents. Sewage and other contaminants long ago eliminated most of these resources, however. Recently, major efforts have been made to revive Chesapeake Bay, America's largest and most productive estuary. These efforts have shown some success, but many challenges remain (see related story "Restoring the Chesapeake" at www.mhhe.com/cunningham10e).

In contrast to the shallow, calm conditions of estuaries, coral reefs, and mangroves, there are violent, wave-blasted shorelines that support fascinating life-forms in **tide pools.** Tide pools are depressions in a rocky shoreline that are flooded at high tide but retain some water at low tide. These areas remain rocky where wave action prevents most plant growth or sediment (mud) accumulation. Extreme conditions, with frigid flooding at high tide and hot, dessicating sunshine at low tide, make life impossible for most species. But the specialized animals and plants that do occur in this rocky intertidal zone are astonishingly diverse and beautiful (fig. 5.16*d*).

**Barrier islands** are low, narrow, sandy islands that form parallel to a coastline (fig. 5.17). They occur where the continental shelf is shallow and rivers or coastal currents provide a steady source of sediments. They protect brackish (moderately salty), inshore lagoons and salt marshes from storms, waves, and tides. One of the world's most extensive sets of barrier islands lines the Atlantic coast from New England to Florida, as well as along the Gulf coast of Texas. Composed of sand that is constantly reshaped by wind and waves, these islands can be formed or removed by a single violent storm. Because they are mostly beach, barrier islands are also popular places for real estate development. About 20 percent of the barrier island surface in the United States has been developed. Barrier islands are also critical to preserving coastal shorelines, settlements, estuaries, and wetlands.

Unfortunately, human occupation often destroys the value that attracts us there in the first place. Barrier islands and beaches are dynamic environments, and sand is hard to keep in place. Wind and wave erosion is a constant threat to beach developments. Walking or driving vehicles over dune grass destroys the stabilizing vegetative cover and accelerates, or triggers, erosion. Cutting roads through the dunes further destabilizes these islands, making them increasingly vulnerable to storm damage. When Hurricane Katrina hit the U.S. Gulf coast in 2005, it caused at least $200 billion in property damage and displaced 4 million people. Thousands of homes were destroyed (fig. 5.18), particularly on low-lying barrier islands.

Because of these problems, we spend billions of dollars each year building protective walls and barriers, pumping sand onto beaches from offshore, and moving sand from one beach area to

**FIGURE 5.17** A barrier island, Assateague, along the Maryland–Virginia coast. Grasses cover and protect dunes, which keep ocean waves from disturbing the bay, salt marshes, and coast at right. Roads cut through the dunes expose them to erosion.

**FIGURE 5.18** Winter storms have eroded the beach and undermined the foundations of homes on this barrier island. Breaking through protective dunes to build such houses damages sensitive plant communities and exposes the whole island to storm sand erosion. Coastal zone management attempts to limit development on fragile sites.

another. Much of this expense is borne by the public. Some planners question whether we should allow rebuilding on barrier islands, especially after they've been destroyed multiple times.

## 5.3 Freshwater Ecosystems

Freshwater environments are far less extensive than marine environments, but they are centers of biodiversity. Most terrestrial communities rely, to some extent, on freshwater environments. In deserts, isolated pools, streams, and even underground water systems, support astonishing biodiversity as well as provide water to land animals. In Arizona, for example, most birds gather in trees and bushes surrounding the few available rivers and streams.

### Lakes have open water

Freshwater lakes, like marine environments, have distinct vertical zones (fig. 5.19). Near the surface a subcommunity of plankton, mainly microscopic plants, animals, and protists (single-celled organisms such as amoebae), float freely in the water column. Insects such as water striders and mosquitoes also live at the air-water interface. Fish move through the water column, sometimes near the surface, and sometimes at depth.

Finally, the bottom, or *benthos,* is occupied by a variety of snails, burrowing worms, fish, and other organisms. These make up the benthic community. Oxygen levels are lowest in the benthic environment, mainly because there is little mixing to introduce oxygen to this zone. Anaerobic bacteria (not using oxygen) may live in low-oxygen sediments. In the littoral zone, emergent plants such as cattails and rushes grow in the bottom sediment. These plants create important functional links between layers of an aquatic ecosystem, and they may provide the greatest primary productivity to the system.

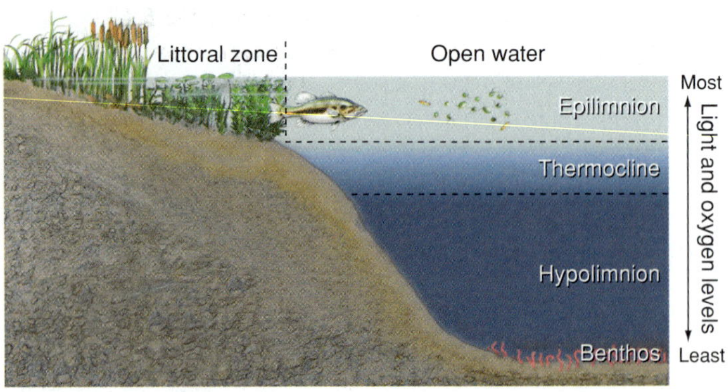


**FIGURE 5.19** The layers of a deep lake are determined mainly by gradients of light, oxygen, and temperature. The epilimnion is affected by surface mixing from wind and thermal convections, while mixing between the hypolimnion and epilimnion is inhibited by a sharp temperature and density difference at the thermocline.

**FIGURE 5.20** The character of freshwater ecosystems is greatly influenced by the immediately surrounding terrestrial ecosystems, and even by ecosystems far upstream or far uphill from a particular site.

Lakes, unless they are shallow, have a warmer upper layer that is mixed by wind and warmed by the sun. This layer is the *epilimnion.* Below the epilimnion is the hypolimnion (*hypo* = below), a colder, deeper layer that is not mixed. If you have gone swimming in a moderately deep lake, you may have discovered the sharp temperature boundary, known as the **thermocline,** between these layers. Below this boundary, the water is much colder. This boundary is also called the mesolimnion.

Local conditions that affect the characteristics of an aquatic community include (1) nutrient availability (or excess) such as nitrates and phosphates; (2) suspended matter, such as silt, that affects light penetration; (3) depth; (4) temperature; (5) currents; (6) bottom characteristics, such as muddy, sandy, or rocky floor; (7) internal currents; and (8) connections to, or isolation from, other aquatic and terrestrial systems (fig. 5.20).

### Wetlands are shallow and productive

**Wetlands** are shallow ecosystems in which the land surface is saturated or submerged at least part of the year. Wetlands have vegetation that is adapted to grow under saturated conditions. These legal definitions are important because although wetlands make up only a small part of most countries, they are disproportionately important in conservation debates and are the focus of continual legal disputes around the world and in North America. Beyond these basic descriptions, defining wetlands is a matter of hot debate. How often must a wetland be saturated, and for how long? How large must it be to deserve legal protection? Answers can vary, depending on political, as well as ecological, concerns.

(a) Swamp, or wooded wetland (b) Marsh (c) Coastal saltmarsh

**FIGURE 5.21** Wetlands provide irreplaceable ecological services, including water filtration, water storage and flood reduction, and habitat. Forested wetlands (a) are often called swamps; marshes (b) have no trees; coastal saltmarshes (c) are tidal and have rich diversity.

These relatively small systems support rich biodiversity, and they are essential for both breeding and migrating birds. Although wetlands occupy less than 5 percent of the land in the United States, the Fish and Wildlife Service estimates that one-third of all endangered species spend at least part of their lives in wetlands. Wetlands retain storm water and reduce flooding by slowing the rate at which rainfall reaches river systems. Floodwater storage is worth \$3 billion to \$4 billion per year in the United States. As water stands in wetlands, it also seeps into the ground, replenishing groundwater supplies. Wetlands filter, and even purify, urban and farm runoff, as bacteria and plants take up nutrients and contaminants in water. They are also in great demand for filling and development. They are often near cities or farms, where land is valuable, and once drained, wetlands are easily converted to more lucrative uses.

Wetlands are described by their vegetation. **Swamps** are wetlands with trees (fig. 5.21*a*). **Marshes** are wetlands without trees (fig. 5.21*b*). **Bogs** are areas of saturated ground, and usually the ground is composed of deep layers of accumulated, undecayed vegetation known as peat. **Fens** are similar to bogs except that they are mainly fed by groundwater, so that they have mineral-rich water and specially adapted plant species. Bogs are fed mainly by precipitation. Swamps and marshes have high biological productivity. Bogs and fens, which are often nutrient-poor, have low biological productivity. They may have unusual and interesting species, though, such as sundews and pitcher plants, which are adapted to capture nutrients from insects rather than from soil.

The water in marshes and swamps usually is shallow enough to allow full penetration of sunlight and seasonal warming (fig. 5.21*c*). These mild conditions favor great photosynthetic activity, resulting in high productivity at all trophic levels. In short, life is abundant and varied. Wetlands are major breeding, nesting, and migration staging areas for waterfowl and shorebirds.

Wetlands may gradually convert to terrestrial communities as they fill with sediment, and as vegetation gradually fills in toward the center. Often this process is accelerated by increased sediment loads from urban development, farms, and roads. Wetland losses are one of the areas of greatest concern among biologists.

## 5.4 Human Disturbance

Humans have become dominant organisms over most of the earth, damaging or disturbing more than half of the world's terrestrial ecosystems to some extent. By some estimates, humans preempt about 40 percent of the net terrestrial primary productivity of the biosphere either by consuming it directly, by interfering with its production or use, or by altering the species composition or physical processes of human-dominated ecosystems. Conversion of natural habitat to human uses is the largest single cause of biodiversity losses.

Researchers from the environmental group Conservation International have attempted to map the extent of human disturbance of the natural world (fig. 5.22). The greatest impacts have been in Europe, parts of Asia, North and Central America, and islands such as Madagascar, New Zealand, Java, Sumatra, and those in the Caribbean. Data from this study are shown in table 5.1.

Temperate broad-leaf forests are the most completely human-dominated of any major biome. The climate and soils that support such forests are especially congenial for human occupation. In eastern North America or most of Europe, for example, only remnants of the original forest still persist. Regions with a Mediterranean climate generally are highly desired for human habitation. Because these landscapes also have high levels of biodiversity, conflicts between human preferences and biological values frequently occur.

Temperate grasslands, temperate rainforests, tropical dry forests, and many islands also have been highly disturbed by human activities. If you have traveled through the American cornbelt states such as Iowa or Illinois, you have seen how thoroughly former prairies have been converted to farmlands. Intensive cultivation of this land exposes the soil to erosion and fertility losses (chapter 9). Islands, because of their isolation, often have high

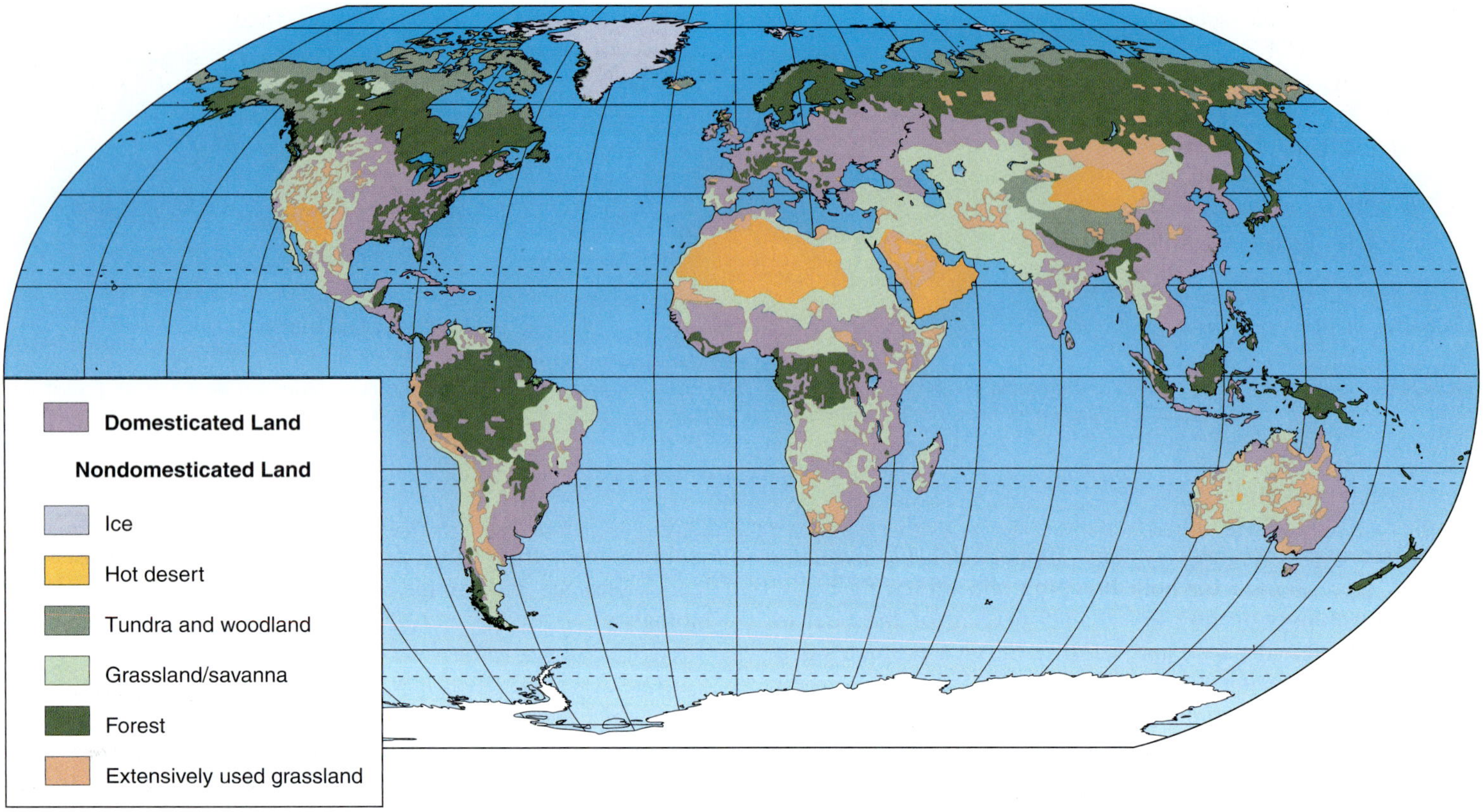


**FIGURE 5.22** Domesticated land has replaced much of the earth's original land cover.
**Source:** United Nations Environment Programme, *Global Environment Outlook,* 1997.

**TABLE 5.1**

**Human Disturbance**

| Biome | Total Area ($10^6$ $km^2$) | % Undisturbed Habitat | % Human Dominated |
|---|---|---|---|
| Temperate broad-leaf forests | 9.5 | 6.1 | 81.9 |
| Chaparral | 6.6 | 6.4 | 67.8 |
| Temperate grasslands | 12.1 | 27.6 | 40.4 |
| Temperate rainforests | 4.2 | 33.0 | 46.1 |
| Tropical dry forests | 19.5 | 30.5 | 45.9 |
| Mixed mountain systems | 12.1 | 29.3 | 25.6 |
| Mixed island systems | 3.2 | 46.6 | 41.8 |
| Cold deserts/semideserts | 10.9 | 45.4 | 8.5 |
| Warm deserts/semideserts | 29.2 | 55.8 | 12.2 |
| Moist tropical forests | 11.8 | 63.2 | 24.9 |
| Tropical grasslands | 4.8 | 74.0 | 4.7 |
| Temperate coniferous forests | 18.8 | 81.7 | 11.8 |
| Tundra and arctic desert | 20.6 | 99.3 | 0.3 |

**Note:** Where undisturbed and human-dominated areas do not add up to 100 percent, the difference represents partially disturbed lands.
**Source:** Hannah, Lee, et al., "Human Disturbance and Natural Habitat: A Biome Level Analysis of a Global Data Set," in *Biodiversity and Conservation,* 1995, Vol. 4:128–55.

numbers of endemic species. Many islands, such as Madagascar, Haiti, and Java have lost more than 99 percent of their original land cover.

Tundra and arctic deserts are the least disturbed biomes in the world. Harsh climates and unproductive soils make these areas unattractive places to live for most people. Temperate conifer forests also generally are lightly populated and large areas remain in a relatively natural state. However, recent expansion of forest harvesting in Canada and Siberia may threaten the integrity of this biome. Large expanses of tropical moist forests still remain in the Amazon and Congo basins but in other areas of the tropics such as West Africa, Madagascar, Southeast Asia, and the Indo-Malaysian peninsula and archipelago, these forests are disappearing at a rapid rate (chapter 12).

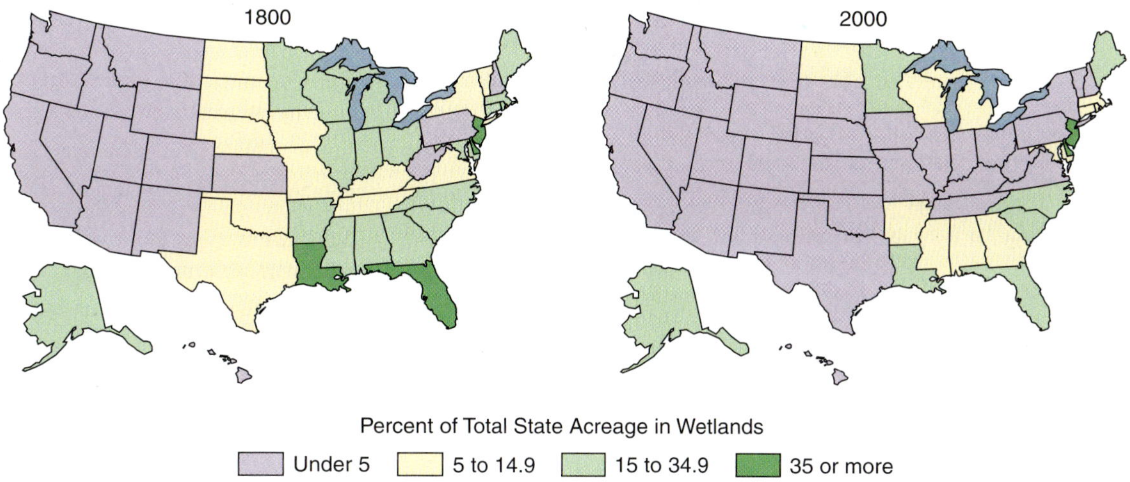


**FIGURE 5.23** Over the past two centuries, more than half of the original wetlands in the lower 48 states have been drained, filled, polluted, or otherwise degraded. Some of the greatest losses have been in midwestern farming states where up to 99 percent of all wetlands have been lost.

As mentioned earlier, wetlands have suffered severe losses in many parts of the world. About half of all original wetlands in the United States have been drained, filled, polluted, or otherwise degraded over the past 250 years. In the prairie states, small potholes and seasonally flooded marshes have been drained and converted to croplands on a wide scale. Iowa, for example, is estimated to have lost 99 percent of its presettlement wetlands (fig. 5.23). Similarly, California has lost 90 percent of the extensive marshes and deltas that once stretched across its central valley. Wooded swamps and floodplain forests in the southern United States have been widely disrupted by logging and conversion to farmland.

Similar wetland disturbances have occurred in other countries as well. In New Zealand, over 90 percent of natural wetlands have been destroyed since European settlement. In Portugal, some 70 percent of freshwater wetlands and 60 percent of estuarine habitats have been converted to agriculture and industrial areas. In Indonesia, almost all the mangrove swamps that once lined the coasts of Java have been destroyed, while in the Philippines and Thailand, more than two-thirds of coastal mangroves have been cut down for firewood or conversion to shrimp and fish ponds.

Slowing this destruction, or even reversing it, is a challenge that we will discuss in chapter 13.

## CONCLUSION

The potential location of biological communities is determined in large part by climate, moisture availability, soil type, geomorphology, and other natural features. Understanding the global distribution of biomes, and knowing the differences in who lives where and why, are essential to the study of global environmental science. Human occupation and use of natural resources is strongly dependent on the biomes found in particular locations. We tend to prefer mild climates and the highly productive biological communities found in temperate zones. These biomes also suffer the highest rates of degradation and overuse.

Being aware of the unique conditions and the characteristics evolved by plants and animals to live in those circumstances can help you appreciate how and why certain species live in unique biomes, such as seasonal tropical forests, alpine tundra, or chaparral shrublands.

Oceans cover nearly three-fourths of the earth's surface, and yet we know relatively little about them. Some marine biomes, such as coral reefs, can be as biologically diverse and productive as any terrestrial biome. People have depended on rich, complex coastal ecosystems for eons, but in recent times rapidly growing human populations, coupled with more powerful ways to harvest resources, have led to damage—and, in some cases, irreversible destruction—of these irreplaceable treasures. Still, there is reason to hope that we'll find ways to live sustainably with nature. The opening case study of this chapter illustrates how, without expensive high technology, a group of local residents protected and restored their coral reef. It gives us optimism that we'll find similar solutions in other biologically rich but endangered biomes.

## Reviewing Learning Outcomes

By now you should be able to explain the following points:

**5.1** Recognize the characteristics of some major terrestrial biomes as well as the factors that determine their distribution.

- Tropical moist forests are warm and wet year-round.
- Tropical seasonal forests have annual dry seasons.
- Tropical savannas and grasslands are dry most of the year.
- Deserts are hot or cold, but always dry.
- Temperate grasslands have rich soils.
- Temperate shrublands have summer drought.
- Temperate forests can be evergreen or deciduous.
- Boreal forests occur at high latitudes.
- Tundra can freeze in any month.

**5.2** Understand how and why marine environments vary with depth and distance from shore.

- Open-ocean communities vary from surface to hadal zones.
- Coastal zones support rich, diverse biological communities.

**5.3** Compare the characteristics and biological importance of major freshwater ecosystems.

- Lakes have open water.
- Wetlands are shallow and productive.

**5.4** Summarize the overall patterns of human disturbance of world biomes.

- Biomes that humans find comfortable and profitable have high rates of disturbance, while those that are less attractive or have limited resources have large pristine areas.

## Practice Quiz

1. Throughout the central portion of North America is a large biome once dominated by grasses. Describe how physical conditions and other factors control this biome.
2. What is taiga and where is it found? Why might logging in taiga be more disruptive than in southern coniferous forests?
3. Why are tropical moist forests often less suited for agriculture and human occupation than tropical deciduous forests?
4. Find out the annual temperature and precipitation conditions where you live (fig. 5.2). Which biome type do you occupy?
5. Describe four different kinds of wetlands and explain why they are important sites of biodiversity and biological productivity.
6. Forests differ according to both temperature and precipitation. Name and describe a biome that occurs in (a) hot, (b) cold, (c) wet, and (d) dry climates (one biome for each climate).
7. How do physical conditions change with depth in marine environments?
8. Describe four different coastal ecosystems.

## Critical Thinking and Discussion Questions

1. What physical and biological factors are most important in shaping your biological community? How do the present characteristics of your area differ from those 100 or 1,000 years ago?
2. Forest biomes frequently undergo disturbances such as fire or flooding. As more of us build homes in these areas, what factors should we consider in deciding how to protect people from natural disturbances?
3. Often humans work to preserve biomes that are visually attractive. What biomes might be lost this way? Is this a problem?
4. Disney World in Florida wants to expand onto a wetland. It has offered to buy and preserve a large nature preserve in a different area to make up for the wetland it is destroying. Is that reasonable? What conditions would make it reasonable or unreasonable?
5. Suppose further that the wetland being destroyed in question 4 and its replacement area both contain several endangered species (but different ones). How would you compare different species against each other? How many plant or insect species would one animal species be worth?
6. Historically, barrier islands have been hard to protect because links between them and inshore ecosystems are poorly recognized. What kinds of information would help a community distant from the coast commit to preserving a barrier island?

## DATA analysis Reading Climate Graphs

As you've learned in this chapter, temperature and moisture are critical factors in determining the distribution and health of ecosystems. But how do you read the climate and precipitation graphs that accompany the description of each biome? To begin, examine the three climate graphs in this box. These graphs show annual trends in temperature and precipitation (rainfall and snow). They also indicate the relationship between potential evaporation, which depends on temperature and precipitation. When evaporation exceeds precipitation, dry conditions result (yellow areas). Extremely wet months are shaded dark blue on the graphs. Moist climates may vary in precipitation rates, but evaporation rarely exceeds precipitation. Months above freezing temperature (shaded brown on the X-axis) have most evaporation. Comparing these climate graphs helps us understand the different seasonal conditions that control plant and animal lives in different biomes.

1. What are the maximum and minimum temperatures in each of the three locations shown?
2. What do these temperatures correspond to in Fahrenheit? (*Hint:* look at the table in the back of your book).
3. Which area has the wettest climate; which is driest?
4. How do the maximum and minimum monthly rainfalls in San Diego and Belém compare?
5. Describe these three climates.
6. What kinds of biomes would you expect to find in these areas?

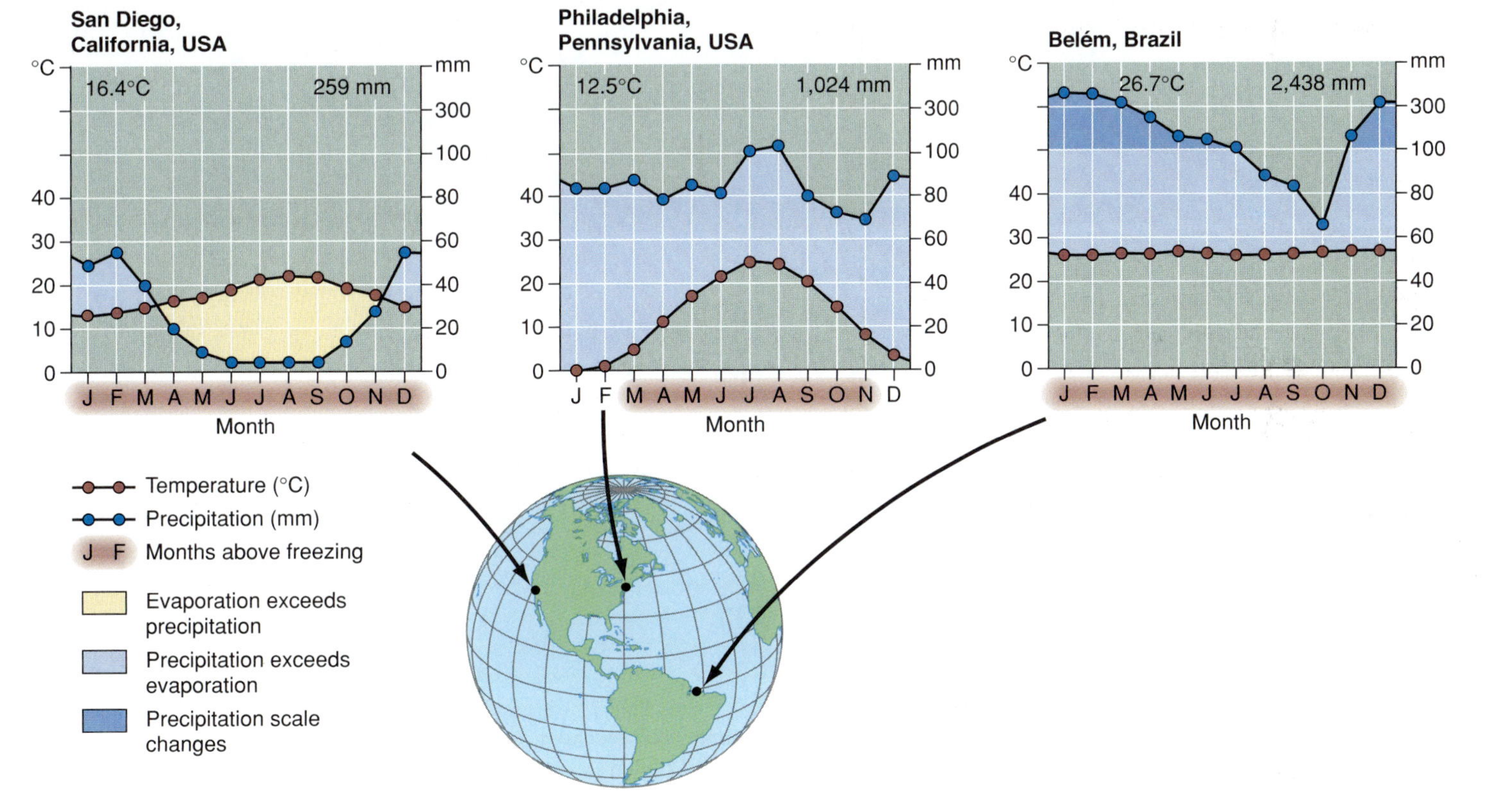


Moisture availability depends on temperature as well as precipitation. The horizontal axis on these climate diagrams shows months of the year; vertical axes show temperature (*left side*) and precipitation (*right*). The number of dry months (*shaded yellow*) and wetter months (*blue*) varies with geographic location. Mean annual temperature (°C) and precipitation (mm) are shown at the top of each graph.

**For Additional Help in Studying This Chapter,** please visit our website at www.mhhe.com/cunningham10e. You will find additional practice quizzes and case studies, flashcards, regional examples, place markers for Google Earth™ mapping, and an extensive reading list, all of which will help you learn environmental science.

Crew members sort fish on a trawler. As the large predators, such as cod, have been exhausted, we turn our attention to smaller prey. Some call this fishing down the food chain.

# CHAPTER 6
# Population Biology

*Nature teaches more than she preaches.*

*—John Burroughs—*

## LEARNING OUTCOMES

After studying this chapter, you should be able to:

**6.1** Describe the dynamics of population growth.

**6.2** Summarize the factors that increase or decrease populations.

**6.3** Compare and contrast the factors that regulate population growth.

**6.4** Identify some applications of population dynamics in conservation biology.

# Case Study How Many Fish in the Sea?

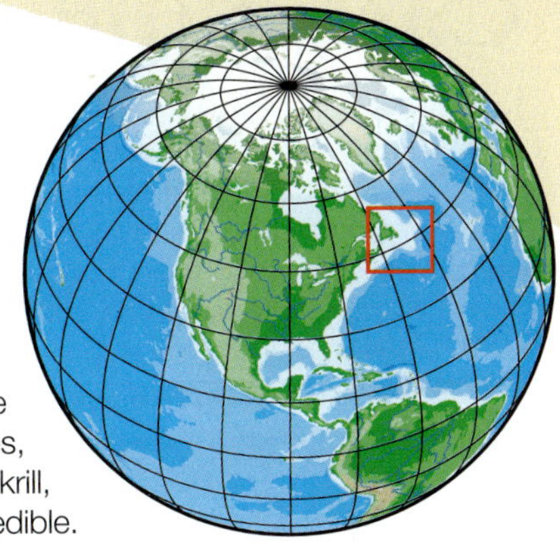

When John Cabot discovered Newfoundland in 1497, cod (*Gadus morhua*) were so abundant that sailors could simply scoop them up in baskets. Growing up to 2 meters (6 feet) long, weighing as much as 100 kg (220 lbs) and living up to 25 years, cod have been a major food resource for Europeans for more than 500 years. Because the firm, white flesh of the cod has little fat, it can be salted and dried to produce a long-lasting food that can be stored or shipped to distant markets.

No one knows how many cod there may have once been in the ocean. Coastal people recognized centuries ago that huge schools would gather to spawn on shoals and rocky reefs from Massachusetts around the North Atlantic to the British Isles. In 1990, Canadian researchers watched on sonar as a school estimated to contain several hundred million fish spawned on the Gorges Bank off Newfoundland. Because a single mature female cod can lay up to 10 million eggs in a spawning, a school like this—only one of many in the ocean—might have produced a quadrillion eggs.

It seems that such an abundant and fecund an animal could never be threatened by humans. In 1883, Thomas Huxley, the eminent biologist and friend of Charles Darwin, said, "I believe that the cod fishery . . . and probably all the great sea fisheries are inexhaustible . . . Nothing we do seriously affects the number of fish." But in Huxley's time, most cod were caught on handlines by fishermen in small wooden dories. He couldn't have imagined the size and efficiency of modern fishing fleets. Following World War II, fishing boats grew larger, more powerful, and more numerous, while their fish-finding and harvesting technology grew tremendously more effective.

Modern trawlers now pull nets with mouths large enough to engulf a dozen jumbo jets at a time. Heavy metal doors, connected by a thick metal chain, hold the net down on the ocean floor, where it crushes bottom-dwelling organisms and reduces habitat to rubble. A single pass of the trawler not only can scoop up millions of fish, it leaves a devastated community that may take decades to repair. Some environmental groups have called for a complete ban on trawling everywhere in the world.

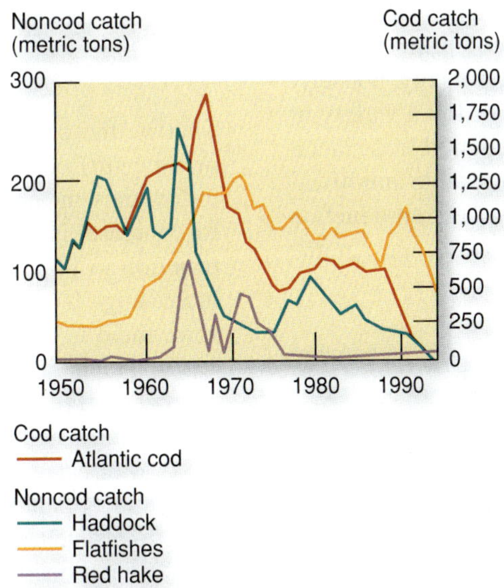


**FIGURE 6.1** Commercial harvests in the Northwest Atlantic of some important ground (bottom) fish, 1950–1995.
**Source:** World Resources Institute, 2000.

It's difficult to know how many fish are in the ocean. We can't see them easily, and often we don't even know where they are. Our estimates of population size often are based on the harvest brought in by fishing boats. Biologists warn that many marine species are overfished and in danger of catastrophic population crashes. Research shows that 90 percent of large predators such as tuna, marlin, swordfish, sharks, cod, and halibut are gone from the ocean.

Fish and seafood (including freshwater species) contribute more than 140 million metric tons of highly valued food every year, and are the main animal protein source for about one-quarter of the world population. Marine biologists note, however, that we're "fishing down the food chain." First we pursued the top predators and ground fish until they were commercially extinct, then we went after smaller fish, such as pilchard, capelin, pollock, and eels. When they became scarce, we turned to squid, skates, and other species once discarded as unwanted by-catch. Finally, we've begun harvesting invertebrates, such as sea cucumbers and krill, that many people regard as inedible.

In 2006, an international team of researchers predicted that all the world's major fish and seafood populations will collapse by 2048 if current trends in overfishing and habitat destruction continue. Marine biodiversity, they found, has declined dramatically, particularly since the 1950s. Three-fourths of all major marine fisheries are reported to be fully exploited, overfished, or severely depleted. About one-third of those species are already in collapse—defined as having catches decline 90 percent from the maximum catch. Nevertheless, scientists say, it's not too late to turn this situation around. Many fish stocks can recover quickly if we change destructive fishing practices.

Some governments already have heeded warnings about declining marine fisheries. In 1972, Iceland unilaterally declared a 200 nautical mile (370 km) exclusive economic zone that excluded all foreign fishing boats. In 2003 the Canadian government, in response to declining populations of prized ground fish (fig. 6.1) banned all trawling in the Gulf of St. Lawrence and in the Atlantic Ocean northeast of Newfoundland and Labrador. More than 40,000 Canadians lost their jobs, and many fishing towns were decimated. Marine scientists have called for similar bans in European portions of the North Atlantic, but governments there have been reluctant to impose draconian regulations. They've closed specific fisheries, such as anchovy harvest in the Bay of Biscay and sand eel fishing off Scotland, but they've only gradually reduced quotas for fisheries, such as cod, despite growing evidence of population declines.

Industry trade groups deny that there's a problem with marine fish populations. If restrictions were lifted, they argue, they could catch plenty of fish. It's true that some cod stocks, including the Barents Sea and the Atlantic around Iceland, are stable or even increasing. Establishing marine preserves, like the one around Apo Island in the Philippines, described in chapter 5, can quickly replenish many species if enough fish are available for breeding.

It's questionable, however, if some areas will ever recover their former productivity. It appears that overharvesting may have irreversibly disrupted marine ecosystems and food webs. On the Gorges Bank off the coast of Newfoundland, trillions of tiny tentacled organisms called hydroids now prey on both the organisms that once fed young cod as well as the juvenile cod themselves. Although hydroids have probably always been present, they once were held in check by adult fish.

# Case Study *continued*

Now not enough fish survive to regulate hydroid populations. Is this a shift to a permanent new state, or just a temporary situation?

This case study illustrates some of the complexities and importance of population biology. How can we predict the impacts of human actions and environmental change on different kinds organisms? What are acceptable harvest limits and minimum viable population sizes? In this chapter, we'll look at some of the factors that affect population dynamics of biological organisms.

**For more information, see**

Worm, B., et al. 2006. Impacts of biodiversity loss on ocean ecosystem services. *Science* 314 (5800):787–90.

## 6.1 Dynamics of Population Growth

Many biological organisms can produce unbelievable numbers of offspring if environmental conditions are right. Consider the common housefly (*Musca domestica*). Each female fly lays 120 eggs (assume half female) in a generation. In 56 days those eggs become mature adults, able to reproduce. In one year, with seven generations of flies being born and reproducing, that original fly would be the proud parent of 5.6 trillion offspring. If this rate of reproduction continued for ten years, the entire earth would be covered in several meters of housefly bodies. Luckily housefly reproduction, as for most organisms, is constrained in a variety of ways—scarcity of resources, competition, predation, disease, accident. The housefly merely demonstrates the remarkable amplification—the **biotic potential**—of unrestrained biological reproduction (fig. 6.2). Population dynamics describes these changes in the number of organisms in a population over time.

**FIGURE 6.2** Reproduction gives many organisms the potential to expand populations explosively. The cockroaches in this kitchen could have been produced in only a few generations. A single female cockroach can produce up to 80 eggs every six months. This exhibit is in the Smithsonian Institute's National Museum of Natural History.

### Growth without limits is exponential

As you learned in chapter 3, a population consists of all the members of a single species living in a specific area at the same time. The growth of the housefly population just described is **exponential,** having no limit and possessing a distinctive shape when graphed over time. An exponential growth rate (increase in numbers per unit of time) is expressed as a constant fraction, or exponent, which is used as a multiplier of the existing population. The mathematical formula for exponential growth is:

$$\frac{dN}{dt} = rN$$

That is, the change in numbers of individuals ($dN$) per change in time ($dt$) equals the rate of growth ($r$) times the number of individuals in the population ($N$). The $r$ term (intrinsic capacity for increase) is a fraction representing the average individual contribution to population growth. If $r$ is positive, the population is increasing. If $r$ is negative, the population is shrinking. If $r$ is zero, there is no change, and $dN/dt = 0$.

A graph of exponential population growth is described as a **J curve** (fig. 6.3) because of its shape. As you can see, the number of individuals added to a population at the beginning of an exponential growth curve can be rather small. But within a very short time, the numbers begin to increase quickly because a fixed percentage leads to a much larger increase as the population size grows.

The exponential growth equation is a very simple model; it is an idealized, simple description of the real world. The same equation is used to calculate growth in your bank account due to compounded interest rates; achieving the potential of your savings depends on you never making a withdrawal. Just as species populations lose individuals and experience reduced biotic potential, not all of your dollars will survive to a ripe old age and contribute fully to your future cash position.

### Carrying capacity relates growth to its limits

In the real world there are limits to growth. Around 1970, ecologists developed the concept of **carrying capacity** to mean the number or biomass of animals that can be supported (without harvest) in a certain area of habitat. The concept is now used more generally to suggest a limit of sustainability that an environment has in relation to the size of a species population. Carrying capacity is helpful in understanding the population dynamics of some species, perhaps even humans.

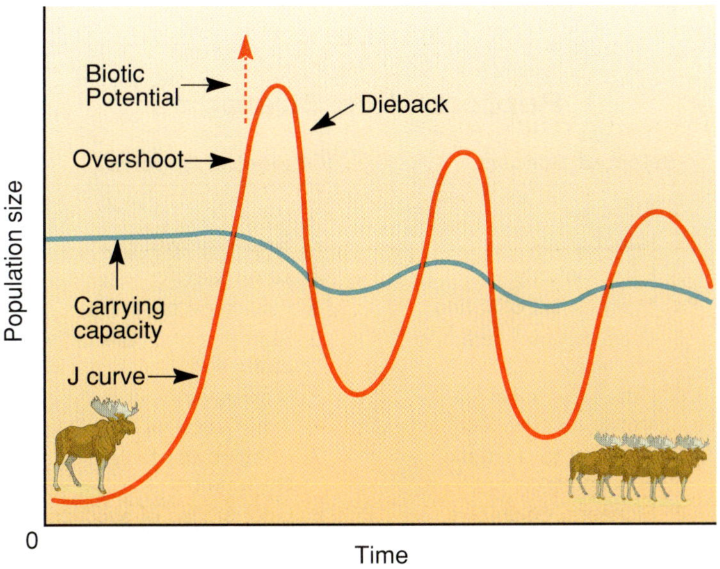


**FIGURE 6.3** J curve, or exponential growth curve, with overshoot of carrying capacity. Exponential growth in an unrestrained population (*left side of curve*) leads to a population crash and oscillations below former levels. After the overshoot, carrying capacity may be reduced because of damage to the resources of the habitat. Moose on Isle Royale in Lake Superior may be exhibiting this growth pattern in response to their changing environment.

When a population **overshoots** or exceeds the carrying capacity of its environment, resources become limited and death rates rise. If deaths exceed births, the growth rate becomes negative and the population may suddenly decrease, a change called a **population crash** or dieback. (see fig. 6.3). Some populations can go through repeated **boom-and-bust cycles** in which they repeatedly overshoot the carrying capacity of their habitat and then crash catastrophically. This oscillation can eventually lower the environmental carrying capacity for an entire food web.

Moose and other browsers or grazers sometimes overgraze their food plants, for example, such that future populations of herbivores in the same habitat find less preferred food to sustain them, at least until the habitat recovers. Some species go through predictable cycles if simple factors are involved, such as the seasonal light- and temperature-dependent bloom of algae in a lake. Cycles can be irregular if complex environmental and biotic relationships exist. Irregular cycles include migratory locust outbreaks in the Sahara, or tent caterpillars in temperate forests—these represent irruptive population growth. Population dynamics are also affected by the emigration of organisms from an overcrowded habitat, or immigration of individuals into new habitat, as occurred in 2005 when owls suddenly invaded the northern United States due to a food shortage in their Canadian habitat.

## Feedback produces logistic growth

Not all biological populations cycle through exponential overshoot and catastrophic dieback. Many species are regulated by both internal and external factors and come into equilibrium with their environmental resources while maintaining relatively stable population sizes. When resources are unlimited, they may even grow exponentially, but this growth slows as the carrying capacity of the environment is approached. This population dynamic is called **logistic growth** because of its constantly changing rate.

Mathematically, this growth pattern is described by the following equation, which adds a feedback term for carrying capacity ($K$) to the exponential growth equation:

$$\frac{dN}{dt} = rN\left(1 - \frac{N}{K}\right)$$

The logistic growth equation says that the change in numbers over time ($dN/dt$) equals the exponential growth rate ($rN$) times the portion of the carrying capacity ($K$) not already taken by the current population size ($N$). The term $(1 - N/K)$ establishes the relationship between population size at any given time step and the number of individuals the environment can support. Depending on whether $N$ is less than or greater than $K$, the rate of growth will be positive or negative.

The logistic growth curve has a different shape than the exponential growth curve. It is a sigmoidal-shaped, or **S curve** (fig. 6.4). It describes a population that decreases if its numbers exceed the carrying capacity of the environment. In the Data Analysis exercise at the end of this chapter, you will learn how the terms in the exponential and logistic equations influence the size of a population at any time.

Population growth rates are affected by external and internal factors. External factors are habitat quality, food availability, and interactions with other organisms. As populations grow, food becomes scarcer and competition for resources more intense. With a larger population, there is an increased risk that disease or para-

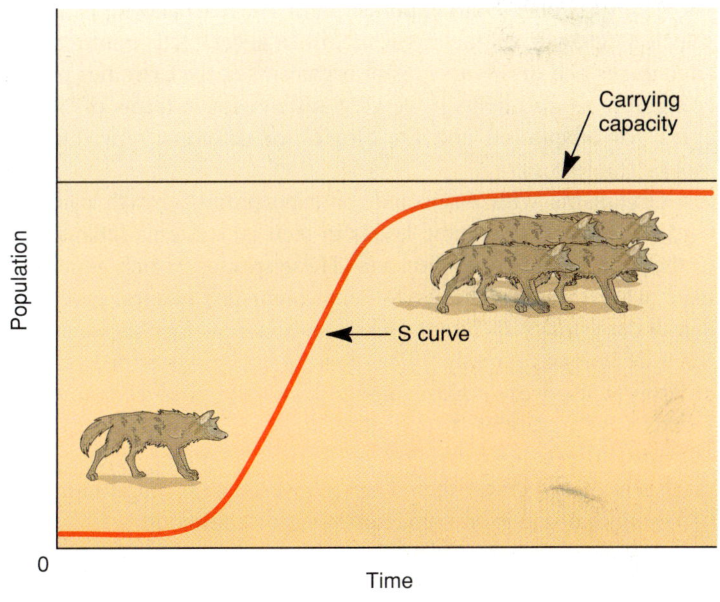


**FIGURE 6.4** S curve, or logistic growth curve, describes a population's changing numbers over time in response to feedback from the environment or its own population density. Over the long run, a conservative and predictable population dynamic may win the race over an exponential population dynamic. Species with this growth pattern tend to be *K*-selected.

sites will spread, or that predators will be attracted to the area. Some organisms become physiologically stressed when conditions are crowded, but other internal factors of maturity, body size, and hormonal status may cause them to reduce their reproductive output. Overcrowded house mice ($>1,600/m^3$), for instance, average 5.1 baby mice per litter, while uncrowded house mice ($<34/m^3$) produce 6.2 babies per litter. All these factors are **density-dependent,** meaning as population size increases, the effect intensifies. With density-independent factors, a population is affected no matter what its size. Drought, an early killing frost, flooding, landslide, or habitat destruction by people—all increase mortality rates regardless of the population size. Density-independent limits to population are often nonbiological, capricious acts of nature.

## Species respond to limits differently: *r*- and *K*-selected species

The story of the race between the hare and the tortoise has parallels to the way that species deal with limiting factors in their environment. Some organisms, such as dandelions and barnacles, depend on a high rate of reproduction and growth ($rN$) to secure a place in the environment. These organisms are called ***r*-selected species** because they employ a high reproductive rate ($r$) to overcome the high mortality of virtually ignored offspring. Without predators or diseases to control their population, those species can overshoot carrying capacity and experience population crashes, but as long as vast quantities of young are produced, a few will survive. Other organisms that reproduce more conservatively—longer generation times, late sexual maturity, fewer young—are referred to as ***K*-selected species,** because their growth slows as the carrying capacity ($K$) of their environment is approached.

Many species blend exponential (*r*-selected) and logistic (*K*-selected) growth characteristics. Still, it's useful to contrast the advantages and disadvantages of organisms at the extremes of the continuum. It also helps if we view differences in terms of "strategies" of adaptation and the "logic" of different reproductive modes (table 6.1).

Organisms with *r*-selected, or exponential, growth patterns tend to occupy low trophic levels in their ecosystems (chapter 3) or they are successional pioneers. These species, which generally have wide tolerance limits for environmental factors, and thus can occupy many different niches and habitats, are the ones we often describe as "weedy." They tend to occupy disturbed or new environments, grow rapidly, mature early, and produce many offspring with excellent dispersal abilities. As individual parents, they do little to care for their offspring or protect them from predation. They invest their energy in producing huge numbers of young and count on some surviving to adulthood.

A female clam, for example, can release up to 1 million eggs in her lifetime. The vast majority of young clams die before reaching maturity, but if even a few survive, the species will continue. Many marine invertebrates, parasites, insects, rodents, and annual plants follow this reproductive strategy. Also included in this group are most invasive and pioneer organisms, weeds, pests, and nuisance species.

**TABLE 6.1**

**Reproductive Strategies**

| *r*-Selected Species | *K*-Selected Species |
|---|---|
| 1. Short life | 1. Long life |
| 2. Rapid growth | 2. Slower growth |
| 3. Early maturity | 3. Late maturity |
| 4. Many, small offspring | 4. Few, large offspring |
| 5. Little parental care and protection | 5. High parental care or protection |
| 6. Little investment in individual offspring | 6. High investment in individual offspring |
| 7. Adapted to unstable environment | 7. Adapted to stable environment |
| 8. Pioneers, colonizers | 8. Later stages of succession |
| 9. Niche generalists | 9. Niche specialists |
| 10. Prey | 10. Predators |
| 11. Regulated mainly by intrinsic factors | 11. Regulated mainly by extrinsic factors |
| 12. Low trophic level | 12. High trophic level |

So-called *K*-selected organisms are usually larger, live long lives, mature slowly, produce few offspring in each generation, and have few natural predators. Elephants, for example, are not reproductively mature until they are 18 to 20 years old. In youth and adolescence, a young elephant belongs to an extended family that cares for it, protects it, and teaches it how to behave. A female elephant normally conceives only once every 4 or 5 years. The gestation period is about 18 months; thus, an elephant herd doesn't produce many babies in any year. Since elephants have few enemies and live a long life (60 or 70 years), this low reproductive rate produces enough elephants to keep the population stable, given good environmental conditions and no poachers.

When you consider the species you recognize from around the world, can you pigeonhole them into categories of *r*- or *K*-selected species? What strategies seem to be operating for ants, bald eagles, cheetahs, clams, dandelions, giraffes, or sharks?

***Think About It***

Which of the following strategies do humans follow: Do we more closely resemble wolves and elephants in our population growth, or does our population growth pattern more closely resemble that of moose and rabbits? Will we overshoot our environment's carrying capacity (or are we already doing so), or will our population growth come into balance with our resources?

# 6.2 Factors that Increase or Decrease Populations

Now that you have seen population dynamics in action, let's focus on what happens *within* populations, which are, after all, made up of individuals. In this section, we will discuss how new members

## What Do You Think?

### Too Many Deer?

A century ago, few Americans had ever seen a wild deer. Uncontrolled hunting and habitat destruction had reduced the deer population to about 500,000 animals nationwide. Some states had no deer at all. To protect the remaining deer, laws were passed in the 1920s and 1930s to restrict hunting, and the main deer predators—wolves and mountain lions—were exterminated throughout most of their former range.

As Americans have moved from rural areas to urban centers, forests have regrown, and deer populations have undergone explosive growth. Maturing at age two, a female deer can give birth to twin fawns every year for a decade or more. Increasing more than 20 percent annually, a deer population can double in just three years, an excellent example of irruptive, exponential growth.

Wildlife biologists estimate that the contiguous 48 states now have a population of more than 30 million white-tailed deer (*Odocoileus virginianus*), probably triple the number present in pre-Columbian times. Some areas have as many as 200 deer per square mile (77/km$^2$). At this density, woodland plant diversity is generally reduced to a few species that deer won't eat. Most deer, in such conditions, suffer from malnourishment, and many die every year of disease and starvation. Other species are diminished as well. Many small mammals and ground-dwelling birds begin to disappear when deer populations reach 25 animals per square mile. At 50 deer per square mile, most ecosystems are seriously impoverished.

A white-tailed deer (*Odocoileus virginianus*)

The social costs of large deer populations are high. In Pennsylvania alone, where deer numbers are now about 500 times greater than a century ago, deer destroy about $70 million worth of crops and $75 million worth of trees annually. Every year some 40,000 collisions with motor vehicles cause $80 million in property damage. Deer help spread Lyme disease, and, in many states, chronic wasting disease is found in wild deer herds. Some of the most heated criticisms of current deer management policies are in the suburbs. Deer love to browse on the flowers, young trees, and ornamental bushes in suburban yards. Heated disputes often arise between those who love to watch deer and their neighbors who want to exterminate them all.

In remote forest areas, many states have extended hunting seasons, increased the bag limit to four or more animals, and encouraged hunters to shoot does (females) as well as bucks (males). Some hunters criticize these changes because they believe that fewer deer will make it harder to hunt successfully and less likely that they'll find a trophy buck. Others, however, argue that a healthier herd and a more diverse ecosystem is better for all concerned.

In urban areas, increased sport hunting usually isn't acceptable. Wildlife biologists argue that the only practical way to reduce deer herds is culling by professional sharpshooters. Animal rights activists protest lethal control methods as cruel and inhumane. They call instead for fertility controls, reintroduction of predators, such as wolves and mountain lions, or trap and transfer programs. Birth control works in captive populations but is expensive and impractical with wild animals. Trapping, also, is expensive, and there's rarely anyplace willing to take surplus animals. Predators may kill domestic animals or even humans.

This case study shows that carrying capacity can be more complex than the maximum number of organisms an ecosystem can support. While it may be possible for 200 deer to survive in a square mile, there's an ecological carrying capacity lower than that if we consider the other species dependent on that same habitat. There's also an ethical carrying capacity if we don't want to see animals suffer from malnutrition and starve to death every winter. And there's a cultural carrying capacity if we consider the tolerable rate of depredation on crops and lawns or an acceptable number of motor vehicle collisions.

If you were a wildlife biologist charged with managing the deer herd in your state, how would you reconcile the different interests in this issue? How would you define the optimum deer population, and what methods would you suggest to reach this level? What social or ecological indicators would you look for to gauge whether deer populations are excessive or have reached an appropriate level?

---

are added to and old members removed from populations. We also will examine the composition of populations in terms of age classes and introduce terminology that will apply in subsequent chapters.

## Natality, fecundity, and fertility are measures of birth rates

**Natality** is the production of new individuals by birth, hatching, germination, or cloning, and is the main source of addition to most biological populations. Natality is usually sensitive to environmental conditions so that successful reproduction is tied strongly to nutritional levels, climate, soil or water conditions, and—in some species—social interactions between members of the species. The maximum rate of reproduction under ideal conditions varies widely among organisms and is a species-specific characteristic. We already have mentioned, for instance, the differences in natality between several different species.

**Fecundity** is the physical ability to reproduce, while **fertility** is a measure of the actual number of offspring produced. Because of lack of opportunity to mate and successfully produce offspring, many fecund individuals may not contribute to population growth. Human fertility often is determined by personal choice of fecund individuals.

## Immigration adds to populations

Organisms are introduced into new ecosystems by a variety of methods. Seeds, spores, and small animals may be floated on winds or water currents over long distances. This is a major route of colonization for islands, mountain lakes, and other remote locations. Sometimes organisms are carried as hitchhikers in the fur, feathers, or intestines of animals traveling from one place to another. They also may ride on a raft of drifting vegetation. Some animals travel as adults—flying, swimming, or walking. In some ecosystems, a population is maintained only by a constant influx of immigrants. Salmon, for instance, must be important predators in some parts of the ocean, but their numbers are maintained only by recruitment from mountain streams thousands of kilometers away.

## Mortality and survivorship measure longevity

An organism is born and eventually it dies; it is mortal. **Mortality,** or death rate, is determined by dividing the number of organisms that die in a certain time period by the number alive at the beginning of the period.

Since the number of survivors is more important to a population than is the number that died, mortality is often better expressed in terms of **survivorship** (the percentage of a cohort that survives to a certain age) or **life expectancy** (the probable number of years of survival for an individual of a given age). Aging has an interesting effect on life expectancy. For each year you survive, your life expectancy increases. If you were one year old in 2000, for example, you could expect to live, on average, 76.7 years more. But if you had already reached age 75 that year, rather than having only 1.7 more years to live, you could expect to survive another 11.2 years, on average. How could this be? At age 1, many in your cohort were likely to die early for one reason or another. By age 75, most of those individuals are already dead, giving the rest of you a longer average life probability. Even at age 100, long past your starting life expectancy, you would still have 2.7 more years to live, on average.

Human life expectancies have risen dramatically nearly everywhere in the world over the past century. In 1900 the world average life expectancy was only about 30 years, which was not much higher than the average lifespan in the Roman Empire 2,000 years earlier. By 2006, the average was 64.3 years (fig. 6.5). Improved nutrition, sanitation, and medical care were responsible for most of that increase. Demographers wonder how much more life expectancies can increase. Notice the great discrepancy in life expectancies between rich and poor countries. Currently, microstates Andorra, San Marino, and Singapore have the world's highest life expectancies (83.5, 82.1, and 81.6 years, respectively). Japan is nearly as high with a countrywide average of 81.5 years. The lowest national life expectancies are in Africa, where diseases, warfare, poverty, and famine cause many early deaths. In Swaziland, Botswana, and Lesotho, for example, the average person lives only 32.6, 33.7, and 34.4 years, respectively. In many African countries AIDS has reduced life expectancies by about 25 percent in the past two decades. This can be seen in the lag in progress in life expectancies between 1980 and 2000 in these countries in fig. 6.5.

Large discrepancies also exist in the United States. While the nation-wide average life expectancy is 77.5 years, Asian American women in Bergen County, New Jersey, live 91 years on average, while Native American men on the Pine Ridge Reservation in South Dakota are reported to typically live only 48 years. Two-thirds of African countries have life expectancies greater than Pine Ridge. Women almost always have higher life expectancies than men. Worldwide, the average difference between sexes is 3 years, but in Russia the difference between men and women is 13 years. Is this because women are biologically superior to men, and thus live longer? Or is it simply that men are generally employed in more hazardous occupations and often engage in more dangerous behaviors (drinking, smoking, reckless driving)?

**Life span** is the longest period of life reached by a given type of organism. The process of living entails wear and tear that eventually overwhelm every organism, but maximum age is dictated

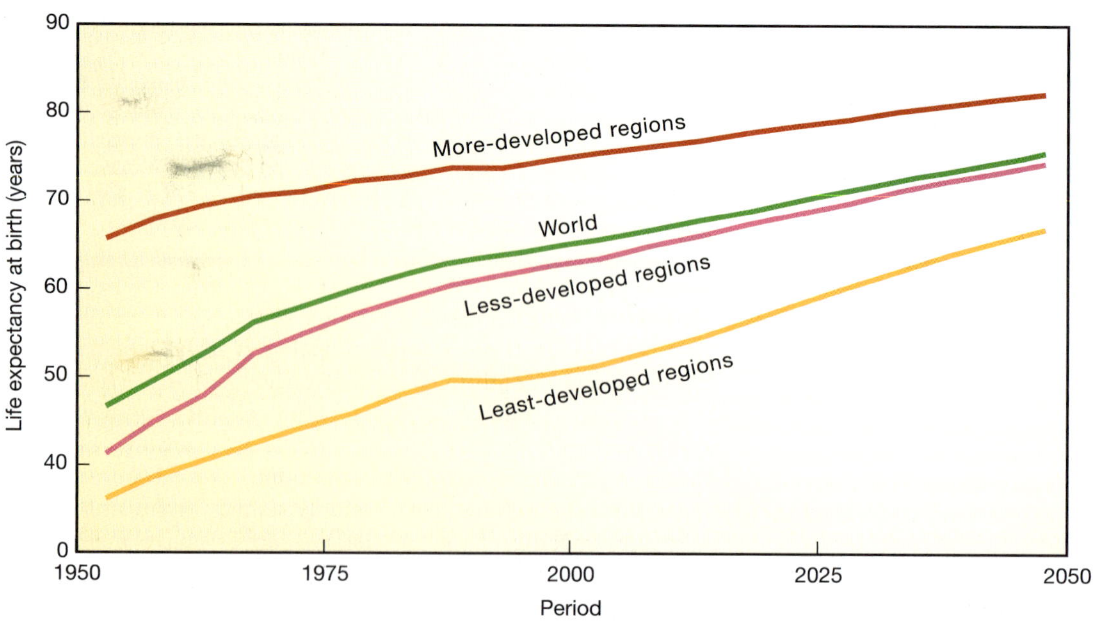


**FIGURE 6.5** Life expectancy has increased nearly everywhere in the world, but the increase has lagged in the least-developed countries.
Data from the Population Division of the United Nations, 2006.

primarily by physiological aspects of the organism itself. There is an enormous difference in life span between different species. Some microorganisms live their whole life cycles in a matter of hours or minutes. Bristlecone pine trees in the mountains of California, on the other hand, have life spans up to 4,600 years. The maximum life span for humans appears to be about 120 years.

Most organisms do not live anywhere near the maximum life span for their species. The major factors in early mortality are predation, parasitism, disease, accidents, fighting, and environmental influences, such as climate and nutrition. Important differences in relative longevity among different species are reflected in the survivorship curves shown in figure 6.6.

Four general patterns of survivorship can be seen in this idealized figure. Curve (*a*) is the pattern of organisms that tend to live their full physiological life span if they reach maturity and then have a high mortality rate when they reach old age. This pattern is typical of many large mammals, such as whales, bears, and elephants (when not hunted by humans), as well as humans in developed countries. Interestingly, some very small organisms, including predatory protozoa and rotifers (small, multicellular, freshwater animals), have similar survivorship curves even though their maximum life spans may be hundreds or thousands of times shorter than those of large mammals. In general, curve (*a*) is the pattern for top consumers in an ecosystem, although many annual plants have a similar survivorship pattern.

Curve (*b*) represents the survivorship pattern for organisms for which the probability of death is generally unrelated to age. Sea gulls, for instance, die from accidents, poisoning, and other factors that act more or less randomly. Their mortality rate is generally constant with age, and their survivorship curve is a straight line.

Curve (*c*) is characteristic of many songbirds, rabbits, members of the deer family, and humans in less-developed countries (see chapter 7). They have a high mortality early in life when they are more susceptible to external factors, such as predation, disease, starvation, or accidents. Adults in the reproductive phase have a high level of survival. Once past reproductive age, they become susceptible again to external factors and the number of survivors falls quite rapidly.

Curve (*d*) is typical of organisms at the base of a food chain or those especially susceptible to mortality early in life. Many tree species, fish, clams, crabs, and other invertebrate species produce a very large number of highly vulnerable offspring, few of which survive to maturity. Those individuals that do survive to adulthood, however, have a very high chance of living most of the maximum life span for the species. Figure 6.7 shows some examples of organisms with each of these survivorship patterns.

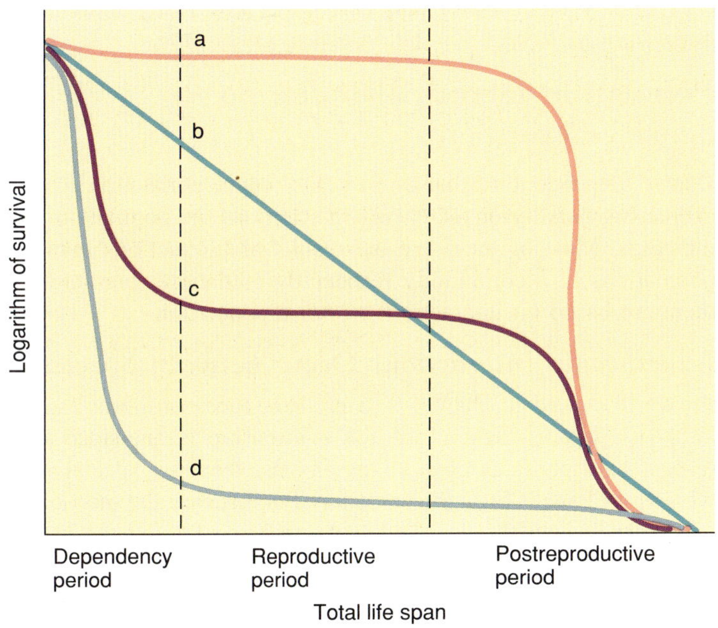


**FIGURE 6.6** Four basic types of survivorship curves for organisms with different life histories. Curve (a) represents organisms such as humans or whales, which tend to live out the full physiological life span if they survive early growth. Curve (b) represents organisms such as sea gulls, which have a fairly constant mortality at all age levels. Curve (c) represents such organisms as white-tailed deer, which have high mortality rates in early and late life. Curve (d) represents such organisms as clams and redwood trees, which have a high mortality rate early in life but live a full life if they reach adulthood.

**FIGURE 6.7** Different survivorship patterns. (a) Most elephants survive to old age. (b) Seagulls die randomly at all ages from accidents. (c) Antelope have high infant mortality, but adults survive well. (d) Redwood trees have very high seedling losses, but mature trees live thousands of years.

**Think About It**

Which of these survivorship patterns best describes humans? Are we more like elephants or deer? Do wealth and modernity have something to do with it? Might people in Bangladesh have different survivorship prospects than you do?

## Emigration removes members of a population

**Emigration,** the movement of members out of a population, is the second major factor that reduces population size. The dispersal factors that allow organisms to migrate into new areas are important in removing surplus members from the source population. Emigration can even help protect a species. For instance, if the original population is destroyed by some catastrophic change in their environment, their genes still are carried by descendants in other places. Many organisms have very specific mechanisms to facilitate migration of one or more of each generation of offspring.

# 6.3 Factors that Regulate Population Growth

So far, we have seen that differing patterns of natality, mortality, life span, and longevity can produce quite different rates of population growth. The patterns of survivorship and age structure created by these interacting factors not only show us how a population is growing but also can indicate what general role that species plays in its ecosystem. They also reveal a good deal about how that species is likely to respond to disasters or resource bonanzas in its environment. But what factors *regulate* natality, mortality, and the other components of population growth? In this section, we will look at some of the mechanisms that determine how a population grows.

Various factors regulate population growth, primarily by affecting natality or mortality, and can be classified in different ways. They can be *intrinsic* (operating within individual organisms or between organisms in the same species) or *extrinsic* (imposed from outside the population). Factors can also be either **biotic** (caused by living organisms) or **abiotic** (caused by nonliving components of the environment). Finally, the regulatory factors can act in a *density-dependent* manner (effects are stronger or a higher proportion of the population is affected as population density increases) or *density-independent* manner (the effect is the same or a constant proportion of the population is affected regardless of population density).

In general, biotic regulatory factors tend to be density-dependent, while abiotic factors tend to be density-independent. There has been much discussion about which of these factors is most important in regulating population dynamics. In fact, it probably depends on the particular species involved, its tolerance levels, the stage of growth and development of the organisms involved, the specific ecosystem in which they live, and the way combinations of factors interact. In most cases, density-dependent and density-independent factors probably exert simultaneous influences. Depending on whether regulatory factors are regular and predictable or irregular and unpredictable, species will develop different strategies for coping with them.

## Population factors can be density-independent

In general, the factors that affect natality or mortality independently of population density tend to be abiotic components of the ecosystem. Often weather (conditions at a particular time) or climate (average weather conditions over a longer period) are among the most important of these factors. Extreme cold or even moderate cold at the wrong time of year, high heat, drought, excess rain, severe storms, and geologic hazards—such as volcanic eruptions, landslides, and floods—can have devastating impacts on particular populations.

Abiotic factors can have beneficial effects as well, as anyone who has seen the desert bloom after a rainfall can attest. Fire is a powerful shaper of many biomes. Grasslands, savannas, and some montane and boreal forests often are dominated—even created—by periodic fires. Some species, such as jack pine and Kirtland's warblers, are so adapted to periodic disturbances in the environment that they cannot survive without them.

In a sense, these density-independent factors don't really regulate population *per se,* since regulation implies a homeostatic feedback that increases or decreases as density fluctuates. By definition, these factors operate without regard to the number of organisms involved. They may have such a strong impact on a population, however, that they completely overwhelm the influence of any other factor and determine how many individuals make up a particular population at any given time.

## Population factors also can be density-dependent

Density-dependent mechanisms tend to reduce population size by decreasing natality or increasing mortality as the population size increases. Most of them are the results of interactions *between* populations of a community (especially predation), but some of them are based on interactions *within* a population.

### *Interspecific Interactions Occur between Species*

As we discussed in chapter 4, a predator feeds on—and usually kills—its prey species. While the relationship is one-sided with respect to a particular pair of organisms, the prey species as a whole may benefit from the predation. For instance, the moose that gets eaten by wolves doesn't benefit individually, but the moose *population* is strengthened because the wolves tend to kill old or sick members of the herd. Their predation helps prevent population overshoot, so the remaining moose are stronger and healthier.

Sometimes predator and prey populations oscillate in a sort of synchrony with each other as is shown in figure 6.8, which shows the number of furs brought into Hudson Bay Company trading posts in Canada between 1840 and 1930. As you can see, the numbers of Canada lynx fluctuate on about a ten-year cycle that is similar to, but slightly out of phase with, the population peaks of snowshoe hares. Although there are some doubts now about

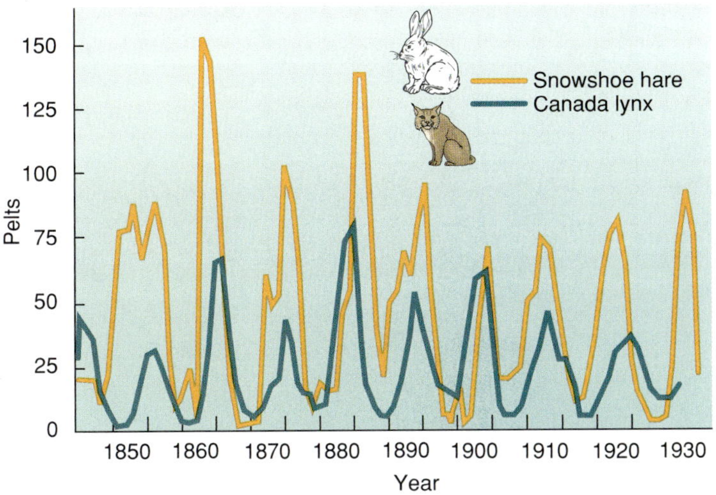


**FIGURE 6.8** Ten-year oscillations in the populations of snowshoe hare and lynx in Canada suggest a close linkage of predator and prey, but may not tell the whole story. These data are based on the number of pelts received by the Hudson Bay Company each year, meaning fur-traders were unwitting accomplices in later scientific research.
**Source:** Data from D. A. MacLulich, *Fluctuations in the Numbers of the Varying Hare* (Lepus americus).Toronto: University of Toronto Press, 1937, reprinted 1974.

how and where these data were collected, this remains a classic example of population dynamics. When prey populations (hares) are abundant, predators (lynx) reproduce more successfully and their population grows. When hare populations crash, so do the lynx. This predator-prey oscillation is known as the Lotka-Volterra model after the scientists who first described it mathematically.

Not all interspecific interactions are harmful to one of the species involved. Mutualism and commensalism, for instance, are interspecific interactions that are beneficial or neutral in terms of population growth (chapter 4).

### *Intraspecific Interactions Occur within Species*

Individuals within a population also compete for resources. When population density is low, resources are likely to be plentiful and the population growth rate will approach the maximum possible for the species, assuming that individuals are not so dispersed that they cannot find mates. As population density approaches the carrying capacity of the environment, however, one or more of the vital resources becomes limiting. The stronger, quicker, more aggressive, more clever, or luckier members get a larger share, while others get less and then are unable to reproduce successfully or survive.

Territoriality is one principal way many animal species control access to environmental resources. The individual, pair, or group that holds the territory will drive off rivals if possible, either by threats, displays of superior features (colors, size, dancing ability), or fighting equipment (teeth, claws, horns, antlers). Members of the opposite sex are attracted to individuals that are able to seize and defend the largest share of the resources. From a selective point of view, these successful individuals presumably represent superior members of the population and the ones best able to produce offspring that will survive.

**FIGURE 6.9** Animals often battle over resources. This conflict can induce stress and affect reproductive success.

### *Stress and Crowding Can Affect Reproduction*

Stress and crowding also are density-dependent population control factors. When population densities get very high, organisms often exhibit symptoms of what is called stress shock or **stress-related diseases.** These terms describe a loose set of physical, psychological, and/or behavioral changes that are thought to result from the stress of too much competition and too close proximity to other members of the same species. There is a considerable controversy about what causes such changes and how important they are in regulating natural populations. The strange behavior and high mortality of arctic lemmings or hares during periods of high population density may be a manifestation of stress shock (fig. 6.9). On the other hand, they could simply be the result of malnutrition, infectious disease, or some other more mundane mechanism at work.

Some of the best evidence for the existence of stress-related disease comes from experiments in which laboratory animals, usually rats or mice, are grown in very high densities with plenty of food and water but very little living space. A variety of symptoms are reported, including reduced fertility, low resistance to infectious diseases, and pathological behavior. Dominant animals seem to be affected least by crowding, while subordinate animals—the ones presumably subjected to the most stress in intraspecific interactions—seem to be the most severely affected.

***Case Study A Plague of Locusts*** *Schistocerca gregarius*, the desert locust, has been called the world's most destructive insect. Throughout recorded human history, locust plagues have periodically swarmed out of deserts and into settled areas. Their impact on human lives has often been so disruptive that records of plagues have taken on religious significance and made their way into sacred and historical texts.

Locusts usually are solitary creatures resembling ordinary grasshoppers. Every few decades, however, when rain comes to the desert and vegetation flourishes, locusts reproduce rapidly until the ground is literally crawling with bugs. High population densities and stress bring ominous changes in these normally innocuous insects. They stop reproducing, grow longer wings, group together in enormous swarms, and begin to move across the desert. Dense clouds of insects darken the sky, moving as much as 100 km per day. Locusts may be small, but they can eat their own body weight of vegetation every day. A single swarm can cover 1,200 $km^2$ and contain 50 to 100 billion individuals. The swarm can strip pastures, denude trees, and destroy crops in a matter of hours, consuming as much food in a day as 500,000 people would need for a year. Eventually, having exhausted their food supply and migrated far from the desert where conditions favor reproduction, the locusts die and aren't seen again for decades.

Huge areas of crops and rangeland in northern Africa, the Middle East, and Asia are within the reach of the desert locust. This small insect, with its voracious appetite, can affect the livelihood of at least one-tenth of the world's population. During quiet periods, called recessions, African locusts are confined to the Sahara Desert, but when conditions are right, swarms invade countries as far away as Spain, Russia, and India. Swarms are even reported to have crossed the Atlantic Ocean from Africa to the Caribbean.

Unusually heavy rains in the Sahara in 2004 created the conditions for a locust explosion. Four generations bred in rapid succession, and swarms of insects moved out of the desert. Twenty-eight countries in Africa and the Mediterranean area were afflicted. Crop losses reached 100 percent in some places, and food supplies for millions of people were threatened. Officials at the United Nations warned that we could be headed toward another great plague. Hundreds of thousands of hectares of land were treated with pesticides, but millions of dollars of crop damage were reported anyway.

This case study illustrates the power of exponential growth and the disruptive potential of a boom-and-bust life cycle. Stress, population density, migration, and intraspecific interactions all play a role in this story. Although desert conditions usually keep locust numbers under control, their biotic potential for reproduction is a serious worry for residents of many countries.

## 6.4 Conservation Biology

Small, isolated populations can undergo catastrophic declines due to environmental change, genetic problems, or stochastic (random or unpredictable) events. A critical question in conservation biology is the minimum population size of a rare and endangered species required for long-term viability. In this section, we'll look at some factors that limit species and genetic diversity. We'll also consider the interaction of collections of subpopulations of species in fragmented habitats.

### Island biogeography describes isolated populations

In a classic 1967 study, R. H. MacArthur and E. O. Wilson asked why it is that small islands far from the mainland generally have far fewer species than larger or nearer islands. Their theory of **island biogeography** explains that diversity in isolated habitats is a balance between colonization and extinction rates. An island far from a population source has a lower colonization rate for terrestrial species because it is harder for organisms to reach (fig. 6.10). At

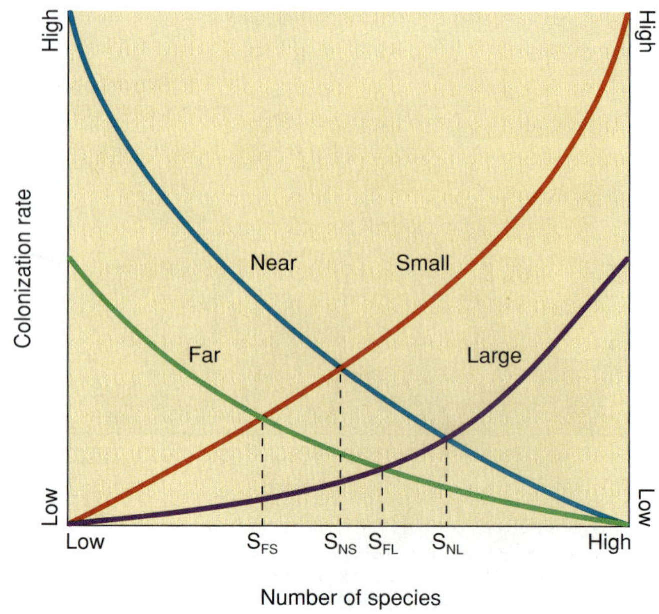


**FIGURE 6.10** Predicted species richness on an island resulting from a balance between colonization (immigration) and extinction by natural causes. This island biogeography theory of MacArthur and Wilson (1967) is used to explain why large islands near a mainland ($S_{NL}$) tend to have more species than small, far islands ($S_{FS}$).
**Source:** Based on MacArthur and Wilson, *The Theory of Island Biogeography,* 1967, Princeton University Press.

the same time, the limited habitat on a small island can support fewer individuals of any given species. This creates a greater probability that a species could go extinct due to natural disasters, diseases, or demographic factors such as imbalance between sexes in a particular generation. Larger islands, or those closer to the mainland, on the other hand, are more likely to be colonized or to retain those species already successfully established. Thus they tend to have greater diversity than smaller, more remote places.

Island biogeographical effects have been observed in many places. Cuba, for instance, is 100 times as large and has about 10 times as many amphibian species as its Caribbean neighbor, Monserrat. Similarly, in a study of bird species on the California Channel Islands, Jared Diamond observed that on islands with fewer than 10 breeding pairs, 39 percent of the populations went extinct over an 80-year period. In contrast, only 10 percent of populations numbering between 10 and 100 pairs went extinct, and no species with over 1,000 pairs disappeared over this time (fig. 6.11). This theory of a balance between colonization and extinction is now seen to explain species dynamics in many small, isolated habitat fragments whether on islands or not.

### Conservation genetics is important in survival of endangered species

Genetics plays an important role in the survival or extinction of small, isolated populations. In large populations, genetic variation

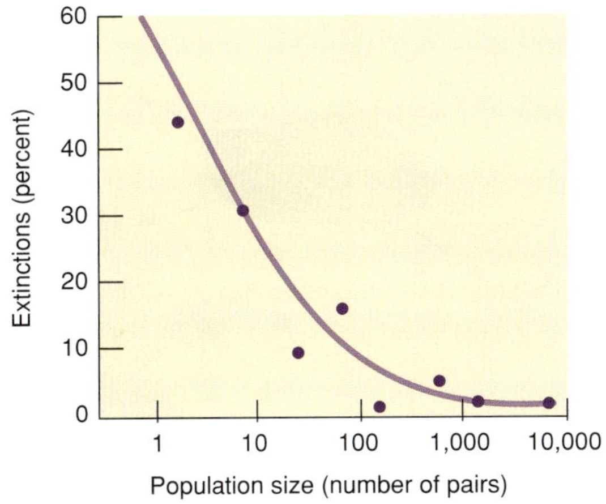


**FIGURE 6.11** Extinction rates of bird species on the California Channel Islands as a function of population size over 80 years.
**Source:** H. L. Jones and J. Diamond, "Short-term-base studies of turnover in breeding bird populations on the California coast island," in *Condor,* vol. 78:526–49, 1976.

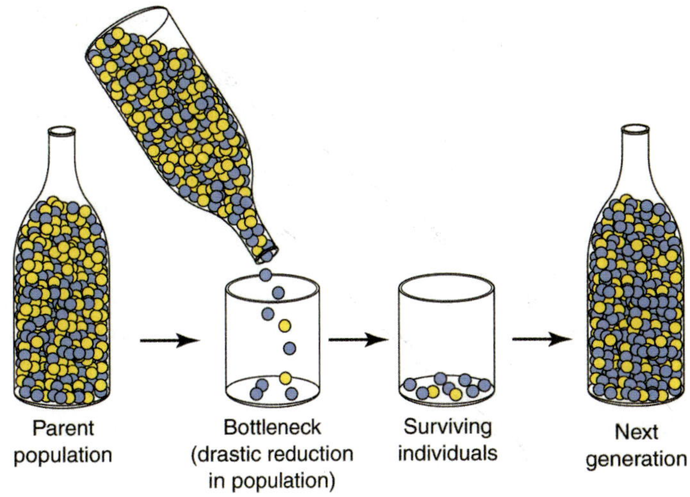


**FIGURE 6.12** Genetic drift: the bottleneck effect. The parent population contains roughly equal numbers of blue and yellow individuals. By chance, the few remaining individuals that comprise the next generation are mostly blue. The bottleneck occurs because so few individuals form the next generation, as might happen after an epidemic or catastrophic storm.

tends to persist in what is called a Hardy-Weinberg equilibrium, named after the scientists who first described why this occurs. If mating is random, no mutations (changes in genetic material) occur, and there is no gene in-flow or selective pressure for or against particular traits, random distribution of gene types (alleles) will occur during gamete formation and sexual reproduction. That is, different alleles will be distributed in the offspring in the same ratio they occur in the parents, and genetic diversity is preserved.

In a large population, these conditions for maintaining genetic equilibrium are generally operative. The addition or loss of a few individuals or appearance of new genotypes makes little difference in the total gene pool, and genetic diversity is relatively constant. In small, isolated populations, however, immigration, mortality, mutations, or chance mating events involving only a few individuals can greatly alter the genetic makeup of the whole population. We call the gradual changes in gene frequencies due to random events **genetic drift.**

For many species, loss of genetic diversity causes a number of harmful effects that limit adaptability, reproduction, and species survival. A **founder effect** or **demographic bottleneck** occurs when just a few members of a species survive a catastrophic event or colonize new habitat geographically isolated from other members of the same species. Any deleterious genes present in the founders will be overrepresented in subsequent generations (fig. 6.12). Inbreeding, mating of closely related individuals, also makes expression of rare or recessive genes more likely.

Some species seem not to be harmed by inbreeding or lack of genetic diversity. The northern elephant seal, for example, was reduced by overharvesting a century ago, to fewer than 100 individuals. Today there are more than 150,000 of these enormous animals along the Pacific coast of Mexico and California. No marine mammal is known to have come closer to extinction and then made such a remarkable recovery. All northern elephant seals today appear to be essentially genetically identical and yet they seem to have no apparent problems. Although interpretations of their situation are controversial, in highly selected populations, where only the most fit individuals reproduce, or in which there are few deleterious genes, inbreeding and a high degree of genetic identity may not be such a negative factor.

Cheetahs, also appear to have undergone a demographic bottleneck sometime in the not-too-distant past. All the male cheetahs throughout the world appear to be nearly genetically identical,

**FIGURE 6.13** Sometime in the past, cheetahs underwent a severe population crash. Now all male cheetahs in the world are nearly genetically identical, and deformed sperm, low fertility levels, and low infant survival are common in the species.

suggesting that they all share a single male ancestor (fig. 6.13). This lack of diversity is thought to be responsible for an extremely low fertility rate, a high abundance of abnormal sperm, and low survival rate for offspring, all of which threatens the survival of the species.

## Population viability analysis calculates chances of survival

Conservation biologists use the concepts of island biogeography, genetic drift, and founder effects to determine **minimum viable population size,** or number of individuals needed for long-term survival of rare and endangered species. A classic example is that of the grizzly bear (*Ursus arctos horribilis*) in North America. Before European settlement, grizzlies roamed from the Great Plains west to California and north to Alaska. Hunting and habitat destruction reduced the number of grizzlies from an estimated 100,000 in 1800 to less than 1,200 animals in six separate subpopulations that now occupy less than 1 percent of the historic range. Recovery target sizes—based on estimated environmental carrying capacities—call for fewer than 100 bears for some subpopulations. Conservation genetics predicts that a completely isolated population of 100 bears cannot be maintained for more than a few generations. Even the 600 bears now in Yellowstone National Park will be susceptible to genetic problems if completely isolated. Interestingly, computer models suggest that translocating only two unrelated bears into small populations every generation (about ten years) could greatly increase population viability.

## Metapopulations are important interconnections

For mobile organisms, separated populations can have gene exchange if suitable corridors or migration routes exist. A **metapopulation** is a collection of populations that have regular or intermittent gene flow between geographically separate units (fig. 6.14). For example, the Bay checkerspot butterfly (*Euphydrays editha bayensis*) in California exists in several distinct habitat patches. Individuals occasionally move among these patches, mating with existing animals or recolonizing empty habitats. Thus, the apparently separate groups form a functional metapopulation.

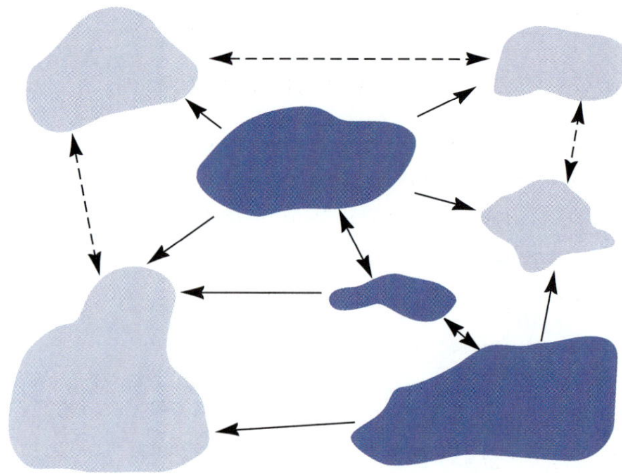

**FIGURE 6.14** A metapopulation is composed of several local populations linked by regular (*solid arrows*) or occasional (*dashed lines*) gene flows. Source populations (*dark*) provide excess individuals, which emigrate to and colonize sink habitats (*light*).

A "source" habitat, where birth rates are higher than death rates, produces surplus individuals that can migrate to new locations within a metapopulation. "Sink" habitats, on the other hand, are places where mortality exceeds birth rates. Sinks may be spatially larger than sources but because of unfavorable conditions, the species would disappear in the sink habitat if it were not periodically replenished from a source population. In general, the larger a reserve is, the better it is for endangered species. Sometimes, however, adding to a reserve can be negative if the extra area is largely sink habitat. Individuals dispersing within the reserve may settle in unproductive areas if better habitat is hard to find. Recent studies using a metapopulation model for spotted owls predict just such a problem for this species in the Pacific Northwest.

Some conservation biologists argue that we ought to try to save every geographically distinct population or "evolutionarily significant unit (ESU)" possible in order to preserve maximum genetic diversity. Paul Ehrlich and Gretchen Daily estimate there may be an average of 220 ESU for every one of the 10 to 50 billion species in the world. Saving all of them would be a gargantuan task.

# Conclusion

Given optimum conditions, populations of many organisms can grow exponentially; that is, they can expand at a constant *rate* per unit of time. This biotic potential can produce enormous populations that far surpass the carrying capacity of the environment if left unchecked. Obviously, no population grows at this rate forever. Sooner or later, predation, disease, starvation, or some other factor will cause the population to crash. Not all species follow this boom-and-bust pattern, however. Most top predators have intrinsic factors that limit their reproduction and prevent overpopulation.

Overharvesting of species, habitat destruction, predator elimination, introduction of exotic species, and other forms of human disruption can also drive populations to boom and/or crash. Population dynamics are an important part of conservation biology. Principles, such as island biogeography, genetic drift, demographic bottlenecks, and metapopulation interactions are critical in endangered species protection.

More than 1.1 billion people live in India, more than one-quarter of them in object poverty.

# CHAPTER 7
# Human Populations

*Live simply so that others may simply live.*

*—Mahatma Gandhi—*

## LEARNING OUTCOMES

After studying this chapter, you should be able to:

**7.1** Trace the history of human population growth.
**7.2** Summarize different perspectives on population growth.
**7.3** Analyze some of the factors that determine population growth.
**7.4** Explain how ideal family size is culturally and economically dependent.
**7.5** Describe how a demographic transition can lead to stable population size.
**7.6** Relate how family planning gives us choices.
**7.7** Reflect on what kind of future we are creating.

# Case Study Family Planning in Thailand: A Success Story

Down a narrow lane off Bangkok's busy Sukhumvit Road, is a most unusual café. Called Cabbages and Condoms, it's not only highly rated for its spicy Thai food, but it's also the only restaurant in the world dedicated to birth control. In an adjoining gift shop, baskets of condoms stand next to decorative handicrafts of the northern hill tribes. Piles of T-shirts carry messages, such as, "A condom a day keeps the doctor away," and "Our food is guaranteed not to cause pregnancy." Both businesses are run by the Population and Community Development Association (PDA), Thailand's largest and most influential nongovernmental organization.

The PDA was founded in 1974 by Mechai Viravaidya, a genial and fun-loving former Thai Minister of Health, who is a genius at public relations and human motivation (fig. 7.1). While traveling around Thailand in the early 1970s, Mechai recognized that rapid population growth—particularly in poor rural areas—was an obstacle to community development. Rather than lecture people about their behavior, Mechai decided to use humor to promote family planning. PDA workers handed out condoms at theaters and traffic jams, anywhere a crowd gathered. They challenged governmental officials to condom balloon-blowing contests, and taught youngsters Mechai's condom song: "Too Many Children Make You Poor." The PDA even pays farmers to paint birth control ads on the sides of their water buffalo.

This campaign has been extremely successful at making birth control and family planning, which once had been taboo topics in polite society, into something familiar and unembarrassing. Although condoms—now commonly called "mechais" in Thailand—are the trademark of PDA, other contraceptives, such as pills, spermicidal foam, and IUDs, are promoted as well. Thailand was one of the first countries to allow the use of the injectable contraceptive DMPA, and remains a major user. Free non-scalpel vasectomies are available on the king's birthday. Sterilization has become the most widely used form of contraception in the country. The campaign to encourage condom use has also been helpful in combating AIDS.

In 1974, when PDA started, Thailand's growth rate was 3.2 percent per year. In just fifteen years, contraceptive use among married couples increased from 15 to 70 percent, and the growth rate had dropped to 1.6 percent, one of the most dramatic birth rate declines ever recorded. Now Thailand's growth rate is 0.7 percent, or nearly the same as the United States. The fertility rate (or average number of children per woman) decreased from 7 in 1974 to 1.7 in 2006. The PDA is credited with the fact that Thailand's population is 20 million less than it would have been if it had followed its former trajectory.

**FIGURE 7.1** Mechai Viravaidya (right) is joined by Peter Piot, Executive Director of UNAIDS, in passing out free condoms on family planning and AIDS awareness day in Bangkok.

In addition to Mechai's creative genius and flair for showmanship, there are several reasons for this success story. Thai people love humor and are more egalitarian than most developing countries. Thai spouses share in decisions regarding children, family life, and contraception. The government recognizes the need for family planning and is willing to work with volunteer organizations, such as the PDA. And Buddhism, the religion of 95 percent of Thais, promotes family planning.

The PDA hasn't limited itself to family planning and condom distribution. It has expanded into a variety of economic development projects. Microlending provides money for a couple of pigs, or a bicycle, or a small supply of goods to sell at the market. Thousands of water-storage jars and cement rainwater-catchment basins have been distributed. Larger scale community development grants include road building, rural electrification, and irrigation projects. Mechai believes that human development and economic security are keys to successful population programs.

This case study introduces several important themes of this chapter. What might be the effects of exponential growth in human populations? How might we manage fertility and population growth? And what are the links between poverty, birth rates, and our common environment? Keep in mind, as you read this chapter, that resource limits aren't simply a matter of total number of people on the planet, they also depend on consumption levels and the types of technology used to produce the things we use.

## 7.1 Population Growth

Every second, on average, four or five children are born, somewhere on the earth. In that same second, two other people die. This difference between births and deaths means a net gain of roughly 2.3 more humans per second in the world's population. In mid-2007 the total world population stood at roughly 6.6 billion people and was growing at 1.17 percent per year. This means we are now adding nearly 79 million more people per year, and if this rate persists, our global population will double in about 58 years. Humans are now probably the most numerous vertebrate species on the earth. We also are more widely distributed and manifestly have a greater global environmental impact than any other species. For the families to whom these children are born, this may well be a joyous and long-awaited event (fig. 7.2). But is a continuing increase in humans good for the planet in the long run?

Many people worry that overpopulation will cause—or perhaps already is causing—resource depletion and environmental degradation that threaten the ecological life-support systems on which we all depend. These fears often lead to demands for immediate, worldwide birth control programs to reduce fertility rates and to eventually stabilize or even shrink the total number of humans.

Others believe that human ingenuity, technology, and enterprise can extend the world carrying capacity and allow us to overcome any problems we encounter. From this perspective, more people may be beneficial rather than disastrous. A larger population means a larger workforce, more geniuses, more ideas about what to do. Along with every new mouth comes a pair of hands. Proponents of this worldview—many of whom happen to be economists—argue that continued economic and technological growth can both feed the world's billions and enrich everyone enough to end the population explosion voluntarily. Not so, counter many ecologists. Growth is the problem; we must stop both population and economic growth.

Yet another perspective on this subject derives from social justice concerns. In this worldview, there are sufficient resources for everyone. Current shortages are only signs of greed, waste, and oppression. The root cause of environmental degradation, in this view, is inequitable distribution of wealth and power rather than population size. Fostering democracy, empowering women and minorities, and improving the standard of living of the world's poorest people are what are really needed. A narrow focus on population growth only fosters racism and an attitude that blames the poor for their problems while ignoring the deeper social and economic forces at work.

Whether human populations will continue to grow at present rates and what that growth would imply for environmental quality and human life are among the most central and pressing questions in environmental science. In this chapter, we will look at some causes of population growth as well as how populations are measured and described. Family planning and birth control are essential for stabilizing populations. The number of children a couple decides to have and the methods they use to regulate fertility, however, are strongly influenced by culture, religion, politics, and economics, as well as basic biological and medical considerations. We will examine how some of these factors influence human demographics.

**FIGURE 7.2** A Mayan family in Guatemala with four of their six living children. Decisions on how many children to have are influenced by many factors, including culture, religion, need for old age security for parents, immediate family finances, household help, child survival rates, and power relationships within the family. Having many children may not be in the best interest of society at large, but may be the only rational choice for individual families.

### Human populations grew slowly until relatively recently

For most of our history, humans have not been very numerous compared to other species. Studies of hunting and gathering societies suggest that the total world population was probably only a few million people before the invention of agriculture and the domestication of animals around 10,000 years ago. The larger and more secure food supply made available by the agricultural revolution allowed the human population to grow, reaching perhaps 50 million people by 5000 B.C. For thousands of years, the number of humans increased very slowly. Archaeological evidence and historical descriptions suggest that only about 300 million people were living at the time of Christ (table 7.1).

Until the Middle Ages, human populations were held in check by diseases, famines, and wars that made life short and uncertain for most people (fig. 7.3). Furthermore, there is evidence that many early societies regulated their population size through cultural taboos and practices such as abstinence and infanticide. Among the most destructive of natural population controls were bubonic plagues (or Black Death) that periodically swept across Europe between 1348 and 1650. During the worst plague years (between 1348 and 1350), it is estimated that at least one-third of the European population perished. Notice, however, that this didn't retard population growth for very long. In 1650, at the end of the last great plague, there were about 600 million people in the world.

**TABLE 7.1**

**World Population Growth and Doubling Times**

| Date | Population | Doubling Time |
|---|---|---|
| 5000 B.C. | 50 million | ? |
| 800 B.C. | 100 million | 4,200 years |
| 200 B.C.. | 200 million | 600 years |
| A.D. 1200 | 400 million | 1,400 years |
| A.D. 1700 | 800 million | 500 years |
| A.D. 1900 | 1,600 million | 200 years |
| A.D. 1965 | 3,200 million | 65 years |
| A.D. 2005 | 6,400 million | 40 years |
| A.D. 2050 (estimate) | 8,920 million | 140 years |

**Source:** United Nations Population Division.

As you can see in figure 7.3, human populations began to increase rapidly after A.D. 1600. Many factors contributed to this rapid growth. Increased sailing and navigating skills stimulated commerce and communication between nations. Agricultural developments, better sources of power, and better health care and hygiene also played a role. We are now in an exponential or J curve pattern of growth.

It took all of human history to reach 1 billion people in 1804, but little more than 150 years to reach 3 billion in 1960. To go from 5 to 6 billion took only 12 years. Another way to look at population growth is that the number of humans tripled during the twentieth century. Will it do so again in the twenty-first century? If it does, will we overshoot the carrying capacity of our environment and experience a catastrophic dieback similar to those described in chapter 6? As you will see later in this chapter, there is evidence that population growth already is slowing, but whether we will reach equilibrium soon enough and at a size that can be sustained over the long term remains a difficult but important question.

## 7.2 Perspectives on Population Growth

As with many topics in environmental science, people have widely differing opinions about population and resources. Some believe that population growth is the ultimate cause of poverty and environmental degradation. Others argue that poverty, environmental degradation, and overpopulation are all merely symptoms of deeper social and political factors The worldview we choose to believe will profoundly affect our approach to population issues. In this section, we will examine some of the major figures and their arguments in this debate.

### Does environment or culture control human populations?

Since the time of the Industrial Revolution, when the world population began growing rapidly, individuals have argued about the causes and consequences of population growth. In 1798 Thomas Malthus (1766–1834) wrote *An Essay on the Principle of Population,* changing the way European leaders thought about population growth. Malthus marshaled evidence to show that populations

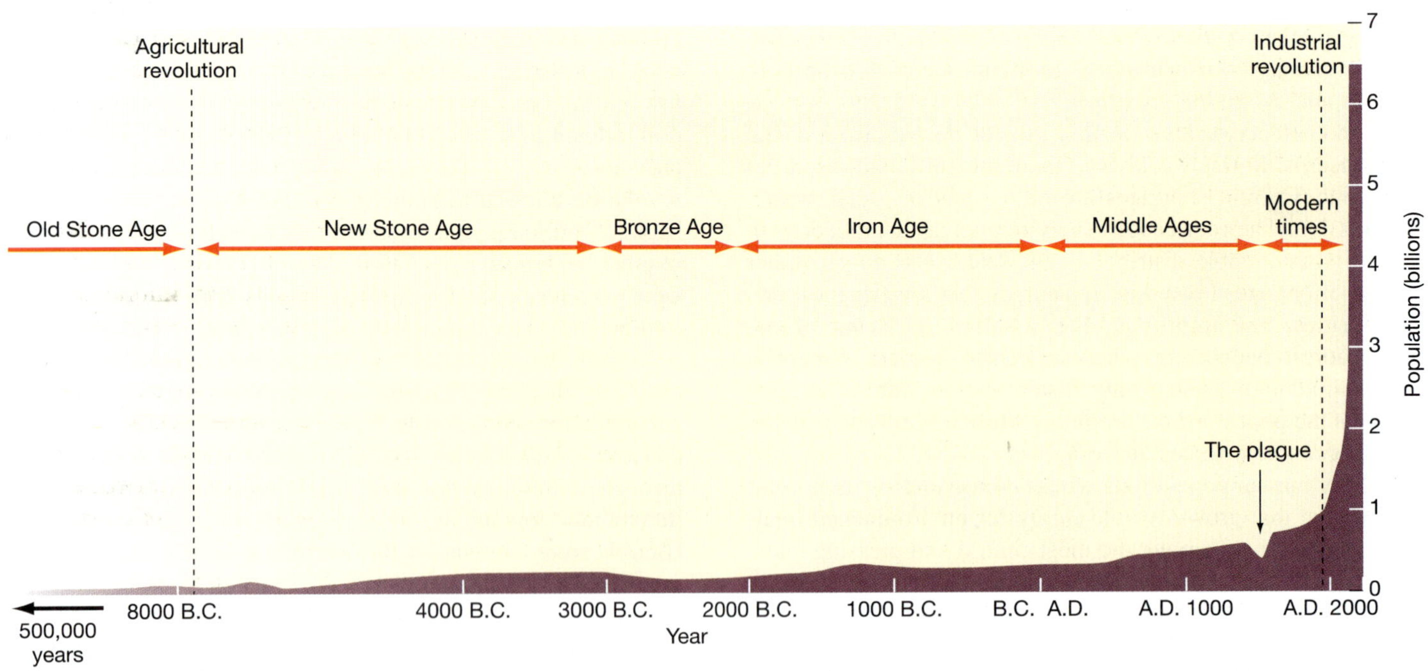


**FIGURE 7.3** Human population levels through history. Since about A.D. 1000, our population curve has assumed a J shape. Are we on the upward slope of a population overshoot? Will we be able to adjust our population growth to an S curve? Or can we just continue the present trend indefinitely?

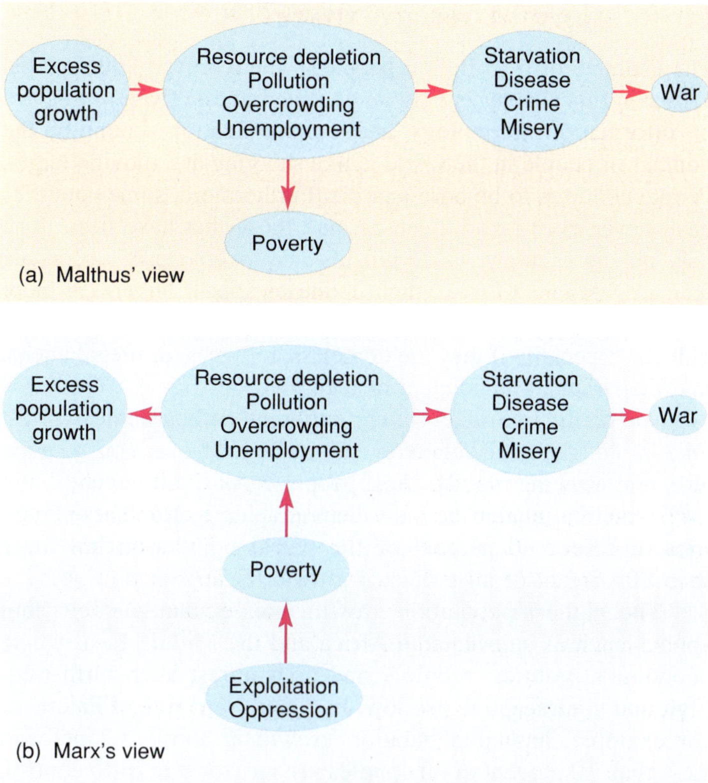


**FIGURE 7.4** (a) Thomas Malthus argued that excess population growth is the ultimate cause of many other social and environmental problems. (b) Karl Marx argued that oppression and exploitation are the real causes of poverty and environmental degradation. Population growth in this view is a symptom or result of other problems, not the source.

tended to increase at an exponential, or compound, rate while food production either remained stable or increased only slowly. Eventually human populations would outstrip their food supply and collapse into starvation, crime, and misery (fig. 7.4*a*). He converted most economists of the day from believing that high fertility increased gross domestic output to believing that per capita output actually fell with rapidly rising population.

In Malthusian terms, growing human populations stop growing when disease or famine kills many, or when constraining social conditions compel others to reduce their birth rates—late marriage, insufficient resources, celibacy, and "moral restraint." Several decades later, the economist Karl Marx (1818–1883) presented an opposing view, that population growth resulted from poverty, resource depletion, pollution, and other social ills. Slowing population growth, said Marx, required that people be treated justly, and that exploitation and oppression be eliminated from social arrangements (fig. 7.4*b*).

Both Marx and Malthus developed their theories about human population growth when understanding of the world, technology, and society were much different than they are today. But these different views of human population growth still inform competing approaches to family planning today. On the one hand, some believe that we are approaching, or may have surpassed, the earth's carrying capacity. Joel Cohen, a mathematical biologist at Rockefeller University, reviewed published estimates of the maximum human population size the planet can sustain. The estimates, spanning 300 years of thinking, converged on a median value of 10–12 billion. We are more than 6.5 billion strong today, and growing, an alarming prospect for some. Cornell University entomologist David Pimental, for example, has said: "By 2100, if current trends continue, twelve billion miserable humans will suffer a difficult life on Earth." In this view, birth control should be our top priority. On the other hand, many scholars agree with Marx that improved social conditions and educational levels can stabilize populations humanely. In this perspective, the earth is bountiful in its resource base, but poverty and high birth rates result from oppressive social relationships that unevenly distribute wealth and resources. Consequently, this position believes, technological development, education, and just social conditions are the means of achieving population control. Mohandas Gandhi stated it succinctly: "There is enough for everyone's need, but not enough for anyone's greed."

## Technology can increase carrying capacity for humans

Optimists argue that Malthus was wrong in his predictions of famine and disaster 200 years ago because he failed to account for scientific and technical progress. In fact, food supplies have increased faster than population growth since Malthus' time. For example, according to the UN FAO Statistics Division, each person on the planet averaged 2,435 calories of food per day in 1970, while in 2000 the caloric intake reached 2,807 calories. Even poorer, developing countries saw a rise, from an average of 2,135 calories per day in 1970 to 2,679 in 2000. In that same period the world population went from 3.7 to more than 6 billion people. Certainly terrible famines have stricken different locations in the past 200 years, but they were caused more by politics and economics than by lack of resources or population size. Whether the world can continue to feed its growing population remains to be seen, but technological advances have vastly increased human carrying capacity so far.

The burst of world population growth that began 200 years ago was stimulated by scientific and industrial revolutions. Progress in agricultural productivity, engineering, information technology, commerce, medicine, sanitation, and other achievements of modern life have made it possible to support thousands of times as many people per unit area as was possible 10,000 years ago. Economist Stephen Moore of the Cato Institute in Washington, D.C., regards this achievement as "a real tribute to human ingenuity and our ability to innovate." There is no reason, he argues, to think that our ability to find technological solutions to our problems will diminish in the future.

Much of our growth and rising standard of living in the past 200 years, however, has been based on easily acquired natural resources, especially cheap, abundant fossil fuels (see chapter 19). Whether rising prices of fossil fuels will constrain that production and result in a crisis in food production and distribution, or in some other critical factor in human society, concerns many people.

However, technology can be a double-edged sword. Our environmental effects aren't just a matter of sheer population size; they also depend on what kinds of resources we use and how we use them. This concept is summarized as the **I = PAT** formula. It says that our environmental impacts (I) are the product of our population size (P) times affluence (A) and the technology (T) used to produce the goods and services we consume. A single American living an affluent lifestyle that depends on high levels of energy and material consumption, and that produces excessive amounts of pollution, probably has a greater environmental impact than a whole village of Asian or African farmers. Ideally, Americans will begin to use nonpolluting, renewable energy and material sources. Better yet, Americans will extend the benefits of environmentally friendly technology to those villages of Asians and Africans so everyone can enjoy the benefits of a better standard of living without degrading their environment.

## Population growth could bring benefits

Think of the gigantic economic engine that China is becoming as it continues to industrialize and its population becomes more affluent. More people mean larger markets, more workers, and efficiencies of scale in mass production of goods. Moreover, adding people boosts human ingenuity and intelligence that will create new resources by finding new materials and discovering new ways of doing things. Economist Julian Simon (1932–1998), a champion of this rosy view of human history, believed that people are the "ultimate resource" and that no evidence suggests that pollution, crime, unemployment, crowding, the loss of species, or any other resource limitations will worsen with population growth. In a famous bet in 1980, Simon challenged Paul Ehrlich, author of *The Population Bomb,* to pick five commodities that would become more expensive by the end of the decade. Ehrlich chose metals that actually became cheaper, and he lost the bet. Leaders of many developing countries share this outlook and insist that, instead of being obsessed with population growth, we should focus on the inordinate consumption of the world's resources by people in richer countries.

***Think About It***

What larger worldviews are reflected in this population debate? What positions do you believe neo-Malthusians and neo-Marxists might take on questions of human rights, resource abundance, or human perfectability? Where do you stand on these issues?

# 7.3 Many Factors Determine Population Growth

**Demography** is derived from the Greek words *demos* (people) and *graphos* (to write or to measure). It encompasses vital statistics about people, such as births, deaths, and where they live, as well as total population size. In this section, we will survey ways human populations are measured and described, and discuss demographic factors that contribute to population growth.

## How many of us are there?

The estimate of 6.55 billion people in the world in 2006 quoted earlier in this chapter is only an educated guess. Even in this age of information technology and communication, counting the number of people in the world is like shooting at a moving target. People continue to be born and die. Furthermore, some countries have never even taken a census, and those that have been done may not be accurate. Governments may overstate or understate their populations to make their countries appear larger and more important or smaller and more stable than they really are. Individuals, especially if they are homeless, refugees, or illegal aliens, may not want to be counted or identified.

We really live in two very different demographic worlds. One is old, rich, and relatively stable. The other is young, poor, and growing rapidly. Most people in Asia, Africa, and Latin America inhabit the latter demographic world. These countries represent 80 percent of the world population but more than 90 percent of all projected growth (fig. 7.5).

The highest population growth rates occur in a few "hot spots," such as sub-Saharan Africa and the Middle East, where economics, politics, religion, and civil unrest keep birth rates high and contraceptive use low. In Niger, Yemen, and Palestine, for example, annual population growth is above 3.2 percent. Less than 10 percent of all couples use any form of birth control, women average more than seven children each, and nearly half the population is less than 15 years old. The world's highest current growth rate is in the United Arab Emirates, where births plus immigration are producing an annual increase of 6.8 percent (the highest immigration rate in the world is responsible for 80 percent of that growth). This means that the UAE is doubling its population size approximately every decade. Obviously, a small country with limited resources (except oil) and almost no fresh water or agriculture, can't sustain that high growth rate indefinitely.

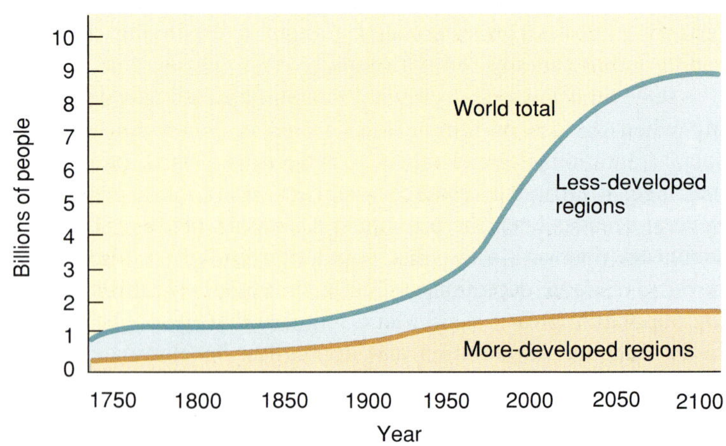


**FIGURE 7.5** Estimated human population growth, 1750–2100, in less-developed and more-developed regions. Almost all growth projected for the twenty-first century is in the less-developed countries.
**Source:** UN Population Division, 2005.

**TABLE 7.2**

**The World's Largest Countries**

| 2006 | | 2050* | |
|---|---|---|---|
| **Country** | **Population (millions)** | **Country** | **Population (millions)** |
| China | 1,311 | India | 1,628 |
| India | 1,122 | China | 1,437 |
| United States | 299 | United States | 420 |
| Indonesia | 225 | Nigeria | 299 |
| Brazil | 187 | Pakistan | 295 |
| Pakistan | 166 | Indonesia | 285 |
| Bangladesh | 147 | Brazil | 260 |
| Russia | 142 | Bangladesh | 231 |
| Nigeria | 135 | Dem. Rep. of Congo | 183 |
| Japan | 128 | Ethiopia | 145 |

*Estimate.

**Source:** Population Reference Bureau, 2006.

Some countries in the developing world have experienced amazing growth rates and are expected to reach extraordinary population sizes by the middle of the twenty-first century. Table 7.2 shows the ten largest countries in the world, arranged by their estimated size in 2006 and projected size in 2050. Note that, while China was the most populous country throughout the twentieth century, India is expected to pass China in the twenty-first century. Nigeria, which had only 33 million residents in 1950, is forecast to have nearly 300 million in 2050. Ethiopia, with about 18 million people 50 years ago, is likely to grow nearly eight-fold over a century. In many of these countries, rapid population growth is a serious problem. Bangladesh, about the size of Iowa, is already overcrowded at 147 million people. Another 84 million people by 2050 will only add to current problems.

The other demographic world is made up of the richer countries of North America, western Europe, Japan, Australia, and New Zealand. This world is wealthy, old, and mostly shrinking. Italy, Germany, Hungary, and Japan, for example, all have negative growth rates. The average age in these countries is now 40, and life expectancy of their residents is expected to exceed 90 by 2050. With many couples choosing to have either one or no children, the populations of these countries are expected to decline significantly over the next century. Japan, which has 128 million residents now, is expected to shrink to about 100 million by 2050. Europe, which now makes up about 12 percent of the world population, will constitute less than 7 percent in 50 years, if current trends continue. Even the United States and Canada would have nearly stable populations if immigration were stopped.

It isn't only wealthy countries that have declining populations. Russia, for instance, is now declining by nearly 1 million people per year as death rates have soared and birth rates have plummeted. A collapsing economy, hyperinflation, crime, corruption, and despair have demoralized the population. Horrific pollution levels left from the Soviet era, coupled with poor nutrition and health care, have resulted in high levels of genetic abnormalities, infertility, and infant mortality. Abortions are twice as common as live births, and the average number of children per woman is now 1.3, one of the lowest in the world. Death rates, especially among adult men, have risen dramatically. Male life expectancy dropped from 68 years in 1990 to 59 years in 2006. Russia, which is the world's largest country geographically, is expected to decline from 142 million people in 2006 to less than 100 million in 2050. It will then have a smaller population than Vietnam, Egypt, or Uganda. Other former Soviet states are experiencing similar declines. Estonia, Bulgaria, Georgia, and Ukraine, for example, now have negative growth rates and are expected to lose about 40 percent of their population in the next 50 years.

The situation is even worse in many African countries, where AIDS and other communicable diseases are killing people at a terrible rate. In Zimbabwe, Botswana, Zambia, and Namibia, for example, up to 39 percent of the adult population have AIDS or are HIV positive. Health officials predict that more than two-thirds of the 15-year-olds now living in Botswana will die of AIDS before age 50. Without AIDS, the average life expectancy is estimated to be 69.7 years. Now, with AIDS, Botswana's life expectancy has dropped to only 31.6 years. The populations of many African countries are now falling because of this terrible disease (fig. 7.6). Altogether, Africa's population is expected to be nearly 200 million lower in 2050 than it would have been without AIDS.

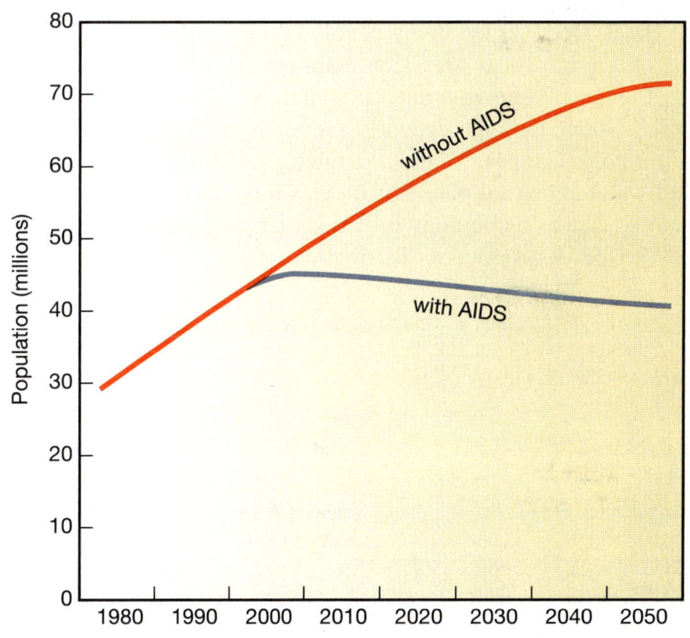


**FIGURE 7.6** Projected population of south Africa with and without AIDS.

**Data source:** UN Population Division, 2006.

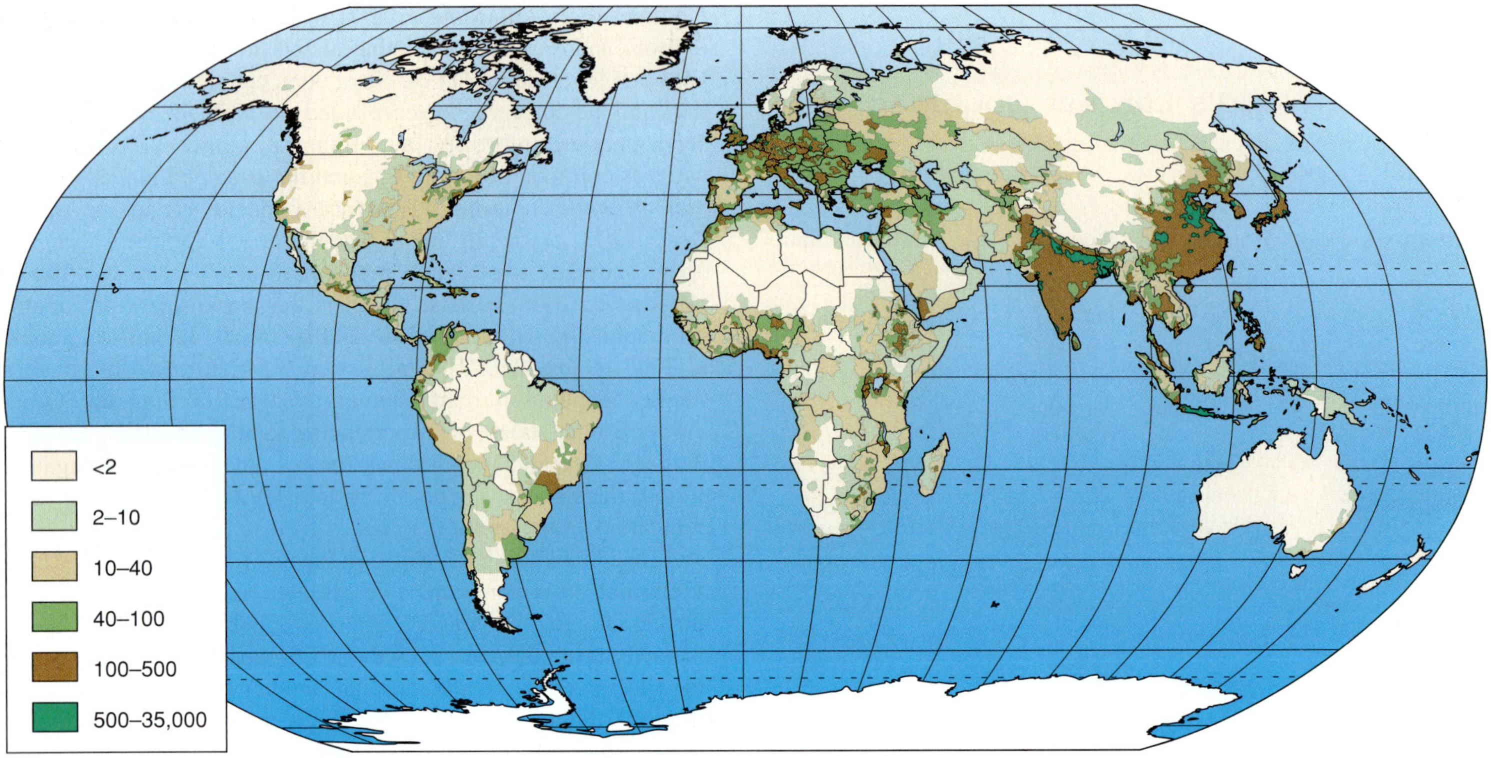


**FIGURE 7.7** Population density in persons per square kilometer.
**Source:** World Bank, 2000.

AIDS is now spreading in Asia. Because of the large population there, Asia is expected to pass Africa in 2020 in total number of deaths. Although a terrible human tragedy, this probably won't affect total world population very much. Remember that the Black Death killed many people in the fourteenth century but had only a transitory effect on demography.

Figure 7.7 shows human population distribution around the world. Notice the high densities supported by fertile river valleys of the Nile, Ganges, Yellow, Yangtze, and Rhine Rivers and the well-watered coastal plains of India, China, and Europe. Historic factors, such as technology diffusion and geopolitical power, also play a role in geographic distribution.

## Fertility measures the number of children born to each woman

As we pointed out in chapter 6, fecundity is the physical ability to reproduce, while fertility describes the actual production of offspring. Those without children may be fecund but not fertile. The most accessible demographic statistic of fertility is usually the **crude birth rate,** the number of births in a year per thousand persons. It is statistically "crude" in the sense that it is not adjusted for population characteristics such as the number of women in reproductive age. Table 7.3 shows birth rates more-developed and less-developed countries in 2006.

**TABLE 7.3**

**Birth and Death Rates in 2006**

| | Population (Millions) | Births per 1,000 | Deaths per 1,000 | Total Fertility Rate | Natural Increase (Percent) |
|---|---|---|---|---|---|
| **World** | 6,555 | 21 | 9 | 2.7 | 1.2 |
| **More-Developed Regions** | 1,216 | 11 | 10 | 1.6 | 0.1 |
| **Less-Developed Regions** | 5,339 | 23 | 8 | 2.9 | 1.5 |

**Source:** Population Reference Bureau, 2006.

The **total fertility rate** is the number of children born to an average woman in a population during her entire reproductive life. Upper-class women in seventeenth- and eighteenth-century England, whose babies were given to wet nurses immediately after birth and who were expected to produce as many children as possible, often had 25 or 30 pregnancies. The highest recorded total fertility rates for working-class people is among some Anabaptist agricultural groups in North America who have averaged up to 12 children per woman. In most tribal or traditional societies, food shortages, health problems, and cultural practices limit total fertility to about six or seven children per woman even without modern methods of birth control.

**Zero population growth (ZPG)** occurs when births plus immigration in a population just equal deaths plus emigration. It takes several generations of replacement level fertility (where people just replace themselves) to reach ZPG. Where infant mortality rates are high, the replacement level may be five or more children per couple. In the more highly developed countries, however, this rate is usually about 2.1 children per couple because some people are infertile, have children who do not survive, or choose not to have children.

Fertility rates have declined dramatically in every region of the world except Africa over the past 50 years (fig. 7.8). Only a few decades ago, total fertility rates above 6 were common in many countries. The average family in Mexico in 1975, for instance, had 7 children. By 2006, however, the average Mexican woman had only 2.4 children. According to the World Health Organization, 61 out of the world's 190 countries are now at or below a replacement rate of 2.1 children per couple, and by 2050, all but a few of the least-developed countries are expected to have reached that milestone. The greatest fertility reduction has been in Southeast Asia, where rates have fallen by more than half. Most of this decrease has occurred in just the past few decades and, contrary to what many demographers expected, some of the poorest countries in the world have been remarkably successful in lowering growth rates. As the opening case study for this chapter shows, Thailand reduced its total fertility rate from 7.0 in 1979 to 1.7 (lower than that in the United States) in 2006.

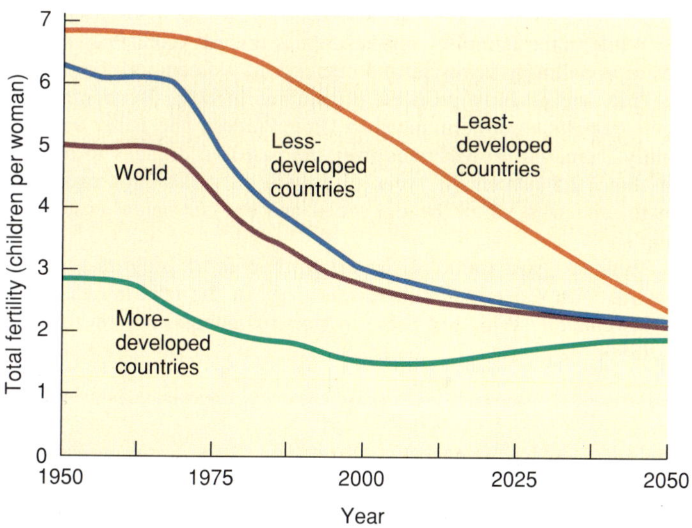


**FIGURE 7.8** Average total fertility rates for less-developed countries fell by more than half over the past 50 years. Much of this dramatic change was due to China's one-child policy. Progress has been slower in the least-developed countries, but by 2050, they should be approaching the replacement rate of 2.1 children per woman of reproductive age.

**FIGURE 7.9** China's one-child-per-family policy, promoted in this billboard, has been remarkably successful in reducing birth rates. It may, however, have created a generation of "little emperors," since parents and grandparents focus all their attention on an only child.

China's one-child-per-family policy decreased the fertility rate from 6 in 1970 to 1.8 in 1990 and 1.6 in 2006 (fig. 7.9). This policy, however, has sometimes resulted in abortions, forced sterilizations, and even infanticide. Another adverse result is that the only children (especially boys) allowed to families may grow up to be spoiled "little emperors" who have an inflated impression of their own importance. (fig. 7.9) Furthermore, there may not be enough workers to maintain the army, sustain the economy, or support retirees when their parents reach old age.

China reports that 119 boys are now being born for every 100 girls. Normal ratios would be about 105 boys to 100 girls. If this imbalance persists, there will be a shortage of brides in another generation. The government is considering easing the one-child policy. Macao, with a total average fertility rate of only 0.9, now has the lowest birth rate in the world.

Although the world as a whole still has an average fertility rate of 2.7, growth rates are now lower than at any time since World War II. If fertility declines like those in Thailand and China were to occur everywhere in the world, global population could begin to decline by 2050, and might be below 6 billion by 2150. Most of Eastern Europe now has fertility levels of 1.2 children per woman. Interestingly, Spain and Italy, although predominately Roman Catholic, have similar fertility rates. Several Indian states have reached zero population growth, but their means of doing so have been very different (What Do You Think? p. 140).

## What Do You Think?

### How to Reduce Population Growth?

Every year, India adds more people to the world's population than any other country. In 2006, having added more than 185 million residents in the previous decade, India had more than 1.1 billion people. By 2050, if current growth rates persist, India will have increased its population by more than 50 percent over current levels, and will have around 1.63 billion residents, making it the most populous country in the world. How will the country, which already has more than a quarter of its population living in abject poverty, feed, house, educate, and employ all those being added each year? And what's the best way to slow this rapid growth? The fierce debate now taking place about how to control India's population has ramifications for the rest of the world as well.

Currently, fewer than half of all Indian women use any form of birth control. How can this percentage be raised? On one side of this issue are those who believe that the best way to reduce the number of children born is poverty eradication and progress for women. Drawing on social justice principles established at the 1994 UN Conference on Population and Development in Cairo, some argue that responsible economic development, a broad-based social welfare system, education and empowerment of women, and high-quality health care—including family planning services—are essential components of population control. Without progress in these areas, they believe, efforts to provide contraceptives or encourage sterilization are futile.

Indian states have taken very different approaches to family planning and human development.

On the other side of this debate are those who contend that, while social progress is an admirable goal, India doesn't have the time or resources to wait for an indirect approach to population control. The government must push aggressively, they argue, to reduce births now or the population will be so huge and its use of resources so great that only mass starvation, class war, crime, and disease will be able to bring it down to a manageable size.

Unable to reach a consensus on population policy, the Indian government decided in 2000 to let each state approach the problem in its own way. Some states have chosen to focus on social justice, while others have adopted more direct, interventionist policies.

The model for the social justice approach is the southern state of Kerala, which achieved population stabilization in the mid-1980s, the first Indian state to do so. Although still one of the poorest places in the world, economically, Kerala's fertility rate is comparable to that of many industrialized nations, including the United States. Both women and men have a nearly 100 percent literacy rate and share affordable and accessible health care, family planning, and educational opportunities; therefore, women have only the number of children they want, usually two. The Kerala experience suggests that increased wealth isn't a prerequisite for zero population growth.

Taking a far different path to birth reduction is the nearby state of Andra Pradesh, which reached a stable growth rate in 2001. Boasting the most dramatic fertility decline of any large Indian state, Andra Pradesh has focused on targeted, strongly enforced sterilization programs. The poor are encouraged—some would say, compelled—to be sterilized after having only one or two children. The incentives include cash payments. You might receive 500 rupees—equivalent to (U.S.) $11 or a month's wages for an illiterate farm worker—if you agree to have "the operation." In addition, participants are eligible for better housing, land, wells, and subsidized loans.

The pressure to be sterilized is overwhelmingly directed at women, for whom the procedure is major abdominal surgery. Sterilizations often are done by animal husbandry staff and carried out in government sterilization camps. This practice raises troubling memories of the 1970s for many people, when then-Prime Minister Indira Gandhi suspended democracy and instituted a program of forced sterilization of poor people. There were reports at the time of people being rounded up like livestock and castrated or neutered against their will.

While many feminists and academics regard Andra Pradesh's policies as appallingly intrusive and coercive to women and the poor, the state has successfully reduced population growth. By contrast, the hugely populous northern states of Uttar Pradesh and Bihar have seen slightly increased growth rates over the past two decades to a current rate above 2.5 percent per year. How will they slow this exponential growth, and what might be the social and environmental costs of not doing so?

What do you think? How do birth control programs in India compare with those in Thailand describes in the opening case study for this chapter? Which of these models for population control would you favor?

## Mortality is the other half of the population equation

A traveler to a foreign country once asked a local resident, "What's the death rate around here?" "Oh, the same as anywhere," was the reply, "about one per person." In demographics, however, **crude death rates** (or crude mortality rates) are expressed in terms of the number of deaths per thousand persons in any given year. Countries in Africa where health care and sanitation are limited may have mortality rates of 20 or more per 1,000 people. Wealthier countries generally have mortality rates around 10 per 1,000. The number of deaths in a population is sensitive to the age structure of the population. Rapidly growing, developing countries such as Libya or Costa Rica have lower crude death rates (4 per 1,000) than do the more-developed, slowly growing countries, such as Denmark (12 per 1,000). This is because there are proportionately more youths and fewer elderly people in a rapidly growing country than in a more slowly growing one.

Crude death rate subtracted from crude birth rate gives the **natural increase** of a population. We distinguish natural increase from the **total growth rate,** which includes immigration and emigration, as well as births and deaths. Both of these growth rates are usually expressed as a percent (number per hundred people) rather than per thousand. A useful rule of thumb is that if you divide 70 by the annual percentage growth, you will get the approximate doubling time in years. Niger, for example, which is growing 3.4 percent per year, is doubling its population every 20 years. The United States, which has a natural increase rate of 0.6 percent per year, would double, without immigration, in 116.7 years. Belgium and Sweden, with natural increase rates of 0.1 percent, are doubling in about 700 years. Ukraine, on the other hand, with a growth rate of –0.8 percent, will lose about 40 percent of its population in the next 50 years. The world growth rate is now 1.2 percent, which means that the population will double in about 58 years if this rate persists.

## Life span and life expectancy describe our potential longevity

Life span is the oldest age to which a species is known to survive. Although there are many claims in ancient literature of kings living a millennium or more, the oldest age that can be certified by written records was that of Jeanne Louise Calment of Arles, France, who was 122 years old at her death in 1997. The aging process is still a medical mystery, but it appears that cells in our bodies have a limited ability to repair damage and produce new components. At some point they simply wear out, and we fall victim to disease, degeneration, accidents, or senility.

**Life expectancy** is the average age that a newborn infant can expect to attain in any given society. It is another way of expressing the average age at death. For most of human history, we believe that average life expectancy in most societies has been about 30 years. This doesn't mean that no one lived past age 40, but rather that so many deaths at earlier ages (mostly early childhood) balanced out those who managed to live longer.

Declining mortality, not rising fertility, is the primary cause of most population growth in the past 300 years. Crude death rates began falling in western Europe during the late 1700s. Most of this advance in survivorship came long before the advent of modern medicine and is due primarily to better food and better sanitation.

The twentieth century has seen a global transformation in human health unmatched in history. This revolution can be seen in the dramatic increases in life expectancy in most places. Worldwide, the average life expectancy has risen from about 30 to 65 years over the past century (see fig. 6.5). Table 7.4 shows gains in some selected countries. Globally, the number of people over 60 years old is expected to triple, increasing from 600 million today to nearly 2 billion in 2050. The oldest old (over 80 years) is projected to grow five-fold to about 400 million in that same period.

**TABLE 7.4**

**Life Expectancy at Birth for Selected Countries in 1900 and 2006**

| Country | 1900 | | 2006 | |
|---|---|---|---|---|
| | Males | Females | Males | Females |
| India | 23 | 23 | 62 | 63 |
| Japan | 42 | 44 | 79 | 86 |
| Russia | 31 | 33 | 59 | 72 |
| Sweden | 57 | 60 | 78 | 83 |
| United States | 46 | 48 | 75 | 80 |

Source: Population Reference Bureau, 2006.

The greatest progress in life expectancy has been in developing countries. Take the case of Nicaragua, for example. In 1900, the average Nicaraguan man could expect to live only 29 years, while the average woman would reach just 33 years. By 2006, although Nicaragua had an annual per capita income of about (U.S.)$3,600, the average life expectancy for both men and women had more than doubled and was close to that of countries with ten times its income level. Longer lives were due primarily to better nutrition, improved sanitation, clean water, and education rather than miracle drugs or high-tech medicine. While the gains were not as great for the already industrialized countries, residents of the United States, Italy, and Japan, for example, now live about half-again as long as they did at the beginning of the twentieth century.

As figure 7.10 shows, there is a good correlation between annual income and life expectancy up to about (U.S.) $4,000 per person. Beyond that level—which is generally enough for adequate food, shelter, and sanitation for most people—life expectancies level out at about 75 years for men and 85 for women.

Large discrepancies in how benefits of modernization and social investment are distributed within countries are revealed in differential longevity of various groups. The greatest life expectancy reported anywhere in the United States is for Asian American women in New Jersey, who live to an average age of 91. By contrast, Native American men on Pine Ridge Indian Reservation in neighboring South Dakota, live, on average, only to age 48. Only a few countries in Africa have a lower life expectancy. The Pine Ridge Reservation is the poorest area in America with an unemployment rate near 75 percent and high rates of poverty, alcoholism, drug use, and cultural alienation. Similarly, African-American men in Washington, D.C., live, on average, only 57.9 years, or less than men in Lesotho or Swaziland.

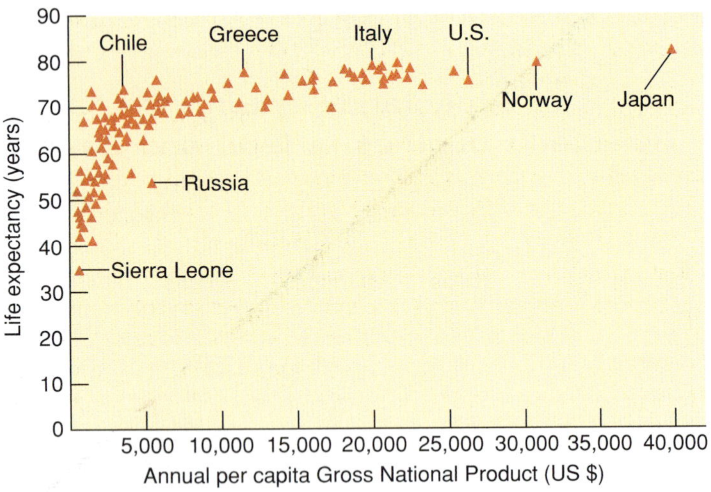


**FIGURE 7.10** As incomes rise, so does life expectancy up to about (U.S.) $4,000. Russia is an exception, with a life expectancy nearly 20 years less than that of Chile, even though their GNP is about the same.
**Source:** The World Bank, *World Development Indicators 1997,* the World Bank, Washington, D.C., 1997.

Some demographers believe that life expectancy is approaching a plateau, while others predict that advances in biology and medicine might make it possible to live 150 years or more. If our average age at death approaches 100 years, as some expect, society will be profoundly affected. In 1970 the median age in the United States was 30. By 2100 the median age could be over 60. If workers continue to retire at 65, half of the population could be unemployed, and retirees might be facing 35 or 40 years of retirement. We may need to find new ways to structure and finance our lives.

## Living longer has demographic implications

A population that is growing rapidly by natural increase has more young people than does a stationary population. One way to show these differences is to graph age classes in a histogram as shown in figure 7.11. In Niger, which is growing at a rate of 3.4 percent per year, 49 percent of the population is in the prereproductive category (below age 15). Even if total fertility rates were to fall abruptly, the total number of births, and population size, would continue to grow for some years as these young people enter reproductive age. This phenomenon is called population momentum.

By contrast, a country with a stable population, like Sweden, has nearly the same number in each age cohort. A population that has recently entered a lower growth rate pattern, such as Singapore, has a bulge in the age classes for the last high-birth-rate generation. Notice that there are more females than males in the older age group in Sweden because of differences in longevity between the sexes.

Both rapidly growing countries and slowly growing countries can have a problem with their **dependency ratio,** or the number of nonworking compared to working individuals in a population. In Mexico, for example, each working person supports

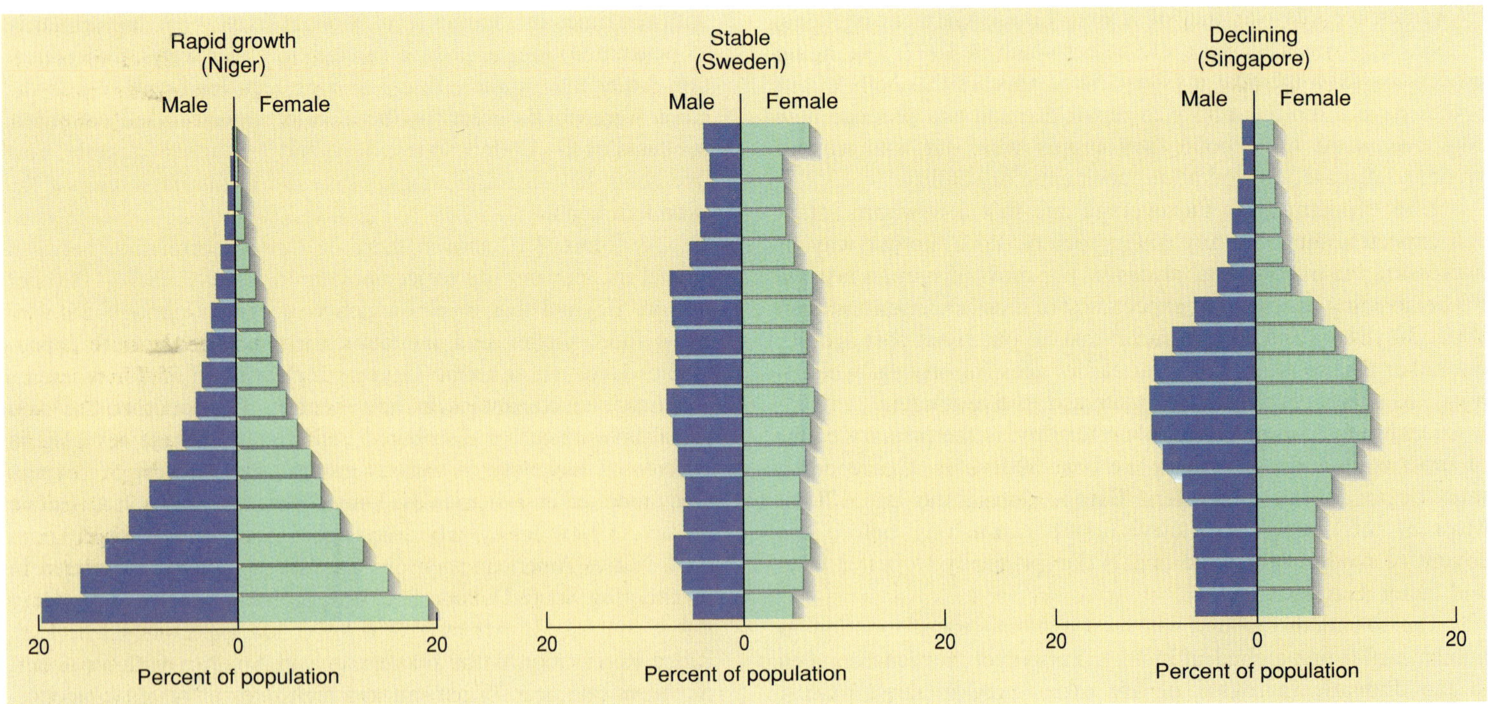


**FIGURE 7.11** Age structure graphs for rapidly growing, stable, and declining populations.
**Source:** U.S. Census Bureau, 2006.

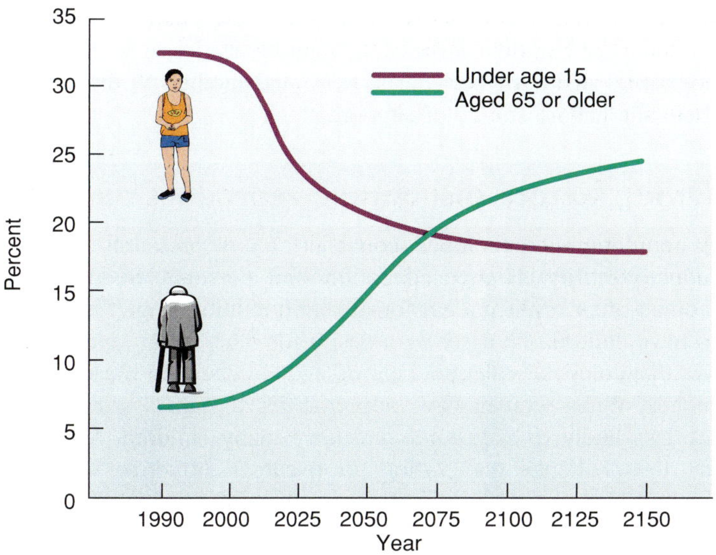


**FIGURE 7.12** By the mid-twenty-first century, children under age 15 will make up a smaller percentage of world population, while people over age 65 will contribute a larger and larger share of the population.

a high number of children. In the United States, by contrast, a declining working population is now supporting an ever larger number of retired persons and there are dire predictions that the social security system will soon be bankrupt. This changing age structure and shifting dependency ratio are occurring worldwide (fig. 7.12). By 2050 the UN predicts there will be two older persons for every child in the world. Many countries are rethinking their population policies and beginning to offer incentives for marriage and child-rearing.

### Emigration and immigration are important demographic factors

Humans are highly mobile, so emigration and immigration play a larger role in human population dynamics than they do in those of many species. Currently, about 800,000 people immigrate legally to the United States each year, but many more enter illegally. Western Europe receives about 1 million applications each year for asylum from economic chaos and wars in former socialist states and the Middle East. The United Nations High Commission on Refugees reported that in 2006 there were 20.8 million refugees who had left their countries for political or economic reasons, while about 25 million more were displaced persons in their own countries, and 175 million migrants had left their homes to look for work, greater freedom, or better opportunities.

The more-developed regions are expected to gain about 2 million immigrants per year for the next 50 years. Without migration, the population of the wealthiest countries would already be declining and would be more than 126 million less than the current 1.2 billion by 2050. In 2006, nearly 42 million U.S. residents (14.5 percent of the total population) classified themselves as Hispanic or Latino. They now constitute the largest U.S. minority.

Immigration is a controversial issue in many countries. "Guest workers" often perform heavy, dangerous, or disagreeable work that citizens are unwilling to do. Many migrants and alien workers are of a different racial or ethnic background than the majority in their new home. They generally are paid low wages and given substandard housing, poor working conditions, and few rights. Local residents often complain, however, that immigrants take away jobs, overload social services, and ignore established rules of behavior or social values. Anti-immigrant groups are springing up in many rich countries.

Some nations encourage, or even force, internal mass migrations as part of a geopolitical demographic policy. In the 1970s, Indonesia embarked on an ambitious "transmigration" plan to move 65 million people from the overcrowded islands of Java and Bali to relatively unpopulated regions of Sumatra, Borneo, and New Guinea. Attempts to turn rainforest into farmland had disastrous environmental and social effects, however, and this plan was greatly scaled back. China has announced a plan to move up to 100 million people to a sparsely populated region along the Amur River in Heilongjang. By some estimates, more than 250 million internal migrants in China have moved from rural areas to the cities to look for work.

## 7.4 Ideal Family Size Is Culturally and Economically Dependent

A number of social and economic pressures affect decisions about family size, which in turn affects the population at large. In this section we will examine both positive and negative pressures on reproduction.

### Many factors increase our desire for children

Factors that increase people's desires to have babies are called **pronatalist pressures.** Raising a family may be the most enjoyable and rewarding part of many people's lives. Children can be a source of pleasure, pride, and comfort. They may be the only source of support for elderly parents in countries without a social security system. Where infant mortality rates are high, couples may need to have many children to be sure that at least a few will survive to take care of them when they are old. Where there is little opportunity for upward mobility, children give status in society, express parental creativity, and provide a sense of continuity and accomplishment otherwise missing from life. Often children are valuable to the family not only for future income, but even more as a source of current income and help with household chores. In much of the developing world, children as young as 6 years old tend domestic animals and younger siblings, fetch water, gather firewood, and help grow crops or sell things in the

**FIGURE 7.13** In rural areas with little mechanized agriculture (a) children are needed to tend livestock, care for younger children, and help parents with household chores. Where agriculture is mechanized (b) rural families view children just as urban families do—helpful, but not critical to survival. This affects the decision about how many children to have.

marketplace (fig. 7.13) Parental desire for children rather than an unmet need for contraceptives may be the most important factor in population growth in many cases.

Society also has a need to replace members who die or become incapacitated. This need often is codified in cultural or religious values that encourage bearing and raising children. In some societies, families with few or no children are looked upon with pity or contempt. The idea of deliberately controlling fertility may be shocking, even taboo. Women who are pregnant or have small children are given special status and protection. Boys frequently are more valued than girls because they carry on the family name and are expected to support their parents in old age. Couples may have more children than they really want in an attempt to produce a son.

Male pride often is linked to having as many children as possible. In Niger and Cameroon, for example, men, on average, want 12.6 and 11.2 children, respectively. Women in these countries consider the ideal family size to be only about one-half that desired by their husbands. Even though a woman might desire fewer children, however, she may have few choices and little control over her own fertility. In many societies, a woman has no status outside of her role as wife and mother. Without children, she has no source of support.

## Other factors discourage reproduction

In more highly developed countries, many pressures tend to reduce fertility. Higher education and personal freedom for women often result in decisions to limit childbearing. The desire to have children is offset by a desire for other goods and activities that compete with childbearing and childbearing for time and money. When women have opportunities to earn a salary, they are less likely to stay home and have many children. Not only are the challenge and variety of a career attractive to many women, but the money that they can earn outside the home becomes an important part of the family budget. Thus, education and socioeconomic status are usually inversely related to fertility in richer countries. In developing countries, however, fertility is likely to increase as educational levels and socioeconomic status rise. With higher income, families are better able to afford the children they want; more money means that women are likely to be healthier, and therefore better able to conceive and carry a child to term.

In less-developed countries where feeding and clothing children can be a minimal expense, adding one more child to a family usually doesn't cost much. By contrast, raising a child in the United States can cost hundreds of thousands of dollars by the time the child is through school and is independent. Under these circumstances, parents are more likely to choose to have one or two children on whom they can concentrate their time, energy, and financial resources.

Figure 7.14 shows U.S. birth rates between 1910 and 2000. As you can see, birth rates have fallen and risen in a complex pattern. The period between 1910 and 1930 was a time of industrialization and urbanization. Women were getting more education

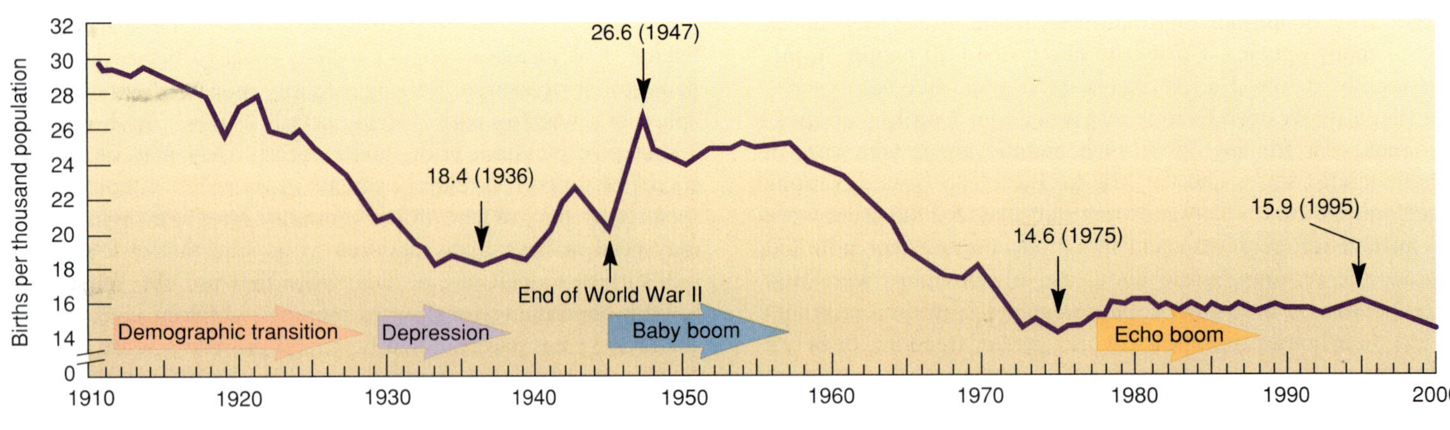


**FIGURE 7.14** Birth rates in the United States, 1910–2000. The falling birth rate from 1910 to 1929 represents a demographic transition from an agricultural to an industrial society. The baby boom following World War II lasted from 1945 to 1965. A much smaller "echo boom" occurred around 1980 when the baby boomers started to reproduce.

**Source:** Data from Population Reference Bureau and U.S. Bureau of the Census.

than ever before and entering the workforce. The Great Depression in the 1930s made it economically difficult for families to have children, and birth rates were low. The birth rate increased at the beginning of World War II (as it often does in wartime). For reasons that are unclear, a higher percentage of boys are usually born during war years.

At the end of the war, there was a "baby boom" as couples were reunited and new families started. This high birth rate persisted through the times of prosperity and optimism of the 1950s, but began to fall in the 1960s. Part of this decline was caused by the small number of babies born in the 1930s. This meant fewer young adults to give birth in the 1960s. Part was due to changed perceptions of the ideal family size. Whereas in the 1950s women typically wanted four children or more, in the 1970s the norm dropped to one or two (or no) children. A small "echo boom" occurred in the 1980s as people born in the 1960s began to have babies, but changing economics and attitudes seem to have permanently altered our view of ideal family size in the United States.

***Think About It***

How many children (if any) do you want to have? Is this number different from that of your parents or grandparents? Why or why not?

## Could we have a birth dearth?

Most European countries now have birth rates below replacement rates, and Italy, Russia, Austria, Germany, Greece, and Spain are experiencing negative rates of natural population increase. Asia, Japan, Singapore, and Taiwan are also facing a "child shock" as fertility rates have fallen well below the replacement level of 2.1 children per couple. There are concerns in all these countries about falling military strength (lack of soldiers), economic power (lack of workers), and declining social systems (not enough workers and taxpayers) if low birth rates persist or are not balanced by immigration. In a sense, the United States is fortunate to have a high influx of immigrants that provides youth and energy to its population.

Economist Ben Wattenberg warns that this "birth dearth" might seriously erode the powers of Western democracies in world affairs. He points out that Europe and North America accounted for 22 percent of the world's population in 1950. By the 1980s, this number had fallen to 15 percent, and by the year 2030, Europe and North America probably will make up only 9 percent of the world's population. Germany, Hungary. Denmark, and Russia now offer incentives to encourage women to bear children. Japan offers financial support to new parents, and Singapore provides a dating service to encourage marriages among the upper classes as a way of increasing population.

On the other hand, since Europeans and North Americans consume so many more resources per capita than most other people in the world, a reduction in the population of these countries will do more to spare the environment than would a reduction in population almost anywhere else.

One reason that birth rates have been falling in many industrialized countries may be that toxins and endocrine hormone disrupters in our environment interfere with sperm production. Sperm numbers and quality (fertilization ability) appear to have fallen by about half over the past 50 years in a number of countries. Widespread chemicals, such as phthalates—common ingredients in plastics—that disrupt sperm production may be responsible for this decline. We'll discuss this further in chapter 8.

# 7.5 A Demographic Transition Can Lead to Stable Population Size

In 1945, demographer Frank Notestein pointed out that a typical pattern of falling death rates and birth rates due to improved living conditions usually accompanies economic development. He called this pattern the **demographic transition** from high birth and death rates to lower birth and death rates. Figure 7.15 shows an idealized model of a demographic transition. This model is often used to explain connections between population growth and economic development.

## Economic and social development influence birth and death rates

Stage I in figure 7.15 represents the conditions in a premodern society. Food shortages, malnutrition, lack of sanitation and medicine, accidents, and other hazards generally keep death rates in such a society around 35 per 1,000 people. Birth rates are correspondingly high to keep population densities relatively constant. As economic development brings better jobs, medical care, sanitation, and a generally improved standard of living in Stage II, death rates often fall very rapidly. Birth rates may actually rise at first as more money and better nutrition allow people to have the children they always wanted. Eventually, in a mature industrial economy (Stage III), birth rates fall as people see that all their children are more likely to survive and that the whole family benefits from concentrating more resources on fewer children. Note that population continues to grow rapidly during this stage because of population momentum (baby boomers reaching reproductive age). Depending on how long it takes to complete the transition, the population may go through one or more rounds of doubling before coming into balance again.

Stage IV in figure 7.15 represents conditions in developed countries, where the transition is complete and both birth rates and death rates are low, often, a third or less than those in the predevelopment era. The population comes into a new equilibrium in this phase, but at a much larger size than before. Most of the countries of northern and western Europe went through a demographic transition in the nineteenth or early twentieth century similar to the curves shown in this figure.

Many of the most rapidly growing countries in the world, such as Kenya, Yemen, Libya, and Jordan, now are in the Stage I of this demographic transition. Their death rates have fallen

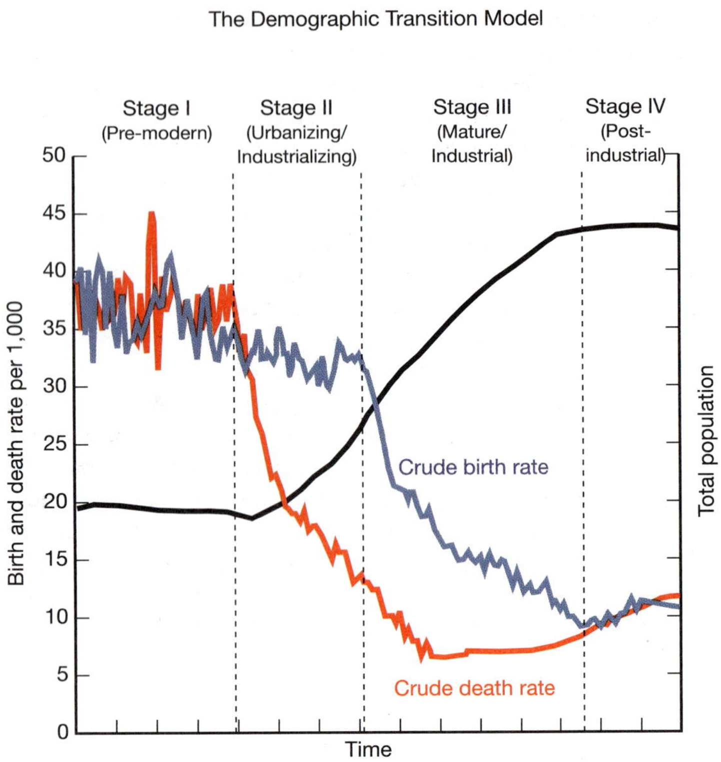


**FIGURE 7.15** Theoretical birth, death, and population growth rates in a demographic transition accompanying economic and social development. In a predevelopment society, birth and death rates are both high, and total population remains relatively stable. During development, death rates tend to fall first, followed in a generation or two by falling birth rates. Total population grows rapidly until both birth and death rates stabilize in a fully developed society.

close to the rates of the fully developed countries, but birth rates have not fallen correspondingly. In fact, both their birth rates and total population are higher than those in most European countries when industrialization began 300 years ago. The large disparity between birth and death rates means that many developing countries now are growing at 3 to 4 percent per year. Such high growth rates in developing countries could boost total world population to 9 billion or more before the end of the twenty-first century. This raises what may be the two most important questions in this entire chapter: Why are birth rates not yet falling in these countries, and what can be done about it?

## There are reasons to be optimistic about population

Four conditions are necessary for a demographic transition to occur: (1) improved standard of living, (2) increased confidence that children will survive to maturity, (3) improved social status of women, and (4) increased availability and use of birth control. As the example of Thailand in the opening case study for this chapter shows, these conditions can be met, even in relatively poor countries.

Some demographers claim that a demographic transition already is in progress in most developing nations. Problems in taking censuses and a normal lag between falling death and birth rates may hide this for a time, but the world population should stabilize sometime in the next century. Some evidence supports this view. As we mentioned earlier in this chapter, fertility rates have fallen dramatically nearly everywhere in the world over the past half century.

Some countries have had remarkable success in population control. In Thailand, Indonesia, Colombia, and Iran, for instance, total fertility dropped by more than half in 20 years. Morocco, Dominican Republic, Jamaica, Peru, and Mexico all have seen fertility rates fall between 30 percent and 40 percent in a single generation. The following factors could contribute to stabilizing populations:

- Growing prosperity and social reforms that accompany development reduce the need and desire for large families in most countries.
- Technology is available to bring advances to the developing world much more rapidly than was the case a century ago, and the rate of technology transfer is much faster than it was when Europe and North America were developing.
- Less-developed countries have historic patterns to follow. They can benefit from our mistakes and chart a course to stability more quickly than they might otherwise do.
- Modern communications (especially television) have caused a revolution of rising expectations that act as a stimulus to spur change and development.

## Many people remain pessimistic about population growth

Economist Lester Brown takes a more pessimistic view. He warns that many of the poorer countries of the world appear to be caught in a "demographic trap" that prevents them from escaping from the middle phase of the demographic transition. Their populations are now growing so rapidly that human demands exceed the sustainable yield of local forests, grasslands, croplands, or water resources. The resulting resource shortages, environmental deterioration, economic decline, and political instability may prevent these countries from ever completing modernization. Their populations may continue to grow until catastrophe intervenes.

Many people argue that the only way to break out of the demographic trap is to immediately and drastically reduce population growth by whatever means are necessary. They argue strongly for birth control education and bold national policies to encourage lower birth rates. Some agree with Malthus that helping the poor will simply increase their reproductive success and further threaten the resources on which we all depend. Author Garret Hardin described this view as lifeboat ethics. "Each rich nation," he said, "amounts to a lifeboat full of comparatively rich people. The poor of the world are in other much more crowded lifeboats. Continuously, so to speak, the poor fall out of their lifeboats and swim for a while, hoping to be admitted to a rich lifeboat, or in some other way to benefit from the goodies on board. . . . We cannot risk the safety of all the passengers by helping others in need. What happens if you share space in a lifeboat? The boat is swamped and everyone drowns. Complete justice, complete catastrophe."

## Social justice is an important consideration

A third view is that **social justice** (a fair share of social benefits for everyone) is the real key to successful demographic transitions. The world has enough resources for everyone, but inequitable social and economic systems cause maldistributions of those resources. Hunger, poverty, violence, environmental degradation, and overpopulation are symptoms of a lack of social justice rather than a lack of resources. Although overpopulation exacerbates other problems, a narrow focus on this factor alone encourages racism and hatred of the poor. A solution for all these problems is to establish fair systems, not to blame the victims. Small nations and minorities often regard calls for population control as a form of genocide. Figure 7.16 expresses the opinion of many people in less-developed countries about the relationship between resources and population.

An important part of this view is that many of the rich countries are, or were, colonial powers, while the poor, rapidly growing countries were colonies. The wealth that paid for progress and security for developed countries was often extracted from colonies, which now suffer from exhausted resources, exploding populations, and chaotic political systems. Some of the world's poorest countries such as India, Ethiopia, Mozambique, and Haiti had rich resources and adequate food supplies before they were impoverished by colonialism. Those of us who now enjoy abundance may need to help the poorer countries not only as a matter of justice but because we all share the same environment.

In addition to considering the rights of fellow humans, we should also consider those of other species. Rather than ask what is the maximum number of humans that the world can possibly support, perhaps we should think about the needs of other creatures. As we convert natural landscapes into agricultural or industrial areas, species are crowded out that may have just as much right to exist as we do. Perhaps we should seek the optimum number of people at which we can provide a fair and decent life for all humans while causing the minimum impact on nonhuman neighbors.

**FIGURE 7.16** Controlling our population and resources—there may be more than one side to the issue.

## Women's rights affect fertility

Opportunities for education and paying jobs are critical factors in fertility rates (fig. 7.17). Child survival also is crucial in stabilizing population. When Infant and child mortality rates are high, as they are in much of the developing world, parents tend to have high numbers of children to ensure that some will survive to adulthood. There has never been a sustained drop in birth rates that was not first preceded by a sustained drop in infant and child mortality. One of the most important distinctions in our demographically divided world is the high infant mortality rates in the less-developed countries. Better nutrition, improved health care, simple oral rehydration therapy, and immunization against infectious diseases (chapter 8) have brought about dramatic reductions in child mortality rates, which have been accompanied in most regions by falling birth rates. It has been estimated that saving 5 million children each year from easily preventable communicable diseases would avoid 20 or 30 million extra births.

Increasing family income does not always translate into better welfare for children since men in many cultures control most financial assets. Often the best way to improve child survival is to ensure the rights of mothers. Land reform, political rights, opportunities to earn an independent income, and improved health status of women often are better indicators of total fertility and family welfare than rising GNP.

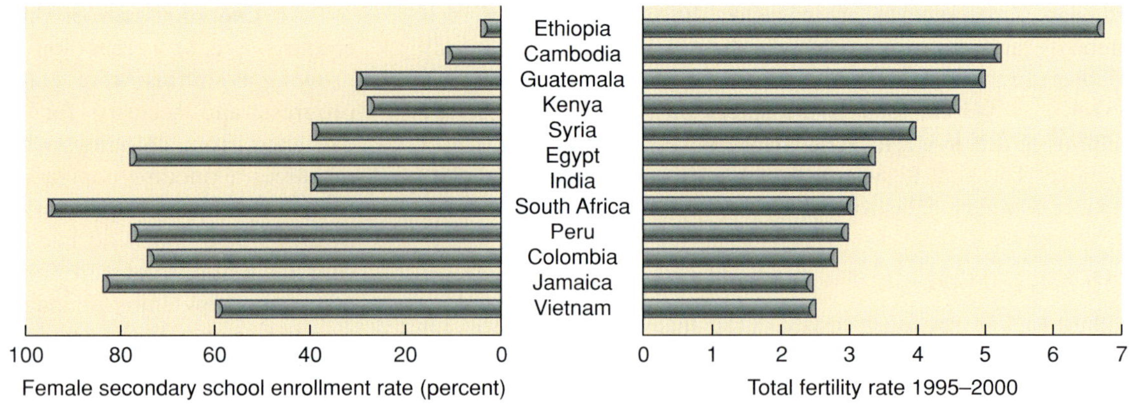


**FIGURE 7.17** Total fertility declines as women's education increases.
**Source:** Worldwatch Institute, 2003.

# 7.6 Family Planning Gives Us Choices

**Family planning** allows couples to determine the number and spacing of their children. It doesn't necessarily mean fewer children—people may use family planning to have the maximum number of children possible—but it does imply that the parents will control their reproductive lives and make rational, conscious decisions about how many children they will have and when those children will be born, rather than leaving it to chance. As the desire for smaller families becomes more common, birth control becomes an essential part of family planning in most cases. In this context, **birth control** usually means any method used to reduce births, including abstinence, delayed marriage, contraception, methods that prevent implantation of embryos, and induced abortions. As the opening case study in this chapter shows, there are many ways to encourage family planning.

## Fertility control has existed throughout history

Evidence suggests that people in every culture and every historic period have used a variety of techniques to control population size. Studies of hunting and gathering people, such as the !Kung or San of the Kalahari Desert in southwest Africa, indicate that our early ancestors had stable population densities, not because they killed each other or starved to death regularly, but because they controlled fertility.

For instance, San women breast-feed children for three or four years. When calories are limited, lactation depletes body fat stores and suppresses ovulation. Coupled with taboos against intercourse while breast-feeding, this is an effective way of spacing children. Other ancient techniques to control population size include abstinence, folk medicines, abortion, and infanticide. We may find some or all of these techniques unpleasant or morally unacceptable, but we shouldn't assume that other people are too ignorant or too primitive to make decisions about fertility.

## Today there are many options

Modern medicine gives us many more options for controlling fertility than were available to our ancestors. The major categories of birth control techniques include (1) avoidance of sex during fertile periods (for example, celibacy or the use of changes in body temperature or cervical mucus to judge when ovulation will occur), (2) mechanical barriers that prevent contact between sperm and egg (for example, condoms, spermicides, diaphragms, cervical caps, and vaginal sponges), (3) surgical methods that prevent release of sperm or egg (for example, tubal ligations in females and vasectomies in males), (4) hormone-like chemicals that prevent maturation or release of sperm or eggs or that prevent embryo implantation in the uterus (for example, estrogen plus progesterone, or progesterone alone, for females: gossypol for males), (5) physical barriers to implantation (for example, intrauterine devices), and (6) abortion.

**TABLE 7.5**

**Some Birth Control Methods and Pregnancy Prevention Rates**

| Method | Number of Women in 100 Who Become Pregnant |
|---|---|
| Sterilization (male, female) | <1 |
| IUD | <1 |
| Oral contraceptive (the Pill) | 1–2 |
| Hormones (implant, patch, injection, etc.) | 1–2 |
| Male condom | 11 |
| Sponge and spermicide | 14–28 |
| Female condom (e.g., cervical cap) | 15–23 |
| Diaphragm together with spermicide | 17 |
| Abstinence during fertile periods | 20 |
| Morning-after-pill (e.g., Preven) | 20 |
| Spermicide alone | 20–50 |
| Actively seeking pregnancy | 85 |

**Source:** U.S. Food and Drug Administration, *Birth Control Guide*, 2003 Revision.

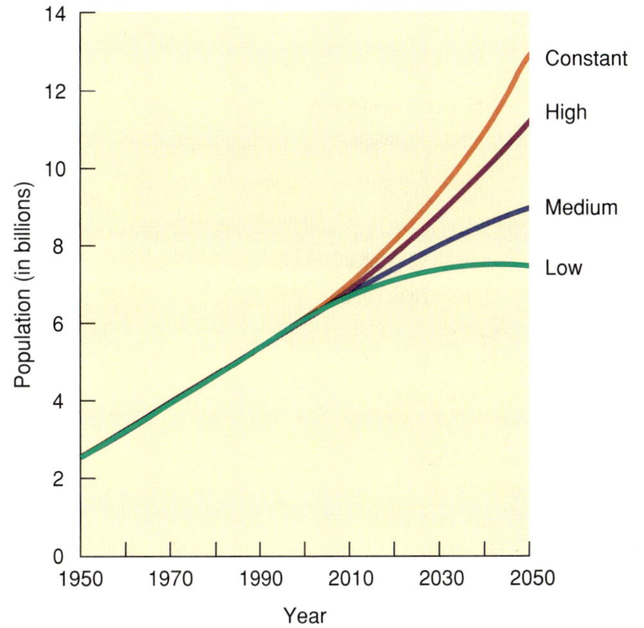


**FIGURE 7.18** Population projections for different growth scenarios. Recent progress in family planning and economic development have led to significantly reduced estimates compared to a few years ago. The medium projection is 8.9 billion in 2050, compared to previous estimates of over 10 billion for that date.
**Source:** UN Population Division, 2004.

Not surprisingly, the most effective birth control methods are also the ones most commonly used (table 7.5). In the United States, the majority of women younger than 30 who eventually want to become pregnant use the Pill. Most women over 35, with their child-bearing years behind them, choose sterilization. Male condom use is more effective than the remaining techniques in the table, and increases in effectiveness when used with a spermicide, Only 2 to 6 women in a hundred become pregnant in a year using this combination method. Condoms have the added advantage of protecting partners against sexually transmitted diseases, including AIDS, if they are made of latex and used correctly. That may partly explain why their use in the United States went from 3.5 million users in 1980 to 8 million in 2000. Condoms are an ancient birth control method; the Egyptians used them some 3,000 years ago.

More than 100 new contraceptive methods are now being studied, and some appear to have great promise. Nearly all are biologically based (e.g., hormonal), rather than mechanical (e.g., condom, IUD). Recently, the U.S. Food and Drug Administration approved five new birth control products. Four of these use various methods to administer female hormones that prevent pregnancy. Other methods are years away from use, but take a new direction entirely. Vaccines for women are being developed that will prepare the immune system to reject the hormone choriononic gonadotropin, which maintains the uterine lining and allows egg implant, or that will cause an immune reaction against sperm. Injections for men are focused on reducing sperm production, and have proven effective in mice. Without a doubt, the contemporary couple has access to many more birth control options than their grandparents had.

## 7.7 What Kind of Future Are We Creating?

How many people will be in the world a century from now? Most demographers believe that world population will stabilize sometime during the next century. The total number of humans, when we reach that equilibrium, is likely to be somewhere around 8 to 10 billion people, depending on the success of family planning programs and the multitude of other factors affecting human populations. Figure 7.18 shows three scenarios projected by the UN Population Division in its 2004 revision. The optimistic (low) projection shows that world population might reach about 7 billion in 2050, and then fall back below 6 billion by 2150. The medium projection suggests that growth might continue to around 8.9 billion in 2050, and then stabilize. The most pessimistic projection assumes a constant rate of growth (no change from present) to 25 billion people by 2150.

Which of these scenarios will we follow? As you have seen in this chapter, population growth is a complex subject. To accomplish a stabilization or reduction of human populations will require substantial changes from business as usual.

An encouraging sign is that worldwide contraceptive use has increased sharply in recent years. About half of the world's married couples used some family planning techniques in 2000, compared to only 10 percent 30 years earlier, but another 100 million couples say they want, but do not have access to, family planning. Contraceptive use varies widely by region. More than

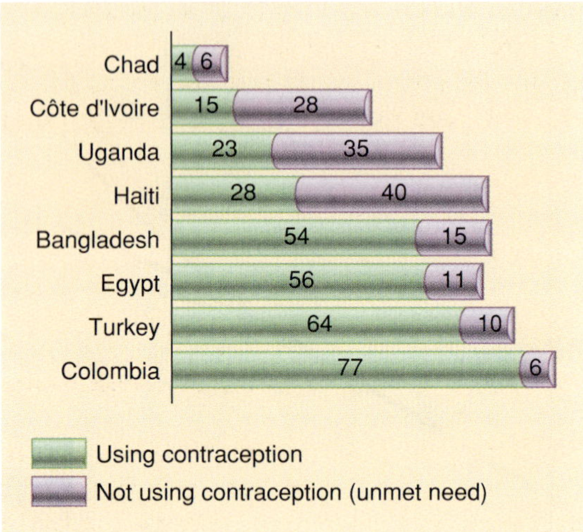


**FIGURE 7.19** Unmet need for family planning in selected countries. Globally, more than 100 million women in developing countries would prefer to avoid pregnancy but do not have access to family planning.
**Source:** Population Reference Bureau, 2003.

70 percent of women in Latin America use some form of birth control, compared to 51 percent in Asia (excluding China), and only 21 percent in Africa.

Figure 7.19 shows the unmet need for family planning among married women in some representative countries. When people in developing countries are asked what they want most, men say they want better jobs, but the first choice for a vast majority of women is family planning assistance. In general, a 15 percent increase in contraceptive use equates to about one fewer birth per woman per lifetime. In Chad, for example, where only 4 percent of all women use contraceptives, the average fertility is 6.6 children per woman. In Columbia, by contrast, where 77 percent of the women who would prefer not to be pregnant use contraceptives, the average fertility is 2.6.

## Religion and politics complicate family planning

In 1994, the United Nations convened a historic meeting in Cairo, Egypt, to discuss women's rights and population. The United States played a lead role in the International Conference on Population and Development (ICPD), which identified links between population growth, economic development, environmental degradation, and the social status of women and girls. To address these issues, 179 countries, including the United States, endorsed the goal of universally available reproductive health services, including family planning, by 2015.

Since 2000, however, the United States has refused to reaffirm the ICPD because it maintained that the document could be interpreted as promoting abortion—even though the ICPD clearly states, "In no case should abortion be promoted as a method of family planning."

In particular, the United States has withheld funds from the United Nations Population Fund (UNFPA) due to claims that, by working in China, the fund tacitly supports the forced abortions reported to be part of that country's one-child policy. A fact-finding team sent to China in 2002 found "no evidence of UNFPA knowledge of or support for such measures," but funding was still halted. Ms. Thoraya Obaid, executive director of the UNFPA, said her organization "does not, and never will condone or support coercive activities of any kind, anywhere."

She also said, "The denial of these funds will, unfortunately, significantly affect millions of women and children worldwide for whom the life-saving services provided by the UNFPA will have to be discontinued." She estimated that the funds withheld by the United States could have prevented 2 million unwanted pregnancies, 800,000 abortions, 4,700 maternal deaths, 60,000 cases of serious maternal illness, and more than 77,000 infant and child deaths.

Many Muslim countries encourage couples to have as many children as possible. Access to birth control is difficult or forbidden outright. Still, some Islamic governments recognize the need for family planning. Iran, for example, decided, in the 1990s, to promote smaller families. It succeeded in cutting birth rates by more than half in ten years.

The World Health Organization estimates that nearly 1 million conceptions occur daily around the world as a result of some 100 million sex acts. At least half of those conceptions are unplanned or unwanted. But there are still places where people desire large families (fig. 7.20).

Deep societal changes are often required to make family planning programs successful. Among the most important of these are (1) improved social, educational, and economic status

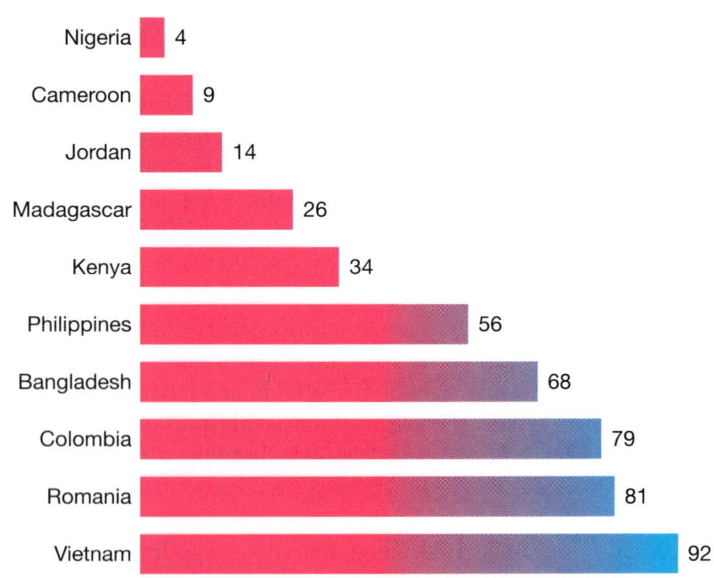


**FIGURE 7.20** Percent of married reproductive-age women with two living children who do not want another child.
**Data source:** UN Population Division, 2006.

for women (birth control and women's rights are often interdependent); (2) improved status for children (fewer children are born as parents come to regard them as valued individuals rather than possessions); (3) acceptance of calculated choice as a valid element in life in general and in fertility in particular (belief that we have no control over our lives discourages a sense of responsibility); (4) social security and political stability that give people the means and the confidence to plan for the future; (5) knowledge, availability, and use of effective and acceptable means of birth control.

## Conclusion

A few decades ago, we were warned that a human population explosion was about to engulf the world. Exponential population growth was seen as a cause or corollary to nearly every important environmental problem. Some people still warn that the total number of humans might grow to 30 or 40 billion by the end of this century. Birth rates have fallen, however, almost everywhere, and most demographers now believe that we will reach an equilibrium around 9 billion people in about 2050. Some claim that if we promote equality, democracy, human development, and modern family planning techniques, population might even decline to below its current level of 6.5 billion in the next 50 years. How we should carry out family planning and birth control remains a controversial issue. Should we focus on political and economic reforms, and hope that a demographic transition will naturally follow; or should we take more direct action (or any action) to reduce births?

Whether our planet can support 9 billion—or even 6 billion—people on a long-term basis remains a vital question. If all those people try to live at a level of material comfort and affluence now enjoyed by residents of the wealthiest nations, using the old, polluting, inefficient technology that we now employ, the answer is almost certain that even 6 billion people is too many in the long run. If we find more sustainable ways to live, however, it may be that 9 billion people could live happy, comfortable, productive lives. If we don't find new ways to live, we probably face a crisis no matter what happens to our population size. We'll discuss pollution problems, energy sources, and sustainability in subsequent chapters of this book.

## Reviewing Learning Outcomes

By now you should be able to explain the following points:

**7.1** Trace the history of human population growth.

- Human populations grew slowly until relatively recently.

**7.2** Summarize different perspectives on population growth.

- Does environment or culture control human populations?
- Technology can increase carrying capacity for humans.
- Population growth could bring benefits.

**7.3** Analyze some of the factors that determine population growth.

- How many of us are there?
- Fertility measures the number of children born to each woman.
- Mortality is the other half of the population equation.
- Life span and life expectancy describe our potential longevity.
- Living longer has demographic implications.
- Emigration and immigration are important demographic factors.

**7.4** Explain how ideal family size is culturally and economically dependent.

- Many factors increase our desire for children.
- Other factors discourage reproduction.
- Could we have a birth dearth?

**7.5** Describe how a demographic transition can lead to stable population size.

- Economic and social development influence birth and death rates.
- There are reasons to be optimistic about population.
- Many people remain pessimistic about population growth.
- Social justice is an important consideration.
- Women's rights affect fertility.

**7.6** Relate how family planning gives us choices.

- Fertility control has existed throughout history.
- Today there are many options.

**7.7** Reflect on what kind of future we are creating.

- Religion and politics complicate family planning.

## Practice Quiz

1. At what point in history did the world population pass its *first* billion? What factors restricted population before that time, and what factors contributed to growth after that point?
2. How might growing populations be beneficial in solving development problems?
3. Why do some economists consider human resources more important than natural resources in determining the future of a country?
4. Where will most population growth occur in the next century? What conditions contribute to rapid population growth in some countries?
5. Define *crude birth rate, total fertility rate, crude death rate,* and *zero population growth.*
6. What is the difference between life expectancy and life span?
7. What is dependency ratio, and how might it affect the United States in the future?
8. What pressures or interests make people want or not want to have babies?
9. Describe the conditions that lead to a demographic transition.
10. Describe some choices in modern birth control.

## Critical Thinking and Discussion Questions

1. What do you think is the optimum human population? The maximum human population? Are the numbers different? If so, why?
2. Some people argue that technology can provide solutions for environmental problems; others believe that a "technological fix" will make our problems worse. What personal experiences or worldviews do you think might underlie these positions?
3. Karl Marx called Thomas Malthus a "shameless sycophant of the ruling classes." Why would the landed gentry of the eighteenth century be concerned about population growth of the lower classes? Are there comparable class struggles today?
4. Try to imagine yourself in the position of a person your age in a developing world country. What family planning choices and pressures would you face? How would you choose among your options?
5. Some demographers claim that population growth has already begun to slow; others dispute this claim. How would you evaluate the competing claims of these two camps? Is this an issue of uncertain facts or differing beliefs? What sources of evidence would you accept as valid?
6. What role do race, ethnicity, and culture play in our immigration and population policies? How can we distinguish between prejudice and selfishness on one hand and valid concerns about limits to growth on the other?

## DATA analysis Telling a Story with Graphs

Graphs are pictorial representations of data designed to communicate information. They are particularly useful in conveying patterns and relationships. Most of us, when confronted by a page of numbers, tend to roll our eyes and turn to something else. A graph can simplify trends and highlight important information. If you read a newspaper or news magazine these days, you will undoubtedly see numerous graphs portraying a wide variety of statistics. Why are they used so widely and why do they make such powerful impressions on us?

Graphs can be used to explain, persuade, or inform. Like photographs or text, however, they also can be used to manipulate or even mislead the viewer. When you create a graph, you inevitably make decisions about what details to display, and which ones to omit. By selecting certain scales or methods of displaying data, you deliberately choose the message you intend to convey. It takes critical thinking skills to understand what a graph is (and isn't) showing.

Central questions addressed in this chapter are: How serious is the world population problem? Does it demand action? Many graphs present information about population growth, yet the impressions each conveys to the reader vary considerably, depending on the type of data used and the design of the graph. Reexamine figure 7.3, which presents a historical perspective on human population growth. What impressions do you draw from it? What conclusions do you suppose the person who originally created this graph intended for you to take away from this presentation? Does it give the impression that human numbers are growing explosively? Does choosing a 5,000-year time horizon emphasize this idea?

Now look at figure 1. It plots world population growth, as does figure 7.3, but gives a very different picture. By choosing only a 100-year time horizon, the author of figure 1 makes it much easier to see precisely what's happening in this century, but provides a very different emotional reaction than figure 7.3. Clearly,

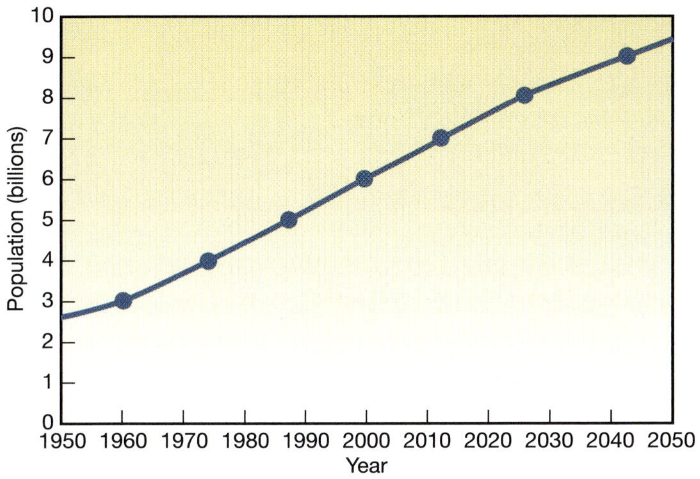


**FIGURE 1** Growth of the world population, 1950–2050.
**Source:** U.S. Census Bureau, 2006.

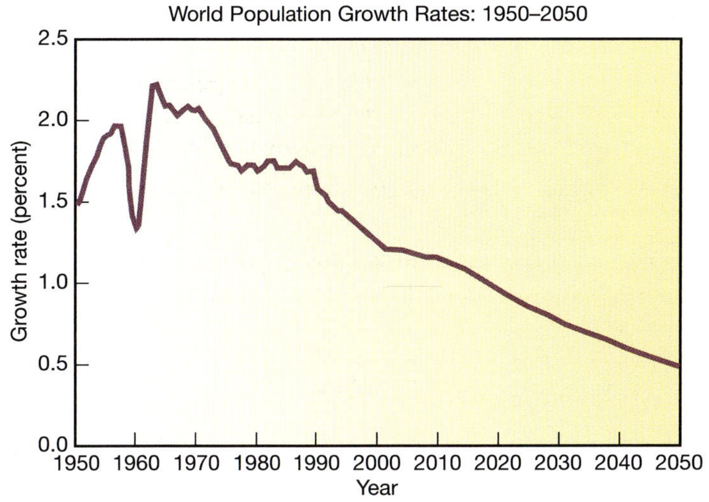


**FIGURE 2** World population growth rates, 1950–2050.
**Source:** U.S. Census Bureau, 2006.

these two graphs were intended to convey different messages. Is one more accurate than the other? Not necessarily; they're just showing different aspects of the same data set. But if you weren't aware that there's more than one way to display information, you might think that only one presentation is valid.

Now study figure 2. Why does it look so different from the other two? Notice that the Y-axis is rate of growth rather than total population. Again, this graph can be derived from the same data set, but it gives yet another message. The rate of population growth is slowing. Fifty years from now, we could be approaching a zero growth rate. This doesn't address what the total population will be at that point. It could well be the 9 billion predicted by figure 1. In fact, if you look closely at the curve between 2040 and 2050 in figure 1, you might detect that the rate of growth is slowing, but when the Y-scale is in billions of people, the difference between a 0.7 and 0.5 percent growth rate is very small.

What do we learn from this exercise? All three graphs are probably correct, but each was designed to communicate different information. It's helpful to ask yourself, when you look at each graph, what did the author intend to say? Are there other ways to display these same data? Would it give a different impression, or tell a different story if you did so?

**For Additional Help in Studying This Chapter,** please visit our website at www.mhhe.com/cunningham10e. You will find additional practice quizzes and case studies, flashcards, regional examples, place markers for Google Earth™ mapping, and an extensive reading list, all of which will help you learn environmental science.

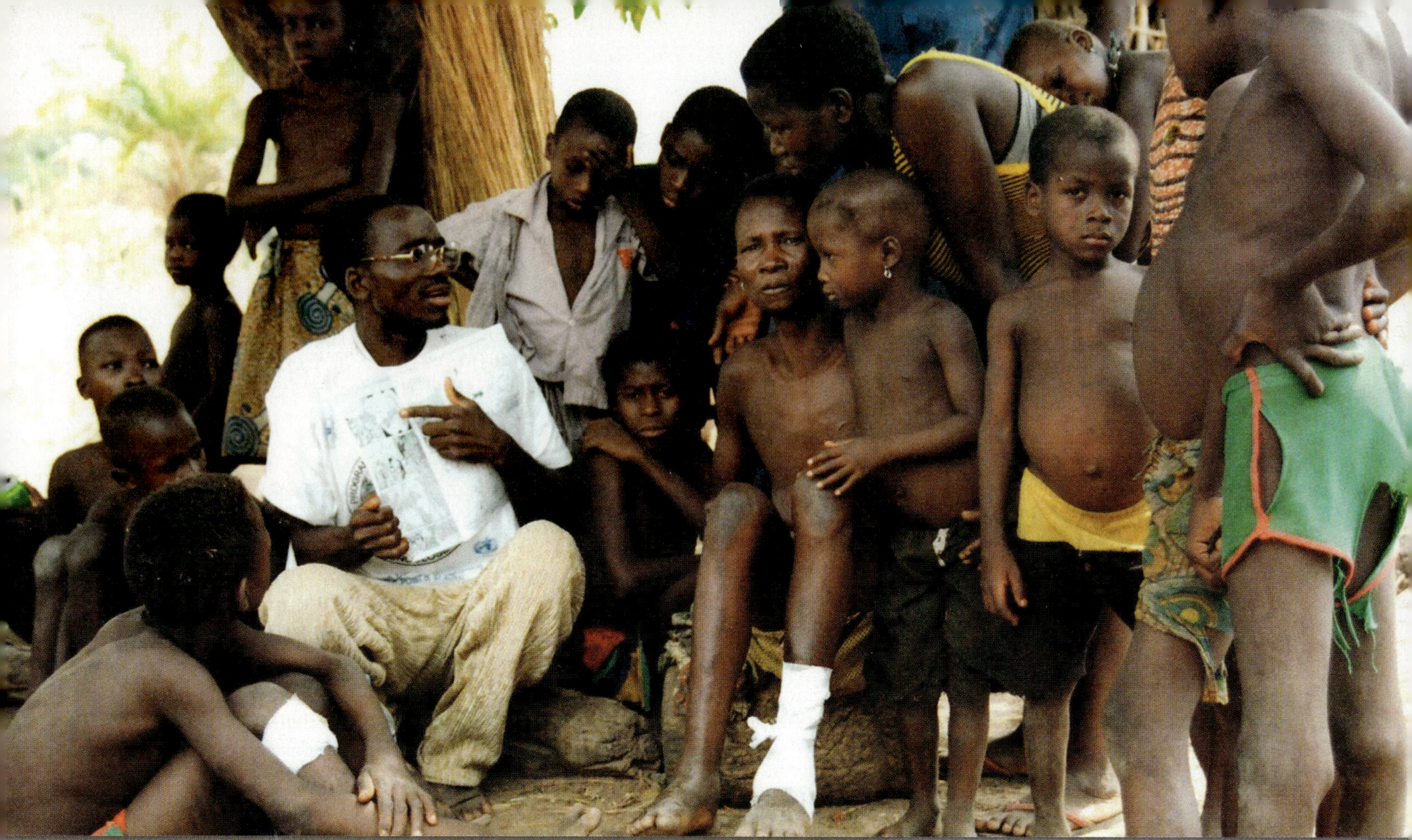

A public health worker explains to villagers that guinea worm is a parasitic disease, not the result of bad behavior or sorcery. Once the most guinea-worm-plagued region in the world, West Africa is now almost totally free of this terrible disease.

CHAPTER 8

# Environmental Health and Toxicology

*To wish to become well is a part of becoming well.*

*—Seneca—*

## LEARNING OUTCOMES

After studying this chapter, you should be able to:

**8.1** Describe health and disease and how global disease burden is now changing.

**8.2** Summarize the principles of toxicology.

**8.3** Discuss the movement, distribution, and fate of toxins in the environment.

**8.4** Characterize mechanisms for minimizing toxic effects.

**8.5** Explain ways we measure and describe toxicity.

**8.6** Evaluate risk assessment and acceptance.

**8.7** Relate how we establish health policy.

# Case Study Defeating the Fiery Serpent

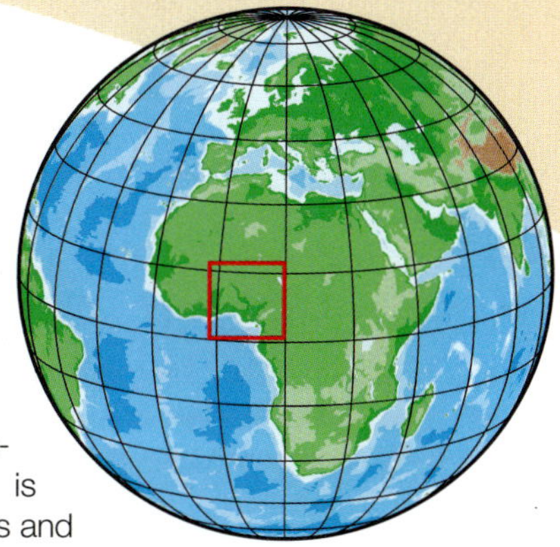

Fighting back tears, an African child cradles her swollen foot as a thin, white worm emerges from an oozing sore. The unfortunate youngster has been invaded by a guinea worm (*Dracunculus medinensis*). The pain is intense. Her whole leg feels like it's on fire. It's difficult to walk or work when you're afflicted with one of these terrible invaders. The disease is known in Africa as "empty granary," because the worms usually erupt during harvest season, making it impossible to work in the fields and harvest crops on which the whole family depends.

The infection cycle starts when someone suffering from the pain of an emerging worm (fig. 8.1) bathes their wound in a local lake or pond. The worm, sensing water, emerges to release thousands of larvae, which are ingested by freshwater copepods ("water fleas"). Inside the water fleas, the larvae develop into the infective stage in about two weeks. When villagers drink the contaminated water, the copepods are digested, but the worms survive and penetrate the wall of the intestine and move to the abdominal cavity. Over the next year, the worm grows to about a meter long (3 feet) and as thick as a spaghetti noodle. When fully grown, the worm migrates to the site where it will erupt, usually in the legs or feet—or even eye sockets—of victims. It takes several weeks for the worm to emerge completely. If you pull too hard on it, the worm breaks, and the part left in your body festers and decays. If the suffering host soaks the lesion in water to soothe the pain, the cycle begins again.

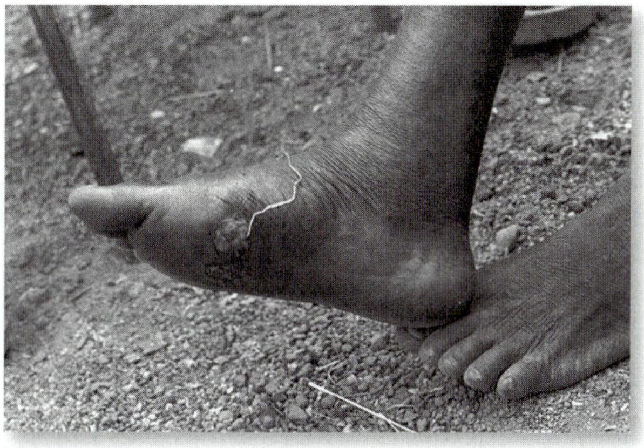


**FIGURE 8.1** A guinea worm emerges from a patient's foot.

This terrible parasite has plagued tropical countries for thousands of years. It has been found in Egyptian mummies and is thought to be the "fiery serpent" described in the Old Testament as torturing the Israelites in the desert. As recently as 1986, at least 3 million people in 16 countries suffered from this affliction, and more than 100 million people were at risk worldwide. But there's a happy ending to this story. In 1988, the world health community, led by former U.S. President Jimmy Carter, began a campaign to eliminate guinea worms.

Eradicating the parasite is fairly simple and relatively inexpensive. There is no medicine to cure the infection once the worm is inside your body, but wells can be drilled to prevent patients from contaminating drinking-water sources. And local ponds can be treated with pesticides that are safe for human consumption but kill the worm larvae and copepods. Furthermore, families can be taught to pour their drinking water through a fine cloth filter to remove any remaining water fleas. The problem is simply to get the proper materials and information to remote villages. Having the prestige of a former American president has been a powerful tool in convincing local officials that the world cares, and that a cure is possible.

Progress has been spectacular. Only a few countries still are afflicted by this horrendous disease. Nigeria is an example of this remarkable success. In 1986, the country was the most Guinea-worm-plagued country in the world, with more than 650,000 cases in 36 states. By 2006, more than 99.9 percent of the infections had been eliminated, and only 120 people still suffered from infections. Worldwide, there now are fewer than 12,000 cases, mostly in Sudan, where civil war has made public health intervention difficult. When guinea worms are finally eradicated, it will be only the second disease ever completely eliminated (smallpox, which was abolished in 1977 was first), and the only human parasite totally exterminated worldwide.

An encouraging outcome of this crusade is the demonstration that public health education and community organization can be effective, even in some of the poorest and most remote areas. Once people understand how the disease spreads and what they need to do to protect themselves and their families, they do change their behavior. And when the campaign is completed and guinea worms are completely vanquished, the health workers and volunteers will be available for further community development projects.

This case study reminds us of the importance of public health and how susceptible humans have always been to diseases and contaminants. In this chapter we'll look at the principles of environmental health to help you understand some of the risks we face and what we might do about them.

## 8.1 Environmental Health

What is health? The WHO defines **health** as a state of complete physical, mental, and social well-being, not merely the absence of disease or infirmity. By that definition, we all are ill to some extent. Likewise, we all can improve our health to live happier, longer, more productive, and more satisfying lives if we think about what we do.

What is disease? A **disease** is an abnormal change in the body's condition that impairs important physical or psychological functions. Diet and nutrition, infectious agents, toxic substances, genetics, trauma, and stress all play roles in **morbidity** (illness)

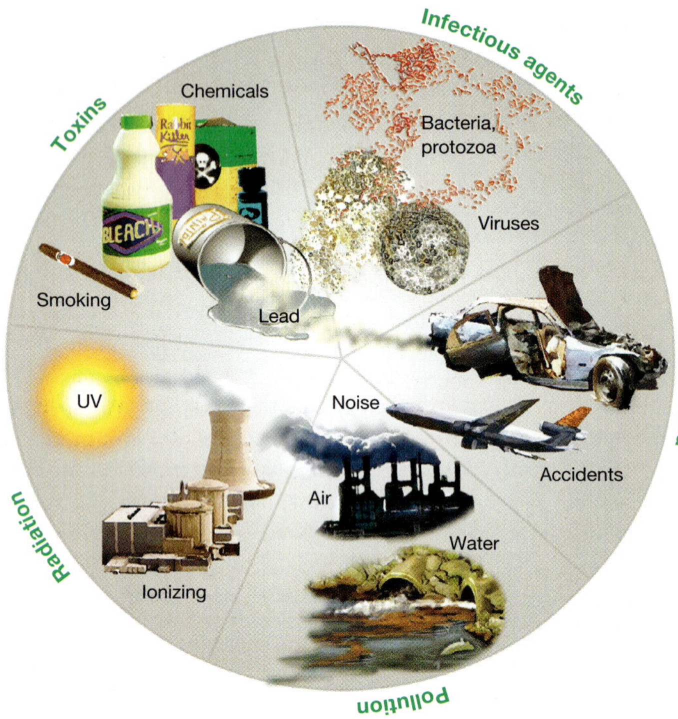


**FIGURE 8.2** Major sources of environmental health risks.

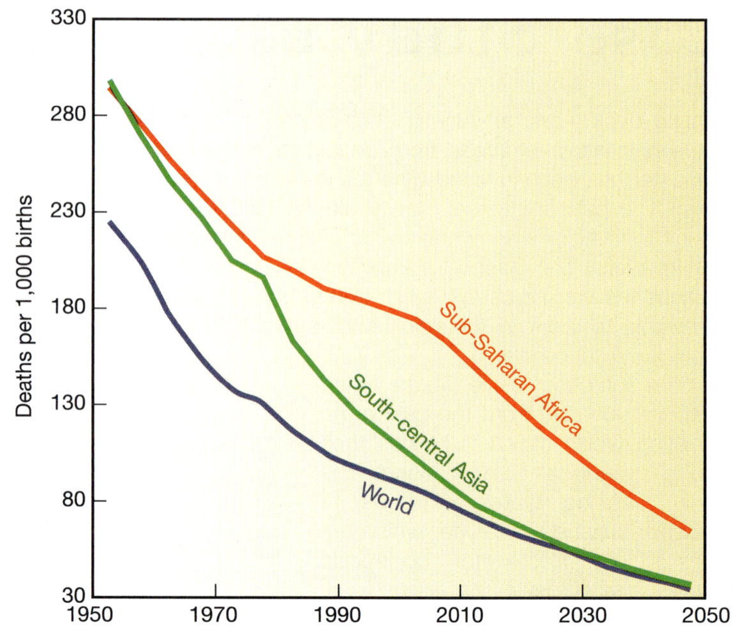


**FIGURE 8.3** Child mortality has fallen dramatically over the past 50 years and is expected to continue this decline in the future. Note that south-central Asia saw a significant improvement in child survival in the 1970s and 80s, while sub-Saharan Africa lagged behind—mainly because of wars and AIDS. Data from the UN Population Division, 2006.

and **mortality** (death). **Environmental health** focuses on external factors that cause disease, including elements of the natural, social, cultural, and technological worlds in which we live. Figure 8.2 shows some major environmental disease agents as well as the media through which we encounter them. Ever since the publication of Rachel Carson's *Silent Spring* in 1962, the discharge, movement, fate, and effects of synthetic chemical toxins have been a special focus of environmental health. Later in this chapter, we'll study these topics in detail. First, however, let's look at some of the major causes of illness worldwide.

## The global disease burden is changing

In addition to tremendous progress in eradicating guinea worms, there are many other public health successes. Smallpox was completely wiped out in 1977. Polio has been eliminated everywhere in the world except for a few remote villages in northern Nigeria. Epidemics of typhoid fever, cholera, and yellow fever that regularly killed thousands of people in North America a century ago are now rarely encountered. AIDS, which once was an immediate death sentence, has become a highly treatable disease. The average HIV-positive person in the United States now lives 24 years after diagnosis if treated faithfully with modern medicines.

A way of demonstrating these advances is to look at how much longer we're living. During the twentieth century, world average life expectancies more than doubled from 30 to 64.3 years. For richer countries, residents can expect to live, on average about three times as long as their great grandparents did a century earlier (see chapters 6 and 7).

A vital component of rising life expectancies is declining child mortality. In 1950, almost one-quarter (224 per 1,000) of all children born worldwide didn't live to their fifth birthday (fig. 8.3). By 2000, this rate had fallen to 86 per 1,000. By 2050, it's expected to be nearly 90 percent less than a century earlier. This is essential if we are to reach zero population growth. Sub-Saharan Africa and South-Central Asia had much higher child mortality than the world average in 1950. Asia has caught up with the rest of the world, but Africa, for a variety of reasons, lags behind.

In the past, health organizations have focused on the leading causes of death as the best summary of world health. Mortality data, however, fail to capture the impacts of nonfatal outcomes of disease and injury, such as dementia or blindness, on human well-being. When people are ill, work isn't done, crops aren't planted or harvested, meals aren't cooked, and children can't study and learn. Health agencies now calculate **disability-adjusted life years (DALYs)** as a measure of disease burden. DALYs combine premature deaths and loss of a healthy life resulting from illness or disability. This is an attempt to evaluate the total cost of disease, not simply how many people die. Clearly, many more years of expected life are lost when a child dies of neonatal tetanus than when an 80-year-old dies of pneumonia. Similarly, a teenager permanently paralyzed by a traffic accident will have many more years of suffering and lost potential than will a senior citizen who has a stroke. According to the WHO, chronic diseases now account for nearly 60 percent of the 56.5 million total deaths worldwide each year and about half of the global disease burden.

**TABLE 8.1**

**Leading Causes of Global Disease Burden**

| Rank | 1990 | Rank | 2020 |
|---|---|---|---|
| 1 | Pneumonia | 1 | Heart disease |
| 2 | Diarrhea | 2 | Depression |
| 3 | Perinatal conditions | 3 | Traffic accidents |
| 4 | Depression | 4 | Stroke |
| 5 | Heart disease | 5 | Chronic lung disease |
| 6 | Stroke | 6 | Pneumonia |
| 7 | Tuberculosis | 7 | Tuberculosis |
| 8 | Measles | 8 | War |
| 9 | Traffic accidents | 9 | Diarrhea |
| 10 | Birth defects | 10 | HIV/AIDS |
| 11 | Chronic lung disease | 11 | Perinatal conditions |
| 12 | Malaria | 12 | Violence |
| 13 | Falls | 13 | Birth defects |
| 14 | Iron anemia | 14 | Self-inflicted injuries |
| 15 | Malnutrition | 15 | Respiratory cancer |

**Source:** World Health Organization, 2002.

The world is now undergoing a dramatic epidemiological transition. Chronic conditions, such as cardiovascular disease and cancer, no longer afflict only wealthy people. Although the traditional killers in developing countries—infections, maternal and perinatal (birth) complications, and nutritional deficiencies—still take a terrible toll, diseases such as depression and heart attacks that once were thought to occur only in rich countries are rapidly becoming the leading causes of disability and premature death everywhere.

In 2020, the WHO predicts heart disease, which was fifth in the list of causes of global disease burden a decade ago, will be the leading source of disability and deaths worldwide (table 8.1). Most of that increase will be in the poorer parts of the world where people are rapidly adopting the lifestyles and diet of the richer countries. Similarly, global cancer rates will increase by 50 percent. By 2020, it's expected that 15 million people will have cancer and 9 million will die from it.

A silent epidemic of diabetes is now sweeping through our population. It's estimated that one-third of all children born today in North America will develop this disease in their lifetime. Obesity, diets high in sugar and fat, lack of exercise, and poverty (which encourages fast-food intake and makes health food unavailable) all play important roles in this disease. Blindness, circulatory problems, and kidney failure are common results of severe, uncontrolled diabetes. Seventy percent of all lower limb amputations are diabetes-related. In some Native American groups, more than half of all adults have this disease. It used to be thought that diabetes only affected rich people, but obesity, sedentary lifestyles, and diabetes are spreading around the world.

Taking disability as well as death into account in our assessment of disease burden reveals the increasing role of mental health as a worldwide problem. WHO projections suggest that psychiatric and neurological conditions could increase their share of the global burden from 10 percent currently to 15 percent of the total load by 2020. Again, this isn't just a problem of the developed world. Depression is expected to be the second largest cause of all years lived with disability worldwide, as well as the cause of 1.4 percent of all deaths. For women in both developing and developed regions, depression is the leading cause of disease burden, while suicide, which often is the result of untreated depression, is the fourth largest cause of female deaths.

Notice in table 8.1 that diarrhea, which was the second leading cause of disease burden in 1990, is expected to be ninth on the list in 2020, while measles and malaria are expected to drop out of the top 15 causes of disability. Tuberculosis, which is becoming resistant to antibiotics and is spreading rapidly in many areas (especially in the former Soviet Union), is the only infectious disease whose ranking is not expected to change over the next 20 years. Traffic accidents are now soaring as more people drive. War, violence, and self-inflicted injuries similarly are becoming much more important health risks than ever before.

Chronic obstructive lung diseases (e.g., emphysema, asthma, and lung cancer) are expected to increase from eleventh to fifth in disease burden by 2020. A large part of the increase is due to rising use of tobacco in developing countries, sometimes called "the tobacco epidemic." Every day about 100,000 young people—most of them in poorer countries—become addicted to tobacco. At least 1.1 billion people now smoke, and this number is expected to increase at least 50 percent by 2020. If current patterns persist, about 500 million people alive today will eventually be killed by tobacco. This is expected to be the biggest single cause of death worldwide (because illnesses such as heart attack and depression are triggered by multiple factors). In 2003, the World Health Assembly adopted a historic tobacco-control convention that requires countries to impose restrictions on tobacco advertising, establish clean indoor air controls, and clamp down on tobacco smuggling. Dr. Gro Harlem Brundtland, former director-general of the WHO, predicted that the convention, if implemented, could save billions of lives.

***Think About It***

What changes could you make in your lifestyle to lessen your risks from the diseases in table 8.1? What would have the greatest impact on your future well-being?

## Infectious and emergent diseases still kill millions of people

Although the ills of modern life have become the leading killers almost everywhere in the world, communicable diseases still are responsible for about one-third of all disease-related mortality. Diarrhea, acute respiratory illnesses, malaria, measles, tetanus, and a few other infectious diseases kill about 11 million children

**FIGURE 8.4** At least 3 million children die every year from easily preventable diseases. This billboard in Guatemala encourages parents to have their children vaccinated against polio, diphtheria, TB, tetanus, pertussis (whooping cough), and scarlet fever.

under age five every year in the developing world. Better nutrition, clean water, improved sanitation, and inexpensive inoculations could eliminate most of those deaths (fig. 8.4).

A wide variety of **pathogens** (disease-causing organisms) afflict humans, including viruses, bacteria, protozoans (single-celled animals), parasitic worms, and flukes (fig. 8.5). The greatest loss of life from an individual disease in a single year was the great influenza pandemic of 1918. Epidemiologists now estimate that at least one-third of all humans living at the time were infected, and that between 50 to 100 million died. Businesses, schools, churches, and sports or entertainment events were shut down for months. Public health authorities warn that if bird flu picks up the ability to spread through the human population as well as it moves from bird to bird, it could sicken 2 billion people, kill up to 150 million, and bring the world economy to a standstill. Influenza is caused by a family of viruses (fig. 8.5*a*) that mutate rapidly and move from wild and domestic animals to humans, making control of this disease very difficult.

Every year there are 76 million cases of foodborne illnesses in the United States, resulting in 300,000 hospitalizations and 5,000 deaths. Both bacteria and intestinal protozoa cause these illnesses (fig. 8.5*b* and *c*). They are spread from feces through food and water. In 2006, a deadly strain of *E. coli* was spread in lettuce that had been fertilized with cattle manure. At any given time, around 2 billion people—one-third of the world population—suffer from worms, flukes, and other internal parasites. Guinea worms (see opening case study) are one example, while people rarely die from parasites, they can be extremely debilitating, and can cause poverty that leads to other, more deadly, diseases.

Malaria is one of the most prevalent remaining infectious diseases. Every year about 500 million new cases of this disease occur, and about 2 million people die from it. The territory infected by this disease is expanding as global climate change allows mosquito vectors to move into new territory. Simply providing insecticide-treated bed nets and a few dollars worth of chloroquine pills could prevent tens of millions of cases of this debilitating disease every year. Tragically, some of the countries where malaria is most widespread tax both bed nets and medicine as luxuries, placing them out of reach for ordinary people.

**Emergent diseases** are those not previously known or that have been absent for at least 20 years. The new strain of bird flu now spreading around the world is a good example. There have been at least 39 outbreaks of emergent diseases over the past two decades, including the extremely deadly Ebola and Marburg fevers, which have afflicted Central Africa in at least six different locations in the past decade. Similarly, cholera, which had been absent from South America for more than a century, reemerged in Peru in 1992 (fig. 8.6). Some other examples include a new drug-resistant form of tuberculosis, now spreading in Russia; dengue fever, which is spreading through Southeast Asia and the Caribbean; malaria, which is reappearing in places where it had

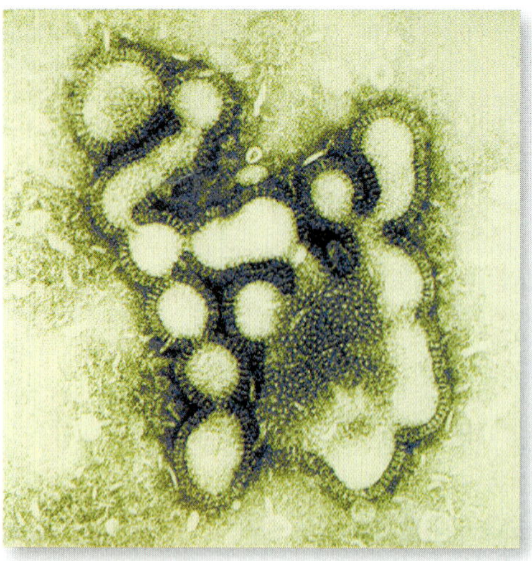
(a) Influenza virus

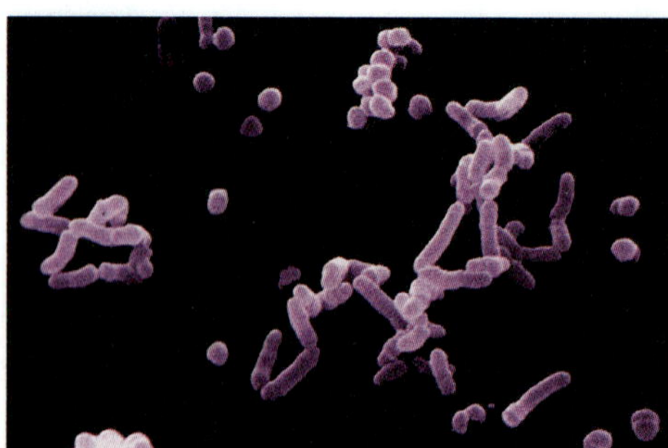
(b) Pathogenic bacteria

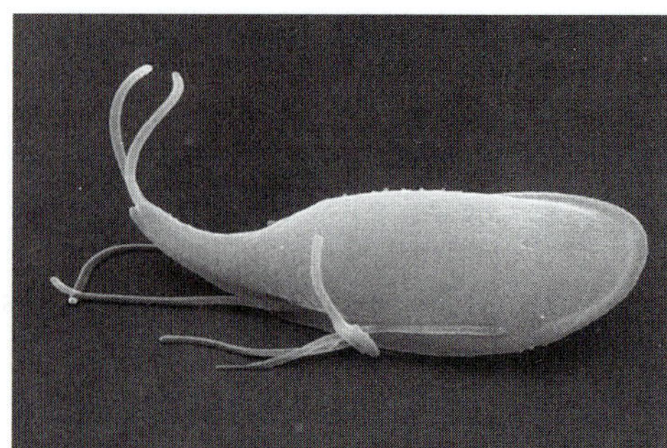
(c) *Giardia*

**FIGURE 8.5** (a) A group of influenza viruses magnified about 300,000 times. (b) Pathogenic bacteria magnified about 50,000 times. (c) *Giardia,* a parasitic intestinal protozoan, magnified about 10,000 times.

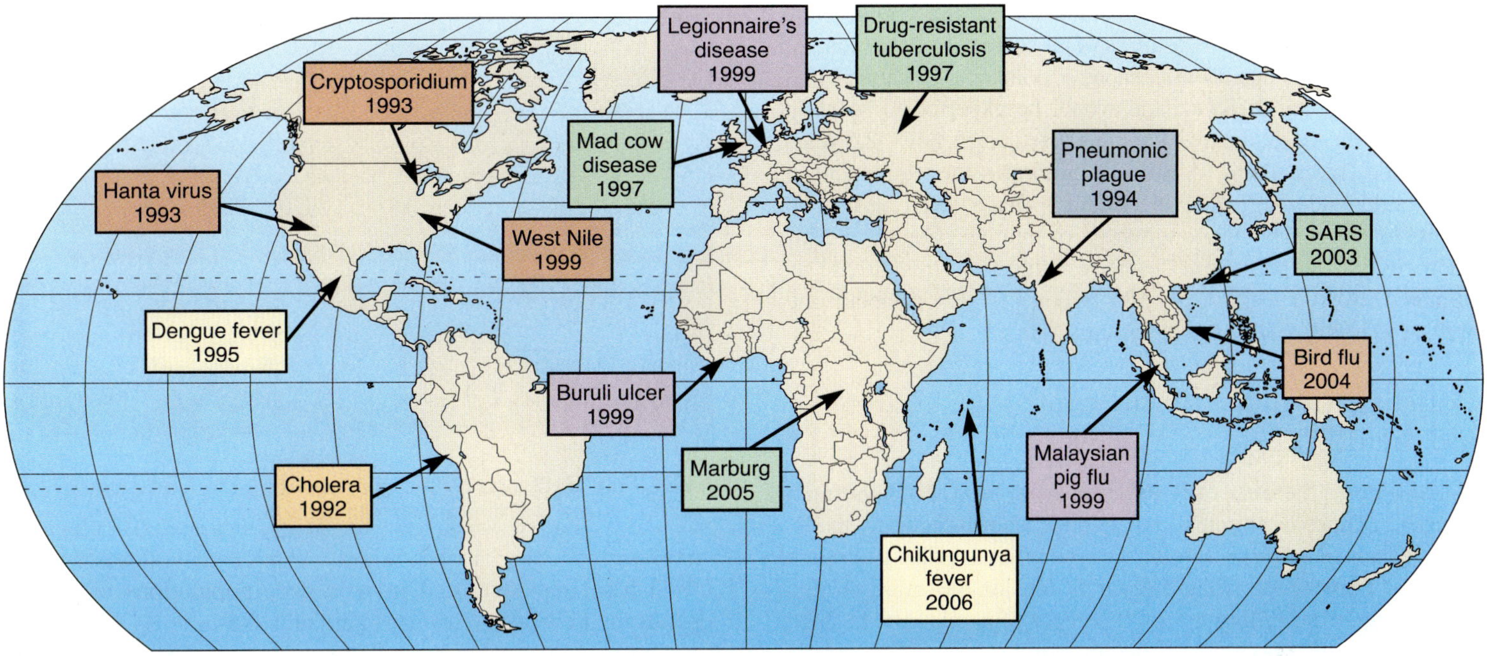


**FIGURE 8.6** Some recent outbreaks of highly lethal infectious diseases. Why are supercontagious organisms emerging in so many different places?
**Source:** Data from U.S. Centers for Disease Control and Prevention.

been nearly eradicated; and a new human lymphotropic virus (HTLV), which is thought to have jumped from monkeys into people in Cameroon who handled or ate bushmeat. Theses HTLV strains are now thought to infect 25 million people.

Growing human populations push people into remote areas where they encounter diseases that may have existed for a long time, but only now are exposed to humans. Rapid international travel makes it possible for these new diseases to spread around the world at jet speed. Epidemiologists warn that the next deadly epidemic is only a plane ride away.

West Nile virus shows how fast new diseases can travel. West Nile belongs to a family of mosquito-transmitted viruses that cause encephalitis (brain inflammation). Although recognized in Africa in 1937, the West Nile virus was absent from North America until 1999, when it apparently was introduced by a bird or mosquito from the Middle East. The disease spread rapidly from New York, where it was first reported, throughout the eastern United States in only two years (fig. 8.7). Now, it's found everywhere in the lower 48 states. The virus infects 230 species of animals, including 130 bird species. By 2006, more than 1,000 people had died from this disease.

The largest recent death toll from an emergent disease is HIV/AIDS. Although virtually unknown 15 years ago, acquired immune deficiency syndrome has now become the fifth greatest cause of contagious deaths. The WHO estimates that 60 million people are now infected with the human immune-deficiency virus, and that 3 million die every year from AIDS complications. Although two-thirds of all current HIV infections are now in sub-Saharan Africa, the disease is spreading rapidly in South and East Asia. Over the next 20 years, there could be an additional 65 million AIDS deaths.

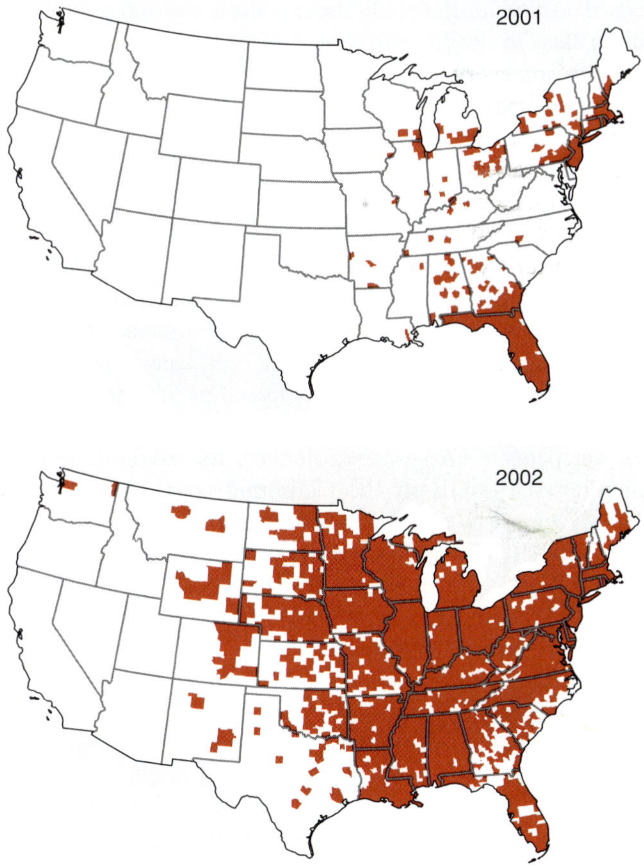


**FIGURE 8.7** The spread of West Nile virus in birds, 2001–2002. By 2003, it occurred throughout the 48 contiguous states.
**Source:** CDC and USGS.

In Swaziland, health officials estimate about one-third of all adults are HIV positive and that two-thirds of all current 15-year-olds will die of AIDS before age 50. As chapter 7 points out, without AIDS, the life expectancy in Swaziland would be expected to be about 65 years. With AIDS, the average life expectancy is now 32.6 years. Worldwide, more than 14 million children—the equivalent of every child under age five in America—have lost one or both parents to AIDS. The economic costs of treating patients and lost productivity from premature deaths resulting from this disease are estimated to be at least (U.S.) $35 billion per year or about one-tenth of the total GDP of sub-Saharan Africa.

## Conservation medicine combines ecology and health care

Humans aren't the only ones to suffer from new and devastating diseases. Domestic animals and wildlife also experience sudden and widespread epidemics, which are sometimes called **ecological diseases.** Ebola hemorrhagic fever is one of the most virulent viruses ever seen, killing up to 90 percent of its victims. In 2002, an outbreak of Ebola fever began killing humans along the Gabon-Congo border. A few months later, researchers found that 221 of the 235 western lowland gorillas they had been studying in this area disappeared in just a few months. Many chimpanzees also died. Although the study team could only find a few of the dead gorillas, 75 percent of those tested positive for Ebola. Altogether, researchers estimate that 5,000 gorillas died in this small area of the Congo. Extrapolating to all of central Africa, it's possible that Ebola has killed one-quarter of all the gorillas in the world. It's thought that the spread of this disease in humans resulted from the practice of hunting and eating primates.

A viral hemorrhagic septicemia much like Ebola is spreading through fish in North America's Great Lakes. It is highly infectious and also has an extremely high fatality rate. The virus infects about 40 species of fish. Shipments of game fish, such as salmon, trout, walleye, and perch, have been banned across much of America in an effort to slow the spread of this disease, which probably was introduced in ballast water of ships from Europe.

A cat parasite (*Toxoplasma gondii*), for example, is killing endangered sea otters off the California coast. The disease is apparently spread by cat owners who empty kitty litter boxes into sewer systems that empty into the ocean. In another human/wildlife connection, 18 percent of all stranded California sea lions have tumors caused by the herpes I virus, which also causes Kaposi's sarcoma in AIDS patients. Presumably, this virus also is spread through sewer systems.

Chronic wasting disease (CWD) is spreading through deer and elk populations in North America (fig. 8.8). Caused by a strange protein called a prion, CWD is one of a family of irreversible, degenerative neurological diseases known as transmissible spongiform encephalopathies (TSE) that include mad cow disease in cattle, scrapie in sheep, and Creutzfelt-Jacob disease in humans. CWD probably started when elk ranchers fed contaminated animal by-products to their herds. Infected animals were sold to other ranches, and now the disease has spread to wild populations. First recognized in 1967 in Saskatchewan, CWD has been identified in wild deer populations and ranch operations in at least eleven American states.

**FIGURE 8.8** Wild elk and deer are widely affected by chronic wasting disease, which originated with domestic herds.

No humans are known to have contracted TSE from deer or elk, but there is a concern that we might see something like the mad cow disaster that inflicted Europe in the 1990s. At least 100 people died, and nearly 5 million European cattle and sheep were slaughtered in an effort to contain that disease.

In 1995, California residents noticed that the oak trees were dying from a fast-spreading new disease known as sudden oak death syndrome (SODS), The disease is caused by funguslike organism known as *Phytophthora ramorum,* part of a group known as water molds. Possibly imported with Asian rhododendrons shipped from European plant nurseries, this organism is a close relative of the pathogen that caused Ireland's potato blight and another species that has wiped out vast swaths of forest in western Australia. Recently, foresters reported that Douglas fir, one of the nation's most economically important timber species, and California coast redwood also are infected with SODS. In addition to being beautiful trees and dominant members of forest communities, each of these species provides commercial products worth more than a billion dollars per year in the United States (fig. 8.9).

One thing that emergent diseases in humans and ecological diseases in natural communities have in common is environmental change that stresses biological systems and upsets normal ecological relationships. We cut down forests and drain wetlands, destroying habitat for native species. Invasive organisms and diseases are accidentally or intentionally introduced into new areas where they can grow explosively. Increasing incursion into former wilderness is spurred by human population growth and ecotourism. In 1950, only about 3 million people a year flew on commercial jets; by 2000, more than 300 million did. Diseases can spread around the globe in mere days as people pass through international travel hubs.

We are coming to recognize that the delicate ecological balances that we value so highly—and disrupt so frequently—are important to our own health. **Conservation medicine** is an emerging discipline that attempts to understand how our environmental changes threaten our own health as well as that of the natural communities on which we depend for ecological services.

FIGURE 8.9 The pathogen that causes sudden oak death also infects redwoods, Douglas fir, and dozens of other species.

While still small, this new field is gaining recognition from mainstream funding sources such as the World Bank, the World Health Organization, and the U.S. National Institutes of Health.

## Resistance to drugs, antibiotics, and pesticides is increasing

Malaria, the most deadly of all insect-borne illnesses, is an example of the return of a disease that once was thought to be nearly vanquished. Malaria now claims about a million lives every year—90 percent are in Africa, and most of them children. With the advent of modern medicines and pesticides, malaria had nearly been wiped out in many places but recently has come roaring back. The protozoan parasite that causes the disease is now resistant to most drugs, while the mosquitoes that transmit it have developed resistance to many insecticides. Spraying of DDT in India and Sri Lanka, for instance, reduced malaria from millions of infections per year to only a few thousand in the 1950s and 1960s. Now South Asia is back to its pre-DDT level of about half a million new cases of malaria every year. Other places that never had cases of malaria now have them as a result of climate change and habitat alteration.

Why have vectors such as mosquitoes, and pathogens such as the malaria parasite become resistant to pesticides and drugs? Part of the answer is natural selection and the ability of many organisms to evolve rapidly. Another factor is the human tendency to use control measures carelessly. When we discovered that DDT and other insecticides could control mosquito populations, we spread them everywhere. This not only harmed wildlife and beneficial insects, but it created perfect conditions for natural selection. Many pests and pathogens were exposed only minimally to control measures, allowing those with natural resistance to survive and spread their genes through the population (fig. 8.10). After repeated cycles of exposure and selection, many microorganisms and their vectors are insensitive to almost all our weapons against them.

As chapter 9 discusses, raising huge numbers of cattle, hogs, and poultry in densely packed barns and feedlots helps spread antibiotic resistance in pathogens. Confined animals are dosed constantly with antibiotics and steroid hormones to keep them disease-free and to make them gain weight faster. More than half of all antibiotics used in the United States each year is fed to livestock. A significant amount of these antibiotics and hormones are excreted in urine and feces, which are spread, untreated, on the land or discharged into surface water where they contribute further to the evolution of supervirulent pathogens.

At least half of the 100 million antibiotic doses prescribed for humans every year in the United States are unnecessary or are the wrong ones. Furthermore, many people who start a course of antibiotic treatment fail to carry it out for the time prescribed. For your own health and that of the people around you, if you are taking an antibiotic, follow your doctor's orders and don't stop taking the medicine as soon as you start feeling better.

## Who should pay for health care?

The heaviest burden of illness is borne by the poorest people who can afford neither a healthy environment nor adequate health care. Women in sub-Saharan Africa, for example, suffer six times the disease burden per 1,000 population as do women in most European countries. The WHO estimates that 90 percent of all disease burden occurs in developing countries where less than one-tenth of all health care dollars is spent. The group Medecins Sans Frontieres (MSF, or Doctors Without Borders) calls this the 10/90 gap. While wealthy nations pursue drugs to treat baldness and obesity,

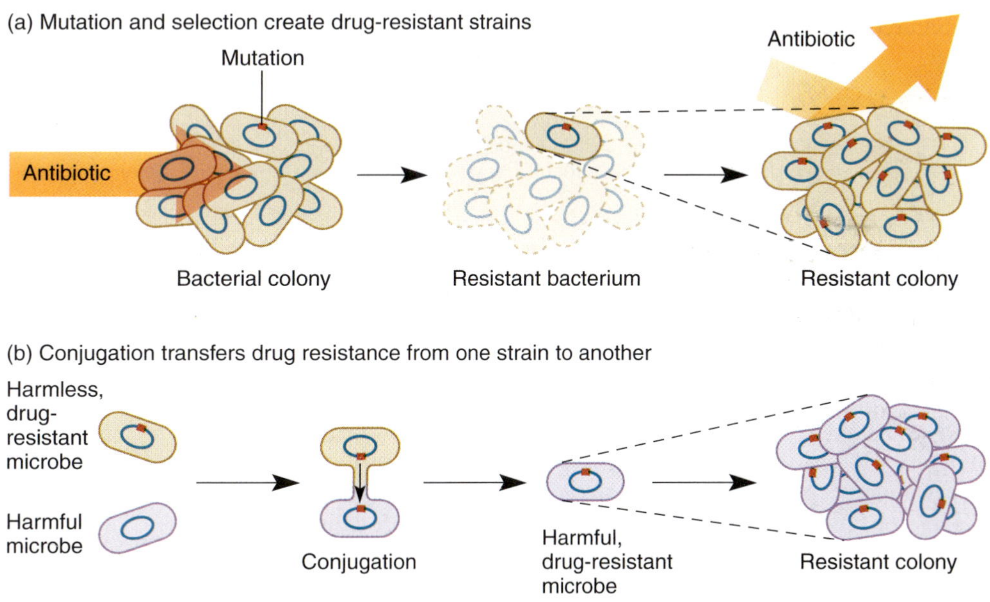


FIGURE 8.10 How microbes acquire antibiotic resistance. Random mutations make a few cells resistant. When challenged by antibiotics, only those cells survive to give rise to a resistant colony. Sexual reproduction (conjugation) or plasmid transfer moves genes from one strain or species to another.

depression in dogs, and erectile dysfunction, billions of people are sick or dying from treatable infections and parasitic diseases to which little attention is paid. Worldwide, only 2 percent of the people with AIDS have access to modern medicines. Every year, some 600,000 infants acquire HIV—almost all of them through mother-to-child transmission during birth or breast-feeding. Antiretroviral therapy costing only a few dollars can prevent most of this transmission. The Bill and Melinda Gates Foundation has pledged $200 million for medical aid to developing countries to help fight AIDS, TB, and malaria.

Dr. Jeffrey Sachs of the Columbia University Earth Institute says that disease is as much a cause as a consequence of poverty and political unrest, yet the world's richest countries now spend just $1 per person per year on global health. He predicts that raising our commitment to about $25 billion annually (about 0.1 percent of the annual GDP of the 20 richest countries) would not only save about 8 million lives each year, but would boost the world economy by billions of dollars. There also would be huge social benefits for the rich countries in not living in a world endangered by mass social instability, the spread of pathogens across borders, and the spread of other ills such as terrorism and drug trafficking caused by social problems. Sachs also argues that reducing disease burden would help reduce population growth. When parents believe their offspring will survive, they have fewer children and invest more in food, health, and education for smaller families.

The United States is the least generous of the world's rich countries, donating only about 12 cents per $100 of GDP on international development aid. Could the U.S. do better? During a time of fear of terrorism and rising anti-American feelings around the globe, it's difficult to interest legislators in international aid, and yet, helping to reduce disease might win the United States more friends and make the nation safer than buying more bombs and bullets. Improved health care in poorer countries may also help prevent the spread of emergent diseases in a globally interconnected world.

***Think About It***

If you were making a case for greater U.S. funding for international health care, what points would you stress? Do we have a moral obligation to help others?

## 8.2 Toxicology

Toxicology is the study of **toxins** (poisons) and their effects, particularly on living systems. Because many substances are known to be poisonous to life (whether plant, animal, or microbial), toxicology is a broad field, drawing from biochemistry, histology, pharmacology, pathology, and many other disciplines. Toxins damage or kill living organisms because they react with cellular components to disrupt metabolic functions. Because of this reactivity, toxins often are harmful even in extremely dilute concentrations. In some cases billionths, or even trillionths of a gram can cause irreversible damage.

All toxins are hazardous, but not all hazardous materials are toxic. Some substances, for example, are dangerous because they're flammable, explosive, acidic, caustic, irritants, or sensitizers. Many of these materials must be handled carefully in large doses or high concentrations, but can be rendered relatively innocuous by dilution, neutralization, or other physical treatment. They don't react with cellular components in ways that make them poisonous at low concentrations.

Environmental toxicology or ecotoxicology specifically deals with the interactions, transformation, fate, and effects of natural and synthetic chemicals in the biosphere including individual organisms, populations, and whole ecosystems. In aquatic systems the fate of the pollutants is primarily studied in relation to mechanisms and processes at interfaces of the ecosystem components. Special attention is devoted to the sediment/water, water/organisms, and water/air interfaces. In terrestrial environments, the emphasis tends to be on effects of metals on soil fauna community and population characteristics.

Table 8.2 lists the top 20 toxins compiled by the U.S. Enviromental Protection Agency from the 275 substances regulated by the Comprehensive Environmental Response, Compensation, and

**TABLE 8.2**

**Top 20 Toxic and Hazardous Substances**

| Material | Major Sources |
|---|---|
| 1. Arsenic | Treated lumber |
| 2. Lead | Paint, gasoline |
| 3. Mercury | Coal combustion |
| 4. Vinyl chloride | Plastics industrial uses |
| 5. Polychlorinated biphenyls (PCBs) | Electric insulation |
| 6. Benzene | Gasoline, industrial use |
| 7. Cadmium | Batteries |
| 8. Benzo(a)pyrene | Waste incineration |
| 9. Polycyclic aromatic hydrocarbons | Combustion |
| 10. Benzo(b)fluoranthene | Fuels |
| 11. Chloroform | Water purification, industry |
| 12. DDT | Pesticide use |
| 13. Aroclor 1254 | Plastics |
| 14. Aroclor 1260 | Plastics |
| 15. Trichloroethylene | Solvents |
| 16. Dibenz(a,h)anthracene | Incineration |
| 17. Dieldrin | Pesticides |
| 18. Chromium, hexavalent | Paints, coatings, welding, anticorrosion agents |
| 19. Chlordane | Pesticides |
| 20. Hexachlorobutadiene | Pesticides |

**Source:** Data from U.S. Environmental Protection Agency, 2003.

**FIGURE 8.11** Some sources of toxic and hazardous substances in a typical home.

Liability Act (CERCLA), commonly known as the Superfund Act. These materials are listed in order of assessed importance in terms of human and environmental health.

## How do toxins affect us?

**Allergens** are substances that activate the immune system. Some allergens act directly as **antigens;** that is, they are recognized as foreign by white blood cells and stimulate the production of specific antibodies (proteins that recognize and bind to foreign cells or chemicals). Other allergens act indirectly by binding to and changing the chemistry of foreign materials so they become antigenic and cause an immune response.

Formaldehyde is a good example of a widely used chemical that is a powerful sensitizer of the immune system. It is directly allergenic and can also trigger reactions to other substances. Widely used in plastics, wood products, insulation, glue, and fabrics, formaldehyde concentrations in indoor air can be thousands of times higher than in normal outdoor air. Some people suffer from what is called **sick building syndrome:** headaches, allergies, chronic fatigue, and other symptoms caused by poorly vented indoor air contaminated by mold spores, carbon monoxide, nitrogen oxides, formaldehyde, and other toxins released from carpets, insulation, plastics, building materials, and other sources (fig. 8.11). The Environmental Protection Agency estimates that poor indoor air quality may cost the United States $60 billion a year in absenteeism and reduced productivity.

Immune system depressants are pollutants that suppress the immune system rather than activate it. Little is known about how this occurs or which chemicals are responsible. Immune system failure is thought to have played a role, however, in widespread deaths of seals in the North Atlantic and of dolphins in the Mediterranean. These dead animals generally contain high levels of pesticide residues, polychlorinated biphenyls (PCBs), and other contaminants that are suspected of disrupting the immune system and making it susceptible to a variety of opportunistic infections.

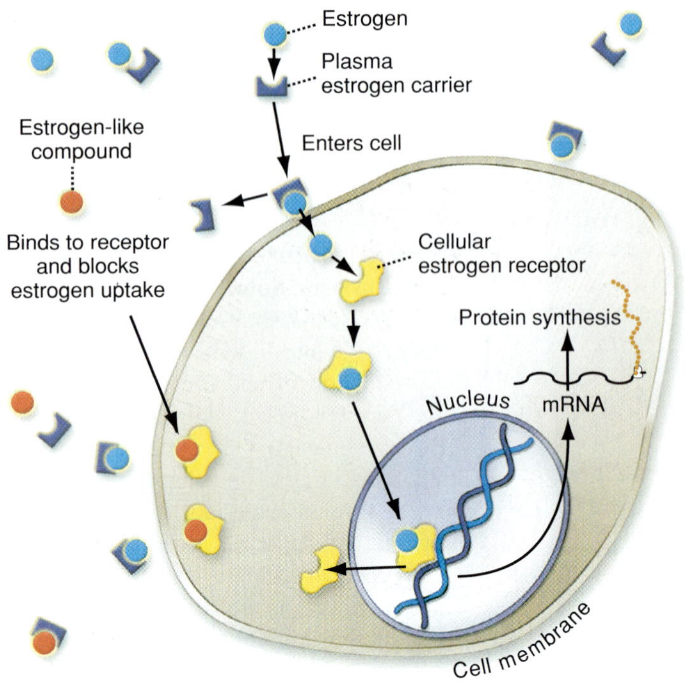


**FIGURE 8.12** Steroid hormone action. Plasma hormone carriers deliver regulatory molecules to the cell surface, where they cross the cell membrane. Intracellular carriers deliver hormones to the nucleus, where they bind to and regulate expression of DNA. Estrogen-like compounds bind to receptors and either block uptake of endogenous hormone, or act as a substitute hormone to disrupt gene expression.

**Endocrine disrupters** are chemicals that disrupt normal hormone functions. Hormones are chemicals released into the blood stream by cells in one part of the body to regulate development and function of tissues and organs elsewhere in the body (fig. 8.12). You undoubtedly have heard about sex hormones and their powerful effects on how we look and behave, but these are only one example of the many regulatory hormones that rule our lives. Some other powerful hormones include thyroxin, insulin, adrenalin, and endorphins.

We now know that some of the most insidious effects of persistent chemicals such as dioxins and PCBs are that they interfere with normal growth, development, and physiology of a variety of animals—including humans—at very low doses. In some cases, picogram concentrations (trillionths of a gram per liter) may be enough to cause developmental abnormalities in sensitive organisms. Because these chemicals often cause sexual dysfunction (reproductive health problems in females or feminization of males, for example), these chemicals are sometimes called environmental estrogens or androgens. They are just as likely, however, to disrupt functions of other important regulatory molecules as they are to obstruct sex hormones.

**Neurotoxins** are a special class of metabolic poisons that specifically attack nerve cells (neurons). The nervous system is so important in regulating body activities that disruption of its activities is especially fast-acting and devastating. Different types of neurotoxins act in different ways. Heavy metals such as lead and mercury kill nerve cells and cause permanent neurological damage. Anesthetics (ether, chloroform, halothane, etc.) and chlorinated hydrocarbons (DDT, Dieldrin, Aldrin) disrupt nerve cell membranes necessary for nerve action. Organophosphates (Malathion, Parathion) and carbamates (carbaryl, zeneb, maneb) inhibit acetylcholinesterase, an enzyme that regulates signal transmission between nerve cells and the tissues or organs they innervate (for example, muscle). Most neurotoxins are both extremely toxic and fast-acting.

**Mutagens** are agents, such as chemicals and radiation, that damage or alter genetic material (DNA) in cells. This can lead to birth defects if the damage occurs during embryonic or fetal growth. Later in life, genetic damage may trigger neoplastic (tumor) growth. When damage occurs in reproductive cells, the results can be passed on to future generations. Cells have repair mechanisms to detect and restore damaged genetic material, but some changes may be hidden, and the repair process itself can be flawed. It is generally accepted that there is no "safe" threshold for exposure to mutagens. Any exposure has some possibility of causing damage.

## What Can You Do?

### Tips for Staying Healthy

- Eat a balanced diet with plenty of fresh fruits, vegetables, legumes, and whole grains. Wash fruits and vegetables carefully; they may well have come from a country where pesticide and sanitation laws are lax.
- Use unsaturated oils such as olive or canola rather than hydrogenated or semisolid fats such as margarine.
- Cook meats and other foods at temperatures high enough to kill pathogens; clean utensils and cutting surfaces; store food properly.
- Wash your hands frequently. You transfer more germs from hand to mouth than any other means of transmission.
- When you have a cold or flu, don't demand antibiotics from your doctor—they aren't effective against viruses.
- If you're taking antibiotics, continue for the entire time prescribed—quitting as soon as you feel well is an ideal way to select for antibiotic-resistant germs.
- Practice safe sex.
- Don't smoke and avoid smoky places.
- If you drink, do so in moderation. Never drive when your reflexes or judgment are impaired.
- Exercise regularly: walk, swim, jog, dance, garden. Do something you enjoy that burns calories and maintains flexibility.
- Get enough sleep. Practice meditation, prayer, or some other form of stress reduction.
- Make a list of friends and family who make you feel more alive and happy. Spend time with one of them at least once a week.

**Teratogens** are chemicals or other factors that specifically cause abnormalities during embryonic growth and development. Some compounds that are not otherwise harmful can cause tragic problems in these sensitive stages of life. Perhaps the most prevalent teratogen in the world is alcohol. Drinking during pregnancy can lead to **fetal alcohol syndrome**—a cluster of symptoms including craniofacial abnormalities, developmental delays, behavioral problems, and mental defects that last throughout a child's life. Even one alcoholic drink a day during pregnancy has been associated with decreased birth weight.

By some estimates, between 300,000 and 600,000 children born every year in the United States are exposed in the womb to unsafe levels of mercury. The effects are subtle, but include reduced intelligence, attention deficit, and behavioral problems. The total cost of these effects is estimated to be $8.7 billion per year.

**Carcinogens** are substances that cause **cancer,** invasive, out-of-control cell growth that results in malignant tumors. Cancer rates rose in most industrialized countries during the twentieth century, and cancer is now the second leading cause of death in the United States, killing more than half a million people in 2002. According to the American Cancer Society, 1 in 2 males and 1 in 3 females in the United States will have some form of cancer in their lifetime. Some authors blame this cancer increase on toxic synthetic chemicals in our environment and diet. Others argue that it is attributable mainly to lifestyle (smoking, sunbathing, alcohol) or simply living longer. The U.S. EPA estimates that 200 million U.S. residents live in areas where the combined lifetime cancer risk from environmental carcinogens exceeds 1 in 100,000, or ten times the risk normally considered acceptable.

## How does diet influence health?

Diet also has an important effect on health. For instance, there is a strong correlation between cardiovascular disease and the amount of salt and animal fat in one's diet.

Fruits, vegetables, whole grains, complex carbohydrates, and dietary fiber (plant cell walls) often have beneficial health effects. Certain dietary components, such as pectins; vitamins A, C, and E; substances produced in cruciferous vegetables (cabbage, broccoli, cauliflower, brussels sprouts); and selenium, which we get from plants, seem to have anticancer effects.

Eating too much food is a significant dietary health factor in developed countries and among the well-to-do everywhere. Sixty percent of all U.S. adults are now considered overweight, and the worldwide total of obese or overweight people is estimated to be over 1 billion. Every year in the United States, 300,000 deaths are linked to obesity.

The U.S. Centers for Disease Control in Atlanta warn that one in three U.S. children will become diabetic unless many more people start eating less and exercising more. The odds are worse for Black and Hispanic children: nearly half of them are likely to develop the disease. And among the Pima tribe of Arizona, nearly 80 percent of all adults are diabetic. More information about food and its health effects is available in chapter 9.

# 8.3 Movement, Distribution, and Fate of Toxins

There are many sources of toxic and hazardous chemicals in the environment and many factors related to each chemical itself, its route or method of exposure, and its persistence in the environment, as well as characteristics of the target organism (table 8.3), that determine the danger of the chemical. We can think of both individuals and an ecosystem as sets of interacting compartments between which chemicals move, based on molecular size, solubility, stability, and reactivity (fig. 8.13). The dose (amount), route of entry, timing of exposure, and sensitivity of the organism all play important roles in determining toxicity. In this section, we will consider some of these characteristics and how they affect environmental health.

## Solubility and mobility determine where and when chemicals move

Solubility is one of the most important characteristics in determining how, where, and when a toxic material will move through the

**TABLE 8.3**

**Factors in Environmental Toxicity**

| Factors Related to the Toxic Agent |
|---|
| 1. Chemical composition and reactivity |
| 2. Physical characteristics (such as solubility, state) |
| 3. Presence of impurities or contaminants |
| 4. Stability and storage characteristics of toxic agent |
| 5. Availability of vehicle (such as solvent) to carry agent |
| 6. Movement of agent through environment and into cells |
| **Factors Related to Exposure** |
| 1. Dose (concentration and volume of exposure) |
| 2. Route, rate, and site of exposure |
| 3. Duration and frequency of exposure |
| 4. Time of exposure (time of day, season, year) |
| **Factors Related to Organism** |
| 1. Resistance to uptake, storage, or cell permeability of agent |
| 2. Ability to metabolize, inactivate, sequester, or eliminate agent |
| 3. Tendency to activate or alter nontoxic substances so they become toxic |
| 4. Concurrent infections or physical or chemical stress |
| 5. Species and genetic characteristics of organism |
| 6. Nutritional status of subject |
| 7. Age, sex, body weight, immunological status, and maturity |

**Source:** U.S. Department of Health and Human Services, 1995.

**FIGURE 8.13** Movement and fate of chemicals in the environment. Toxins also move directly from a source to soil and sediment.

environment or through the body to its site of action. Chemicals can be divided into two major groups: those that dissolve more readily in water and those that dissolve more readily in oil. Water-soluble compounds move rapidly and widely through the environment because water is ubiquitous. They also tend to have ready access to most cells in the body because aqueous solutions bathe all our cells. Molecules that are oil- or fat-soluble (usually organic molecules) generally need a carrier to move through the environment, into, and within, the body. Once inside the body, however, oil-soluble toxins penetrate readily into tissues and cells because the membranes that enclose cells are themselves made of similar oil-soluble chemicals. Once they get inside cells, oil-soluble materials are likely to be accumulated and stored in lipid deposits where they may be protected from metabolic breakdown and persist for many years.

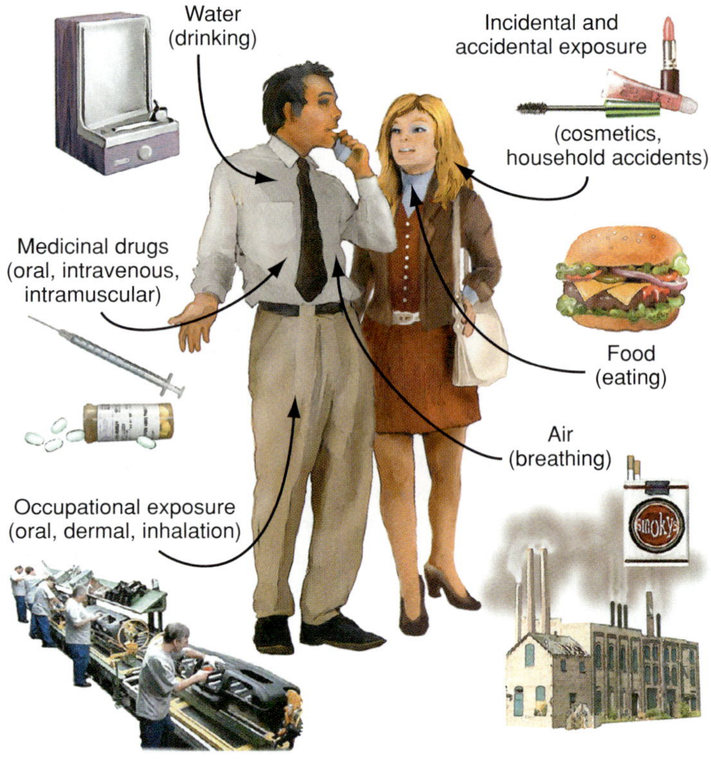


**FIGURE 8.14** Routes of exposure to toxic and hazardous environmental factors.

## Exposure and susceptibility determine how we respond

Just as there are many sources of toxins in our environment, there are many routes for entry of dangerous substances into our bodies (fig. 8.14). Airborne toxins generally cause more ill health than any other exposure source. We breathe far more air every day than the volume of food we eat or water we drink. Furthermore, the lining of our lungs, which is designed to exchange gases very efficiently, also absorbs toxins very well. Epidemiologists estimate that some 3 million people—two-thirds of them children—die each year from diseases caused or exacerbated by air pollution. Still, food, water, and skin contact also can expose us to a wide variety of toxins. The largest exposures for many toxins are found in industrial settings, where workers may encounter doses thousands of times higher than would be found anywhere else. The European Agency for Safety and Health at Work warns that 32 million people (20 percent of all employees) in the European Union are exposed to unacceptable levels of carcinogens and other toxins in their workplace.

Condition of the organism and timing of exposure also have strong influences on toxicity. Healthy adults, for example, may be relatively insensitive to doses that would be very dangerous for young children or for someone already weakened by disease (What Do You Think? p. 167). Similarly, exposure to a toxin

## What Do You Think?

### Protecting Children's Health

Increasing evidence shows that children are much more vulnerable than adults to environmental health hazards. Pound for pound, children drink more water, eat more food, and breathe more air than do adults. Putting fingers, toys, and other objects into their mouths increases children's exposure to toxins in dust or soil. Compared to adults, children generally have less-developed immune systems or processes to degrade or excrete toxins. The brain is especially sensitive to disruption. From before birth to adolescence, the brain undergoes a highly complex series of changes. Obviously, anything that interferes with brain development can have tragic long-term results.

In 2006, an international team of epidemiologists, led by Drs. Phillippe Grandjean, of Harvard, and Philip Landrigan, of the Mount Sinai School of Medicine, both of whom are pioneers in studying long-term effects of toxins on children, surveyed toxicity data on common industrial chemicals. They identified 202 toxic substances—about half of them used in very large volumes—known to damage human brains. They found that effects of these chemicals on c hildren have generally been neglected.

"The human brain is a precious and vulnerable organ. And because optimal brain function depends on the integrity of the organ, even limited damage may have serious consequences," says Dr. Grandjean. He estimates that one out of every six children in America has a developmental disability, usually involving the nervous system. Among the neurodevelopmental disorders that may be associated with toxic chemical exposure are autism, mental retardation, attention deficit disorder, shortened attention spans, slowed motor coordination, increased aggressiveness, and diminished social skills.

Despite years of evidence that many industrial substances are dangerous, few have been studied thoroughly for their effects on children. Lead was the first toxicant identified as having effects at low levels on developing brains of children. Banning leaded gasoline and paint was one of the most successful steps ever taken to protect children's health. Between 1920 and 1970, virtually all children in industrialized countries were exposed to lead from house paint and leaded gasoline. Before these sources were banned in the 1970s, at least 4.4 million American children had more than 10 $\mu g$ of lead per dl of blood, (the level considered safe by the U.S. EPA), and hundreds of children died each year from acute lead poisoning. Epidemiologists now estimate that for every 10 $\mu g$ of lead per dl of blood, IQ scores drop 4.6 points. Grandjean and Landrigan suggest that high lead exposures in the United States reduced IQ scores above 130 (considered superior intelligence) by half, while the number of scores less than 70 increased comparably.

By 2006, the number of children with elevated blood lead levels had fallen more than 90 percent (fig. 1). This is a remarkable success story, but about 430,000 children—especially in poor urban areas—still have unsafe lead levels. Recent evidence suggests that even 10 $\mu g$/dl represents a risk of subtle but important effects in growing children's brains, including diminished learning ability, attention deficit disorder, and hyperactivity. Many public health experts argue that blood lead levels in all children should be below 2.5 $\mu g$/dl (fig. 2).

Grandjean and Landrigan point out that, in addition to lead, only four other toxins—methylmercury, arsenic, PDBs and toluene—are specifically regulated to protect children. Because industrialization has now become global, they warn that this represents a "silent pandemic" in which millions of children may be suffering reduced intelligence. They suggest that to protect children from industrial chemicals that cause mental retardation and developmental disabilities, we should adopt a precautionary approach to chemical testing and control. This philosophy is already being applied in Europe, where industry is required to show that a chemical is safe before it is introduced, rather than requiring regulatory agencies to prove that it is unsafe in order to restrict its use.

What do you think? How important is it to protect children's mental development? What does it cost society to have large numbers of slow learners and behavioral problems? Industry claims that more rigorous testing and stricter regulations on industrial chemicals will drive up consumer prices and limit consumer choices. Would you support stronger regulations now, which could be relaxed later if hazards are less than anticipated, or the current regulations that require a high level of proof before restricting use of toxic products?.

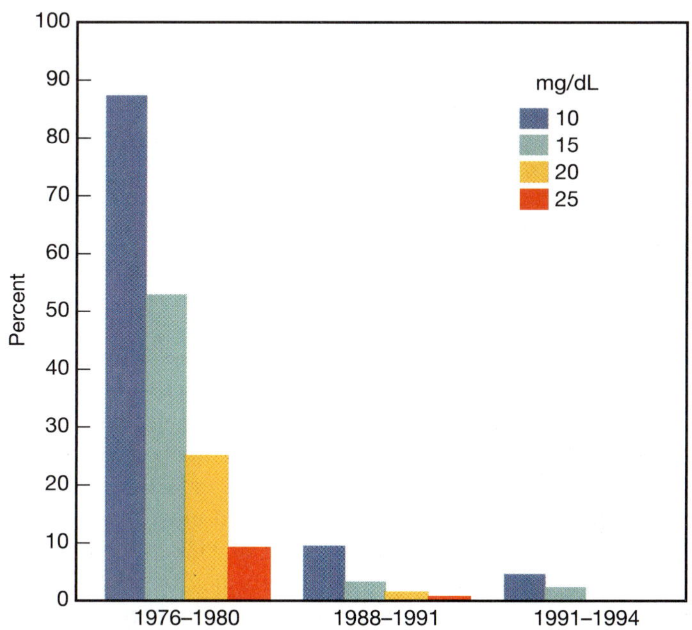


**FIGURE 1** Blood lead levels in U.S. preschoolers fell 90 percent between the 1970s and 1990s. This is one of our greatest environmental health successes ever.
**Source:** J. Pickle et al. 1998.

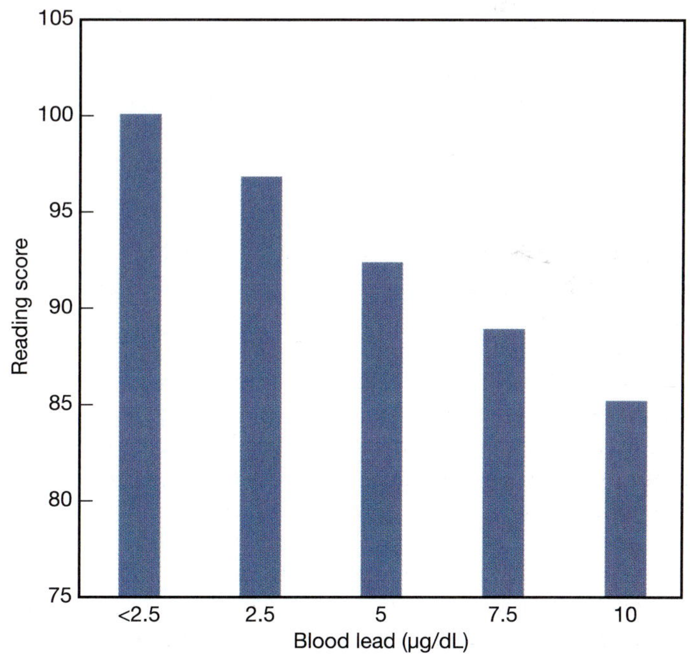


**FIGURE 2** Reading score of U.S. children are inversely related to blood lead levels even below 10 µg/dL, the amount considered safe by the U.S. EPA.
**Source:** J. Pickle et al. 1998.

may be very dangerous at certain stages of developmental or metabolic cycles, but may be innocuous at other times. A single dose of the notorious teratogen thalidomide, for example, taken in the third week of pregnancy (a time when many women aren't aware they're pregnant) can cause severe abnormalities in fetal limb development. A complication in measuring toxicity is that great differences in sensitivity exist between species. Thalidomide was tested on a number of laboratory animals without any deleter-ious effects. Unfortunately, however, it is a powerful teratogen in humans.

## Bioaccumulation and biomagnification increase concentrations of chemicals

Cells have mechanisms for **bioaccumulation,** the selective absorption and storage of a great variety of molecules. This allows them to accumulate nutrients and essential minerals, but at the same time, they also may absorb and store harmful substances through these same mechanisms. Toxins that are rather dilute in the environment can reach dangerous levels inside cells and tissues through this process of bioaccumulation.

The effects of toxins also are magnified in the environment through food webs. **Biomagnification** occurs when the toxic burden of a large number of organisms at a lower trophic level is accumulated and concentrated by a predator in a higher trophic level. Phytoplankton and bacteria in aquatic ecosystems, for instance, take up heavy metals or toxic organic molecules from water or sediments (fig. 8.15). Their predators—zooplankton and small fish—collect and retain the toxins from many prey organisms, building up higher concentrations of toxins. The top carnivores in the food chain—game fish, fish-eating birds, and humans—can accumulate such high toxin levels that they suffer adverse health effects (chapter 18). One of the first known examples of bioaccumulation and biomagnification was DDT, which accumulated through food chains so that by the 1960s it was shown to be interfering with reproduction of peregrine falcons, brown pelicans, and other predatory birds at the top of their food chains (chapter 10).

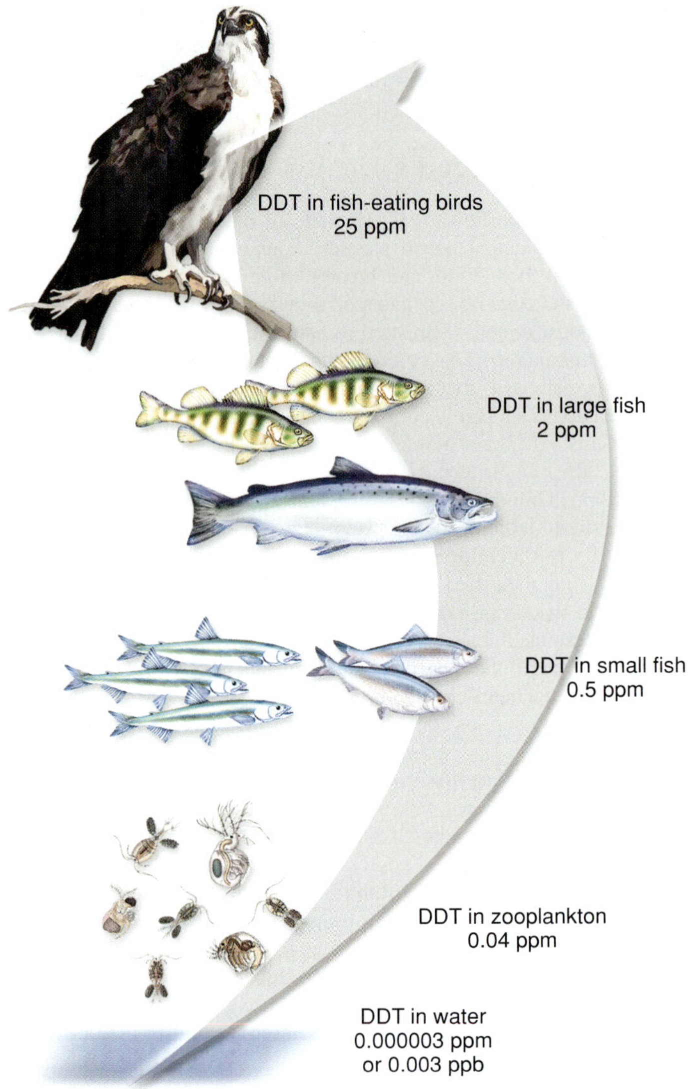


**FIGURE 8.15** Bioaccumulation and biomagnification. Organisms lower on the food chain take up and store toxins from the environment. They are eaten by larger predators, who are eaten, in turn, by even larger predators. The highest members of the food chain can accumulate very high levels of the toxin.

## Persistence makes some materials a greater threat

Some chemical compounds are very unstable and degrade rapidly under most environmental conditions so that their concentrations decline quickly after release. Most modern herbicides and pesticides, for instance, quickly lose their toxicity. Other substances are more persistent and last for years or even centuries in the environment. Metals—such as lead—PVC plastics, chlorinated hydrocarbon pesticides, and asbestos are valuable because they are resistant to degradation. This stability, however, also causes problems because these materials persist in the environment and have unexpected effects far from the sites of their original use.

Some **persistent organic pollutants (POPs)** have become extremely widespread, being found now from the tropics to the Arctic. They often accumulate in food webs and reach toxic concentrations in long-living top predators such as humans, sharks, raptors, swordfish, and bears. Some of greatest current concerns are:

- Polybrominated diphenyl ethers (PBDE). Widely used as flame-retardants in textiles, foam in upholstery, and plastic in appliances and computers, these chemicals are now found in humans and other species everywhere in the world. Nearly 150 million metric tons (330 million lbs) of PBDEs are used every year worldwide. The toxicity and environmental persistence of PBDE is much like that of PCBs, to which it is closely related chemically. Relatively low exposures in the womb or shortly after birth can irreparably harm children's reproductive and nervous systems. The European Union has already banned this compound.

- Perfluorooctane sulfonate (PFOS) and perfluorooctanoic acid (PFOA, also known as C8) are members of a chemical family used to make nonstick, waterproof, and stain-resistant products such as Teflon, Gortex, Scotchguard, and Stainmaster. Industry makes use of their slippery, heat-stable properties to manufacture everything from airplanes and computers to cosmetics and household cleaners. Now these chemicals—which are reported to be infinitely persistent in the environment—are found throughout the world, even the most remote and seemingly pristine sites. Almost all Americans have one or more perfluorinated compounds in their blood. Heating some nonstick cooking pans above 500°F (260°C) can release enough PFOA to kill pet birds. This chemical family has been shown to cause liver damage as well as various cancers and reproductive and developmental problems in rats. Exposure may be especially dangerous to women and girls, who may be 100 times more sensitive than men to these chemicals. In 2005, the EPA announced the start of a study of human health effects of these chemicals.
- Phthalates (pronounced *thalates*) are found in cosmetics, deodorants, and many plastics (such as soft polyvinyl chloride or PVC) used for food packaging, children's toys, and medical devices. Some members of this chemical family are known to be toxic to laboratory animals, causing kidney and liver damage and possibly some cancers. In addition, many phthalates act as endocrine hormone disrupters, and have been linked to reproductive abnormalities and decreased fertility in humans. A correlation has been found between phthalate levels in urine and low sperm numbers and decreased sperm motility in men. Nearly everyone in the United States has phthalates in their body at levels reported to cause these problems. While not yet conclusive, these results could help explain a 50-year decline in semen quality in most industrialized countries.
- Perchlorate is a waterborne contaminant left over from propellants and rocket fuels. About 12,000 sites in the United States were used by the military for live munition testing and are contaminated with perchlorate. Polluted water used to irrigate crops such as alfalfa and lettuce has introduced the chemical into the human food chain. Tests of cow's milk and human breast milk detected perchlorate in nearly every sample from throughout the United States. Perchlorate can interfere with iodine uptake in the thyroid gland, disrupting adult metabolism and childhood development.
- Bisphenol A (BPA), a prime ingredient in polycarbonate plastic (commonly used for products ranging from water bottles to tooth-protecting sealants), has been widely found in humans, even those who have no known exposure to these chemicals. So far, there is little direct evidence linking BPA exposure to human health risks, but studies in animals have found that the chemical can cause abnormal chromosome numbers, a condition called aneuploidy, which is the leading cause of miscarriages and several forms of mental retardation. It also is an environmental estrogen and may alter sexual development in both males and females.
- Atrazine is the most widely used herbicide in America. More than 60 million pounds of this compound are applied per year, mainly on corn and cereal grains, but also on golf courses, sugar cane, and Christmas trees. It has long been known to disrupt endocrine hormone functions in mammals, resulting in spontaneous abortions, low birth weights, and neurological disorders. Studies of families in corn-producing areas in the American Midwest have found higher rates of developmental defects among infants, and certain cancers in families with elevated atrazine levels in their drinking water. University of California professor Tyrone Hayes has shown that atrazine levels as low as 0.1 ppb (30 times less than the EPA maximum contaminant level) caused severe reproductive effects in amphibians, including abnormal gonadal development and hermaphroditism. Atrazine now is found in rain and surface waters nearly everywhere in the United States at levels that could cause abnormal development in frogs. In 2003, the European Union withdrew regulatory approval for this herbicide, and several countries banned its use altogether. Some toxicologists have suggested a similar rule in the United States.

Bills to ban Bisphenol A and certain phthalates—particularly in children's toys and feeding products—were introduced in several states in 2005. None of them passed, but the city of San Francisco did pass a similar measure. All these compounds already are regulated in Europe.

Everyone of us has dozens, if not hundreds, of persistent toxins in our body. This accumulation is called our **body burden.** We acquire it from our air, water, diet, and surroundings. Many of these toxins are present in parts per billion, or even parts per trillion. We don't know how dangerous this persistent burden is, but its presence is a matter of concern. If we're anything like the frogs that Tyrone Hayes studies, this accumulated dose of toxins may be a serious problem. Further discussion of POPs can be found in chapter 10.

## Chemical interactions can increase toxicity

Some materials produce *antagonistic* reactions. That is, they interfere with the effects or stimulate the breakdown of other chemicals. For instance, vitamins E and A can reduce the response to some carcinogens. Other materials are *additive* when they occur together in exposures. Rats exposed to both lead and arsenic show twice the toxicity of only one of these elements. Perhaps the greatest concern is synergistic effects. **Synergism** is an interaction in which one substance exacerbates the effects of another. For example, occupational asbestos exposure increases lung cancer rates 20-fold. Smoking increases lung cancer rates by the same amount. Asbestos workers who also smoke, however, have a 400-fold increase in cancer rates. How many other toxic chemicals are we exposed to that are below threshold limits individually but combine to give toxic results?

## 8.4 Mechanisms for Minimizing Toxic Effects

A fundamental concept in toxicology is that every material can be poisonous under some conditions, but most chemicals have some safe level or threshold below which their effects are undetectable or insignificant. Each of us consumes lethal doses of many chemicals over the course of a lifetime. One hundred cups of strong coffee, for instance, contain a lethal dose of caffeine. Similarly, one hundred aspirin tablets, or 10 kilograms (22 lbs) of spinach or rhubarb, or a liter of alcohol would be deadly if consumed all at once. Taken in small doses, however, most toxins can be broken down or excreted before they do much harm. Furthermore, damage they cause can be repaired. Sometimes, however, mechanisms that protect us from one type of toxin or at one stage in the life cycle become deleterious with another substance or in another stage of development. Let's look at how these processes help protect us from harmful substances as well as how they can go awry.

### Metabolic degradation and excretion elimimate toxins

Most organisms have enzymes that process waste products and environmental poisons to reduce their toxicity. In mammals, most of these enzymes are located in the liver, the primary site of detoxification of both natural wastes and introduced poisons. Sometimes, however, these reactions work to our disadvantage. Compounds, such as benzopyrene, for example, that are not toxic in their original form are processed by these same enzymes into cancer-causing carcinogens. Why would we have a system that makes a chemical more dangerous? Evolution and natural selection are expressed through reproductive success or failure. Defense mechanisms that protect us from toxins and hazards early in life are "selected for" by evolution. Factors or conditions that affect postreproductive ages (like cancer or premature senility) usually don't affect reproductive success or exert "selective pressure."

We also reduce the effects of waste products and environmental toxins by eliminating them from our body through excretion. Volatile molecules, such as carbon dioxide, hydrogen cyanide, and ketones are excreted via breathing. Some excess salts and other substances are excreted in sweat. Primarily, however, excretion is a function of the kidneys, which can eliminate significant amounts of soluble materials through urine formation. Accumulation of toxins in the urine can damage this vital system, however, and the kidneys and bladder often are subjected to harmful levels of toxic compounds. In the same way, the stomach, intestine, and colon often suffer damage from materials concentrated in the digestive system and may be afflicted by diseases and tumors.

### Repair mechanisms mend damage

In the same way that individual cells have enzymes to repair damage to DNA and protein at the molecular level, tissues and organs that are exposed regularly to physical wear-and-tear or to toxic or hazardous materials often have mechanisms for damage repair. Our skin and the epithelial linings of the gastrointestinal tract, blood vessels, lungs, and urogenital system have high cellular reproduction rates to replace injured cells. With each reproduction cycle, however, there is a chance that some cells will lose normal growth controls and run amok, creating a tumor. Thus any agent, such as smoking or drinking, that irritates tissues is likely to be carcinogenic. And tissues with high cell-replacement rates are among the most likely to develop cancers.

**_Think About It_**

Some of the mechanisms that help repair wounds or fight off infections can result in cancer when we're old. Why hasn't evolution eliminated these processes?

*Hint:* Do conditions of postreproductive age affect natural selection?

## 8.5 Measuring Toxicity

Almost 500 years ago, the Swiss scientist Paracelsus said "the dose makes the poison," by which he meant that almost everything is toxic at some level. This remains the most basic principle of toxicology. Sodium chloride (table salt), for instance, is essential for human life in small doses. If you were forced to eat a kilogram of salt all at once, however, it would make you very sick. A similar amount injected into your bloodstream would be lethal. How a material is delivered—at what rate, through which route of entry, and in what medium—plays a vitally important role in determining toxicity.

This does not mean that all toxins are identical, however. Some are so poisonous that a single drop on your skin can kill you. Others require massive amounts injected directly into the blood to be lethal. Measuring and comparing the toxicity of various materials is difficult because not only do species differ in sensitivity, but individuals within a species respond differently to a given exposure. In this section, we will look at methods of toxicity testing and at how results are analyzed and reported.

### We usually test toxins on lab animals

The most commonly used and widely accepted toxicity test is to expose a population of laboratory animals to measured doses of a specific substance under controlled conditions. This procedure is expensive, time-consuming, and often painful and debilitating to the animals being tested. It commonly takes hundreds—or even thousands—of animals, several years of hard work, and hundreds of thousands of dollars to thoroughly test the effects of a toxin at very low doses. More humane toxicity tests using computer simulation of model reactions, cell cultures, and other substitutes for whole living animals are being developed. However, conventional large-scale animal testing is the method in which we have the most confidence and on which most public policies about pollution and environmental or occupational health hazards are based.

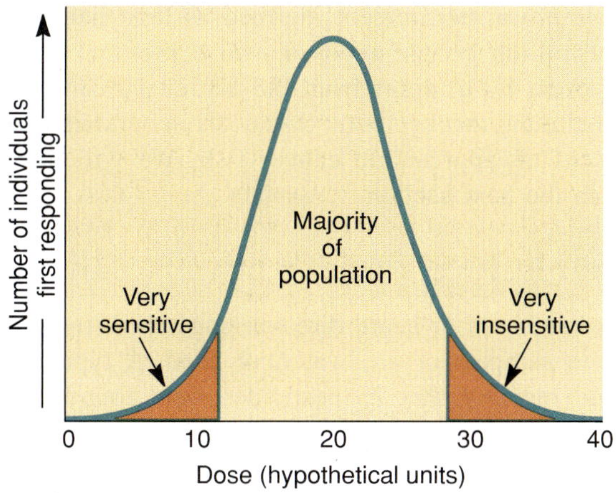


**FIGURE 8.16** Probable variations in sensitivity to a toxin within a population. Some members of a population may be very sensitive to a given toxin, while others are much less sensitive. The majority of the population falls somewhere between the two extremes.

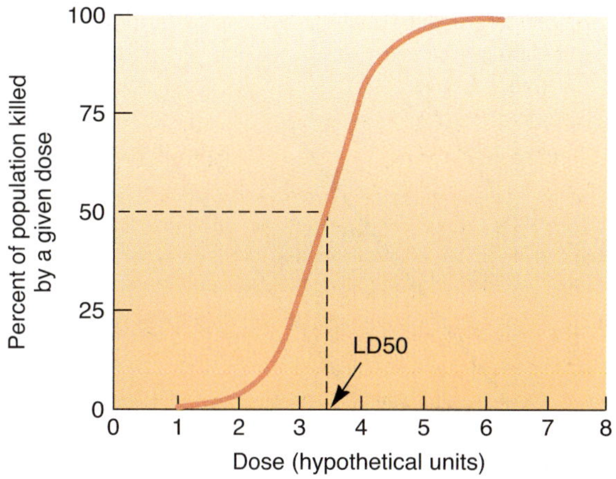


**FIGURE 8.17** Cumulative population response to increasing doses of a toxin. The LD50 is the dose that is lethal to half the population.

In addition to humanitarian concerns, there are several problems in laboratory animal testing that trouble both toxicologists and policymakers. One problem is differences in sensitivity to a toxin of the members of a specific population. Figure 8.16 shows a typical dose/response curve for exposure to a hypothetical toxin. Some individuals are very sensitive to the toxin, while others are insensitive. Most, however, fall in a middle category forming a bell-shaped curve. The question for regulators and politicians is whether we should set pollution levels that will protect everyone, including the most sensitive people, or only aim to protect the average person. It might cost billions of extra dollars to protect a very small number of individuals at the extreme end of the curve. Is that a good use of resources?

Dose/response curves are not always symmetrical, making it difficult to compare toxicity of unlike chemicals or different species of organisms. A convenient way to describe toxicity of a chemical is to determine the dose to which 50 percent of the test population is sensitive. In the case of a lethal dose (LD), this is called the **LD50** (fig. 8.17).

Unrelated species can react very differently to the same toxin, not only because body sizes vary but also because of differences in physiology and metabolism. Even closely related species can have very dissimilar reactions to a particular toxin. Hamsters, for instance, are nearly 5,000 times less sensitive to some dioxins than are guinea pigs. Of 226 chemicals found to be carcinogenic in either rats or mice, 95 caused cancer in one species but not the other. These variations make it difficult to estimate the risks for humans since we don't consider it ethical to perform controlled experiments in which we deliberately expose people to toxins.

## There is a wide range of toxicity

It is useful to group materials according to their relative toxicity. A moderate toxin takes about one gram per kilogram of body weight (about two ounces for an average human) to make a lethal dose. Very toxic materials take about one-tenth that amount, while extremely toxic substances take one-hundredth as much (only a few drops) to kill most people. Supertoxic chemicals are extremely potent; for some, a few micrograms (millionths of a gram—an amount invisible to the naked eye) make a lethal dose. These materials are not all synthetic. One of the most toxic chemicals known, for instance, is ricin, a protein found in castor bean seeds. It is so toxic that 0.3 billionths of a gram given intravenously will generally kill a mouse. If aspirin were this toxic, a single tablet, divided evenly, could kill 1 million people.

Many carcinogens, mutagens, and teratogens are dangerous at levels far below their direct toxic effect because abnormal cell growth exerts a kind of biological amplification. A single cell, perhaps altered by a single molecular event, can multiply into millions of tumor cells or an entire organism. Just as there are different levels of direct toxicity, however, there are different degrees of carcinogenicity, mutagenicity, and teratogenicity. Methanesulfonic acid, for instance, is highly carcinogenic, while the sweetener saccharin is a suspected carcinogen whose effects may be vanishingly small.

## Acute and chronic doses and effects differ

Most of the toxic effects that we have discussed so far have been **acute effects.** That is, they are caused by a single exposure to the toxin and result in an immediate health crisis of some sort. Often, if the individual experiencing an acute reaction survives this immediate crisis, the effects are reversible. **Chronic effects,** on the other hand, are long lasting, perhaps even permanent. A chronic effect can result from a single dose of a very toxic substance, or it can be the result of a continuous or repeated sublethal exposure.

We also describe long-lasting *exposures* as chronic, although their effects may or may not persist after the toxin is removed. It usually is difficult to assess the specific health risks of chronic exposures because other factors, such as aging or normal diseases, act simultaneously with the factor under study. It often requires

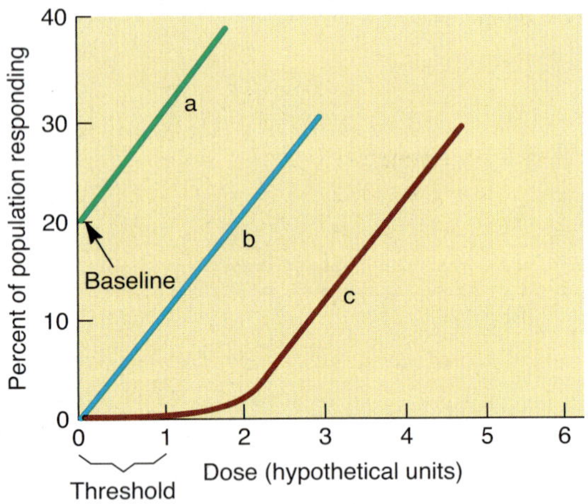


**FIGURE 8.18** Three possible dose-response curves at low doses. (a) Some individuals respond, even at zero dose, indicating that some other factor must be involved. (b) Response is linear down to the lowest possible dose. (c) Threshold must be passed before any response is seen.

very large populations of experimental animals to obtain statistically significant results for low-level chronic exposures. Toxicologists talk about "megarat" experiments in which it might take a million rats to determine the health risks of some supertoxic chemicals at very low doses. Such an experiment would be terribly expensive for even a single chemical, let alone for the thousands of chemicals and factors suspected of being dangerous.

An alternative to enormous studies involving millions of animals is to give massive amounts—usually the maximum tolerable dose—of a toxin being studied to a smaller number of individuals and then to extrapolate what the effects of lower doses might have been. This is a controversial approach because it is not clear that responses to toxins are linear or uniform across a wide range of doses.

Figure 8.18 shows three possible results from low doses of a toxin. Curve (*a*) shows a baseline level of response in the population, even at zero dose of the toxin. This suggests that some other factor in the environment also causes this response. Curve (*b*) shows a straight-line relationship from the highest doses to zero exposure. Many carcinogens and mutagens show this kind of response. Any exposure to such agents, no matter how small, carries some risks. Curve (*c*) shows a threshold for the response where some minimal dose is necessary before any effect can be observed. This generally suggests the presence of some defense mechanism that prevents the toxin from reaching its target in an active form or repairs the damage that it causes. Low levels of exposure to the toxin in question may have no deleterious effects, and it might not be necessary to try to keep exposures to zero.

Which, if any, environmental health hazards have thresholds is an important but difficult question. The 1958 Delaney Clause to the U.S. Food and Drug Act forbids the addition of *any* amount of known carcinogens to food and drugs, based on the assumption that any exposure to these substances represents unacceptable risks. This standard was replaced in 1996 by a *no reasonable harm* requirement, defined as less than one cancer for every million people exposed over a lifetime. This change was supported by a report from the National Academy of Sciences concluding that synthetic chemicals in our diet are unlikely to represent an appreciable cancer risk. We will discuss risk analysis in the next section.

### Detectable levels aren't always dangerous

You may have seen or heard dire warnings about toxic materials detected in samples of air, water, or food. A typical headline announced recently that 23 pesticides were found in 16 food samples. What does that mean? The implication seems to be that any amount of dangerous materials is unacceptable and that counting the numbers of compounds detected is a reliable way to establish danger. We have seen, however, that the dose makes the poison. It matters not only what is there, but how much, where it is located, how accessible it is, and who is exposed. At some level, the mere presence of a substance is insignificant.

Toxins and pollutants may seem to be more widespread now than in the past, and this is surely a valid perception for many substances. The daily reports we hear of new materials found in new places, however, are also due, in part, to our more sensitive measuring techniques. Twenty years ago, parts per million were generally the limits of detection for most chemicals. Anything below that amount was often reported as zero or absent rather than more accurately as undetected. A decade ago, new machines and techniques were developed to measure parts per billion. Suddenly, chemicals were found where none had been suspected. Now we can detect parts per trillion or even parts per quadrillion in some cases. Increasingly sophisticated measuring capabilities may lead us to believe that toxic materials have become more prevalent. In fact, our environment may be no more dangerous; we are just better at finding trace amounts.

## 8.6 Risk Assessment and Acceptance

**Risk** is the possibility of suffering harm or loss. **Risk assessment** is the scientific process of estimating the threat that particular hazards pose to human health. This process includes risk identification, dose response assessment, exposure appraisal, and risk characterization. In hazard identification, scientists evaluate all available information about the effects of a toxin to estimate the likelihood that a chemical will cause a certain effect in humans. The best evidence comes from human studies, such as physician case reports. Animal studies are also used to assess health risks. Risk assessment for identified toxicity hazards (for example, lead) includes collection and analysis of site data, development of exposure and risk calculations, and preparation of human health and ecological impact reports. Exposure assessment is the estimation or determination of the magnitude, frequency, duration, and route of exposure to a possible toxin. Toxicity assessment weighs all available evidence and estimates the potential for adverse health effects to occur.

## Risk perception isn't always rational

A number of factors influence how we perceive relative risks associated with different situations.

- People with social, political, or economic interests—including environmentalists—tend to downplay certain risks and emphasize others that suit their own agendas. We also tend to tolerate risks that we choose—such as driving, smoking, or overeating—while objecting to risks we cannot control—such as potential exposure to slight amounts of toxic substances.
- Most people have difficulty understanding and believing probabilities. We feel that there must be patterns and connections in events, even though statistical theory says otherwise. If the coin turned up heads last time, we feel certain that it will turn up tails next time. In the same way, it is difficult to understand the meaning of a 1-in-10,000 risk of being poisoned by a chemical.
- Our personal experiences often are misleading. When we have not personally experienced a bad outcome, we feel it is more rare and unlikely to occur than it actually may be. Furthermore, the anxieties generated by life's gambles make us want to deny uncertainty and to misjudge many risks (fig. 8.19).
- We have an exaggerated view of our own abilities to control our fate. We generally consider ourselves above-average drivers, safer than most when using appliances or power tools, and less likely than others to suffer medical problems, such as heart attacks. People often feel they can avoid hazards because they are wiser or luckier than others.
- News media give us a biased perspective on the frequency of certain kinds of health hazards, overreporting some accidents or diseases while downplaying or underreporting others. Sensational, gory, or especially frightful causes of death like murders, plane crashes, fires, or terrible accidents occupy a disproportionate amount of attention in the public media. Heart diseases, cancer, and stroke kill nearly 15 times as many people in the United States as do accidents and 75 times as many people as do homicides, but the emphasis placed by the media on accidents and homicides is nearly inversely proportional to their relative frequency compared to either cardiovascular disease or cancer. This gives us an inaccurate picture of the real risks to which we are exposed.
- We tend to have an irrational fear or distrust of certain technologies or activities that leads us to overestimate their dangers. Nuclear power, for instance, is viewed as very risky, while coal-burning power plants seem to be familiar and relatively benign; in fact, coal mining, shipping, and combustion cause an estimated 10,000 deaths each year in the United States. An old, familiar technology seems safer and more acceptable than does a new, unknown one (see Data Analysis p. 177).

***Think About It***

Why might you and your mother rank some risks differently? List some activities on which the two of you might disagree.

## Risk acceptance depends on many factors

How much risk is acceptable? How much is it worth to minimize and avoid exposure to certain risks? Most people will tolerate a higher probability of occurrence of an event if the harm caused by that event is low. Conversely, harm of greater severity is acceptable only at low levels of frequency. A 1-in-10,000 chance of being killed might be of more concern to you than a 1-in-100 chance of being injured. For most people, a 1-in-100,000 chance of dying from some event or some factor is a threshold for changing what we do. That is, if the chance of death is less than 1 in 100,000, we are not likely to be worried enough to change our ways. If the risk is greater, we will probably do something about it. The Environmental Protection Agency generally assumes that a risk of 1 in 1 million is acceptable for most environmental hazards. Critics of this policy ask, acceptable to whom?

**FIGURE 8.19** How dangerous are motorcycles? Many parents regard them as extremely risky, while many students—especially males—believe the risks (which are about the same as your chances of dying from surgery or other medical care) are acceptable. Perhaps the more important question is whether the benefits outweigh the risks.

For activities that we enjoy or find profitable, we are often willing to accept far greater risks than this general threshold. Conversely, for risks that benefit someone else we demand far higher protection. For instance, your chance of dying in a motor vehicle accident in any given year is about 1 in 5,000, but that doesn't deter many people from riding in automobiles. Your lifetime chance of dying from lung cancer if you smoke one pack of cigarettes per day is about 1 in 4. By comparison, the risk from drinking water with the EPA limit of trichloroethylene is about 1 in 10 million. Strangely, many people demand water with zero levels of trichloroethylene, while continuing to smoke cigarettes.

Table 8.4 lists lifetime odds of dying from a few leading diseases and accidents. These are statistical averages, of course, and there clearly are differences in where one lives or how one behaves that affect the danger level of these activities. Although the average lifetime chance of dying in an automobile accident is 1 in 100, there clearly are things you can do—like wearing a seatbelt, following safety rules, and avoiding risky situations—that improve your odds. Still, it is interesting how we readily accept some risks while shunning others.

Our perception of relative risks is strongly affected by whether risks are known or unknown, whether we feel in control of the outcome, and how dreadful the results are. Risks that are unknown or unpredictable and results that are particularly gruesome or disgusting seem far worse than those that are familiar and socially acceptable.

**TABLE 8.4**

**Lifetime Chances of Dying in the U.S.**

| Source | Odds (1 in *x*) |
|---|---|
| Heart disease | 2 |
| Cancer | 3 |
| Smoking | 4 |
| Lung disease | 15 |
| Pneumonia | 30 |
| Automobile accident | 100 |
| Suicide | 100 |
| Falls | 200 |
| Firearms | 200 |
| Fires | 1,000 |
| Airplane accident | 5,000 |
| Jumping from high places | 6,000 |
| Drowning | 10,000 |
| Lightning | 56,000 |
| Hornets, wasps, bees | 76,000 |
| Dog bite | 230,000 |
| Poisonous snakes, spiders | 700,000 |
| Botulism | 1 million |
| Falling space debris | 5 million |
| Drinking water with EPA limit of trichloroethylene | 10 million |

**Source:** U.S. National Safety Council, 2003.

Studies of public risk perception show that most people react more to emotion than statistics. We go to great lengths to avoid some dangers while gladly accepting others. Factors that are involuntary, unfamiliar, undetectable to those exposed, catastrophic, or that have delayed effects or are a threat to future generations are especially feared while those that are voluntary, familiar, detectable, or immediate cause less anxiety. Even though the actual number of deaths from automobile accidents, smoking, or alcohol, for instance, are thousands of times greater than those from pesticides, nuclear energy, or genetic engineering, the latter preoccupy us far more than the former.

## 8.7 Establishing Health Policy

Risk management combines principles of environmental health and toxicology together with regulatory decisions based on socio-economic, technical, and political considerations (fig. 8.20). The biggest problem in making regulatory decisions is that we are usually dealing with many sources of harm to which we are exposed, often without being aware of them. It is difficult to separate the effects of all these different hazards and to evaluate their risks accurately, especially when the exposures are near the threshold of measurement and response. In spite of often vague and contradictory data, public policymakers must make decisions.

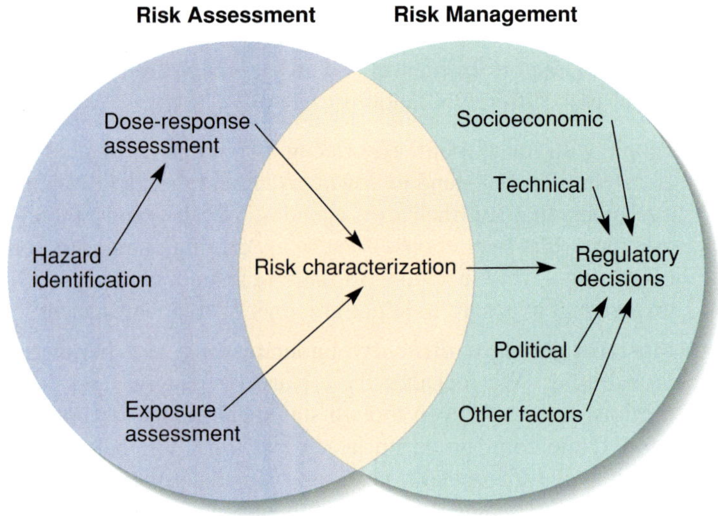


**FIGURE 8.20** Risk assessment organizes and analyzes data to determine relative risk. Risk management sets priorities and evaluates relevant factors to make regulatory decisions.

The case of the sweetener saccharin is a good example of the complexities and uncertainties of risk assessment in public health. Studies in the 1970s at the University of Wisconsin and the Canadian Health Protection Branch suggested a link between saccharin and bladder cancer in male rats. Critics of these studies pointed out that humans would have to drink 800 cans of diet soda *per day* to get a saccharin dose equivalent to that given to the rats. Furthermore, they argued this response may be unique to male rats. In 2000, the U.S. Department of Health concluded a study that found no association between saccharin and cancer in humans. Congress ordered that all warnings be removed from saccharin-containing products. Still, some groups like the Center for Science in the Public Interest consider this sweetener dangerous and urge us to avoid it if possible.

In setting standards for environmental toxins, we need to consider (1) combined effects of exposure to many different sources of damage, (2) different sensitivities of members of the population, and (3) effects of chronic as well as acute exposures. Some people argue that pollution levels should be set at the highest amount that does *not* cause measurable effects. Others demand that pollution be reduced to zero if possible, or as low as is technologically feasible. It may not be reasonable to demand that we be protected from every potentially harmful contaminant in our environment, no matter how small the risk. As we have seen, our bodies have mechanisms that enable us to avoid or repair many kinds of damage so that most of us can withstand some minimal level of exposure without harm (fig. 8.21).

On the other hand, each challenge to our cells by toxic substances represents stress on our bodies. Although each individual stress may not be life-threatening, the cumulative effects of all

**FIGURE 8.21** "Do you want to stop reading those ingredients while we're trying to eat?"

the environmental stresses, both natural and human-caused, to which we are exposed may seriously shorten or restrict our lives. Furthermore, some individuals in any population are more susceptible to those stresses than others. Should we set pollution standards so that no one is adversely affected, even the most sensitive individuals, or should the acceptable level of risk be based on the average member of the population?

Finally, policy decisions about hazardous and toxic materials also need to be based on information about how such materials affect the plants, animals, and other organisms that define and maintain our environment. In some cases, pollution can harm or destroy whole ecosystems with devastating effects on the life-supporting cycles on which we depend. In other cases, only the most sensitive species are threatened. Table 8.5 shows the Environmental Protection Agency's assessment of relative risks to human welfare. This ranking reflects a concern that our exclusive focus on reducing pollution to protect human health has neglected risks to natural ecological systems. While there have been many benefits from a case-by-case approach in which we evaluate the health risks of individual chemicals, we have often missed broader ecological problems that may be of greater ultimate importance.

**TABLE 8.5**

**Relative Risks to Human Welfare**

| **Relatively High-Risk Problems** |
|---|
| Habitat alteration and destruction |
| Species extinction and loss of biological diversity |
| Stratospheric ozone depletion |
| Global climate change |
| **Relatively Medium-Risk Problems** |
| Herbicides/pesticides |
| Toxics and pollutants in surface waters |
| Acid deposition |
| Airborne toxics |
| **Relatively Low-Risk Problems** |
| Oil spills |
| Groundwater pollution |
| Radionuclides |
| Thermal pollution |

**Source:** Environmental Protection Agency.

## Conclusion

We have made marvelous progress in reducing some of the worst diseases that have long plagued humans. Smallpox is the first major disease to be completely eliminated. Guinea worms and polio are nearly eradicated worldwide; typhoid fever, cholera, yellow fever, tuberculosis, mumps, and other highly communicable diseases are rarely encountered in advanced countries. Childhood mortality has decreased 90 percent globally, and people almost everywhere are living twice as long, on average, as they did a century ago.

But the technological innovations and affluence that have diminished many terrible diseases, have also introduced new risks. Chronic conditions, such as cardiovascular disease, cancer, depression, dementia, diabetes, and traffic accidents, that once were confined to richer countries, now have become leading health problems nearly everywhere. Part of this change is that we no longer die at an early age of infectious disease, so we live long enough to develop the infirmities of old age. Another factor is that affluent lifestyles, lack of exercise, and unhealthy diets aggravate these chronic conditions.

New, emergent diseases are appearing at an increasing rate. With increased international travel, diseases can spread around the globe in a few days. Epidemiologists warn that the next deadly epidemic may be only a plane ride away. In addition, modern industry is introducing thousands of new chemical substances every year, most of which aren't studied thoroughly for health effects. Endocrine disrupters, neurotoxins, carcinogens, mutagens, teratogens, and other toxins can have tragic outcomes. The effects of lead on children's mental development is an example of both how we have introduced materials with unintended consequences, and a success story of controlling a serious health risk. Many other industrial chemicals could be having similar harmful effects.

Determining what levels of environmental health risk are acceptable is difficult. We are exposed to many different health threats simultaneously. Furthermore, people consider some dangers tolerable, but dread others—especially those that are new, involuntary, difficult to detect, and whose effects are unknown to science. The situation is complicated by the fact that news media gives us a biased perspective on some hazards, while our personal experiences and our sense of our own abilities are often misleading.

There are many steps that each of us can take to protect our health. Eating a healthy diet, exercising regularly, drinking in moderation, driving prudently, and practicing safe sex are among the most important.

## Reviewing Learning Outcomes

By now you should be able to explain the following points:

**8.1** Describe health and disease and how global disease burden is now changing.

- The global disease burden is changing.
- Infectious and emergent diseases still kill millions of people.
- Conservation medicine combines ecology and health care.
- Resistance to drugs, antibiotics, and pesticides is increasing.
- Who should pay for health care?

**8.2** Summarize the principles of toxicology.

- How do toxins affect us?
- How does diet influence health?

**8.3** Discuss the movement, distribution, and fate of toxins in the environment.

- Solubility and mobility determine where and when chemicals move.
- Exposure and susceptibility determine how we respond.
- Bioaccumulation and biomagnification increase concentrations of chemicals.
- Persistence makes some materials a greater threat.
- Chemical interactions can increase toxicity.

**8.4** Characterize mechanisms for minimizing toxic effects.

- Metabolic degradation and excretion eliminate toxins.
- Repair mechanisms mend damage.

**8.5** Explain ways we measure and describe toxicity.

- We usually test toxins on lab animals.
- There is a wide range of toxicity.
- Acute and chronic doses and effects differ.
- Detectable levels aren't always dangerous.

**8.6** Evaluate risk assessment and acceptance.

- Risk perception isn't always rational.
- Risk acceptance depends on many factors.

**8.7** Relate how we establish health policy.

## Practice Quiz

1. What are guinea worms, and how are they acquired?
2. Define the terms *disease* and *health.*
3. What were some of the most serious diseases in the world in 1990? How is this list expected to change in the next 20 years?
4. What are emergent diseases? Give a few examples, and describe their cause and effects.
5. How do bacteria acquire antibiotic resistance? How might we prevent this?
6. What is the difference between toxic and hazardous? Give some examples of materials in each category.
7. How do the physical and chemical characteristics of materials affect their movement, persistence, distribution, and fate in the environment?
8. What is the difference between acute and chronic toxicity?
9. Define *carcinogenic, mutagenic, teratogenic,* and *neurotoxic.*
10. What are the relative risks of smoking, driving a car, and drinking water with the maximum permissible levels of trichloroethylene? Are these relatively equal risks?

## Critical Thinking and Discussion Questions

1. What consequences (positive or negative) do you think might result from defining health as a state of complete physical, mental, and social well-being? Who might favor or oppose such a definition?
2. Do rich countries bear any responsibilities if the developing world adopts unhealthy lifestyles or diets? What could (or should) we do about it?
3. Why do we spend more money on heart or cancer research than childhood illnesses?
4. What are the premises in the discussion of assessing risk? Could conflicting conclusions be drawn from the facts presented in this section? What is your perception of risk from your environment?

5. Should pollution levels be set to protect the average person in the population or the most sensitive? Why not have zero exposure to all hazards?

6. What level of risk is acceptable to you? Are there some things for which you would accept more risk than others?

## DATA analysis Graphing Multiple Variables

Is it possible to show relationships between two dependent variables on the same graph? Sometimes that's desirable when you want to make comparisons between them. The graph shown here does just that. It's a description of how people perceive different risks. We judge the severity of risks based on how familiar they are and how much control we have over our exposure. The Y-axis represents how mysterious, unknown or delayed the risk seems to be. Things that are unobservable, unknown to those exposed, delayed in their effects, and unfamiliar or unknown to science tend to be more greatly feared than those that are observable, known, immediate, familiar, and known to science. The X-axis represents a measure of dread, which combines how much control we feel we have over the risk, how terrible the results could potentially be, and how equitably the risks are distributed. The size of the symbol for each risk indicates the combined effect of these two variables. Notice that things such as DNA technology or nuclear waste, which have high levels of both mystery and dread, tend to be regarded with the greatest fear, while familiar, voluntary, personally rewarding behaviors such as riding in automobiles or on bicycles, or drinking alcohol are thought to be relatively minor risks. Actuarial experts (statisticians who gather mortality data) would tell you that automobiles, bicycles, and alcohol have killed far more people (so far) than DNA technology or radioactive waste. But this isn't just a question of data. It's a reflection of how much we fear various risks. Notice that this is a kind of scatter plot mapping categories of data that have no temporal sequence. Still, you can draw some useful inferences from this sort of graphic presentation.

Compare the graph with table 8.4.

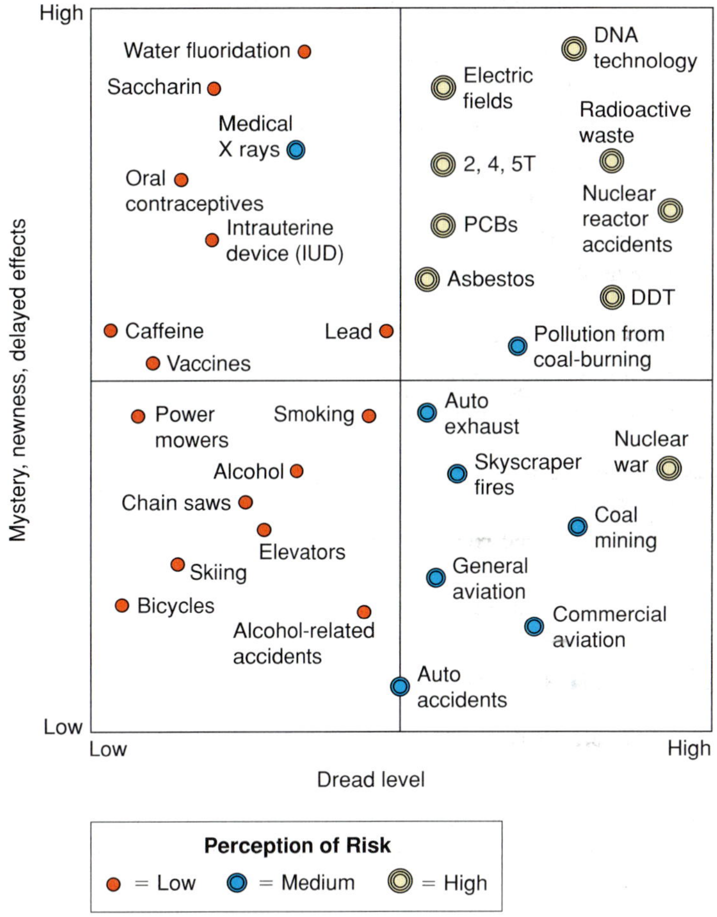


Public perception of risk depending on the familiarity, apparent potential for harm, and personal control over the risk.

**Source:** Data from Slovic, Paul, 1987. Perception of Risk, *Science,* 236 (4799): 286–90.

1. What is the highest risk factor in table 8.4 that also appears on the perception of risk graph?
2. How do the lifetime risk for smoking and auto accidents calculated by the National Safety Council correspond to the perception of risk in this graph?
3. How do airplane accidents compare in these two assessments?
4. Why do DNA technology and radioactive waste appear to generate so much dread, but don't even appear in the lifetime chances of dying list?
5. Thousands of people die every year from alcohol-related accidents, yet alcohol ranks very low in the two scales shown in this graph. On the other hand, no one has ever died—as far as we know—from DNA technology. Why the great discrepancy in rankings?

**For Additional Help in Studying This Chapter,** please visit our website at www.mhhe.com/cunningham10e. You will find additional practice quizzes and case studies, flashcards, regional examples, place markers for Google Earth™ mapping, and an extensive reading list, all of which will help you learn environmental science.

Enormous farms have been carved out of Brazil's Cerrado (savanna), which once was the most biodiverse grassland and open tropical forest complex in the world.

CHAPTER 9

# Food and Agriculture

*We abuse the land because we regard it as a commodity belonging to us. When we see land as a community to which we belong, we may begin to use it with love and respect.*

*—Aldo Leopold—*

## LEARNING OUTCOMES

After studying this chapter, you should be able to:

**9.1** Describe world food supplies and nutritional requirements.

**9.2** Compare key food sources.

**9.3** Understand how farm policy can help or hurt soil conservation.

**9.4** Illustrate why soil is a renewable resource.

**9.5** Summarize the ways we use and abuse soil.

**9.6** Identify other agricultural resources.

**9.7** Explain new crops and genetic engineering.

**9.8** Appraise sustainable agriculture.

# Case Study Farming the Cerrado

A soybean boom is sweeping across South America. Inexpensive land, availability of new crop varieties, and government policies that favor agricultural expansion have made South America the fastest growing agricultural area in the world. The center of this rapid expansion is the Cerrado, a huge area of grassland and tropical forest stretching from Bolivia and Paraguay across the center of Brazil almost to the Atlantic Ocean (fig. 9.1). Biologically, this rolling expanse of grasslands and tropical woodland is the richest savanna in the world, with at least 130,000 different plant and animal species, many of which are threatened by agricultural expansion.

Until recently, the Cerrado, which is roughly equal in size to the American Midwest, was thought to be unsuitable for cultivation. Its red iron-rich soils are highly acidic and poor in essential plant nutrients. Furthermore, the warm, humid climate harbors many destructive pests and pathogens. For hundreds of years, the Cerrado was primarily cattle country with many poor-quality pastures producing low livestock yields.

In the past few decades, however, Brazilian farmers have learned that modest applications of lime and phosphorus can quadruple yields of soybeans, maize, cotton, and other valuable crops. Researchers have developed more than 40 varieties of soybeans—mostly through conventional breeding, but some created with molecular techniques—specially adapted for the soils and climate of the Cerrado. Until about 30 years ago, soybeans were a relatively minor crop in Brazil. Since 1975, however, the area planted with soy has doubled about every four years, reaching more than 25 million ha (60 million acres) in 2006. Although that's a large area, it represents only one-eighth of the Cerrado, more than half of which is still occupied by pasture.

**FIGURE 9.1** Brazil's Cerrado, 2 million ha of savanna (grassland) and open woodland, is the site of the world's fastest growing soybean production. Cattle ranchers and agricultural workers, displaced by mechanized crop production, are moving northward into the "arc of destruction" at the edge of the Amazon rainforest, where the continent's highest rate of forest clearing is occurring.

Brazil is now the world's top soy exporter, shipping some 50 million metric tons per year, or about 10 percent more than the United States. With two crops per year, cheap land, low labor costs, favorable tax rates, and yields per hectare equal to those in the American Midwest, Brazilian farmers can produce soybeans for less than half the cost in America. Agricultural economists predict that, by 2020, the global soy crop will double from the current 160 million metric tons per year, and that South America could be responsible for most of that growth. In addition to soy, Brazil now leads the world in beef, corn (maize), oranges, and coffee exports. This dramatic increase in South American agriculture helps answer the question of how the world may feed a growing human population.

A major factor in Brazil's current soy expansion is rising income in China. With more money to spend, the Chinese are consuming more soy, both directly as tofu and other soy products, and indirectly as animal feed. A decade ago, China was self-sufficient in soy production. Now, China imports about 30 million tons of soy annually. About half that amount comes from Brazil, which passed the United States in 2007 as the world's leading soy exporter. In 1997, Brazil shipped only 2 million tons of soy. A decade later, exports reached 28 million tons.

The outbreak of mad cow disease (or BSE, see chapter 8) in Europe, Canada, and Japan has fueled increased worldwide demand for Brazilian beef. With 175 million free-range, grass-fed (and presumably BSE-free) cattle, Brazil has become the world's largest beef exporter.

Increasing demand for both soybeans and beef create land conflicts in Brazil. The pressure for more cropland and pasture is a leading cause of deforestation and habitat loss, most of which is occurring in the "arc of destruction" between the Cerrado and the Amazon. Small family farms are being gobbled up, and farm workers, displaced by mechanization, often migrate either to the big cities or to frontier forest areas. Increasing conflicts be-tween poor farmers and big landowners have led to violent confrontations. The Landless Workers Movement claims that 1,237 rural workers died in Brazil between 1985 and 2000 as a result of assassinations and clashes over land rights. In 2005, a 74-year-old Catholic nun, Sister Dorothy Stang, was shot by gunmen hired by ranchers who resented her advocacy for native people, workers, and environmental protection. Over the past 20 years, Brazil claims to have resettled 600,000 families from the Cerrado. Still, tens of thousands of landless farm workers and displaced families live in unauthorized squatter camps and shanty-towns across the country awaiting relocation.

# Case Study *continued*

As you can see, rapid growth of beef and soy production in Brazil have both positive and negative aspects. On one hand, more high-quality food is now available to feed the world. The 2 million $km^2$ of the Cerrado represents one of the world's last opportunities to open a large area of new, highly productive cropland. On the other hand, the rapid expansion and mechanization of agriculture in Brazil is destroying biodiversity and creating social conflicts as people move into formerly pristine lands. The issues raised in this case study illustrate many of the major themes in this chapter. Will there be enough food for everyone in the world? What will be the environmental and social consequences of producing the nutrition we need? In this chapter, we'll look at world food supplies, agricultural inputs, and sustainable approaches that can help solve some of the difficult questions we face.

## 9.1 World Food and Nutrition

Despite dire predictions that runaway population growth would soon lead to terrible famines (chapter 7), world food supplies have more than kept up with increasing human numbers over the past two centuries. The past 40 years have seen especially encouraging strides in reducing world hunger. While population growth averaged 1.7 percent per year during that time, world food production increased an average of 2.2 percent. Increased use of irrigation, improved crop varieties, more readily available fertilizers, and distribution systems to transport food from regions with surpluses to those in need have brought improved nutrition to billions of people. In this section, we'll look at the causes and effects of remaining chronic and acute food shortages, as well as recommendations for a balanced, healthful diet.

### Millions of people don't have enough to eat

In 1960, nearly 60 percent of the residents of developing countries were considered **chronically undernourished,** meaning their diet didn't provide the 2,200 kcal per day, on average, considered necessary for a healthy productive life. Today, despite the fact that their population has doubled over the past 46 years, the proportion in these countries suffering from chronic caloric deficiency has fallen to less than 14 percent.

The UN Food and Agriculture Organization (FAO) expects agricultural production to continue to grow over the next few decades. Where the current world food supply would be sufficient, if equitably distributed, to provide an average of 2,800 kcal per person per day, the FAO predicts that by 2030, there will be enough food available to supply 3,050 kcal per day to everyone, or about 30 percent more than most of us need. In countries such as the United States, the problem already has become what to do with surplus food. Farmers in these countries are paid billions of dollars per year not to grow crops.

Still, in a world of surplus food, at least 850 million people don't have enough to eat. Figure 9.2 shows countries with the highest risk of food shortages. As you can see, most of sub-Saharan Africa, South and Southeast Asia, and parts of Latin America fall in this category. Ninety-five percent of the chronically undernourished are in developing countries, but an increasing number are in transition countries (primarily states of the former Soviet Union and its allies undergoing a change from socialism to market economies) where bad weather, poor management, and social crises have resulted in sharply falling agricultural production (fig. 9.3). Even in the richest countries, where excess calories are the greatest problem for the majority, some 11 million people don't have enough to eat.

Poverty is the greatest threat to **food security,** or the ability to obtain sufficient food on a day-to-day basis. The 1.5 billion people in the world who live on less than $1 per day all too often can't buy the food they need and don't have access to resources to grow it for themselves. Food security occurs at multiple scales. In the poorest countries, hunger may affect nearly everyone. In other countries, although the average food availability may be satisfactory, some individual communities or families may not have enough to eat. And within families, males often get both the largest share as well as the most nutritious food, while women and children—who need food most—all too often get the poorest diet. At least 6 million children under 5 years old die every year (one every 5 seconds) from hunger and malnutrition. Providing a healthy diet might eliminate as much as 60 percent of all premature deaths worldwide.

Hungry people can't work their way out of poverty, the Nobel Prize-winning economist Robert Fogel points out. He estimates that in 1790, about 20 percent of the population of England and France was effectively excluded from the labor force because they were too weak and hungry to work. Improved nutrition, he calculates, accounted for about half of all European economic growth during the nineteenth century. Since many developing countries are as poor now (in relative terms) as Britain and France were in 1790, his analysis suggests that reducing hunger could yield more than (U.S.) $120 billion in economic growth produced by longer, healthier, more productive lives for several hundred million people.

The 2003 UN World Food Summit reaffirmed the goal set by previous conventions of reducing the number of chronically undernourished people from 850 million to 400 million by 2015. We aren't on track to meet that goal, but some countries have made impressive progress. China, alone, has reduced its number of undernourished people by 74 million over the past decade. Indonesia, Vietnam, Thailand, Nigeria, Ghana, and Peru each reduced chronic hunger by about 3 million people. In 47 other countries, however, the numbers of chronically underfed

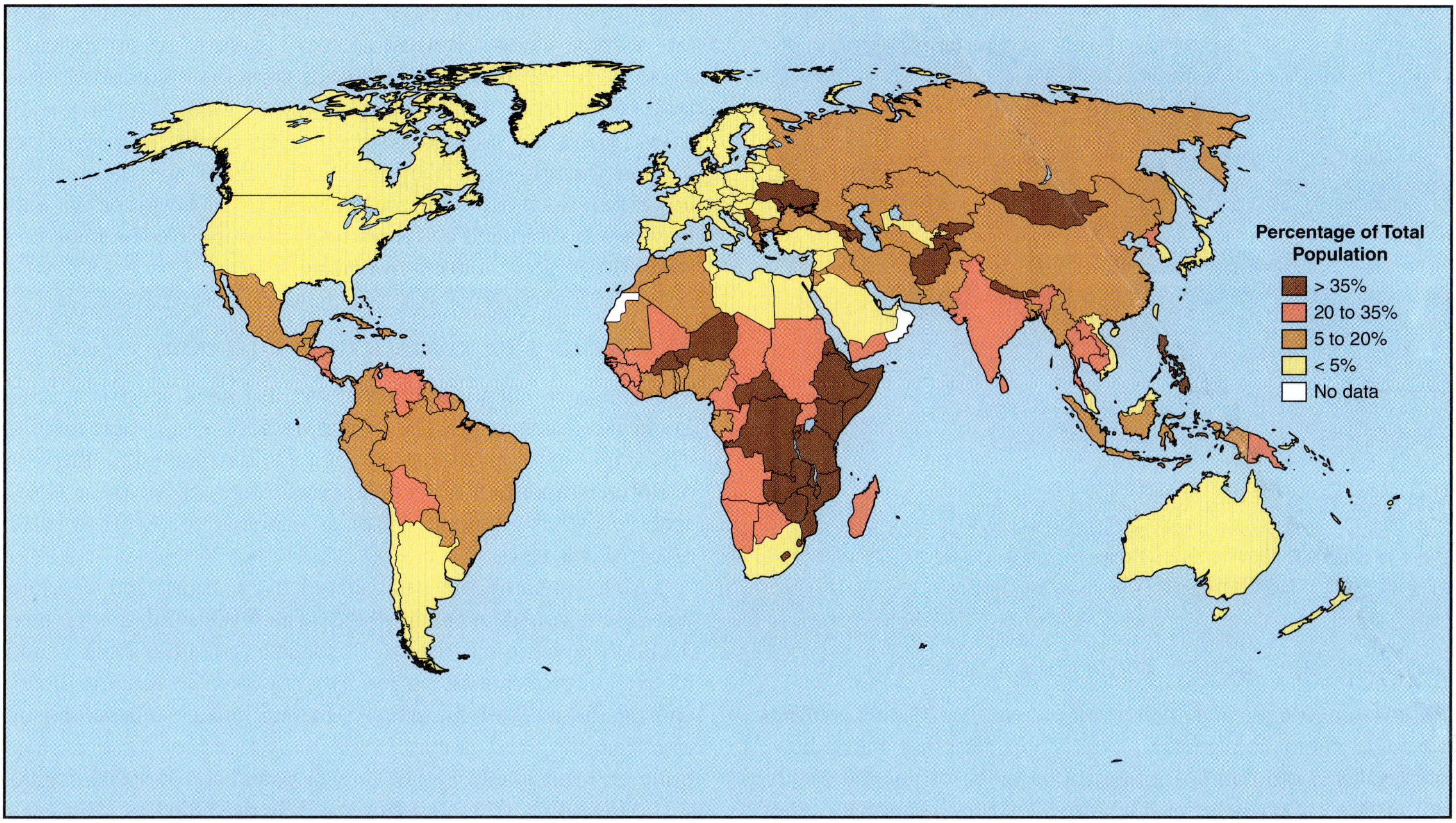


**FIGURE 9.2** Hunger around the world. In 2005, the United Nations reported that 852 million people—815 million of them in developing countries—suffered from chronic hunger and malnutrition. Africa has the largest number of countries with food shortages.

people have increased. India now has by far the largest number of persistently hungry people in the world, while central Africa has the highest percentage (fig. 9.4).

Recognizing the role of women in food production is an important step toward food security for all. Throughout the developing world, women do 50 to 70 percent of all farm work but control only a tiny fraction of the land and rarely have access to capital or developmental aid. In Nigeria, for example, home gardens occupy only 2 percent of all cropland, but provide half the food families eat. Making land, credit, education, and access to markets available to women could contribute greatly to family nutrition.

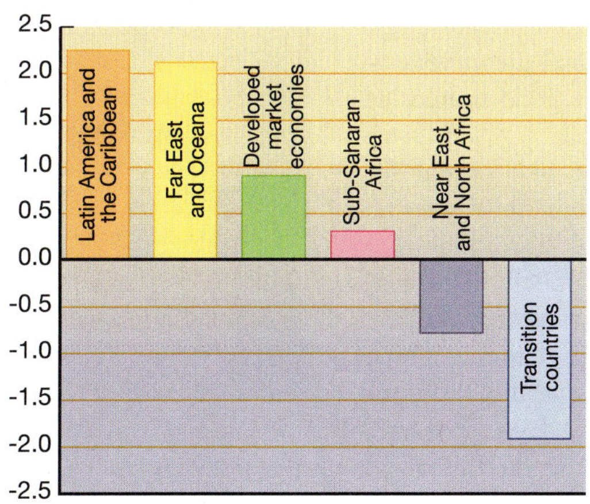


**FIGURE 9.3** Changes in agricultural production in 2000. Transition nations are states from the former Soviet Union undergoing a change from socialism to capitalism.
**Source:** Food and Agriculture Organization (FAO), 2002.

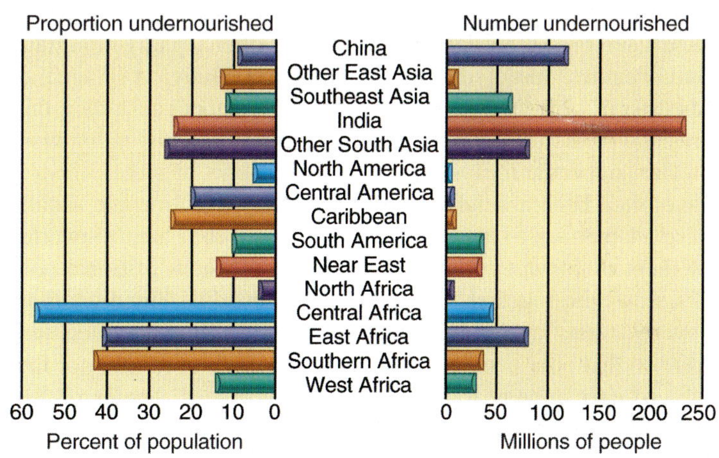


**FIGURE 9.4** Number and percent of population chronically undernourished by region in the developing world.
**Source:** Food and Agriculture Organization (FAO), 2002

**FIGURE 9.5** Children wait for their daily ration of porridge at a feeding station in Somalia. When people are driven from their homes by hunger or war, social systems collapse, diseases spread rapidly, and the situation quickly becomes desperate.

## Famines are acute food emergencies

The chronic hunger and malnutrition described in the previous section is silent and generally invisible, affecting individuals, families, and communities, often in the midst of general plenty, but **famines** are characterized by large-scale food shortages, massive starvation, social disruption, and economic chaos. Starving people eat their seed grain and slaughter their breeding stock in a desperate attempt to keep themselves and their families alive. Even if better conditions return, they have sacrificed their productive capacity and will take a long time to recover. Famines are characterized by mass migrations as starving people travel to refugee camps in search of food and medical care (fig. 9.5). Many die on the way or fall prey to robbers.

In 2006, the FAO reported that 58 million people in 36 countries (two-thirds of them in sub-Saharan Africa) needed emergency food aid. What causes these emergencies? Droughts, earthquakes, severe storms, and other natural disasters are usually the immediate trigger, but politics and economics are often equally important. Bad weather, insect outbreaks (see Case Study: A Plague of Locusts, p. 125), and other environmental factors cause crop failures and create food shortages. But the Nobel Prize-winning work of Harvard economist Amartya K. Sen shows that these factors have often been around for a long time, and local people usually have adaptations to get through hard times if they aren't thwarted by inept or corrupt governments and greedy elites. National politics, however, together with commodity hoarding, price gouging, poverty, wars, landlessness, and other external factors often conspire so that poor people can neither grow their own food nor find jobs to earn money to buy the food they need. Professor Sen points out that armed conflict and political oppression almost always are at the root of famine. No democratic country with a relatively free press, he says, has ever had a major famine.

The aid policies of rich countries often serve more to get rid of surplus commodities and make us feel good about our generosity than to get at the root causes of starvation. But herding people into feeding camps often is the worst thing to do for them. The stress of getting there kills many of them, and the crowding and lack of sanitation in the camps exposes them to epidemic diseases. There are no jobs in the refugee camps, so people can't support themselves if they try. Social chaos and family breakdown expose those who are weakest to robbery and violence. Having left their land and tools behind, people can't replant crops when the weather returns to normal.

## We need the right kinds of food

In addition to energy (calories), we also need specific nutrients in our diet, such as proteins, vitamins, and certain trace minerals. You might have more than enough calories and still suffer from **malnourishment,** a nutritional imbalance caused by a lack of specific dietary components or an inability to absorb or utilize essential nutrients.

Many poor people can't afford meat, fruits, and vegetables that would provide a balanced diet. The FAO estimates that nearly 3 billion people (half the world population) suffer from vitamin, mineral, or protein deficiencies. This results in devastating illnesses and deaths as well as reduced mental capacity, developmental abnormalities, and stunted growth. Altogether, these problems bring an incalculable loss of human potential and social capital.

**Anemia** (low hemoglobin levels in the blood, usually caused by dietary iron deficiency) is the most common nutritional problem in the world. According to the FAO, more than 2 billion people (52 percent are pregnant women and 39 percent are children under age 5) suffer from iron deficiencies. The problem is most severe in India, where it's estimated that 80 percent of all pregnant women are anemic. Anemia increases the risk of maternal deaths from hemorrhage in childbirth and affects childhood development. Red meat, eggs, legumes, and green vegetables all are good sources of dietary iron.

Iodine is essential for synthesis of thyroxin, an endocrine hormone that regulates metabolism and brain development, among other things. Chronic iodine deficiency causes goiter (a swollen thyroid gland, fig. 9.6), stunted growth, and reduced mental ability. The FAO estimates that 740 million people—mainly in South and Southeast Asia—suffer from iodine

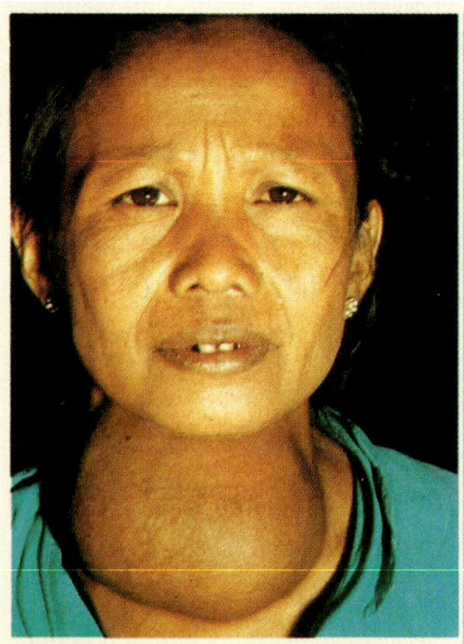

**FIGURE 9.6** Goiter, a swelling of the thyroid gland at the base of the neck, is often caused by an iodine deficiency. It is a common problem in many parts of the world, particularly where soil iodine is low and seafood is unavailable.

deficiency and that 177 million children have stunted growth and development. Adding a few pennies worth of iodine to our salt has largely eliminated this problem in developed countries.

Starchy foods like maize (corn), polished rice, and manioc (tapioca), which form the bulk of the diet for many poor people, tend to be low in several essential vitamins as well as minerals. According to the FAO, vitamin A deficiencies affect 100–140 million children at any given time. At least 350,000 go blind every year from the effects of this vitamin shortage. Folic acid (found in dark green, leafy vegetables) is essential for early fetal development. Folic acid deficiencies have been linked to neurological problems in babies including microencephaly (an abnormally small head) or even anencephaly (lacking a brain).

Protein also is essential for normal growth and development. The two most widespread human protein deficiency diseases are kwashiorkor and marasmus. **Kwashiorkor** is a West African word meaning "displaced child." (A young child is displaced—and deprived of nutritious breast milk—when a new baby is born.) This condition most often occurs in young children who eat mainly cheap starchy food and don't get enough good-quality protein. Children with kwashiorkor often have reddish-orange hair, puffy, discolored skin, and a bloated belly. **Marasmus** (from the Greek "to waste away") is caused by a diet low in both calories and protein. A child suffering from severe marasmus is generally thin and shriveled, like a tiny, very old starving person (fig. 9.7). Children with both these deficiencies have low resistance to infections and are likely to suffer from stunted growth, mental retardation, and other developmental problems. Altogether, the FAO estimates that the annual losses due to deaths and diseases caused by calorie and nutrient deficiencies are equivalent to 46 million years of productive life (see more in chapter 8 on disability-adjusted life years).

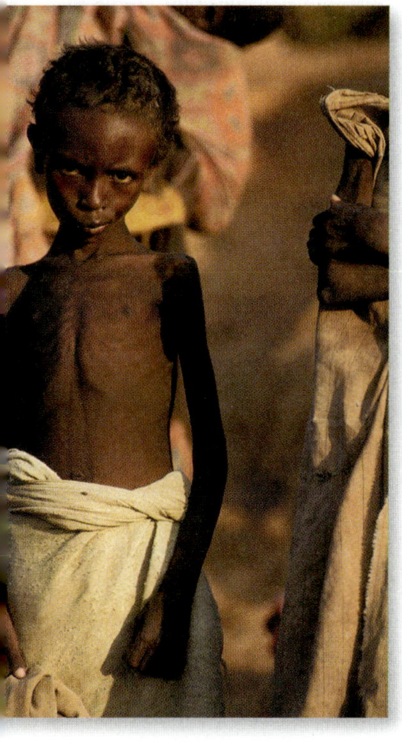

**FIGURE 9.7** Marasmus is caused by combined energy (calorie) and protein deficiencies. Children with marasmus have the wizened look and dry, flaky skin of an old person.

## Eating a balanced diet is essential for good health

Those of us in richer countries often eat too much meat, salt, sugar, and saturated fat, and too little fiber, vitamins, trace minerals, and other components lost from highly processed foods. On average, we consume about one-third more calories than we need and get too little exercise. Food provides solace in stressful lives, and we're constantly bombarded with advertising tempting us to consume more. Ironically, there are now probably more people in the world (1 billion) who suffer from too many calories, than there are who suffer from too few. According to the U.S. Surgeon General, 62 percent of all adult Americans are overweight, up from 40 percent only a decade ago, and about one-third of us are seriously overweight or **obese**—generally considered to be a body mass greater than 30 kg/m$^2$, or roughly 30 pounds above normal for an average person.

Being overweight substantially raises your risk of hypertension, diabetes, heart attacks, stroke, gallbladder disease, osteoarthritis, respiratory problems, and certain cancers. Every year, about 300,000 in the United States die from illnesses related to obesity. This number is close to the 400,000 annual deaths from diseases related to smoking. Paradoxically, food insecurity and poverty can contribute to obesity. In one study, more than half of the women who reported not having enough to eat were overweight compared to one-third of food-secure women. Lack of time for cooking and access to health food choices along with ready availability of fast-food snacks and calorie-laden drinks lead to dangerous dietary imbalances for many people.

This trend isn't limited to richer countries. Obesity is spreading around the world (fig. 9.8). As chapter 8 points out, western diets and lifestyles are being adopted by many in the developing world, and diseases such as heart attack, stroke, diabetes, and depression that once were thought to inflict only wealthy nations are now becoming the most prevalent causes of death and disability everywhere.

What's the best way to be sure you're getting a healthy diet? Generally, it isn't necessary to take synthetic dietary supplements. Eating a good variety of foods should give you all the nutrients you need. For years, Americans were advised to eat daily servings of four major food groups: meat, dairy products, grains, and fruits and vegetables. These recommendations were revised in 1992 to emphasize only sparing servings of meat, dairy, fats, and sweets.

Some nutritionists believe that the 1992 recommendations for 6 to 11 servings a day of bread, cereal, rice, and pasta still provides too many simple sugars (or starches that are quickly converted to

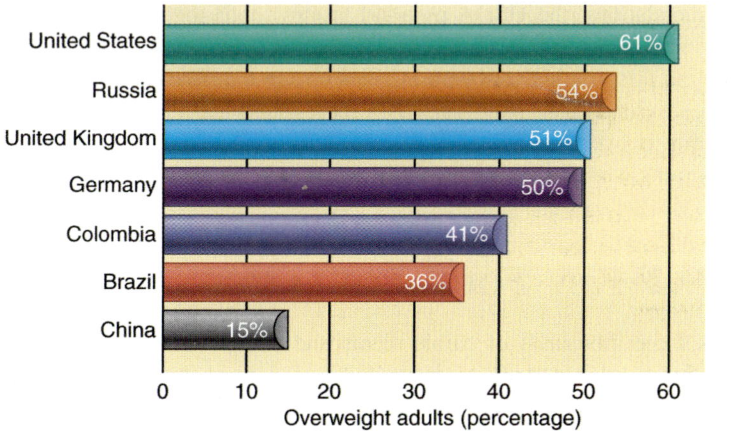


**FIGURE 9.8** While nearly a billion people are chronically undernourished, people in wealthier countries are at risk from eating too much. **Source:** Worldwatch Institute, 2001.

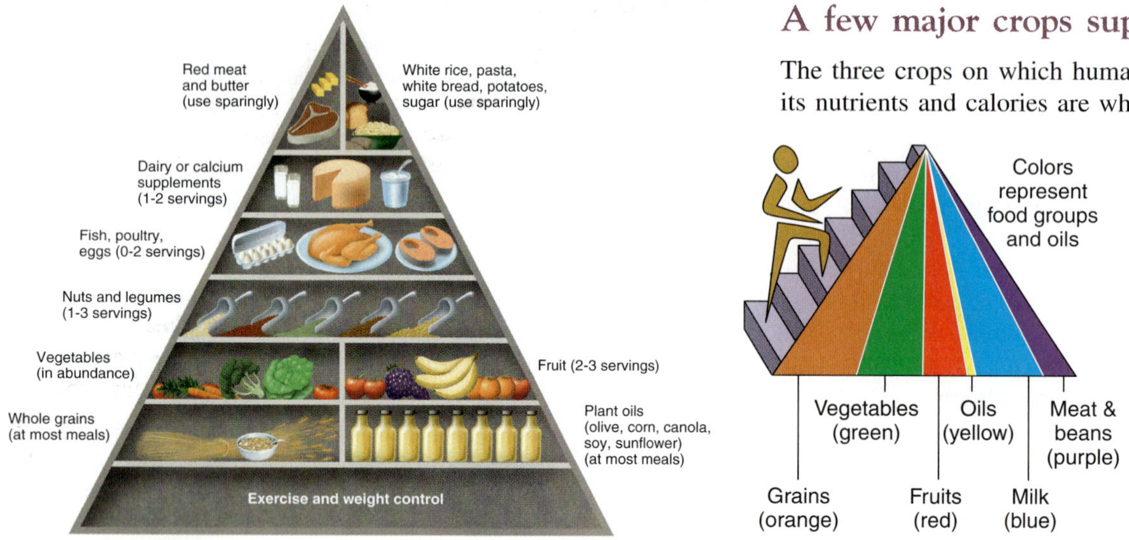


(a) Harvard food pyramid

(b) USDA recommendations

**FIGURE 9.9** The Harvard food pyramid (a) emphasizes fruits, vegetables, and whole grains as the basis of a healthy diet. Red meat, white rice, pasta, and potatoes should be used sparingly. The new USDA food pyramid (b) is confusing to many people and seems to encourage more consumption of meat and milk.
**Source:** (a) Data from Willett and Stampfer, 2002 (b) USDA, 2005.

sugar). Based on observations of health effects of Mediterranean diets as well as a long-term study of 140,000 U.S. health professionals, Drs. Walter Willett and Meir Stampfer of Harvard University have recommended a new dietary pyramid (fig. 9.9*a*). Both red meat and starchy food such as white rice, white bread, potatoes, and pasta should be eaten sparingly. Nuts, legumes (beans, peas, and lentils), fruit, vegetables, and whole grain foods form the basis of this diet. Unsaturated plant oils should make up 30 to 40 percent of dietary calories, according to this view. Trans fat (the kind found in hydrogenated margarine), on the other hand, is not recommended at all. Combined with regular, moderate exercise, this food selection should provide most people with all the nutrition they need.

In 2005, the U.S. Department of Agriculture unveiled an alternative food pyramid (fig. 9.9*b*). Rather than give specific recommendations, the USDA pyramid groups foods together and shows some consumption of each food group at every level. Many people find this graphic confusing. Critics charge that it was developed by consultants who simultaneously represented the food industry, and that it overemphasizes meat and dairy while downplaying whole grains, vegetables, fruits, and unsaturated oils.

## 9.2 Key Food Sources

Of the thousands of edible plants and animals in the world, only about a dozen types of seeds and grains, three root crops, twenty or so common fruits and vegetables, six mammals, two domestic fowl, and a few fish and other forms of marine life make up almost all of the food humans eat. Table 9.1 shows annual production of some important foods in human diets. In this section, we will highlight the characteristics of some important food sources.

### A few major crops supply most of our food

The three crops on which humanity depends for the majority of its nutrients and calories are wheat, rice, and maize (called corn in the United States). Together, more than 1,900 million metric tons of these three grains are grown each year. Wheat and rice are especially important since they are the staple foods for most of the 5.5 billion people in the developing countries of the world. These two grass species supply around 60 percent of the calories consumed directly by humans.

Potatoes, barley, oats, and rye are staples in mountainous regions and high latitudes (northern Europe, north Asia) because they grow well in cool, moist climates. Cassava, sweet potatoes, and other roots and tubers grow well in warm, wet areas and are staples in Amazonian, Africa, Melanesia, and the South Pacific. Barley, oats, and rye can grow in cool, short-season climates. Sorghum and millet are drought resistant and are staples in the dry regions of Africa.

Fruits, vegetables, and vegetable oils make a surprisingly large contribution to human diets. They are especially welcome because they typically contain high levels of vitamins, minerals, dietary fiber, and complex carbohydrates.

**TABLE 9.1**

**Some Important Food Sources**

| Crop | 2006 Yield (Million Metric Tons) |
|---|---|
| Wheat | 624 |
| Rice (paddy) | 635 |
| Maize (corn) | 683 |
| Potatoes | 320 |
| Coarse grains* | 1,009 |
| Soybeans | 222 |
| Cassava and sweet potato | 550 |
| Sugar (cane and beet) | 150 |
| Pulses (beans, peas) | 61 |
| Oil seeds | 397 |
| Vegetables and fruits | 1,220 |
| Meat and milk | 927 |
| Fish and seafood | 150 |

**Barley, oats, sorghum, rye, millet.*

**Source:** Food and Agriculture Organization (FAO), 2006.

**FIGURE 9.10** Meat and dairy consumption have quadrupled in the past 40 years, and China represents about 40 percent of that increased demand.

## Meat and dairy are important protein sources

Protein-rich foods are prized by people nearly everywhere. In the past, the rich countries consumed a vast majority of the meat, dairy, and high-quality seafood traded internationally. The four-fifths of the world's people in less-developed countries generally raised 60 percent of the 3 billion domestic ruminants and 20 billion poultry in the world but consumed only one-fifth of all commercial animal products. The FAO reports, however, that as incomes rise in developing countries, food choices throughout the world are shifting toward higher-quality and more expensive foods. Meat consumption in developing countries, for example, has risen from only 10 kg per person annually in the 1960s to 26 kg currently (fig. 9.10). By 2030, average annual meat consumption in those countries is expected to be nearly 40 kg per person.

Much of that increase in meat production could be provided by improved productivity in tropical areas. As the opening Case Study for this chapter shows, there has been a huge increase in beef and soy production in South America. Modern breeding techniques that combine the best qualities of Indian zebu (humped) cattle and beef-producing varieties from Europe and North America have provided animals that resist heat and tropical diseases but have meat characteristics preferred by consumers in developed countries (fig. 9.11*a*). Over the past 40 years, the area devoted to soybeans in Brazil has increased more than 20-fold (fig. 9.11*b*.), while the yield per hectare has doubled. Carving farmland and pasture out of tropical forest and savanna is helping to feed the world, but it is also reducing biological diversity and causing social conflict, as small subsistence farms are bought up by large-scale industrial agriculture, and landless farm workers are forced off the land into shantytowns (fig. 9.11*c*).

Soy farmers claim they aren't causing deforestation because they're only turning low-quality pasture, which was cleared years ago for ranching, into high-production row cropping. In 2006, Brazil's grain and processing industry declared a two-year moratorium on purchasing soybeans from farmers who deforest land in the Amazon. It's true that the land is more productive under intensive farming, but as grazing lands are turned into farms, ranchers and displaced rural families often move further into the forest in search of new land. In 2006, there were 1,690 violent clashes and 93 people were killed in disputes over landownership in Brazil, mostly involving indigenous or local people and large landowners.

Most of the livestock grown in North America are confined in **concentrated animal feeding operations (CAFOs)**

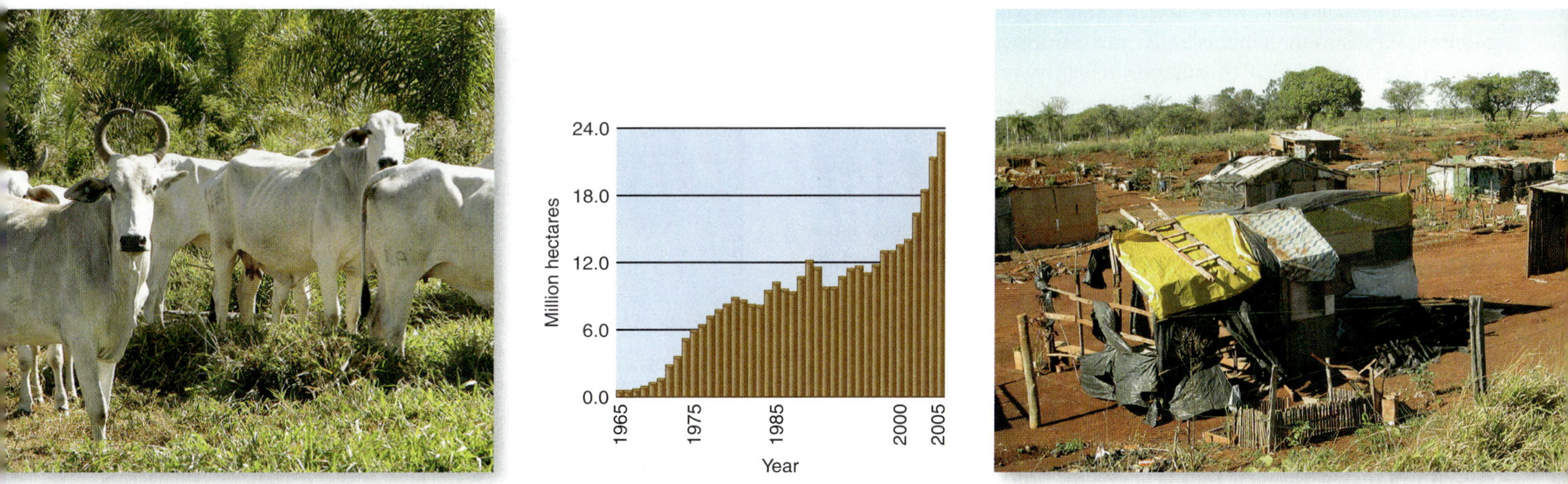


(a) Brazilian zebu cattle (b) Brazil soy area (c) Favela

**FIGURE 9.11** Beef and soy production in tropical countries, such as Brazil, could help satisfy the growing demand for meat in the developing world. Beef production and soy yields in Brazil have doubled in the past 40 years, while the area devoted to soybeans has increased more than 20-fold. This rapid expansion has had severe environmental costs, however. And thousands of rural families have been displaced to live in shanty towns called *favelas*.
**Sources:** (b) Data from USDA, 2005.

**FIGURE 9.12** Concentrated feeding operations fatten animals quickly and efficiently, but create enormous amounts of waste and expose livestock to unhealthy living conditions.

for part or all of their lifetime. A diet rich in grain, oil, and protein fattens animals quickly and produces meat preferred by many consumers. Globally, some 680 million metric tons of cereals are used as livestock feed each year, representing more than one-third of the total maize, soy, and coarse grain production. It's often pointed out that we could feed about ten times as many people if we ate grain directly rather than feeding it to livestock. Surprisingly, the FAO claims that using cereals as animal feed does not contribute to hunger and under nutrition. Given current agricultural economics, if these cereals were not used as animal feed, they would probably not be produced at all, and thus would not be available as food. According to this view, if everyone became a vegetarian, the lack of demand for cereals for livestock production would simply lead to lower crop production, not more food for the hungry.

The rapid proliferation of CAFOs raises a number of social and environmental concerns. Housing up to a million animals in a single facility can cause serious local air and water pollution. Altogether, CAFOs in the United States are estimated to generate some 500 million metric tons of manure annually, or roughly 25 times all the human waste produced in the country each year (fig. 9.12). Storage and disposal of this waste is a serious problem. CAFOs are almost never served by sewage treatment plants. Land disposal of waste overloads soil with nutrients, and can contaminate food and groundwater supplies. In 2006, more than 200 people were sickened and four died from eating fresh spinach contaminated by run-off from a nearby cattle feed lot.

Local residents often find the stench of so many animals and so much waste overwhelming. A study by the University of North Carolina found that people living near large CAFOs suffer high levels of respiratory illness. Many communities have tried to restrict or regulate CAFOs, but find it difficult to do.

Animal wastes, particularly from hog farms, often are stored in huge open lagoons that can leak or rupture, poisoning local surface waters. In 1999, for example, Hurricane Floyd dumped torrential rains on North Carolina. An estimated 10 million $m^3$ (2.5 billion gal) of hog and poultry waste overflowed from storage lagoons into local rivers. Bacterial growth, which was stimulated by the manure, deoxygenated the water, killed millions of fish, and created a "dead zone" in Pamlico Sound that lasted for months.

The high density of animals in these facilities and the rich diet they're fed to speed weight gain requires a constant use of antibiotics and growth hormones. About 25 million pounds of antibiotics are added to animal feed annually in the United States, mostly in CAFOs. This is eight times as much as is used in human therapy. A big part of the rapid rise in antibiotic resistant pathogens is due to this massive and constant dosing of domestic livestock with drugs.

## Seafood is another important protein source

We currently harvest about 95 million metric tons of wild fish and seafood every year, but we directly eat only about two-thirds of that amount. The rest is used as feed for captive-raised aquatic species, which now make up about half of the seafood we eat. Seafood is the main animal protein source for about 1.5 billion people in developing countries and is an important nutritional source for billions of others. Unfortunately, overharvesting and habitat destruction threaten most of the world's wild fisheries. Annual catches of ocean fish rose by about 4 percent annually between 1950 and 1988. Since 1989, however, 13 of 17 major marine fisheries have declined dramatically or become commercially unsustainable (see Data Analysis p. 205). An international team of marine biologists warn that if current trends continue, all the world's major fisheries will be exhausted by 2048.

The problem is too many boats using efficient but destructive technology to exploit a dwindling resource base. Boats as big as ocean liners travel thousands of kilometers and drag nets large enough to scoop up a dozen jumbo jets, sweeping a large patch of ocean clean of fish in a few hours. Long-line fishing boats set cables up to 10 km long with hooks every 2 meters that catch birds, turtles, and other unwanted "by-catch" along with targeted species. Trawlers drag heavy nets across the bottom, scooping up everything indiscriminately and reducing broad swaths of habitat to rubble. In some operations, up to 15 kg of dead and dying by-catch is dumped back into the ocean for every kilogram of marketable food. The FAO estimates that operating costs for the 4 million boats now harvesting wild fish exceed sales by (U.S.) \$50 billion per year. Countries subsidize fishing fleets to preserve jobs and to ensure access to this valuable resource.

### *Think About It*

Americans spend a smaller proportion of their income on food than any other industrial nation. A major reason for this is that we've industrialized our food supply in the same way that we produce cars, shoes, and television sets. Are there reasons that we should treat food differently from other commodities? What would be your ideal food production system? What could you do to attain that ideal?

**Aquaculture** (growing aquatic species in net pens or tanks) is of increasing importance in our diets. Fish can be grown in farm ponds that take relatively little space but are highly produc-

**FIGURE 9.13** Pens for fish-rearing in Thailand.

tive. Cultivation of high-value carnivorous species, however, such as salmon, sea bass, and tuna threaten wild stocks exploited to stock captive operations or to provide fish food. Building coastal fish-rearing ponds causes destruction of hundreds of thousands of hectares of mangrove forests and wetlands that serve as irreplaceable nurseries for marine species. Net pens anchored in nearshore areas allow spread of diseases, escape of exotic species, and release of feces, uneaten food, antibiotics, and other pollutants into surrounding ecosystems (fig. 9.13).

Polyculture systems of mixed species of herbivores or filter feeders can alleviate many aquaculture problems. Raising species in enclosed, land-based ponds or warehouses can eliminate the pollution problems associated with net pens in lakes or oceans. In China, for example, most fish are raised in ponds or rice paddies. One ecologically balanced system uses four carp species that feed at different levels of the food chain. The grass carp, as its name implies, feeds largely on vegetation, while the common carp is a bottom feeder, living on detritus that settles to the bottom. Silver carp and bighead carp are filter feeders that feed on phytoplankton and zooplankton, respectively. Agricultural wastes such as manure, dead silkworms, and rice straw fertilize ponds and encourage phytoplankton growth. All these carp species are considered dangerous invasive species in North America. Still, these integrated polyculture systems typically boost fish yields per hectare by 50 percent or more compared with monoculture farming. Of the top ten seafood species recommended by the Monterey Aquarium as ecologically acceptable, six are farmed.

## 9.3 Farm Policy

Much of the increase in food production over the past 50 years has been fueled by public support for agricultural education, research, and development projects such as irrigation systems, transportation networks, and crop insurance. While helping local farmers, agricultural subsidies also can distort markets and cripple production in developing countries. The World Bank estimates that rich countries pay their own farmers $350 billion per year, or nearly six times as much as all developmental aid to poor countries. A typical cow in Europe enjoys annual subsidies three times the average yearly income for most African farmers. In the past, European countries have had the highest aid per farm, but recent policy reforms have promised to decouple farm support from crop production.

Powerful political and economic interests protect agricultural assistance in many countries. Over the past decade, the United States, for example, has spent $143 billion in farm support. This aid is distributed very inequitably. According to the Environmental Working Group, 72 percent of all aid goes to the top 10 percent of recipients. One giant operation in Arkansas, for example, received $38 million over a five-year period. Aid also is concentrated geographically. Just 22 of the nation's 435 congressional districts (5 percent) collect more than half of all agricultural payments. Most of this aid goes to a few selected crops, such as corn, wheat, soybeans, rice, and cotton along with certain specially protected commodities including milk, sugar, and peanuts. Legislators claim that crop supports are intended to preserve "family farms," but critics claim this program really amounts to political payoff and corporate welfare. There have been repeated efforts to roll back agricultural payments, but Congress has been more inclined to slash conservation funds and food assistance for the poor rather than reduce rewards for their friends.

One unfortunate effect of these market interventions is that they tend to encourage unhealthy diets. High-calorie, fat- and sugar-rich, processed food is often cheaper and more readily available than fresh fruits and vegetables that would be better for us. Subsidies also allow American farmers to sell their products overseas at as much as 20 percent below the actual cost of production. When cheap commodities flood markets in developing countries, it drives local farmers out of business and destabilizes indigenous food production. The FAO argues that ending distorting financial support in the richer countries would have far more positive impact on local food supplies and livelihoods in the developing world than any aid program.

Every year millions of tons of topsoil and agricultural chemicals wash from U.S. farm fields into rivers, lakes, and, eventually, the ocean. Farmers know that erosion both impoverishes their land and pollutes water, but they're caught in a bind. For every $1 the U.S. government pays farmers to conserve soil and manage nutrients, it pays $7 to support commodities that promote soil loss and chemical runoff. The USDA estimates that farmers would shift 2.5 million ha (6 million acres) of row crops, which require intensive cultivation, into pasture grass and other perennial crops that minimize erosion if federal subsidies didn't favor certain row crops.

The United States already has an effective Conservation Reserve Program (CRP) that pays farmers to take highly erodible land out of production. Contracts expired in 2007 for more than 11 million ha (28 million acres) of this land. The USDA reports that CRP lands prevent the annual loss of 450 million tons of soil every year, protect 270,000 km (170,000 miles) of streams, and store 48 million tons of carbon per year. Just in the upper Midwest, CRP contracts will end on 6.5 million ha of critical wildlife habitat.

Many agronomists suggest that we should have more, not less CRP land. The United States could gradually shift payments from production subsidies to conservation programs that would truly support family farms while also protecting the environment. This could become a "whole farm" safety net that allows farmers to plant whatever crops are appropriate to market and environmental conditions rather than a few specific commodities. Farmers might find it more profitable to grow fruits and vegetables for local consumption rather than low-value grains. Nutrient management could be required as a condition of receiving crop aid.

We'll discuss soil conservation and low-input agriculture further in the next section of this chapter. In the meantime, please keep this question in mind: How can we support family farms and rural economies while also protecting our environment and helping our global neighbors feed themselves?

## 9.4 Soil: A Renewable Resource

Growing the food and fiber needed to support human life is a complex enterprise that requires knowledge from many different fields and cooperation from many different groups of people. In this section, we'll survey some of the principles of soil science and look at some of the inputs necessary for continued agricultural production.

Of all the earth's crustal resources, the one we take most for granted is soil. We are terrestrial animals and depend on soil for life, yet most of us think of it only in negative terms. English is unique in using "soil" as an interchangeable word for earth and excrement. "Dirty" has a moral connotation of corruption and impurity. Perhaps these uses of the word enhance our tendency to abuse soil without scruples; after all, it's only dirt.

The truth is that **soil** is a marvelous substance, a living resource of astonishing beauty, complexity, and frailty. It is a complex mixture of weathered mineral materials from rocks, partially decomposed organic molecules, and a host of living organisms. It can be considered an ecosystem by itself. Soil is an essential component of the biosphere, and it can be used sustainably, or even enhanced, under careful management.

There are at least 15,000 different soil series or types in the United States and many thousands more worldwide. They vary because of the influences of parent material, time, topography, climate, and organisms on soil formation. There are young soils that, because they have not weathered much, are rich in soluble nutrients. There are old soils, like the red soils of the tropics, from which rainwater has washed away most of the soluble minerals and organic matter, leaving behind clay and rust-colored oxides.

To understand the potential for feeding the world on a sustainable basis we need to know how soil is formed, how it is being lost, and what can be done to protect and rebuild good agricultural soil. With careful husbandry, soil can be replenished and renewed indefinitely. Many farming techniques deplete soil nutrients, however, and expose the soil to the erosive forces of wind and moving water. As a result, in many places we are essentially mining this resource and using it much faster than it is being replaced.

Building good soil is a slow process. Under the best circumstances, good topsoil accumulates at a rate of about 10 tons per hectare (2.5 acres) per year—enough soil to make a layer about 1 mm deep when spread over a hectare. Under poor conditions, it can take thousands of years to build that much soil. Perhaps one-third to one-half of the world's current croplands are losing topsoil faster than it is being replaced. In some of the worst spots, erosion carries away about 2.5 cm (1 in.) of topsoil per year. With losses like that, agricultural production has already begun to fall in many areas.

### Soil is a complex mixture

Most soil is about half mineral. The rest is plant and animal residue, air, water, and living organisms. The mineral particles are derived either from the underlying bedrock or from materials transported and deposited by glaciers, rivers, ocean currents, windstorms, or landslides. The weathering processes that break rocks down into soil particles are described in chapter 14.

Particle sizes affect the characteristics of the soil (fig. 9.14). The spaces between sand particles give sandy soil good drainage and usually allow it to be well aerated, but also cause it to dry out quickly when rains are infrequent. Tight packing of small particles in silty or clay soils makes them less permeable to air and water than sandy soils. Tiny capillary spaces between the particles, on the other hand, store water and mineral ions better than more porous soils. Because clay particles have a proportionately large surface area and a high ionic charge, they stick together tenaciously, giving clay its slippery plasticity, cohesiveness, and impermeability. Soils

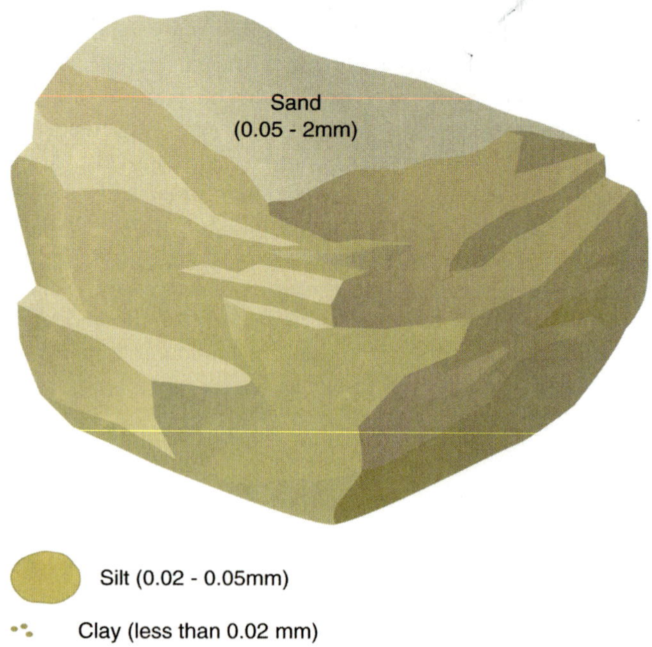


**FIGURE 9.14** Relative sizes of soil particles magnified about 100-fold.

with a high clay content are called "heavy soils," in contrast to easily worked "light soils" that are composed mostly of sand or silt. Varying proportions of these mineral particles occur in each soil type (fig. 9.15). Farmers usually consider sandy loam the best soil type for cultivating crops.

The organic content of soil can range from nearly zero for pure sand, silt, or clay, to nearly 100 percent for peat or muck, which is composed mainly of partly decomposed plant material. Much of the organic material in soil is **humus,** a sticky, brown, insoluble residue from the partially decomposed bodies of dead plants and animals. Humus is the most significant factor in the development of "structure," a description of how the soil particles clump together. Humus coats mineral particles and holds them together in loose crumbs, giving the soil a spongy texture that holds water and nutrients needed by plant roots, and maintains the spaces through which delicate root hairs grow.

## Living organisms create unique properties of soil

Without soil organisms, the earth would be covered with sterile mineral particles far different from the rich, living soil ecosystems on which we depend for most of our food. The activity of the myriad organisms living in the soil help create structure, fertility, and tilth (condition suitable for tilling or cultivation) (fig. 9.16).

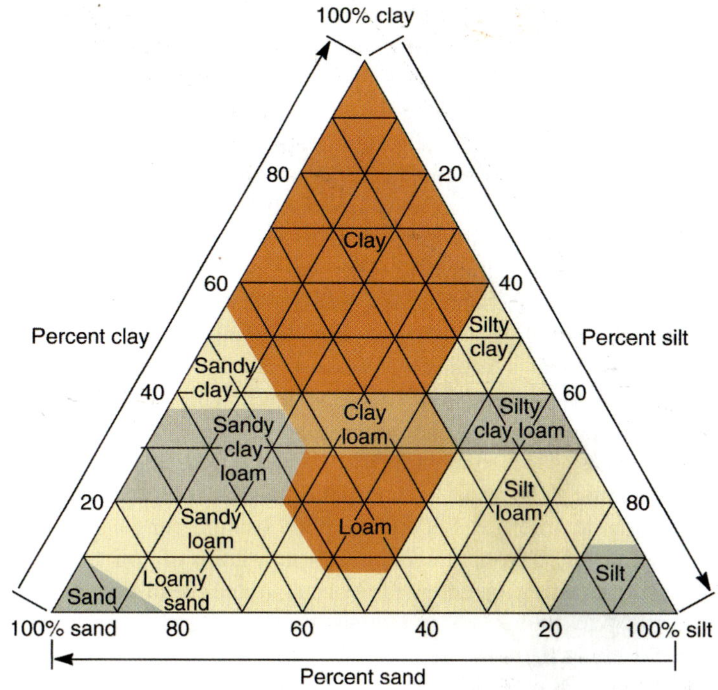


**FIGURE 9.15** Soil texture is determined by the percentages of clay, silt, and sand particles in the soil. Soils with the best texture for most crops are loams, which have enough larger particles (sand) to be loose, yet enough smaller particles (silt and clay) to retain water and dissolved mineral nutrients.

**FIGURE 9.16** Soil ecosystems include numerous consumer organisms, as depicted here: (1) snail, (2) termite, (3) nematodes and nematode-killing constricting fungus, (4) earthworm, (5) wood roach, (6) centipede, (7) carabid (ground) beetle, (8) slug, (9) soil fungus, (10) wireworm (click beetle larva), (11) soil protozoan, (12) sow bug, (13) ant, (14) mite, (15) springtail, (16) pseudoscorpion, and (17) cicada nymph.

Soil organisms usually stay close to the surface, but that thin living layer can contain thousands of species and billions of individual organisms per hectare. Algae live on the surface, while bacteria and fungi flourish in the top few centimeters of soil. A single gram of soil (about one-half teaspoon) can contain hundreds of millions of these microscopic cells. Algae and blue-green bacteria capture sunlight and make new organic compounds. Bacteria and fungi decompose organic detritus and recycle nutrients that plants can use for additional growth. The sweet aroma of freshly turned soil is caused by actinomycetes, bacteria that grow in funguslike strands and give us the antibiotics streptomycin and tetracyclines.

**Micorrhizal symbiosis** is an association between the roots of most plant species and certain fungi. The plant provides organic compounds to the fungus, while water and inorganic nutrients are absorbed from the soil by the fungus and transferred to the plant. As a result, micorrhizal plants often grow better—especially in infertile soil—than those lacking fungal partners. There can be 20 m of fungal strands in a gram of soil.

Roundworms, segmented worms, mites, and tiny insects swarm by the thousands in that same gram of soil from the surface. Some of them are herbivorous, but many of them prey upon one another. Soil roundworms (nematodes) attack plant rootlets and can cause serious crop damage. A carnivorous fungus snares nematodes with tiny loops of living cells that constrict like a noose when a worm blunders into it. Burrowing animals, such as gophers, moles, insect larvae, and worms, tunnel deeper in the soil, mixing and aerating it. Plant roots also penetrate lower soil levels, drawing up soluble minerals and secreting acids that decompose mineral particles. Fallen plant litter adds new organic material to the soil, returning nutrients to be recycled.

## Soils are layered

Most soils are stratified into horizontal layers called **soil horizons** that reveal much about the history and usefulness of the soil. The thickness, color, texture, and composition of each horizon are used to classify the soil. Together these horizons make up a **soil profile** (fig. 9.17).

The soil surface is often covered with a layer of leaf litter, crop residues, or other fresh or partially decomposed organic material. This organic layer is known as the O horizon. Below this layer is the A horizon or **topsoil,** composed of mineral particles mixed with organic material. The A horizon can range from several meters thick under virgin prairie to almost nothing in dry deserts. The O and A horizons contain most of the living organisms and organic material in the soil, and it is in these layers that most plants spread their roots to absorb water and nutrients. An eluviated (leached) E horizon often lies below the A horizon. This layer is depleted of clays and soluble nutrients, which are removed by rainwater seeping down through the soil. These clays and nutrients generally accumulate in the B horizon, or **subsoil.** The B horizon may have a dense or clayey texture because of the accumulated clays. In desert regions, a dense, impermeable "hardpan" layer of accumulated minerals or salts may develop on the B horizon. This hardpan blocks plant root growth and prevents water from draining properly.

**FIGURE 9.17** Soil profile showing possible soil horizons. The actual number, composition, and thickness of these layers varies in different soil types.

Beneath the subsoil is parent material, called the C horizon or **regolith,** made of weathered rock fragments with very little organic material. Weathering in the C horizon, largely accomplished by rain water mixed with organic compounds from plants, produces new soil particles and allows downward expansion of the soil profile. Parent material can be sand, solid rock, or other material. About 70 percent of the parent material in the United States was transported to its present site by glaciers, wind, or water and is not directly related to the underlying bedrock.

## Soils are classified according to their structure and composition

In the United States, soils are classified into 12 *soil orders.* The richest farming soils, with thick, organic-rich A horizons are *mollisols* (formed under grasslands) and *alfisols* (formed under deciduous forests). Both of these develop where rainfall and temperatures are moderate. *Spodosols* also develop in temperate climates, but usually under pine forests, where acidic needle litter causes a characteristic whitish, ashy-looking E horizon. In hot and rainy environments, *oxisols* and *ultisols* develop. These soils are severely depleted of nutrients and tend to be reddish because iron-rich minerals make up much of the remaining soil material. *Aridosols* form in arid environments, have little organic material, and often contain accumulated salt or hardpan layers. Some soil types are defined mainly by parent material: *andisols* develop from volcanic material, and *vertisols* develop from clay-rich material such as lake beds or shale bedrock. *Histosols* are formed of waterlogged, incompletely decayed plant material. Other soils are largely defined by degree of development; *entisols* and *inceptisols* have little or no horizon development. This may be because these soils are on recently exposed parent material (such as glacial debris) or because they are in an environment where soil development is extremely slow because of slight rainfall or organic activity. A relatively new soil order is *gelisols,* soils in permafrost areas. These soils occur only at high latitudes or high altitudes.

**FIGURE 9.18** In many areas, soil or climate constraints limit agricultural production. These hungry goats in Sudan feed on a solitary *Acacia* shrub.

## 9.5 Ways We Use and Abuse Soil

Only about 12.5 percent of the earth's land area (16.6 million $km^2$ out of a total of 132.4 million $km^2$) is currently in agricultural production. Perhaps four times as much land could potentially be converted to cropland, but much of this land serves as a refuge for cultural or biological diversity or suffers from constraints, such as steep slopes, shallow soils, poor drainage, tillage problems, low nutrient levels, metal toxicity, or excess soluble salts or acidity, that limit the types of crops that can be grown there (fig. 9.18). Agronomists worry that global climate change (chapter 15) could alter weather patterns and flood low-lying coastal areas so that world food production could be seriously reduced.

### Arable land is unevenly distributed

In parts of Canada, the United States, and Europe, temperate climates, abundant water, and high soil fertility produce high crop yields that contribute to high standards of living. Other parts of the world, although rich in land area, lack suitable soil, topography, water, or climate to sustain our levels of productivity.

Worldwide, the cropland available for agriculture is shrinking. In 1970, there was a global average of 0.38 ha per person. By 2002, this had declined to 0.21 ha per person. The FAO forecasts that by 2030 the land supply will shrink to 0.16 ha per person. In Asia, cropland will be even more scarce—0.09 ha (0.22 acre) per person—30 years from now. If you live on a typical quarter-acre suburban lot, look at your yard and imagine feeding yourself for a year on what you could produce there.

In the developed countries, 95 percent of recent agricultural growth in the twentieth century has come from improved crop varieties or increased fertilization, irrigation, and pesticide use, rather than from bringing new land into production. In fact, less land is being cultivated now than 100 years ago in North America, or 600 years ago in Europe. As more effective use of labor, fertilizer, and water and improved seed varieties have increased in the more developed countries, productivity per unit of land has increased, and marginal land has been retired, mostly to forests and grazing lands. In many developing countries, land continues to be cheaper than other resources, and new land is still being brought under cultivation, mostly at the expense of forests and grazing lands. Still, at least two-thirds of recent production gains have come from new crop varieties and more intense cropping rather than expansion into new lands.

The largest increases in cropland over the last 30 years occurred in South America and Oceania where forests and grazing lands—such as Brazil's Cerrado, described in this chapter's opening case study—are rapidly being converted to farms. Many developing countries are reaching the limit of lands that can be exploited for agriculture without unacceptable social and environmental costs, but others still have considerable potential for opening new agricultural lands. East Asia, for instance, already uses about three-quarters of its potentially arable land. Most of its remaining land has severe restrictions for agricultural use. Further increases in crop production will probably have to come from higher yields per hectare. Latin America, by contrast, uses only about one-fifth of its potential land, and Africa uses only about one-fourth of the land that theoretically could grow crops. However, there would be serious ecological trade-offs in putting much of this land into agricultural production.

### Land degradation reduces agricultural potential

Agriculture both causes and suffers from environmental degradation. The International Soil Reference and Information Centre in the Netherlands estimates that every year 3 million ha (7.4 million acres) of cropland are ruined by erosion, 4 million ha are turned into deserts, and 8 million ha are converted to nonagricultural uses such as homes, highways, shopping centers, factories, reservoirs, etc.

Over the past 50 years, some 1.9 *billion* ha of agricultural land (an area greater than that now in production) have been degraded to some extent. About 300 million ha of this land are strongly degraded (deep gullies, severe nutrient depletion, crops grow poorly, restoration is difficult and expensive), while 910 million ha—about the size of China—are moderately degraded. Nearly 9 million ha of former croplands are so degraded that they no longer support any crop growth at all. The causes of this extreme degradation vary: In Ethiopia it is water erosion, in Somalia it is wind, and in Uzbekistan salt and toxic chemicals are responsible.

Definitions of degradation are based on both biological productivity and our expectations about what the land should be like. Often this is a subjective judgment and it is difficult to distinguish between human-caused deterioration and natural processes like drought. We generally consider the land degraded when the soil is impoverished or eroded, water runs off or is contaminated more than is normal, vegetation is diminished, biomass production is decreased, or wildlife diversity diminishes. On farmlands this results in lower crop yields. On ranchlands it means fewer livestock can be supported per unit area. On nature reserves it means fewer species.

The amount and degree of land degradation varies by region and country. About 20 percent of land in Africa and Asia is degraded, but most is in either the light or moderate category. In Central America and Mexico, by contrast, 25 percent of all vegetated land suffers moderate to extreme degradation. Figure 9.19 shows some areas of greatest concern for soil degradation in the United States, as well as the prime mechanisms for this problem. Water and wind erosion provide the motive force for the vast majority of all soil degradation, worldwide. Chemical deterioration includes nutrient depletion, salinization (salt accumulation), acidification, and pollution. Physical deterioration includes compaction by heavy machinery or trampling by cattle, waterlogging—water accumulation—from excess irrigation and poor drainage, and laterization—solidification of iron and aluminum-rich tropical soil when exposed to sun and rain.

## Soil erosion is widespread

Erosion is an important natural process, resulting in the redistribution of the products of geologic weathering, and is part of both soil formation and soil loss. The world's landscapes have been sculpted by erosion. When the results are spectacular enough, we enshrine them in national parks as we did with the Grand Canyon. Where erosion has worn down mountains and spread soil over the plains, or deposited rich alluvial silt in river bottoms, we gladly farm it. Erosion is a disaster only when it occurs in the wrong place at the wrong time.

In some places, erosion occurs so rapidly that anyone can see it happen. Deep gullies are created where water scours away the soil, leaving fenceposts and trees sitting on tall pedestals as the land erodes away around them. In most places, however, erosion is more subtle. It is a creeping disaster that occurs in small increments. A thin layer of topsoil is washed off fields

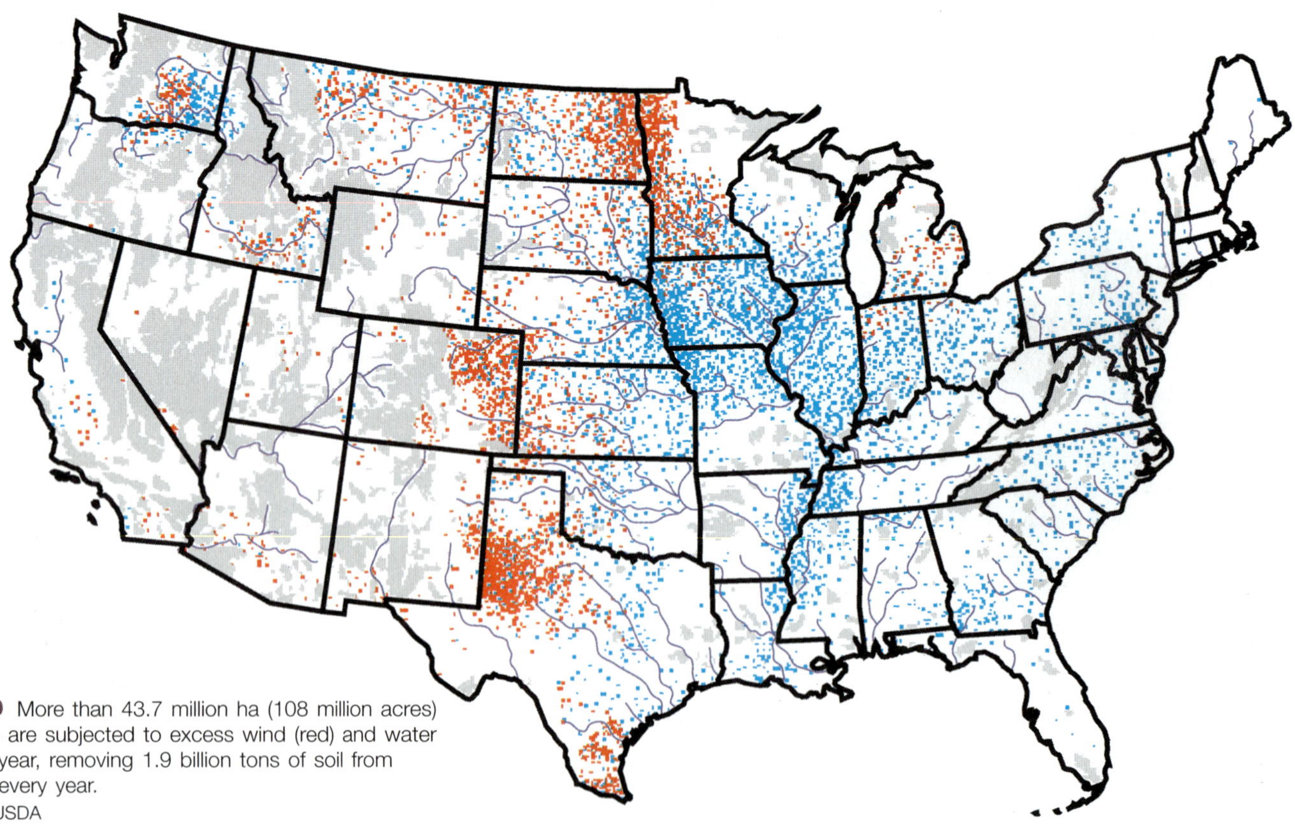

**FIGURE 9.19** More than 43.7 million ha (108 million acres) in the United States are subjected to excess wind (red) and water (blue) erosion each year, removing 1.9 billion tons of soil from fields and pastures every year.
**Source:** Data from USDA

year after year until eventually nothing is left but poor-quality subsoil that requires more and more fertilizer and water to produce any crop at all.

The total annual soil loss from croplands is thought to be 25 billion metric tons. About twice that much soil is lost from rangelands, forests, and urban construction sites each year. The net effect, worldwide, of this general, widespread topsoil erosion is a reduction in crop production equivalent to removing about 1 percent of world cropland each year. Many farmers are able to compensate for this loss by applying more fertilizer and by bringing new land into cultivation.

In addition to reduced land fertility, this erosion results in sediment-loading of rivers and lakes, siltation of reservoirs, smothering of wetlands and coral reefs, and clogging of water intakes and waterpower turbines.

## Wind and water are the main agents that move soil

A thin layer taken off the land surface is called **sheet erosion.** When little rivulets of running water gather together and cut small channels in the soil, the process is called **rill erosion** (fig. 9.20*a*). When rills enlarge to form bigger channels or ravines that are too large to be removed by normal tillage operations, we call the process **gully erosion** (fig. 9.20*b*). Streambank erosion refers to the washing away of soil from the banks of established streams, creeks, or rivers, often as a result of removing trees and brush along streambanks and by cattle damage to the banks.

Most soil loss on agricultural land is sheet or rill erosion. Large amounts of soil can be transported a little bit at a time without being very noticeable. A farm field can lose 20 metric tons of soil per hectare during winter and spring runoff in rills so small that they are erased by the first spring cultivation. That represents a loss of only a few millimeters of soil over the whole surface of the field, hardly apparent to any but the most discerning eye. But it doesn't take much mathematical skill to see that if you lose soil twice as fast as it is being replaced, eventually it will run out.

Wind can equal or exceed water in erosive force, especially in a dry climate and on relatively flat land. When plant cover and surface litter are removed from the land by agriculture or grazing, wind lifts loose soil particles and sweeps them away. In extreme conditions, windblown dunes encroach on useful land and cover roads and buildings (fig. 9.20*c*). Over the past 30 years, China has lost 93,000 $km^2$ (about the size of Indiana) to **desertification,** or conversion of productive land to desert. Advancing dunes from the Gobi desert are now only 160 km (100 mi) from Beijing. Every year more than 1 million tons of sand and dust blow from Chinese drylands, often traveling across the Pacific Ocean to the West Coast of North America.

Some of the highest erosion rates in the world occur in the United States and Canada. The U.S. Department of Agriculture reports that 69 million hectares (170 million acres) of U.S. farmland and range are eroding at rates that reduce long-term productivity. Five tons per acre (11 metric tons per hectare) is generally considered the maximum tolerable rate of soil loss because that is generally the highest rate at which soil forms under optimum conditions. Some farms lose soil at twice that rate or more.

Intensive farming practices are largely responsible for this situation. Row crops, such as corn and soybeans, leave soil exposed for much of the growing season (fig. 9.21). Deep plowing and heavy herbicide applications create weed-free fields that look neat but are subject to erosion. Because big machines cannot easily follow contours, they often go straight up and down the hills, creating ready-made gullies for water to follow. Farmers sometimes plow through grass-lined watercourses and have pulled out windbreaks and fencerows to accommodate the large machines and to get every last square meter into production. Consequently, wind and water carry away the topsoil.

(a) Sheet and rill erosion (b) Gullying (c) Wind erosion and desertification

**FIGURE 9.20** Land degradation affects more than 1 billion ha yearly, or about two-thirds of all global cropland. Water erosion (a) and (b) account for about half that total. Wind erosion affects a nearly equal area (c).

**FIGURE 9.21** Annual row crops leave soil bare and exposed to erosion for most of the year, especially when fields are plowed immediately after harvest, as this one always is.

Pressed by economic conditions, many farmers have abandoned traditional crop rotation patterns and the custom of resting land as pasture or fallow every few years. Continuous monoculture cropping can increase soil loss tenfold over other farming patterns. A soil study in Iowa showed that a three-year rotation of corn, wheat, and clover lost an average of about 6 metric tons per hectare. By comparison, continuous wheat production on the same land caused nearly four times as much erosion, and continuous corn cropping resulted in seven times as much soil loss as the rotation with wheat and clover. The Mississippi River carries enough topsoil and fertilizer every year to create a "dead zone" in the Gulf of Mexico that can be as large as 57,000 $km^2$. Algal growth stimulated by high nitrogen in runoff from farms and cities depletes oxygen within this zone to levels that are lethal for most marine life. A task force recommended a 20 to 30 percent decrease in nitrogen loading to reduce the size and effects of this zone. Similar hypoxic zones occur near mouths of many other rivers that drain agricultural areas (chapter 18).

## Deserts are spreading around the world

According to the United Nations, about one-third of the earth's surface and the livelihoods of at least one billion people are threatened by **desertification** (conversion of productive lands to desert), which contributes to food insecurity, famine, and poverty. UN Secretary General Kofi Annan calls this a "creeping catastrophe" that creates millions of environmental refugees every year. Forced by economic circumstances to overcultivate and overgraze their land, poor people often are both the agents and victims of desertification.

Rangelands and pastures, which generally are too dry for cultivation, are highly susceptible to desertification. According to the UN, 80 percent of the world's grasslands are suffering from overgrazing and soil degradation, and three-quarters of that area has undergone some degree of desertification. The world's 3 billion domestic grazing animals provide livelihood and food for many people, but can have severe environmental effects.

Two areas of particular concern are Africa and China. Arid lands, where rains are sporadic and infrequent and the economy is based mainly on crop and livestock raising, make up about two-thirds of Africa. Nearly 400 million people live around the edges of these deserts. Rapid population growth and poverty create unsustainable pressures on the fragile soils of these areas. Stripping trees and land cover for fodder and firewood exposes the soil to erosion and triggers climate changes that spread desertification, which now affects nearly three-quarters of the arable land in Africa to some extent. The fringes of the two great African deserts, the Sahara and the Kalahari, are particularly vulnerable. Nearly every year there is drought somewhere in Africa. About one-third of the 60 million people who required food aid in 2005 were victims of drought and desertification.

In China, large areas of natural woodland were cleared in the twentieth century to create farmland and provide timber for industry. Especially in the 1950s, trees were cut to fuel backyard iron smelters. This rampant deforestation has been blamed for disastrous summer floods on the Yangtze and Yellow Rivers and severe spring sandstorms that threaten the health of millions of people. Overgrazing and cultivation of fragile lands also contribute to desertification. The advancing Gobi Desert is now within 150 km (100 mi) from Beijing; and the Taklimakan, Asia's largest desert, is spreading eastward to unite with the Gobi. Sandstorms from these deserts have grown from about 5 per year in the 1960s to 24 per year in the 1990s.

China is trying to fight the spread of deserts with an ambitious ecological restoration program. Since 1985, more than 50 billion trees have been planted over an area the size of Germany. Every Chinese citizen between the ages of 11 and 60 is required to plant 3 to 5 trees per year. Many saplings aren't watered, however, and survival rates are reported to be low. Eventually this gargantuan program plans to create a "Green Wall" thousands of kilometers long that will stabilize soils, clear the air, and stop the advancing deserts. China is especially anxious to improve air quality in Beijing before the 2008 Olympics, as we discussed in chapter 1. We'll look at other methods to control soil erosion in our discussion of sustainable agriculture.

# 9.6 Other Agricultural Resources

Soil is only part of the agricultural resource picture. Agriculture is also dependent upon water, nutrients, favorable climates to grow crops, productive crop varieties, and upon the mechanical energy to tend and harvest them.

## All plants need water to grow

Agriculture accounts for the largest single share of global water use. About two-thirds of all fresh water withdrawn from rivers, lakes, and groundwater supplies is used for irrigation (chapter 17). Although estimates vary widely (as do definitions of irrigated land), about 15 percent of all cropland, worldwide, is irrigated.

Some countries are water rich and can readily afford to irrigate farmland, while other countries are water poor and must use water very carefully. The efficiency of irrigation water use is rather low in most countries. High evaporative and seepage losses from unlined and uncovered canals often mean that in some

**FIGURE 9.22** Downward-facing sprinklers on this center-pivot irrigation system deliver water more efficiently than upward-facing sprinklers.

places, up to 80 percent of water withdrawn for irrigation never reaches its intended destination (chapter 17). Farmers often tend to over-irrigate because water prices are relatively low and because they lack the technology to meter water and distribute just the amount needed. In the United States and Canada, many farmers are adopting water-saving technologies such as drip irrigation or downward-facing sprinklers (fig. 9.22).

Excessive use not only wastes water; it often results in **waterlogging.** Waterlogged soil is saturated with water, and plant roots die from lack of oxygen. **Salinization,** in which mineral salts accumulate in the soil, occurs particularly when soils in dry climates are irrigated with saline water. As the water evaporates, it leaves behind a salty crust on the soil surface that is lethal to most plants. Flushing with excess water can wash away this salt accumulation but the result is even more saline water for downstream users.

Worldwide, irrigation problems are a major source of land degradation and crop losses. The Worldwatch Institute reports that 60 million ha (150 million acres) of cropland have been damaged by salinization and waterlogging. Water conservation techniques can greatly reduce problems arising from excess water use. Conservation also makes more water available for other uses or for expanded crop production where water is in short supply (chapter 17).

Many agronomists worry that global climate change (chapter 15) could reduce food production in many countries. It's expected that global warming will bring too little rain to some areas and too much to others. Warmer temperatures could extend growing seasons at high latitudes, but dry soils and stress crops elsewhere.

## Plants need fertilizer

In addition to water, sunshine, and carbon dioxide, plants need small amounts of inorganic nutrients for growth. The major elements required by most plants are nitrogen, potassium, phosphorus, calcium, magnesium, and sulfur. Calcium and magnesium often are limited in areas of high rainfall and must be supplied in the form of lime. Lack of nitrogen, potassium, and phosphorus even more often limits plant growth. Adding these elements in fertilizer usually stimulates growth and greatly increases crop yields. A good deal of the doubling in worldwide crop production since 1950 has come from increased inorganic fertilizer use. In 1950, the average amount of fertilizer used was 20 kg per hectare. In 1990, this had increased to an average of 91 kg per hectare worldwide.

Farmers may overfertilize because they are unaware of the specific nutrient content of their soils or the needs of their crops. While European farmers use more than twice as much fertilizer per hectare as do North American farmers, their yields are not proportionally higher. Phosphates and nitrates from farm fields and cattle feedlots are a major cause of aquatic ecosystem pollution. Nitrate levels in groundwater have risen to dangerous levels in many areas where intensive farming is practiced. Young children are especially sensitive to the presence of nitrates. Using nitrate-contaminated water to mix infant formula can be fatal for newborns.

What are some alternative ways to fertilize crops? Manure and green manure (crops grown specifically to add nutrients to the soil) are important natural sources of soil nutrients. Nitrogen-fixing bacteria living symbiotically in root nodules of legumes are valuable for making nitrogen available as a plant nutrient (chapter 3). Interplanting or rotating beans or some other leguminous crop with such crops as corn and wheat are traditional ways of increasing nitrogen availability.

There is considerable potential for increasing world food supply by increasing fertilizer use in low-production countries if ways can be found to apply fertilizer more effectively and reduce pollution. Africa, for instance, uses an average of only 19 kg of fertilizer per hectare, or about one-fourth of the world average. It has been estimated that the developing world could at least triple its crop production by raising fertilizer use to the world average.

## Farming consumes energy

Farming as it is generally practiced in the industrialized countries is highly energy-intensive. Fossil fuels supply almost all of this energy. Between 1920 and 1980, direct energy use on farms rose as gasoline and diesel fuels were consumed by increasing mechanization of agriculture. An even greater increase in indirect energy use, in the form of synthetic fertilizers, pesticides, and other agricultural chemicals, also occurred, especially after World War II.

After crops leave the farm, additional energy is used in food processing, distribution, storage, and cooking. It has been estimated that the average food item in the American diet travels 2,400 km between the farm that grew it and the person who consumes it. The energy required for this complex processing and distribution system may be five times as much as is used directly in farming. Altogether the food system in the United States consumes about 16 percent of the total energy we use. Most of our foods require more energy to produce, process, and get to market than they yield when we eat them. A British study concluded that eating locally grown food has less environmental impact—even if produced with conventional farming—than organic food from far away.

**FIGURE 9.23** Biofuel crops, such as these sunflowers, can provide a good cash crop for farmers and contribute to national energy supplies.

**FIGURE 9.24** Winged beans bear fruit year-round in tropical climates and are resistant to many diseases that prohibit growing other bean species. Whole pods can be eaten when they are green, or dried beans can be stored for later use. It is a good protein source in a vegetarian diet.

Farmers could assist in moving to a renewable energy future by growing energy crops that can be converted into biofuels. Many Midwestern states have encouraged construction of corn- or soy-based ethanol factories in recent years. Mixing ethanol with gasoline helps reduce air pollution, and having a new market helps struggling farmers. There has been much debate about whether corn-based ethanol produces more energy than is required to grow and process the crop. State-of-the-art facilities do, in fact, appear to have a 150 percent net energy yield. Plant oils from sunflowers, soybeans, corn, and other oil seed crops can be burned directly in most diesel engines, and may be closer than ethanol to a net energy gain (fig. 9.23). Europe plans to produce 5.7 percent of its transportation energy from biofuels by 2010. Fast-growing trees, kenaf, switch grass, cattails, and other energy crops may be grown on marginal land and burned directly in power plants or used to create ethanol. Many American farmers are finding that their most lucrative crop may be the wind (chapter 20).

## 9.7 New Crops and Genetic Engineering

Although at least 3,000 species of plants have been used for food at one time or another, most of the world's food now comes from only 16 species. Many new or unconventional varieties might be valuable human food supplies, however, especially in areas where conventional crops are limited by climate, soil, pests, or other problems. The FAO predicts that 70 percent of future world production growth will come from higher yields and new crop varieties rather than expansion of arable lands. Among the plants now being investigated as potential additions to our crop roster is the winged bean (fig. 9.24), a perennial plant that grows well in hot climates where other legumes will not grow. It is totally edible (pods, mature seeds, shoots, flowers, leaves, and tuberous roots), resistant to diseases, and enriches the soil. Another promising crop is tricale, a hybrid between wheat (*Triticum*) and rye (*Secale*) that grows in light, sandy, infertile soil. It is drought resistant, has nutritious seeds, and is being tested for salt tolerance for growth in saline soils or irrigation with seawater. Some traditional crop varieties grown by Native Americans, such as tepary beans, amaranth, and Sonoran panicgrass are being collected by seed conservator Gary Nabhan both as a form of cultural revival for native people and as a possible food crop for harsh environments.

### The "green revolution" produced dramatic increases in crop yields

So far, the major improvements in farm production have come from technological advances and modification of a few well-known species. Yield increases often have been spectacular. A century ago, when all maize (corn) in the United States was open-pollinated, average yields were about 25 bushels per acre. In 1999, average yields from hybrid maize were around 130 bushels per acre, and under optimum conditions, 250 bushels per acre are possible. Most of this gain was accomplished

**FIGURE 9.25** Semi-dwarf wheat (*right*), bred by Norman Borlaug, has shorter, stiffer stems and is less likely to lodge (fall over) when wet than its conventional cousin (*left*). This "miracle" wheat responds better to water and fertilizer, and has played a vital role in feeding a growing human population.

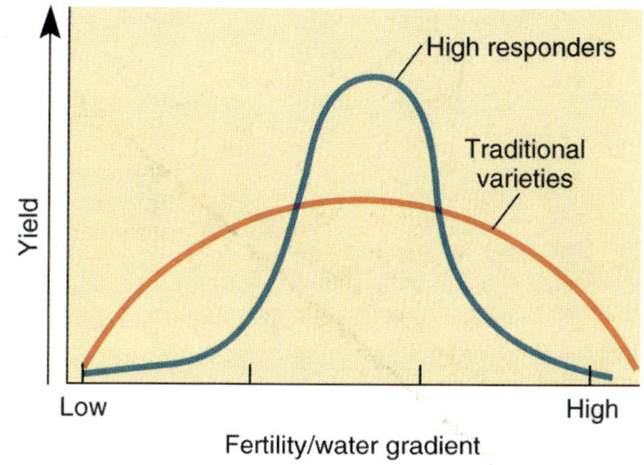


**FIGURE 9.26** Green revolution miracle crops are really high responders, meaning that they have excellent yields under optimum conditions. For poor farmers who can't afford the fertilizer and water needed by high responders, traditional varieties may produce better yields.

by conventional plant breeding: Geneticists laboriously hand-pollinating plants, moving selected genes from one variety to another.

Starting about 50 years ago, agricultural research stations began to breed tropical wheat and rice varieties that would provide food for growing populations in developing countries. The first of the "miracle" varieties was a dwarf, high-yielding wheat developed by Norman Borlaug (who received a Nobel Peace Prize for his work) at a research center in Mexico (fig. 9.25). At about the same time, the International Rice Institute in the Philippines developed dwarf rice strains with three or four times the production of varieties in use at the time. The dramatic increases obtained as these new varieties spread around the world has been called the **green revolution.** It is one of the main reasons that world food supplies have more than kept pace with the growing human population over the past few decades.

Most green revolution breeds really are "high responders," meaning that they yield more than other varieties if given optimum levels of fertilizer, water, and protection from pests and diseases (fig. 9.26). Under suboptimum conditions, on the other hand, high responders may not produce as well as traditional varieties. Poor farmers who can't afford the expensive seed, fertilizer, and water required to become part of this movement, usually are left out of the green revolution. In fact, they may be driven out of farming altogether as rising land values and falling commodity prices squeeze them from both sides.

The first advances under the green revolution were for grains, such as wheat, corn, and rice, grown in temperate climates of wealthy countries. Little research was devoted to tropical crops. Recently, however, the Gates and Rockefeller Foundations announced a new initiative for research on uniquely African crops and agricultural technology. One aim of this research is to create hardy "climate-ready" crops in anticipation of global climate change.

## Genetic engineering uses molecular techniques to produce new crop varieties

**Genetic engineering** involves removing genetic material from one organism and splicing it into the chromosomes of another (fig. 9.27). This new technology has the potential to greatly increase both the quantity and quality of our food supply. It is now possible to build entirely new genes, and even new organisms, taking a bit of DNA from here, a bit from there, even synthesizing artificial DNA sequences to create desired characteristics in **genetically modified organisms (GMOs).**

Proponents predict dramatic benefits from genetic engineering. Research is now underway to improve yields and create crops that resist drought, frost, or diseases. Other strains are being developed to tolerate salty, waterlogged, or low-nutrient soils, allowing degraded or marginal farmland to become productive. All of these could be important for reducing hunger in developing countries. Plants that produce their own pesticides might reduce the need for toxic chemicals; while engineering for improved protein or vitamin content could make our food more nutritious. Attempts to remove specific toxins or allergens from crops also could make our food safer. Crops such as bananas and potatoes have been altered to contain oral vaccines that can be grown in developing countries where refrigeration and sterile needles are unavailable. Plants have been engineered to make industrial oils and plastics. Animals, too, are being genetically modified to grow faster, gain weight on less food, and produce pharmaceuticals such as insulin in their milk. It may soon be possible to create animals with human cell-recognition factors that could serve as organ donors.

Opponents worry that moving genes willy-nilly could create a host of problems, some of which we can't even imagine. GMOs themselves might escape and become pests or they might interbreed with wild relatives. In either case, we may create superweeds or reduce native biodiversity. Constant presence of pesticides in plants could accelerate pesticide resistance in insects or leave toxic residues in soil or our food. Genes for toxicity or

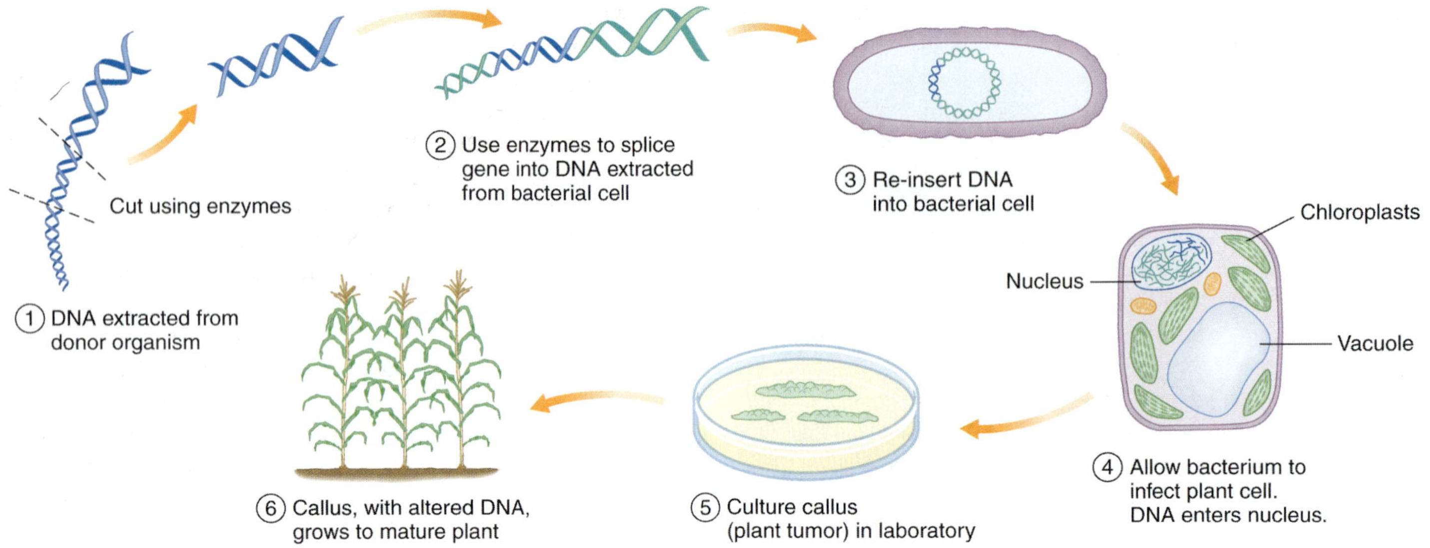


**FIGURE 9.27** One method of gene transfer, using an infectious, tumor-forming bacterium such as agrobacterium. Genes with desired characteristics are cut out of donor DNA and spliced into bacterial DNA using special enzymes. The bacteria then infect plant cells and carry altered DNA into cells' nuclei. The cells multiply, forming a tumor, or callus, which can grow into a mature plant.

allergies could be transferred along with desirable genes, or novel toxins could be created as genes are mixed together. This technology may be available only to the richest countries or the wealthiest corporations, making family farms uncompetitive and driving developing countries even further into poverty.

Both the number of GMO crops and the acreage devoted to growing them is increasing rapidly. Between 1990 and 2006, more than 11,373 field tests for some 750 different crop varieties were carried out in the United States (fig. 9.28). Hawaii is the most popular site for GMO testing, with about 1,500 field releases over the past 15 years. Illinois, Iowa, California, and Puerto Rico each have had more than 1,000 tests during that time. Worldwide, 90 million ha (about 25 percent of all cultivated land) were planted with GMO crops in 2006. The United States accounted for 56 percent of that acreage, followed by Argentina with 19 percent. Canada, Brazil, and China together make up 21 percent of all transgenic cropland. In 2005, China approved GMO rice for commercial production. This is the first GMO cereal grain approved for direct human consumption, and could move China into the forefront of GMO crop production.

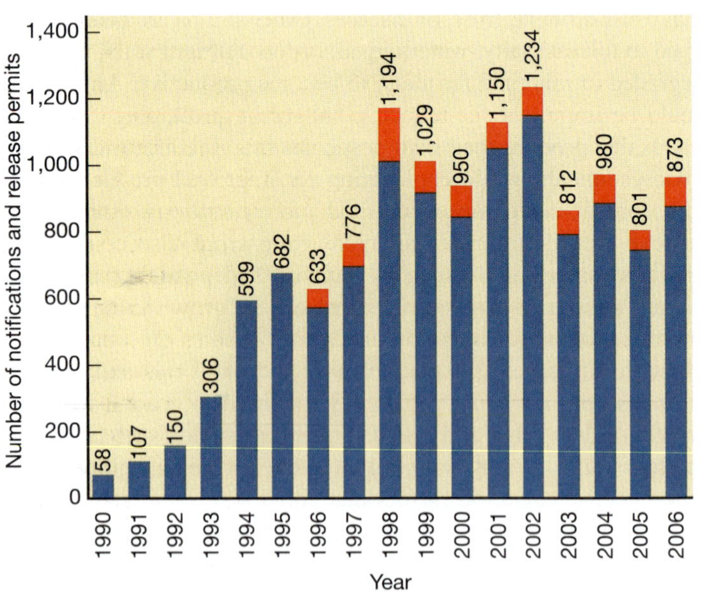


**FIGURE 9.28** Transgenic crop field releases in the United States 1990 to 2006. Approved or acknowledged (*blue*) versus denied or withdrawn (*red*).

**Source:** Data from Information Systems for Biotechnology, Virginia Tech University.

Some transgenic crops have reached mainstream status. Currently about 82 percent of all soybeans, 71 percent of the cotton, and one-quarter of all maize (corn) grown in the United States are GMOs. You've already probably eaten some genetically modified food. Estimates are that at least 60 percent of all processed food in America contains GMO ingredients. Since the United States, Argentina, and Brazil account for 90 percent of international trade in maize and soybeans (in which GMOs often are mixed with other grains), a large fraction of the world most likely has been exposed to some GMO products.

## Most GMOs have been engineered for pest resistance or weed control

Biotechnologists recently have created plants with genes for endogenous insecticides. *Bacillus thuringiensis* (Bt), a bacterium, makes toxins lethal to Lepidoptera (butterfly family) and Coleoptera (beetle family). The genes for some of these toxins have been transferred into crops such as maize (to protect against European cut worms), potatoes (to fight potato beetles), and cotton (for protection against boll weevils). This allows farmers to reduce insecticide spraying. Arizona cotton farmers, for example, report reducing their use of chemical insecticides by 75 percent. Small farms in India report an 80 percent yield increase with Bt cotton compared to neighboring plots growing conventional cotton.

Entomologists worry that Bt plants churn out toxin throughout the growing season, regardless of the level of infestation, creating perfect conditions for selection of Bt resistance in pests. The effectiveness of this natural pesticide—one of the few available to organic growers—is likely to be destroyed within a few years. One solution is to plant at least a part of every field in non-Bt crops that will act as a refuge for nonresistant pests. The hope is that interbreeding between these "wild-type" bugs and those exposed to Bt will dilute out recessive resistance genes. Deliberately harboring pests and letting them munch freely on crops is something that many farmers find hard to do. In addition, devoting a significant part of their land to nonproductive crops lowers the total yield and counteracts the profitability of engineered seed.

There also is a concern about the effects on nontarget species. In laboratory tests, about half of a group of monarch butterfly caterpillars died after being fed on plants dusted with pollen from Bt corn. Under field conditions, however, it was difficult to show any harm to butterflies. Critics still worry, however, that there may be unexpected problems associated with this new technology.

The other major group of transgenic crops are engineered to tolerate high doses of herbicides. Currently these crops occupy 20 million hectares worldwide, or about three-quarters of all genetically engineered acreage. The two main products in this category are Monsanto's "Roundup Ready" crops—so-called because they can withstand treatment with Monsanto's best-selling herbicide, Roundup (glyphosate)—and AgrEvo's "Liberty Link" crops, which resist that company's Liberty (glufosinate) herbicide. Because crops with these genes can grow in spite of high herbicide doses, farmers can spray fields heavily to exterminate weeds. This practice allows for conservation tillage and leaving more crop residue on fields to protect topsoil from erosion, which are both good ideas, but it also means using much more herbicide in higher doses than might otherwise be done.

## Is genetic engineering safe?

Ever since scientists discovered how to move genes from one organism to another, critics have worried about irresponsible use of this technology or unforeseen consequences arising from novel combinations of genetic material. Environmental and consumer groups have campaigned against transgenic organisms, labeling them "Frankenfoods." Industry groups, on the other hand, describe their critics as Neo-Luddites, who blindly oppose any new technology. In 2002, while millions of its people faced famine, Zambia's government refused to accept thousands of tons of genetically modified maize from the United States, claiming that it might be unsafe for human consumption. Most European nations have bans on genetically engineered crops. The United States has filed a suit at the World Trade Organization claiming that European policies constitute an unwarranted restraint on trade. How can we decide what to believe in this welter of claims and counterclaims?

At the base of some people's unease with genetic modification is a feeling that it simply isn't right to mess with nature. It doesn't seem proper to recombine genes any way we want. Doing so raises specters of gruesome monsters stitched together from spare parts, like Mary Shelley's Frankenstein. Is this merely a fear of science, or is it a valid ethical issue?

The U.S. Food and Drug Administration declined to require labeling of foods containing GMOs, saying that these new varieties are "substantially equivalent" to related varieties bred via traditional practices. After all, proponents say, we have been moving genes around for centuries through plant and animal breeding. All domesticated organisms should be classified as genetically modified, some people argue. Biotechnology may be a more precise way of creating novel organisms than normal breeding procedures. We're moving only a few genes—or even part of one gene—at a time with biotechnology, rather than hundreds of unknown genes through classical techniques.

What if GMOs escape and interbreed with native species? In a ten-year study of genetically modified crops, a group from Imperial College, London, concluded that GMOs tested so far do not survive well in the wild, and are no more likely to invade other habitats than their unmodified counterparts. Other scientists counter that some GMO crops already have been shown to spread their genes to nearby fields. Normal rapeseed (canola oil) varieties in Canada were found to contain genes from genetically modified varieties in nearby fields. It isn't clear, however, if the genes involved will have any adverse effect or whether they will persist for many generations. In 2001, researchers reported finding traces of genetically modified corn in wild relatives in Mexico, which supposedly has banned planting of GMOs, especially in regions where species were thought to have originated. This report was later withdrawn due to criticism of the techniques used to detect genetic markers, but worries that GMO genes could spread to wild relatives continues to be among the greatest concerns about this technology.

The first genetically modified animal designed to be eaten by humans is an Atlantic salmon containing extra growth hormone genes from an oceanic pout. The greatest worry from this fish is not that it will introduce extra hormones into our diet—that's already being done by chickens and beef that get extra growth hormone via injections or their diet—but rather the ecological effects if the fish escape from captivity. The transgenic fish grow seven times faster and are more attractive to the opposite sex than a normal salmon. If they escape from captivity, they may outcompete already endangered wild relatives for food, mates, and habitat. Fish farmers say they will grow only sterile females and will keep them in secure net pens. Opponents point out that salmon frequently escape from aquaculture operations, and that if just a few fertile transgenic fish break out it could be catastrophic for wild stocks.

Many people argue that we should take a better-safe-than-sorry "precautionary approach" that errs—if at all—on the side of safety. The 5,000-member Society of Toxicology has concluded that, "Based on available tests, there's no reason to suspect that transgenic plants differ in any substantive way from traditional varieties." On the other hand, a panel of biologists and agricultural scientists convened by the U.S. National Academy of Sciences

## What Do You Think?

### Shade-Grown Coffee and Cocoa

Has it ever occurred to you that your purchases of coffee and chocolate may be contributing to the protection or destruction of tropical forests? Both coffee and cocoa are examples of food products grown exclusively in developing countries but consumed almost entirely in the wealthy nations (vanilla and bananas are some other examples). Coffee grows in cool, mountain areas of the tropics, while cocoa is native to the warm, moist lowlands. Both are small trees of the forest understory, adapted to low light levels.

Until a few decades ago, most of the world's coffee and cocoa were grown under a canopy of large forest trees. Recently, however, new varieties of both crops have been developed that can be grown in full sun. Yields for sun-grown crops are higher because more coffee or cocoa trees can be crowded into these fields, and they get more solar energy than in a shaded plantation.

There are costs, however, in this new technology. Sun-grown trees die earlier from the stress and diseases common in these fields. Furthermore, ornithologists have found that the number of bird species can be cut in half in full-sun plantations, and the number of individual birds may be reduced by 90 percent. Shade-grown coffee and cocoa generally require fewer pesticides (or sometimes none) because the birds and insects residing in the forest canopy eat many of the pests. Shade-grown plantations also need less chemical fertilizer because many of the plants in these complex forests add nutrients to the soil. In addition, shade-grown crops rarely need to be irrigated because heavy leaf fall protects the soil, while forest cover reduces evaporation.

Cocoa pods grow directly on the trunk and large branches of cocoa trees.

Currently, about 40 percent of the world's coffee and cocoa plantations have been converted to full-sun varieties and another 25 percent are in process. Traditional techniques for coffee and cocoa production are worth preserving. Thirteen of the world's 25 biodiversity hot spots occur in coffee or cocoa regions. If all the 20 million ha (49 million acres) of coffee and cocoa plantations in these areas are converted to monocultures, an incalculable number of species will be lost.

The Brazilian state of Bahia is a good example of both the ecological importance of these crops and how they might help preserve forest species. At one time, Brazil produced much of the world's cocoa, but in the early 1900s, the crop was introduced into West Africa. Now Côte d'Ivoire alone grows more than 40 percent of the world total, and the value of Brazil's harvest has dropped by 90 percent. Côte d'Ivoire is aided in this competition by a labor system that reportedly includes widespread child slavery. Even adult workers in Côte d'Ivoire get only about $165 per year (if they get paid at all) compared to a minimum wage of $850 per year in Brazil. As African cocoa production ratchets up, Brazilian landowners are converting their plantations to pastures or other crops.

The area of Bahia where cocoa was once king is part of Brazil's Atlantic forest, one of the most threatened forest biomes in the world. Only 8 percent of this forest remains undisturbed. Although cocoa plantations don't represent the full diversity of intact forests, they protect a surprisingly large sample of what once was there. And shade-grown cocoa can provide an economic rationale for preserving that biodiversity. Brazilian cocoa will probably never compete with that from other areas for lowest cost. There is room in the market, however, for specialty products. If consumers were willing to pay a small premium for organic, fair-trade, shade-grown chocolate and coffee it might provide the incentive needed to preserve biodiversity. Wouldn't you like to know that your chocolate or coffee wasn't grown with child slavery, and is helping protect plants and animal species that might otherwise go extinct?

---

urged the government to more carefully, and more publically, review the potential environmental impacts of genetically modified plants and animals before approving them for commercial use.

Finally, there are social and economic implications of GMOs. Will they help feed the world or will they lead to a greater consolidation of corporate power and economic disparity? Might higher yields and fewer losses to pests and diseases allow poor farmers in developing countries to stop using marginal land and avoid cutting down forests to create farmland? Is this simply a technological fix or could it help promote agricultural sustainability? Critics suggest that there are simpler and cheaper ways other than high-tech crop varieties to provide nutrients to people in developing countries or to increase the income of poor, rural families. Adding a cow or a fishpond or training people in water harvesting or regenerative farming techniques (as we'll discuss in the next section) may have a longer-lasting impact than selling them expensive new seeds.

On the other hand, if we hope to reduce malnutrition and feed 9 billion people in 50 years, maybe we need all the tools we can get. Where do you stand in this debate? What additional information would you need to reach a reasoned judgment about the risks and benefits of this new technology?

***Think About It***

Suppose your grandmother asks you, "what's all this controversy about GMOs? Are they safe or not?" Could you summarize the arguments for and against genetic engineering in a few sentences? Which are the most important issues in this debate, in your opinion?

## 9.8 Sustainable Agriculture

How, then, shall we feed the world? Can we make agriculture compatible with sustainable ecological and social systems? Having discussed some of the problems that beset modern agriculture, we will now consider some suggested ways to overcome problems and make farming and food production just and lasting enterprises. This goal is usually termed **sustainable agriculture,** regenerative farming, or agroecology, all of which aim to produce food and fiber on a sustainable basis and repair the damage caused by destructive practices. Some alternative methods are developed through scientific research; others are discovered in traditional cultures and practices nearly forgotten in our mechanization and industrialization of agriculture.

### Soil conservation is essential

With careful husbandry, soil is a renewable resource that can be replenished and renewed indefinitely. Since agriculture is the area in which soil is most essential and also most often lost through erosion, agriculture offers the greatest potential for soil conservation and rebuilding. Some rice paddies in Southeast Asia, for instance, have been farmed continuously for a thousand years without any apparent loss of fertility. The rice-growing cultures that depend on these fields have developed management practices that return organic material to the paddy and carefully nurture the soil's ability to sustain life.

While American agriculture hasn't reached that level of sustainability, there is evidence that soil conservation programs are having a positive effect. In one Wisconsin study, erosion rates in one small watershed were 90 percent less in 1975–1993 than they were in the 1930s. Among the most important elements in soil conservation are land management, ground cover, climate, soil type, and tillage system.

#### *Managing Topography*

Water runs downhill. The faster it runs, the more soil it carries off the fields. Comparisons of erosion rates in Africa have shown that a 5 percent slope in a plowed field has three times the water runoff volume and eight times the soil erosion rate of a comparable field with a 1 percent slope. Water runoff can be reduced by leaving grass strips in waterways and by **contour plowing,** that is, plowing across the hill rather than up and down. Contour plowing is often combined with **strip farming,** the planting of different kinds of crops in alternating strips along the land contours (fig. 9.29). When one crop is harvested, the other is still present to protect the soil and keep water from running straight downhill. The ridges created by cultivation make little dams that trap water and allow it to seep into the soil rather than running off. In areas where rainfall is very heavy, intersecting, or "tied," ridges are often useful. This method involves a series of ridges running at right angles to each other, so that water runoff is blocked in all directions and is encouraged to soak into the soil.

**Terracing** involves shaping the land to create level shelves of earth to hold water and soil. The edges of the terrace are planted with soil-anchoring plant species. This is an expensive procedure, requiring either much hand labor or expensive machinery, but makes it possible to farm very steep hillsides. Rice terraces in Asia create beautiful landscapes as well as highly productive and sustainable agroecosystems (fig. 9.30).

**FIGURE 9.29** Contour plowing and strip cropping help prevent soil erosion on hilly terrain as well as create a beautiful landscape.

**FIGURE 9.30** Terracing, as in these Balinese rice paddies, can control erosion and make steep hillsides productive.

Planting **perennial species** (plants that grow for more than two years) is the only suitable use for some lands and some soil types. Establishing forest, grassland, or crops such as tea, coffee, or other crops that do not have to be cultivated every year may be necessary to protect certain unstable soils on sloping sites or watercourses (low areas where water runs off after a rain).

### *Providing Ground Cover*

Annual row crops such as corn or beans generally cause the highest erosion rates because they leave soil bare for much of the year (table 9.2). Often, the easiest way to provide cover that protects soil from erosion is to leave crop residues on the land after harvest. They not only cover the surface to break the erosive effects of wind and water, but they also reduce evaporation and soil temperature in hot climates and protect ground organisms that help aerate and rebuild soil. In some experiments, 1 ton of crop residue per acre (0.4 ha) increased water infiltration 99 percent, reduced runoff 99 percent, and reduced erosion 98 percent. Leaving crop residues on the field also can increase disease and pest problems, however, and may require increased use of pesticides and herbicides.

Where crop residues are not adequate to protect the soil or are inappropriate for subsequent crops or farming methods, such **cover crops** as rye, alfalfa, or clover can be planted immediately after harvest to hold and protect the soil. These cover crops can be plowed under at planting time to provide green manure. Another method is to flatten cover crops with a roller and drill seeds through the residue to provide a continuous protective cover during early stages of crop growth.

In some cases, interplanting of two different crops in the same field not only protects the soil but also is more efficient use of the land, providing double harvests. Native Americans and pioneer farmers, for instance, planted beans or pumpkins between the corn rows. The beans provided nitrogen needed by the corn, pumpkins crowded out weeds, and both crops provided foods that nutritionally balance corn. Traditional swidden (slash-and-burn) cultivators in Africa and South America often plant as many as 20 different crops together in small plots. The crops mature at different times so that there is always something to eat, and the soil is never exposed to erosion for very long.

**Mulch** is a general term for a protective ground cover that can include manure, wood chips, straw, seaweed, leaves, and other natural products. For some high-value crops, such as tomatoes, pineapples, and cucumbers, it is cost-effective to cover the ground with heavy paper or plastic sheets to protect the soil, save water, and prevent weed growth. Israel uses millions of square meters of plastic mulch to grow crops in the Negev Desert.

### *Reduced Tillage*

Farmers have traditionally used a moldboard plow to till the soil, digging a deep trench and turning the topsoil upside down. In the 1800s, it was shown that tilling a field fully—until it was "clean"—increased crop production. It helped control weeds and pests, reducing competition; it brought fresh nutrients to the surface, providing a good seedbed; and it improved surface drainage and aerated the soil. This is still true for many crops and many soil types, but it is not always the best way to grow crops. We are finding that less plowing and cultivation often makes for better water management, preserves soil, saves energy, and increases crop yields.

There are several major **reduced tillage systems.** *Minimum till* involves reducing the number of times a farmer disturbs the soil by plowing, cultivating, etc. This often involves a disc or chisel plow rather than a traditional moldboard plow. A chisel plow is a curved chisel-like blade that doesn't turn the soil over but creates ridges on which seeds can be planted. It leaves up to 75 percent of plant debris on the surface between the rows, preventing erosion (fig. 9.31). *Conserv-till* farming uses a coulter, a

**TABLE 9.2**

**Soil Cover and Soil Erosion**

| Cropping System | Average Annual Soil Loss (Tons/Hectare) | Percent Rainfall Runoff |
|---|---|---|
| Bare soil (no crop) | 41.0 | 30 |
| Continuous corn | 19.7 | 29 |
| Continuous wheat | 10.1 | 23 |
| Rotation: corn, wheat, clover | 2.7 | 14 |
| Continuous bluegrass | 0.3 | 12 |

**Source:** Based on 14 years' data from Missouri Experiment Station, Columbia, MO.

**FIGURE 9.31** In ridge tilling, a chisel plow is used to create ridges on which crops are planted and shallow troughs filled with crop residue. Less energy is used in plowing and cultivation, weeds are suppressed, and moisture is retained by the ground cover, or crop residue, left on the field.

sharp disc like a pizza cutter, which slices through the soil, opening up a furrow or slot just wide enough to insert seeds. This disturbs the soil very little and leaves almost all plant debris on the surface. *No-till* planting is accomplished by drilling seeds into the ground directly through mulch and ground cover. This allows a cover crop to be interseeded with a subsequent crop.

Farmers who use these conservation tillage techniques often must depend on pesticides (insecticides, fungicides, and herbicides) to control insects and weeds. Increased use of toxic agricultural chemicals is a matter of great concern. Massive use of pesticides is not, however, a necessary corollary of soil conservation. It is possible to combat pests and diseases with integrated pest management that combines crop rotation, trap crops, natural repellents, and biological controls (chapter 10).

## Low-input agriculture can be good for farmers and their land

In contrast to the trend toward industrialization and dependence on chemical fertilizers, pesticides, antibiotics, and artificial growth factors common in conventional agriculture, some farmers are going back to a more natural, agroecological farming style. Finding that they can't—or don't want to—compete with factory farms, these folks are making money and staying in farming by returning to small-scale, low-input agriculture. The Minar family, for instance, operate a highly successful 150-cow dairy operation on 97 ha (240 acres) near New Prague, Minnesota. No synthetic chemicals are used on their farm. Cows are rotated every day between 45 pastures or paddocks to reduce erosion and maintain healthy grass. Even in the winter, livestock remain outdoors to avoid the spread of diseases common in confinement (fig. 9.32). Antibiotics are used only to fight diseases. Milk and meat from this operation are marketed through co-ops and a community-supported agriculture (CSA) program. Sand Creek, which flows across the Minar land, has been shown to be cleaner when leaving the farm than when it enters.

**FIGURE 9.32** On the Minar family's 230-acre dairy farm near New Prague, Minnesota cows and calves spend the winter outdoors in the snow, bedding down on hay. Dave Minar is part of a growing counterculture that is seeking to keep farmers on the land and bring prosperity to rural areas.

Research at Iowa State University has shown that raising animals on pasture grass rather than grain reduces nitrogen runoff by two-thirds while cutting erosion by more than half. If more Midwestern farmers followed the Minar's example, we could easily eliminate the Gulf of Mexico "Dead Zone" (chapter 18).

Similarly, the Franzen family, who raise livestock on their organic farm near Alta Vista, Iowa, allow their pigs to roam in lush pastures, where they can supplement their diet of corn and soybeans with grasses and legumes. Housing for these happy hogs is in spacious, open-ended hoop structures. As fresh layers of straw are added to the bedding, layers of manure beneath are composted, breaking down into odorless organic fertilizer.

Low-input farms such as these typically don't turn out the quantity of meat or milk that their intensive agriculture neighbors do, but their production costs are lower, and they get higher prices for their crops, so that the all-important net gain is often higher. The Franzens, for example, calculate that they pay 30 percent less for animal feed, 70 percent less for veterinary bills, and half as much for buildings and equipment as their neighboring confinement operations. And on the Minar's farm, erosion after an especially heavy rain was measured to be 400 times lower than a conventional farm nearby.

Preserving small-scale, family farms also helps preserve rural culture. As Marty Strange of the Center for Rural Affairs in Nebraska asks, "Which is better for the enrollment in rural schools, the membership of rural churches, and the fellowship of rural communities—two farms milking 1,000 cows each or twenty farms milking 100 cows each?" Family farms help keep rural towns alive by purchasing machinery at the local implement dealer, gasoline at the neighborhood filling station, and groceries at the mom-and-pop grocery store.

## Consumers' choices play an important role

Adopting a vegetarian, organic diet can help reduce environmental impacts of the food you consume, but an even more environmentally friendly choice may be to become a **locavore** (that is, one who eats locally grown, seasonal food). Buying locally supports family farms and local economies. Where conventional foods were shipped an average of 2,400 km (1,500 mi) to markets in one frequently cited study, the average food item at a farmers' market traveled only 72 km (45 miles). Having any kind of food we want whenever we want it is gratifying, but the industrialized, global agriculture on which we have come to depend requires huge amounts of energy for fertilizer, fuel for farm equipment and shipping, plastics for food packaging, and climate control during shipping, storing, and ripening. While payments at a farmers' market go directly to many individual producers, most profits from conventional foods are retained by a tiny number of giant food corporations (the top three or four corporations in each commodity group typically control between 60 to 80 percent of the U.S. market—and often the same handful of companies, or their affiliates, dominate most food categories).

Co-ops often make an effort to carry locally grown and processed food. An even better way to know where your food comes from and how it's produced is to join a **community-supported agriculture (CSA)** program in which you make an annual contribution to a local farm in return for weekly deliveries of a "share" of whatever the farm produces. CSA farms generally practice organic or low-input agriculture, and they often invite members to visit and learn how their food is grown. A share often provides most fresh fruits and vegetables that an average family would consume during the growing season at a far lower price than similar food (especially if it's organic) would cost at a co-op or conventional grocery store. Much of America's most fertile land is around major cities. Farm failure fuels sprawl. Small-scale, artesinal farms are one of the best ways to preserve rural landscapes around metropolitan areas. There are now at least 1,500 CSA farms in the United States. You should be able to find one in your area.

**Agroecology,** or sustainable farming using ecological knowledge, also can be applied effectively in poorer countries. Brian Halweil, of the Worldwatch Institute, reports that farmers in the Guatemalan Highlands raised their crop yields tenfold without the use of chemical fertilizers or pesticides. Encouraged by World Neighbors, a nongovernmental development organization, to find innovative, local solutions to their problems, the farmers came up with low-cost techniques such as planting cover crops and grass strips to reduce erosion, or rotating corn with beans, peas, and other legumes to add nitrogen to the soil. Corn harvests increased from 0.4 tons to 4.5 tons per hectare, yields comparable to many richer countries.

One of the largest experiments in low-input farming is currently taking place in Cuba, where loss of aid from the former Soviet Union coupled with a trade embargo imposed by the United States have forced farmers to turn to organic, nonmechanized agriculture (chapter 10).

## Conclusion

World food supplies have increased dramatically over the past half century. Despite the fact that human populations have grown nearly threefold during that time, there's now more than enough food available for everyone. Nevertheless, at least 850 million people don't have enough to eat on a daily basis, and many more don't have the right kind of food. Whether we can meet the United Nations Millennium goal of cutting world hunger in half by 2020 remains to be seen. Declining fossil fuel supplies, soil erosion, and global climate change may make it difficult to continue to feed everyone. Ironically, there are now probably more obese people (1 billion) in the world than there are who have too few calories.

Billions of hectares of agricultural and pasture lands suffer soil degradation caused by erosion, nutrient depletion, salinization, and other forms of abuse. It's estimated that 25 billion metric tons of soil are lost from croplands every year. This not only reduces productivity, but also causes air and water pollution. Better farming techniques, including terracing, contour plowing, strip cropping, reduced tillage, providing ground cover, and planting perennial species could conserve our precious soil resources.

Some encouraging progress in low-input farming, community-supported agriculture, and agroecology show us how we can live more harmoniously with the land while still producing the food and fiber that we need. As pioneering ecologist Aldo Leopold wrote in *A Sand County Almanac,* "A thing is right only when it tends to preserve the integrity, stability and beauty of the community; and the community includes the soil, water, fauna and flora, as well as the people."

## Reviewing Learning Outcomes

By now you should be able to explain the following points:

**9.1** Describe world food supplies and nutritional requirements.

- Millions of people don't have enough to eat.
- Famines are acute food emergencies.
- We need the right kinds of food.
- Eating a balanced diet is essential for good health.

**9.2** Compare key food sources.

- A few major crops supply most of our food.
- Meat and dairy are important protein sources.
- Seafood is another important protein source.

**9.3** Understand how farm policy can help or hurt soil conservation.

**9.4** Illustrate why soil is a renewable resource.

- Soil is a complex mixture.
- Living organisms create unique properties of soil.
- Soils are layered.
- Soils are classified according to their structure and composition.

**9.5** Summarize the ways we use and abuse soil.

- Arable land is unevenly distributed.
- Land degradation reduces agricultural potential.
- Soil erosion is widespread.
- Wind and water are the main agents that move soil.
- Deserts are spreading around the world.

**9.6** Identify other agricultural resources.

- All plants need water to grow.
- Plants need fertilizer.
- Farming consumes energy.

**9.7** Explain new crops and genetic engineering.

- The "green revolution" produced dramatic increases in crop yields.
- Genetic engineering uses molecular techniques to produce new crop varieties.
- Most GMOs have been engineered for pest resistance or weed control.
- Is genetic engineering safe?
- How could shade-grown coffee and cocoa save wildlife?

**9.8** Appraise sustainable agriculture.

- Soil conservation is essential.
- Low-input agriculture can be good for farmers and their land.
- Consumers' choices play an important role.

## Practice Quiz

1. How many people in the world are chronically undernourished and how many die each year from starvation and nutritionally related diseases?
2. What is the relation between poverty and food security?
3. Define *malnutrition* and *obesity*. How many Americans are considered overweight?
4. How have recommendations for a balanced diet changed over the years?
5. What do we mean by the "green revolution"?
6. Why are wild fish populations endangered? How might aquaculture contribute to our food supply?
7. What is the composition of soil? What is humus? Why are soil organisms so important?
8. What are four kinds of erosion? Why is erosion a problem?
9. What is genetic engineering, or biotechnology, and how might they help or hurt agriculture and the environment?
10. What is sustainable agriculture?

## Critical Thinking and Discussion Questions

1. Promoters claim that South American lands that are being converted to soybean farms had already been transformed from native savanna to pasture long ago. Critics argue that ranching is simply shifting to virgin lands, and that biodiversity will suffer. How would you weigh the need of growing populations for high-quality protein with the value of pristine nature?
2. Suppose that an agribusiness wants to start a CAFO near your home. What regulations or safeguards would you want to see imposed on its operation? How would you weigh the possible costs and benefits of this operation?
3. Debate the claim that famines are caused more by human actions (or inactions) than by environmental forces. What is the critical element or evidence in this debate?
4. Should farmers be forced to use ecologically sound techniques that serve farmers' best interests in the long run, regardless of short-term consequences? How could we mitigate hardships brought about by such policies?
5. Should we try to increase food production on existing farmland, or should we sacrifice other lands to increase farming areas?
6. Some rice paddies in Southeast Asia have been cultivated continuously for a thousand years or more without losing fertility. Could we, and should we, adapt these techniques to our own country?
7. Outline the arguments you would make to your family and friends for why they should buy shade-grown, fair-trade coffee and cocoa. How much of a premium would you pay for these products?

## DATA analysis Using Relative Values

The graph in this box shows changes in abundance of marine species in six oceanic regions between 1970 and 2000. Look at the scales on the axes. Do you see anything unusual about them? The X-axis is simply five-year intervals. Nothing unusual there. But what are the units for the Y-axis? What is a species index, and why do all the curves start at 100? Essentially, the species index is a percent change from the abundance of species in each region in 1970. Why use an index rather than absolute numbers? We do this because the abundance values vary so much between regions. It's difficult to fit the full range of real numbers of species on the Y-axis scale, and if we did so, the curves would be so far apart that comparison would be difficult. What we're mostly interested in here are the comparative changes and the trends in different areas. A relative scale is useful because it makes it easy to see that species abundance in most areas has declined sharply over this 30-year period as well as where the losses have been greatest.

1. In which regions have species abundance remained relatively constant?
2. What do you think may be responsible for the sharp declines shown here?
3. Can you think of any reason that these regions have experienced less decline than other places?
4. How much has marine abundance declined in the South Atlantic during the period shown?
5. What is the significance of the thick gray line?
6. Why are some of the curves dotted at their ends?
7. How would you describe the marine biodiversity history of the North Atlantic over the time shown?
8. How do you think the agenda of the WWF, which created this graph, might be furthered by the way it's presented?
9. Might some other organization be likely to present the data in another way?

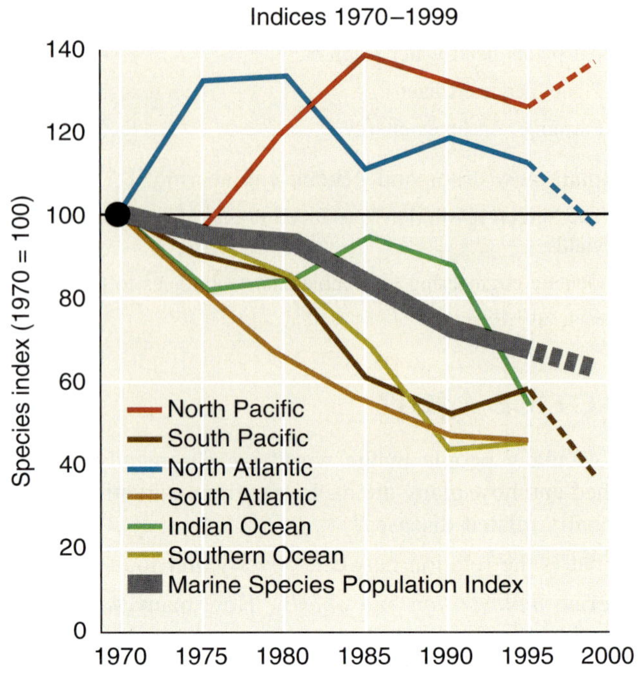


Changes in abundance of 217 species of marine fish, birds, mammals, and reptiles, from 1970 to 1999, by continent.
**Source:** World Wide Fund for Nature (WWF), 2000.

**For Additional Help in Studying This Chapter,** please visit our website at www.mhhe.com/cunningham10e. You will find additional practice quizzes and case studies, flashcards, regional examples, place markers for Google Earth™ mapping, and an extensive reading list, all of which will help you learn environmental science.

Pollinators, such as honeybees, provide services worth billions of dollars per year to farmers and orchardists. Pesticides can help protect pollinators from parasites, but can also have unexpected side effects.

CHAPTER 10

# Pest Control

*It ain't the things we know that cause all the trouble; it's the things we think we know that ain't so.*

*—Will Rogers—*

## LEARNING OUTCOMES

After studying this chapter, you should be able to:

**10.1** Define the major types of pesticides and describe the pest they are meant to manage.

**10.2** Identify some important pesticide benefits.

**10.3** Illustrate some important pesticide problems.

**10.4** Summarize some alternatives to current pesticide uses.

**10.5** Discuss how we might reduce pesticide exposure.

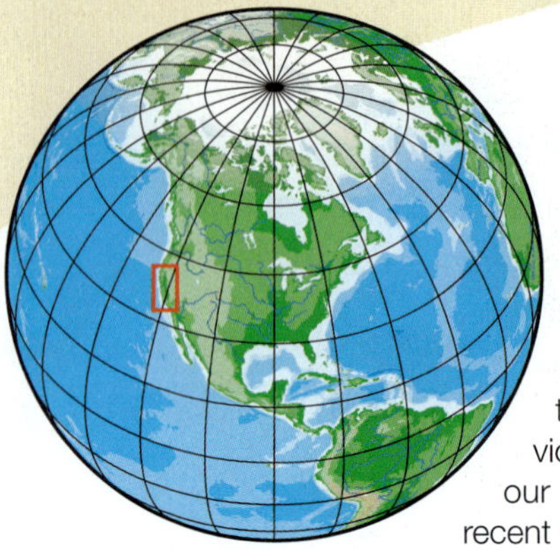

# Case Study The Forgotten Pollinators

Have you thanked an insect today? For many people, pollen means allergies, and insects mean bites or stings. Few understand the importance of pollinators and the service they provide in maintaining our diet and our environment. A recent biodiversity inventory suggests that about 90 percent of all flowering plants depend on insects, birds, bats, and other animal species to assist in delivering the pollen that they need to create seeds. At least one-third of the crops we eat rely on animal pollinators for successful reproduction. The annual value of this ecological service is estimated to be (U.S.) $40 billion. At least 200,000 kinds of animal play a role in assisting plant reproduction, and the vast majority are wild species, many of which play roles unrecognized by science.

For decades, biologists have been worried about declining populations of both wild and domesticated pollinators. Many factors contribute to pollinator disappearance, including habitat loss, climate change, introduction of exotic species, and the spread of diseases, but one of the most important threats is the indiscriminate use of pesticides that eliminate beneficial species along with the pests they are intended to target. Few people realize that the United States now applies twice the amount of pesticides used when Rachel Carson wrote *Silent Spring* in 1962.

A great deal of attention has been focused recently on honeybees. These industrious little insects not only make honey, but they play a vital role in fertilizing crops. In fact, their importance in spreading pollen is worth at least 100 times the value of the honey they make. Bee-pollinated foods include squash, tomatoes, peppers, apples, and pears. Farmers pay commercial beekeepers to bring hives to their fields to pollinate crops. The $1.6 billion annual California almond crop, for example, is totally dependent on rented honeybee colonies to make a successful crop. Unfortunately, there's a shortage of honeybees. The United States has fewer than half as many honeybee colonies now as it did in the 1940s. In 2005, there was a severe shortage of honeybees in California. Beekeepers traveled from as far away as Florida and North Carolina to service California's almond groves. For the first time in 50 years, the U.S. borders were opened to honeybees from New Zealand and Australia.

Native pollinators are just as important for crops as domesticated ones. In the 1970s, the Canadian government began spraying forests in New Brunswick in an attempt to control spruce budworm outbreaks. Blueberry farmers noticed that their crop yields decreased about 1 million pounds per year compared to before spraying started. In the late 1970s, forest spraying was stopped, and blueberry harvests gradually returned to normal. At about the same time, ranchers in the western United States noticed that alfalfa crops (a staple food for dairy cows and other livestock) were failing. Research showed that alkali bees, a wild, solitary, ground-nesting species, were pollinating the alfalfa. Farmers believed that a shift in pesticide use on neighboring crops did the bees in, or perhaps the alfalfa farmer's own use of insecticides to combat pests was responsible. In any case, almost all the $5 billion annual U.S. alfalfa crop depends on leafcutter bees, which are imported from Canada.

This case study introduces some important themes for this chapter. Pesticides are an important tool in disease prevention, food preservation, and property protection. On the other hand, our battle to protect ourselves from biological enemies often has unintended consequences. In many cases, we need more thoughtful, sustainable methods of accomplishing our goals. In this chapter, we'll look at both some benefits and costs of using powerful pesticides, as well as some new ways to use them more effectively.

**For more information see**

Klein, Alexandra-Maria, et al. 2006. Importance of pollinators in changing landscapes for world crops. *Proceedings of the Royal Society B* 3721 27 Oct. 2006, Page FirstCite, DOI 10.1098/rspb.2006.3721, URL http://dx.doi.org/10.1098/rspb.2006.3721.

## 10.1 Pests and Pesticides

A pest is something or someone that annoys us, detracts from some resource that we value, or interferes with a pursuit that we enjoy. In this chapter, we will concentrate on **biological pests,** organisms that reduce the availability, quality, or value of resources useful to humans. What's annoying or undesirable depends, of course, on your perspective. The mosquitoes that swarm in clouds over a marsh in the summer may be irritating to us, but they are an essential food source for birds and bats that feed on them. You may regard dandelions in your yard as tenacious and obnoxious weeds, but in some countries dandelions are cultivated as beautiful flowers and as a food source. Of the millions of species of organisms only about 100 plants, animals, fungi, and microbes cause 90 percent of all crop damage worldwide.

Insects tend to be the most frequent pests, in part, because they make up at least three-quarters of all species on the earth. Most pest organisms tend to be generalists, the opportunistic species that reproduce rapidly, migrate quickly into disturbed areas, and that are pioneers in ecological succession. They compete aggressively against more specialized endemic species and can often take over a biotic community, especially where humans have disrupted natural conditions and created an opening into which they can slip. Most Americans are familiar with dandelions, ragweed, English sparrows, starlings, European pigeons, and other "weedy" species that survive well in urban habitats. Chapter 11 describes how exotic aliens brought in by humans are crowding out native species in many places.

A **pesticide** is a chemical that kills pests. We generally think of toxic substances in this category, but chemicals that drive away pests or prevent their development are sometimes included as well.

**FIGURE 10.1** Broad-spectrum toxins can eliminate pests quickly and efficiently, but what are the long-term costs to us and to our environment?

Pest control can also include activities such as killing pests by burning crop residues or draining wetlands to eliminate breeding sites. A broad-spectrum pesticide that kills a wide range of living organisms is called a **biocide** (fig. 10.1). Fumigants, such as ethylene dibromide or dibromochloropropane, used to protect stored grain or sterilize soil fall into this category. Generally, we prefer narrower spectrum agents that attack a specific type of pest: **herbicides** kill plants; **insecticides** kill insects; **fungicides** kill fungi; acaricides kill mites, ticks, and spiders; nematicides kill nematodes (microscopic roundworms); rodenticides kill rodents; and avicides kill birds. Pesticides can also be defined by their method of dispersal (fumigation, for example) or their mode of action, such as an ovicide, which kills the eggs of pests. In a sense, the antibiotics used in medicine to fight infections are pesticides as well.

## People have always known of ways to control pests

Using chemicals to control pests may well have been among our earliest forms of technology. People in every culture have known that salt, smoke, and insect-repelling plants can keep away bothersome organisms and preserve food. The Sumerians controlled insects and mites with sulfur 5,000 years ago. Chinese texts 2,500 years old describe mercury and arsenic compounds used to control body lice and other pests. Greeks and Romans used oil sprays, ash and sulfur ointments, lime, and other natural materials to protect themselves, their livestock, and their crops from a variety of pests.

In addition to these metals and inorganic chemicals, people have used organic compounds, biological controls, and cultural practices for a long time. Alcohol from fermentation and acids in pickling solutions prevent growth of organisms that would otherwise ruin food. Spices were valued both for their flavors and because they deterred spoilage and pest infestations. Romans burned fields and rotated crops to reduce crop diseases. They also employed cover crops to reduce weeds. The Chinese developed plant-derived insecticides and introduced predatory ants in orchards to control caterpillars 1,200 years ago.

## Modern pesticides provide benefits, but also create problems

The era of synthetic organic pesticides began in 1939 when Swiss chemist Paul Müller discovered the powerful insecticidal properties of dichloro-diphenyl-trichloroethane (DDT). Inexpensive, stable, easily applied, and highly effective, this compound seemed ideal for crop protection and disease prevention. DDT is remarkably lethal to a wide variety of insects but relatively nontoxic to

**FIGURE 10.2** Before we realized the toxicity of DDT, it was sprayed freely on people to control insects as shown here at Jones Beach, New York, in 1948.

mammals. Mass production of DDT started during World War II, when Allied armies used it to protect troops from insect-borne diseases. Manufacture of the compound soared from a few kilograms to thousands of metric tons per year in less than a decade. It was sprayed on crops and houses, dusted on people and livestock, and used to combat insects nearly everywhere (fig. 10.2).

By the 1960s, however, evidence began to accumulate that indiscriminate use of DDT and other long-lasting industrial toxins was having unexpected effects on wildlife. Peregrine falcons, bald eagles, brown pelicans, and other carnivorous birds were disappearing from former territories in eastern North America. Studies revealed that egg shells were thinning in these species as DDT and its breakdown products were concentrated through food chains until it reached endocrine hormone-disrupting levels in top predators (see fig. 8.15). In 1962, biologist Rachel Carson published *Silent Spring,* warning that persistent organic pollutants, such as DDT, pose a threat to wildlife, and perhaps to humans. Most uses of DDT were banned in developed countries in the late 1960s, but it continues to be used in developing countries and remains the most prevalent contaminant on food imported to the United States.

Since the 1940s, many new synthetic pesticides have been invented. Many of them, like DDT, have proven to have unintended consequences on nontarget species. Assessing the relative costs and benefits of using these compounds continues to be a contentious topic, especially when unexpected complications arise, such as increasing pest resistance or damage to beneficial insects as described in the opening case study for this chapter.

According to the EPA, total pesticide use in the United States amounts to about 5.3 billion pounds (2.4 million metric tons) per year. Roughly half of that amount is chlorine and hypochlorites (bleach) used for water purification (fig. 10.3). Eliminating pathogens from drinking water prevents a huge number of infections and deaths, but there's concern that using so much chlorine and hypochlorite to do so may be creating other chronic health risks. The next largest category is conventional pesticides: primarily insecticides, herbicides, and fungicides.

Specialty biocides, such as preservatives used in adhesives and sealants, paints and coatings, leather, petroleum products, and

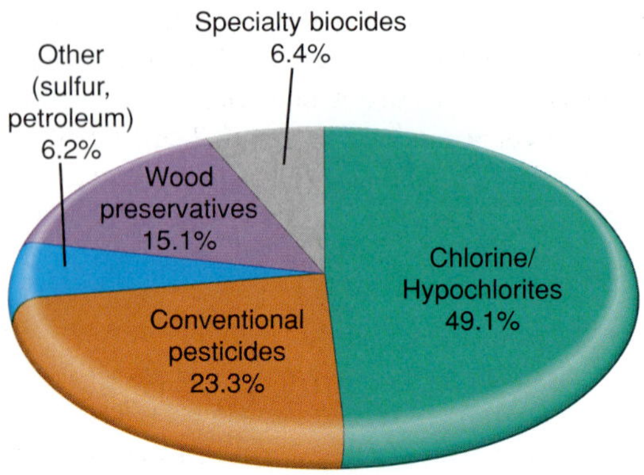


**FIGURE 10.3** Of the 5.3 billion pounds of pesticides used in the United States each year, nearly half is chlorine/hypochlorite disinfectant. Specialty biocides include other antiseptics and sanitizers.
**Source:** U.S. EPA, 2000.

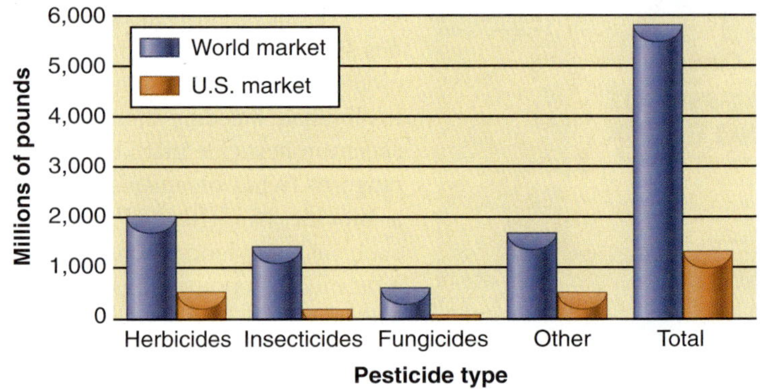


**FIGURE 10.4** World and U.S. "conventional" pesticide use in 1999 in millions of pounds of active ingredients. U.S. consumption of these chemicals is about 20 percent of world total.
**Source:** U.S. EPA, 2002.

plastics as well as recreational and industrial water treatment amount to some 300 million pounds per year, although they represent only about 6 percent of total pesticide use. The "other" category in figure 10.3 includes sulfur, oil, and chemicals used for insect repellents (such as DEET) and moth control. Wood preservatives represent just 15 percent of total pesticide consumption, but can be especially dangerous to our health because they tend to be both highly toxic and very long lasting. We'll discuss some of the concerns about these chemicals later in this chapter.

Roughly 80 percent of all conventional pesticides applied in the United States are used in agriculture or food storage and shipping. According to CropLife America, some 90 million ha of crops in the United States—including 96 percent of all corn (maize) and one-third of all soybeans—are treated with herbicides every year. In addition, 25 million ha of agricultural fields and 7 million ha of parks, lawns, golf courses, and other lands are treated with insecticides and fungicides. By some accounts, cotton has the highest rate of insecticide application of any crop, while golf courses often have higher rates of application of all conventional pesticides per unit area than any farm fields.

Homes and gardens account for only about 8 percent of total annual pesticide use in the United States, but three-quarters of all American homes use some type of pesticide, amounting to 20 million applications per year. Often people use much larger quantities of chemicals in their homes, yards, or gardens than farmers would to eradicate the same species in their fields. Storage and accessibility of toxins in homes also can be a problem. As we'll discuss later in this chapter, children's exposure to toxins in their home may be of greater concern than pesticide residues in food.

Information on global pesticide use is spotty and uncertain. The U.S. EPA estimates that world usage of conventional pesticides alone amounts to some 5.7 billion pounds (2.6 million metric tons) per year. In addition, millions of metric tons of so-called "inert" ingredients are added to pesticide formulations as carriers, stabilizers, emulsifiers, etc. Some of these materials are dangerous in their own right, and should be regulated more carefully than they often are. The 1.24 billion pounds of conventional pesticides used annually in the United States accounts for about 20 percent of total world consumption (fig. 10.4). Good data for water purification compounds, wood preservatives, and specialty biocides are lacking for many countries.

## There are many types of pesticides

One way to classify pesticides is by their chemical structure. This is useful because environmental properties—such as stability, solubility, and mobility—and toxicological characteristics of members of a particular chemical group are often similar.

**Inorganic pesticides** include compounds of arsenic, sulfur, copper, lead, and mercury. These broad-spectrum poisons are generally highly toxic and essentially indestructible, remaining in the environment forever. Seeds are sometimes coated with a mercury or arsenic powder to deter insects and rodents during storage or after planting. Handling such seeds with bare hands can be very dangerous for farmers or gardeners. They are generally neurotoxins and even a single dose can cause permanent damage.

**Natural organic pesticides,** or "botanicals," generally are extracted from plants. Some important examples are nicotine and nicotinoid alkaloids from tobacco; rotenone from the roots of derris and cubé plants; pyrethrum, a complex of chemicals extracted from the daisylike *Chrysanthemum cinerariaefolium* (fig. 10.5); and turpentine, phenols, and other aromatic oils from conifers. All are toxic to insects, but nicotine is also toxic to a broad spectrum of organisms including humans. Rotenone is commonly used to kill fish. Turpentine, phenols, and other natural hydrocarbons are effective pesticides, but synthetic forms such as pentachlorophenol are more stable and more toxic than natural forms. They penetrate surfaces well and are used to prevent wood decay, but they have been banned for most uses in many countries because of their persistence and toxicity.

**Fumigants** are generally small molecules such as carbon tetrachloride, carbon disulfide, ethylene dichloride, ethylene dibromide, methylene bromide, and dibromochloropropane that gasify easily and penetrate rapidly into a variety of materials. They are used to sterilize soil and prevent decay or rodent and insect infestation of stored grain. Because these compounds are extremely dangerous for workers who apply them, many have been restricted or banned altogether.

**FIGURE 10.5** Chrysanthemum flowers are a source of pyrethrum, a natural insecticide.

**Chlorinated hydrocarbons,** or organochlorines such as DDT, chlordane, aldrin, dieldrin, toxaphene, paradichlorobenzene (mothballs), and lindane, are synthetic organic insecticides that inhibit nerve membrane ion transport and block nerve signal transmission. They are fast acting and highly toxic in sensitive organisms. Toxaphene, for instance, kills goldfish at 5 parts per billion (5 μg/liter), one of the highest toxicities for any compound in any organism. Chlorinated hydrocarbons may persist in soil for decades, become concentrated through food chains, and are stored in fatty tissues of a variety of organisms. The chloriphenoxy herbicides 2,4 D and 2,4,5 T have hormone-like growth-regulating properties and are selective for broad-leaf flowering plants. Because these compounds are so toxic and so long-lasting, many chlorinated hydrocarbons have been banned for most uses over much of the world.

**Organophosphates,** such as parathion, malathion, dichlorvos, chlorpyrifos, dimethyldichlorovinylphosphate (DDVP), and tetraethylpyrophosphate (TEPP), are an outgrowth of nerve gas research during World War II. They generally inhibit cholinesterase, an enzyme essential for removing excess neurotransmitter from synapses in the peripheral nervous system. Many organophosphates are extremely toxic to mammals, birds, and fish (generally 10 to 100 times more poisonous than most chlorinated hydrocarbons). A single drop of TEPP on your skin, for example, can be lethal. Because they are quickly degraded, they are much less persistent in the environment than organochlorines, generally lasting only a few hours or a few days. These compounds are very dangerous for workers who often are sent into fields too soon after they have been sprayed (fig. 10.6).

Glyphosate (N-phosphonomethylglycine) is a special organophosphate that does not react with the central nervous system as most others in this group do. Instead, it is a broad spectrum, nonselective herbicide used to kill weeds. Monsanto markets glyphosate as Roundup, and has genetically engineered soy beans, corn, wheat, and other crops to be Roundup resistant. This allows no-till agriculture, but encourages use of the herbicide.

**Carbamates,** or urethanes, such as carbaryl (Sevin), aldicarb (Temik), aminocarb (Zineb), carbofuran (Baygon), and Mirex share many organophosphate properties, including mode of action, toxicity, and lack of environmental persistence and low bioaccumulation. Carbamates generally are extremely toxic to bees and must be used carefully to prevent damage to these beneficial organisms.

*Halogenated pyrroles* are a new class of compounds based on a naturally occurring microbial toxin. The first of these compounds to be released commercially is chlorfenapyr, which is marketed as Pirate. It was hailed as a replacement for organophosphates like malathion, but has been shown to reduce egg laying and egg survival in ducks at very low concentrations. Opponents argue that it may be the new DDT and call for its withdrawal.

**Microbial agents** and **biological controls** are living organisms or toxins derived from them used in place of pesticides. Bacteria such as *Bacillus thuringiensis* or *Bacillus popilliae* kill caterpillars or beetles by producing a toxin that ruptures the digestive tract lining when eaten. Parasitic wasps such as the tiny *Trichogramma* genus attack moth caterpillars and eggs, while lacewings and ladybugs control aphids. Viral diseases also have been used against specific pests.

## 10.2 Pesticide Benefits

Like all organisms, humans compete with other species for food and shelter and struggle to protect ourselves from diseases and predators. Pesticides are important weapons in this fight for survival.

### We have made progress in controlling many insect-borne diseases

Insects and ticks serve as vectors in the transmission of a number of disease-causing pathogens and parasites. Consider malaria, for example. About 300 million people suffer from this disease at any given time, and about 1 million die each year from *Plasmodium* protozoa

**FIGURE 10.6** The United Farm Workers of America claims that 300,000 farmworkers in the United States suffer from pesticide-related illnesses each year. Worldwide, the WHO estimates that 25 million people suffer from pesticide poisoning and 20,000 die each year from improper use or storage of pesticides.

**FIGURE 10.7** Malaria, spread by the *Anopheles* mosquito, is one of the largest causes of human disease and premature death in the world. By controlling mosquitoes, pesticides save millions of lives per year in tropical countries.

spread to humans by *Anopheles* mosquitoes (fig. 10.7). It is estimated that insecticidal mosquito control has prevented at least 50 million deaths from malaria over the past 50 years. Sri Lanka is a classic example of pesticide benefits. In the early 1950s, more than 2 million cases of malaria were reported in Sri Lanka each year. After DDT spraying began in 1954, new malaria cases almost completely disappeared. When DDT spraying was discontinued in 1964, however, malaria reappeared almost immediately.

Within three years the annual incidence was more than 1 million cases per year. The Sri Lankan government resumed DDT spraying in 1968 and continues limited use of this insecticide despite environmental concerns. They concluded that the reductions in medical expenses, social disruption, pain and suffering, and lost work resulting from malaria outweighed the direct costs of insecticide spraying by a thousand to one. Many people advocate resumed use of DDT in other malaria-prone countries as well.

Some other diseases spread by biting insects include yellow fever and related viral diseases such as encephalitis and West Nile, also carried by mosquitoes, and trypanosomiasis or sleeping sickness caused by protozoa transmitted by the tsetse fly. Onchocerciasis (river blindness) and filariasis (one form of which is elephantiasis, fig. 10.8) are caused by tiny worms spread by biting flies that afflict hundreds of millions of people in tropical countries.

**FIGURE 10.8** Elephantiasis is caused by parasitic worms (filaria) that block lymph vessels and cause fluid accumulation in various parts of the body.

All of these terrible diseases can be reduced by judicious use of pesticides. If you were faced with a choice of going blind before age 30 because of masses of worms accumulating in your eyeballs or a small chance of cancer due to pesticide exposures if you live to age 50 or 60, which would you choose?

### Without pesticides, we might lose two-thirds of conventional crops

Although reliable data on crop losses are difficult to obtain, it is thought that plant diseases, insect and bird predation, and competition by weeds reduce crop yields worldwide by at least one-third. Postharvest losses to rodents, insects, and fungi may be as high as another 20 to 30 percent. Without modern chemical pesticides, these losses might be much higher. Although we said in chapter 9 that there is more than enough food in the world to adequately feed everyone now living—if food were equitably distributed—this most certainly would not be true, given current intensive farming practices, if modern chemical pesticides were unavailable.

A commonly quoted estimate is that farmers save \$3 to \$5 for every \$1 spent on pesticides. This means lower costs and generally more reliable quality for consumers. In some cases, insects and fungal diseases cause only small losses in terms of the total crop quantity, but the cosmetic damage they cause greatly reduces the economic value of crops. For example, although codling moth larvae consume very little of the apples they infest, the brown trails they leave as they crawl through the fruit and the possibility of biting into a living worm often make an unsprayed crop unsalable. Which is the more worrisome risk for you—consuming a bug or consuming pesticides?

CropLife America calculates that prohibiting pesticides in the United States would result in a \$21 billion per year loss in food and fiber production. Without herbicides, they conclude, crop yields would be reduced up to 67 percent, soil erosion would increase by at least 1 billion metric tons per year (because of more cultivation to remove weeds), and it would take up to 70 million additional farm laborers to remove weeds by hand. This assumes, however, that we would continue growing the same single-species crops year after year without crop rotation or other sustainable agriculture practices.

**Think About It**

Which of the benefits of pesticides described in this section is most persuasive for you? Do these benefits change your overall attitude toward pesticides? You should be aware that the claims quoted here come from pesticide proponents. Compare them to problems described in the next section from the perspective of pesticide opponents.

## 10.3 Pesticide Problems

While synthetic chemical pesticides have brought us valuable economic and social benefits, they also cause a number of serious problems. In this section we will examine some of the worst of those problems.

**FIGURE 10.9** This machine sprays insecticide on orchard trees—and everything nearby. Up to 90 percent of pesticides applied in this fashion never reach target organisms.

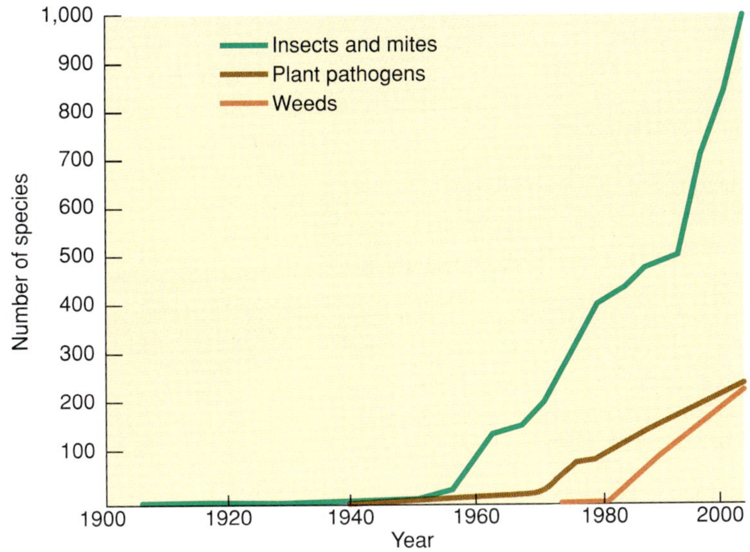


**FIGURE 10.10** Many pests have developed resistance to pesticides. Because insecticides were the first class of pesticide to be used widely, selection pressures led insects to show resistance early. More recently, plant pathogens and weeds are also becoming insensitive to pesticides.
**Source:** Worldwatch Institute, 2003.

## Pesticides often poison nontarget species

It is estimated that in many cases, up to 90 percent of the pesticides we use never reach their intended targets (fig. 10.9). Many beneficial organisms are poisoned unintentionally as a result, as described in the opening case study for this chapter.

In some cases, the effects of poisoning nontarget species are immediate and unmistakable. In one episode in 1972, a single application of the insecticide Azodrin to combat potato aphids on a farm in Dade County, Florida, killed 10,000 migrating robins in three days. Similarly, a 1991 derailment of a Southern Pacific tanker car on a tricky canyon bridge just north of Dunsmuir, California, dumped 75,000 liters (20,000 gallons) of highly toxic metam sodium herbicide into the Sacramento River. The entire river ecosystem—including aquatic plants, insects, amphibians, and at least 100,000 trout—was completely wiped out for 45 kilometers (27 miles) downstream.

In other cases, the effects are more difficult to pin down, although not less serious. Insecticide spraying in Canadian forests has been linked to dramatic declines (up to 77 percent) in Atlantic salmon. The chemical suspected in this case is 4-nonylphenol, a powerful endocrine-hormone disrupter (Exploring Science, p. 214).

## Pesticide resistance is often rapid and widespread

The resistance evolved by Varroa mites is a common phenomenon. Pesticides almost never kill 100 percent of a target species even under the most ideal conditions. As we discussed in chapter 4, every population contains some diversity in tolerance to adverse environmental factors. The most resistant members of a population survive pesticide treatment and produce more offspring like themselves with genes that enable them to withstand further chemical treatment. Because most pests propagate rapidly and produce many offspring, the population quickly rebuilds with pesticide-resistant individuals. We call this phenomenon **pest resurgence** or rebound. The Worldwatch Institute reports that at least 1,000 insect pest species and another 550 or so weeds and plant pathogens worldwide have developed chemical resistance (fig. 10.10). Of the 25 most serious insect pests in California, three-quarters are now resistant to one or more insecticides. This resistance means that it takes constantly increasing doses to get the same effect or that farmers who are caught on a **pesticide treadmill** must constantly try newer and more toxic chemicals in an attempt to stay ahead of the pests. Cornell University entomologist David Pimentel reports that a larger percentage of crops are lost now to insects, diseases, and weeds than in 1944 despite the use of millions of tons more pesticides. Some people worry that we are losing the battle to control pests.

One of the most ominous developments in this race to find effective pesticides is that we increasingly find pests that are resistant to chemicals to which they have never been exposed. Apparently, genes for pesticide resistance are being transferred from one species to another by means of vectors such as viruses and plasmids (naked pieces of virus-like DNA). Often a whole cluster of genes jump between species so that multiple chemical-tolerance is inherited before a particular pest is ever exposed to any of the chemicals. Widespread use of crops genetically engineered to produce pesticides endogenously (chapter 9) is likely to cause even more pesticide resistance.

In a way, every pesticide has a very limited useful life span before target species become resistant to it or the pesticide builds up intolerable environmental concentrations. We may have made a big mistake in broadcasting pesticides recklessly and extravagantly. DDT, for instance, is such a helpful insecticide that we

# *Exploring* SCIENCE

## Endocrine Disrupters

What might alligators in Florida, seals in the North Sea, salmon in the Great Lakes, and you have in common? All are at the top of their respective food chains and all appear to be accumulating threatening levels of toxic environmental chemicals in their body tissues. One of the most serious effects of those chemicals is that many disrupt endocrine hormones that regulate many important bodily functions, including reproductive and immune systems. Evidence for this seems quite convincing in some wildlife populations, but whether it also is true for humans is one of the most contentious and important questions in environmental toxicology today.

A shocking decrease in fertility and hatchling survival of alligators in Lake Apopka has been linked to pesticide pollution. Are humans affected by similar toxins?

One of the first examples of hormone-disrupting chemicals in the environment was a dramatic decline in alligators two decades ago in Florida's Lake Apopka just north of Orlando. Surveys showed that 90 percent of the alligator eggs laid each year were infertile and that of the few that hatched, only about half survived more than two weeks. Male hatchlings had shrunken penises and unusually low levels of the male hormone testosterone. Female alligators, meanwhile, had highly elevated estrogen levels and abnormal ovaries. The explanation seems to be that a DDT spill in the lake in the 1980s, along with pesticide-laden runoff from adjacent farm fields, has led to high levels of DDE (a persistent breakdown product of DDT) in the reptiles' tissues and eggs. Because of a similarity in chemical structure, DDE appears to interfere with the action of androgens and estrogens, the normal sex hormones (see fig. 8.12).

Researchers have begun to suspect that mysterious outbreaks of health and reproductive problems and immune system failures in other wildlife populations may have similar origins. A USGS survey of 139 U.S. rivers, for instance, found that 40 percent had hormonally active industrial chemicals. In many rivers, fish that appeared superficially to be females were actually genetically males. In some rivers and lakes, all male fish are feminized. Recent evidence suggests that atrazine converts male frogs into hermaphrodites and makes both sexes more susceptible to parasites and limb deformities.

Humans may be affected as well. DDE and other compounds have been correlated with elevated cases of estrogen-sensitive breast and vaginal cancers in women, with developmental disorders in children, and possibly with falling sperm counts and low fertility in men.

Good evidence exists from controlled laboratory experiments that rats and mice exposed *in utero* or through mother's milk to very low levels of estrogen-like compounds develop physical, reproductive, and behavioral problems. Among some pesticides shown to have hormone-disrupting effects are the herbicides alachlor and atrazine, the fungicides mancozeb and benomyl, and the insecticides carbaryl, clicofol, endosulfan, methomyl, methoxychlor, parathion, linuron, chlorfenapyr, and sythetic pyrethroids.

We know that some of these chemicals act as synthetic hormones, others are antagonists that block normal hormone function. Furthermore, there can be striking synergy between some compounds. When endosulfan and DDT or chlordane are applied together, for example, the combination is 1,600 times more estrogenic than either chemical alone.

The question is whether specific chemical exposures are linked to particular human health problems. Many of these compounds are hundreds or thousands of times less active than normal hormones, leading skeptics to doubt that they have any noticeable effects except in people exposed to extremely high levels. Furthermore, we may have protective mechanisms that are lacking in highly inbred laboratory rodents, and we can eat a highly varied diet that includes protective factors as well as toxins.

A wide variety of industrial products—including phthalates in plastics and cosmetics, polychlorinated biphenyls (PCBs), dioxins, chlorinated solvents, and metals such as lead, mercury, arsenic, and cadmium—also can disrupt hormone actions. Every human in the world now carries dozens of these chemicals in his or her body from before birth to death. Determining which of these toxins—or which combination—might be responsible for a particular disease or developmental abnormality may never be possible.

The bottom line is that we don't know (and we may never know for sure) whether falling sperm counts, increasing cancers, birth defects, immune diseases, and behavioral disorders in humans are caused by hormonally active, bioaccumulative, environmental chemicals. In 1996, the EPA ordered pesticide manufacturers to begin testing for disrupting effects. Given the continuing uncertainty about the dangers we face, what more do you think we should do? Is this threat serious enough to warrant drastic steps to reduce our risk? If you were head of the Environmental Protection Agency or the Food and Drug Administration, how much certainty would you demand before acting to protect our environment and ourselves from this ominous threat?

should have used it sparingly and carefully, so that it would still be effective against the worst insect pests. It has been spread so widely that 50 of the 60 malaria-carrying mosquitoes are now resistant to it, and the environmental side effects outweigh its benefits. You can think of the useful life of DDT as a nonrenewable resource that we squandered.

## Pesticide misuse can create new pests

Often the worst side effect of broadcast spraying a pesticide is that we kill beneficial predators that previously kept a number of pests under control. Many of the agricultural pests that we want to eliminate, for instance, are herbivores such as aphids, grasshoppers, or moth larvae that eat crop plants. Under natural conditions, their populations are kept under control by predators such as wasps, ladybugs, and praying mantises. As we discussed in chapter 3, there are generally far fewer predators in a food web than the species they prey upon. When we use a broad-spectrum pesticide, higher trophic levels are more likely to be knocked out than lower ones. This means that species that previously were insignificant can be released from natural controls and suddenly become major pests.

Consider the case of the Canete Valley in Peru. Before DDT was introduced in 1949, cotton yields were about 500 kg per hectare. By 1952 yields had risen to nearly 750 kg per hectare, but DDT-resistant boll weevils also had appeared (fig. 10.11). Toxaphene replaced DDT, but within two years it also became ineffective against boll weevils, which rebounded to higher levels than ever. Even worse, *Heliothis worms*—which had not previously been a problem—began increasing rapidly. The wasps that earlier had kept both organisms in check were poisoned by the increasing pesticide doses. By 1955, cotton yields were down to 330 kg per hectare, one-third less than before any pesticides were used.

Similarly, in California in the 1990s, cotton yields plunged by 20 percent despite a doubling of the number of insecticide applications and a 43 percent increase in pesticide costs. With the advent of chemical pest controls, farmers tend to abandon mixed crops, rotation regimes, and other traditional methods of management. This creates a greater potential for pest damage and pest resistance and thus makes it economical to use even higher levels of pesticide treatment. The application of large amounts of pesticides over wide expanses of land, together with changes in the ways the land is managed, have undermined the biological and ecological forces and interactions that previously governed population dynamics among species. A lack of understanding of the ways pesticides adversely affect beneficial organisms is one reason pest managers stick with chemical-based systems despite declining pesticide efficiency.

## Some persistent pesticides can move long distances in the environment

The qualities that make DDT and other chlorinated hydrocarbons so effective—stability, high solubility, and high toxicity—also make them environmental nightmares. Because they persist for years, even decades in some cases, and move freely through air, water, and soil, they often show up far from the point of original application. Some of these compounds have been discovered far from any possible source and long after they most likely were used. Because they have an affinity for fat, many chlorinated hydrocarbons are bio-concentrated and stored in the bodies of predators—such as porpoises, whales, polar bears, trout, eagles, ospreys, and humans—that feed at the top of food webs. In a study of human pesticide uptake and storage, Canadian researchers found the levels of chlorinated hydrocarbons in the breast milk of Inuit mothers living in remote arctic villages were five times that of women from Canada's industrial region some 2,500 km (1,600 mi) to the south. Inuit people have the highest levels of these persistent pollutants of any human population except those contaminated by industrial accidents.

These compounds accumulate in polar regions by what has been called the "grasshopper effect," in which they evaporate from water and soil in warm areas and then condense and precipitate in colder regions. In a series of long-distance hops, they eventually collect in polar regions, where they accumulate in top predators. Polar bears, for instance, have been shown to have concentrations of certain chlorinated compounds 3 billion times greater than the seawater around them. In Canada's St. Lawrence estuary, beluga (white whales), which suffer from a wide range of infectious diseases and tumors thought to be related to environmental toxins, have such high levels of chlorinated hydrocarbons that their carcasses must be treated as toxic waste.

Although DDT hasn't been used legally in the United States or Canada for more than 20 years, in 1999 researchers reported discovery of p, p′ -DDE, a DDT breakdown by-product, in the amniotic fluid of 30 percent of a sample of pregnant women in Los Angeles, California. A growing baby is surrounded by amniotic fluid during its nine months in the mother's womb. Retrospective studies have shown that mothers with high DDT or DDE levels in their blood are more likely to have premature or low-birthweight babies, and that these children have lower fertility rates as adults than those not exposed to pesticides in the womb.

**FIGURE 10.11** The cotton boll weevil caused terrible devastation in southern U.S. cotton fields in the 1930s and 1940s. DDT and other insecticides controlled this pest briefly, but the beetle quickly became resistant.

Atrazine and alochlor are the most widely used herbicides in North America. More than 96 percent of the maize (corn) and soybean crops in the United States are treated with these compounds. The 31 million kg (70 million lbs) of atrazine used every year in the United States makes it the biggest commercial pesticide. Surveys of wells have found that up to 30 percent of all community wells and as much as 60 percent of all private wells in the Midwestern U.S. corn belt are contaminated with atrazine and alochlor. University of California researcher Tyrone Hayes has shown that extremely low levels of atrazine cause sexual abnormalities in frogs. Although the chemical industry has tried to discredit his research and suppress its publication, others have confirmed the estrogenic effects of this herbicide. In 2003, the European Union withdrew regulatory approval for atrazine, citing its human and environmental health effects.

Because **persistent organic pollutants (POPs)** are so long-lasting and so dangerous, 127 countries agreed in 2001 to a global ban on the worst of them, including aldrin, clordane, dieldrin, DDT, endrin, hexachlorobenzene, neptachlor, mirex, toxaphene, polychlorinated biphenyls (PCBs), dioxins, and furans. Most of this "dirty dozen" had been banned or severely restricted in developed countries for years. However, their production continued. Between 1994 and 1996, U.S. ports shipped more than 100,000 tons of POPs each year. Most of this was sent to developing countries where regulations were lax. Ironically, many of these pesticides returned to the United States on bananas and other imported crops. According to the 2001 POPs treaty, eight of the dirty dozen were banned immediately; PCBs, dioxins, and furans will be phased out; and use of DDT, still allowed for limited uses such as controlling malaria, must be publically registered in order to permit monitoring. The POPs treaty has been hailed as a triumph for environmental health and international cooperation. Unfortunately, other compounds—perhaps just as toxic—have been introduced to replace POPs.

**FIGURE 10.12** Handling pesticides requires protective clothing and an effective respirator. Pesticide applicators in tropical countries, however, often can't afford these safeguards or can't bear to wear them because of the heat.

## Many pesticides cause human health problems

Pesticide effects on human health can be divided into two categories: (1) acute effects, including poisoning and illnesses caused by relatively high doses and accidental exposures, and (2) chronic effects suspected to include cancer, birth defects, immunological problems, endometriosis, neurological problems, Parkinson's disease, and other chronic degenerative diseases.

The World Health Organization (WHO) estimates that 25 million people suffer pesticide poisoning and at least 20,000 die each year (fig. 10.12). At least two-thirds of this illness and death results from occupational exposures in developing countries where people use pesticides without proper warnings or protective clothing. A tragic example of occupational pesticide exposure is found among workers in the Latin American flower industry. Fueled by the year-round demand in North America for fresh vegetables, fruits, and flowers, a booming export trade has developed in countries such as Guatemala, Colombia, Chile, and Ecuador. To meet demands in North American markets for perfect flowers, table grapes and other produce, growers use high levels of pesticides, often spraying daily with fungicides, insecticides, nematicides, and herbicides. Working in warm, poorly ventilated greenhouses with little protective clothing, the workers—70 to 80 percent of whom are women—find it hard to avoid pesticide contact. Nearly two-thirds of nearly 9,000 workers surveyed in Colombia experienced blurred vision, nausea, headaches, conjunctivitis, rashes, and asthma. Although harder to document, they also reported serious chronic effects such as stillbirths, miscarriages, and neurological problems.

Long-term health effects caused by chronic pesticide exposures are difficult to document conclusively. As we discussed in chapter 8, isolating a specific pesticide-related disease from the myriad of other risks we encounter is complex. For example, numerous studies have shown that farmers who use pesticides have elevated rates of prostate cancer, non-Hodgkin's lymphoma, Parkinson's, and several other chronic diseases, but distinguishing between pesticide effects, smoking, exposure to gasoline and solvents, and other health risks is complex. Similarly, children of farmers who apply herbicides and fungicides have higher rates of birth defects, but it's difficult to know which of the mixture of chemicals used by these farmers may be responsible.

In an ongoing study of prenatal exposure to PCBs and other persistent contaminants, researchers from Wayne State University in Detroit, Michigan, have documented significant learning and attention problems in children whose mothers ate Lake Michigan fish regularly in the years prior to pregnancy. At age 11, the most highly exposed children had difficulty paying attention, suffered from poorer short- and long-term memory, were twice as likely to be below average in reading comprehension, and were three times as likely to have low IQ scores as their age-matched peers.

Researchers in a similar study in Mexico showed striking differences in behavior and motor skills in children exposed to pesticides compared to peers with minimal pesticide risk. Two groups of Yaqui children from northwestern Mexico were tested in this study. The children were similar in every respect except for their pesticide exposure. Children who lived in the foothills come from ranches where pesticides are used sparingly, if at all. Children from the Yaqui valley, live in a farming area where pesticide use has been heavy for the past 50 years. Samples of mother's milk and umbilical cord blood taken from valley families contained high amounts of several persistent pesticides including aldrin, endrin, dieldrin, heptachlor, and DDE. In tests designed to measure growth and development, the valley children fell far behind their foothill-dwelling peers (fig. 10.13). Farm children also exhibited diminished memory, decreased physical stamina, higher irritability, a greater tendency to fight, and poorer eye-hand coordination.

Farm families aren't the only ones to suffer from pesticide exposure. The 63 million pounds of pesticides used annually in American homes and gardens pose a threat to families, especially those with young children. In California, for example, children from homes fumigated by professional bug exterminators have been found to be three times as likely to have acute lymphocytic leukemia as those from homes where no pesticides were used (fig. 10.14). One recent study concluded that more than 1 million children under age five in the United States may exceed the safe daily dose of organophosphate insecticides. Children are much more susceptible to toxic chemicals than adults because—pound for pound—they eat more food, drink more water, and breathe more air. Children spend more time playing on carpets, grass, or in dirt where pesticides may accumulate. They put their fingers in their mouths more often, and children's rapid growth and development make them vulnerable to toxic effects (see What Do You Think? p. 221). Many health professionals believe that the sudden rise in autism and hyperactivity/attention deficit disorder in America may be linked to environmental toxins.

**FIGURE 10.13** Representative examples of drawings of people by 4- and 5-year-old Yaqui children relatively unexposed to pesticides (foothills) and those heavily exposed (valley).

**FIGURE 10.14** Our approach to nature is to beat it into submission, but at what cost?



All of us now have traces of persistent pollutants in our bodies. In its "Body Burden Study," the Environmental Working Group found detectable levels of 91 different industrial chemicals per person on average (161 different chemicals in all) in a sample of people none of whom had any known occupational exposure. Of these chemicals, 76 are suspected to be cancer-causing, 94 are known to be neurotoxins in laboratory animals, and 79 are suspected of causing birth defects (the total adds up to more than 161 because some have multiple effects). Of course, some materials can be detected at levels far below those known to have toxic effects, but the presence of so many different compounds in ordinary people suggests the ubiquity of these chemicals in our environment. Still, there are steps you can take to reduce your exposure to pesticides (What Can You Do? p. 218).

## What Can You Do?

### Controlling Pests

Based on the principles of Integrated Pest Management, the U.S. EPA has released a helpful and informative guide to pest control. Among their recommendations:

1. *Identify the pest problem.* What kinds of pests and how many do you have? Many free resources are available from your library or County Cooperative Extension Service to help you understand what you face and how best to deal with it.
2. *Decide how much pest control is necessary.* Does your lawn really need to be totally weed free? Could you tolerate some blemished fruits and vegetables, or could you replace some of the plants you now grow with ones less sensitive to pests?
3. *Eliminate pest sources.* Remove food, water, and habitat that encourage pest growth from your house or yard. Block off hiding places or routes of entry into your home. Don't plant the same crops in the same spot year after year.
4. *Develop a healthy, weed-resistant yard.* Make sure your soil has the right pH balance, key nutrients, and good texture. Grow a type of grass or cover crop that does well in your climate. Set realistic weed and pest controls.
5. *Use biological controls.* Beneficial predators such as purple martins and bats eat insects; lady beetles (ladybugs) and their larvae eat aphids, mealybugs, whiteflies, and mites. Other beneficial predators include spiders, centipedes, dragonflies, wasps, and ants. Planting a border of marigolds may keep insects or herbivores away from your crops or flowers.
6. *Use simple manual methods.* Cultivate your garden and hand-pick weeds and pests from your garden. Use a flyswatter. Set traps to control rats, mice, and some insects. Mulch to reduce weed growth.
7. *Use chemical pesticides carefully.* If you decide that the best solution is chemical, choose the right pesticide product, read safety warnings and handling instructions, buy the amount you need, store the product safely, and dispose of excess properly.
8. *Evaluate the results.* Compare pretreatment and posttreatment conditions. Is there clear evidence of pest reduction? Are the benefits of short-term chemical pesticide control worth the risks? Would other treatments, including prevention of pest buildup work as well?

**Source:** Citizen's Guide to Pest Control and Pesticide Safety: EPA 730-K-95-001.

## 10.4 Alternatives to Current Pesticide Uses

Can we avoid using toxic pesticides or lessen their environmental and human-health impacts? In many cases, improved management programs can cut pesticide use between 50 and 90 percent without reducing crop production or creating new diseases. Some of these techniques are relatively simple and save money while maintaining disease control and yielding crops with just as high quality and quantity as we get with current methods. In this section, we will examine behavioral changes, biological controls, and integrated pest management systems that could substitute for current pest-control methods.

**FIGURE 10.15** The praying mantis looks ferocious and is an effective predator against garden pests, but it is harmless to humans. They can even make interesting and useful pets.

### We can change our behavior

Crop rotation (growing a different crop in a field each year in a two- to six-year cycle) keeps pest populations from building up. For instance, a soybean/corn/hay rotation is effective and economical against white-fringed weevils. Mechanical cultivation can substitute for herbicides but increases erosion. Flooding fields before planting or burning crop residues and replanting with a cover crop can suppress both weeds and insect pests. Habitat diversification, such as restoring windbreaks, hedgerows, and ground cover on watercourses, not only prevents soil erosion but also provides perch areas and nesting space for birds and other predators that eat insect pests. Growing crops in areas where pests are absent makes good sense. Adjusting planting times can avoid pest outbreaks, while switching from huge monoculture fields to mixed polyculture (many crops grown together) makes it more difficult for pests to find the crops they like. Tillage at the right time can greatly reduce pest populations. For instance, spring or fall plowing can help control overwintering corn earworms.

### Useful organisms can help us control pests

Biological controls such as predators (wasps, ladybugs, praying mantises; fig. 10.15) or pathogens (viruses, bacteria, fungi) can control many pests more cheaply and safely than broad-spectrum, synthetic chemicals. *Bacillus thuringiensis* or Bt, for example, is a naturally occurring bacterium that kills the larvae of lepidopteran (butterfly and moth) species but is generally harmless to mammals. A number of important insect pests such as tomato hornworm, corn rootworm, cabbage loopers, and others can be controlled by spraying bacteria on crops. Larger species are effective as well. Ducks, chickens, and geese, among other species, are used to rid fields of both insect pests and weeds. These biological organisms are self-reproducing and often have wide prey tolerance. A few mantises or ladybugs released in your garden in the spring will keep producing

FIGURE 10.16 A Nigerian woman examines a neem tree, the leaves, seeds, and bark of which provide a natural insecticide.

offspring and protect your fruits and vegetables against a multitude of pests for the whole growing season.

Herbivorous insects also have been used to control weeds. For example, the prickly pear cactus was introduced to Australia about 150 years ago as an ornamental plant. This hardy cactus escaped from gardens and found an ideal home in the dry soils of the outback. It quickly established huge, dense stands that dominated 25 million ha (more than 60 million acres) of grazing land. A natural predator from South America, the cactoblastis moth, was introduced into Australia in 1935 to combat the prickly pear. Within a few years, cactoblastis larvae had eaten so much prickly pear that the cactus has become rare and is no longer economically significant.

Some plants make natural pesticides and insect repellents. The neem tree (*Azadirachta indica*) is native to India, but is now grown in many tropical countries (fig. 10.16). The leaves, bark, roots and flowers all contain compounds that repel insects and can be used to combat a number of crop pests and diseases.

Genetics and bioengineering can help in our war against pests. Traditional farmers have long known to save seeds of disease-resistant crop plants or to breed livestock that tolerate pests well. Modern science can speed up this process through selection regimes or by using biotechnology to transfer genes between closely related or even totally unrelated species. (Often we don't know, however what unintended consequences, such as super weeds or new pests, might result from these activities.) Insect pest reproduction has sometimes been reduced by releasing sterile males. Screwworms, for example, are the flesh-eating larvae of flies that lay their eggs in scratches or skin wounds of livestock. They were a terrible problem for ranchers in Texas and Florida in the 1950s, but release of massive numbers of radiation-sterilized males disrupted reproduction and eliminated this pest in Florida. In Texas, where flies continue to cross the border from Mexico, control has been more difficult, but continual vigilance keeps the problem manageable.

Sometimes we can change the biology or behavior of species that are being attacked by pests. Researchers at the University of Minnesota, for example, are fighting Varroa mites by raising honeybees with an unusual behavior. These bees recognize mites and remove them from the hive. When bee-keepers obtain enough honeybees with this behavior, they can save their hives while also reducing reliance on pesticides.

FIGURE 10.17 Different strategies can be used to control pests at various stages of their life cycles. *Bacillus thuringiensis* (*Bt*) kills caterpillars when they eat leaves with these bacteria on the surface. Releasing juvenile hormone in the environment prevents maturation of pupae. Predators attack at all stages.

Other promising approaches are to use hormones that upset development or sex attractants to bait traps containing toxic pesticides. Many municipalities control mosquitoes with these techniques rather than aerial spraying of insecticides because of worries about effects on human health. Briquettes saturated with insect juvenile hormone are scattered in wetlands where mosquitoes breed. The presence of even minute amounts of this hormone prevent larvae from ever turning into biting adults (fig. 10.17). Unfortunately, many beneficial insects as well as noxious ones are affected by the hormone, and birds or amphibians that eat insects may be adversely affected when their food supply is reduced. Some communities that formerly controlled mosquitoes have abandoned these programs, believing that having naturally healthy wetlands is worth getting a few bites in the summer.

## Integrated pest management uses a combination of techniques to fight pests

**Integrated pest management (IPM)** is a flexible, ecologically based strategy that is applied at specific times and aimed at specific crops and pests. It often uses mechanical cultivation and techniques such as vacuuming bugs off crops as an alternative to chemical application (fig. 10.18). IPM doesn't give up chemical pest controls entirely but rather tries to use the minimum amount necessary only as a last resort and avoids broad-spectrum, ecologically disruptive products. IPM relies on preventive practices that encourage growth

**FIGURE 10.18** This machine, nicknamed the "salad vac," vacuums bugs off crops as an alternative to treating them with toxic chemicals.

and diversity of beneficial organisms and enhance plant defenses and vigor. Careful, scientific monitoring of pest populations to determine **economic thresholds,** the point at which potential economic damage justifies pest control expenditures, and the precise time, type, and method of pesticide application is critical in IPM.

Trap crops, small areas planted a week or two earlier than the main crop, are also useful. This plot matures before the rest of the field and attracts pests away from other plants. The trap crop then is sprayed heavily with enough pesticides so that no pests are likely to escape. The trap crop is destroyed so that workers will not be exposed to the pesticide and consumers will not be at risk. The rest of the field should be mostly free of both pests and pesticides.

IPM programs are already in use all over the United States on a variety of crops. Massachusetts apple growers who use IPM have cut pesticide use by 43 percent in the past ten years while maintaining per-acre yields of marketable fruit equal to that of farmers who use conventional techniques. Some of the most dramatic IPM success stories come from the developing world. Cuba, for example (What Do You Think? p. 221), has turned almost entirely to organic farming. In Brazil, pesticide use on soybeans has been reduced up to 90 percent with IPM. In Costa Rica, use of IPM on banana plantations has eliminated pesticides altogether in one region. In Africa, mealybugs were destroying up to 60 percent of the cassava crop (the staple food for 200 million people) before IPM was introduced in 1982. A tiny wasp that destroys mealybug eggs was discovered and now controls this pest in over 65 million ha (160 million acres) in 13 countries.

A successful IPM program that could serve as a model for other countries is found in Indonesia, where brown planthoppers developed resistance to virtually every insecticide and threatened the country's hard-won self-sufficiency in rice. In 1986, President Suharto banned 56 of 57 pesticides previously used in Indonesia and declared a crash program to educate farmers about IPM and the dangers of pesticide use. Researchers found that farmers were spraying their fields habitually—sometimes up to three times a week—regardless of whether fields were infested. By allowing natural predators to combat pests and spraying only when absolutely necessary with chemicals specific for planthoppers, Indonesian farmers using IPM have had higher yields than their neighbors using normal practices and they cut pesticide costs by 75 percent. In 1988, only two years after its initiation, the program was declared a success. It has been extended throughout the whole country. Since nearly half the people in the world depend on rice as their staple crop, this example could have important implications elsewhere (fig. 10.19).

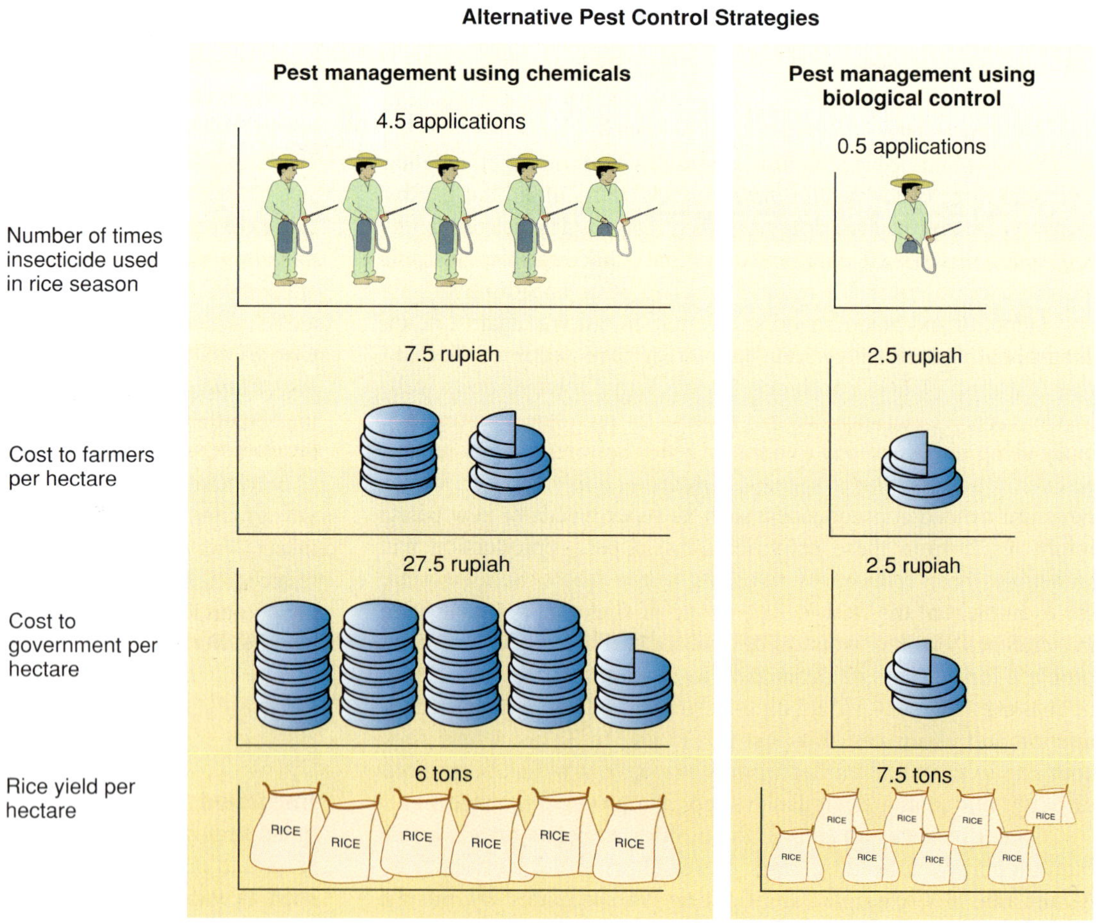


**FIGURE 10.19** Indonesia has one of the world's most successful integrated pest management (IPM) programs. Switching from toxic chemicals to natural pest predators has saved money while also increasing rice production. **Source:** Tolba, et al., *World Environment,* 1972–1992, p. 307, Chapman & Hall, 1992 United Nations Environment Programme.

# What Do You Think?

## Organic Farming in Cuba

The biggest experiment in low-input, sustainable agriculture in world history is occurring now in Cuba. The sudden collapse of the socialist bloc, upon which Cuba had been highly dependent for trade and aid, has forced an abrupt and difficult conversion from conventional agriculture to organic farming on a nationwide scale.

Between the Cuban revolution in 1959 and the breakdown of trading relations with the Soviet Union in 1989, Cuba experienced rapid modernization, a high degree of social equity and welfare, and a strong dependence on external aid. Cuba's economy was supported during this period by the most modern agricultural system in Latin America. Farming techniques, levels of mechanization, and output often rivaled those in the United States. The main crop was sugarcane, almost all of which was grown on huge state farms and sold to the former Soviet Union at premium prices. More than half of all food eaten by Cubans came from abroad, as did most fertilizers, pesticides, fuel, and other farm inputs on which agricultural production depended.

This organic, communal garden in Havana, Cuba, provides most of the fresh produce for the housing complex surrounding it.

Under the theory of comparative advantage, it seemed reasonable for Cuba to rely on international trade. With the collapse of the socialist bloc, however, Cuba's economy also fell apart. In 1990, wheat and grain imports decreased by half and other foodstuffs declined even more. At the same time, fertilizer, pesticide, and petroleum imports were down 60 to 80 percent. Farmers faced a dual challenge: how to produce twice as much food using half the normal inputs.

The crisis prompted a sudden turn to a new model of agriculture. Cuba was forced to adopt sustainable, organic farming practices based on indigenous, renewable resources. Typically, it takes three to five years for a farmer in the United States to make the change from conventional to organic farming profitable. Cuba, however, didn't have that long; it needed food immediately.

Cuba's agricultural system is based on a combination of old and new ideas. Broad community participation and use of local knowledge is essential. Scientific, adaptive management is another key. Diverse crops suitable to local microclimates, soil types, and human nutritional needs have been adopted. Natural, renewable energy sources such as wind, solar, and biomass fuels are being substituted for fossil fuels. Oxen and mules have replaced some 500,000 tractors idled by lack of fuel.

Soil management is vital for sustainable agriculture. Organic fertilizers substitute for synthetic chemicals. Livestock manure, green manure crops, composted municipal garbage, and industrial-scale cultivation of high-quality humus in earthworm farms all replenish soil fertility. In 1995 more than 100,000 metric tons of worm compost were produced and spread on fields.

Pests are suppressed by crop rotation and biological controls rather than chemical pesticides. For example, the parasitic fly (*Lixophaga diatraeae*) controls sugarcane borers; wasps in the genus *Trichogramma* feed on the eggs of grain weevils; while the predatory ant (*Pheidole megacephala*) attacks sweet potato weevils. Pest control also involves innovative use of biopesticides, such as *Bacillis thuringiensis,* that are poisonous or repellent to crop pests. Finally, integrated pest management includes careful monitoring of crops and measures to build populations of native beneficial organisms and to enhance the vigor and defenses of crop species.

Worker brigades from schools and factories help provide farm labor during harvest season. In addition to state farms and rural communes, urban gardening provides a much-needed supplement to city diets. Individual gardens are encouraged, but community or institutional gardens—schools, factories, and mass organizations—also produce large amounts of food.

Although food supplies in Cuba still are limited and diets are austere, the crisis wasn't as bad as many feared. In some ways, this draconian transition is fortunate. Cuba is now a world leader in sustainable agriculture. What do you think; is Cuba's experiment with organic agriculture a model that could be exported to other countries? Does the centrally planned economy of Cuba create special conditions for this project? Which of the aspects of Cuban farming might work in your country?

While IPM can be a good alternative to chemical pesticides, it also presents environmental risks in the form of exotic organisms. Wildlife biologist George Boettner of the University of Massachusetts reported in 2000 that biological controls of gypsy moths, which attack fruit trees and ornamental plants, have also decimated populations of native North American moths. *Compsilura* flies, introduced in 1905 to control the gypsy moths, have a voracious appetite for other moth caterpillars as well. One of the largest North American moths, the Cecropia moth (*Hyalophora cecropia*), with a 15 cm wingspan, was once ubiquitous in the eastern United States, but it is now rare in regions where *Compsilura* flies were released.

## 10.5 Reducing Pesticide Exposure

The total numbers and amounts of different chemicals to which we are potentially exposed is overwhelming. The 2.4 million metric tons of pesticides used in the United States every year contain about 600 active ingredients combined with more than 1,200 presumably inactive carriers, solvents, preservatives, and other ingredients in about 25,000 commercial products. Less than 10 percent of the active pesticide ingredients have been subjected to a full battery of ten chronic health-effect tests, and studies of the inactive ingredients have started only recently (fig. 10.20). Of the 321 pesticides screened so far, the EPA reports that 146 are probable human carcinogens. Since 1972, only 40 pesticides have been banned. Critics charge that this slow pace puts all of us at risk and that assessment and regulation must proceed more quickly.

### Who regulates pesticides?

Three federal agencies share responsibility for regulating pesticides used in food production in the United States: the Environmental Protection Agency (EPA), the Food and Drug Administration (FDA), and the Department of Agriculture (USDA). The EPA regulates the sale and use of pesticides under the Federal Insecticide, Fungicide, and Rodenticide Act (FIFRA), which mandates the "registration" (licensing) of all pesticide products. Based on scientific studies, the EPA determines which pesticides will not pose significant risks to human health or the environment when used according to label directions. Under the Federal Food, Drug, and Cosmetic Act (FFDCA), the EPA also sets "tolerance levels" or limits for the amount of pesticide residues that lawfully may remain in or on foods marketed in the United States, whether grown domestically or imported from abroad. The FDA and USDA enforce pesticide use and tolerance levels set by the EPA. These agencies can seize and destroy food shipments found to contain pesticide residues in violation of limits set by the EPA.

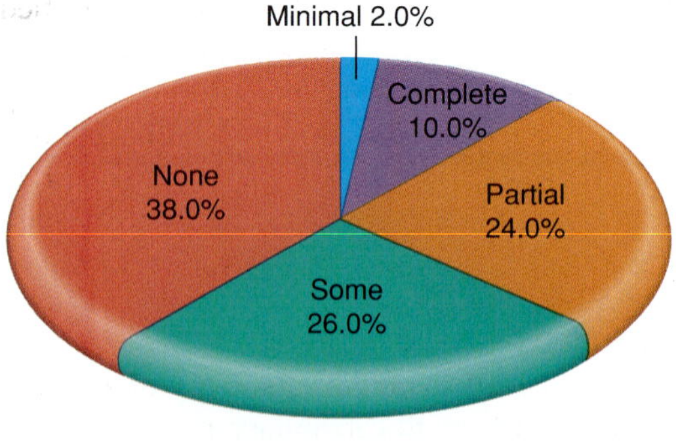


**FIGURE 10.20** Level of testing for health effects of 3,350 active and inert pesticide ingredients. Complete testing includes a battery of ten tests for both chronic and acute toxicity and genetic effects.
**Source:** National Academy of Sciences.

### *Think About It*

Pesticide residues in food are a major concern for many people, but most of us also use toxic chemicals in other aspects of our life as well. Look around your home, how many different toxic products can you find? Are they all necessary? Would you have alternatives to these products?

In Canada, the Pest Control Products Act and Regulations, administered by Agriculture Canada, performs similar functions. Individual Canadian provinces also set local standards. The Ontario Ministry of Agriculture and Food, for instance, introduced a comprehensive program in 1988 to reduce pesticide use by 50 percent. Nonchemical pest control methods, on-farm education, and biotechnology and the development of pest-resistant crop varieties all played a role in this program. Savings on chemical costs are about (C)$100 million per year.

In 1996, as a result of growing evidence of pesticide risks to children, Congress passed the Food Quality Protection Act. Key provisions of this act direct the EPA to set pesticide tolerance levels based on potential aggregate pesticide exposures. Benefits of pesticides can be considered when evaluating tolerances for carcinogens in adults, but not in calculating risks on reproduction and prenatal development or on exposure levels for infants and children. This act requires that by 2006 the EPA reassess the risks of over 9,700 pesticides, reevaluate the allowable levels of pesticides that can remain in or on food, and apply a safety factor of 10 where comprehensive and complete information is not available regarding the cumulative impacts on children.

For the first time, under this act, the EPA is required to examine exemptions given to "inert" pesticide ingredients used to dilute or carry the active agents. Of the 2,500 substances added to pesticides but not usually named on product labels, more than 650 have been identified as hazardous by federal, state, or international agencies. Nearly half of these "inert" ingredients have been classified as carcinogens, occupation hazards, or air and water pollutants. Naphthalene, for instance, which is commonly described as an inert ingredient, is designated a hazardous air pollutant under the Clean Air Act and a priority pollutant under the Clean Water Act.

Based on these new rules, the EPA recently banned the use of methyl parathion on all fruits and many vegetables. It also limited the quantity of azinphos methyl that can be used on foods common in children's diets, such as apples, peaches, and pears. Similarly, the EPA has prohibited most uses of chlorpyrifos, commonly known by its trade name Dursban. Found in more than 1,000 commercial products, this organophosphate represented 20 percent of all residential insecticide applications. It was shown, however, to cause brain damage in developing rat pups, and was judged an unacceptable risk to children. Some 36 other organophosphates are still being reviewed. The EPA is also studying hormone-disrupting chemicals and will probably regulate products on this basis in the near future.

Wood preservatives are another example of pesticide use especially dangerous to children. Thirty years ago, the principal wood preservatives were petroleum-based compounds like tar and creosote. These highly toxic biocides prevented growth of molds, invasion by termites, and just about all other living organisms, but also were dangerous to humans. Many Superfund sites are contaminated with these wood preservatives. As understanding of the lingering toxicity of creosote and its chemical cousins emerged, consumers switched to CCA (chromated copper arsenate) pressure-treated lumber. We now know that the arsenic in CCA-treated wood is a serious hazard, especially to workers exposed to sawdust or children who come in contact with the lumber in play equipment, decks, sidewalks, and other outdoor settings (see What Do You Think? p. 221). CCA-treated wood was banned for virtually all residential uses in the United States in 2003, but are the substitutes safer?

With globalization, our food comes increasingly from other countries where rules governing food safety may be less strict than we would like. The U.S. General Accounting Office reports that the volume of imported food has doubled over the last five years, while food inspections have decreased during that same time. Currently, about 38 percent of the fruits and 12 percent of the vegetables consumed by Americans are imported. The Food and Drug Administration (FDA) inspects only about 2 percent of all food shipments. Less than 0.2 percent of imported fruits and vegetables are examined for microbial contamination, and even a smaller fraction is tested for pesticides. In studies of a wide range of foods collected by the USDA, the State of California, and the Consumers Union between 1994 and 2000, 73 percent of conventionally grown food had residue from at least one pesticide and were six times as likely as organic foods to contain multiple pesticide residues. Only 23 percent of the organic samples of the same groups had any residues. Using these data, the Environmental Working Group has assembled a list of the fruits and vegetables most commonly contaminated with pesticides (table 10.1).

**TABLE 10.1**

**The Twelve Most Contaminated Foods**

| Rank | Food |
|---|---|
| 1. | Strawberries |
| 2. | Bell peppers |
| 3. | Spinach |
| 4. | Cherries (U.S.) |
| 5. | Peaches |
| 6. | Cantaloupe (Mexican) |
| 7. | Celery |
| 8. | Apples |
| 9. | Apricots |
| 10. | Green beans |
| 11. | Grapes (Chilean) |
| 12. | Cucumbers |

**Source:** Environmental Working Group, 2002.

In 2001, the Canadian Supreme Court ruled that pesticides can be excluded from residential areas. In response, the province of Quebec announced that it would immediately ban the use of 30 "cosmetic" pesticides on public lands including parks, schools, daycare centers and hospitals. By 2005, the province will prohibit the use of most nonfarm pesticides. "People's health is more important than a perfect lawn," said Quebec Environment Minister Andre Boisclair. "I enjoin Quebecers to no longer use pesticides."

## Is organic the answer?

Many farmers and consumers are turning to organic agriculture as a way to reduce pesticide exposures. Numerous studies have shown that organic, sustainable agriculture is more eco-friendly and leaves soils healthier than intensive, chemical-based monoculture cropping. A Swiss study spanning two decades found that average yields on organic plots were 20 percent less than adjacent fields farmed by conventional methods. Costs also were lower and prices paid for organic produce were higher, so that net returns were actually higher with organic crops. Energy use, for example, was 56 percent less per unit of yield in organic farming than conventional approaches. In addition, root fungi were 40 percent higher, earthworms were three times as abundant, and spiders and other pest-eating predators were doubled in the organic plots. Both the organic farmers and their families reported better health and greater satisfaction than their conventional neighbors. A study of food quality in Sweden reported that organic food contained more cancer-fighting polyphenolics and antioxidants than pesticide-treated produce. As mentioned in chapter 9, farms using sustainable techniques can have up to 400 times less erosion after heavy rains than monoculture row crops.

Currently, less than 1 percent of all American farmland is devoted to organic growing, but the market for organic products may stimulate more conversion to this approach in the future. Organic food is much more popular in Europe than in North America. Tiny Liechtenstein is probably the leader among industrialized nations with 18 percent of its land in certified organic agriculture. Sweden is second with 11 percent of its land in organic production.

Considerable controversy and confusion have existed in the United States over what constitutes organic agriculture. How do you know if you can believe claims that food has been grown safely (fig. 10.21)? Walmart, for example, has vowed to become the top seller of organic products in the United States. Will this mean that much of our organic food will come from overseas? Already, more than 2,000 farms in China and India are certified "organic," but what does that really mean?

With the market for organic food generating $11 billion per year, some farmers and marketers are likely to try to pass off foods grown inexpensively with pesticides as more valuable organic produce.

**FIGURE 10.21** These strawberries were grown organically, but how could you tell? The USDA reports more pesticides in commercial strawberries than any other fruit.

**FIGURE 10.22** Your local farmers' market is a good source of locally grown, organic produce.

According to U.S. Department of Agriculture rules, products labeled "100 percent organic" must be produced without hormones, antibiotics, pesticides, synthetic fertilizers, or genetic modification. "Organic" means that at least 95 percent of the ingredients must be organic. "Made with organic ingredients" must contain at least 70 percent organic contents. Products containing less than 70 percent organic ingredients can list them individually. Organic animals must be raised on organic feed, given access to the outdoors, given no steroidal growth hormones, and treated with antibiotics only to treat diseases.

Many who endorse the concept of organic food are disappointed by the limited scope of these definitions. They point out that there are three goals for organic, sustainable agriculture. One is growing food in harmony with nature—a nonindustrial approach to food production that treats animals humanely and avoids the use of chemical pesticides. Another is a food distribution system based on co-ops, farmers' markets, community supported agriculture, and local production rather than crops flown and trucked from far-away places where consumers have no connection with producers (fig. 10.22). The third goal is simple, wholesome, nutritious food. It's true that the USDA rules eliminate many of the pesticides, but a myriad of highly processed ingredients are allowed in organic food. You might find yourself someday buying organic Coca-Cola to go with your organic TV dinner. You can already buy organic, intercontinental grapes in which thousands of calories of jet and diesel fuel were consumed to transport every calorie of food energy from Chile to your supermarket. Some of the best farmers are refusing to apply for organic certification, regarding it as a meaningless term under these rules.

Other people doubt that organic growers can produce enough—even within these limited definitions—to feed everyone. We need the efficiency of chemical pesticides and largescale, chemical-intensive farming to provide food for the 8 to 9 billion people expected by the middle of this century. According to Vaclav Smil, a Canadian geographer, without synthetic fertilizer and pesticides we could feed only 2 to 3 billion people. Dennis Avery, director of global food issues for the conservative Hudson Institute calculates that if we were to depend entirely on animal manure to fertilize crops, America would need to increase its cattleherd ninefold and convert almost the entire U.S. landmass to pastureland to support all those animals.

Furthermore, eating organic food doesn't protect you from all risks. Dr. Elsa Murano, undersecretary for food safety in the U.S. Department of Agriculture warns that consumers should be wary of organic food. "We must remember that bacteria and parasites are also all-natural," she said. "As a microbiologist, I know that preservatives are used in foods for a reason . . . to preserve food against the growth of microorganisms." Supporting her point, a Danish research group found more campylobacteria—one of the most common causes of diarrhea—in organic chickens than in standard chickens. Some natural pesticides allowed in organic farming have been linked to health problems in both producers or consumers. Rotenone, for example, has been associated with Parkinson's disease in farmers, while *Bacillus thuringiensis* (Bt) toxin has been found to cause allergies and asthma in crop pickers and handlers.

Biochemist Bruce Ames, who is famous for developing the most widely used cancer screening test, argues that pesticides confer a public-health advantage that far outweighs any risks posed by their toxicity or hormone-disrupting effects. "If the EPA eliminates all pesticides," he says, "all it's going to do is increase the price of fruits and vegetables, and there will be more cancer." It's not that he argues that pesticides are safe, but rather that we're ignoring the real risks we face like smoking, poor nutrition, lack of health care, and accidents. We could save more lives, he claims, by focusing on the factors that kill the vast majority of us than by spending so much time and money worrying about relatively minor hazards.

## You can reduce your own risks

Even though pesticide exposure may not be the greatest threat that most of us face, why not take reasonable steps to reduce your risks? There are many things each of us can do to minimize pesticides in our food and to encourage sustainable agriculture.

- Wash and scrub all fresh fruits and vegetables thoroughly under running water.
- Peel fruits and vegetables when possible. Throw away the outer leaves of leafy vegetables such as lettuce and cabbage.
- Store food carefully so it doesn't get moldy or pick up contaminants from other foods. Use food as soon as possible to ensure freshness.
- Cook or bake foods that you suspect have been treated with pesticides to break down chemical residues.
- Trim the fat from meat, chicken, and fish. Eat lower on the food chain where possible to reduce bioaccumulated chemicals.
- Don't pick and eat berries or other wild foods that grow on the edges of roadsides where pesticides may have been sprayed.
- Grow your own fruits and vegetables without using pesticides or with minimal use of dangerous chemicals.
- Ask for organically grown food at your local grocery store or shop at a farmers' market or co-op where you can get such food.

# Conclusion

Pesticides bring us many benefits, but they have costs as well. Indiscriminate and profligate use of a variety of powerful chemicals kills beneficial organisms, such as honeybees, and often does as much harm as good. Often highly persistent and mobile in the environment, many pesticides move through air, water, and soil to bioaccumulate in food chains and cause serious ecological and human health problems.

Several good alternatives can reduce our dependence on the most dangerous pesticides. Among these are behavioral changes, such as crop rotation, use of biological controls, mechanical cultivation, and planting mixed polycultures rather than vast monocultures. Integrated pest management is a flexible, ecologically based strategy that carefully monitors pest populations to apply just the right kind and amount of pesticide at just the right time to do the most good. It can result in dramatic reductions in pesticide use and unintended consequences while minimizing economic losses. Consumers may have to learn to accept less than perfect fruits and vegetables if they want to reduce their pesticide exposure. In the end, both food producers and consumers will need to be wiser, more perceptive, and, perhaps, more cautious about how we try to defend ourselves against pests.

# Reviewing Learning Outcomes

By now you should be able to explain the following points:

**10.1** Define the major types of pesticides and describe the pest they are meant to manage.

- People have always known of ways to control pests.
- Modern pesticides provide benefits, but also create problems.
- There are many types of pesticides.

**10.2** Identify some important pesticide benefits.

- We have made progress in controlling many insect-borne diseases.
- Without pesticides, we might lose two-thirds of conventional crops.

**10.3** Illustrate some important pesticide problems.

- Pesticides often poison nontarget species.
- Pesticide resistance is often rapid and widespread.
- Pesticide misuse can create new pests.
- Some persistent pesticides can move long distances in the environment.
- Many pesticides cause human health problems.

**10.4** Summarize some alternatives to current pesticide uses.

- We can change our behavior.
- Useful organisms can help us control pests.
- Integrated pest management uses a combination of techniques to fight pests.

**10.5** Discuss how we might reduce pesticide exposure.

- Who regulates pesticides?
- Is organic the answer?
- You can reduce your own risks.

## Practice Quiz

1. What is a pest and what are pesticides? What is the difference between a biocide, a herbicide, an insecticide, and a fungicide?
2. What is DDT and why was it considered a "magic bullet"? Why was it listed among the "dirty dozen" persistent organic pollutants (POPs)?
3. How much pesticide is used worldwide and which of the general categories accounts for the greatest use? Why are "inert" ingredients of concern?
4. Describe fumigants, botanicals, chlorinated hydrocarbons, organophosphates, carbamates, and microbial pesticides.
5. What are endocrine disrupters and why are they dangerous?
6. Explain why pests often resurge or rebound after treatment with pesticides and how they become pesticide resistant. What is a pesticide treadmill?
7. Identify three major categories of alternatives to synthetic pesticides and describe, briefly, how each one works.
8. What is IPM, and how is it used in pest control?
9. Why are children more susceptible to pesticides than adults are? What is being done to protect children?
10. List eight things you could do to reduce your dietary exposure to pesticides.

## Critical Thinking and Discussion Questions

1. Paul Müller, the discoverer of DDT, received a Nobel Prize for his work. What do you think; was it a gift or curse for the world?
2. If you were a public health official in a country in which malaria, filariasis, or onchocerciasis were rampant, would you spray DDT to eradicate vector organisms? Would you spray it in your own house?
3. Pesticide treadmill, "dirty dozen" pesticides, and environmental estrogens are all highly emotional terms. Why would some people choose to use or not use these terms? Can you suggest alternative terms for the same phenomena that convey different values?
4. Many farmworkers who suffer from pesticide poisoning are migrants or minorities. Is this evidence of environmental racism? What evidence would you look for to determine whether environmental justice is being served?
5. Suppose that a developing country believes that it needs a pesticide banned in the United States or Canada to feed or protect the health of its people. Are we right to refuse to sell that pesticide?
6. Do you believe we could grow enough food for everyone if we switched to organic, low-input, sustainable agriculture? How would you design a research program to test this question?
7. What would you personally consider a "negligible" risk? Would you eat grapes if you knew they had a measurable amount of some pesticide? How small would the amount have to be?

## DATA analysis **Assessing Health Risks**

As we discussed in chapter 8, children are much more vulnerable than adults to environmental health hazards. Pound for pound, children drink more water, eat more food, and breathe more air than do adults. Putting fingers, toys, and other objects into their mouths increases children's exposure to toxins in dust or soil. And compared to adults, children generally have less-developed immune systems or processes to degrade or excrete toxins. Because children spend a significant portion of time at school, pesticides used in schools may be an important contributor to overall exposure. There is no federal mandate to collect data on pesticide use in schools, and few states require reporting of this information, so we know little about how pesticides are used in the nation's 110,000 schools.

In 1999 the state of Minnesota conducted a survey on pesticide use in schools. This study focused on indoor uses of pesticides, because pesticide residues can be persistent indoors, and because children spend most of their time indoors when they are at school. The graph below presents results from this study. It presents information about the frequency of spraying and where these pesticides were used. These measures are only a surrogate for exposure. They don't provide information about the toxicity of pesticides used or details about how they were applied and thus cannot provide a complete representation of the risk of adverse effects following exposure. Nevertheless, the frequency of exposure is valuable information about the extent of risk.

After studying the graph, answer the following questions:

1. Do the numbers for all categories add up to 100 percent?
2. Why does this matter?
3. Which areas are sprayed most "as needed"?
4. Why might this be important?
5. Which areas are sprayed most often overall?
6. How might you find an answer for question 5 other than adding up the first three groups of bars?

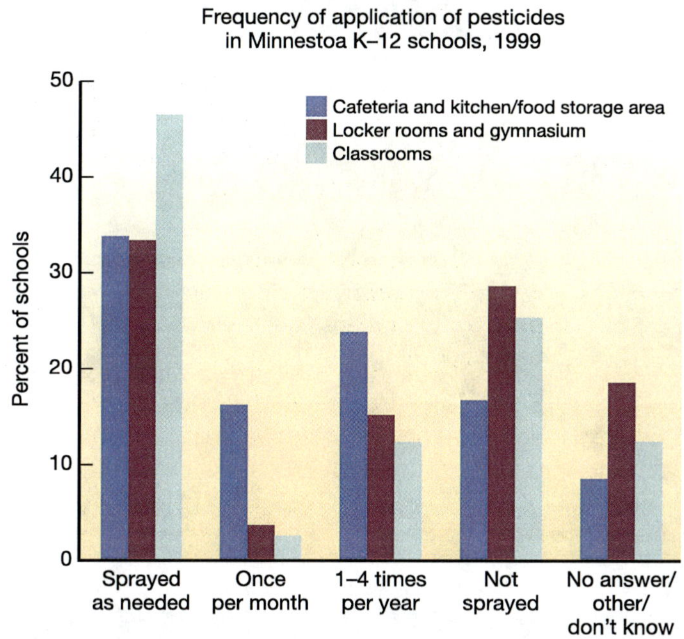


Pesticide use in Minnesota schools.
Data from U.S. EPA, 2003.

**For Additional Help in Studying This Chapter,** please visit our website at www.mhhe.com/cunningham10e. You will find additional practice quizzes and case studies, flashcards, regional examples, place markers for Google Earth™ mapping, and an extensive reading list, all of which will help you learn environmental science.

Now closed and undergoing reclamation, the Beartrack Mine was an open-pit, cyanide heap-leach gold mine.

CHAPTER 14

# Geology and Earth Resources

*Learn geology or die.*

*—Louis Agassiz—*

## LEARNING OUTCOMES

After studying this chapter, you should be able to:

**14.1** Summarize the processes that shape the earth and its resources.

**14.2** Explain how rocks and minerals are formed.

**14.3** Think critically about economic geology and mineralogy.

**14.4** Critique the environmental effects of resource extraction.

**14.5** Discuss ways we could conserve geological resources.

**14.6** Describe geological hazards.

# Case Study **Leaching gold**

The Beartrack Mine is an open-pit, cyanide heap-leach gold mine (see photo opposite page) located within the Salmon National Forest of Idaho (fig. 14.1). The mine is owned and operated by the Meridian Gold Company. Mine construction began in 1994, and the first gold was produced in 1995. At its peak, Beartrack was the single largest gold mine in Idaho's history. In five years of operation, it produced 550,000 ounces of gold worth about $220 million at prevailing prices.

The ore in this deposit contained only about 1 g of gold per ton of rock. This wouldn't have been economical to extract using conventional underground mining and smelting, but with surface mining techniques and the cyanide leaching technology that became popular in the 1970s, gold was extracted from the Beartrack at a cost of about $190 per ounce. The mine consisted of two large open pits (upper right in photo), a 40-million-ton waste rock disposal site in a nearby valley, and heap-leaching facilities (center of photo). During operations, ore excavated from the pits was crushed and piled in 20 ft thick layers on the heap-leach pad, (large tan square), which is underlain by a thick plastic liner. The heap covers about 40 ha to a maximum depth of about 30 m, and contains about 15 million metric tons of ore.

Gold was separated from the ore by spraying a sodium cyanide solution over the top of the heap. This solution bonds with the gold in the ore, percolates through the heap, and drains through a ditch (lower right) into catch basins (large blue squares). The gold-bearing cyanide solution is pumped from the catch basins to a processing plant (white buildings) where the gold is absorbed by activated carbon in large tanks. When the carbon is saturated, the gold is removed by zinc precipitation. The "used" sodium cyanide is sprayed again on the pile (see fig. 14.14).

The mine closed in 2000 when falling gold prices made production uneconomical. Restoration began shortly thereafter and continues today. The reclamation goal is to restore the site to allow essentially the same land uses as existed prior to mining. Ultimately, this will require removal of buildings, impoundments, leach pads, and other mining structures, and the contouring, covering, and revegetation of waste rock piles and other disturbed areas. The photo on the opposite page was taken six years after reclamation began, and shows few changes from when mining ceased. One reason that reclamation is proceeding slowly is that about 4,000 ounces of gold per year continue to be rinsed out of the heap-leach pile.

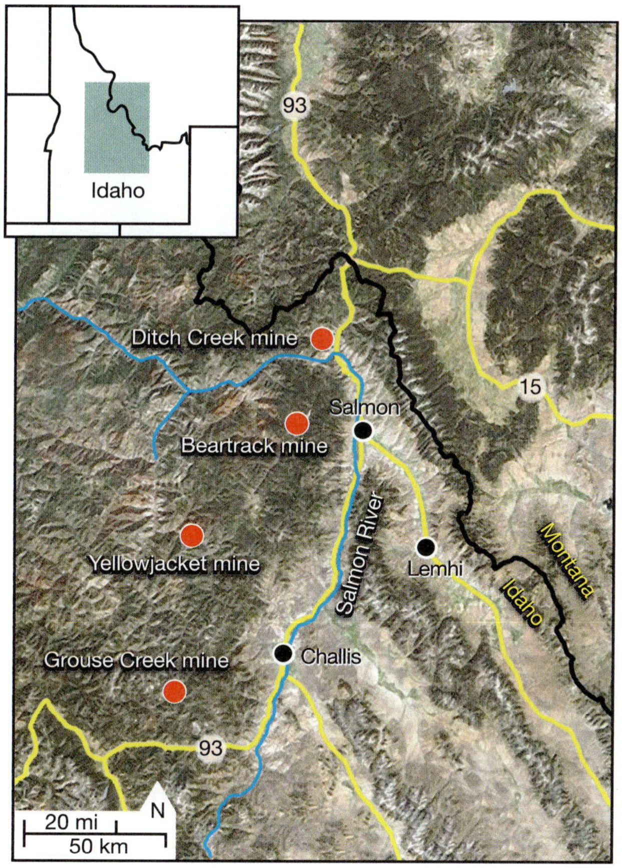


**FIGURE 14.1** The Beartrack Mine is near the historic mining town of Leesburg in Lemhi County of east-central Idaho. The wild and scenic Salmon River runs nearby.

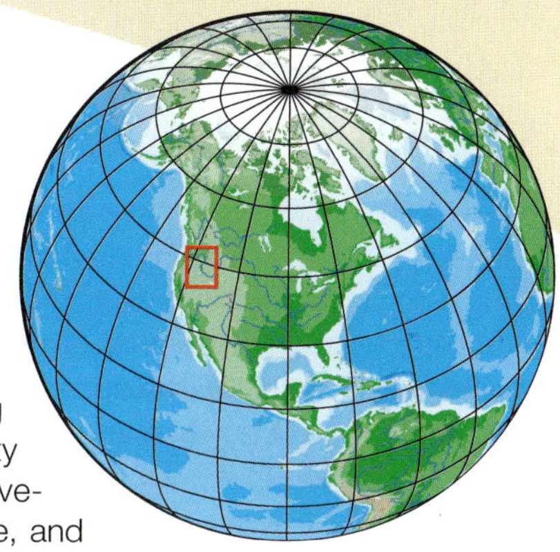

Nevertheless, in 2006, the Bureau of Land Management awarded Meridian Gold the Hardrock Mineral Environmental Award for being an "outstanding example of a mining project with strong community relations, recognizable achievements in reclamation of the site, and use of cutting-edge technology to bring an environmentally sound closure to a mining project." At the same time the Environmental Protection Agency fined Meridian $30,000 for violating its wastewater discharge permit at Beartrack.

Some Idaho residents are understandably concerned about what will happen to the Beartrack Mine, because wastewater from the facility is discharged into Napias Creek, which empties into the Salmon River. Known as the "River of No Return" from the adventures of Lewis and Clark, the Salmon is one of the premier trout streams—and one of the initial seven Wild and Scenic Rivers—of the United States. Every year, thousands of anglers, backpackers, and white water rafters visit the area and make up an important part of Idaho's $2 billion annual tourism industry.

Cyanide heap-leach operations in other parts of the world have had poor environmental records. At about the same time that Beartrack closed, a cyanide leak at a similar facility not far away spilled about 20,000 gallons of cyanide into the Yankee Fork of the Salmon River. The mine owner charged that vandals must have cut the plastic liner. Others believe that ice tore the plastic. Some engineers believe all heap-leach liners eventually tear and leak. Despite industry attempts to block it, a citizen's initiative to phase out open-pit, cyanide-leach mining passed in 1998 in Montana and is now state law.

This case study illustrates some of the dilemmas we face in using geological resources. We need materials from the earth to sustain our modern lifestyle, but often the methods we use to get those materials have severe environmental consequences. How can we supply the resources we need without harming our environment? In this chapter, we'll look at the processes that shape the earth and how rocks and minerals are formed, as well as how we obtain minerals. We'll also look at geological hazards that threaten us and what we might do to reduce our risks.

## 14.1 Earth Processes Shape Our Resources

Every one of us shares the benefits of geological resources. Right now you are probably wearing several geologic products; plastics, including glasses and synthetic fabric, are made from oil; iron, copper, and aluminum mines produced your snaps, zippers, and the screws in your glasses; silver, gold, and diamond mines may have produced your jewelry. All of us also share responsibility for the environmental and social devastation that often results from mining and drilling. Fortunately, there are many promising solutions to reduce these costs, including recycling and alternative materials. The question is whether voters will demand that we use these technologies—and whether consumers will share the costs of responsible production.

Why are these resources distributed as they are? To understand how and where earth resources are created, we must examine the earth's structure and the processes that shape it.

### Earth is a dynamic planet

Although we think of the ground under our feet as solid and stable, the earth is a dynamic and constantly changing structure. Titanic forces inside the earth cause continents to split, move apart, and then crash into each other in slow but inexorable collisions.

The earth is a layered sphere. The **core,** or interior, is composed of a dense, intensely hot mass of metal—mostly iron—thousands of kilometers in diameter (fig. 14.2). Solid in the center but more fluid in the outer core, this immense mass generates the magnetic field that envelops the earth.

Surrounding the molten outer core is a hot, pliable layer of rock called the **mantle.** The mantle is much less dense than the core because it contains a high concentration of lighter elements, such as oxygen, silicon, and magnesium.

The outermost layer of the earth is the cool, lightweight, brittle rock **crust.** The crust below oceans is relatively thin (8–15 km), dense, and young (less than 200 million years old) because of constant recycling. Crust under continents is relatively thick (25–75 km), light, and as old as 3.8 billion years, with new material being added continually. It also is predominantly granitic, while oceanic crust is mainly dense basaltic rock. Table 14.1 compares the composition of the whole earth (dominated by the dense core) and the crust.

### Tectonic processes reshape continents and cause earthquakes

The huge convection currents in the mantle are thought to break the overlying crust into a mosaic of huge blocks called **tectonic plates** (fig. 14.3). These plates slide slowly across the earth's surface like wind-driven ice sheets on water, in some places breaking up into smaller pieces, in other places crashing ponderously into each other to create new, larger landmasses. Ocean basins form where continents crack and pull apart. The Atlantic Ocean, for example, is growing slowly as Europe and Africa move away from the Americas. **Magma** (molten rock) forced

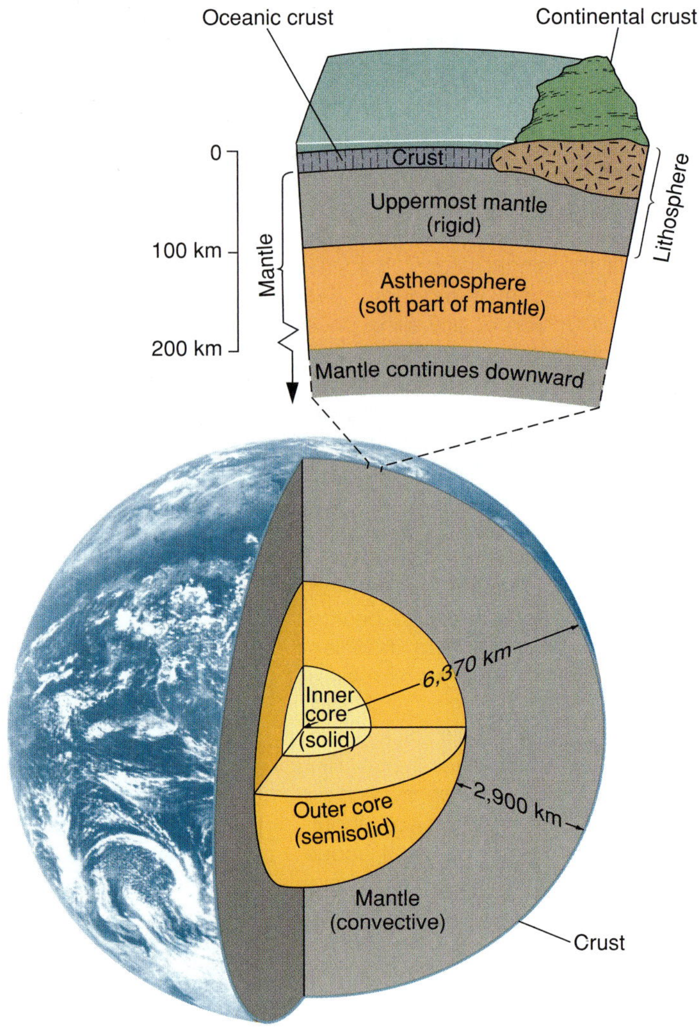


**FIGURE 14.2** Earth's cross-section. Slow convection in the mantle causes the thin, brittle crust to move.

**TABLE 14.1**

**Eight Most Common Chemical Elements (Percent)**

| Whole Earth | | Crust | |
|---|---|---|---|
| Iron | 33.3 | Oxygen | 45.2 |
| Oxygen | 29.8 | Silicon | 27.2 |
| Silicon | 15.6 | Aluminum | 8.2 |
| Magnesium | 13.9 | Iron | 5.8 |
| Nickel | 2.0 | Calcium | 5.1 |
| Calcium | 1.8 | Magnesium | 2.8 |
| Aluminum | 1.5 | Sodium | 2.3 |
| Sodium | 0.2 | Potassium | 1.7 |

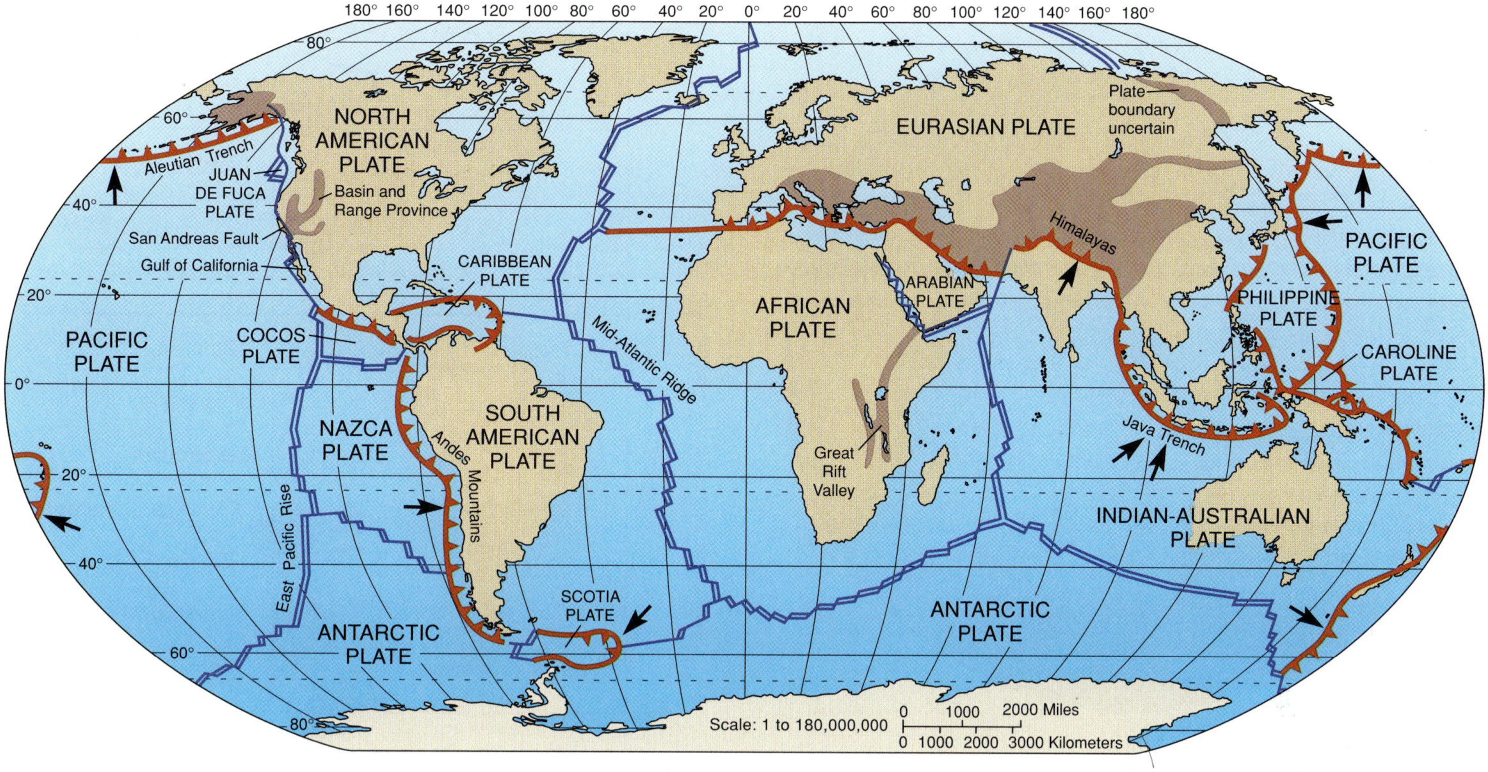


**FIGURE 14.3** Map of tectonic plates. Plate boundaries are dynamic zones, characterized by earthquakes and volcanism and the formation of great rifts and mountain ranges. Arrows indicate direction of subduction where one plate is diving beneath another. These zones are sites of deep trenches in the ocean floor and high levels of seismic and volcanic activity.

up through the cracks forms new oceanic crust that piles up underwater in **mid-ocean ridges.** Creating the largest mountain range in the world, these ridges wind around the earth for 74,000 km (46,000 mi) (see fig. 14.3). Although concealed from our view, this jagged range boasts higher peaks, deeper canyons, and sheerer cliffs than any continental mountains. Slowly spreading from these fracture zones, ocean plates push against continental plates.

Earthquakes are caused by grinding and jerking as plates slide past each other. Mountain ranges like those on the west coast of North America and in Japan are pushed up at the margins of colliding continental plates. The Himalayas are still rising as the Indian subcontinent grinds slowly into Asia. Southern California is slowly sailing north toward Alaska. In about 30 million years, Los Angeles will pass San Francisco, if both still exist by then.

When an oceanic plate collides with a continental landmass, the continental plate usually rides up over the seafloor, while the oceanic plate is **subducted,** or pushed down into the mantle, where it melts and rises back to the surface as magma (fig. 14.4). Deep ocean trenches mark these subduction zones, and volcanoes form where the magma erupts through vents and fissures in the

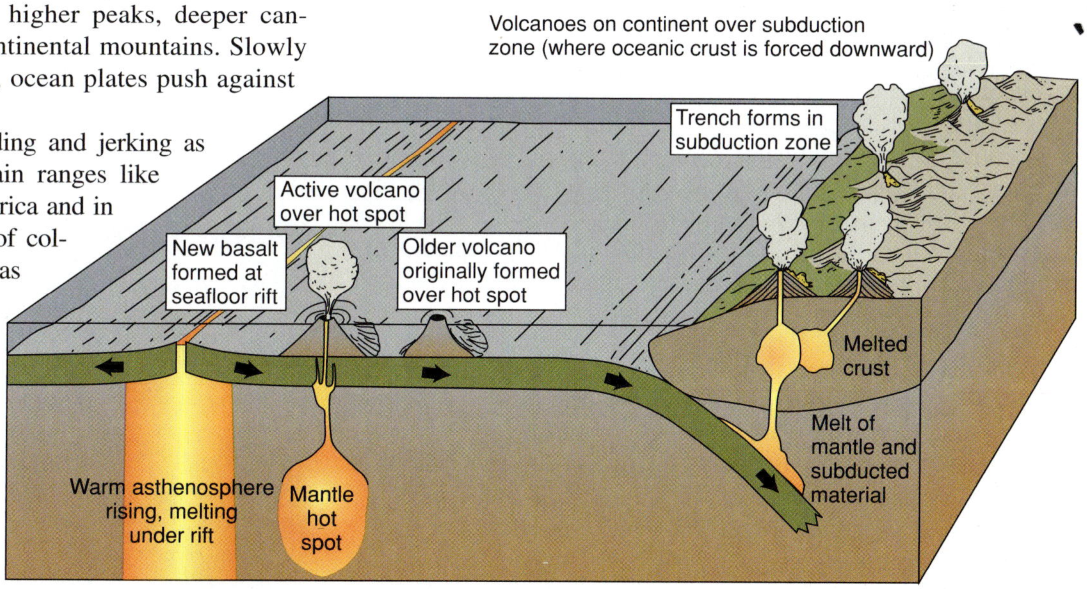


**FIGURE 14.4** Tectonic plate movement. Where thin, oceanic plates diverge, upwelling magma forms mid-ocean ridges. A chain of volcanoes, like the Hawaiian Islands, may form as plates pass over a "hot spot." Where plates converge, melting can cause volcanoes, such as the Cascades.

**FIGURE 14.5** Pangaea, an ancient supercontinent of 200 million years ago, combined all the world's continents in a single landmass. Continents have combined and separated repeatedly.

overlying crust. Trenches and volcanic mountains ring the Pacific Ocean rim from Indonesia to Japan to Alaska and down the west coast of the Americas, forming a so-called ring of fire where oceanic plates are being subducted under the continental plates. This ring is the source of more earthquakes and volcanic activity than any other region on the earth.

Over millions of years, continents can drift long distances. Antarctica and Australia once were connected to Africa, for instance, somewhere near the equator and supported luxuriant forests. Geologists suggest that several times in the earth's history most or all of the continents have gathered to form supercontinents, which have ruptured and re-formed over hundreds of millions of years (fig. 14.5). The redistribution of continents has profound effects on the earth's climate and may help explain the periodic mass extinctions of organisms marking the divisions between many major geologic periods (fig. 14.6).

## 14.2 Rocks and Minerals

A **mineral** is a naturally occurring, inorganic, solid element or compound with a definite chemical composition and a regular internal crystal structure. Naturally occurring means not created by humans (or synthetic). Organic materials, such as coal, produced by living organisms or biological processes are generally not minerals. The two fundamental characteristics of a mineral that distinguish it from all other minerals are its chemical composition and its crystal structure. No two minerals are identical in both respects. Once purified, metals such as iron, aluminum, or copper lack a crystal structure, and thus are not minerals. The ores from which they are extracted, however, are minerals and make up an important part of economic mineralogy.

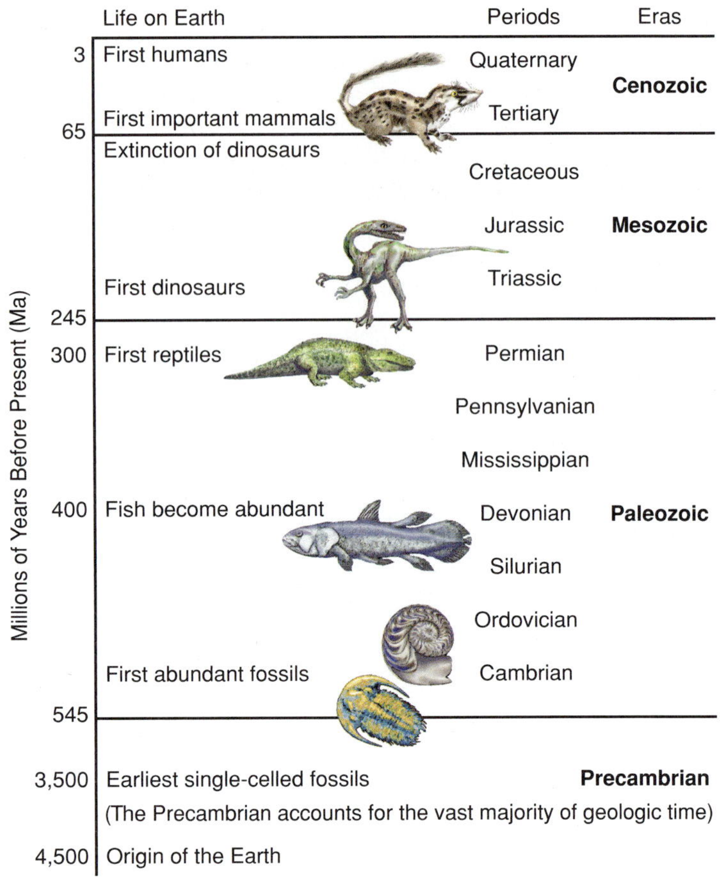


**FIGURE 14.6** Periods and eras in geological time, and major life-forms that mark some periods.

A **rock** is a solid, cohesive, aggregate of one or more minerals. Within the rock, individual mineral crystals (or grains) are mixed together and held firmly in a solid mass (fig. 14.7). The grains may be large or small, depending on how the rock was

**FIGURE 14.7** Crystals of different minerals create beautifully colored patterns in a rock sample seen in a polarizing microscope.

formed, but each grain retains its own unique mineral qualities. Each rock type has a characteristic mixture of minerals (and therefore of different chemical elements), grain sizes, and ways in which the grains are mixed and held together. Granite, for example, is a mixture of quartz, feldspar, and mica crystals. Different kinds of granite have distinct percentages of these minerals and particular grain sizes depending on how quickly the rock solidified. These minerals, in turn, are made up of a few elements such as silicon, oxygen, potassium, and aluminum.

## The rock cycle creates and recycles rocks

What could be harder and more permanent than rocks? Like the continents they create, rocks are also part of a relentless cycle of formation and destruction. They are made and then torn apart, cemented together by chemical and physical forces, crushed, folded, melted, and recrystallized by dynamic processes related to those that shape the large-scale features of the crust. We call this cycle of creation, destruction, and metamorphosis the **rock cycle** (fig. 14.8). Understanding something of how this cycle works helps explain the origin and characteristics of different types of rocks, as well as how they are shaped, worn away, transported, deposited, and altered by geological forces.

There are three major rock classifications: igneous, sedimentary, and metamorphic. In this section, we will look at how they are made and some of their properties.

### *Igneous Rocks*

The most common rock-type in the earth's crust is solidified from magma, welling up from the earth's interior. These rocks are classed as **igneous rocks** (from *igni,* the Latin word for fire). Magma

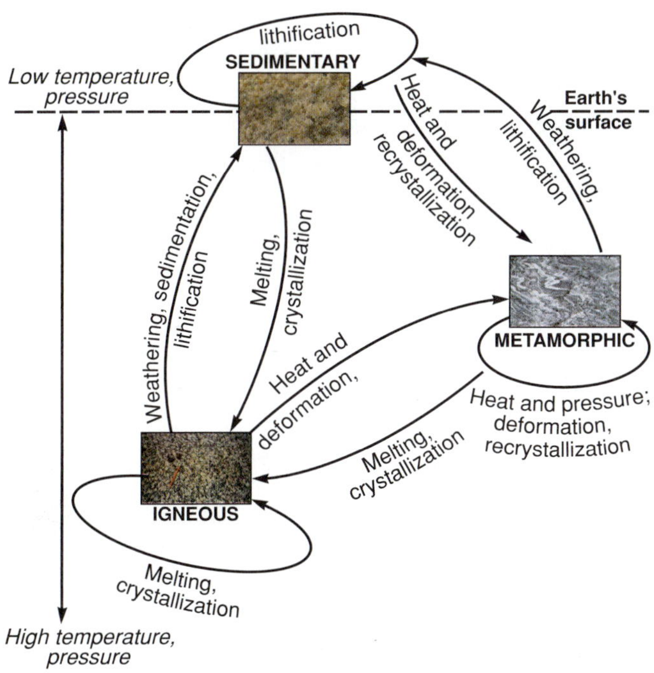


**FIGURE 14.8** The rock cycle includes a variety of geological processes that can transform any rock.

**FIGURE 14.9** Weathering slowly reduces an igneous rock to loose sediment. Here, exposure to moisture expands minerals in the rock, and frost may also force the rock apart.

extruded to the surface from volcanic vents cools quickly to make basalt, rhyolite, andesite, and other fine-grained rocks. Magma that cools slowly in subsurface chambers or is intruded between overlying layers makes granite, gabbro, or other coarse-grained crystalline rocks, depending on its specific chemical composition.

### *Metamorphic Rocks*

Preexisting rocks can be modified by heat, pressure, and chemical agents to create new forms called **metamorphic rock.** Deeply buried strata of igneous, sedimentary, and metamorphic rocks are subjected to great heat and pressure by deposition of overlying sediments or while they are being squeezed and folded by tectonic processes. Chemical reactions can alter both the composition and structure of the rocks as they are metamorphosed. Some common metamorphic rocks are marble (from limestone), quartzite (from sandstone), and slate (from mudstone and shale). Metamorphic rocks are often the host rock for economically important minerals such as talc, graphite, and gemstones.

## Weathering and sedimentation wear down rocks

Most of these crystalline rocks are extremely hard and durable, but exposure to air, water, changing temperatures, and reactive chemical agents slowly breaks them down in a process called **weathering** (fig. 14.9). *Mechanical weathering* is the physical breakup of rocks into smaller particles without a change in chemical composition of the constituent minerals. You have probably seen mountain valleys scraped by glaciers or river and shoreline pebbles that are rounded from being rubbed against one another as they are tumbled by waves and currents. *Chemical weathering* is the selective removal or alteration of specific components that leads to weakening and disintegration of rock. Among the more important chemical weathering processes are oxidation (combination of oxygen with an element to form an oxide or hydroxide mineral) and hydrolysis (hydrogen atoms from water molecules combine with other chemicals to form acids). The products of these reactions are more susceptible to both mechanical weathering and to dissolving in water. For instance, when carbonic acid

**FIGURE 14.10** Different colors of soft sedimentary rocks deposited in ancient seas during the Tertiary period 63 to 40 million years ago have been carved by erosion into the fluted spires and hoodoos of the Pink Cliffs of Bryce Canyon National Park.

(formed when $CO_2$ and $H_2O$ combine) percolates through porous limestone layers in the ground, it dissolves the calcium carbonate (limestone) and creates caves.

Particles of rock are transported by wind, water, ice, and gravity until they come to rest again in a new location. The deposition of these materials is called **sedimentation.** Waterborne particles from sediments cover ocean continental shelves and fill valleys and plains. Most of the American Midwest, for instance, is covered with a layer of sedimentary material hundreds of meters thick in the form of glacierborne till (rock debris deposited by glacial ice), windborne loess (fine dust deposits), riverborne sand and gravel, and ocean deposits of sand, silt, and clay. Deposited material that remains in place long enough, or is covered with enough material to compact it, may once again become rock. Some examples of **sedimentary rock** are shale (compacted mud), sandstone (cemented sand), tuff (volcanic ash), and conglomerates (aggregates of gravel, sand, silt, and clay).

Sedimentary rocks are also formed from crystals that precipitate out of, or grow from a solution. An example is rock salt, made of the mineral halite, which is the name for ordinary table salt (sodium chloride). Salt deposits often form when a body of salt water dries up and salt crystals are left behind.

Sedimentary formations often have distinctive layers that show different conditions when they were laid down. Sedimentary rocks such as sandstone and limestone can be shaped by erosion into striking features (fig. 14.10). Geomorphology is the study of the processes that shape the earth's surface and the structures they create.

Humans have become a major force in shaping landscapes. Geomorphologist Roger Hooke, of the University of Maine, looking only at housing excavations, road building, and mineral production, estimates that we move somewhere around 30 to 35 gigatons (billion tons or Gt) per year worldwide. When combined with the 10 Gt each year that we add to river sediments through erosion, our earth-moving prowess is comparable to, or greater than, any other single geomorphic agent except plate tectonics.

## 14.3 Economic Geology and Mineralogy

Economic mineralogy is the study of minerals that are valuable for manufacturing and are, therefore, an important part of domestic and international commerce. Most economic minerals are metal-bearing ores, such as the gold-bearing deposit at the Beartrack Mine (table 14.2). Nonmetallic economic minerals are mostly graphite, some feldspars, quartz crystals, diamonds, and many other crystals that are valued for their usefulness, beauty, and/or rarity. Metals have been so important in human affairs that major epochs of human history are commonly known by the dominant materials and the technology to use them (Stone Age, Bronze Age, Iron Age, etc.). The mining, processing, and distribution of these materials have broad and varied implications both for culture and our environment. Most economically valuable crustal resources exist everywhere in small amounts; the important thing is to find them concentrated in economically recoverable levels.

Public policy in the United States has encouraged mining on public lands as a way of boosting the economy and utilizing natural resources. Today many people think these laws seem outmoded and in need of reform (What Do You Think? p. 309).

### Metals are essential to our economy

How has the quest for minerals and metals affected global development? We will focus first on world use of metals, earth resources that always have received a great deal of human attention. The availability of metals and the methods to extract and use them have determined technological developments, as

**TABLE 14.2**

**Primary Uses of Some Major Metals Consumed in the United States**

| Metal | Use |
|---|---|
| Aluminum | Packaging foods and beverages (38%), transportation, electronics |
| Chromium | High-strength steel alloys |
| Copper | Building construction, electric and electronic industries |
| Iron | Heavy machinery, steel production |
| Lead | Leaded gasoline, car batteries, paints, ammunition |
| Manganese | High-strength, heat-resistant steel alloys |
| Nickel | Chemical industry, steel alloys |
| Platinum-group | Automobile catalytic converters, electronics, medical uses |
| Gold | Medical, aerospace, electronic uses; accumulation as monetary standard |
| Silver | Photography, electronics, jewelry |

## What Do You Think?

### Should We Revise Mining Laws?

In 1872, the U.S. Congress passed the General Mining Law intended to encourage prospectors to open up the public domain and promote commerce. This law, which has been in effect more than a century, allows miners to stake an exclusive claim anywhere on public lands and to take—for free—any minerals they find. Three-forths of the Beartrack Mine, for example, is on public land. Claim holders can "patent" (buy) the land for $2.50 to $5 per acre (0.4 hectares) depending on the type of claim. Once the patent fee is paid, the owners can do anything they want with the land, just like any other private property. Although $2.50 per acre may have been a fair market value in 1872, many people regard it as ridiculously low today, amounting to a scandalous give-away of public property.

In Nevada, for example, a mining company paid $9,000 for federal land that contains an estimated $20 billion worth of precious metals. Similarly, Colorado investors bought about 7,000 ha (17,000 acres) of rich oil-shale land in 1986 for $42,000 and sold it a month later for $37 million. You don't actually have to find any minerals to patent a claim. A Colorado company paid a total of $400 for 65 ha (160 acres) it claimed would be a gold mine. Almost 20 years later, no mining has been done, but the property—which just happens to border the Keystone Ski Area—is being subdivided for condos and vacation homes.

According to the Bureau of Land Management (BLM), some $4 billion in minerals are mined each year on U.S. public lands. Under the 1872 law, mining companies don't pay a penny for the ores they take. Furthermore, they can deduct a depletion allowance from taxes on mineral profits. Former Senator Dale Bumpers of Arkansas, who calls the 1872 mining law "a license to steal," has estimated that the government could derive $320 million per year by charging an 8 percent royalty on all minerals and probably could save an equal amount by requiring larger bonds to be posted to clean up after mining is finished. The Meridian Gold Company, for example, has posted a $2 million bond for cleaning up the Beartrack Mine (a larger than normal amount). Reclamation, however, is expected to cost 15 times that amount. Chapter 13 has more information on how reclamation and restoration can return damaged sites to beneficial uses.

Thousands of abandoned mines on public lands poison streams and groundwater with acid, metal-laced drainage. This old mine in Montana drains into the Blackfoot River, the setting of Norman Maclean's book. *A River Runs Through It.*

On the other hand, mining companies argue they would be forced to close down if they had to pay royalties or post larger bonds. Many people would lose jobs and the economies of western mining towns would collapse if mining becomes uneconomic. We provide subsidies and economic incentives to many industries to stimulate economic growth. Why not mining for metals essential for our industrial economy? Mining is a risky and expensive business. Without subsidies, mines would close down and we would be completely dependent on unstable foreign supplies.

Mining critics respond that other resource-based industries have been forced to pay royalties on materials they extract from public lands. Coal, oil, and gas companies pay 12.5 percent royalties on fossil fuels obtained from public lands. Timber companies—although they don't pay the full costs of the trees they take—have to bid on logging sales and clean up when they are finished. Even gravel companies pay for digging up the public domain. Ironically, we charge for digging up gravel, but give gold away free.

Over the past decade, numerous mining bills have been introduced in Congress. Those supported by environmental groups generally would require companies mining on federal lands to pay a higher royalty on their production. They also would eliminate the patenting process, impose stricter reclamation requirements, and give federal managers authority to deny inappropriate permits. In contrast, bills offered by Western legislators, and enthusiastically backed by mining supporters, tend to leave most provisions of the 1872 bill in place. They would charge a 2 percent royalty, but only after exploration, production, and other costs were deducted. Permit processes would consider local economic needs before environmental issues in this version. What do you think we should do about this mining law? How could we separate legitimate public-interest land use from private speculation and profiteering? Are current subsidies necessary and justifiable or are they just a form of corporate welfare?

well as economic and political power for individuals and nations. We still are strongly dependent on the unique lightness, strength, and malleability of metals.

The metals consumed in greatest quantity by world industry include iron (740 million metric tons annually), aluminum (40 million metric tons), manganese (22.4 million metric tons), copper and chromium (8 million metric tons each), and nickel (0.7 million metric tons). Most of these metals are consumed in the United States, Japan, and Europe, in that order. They are produced in many countries, often in mountainous areas, where heat and pressure of mountain building tend to concentrate ores (fig. 14.11). The worldwide mineral trade has become crucially important to the economic and social stability of all nations involved.

Nonmetal minerals are a broad class that covers resources from silicate minerals (gemstones, mica, talc, and asbestos) to sand, gravel, salts, limestone, and soils. Durable, highly valuable, and easily portable, gem stones and precious metals have long been a way to store and transport wealth. Unfortunately, these valuable materials also have bankrolled despots, criminal gangs, and terrorism in many countries. In recent years, brutal civil wars

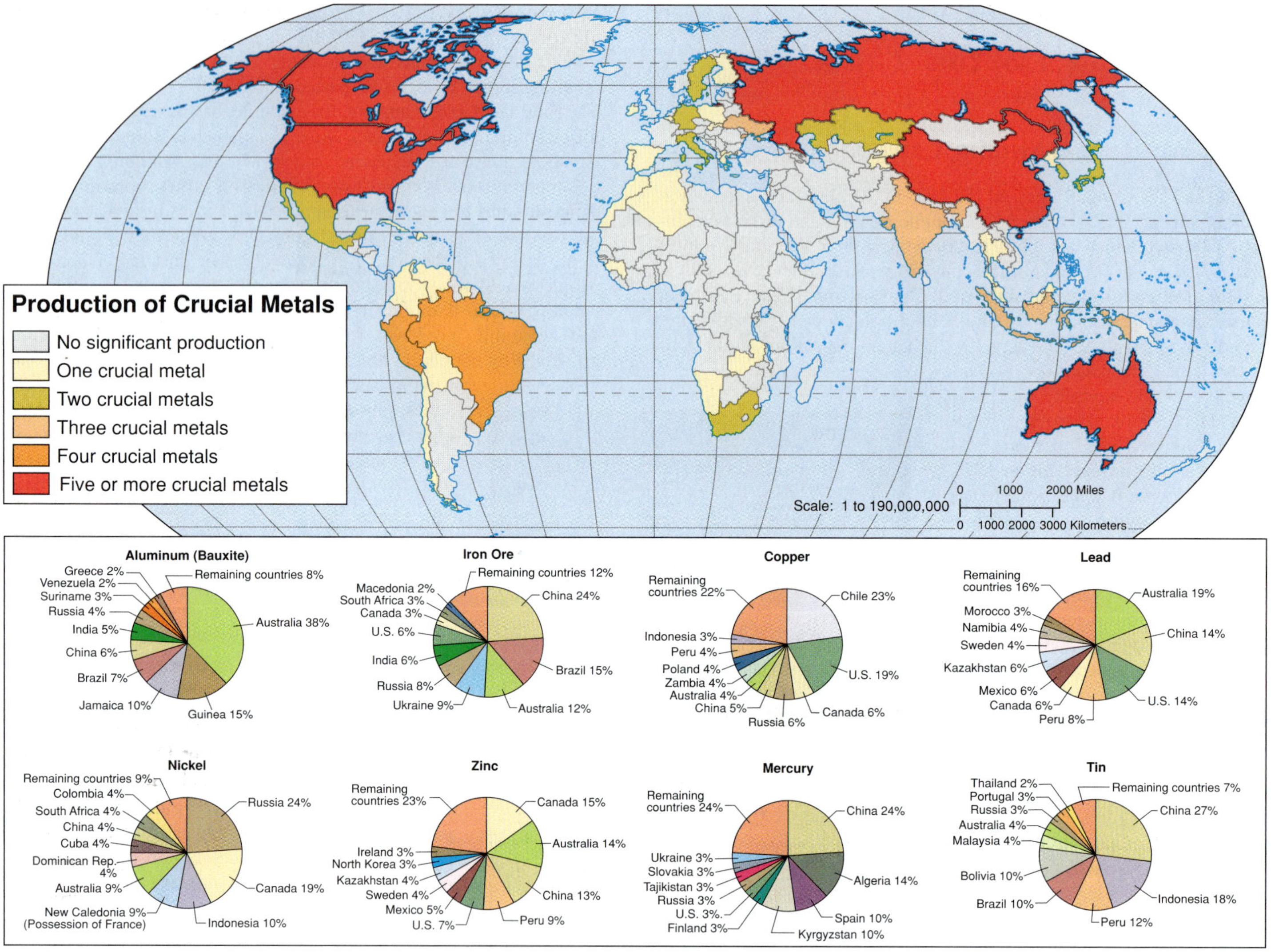


**FIGURE 14.11** World production of metals most essential for an industrial economy. Principal consumers are the United States, Western Europe, Japan, and China.

in Africa have been financed—and often motivated by—gold, diamonds, tantalum ore, and other high-priced commodities. Much of this illegal trade ends up in the $100-billion-per-year global jewelry trade, two-thirds of which sells in the United States. Many people who treasure a diamond ring or a gold wedding band as a symbol of love and devotion are unaware that it may have been obtained through inhumane labor conditions and environmentally destructive mining and processing methods. Civil rights organizations are campaigning to require better documentation of the origins of gems and precious metals to prevent their use as financing for crimes against humanity (see "Conflict Diamonds" at www.mhhe.com/environmentalscience).

In 2004, a group of Nobel Peace Laureates called on the World Bank to overhaul its policies on lending for resource extractive industries. "War, poverty, climate change, and ongoing violations of human rights—all of these scourges are all too often linked to the oil and mining industries" wrote Archbishop Desmond Tutu, winner of the 1984 Nobel Peace Prize for helping to eliminate apartheid in South Africa. In response, the World Bank appointed an Extractive Industries Review headed by former Indonesian environment minister, Emil Salim. In its final report, the committee recommended that some areas of exceptionally high biodiversity value should be "no-go" zones for extractive industries, and that the rights of those affected by extractive projects need better protection.

## Nonmetal minerals include gravel, clay, sand, and salts

You might be surprised to learn that sand and gravel production comprise by far the greatest volume and dollar value of all nonmetal mineral resources and a far greater volume than all metal ores. Sand and gravel are used mainly in brick and concrete construction, paving, as loose road filler, and for sandblasting. High-purity silica sand is our source of glass. These materials usually are retrieved from surface pit mines and quarries, where they have been deposited by glaciers, winds, or ancient oceans.

Limestone, like sand and gravel, is mined and quarried for concrete and crushed for road rock. It also is cut for building stone, pulverized for use as an agricultural soil additive that neutralizes acidic soil, and roasted in lime kilns and cement plants to make plaster (hydrated lime) and cement.

Evaporites (materials deposited by evaporation of chemical solutions) are mined for halite, gypsum, and potash. These are often found at or above 97 percent purity. Halite, or rock salt, is used for water softening and melting ice on winter roads in some northern areas. Refined, it is a source of table salt. Gypsum (calcium sulfate) now makes our plaster wallboard, but it has been used for plaster ever since the Egyptians plastered the walls of their frescoed tombs along the Nile River some 5,000 years ago. Potash is an evaporite composed of a variety of potassium chlorides and potassium sulfates. These highly soluble potassium salts have long been used as a soil fertilizer.

Sulfur deposits are mined mainly for sulfuric acid production. In the United States, sulfuric acid use amounts to more than 200 lbs per person per year, mostly because of its use in industry, car batteries, and some medicinal products.

## 14.4 Environmental Effects of Resource Extraction

Each of us depends daily on geologic resources mined from sites around the world. We use scores of metals and minerals, many of which we've never even heard of, in our lights, computers, watches, fertilizers, and cars. Mining and purifying all these resources can have severe environmental and social consequences. The most obvious effect of mining is often the disturbance or removal of the land surface. Farther-reaching effects, though, include air and water pollution. The EPA lists more than 100 toxic air pollutants, from acetone to xylene, released from U.S. mines every year. Nearly 80,000 metric tons of particulate matter (dust) and 11,000 tons of sulfur dioxide are released from nonmetal mining alone. Chemical- and sediment-runoff pollution is a major problem in many local watersheds.

Resource extraction can affect water quality for several reasons. Gold and other metals are often found in sulfide ores, as is the case at the Beartrack Mine. Sulfide minerals often occur in hydrothermal (hot water) deposits, where sulfur and other elements are mobilized, then deposited as they are heated by deep magma chambers. When sulfur-bearing minerals are exposed to air and water, they produce sulfuric acid, which is highly mobile and strongly acidic. In addition, metal elements often occur in very low concentrations—10 to 20 parts per billion may be economically extractable for gold, platinum, and other metals. Consequently, vast quantities of ore must be crushed and washed to extract metals, as the opening case study shows. A great deal of water is used in cyanide heap-leaching and other washing techniques. In arid Nevada, the USGS estimates that mining consumes about 230,000 $m^3$ (60 million gal) of fresh water per day. After use in ore processing, much of this water contains sulfuric acid, arsenic, heavy metals, and other contaminants. Mine runoff leaking into lakes and streams damages or destroys aquatic ecosystems.

### Mining can have serious environmental impacts

There are many techniques for extracting geological materials. The most common methods are open-pit mining, strip mining, and underground mining. An ancient method of accumulating gold, diamonds, and coal is placer mining, in which pure nuggets are washed from stream sediments. Since the California gold rush of 1849, placer miners have used water cannons to blast away hillsides. This method, which chokes stream ecosystems with sediment, is still used in Alaska, Canada, and many other regions. Another ancient, and much more dangerous, method is underground mining. Ancient Roman, European, and Chinese miners tunneled deep into tin, lead, copper, coal, and other mineral seams. Mine tunnels occasionally collapse, and natural gas in coal mines can explode. Water seeping into mine shafts also dissolves toxic minerals. Contaminated water seeps into groundwater; it is also pumped to the surface, where it enters streams and lakes.

In underground coal mines, another major environmental risk is fires. Hundreds of coal mines smolder in the United States, China, Russia, India, South Africa, and Europe. The inaccessibility and size of these fires make many impossible to extinguish or control. One mine fire in Centralia, Pennsylvania, has been burning since 1962; control efforts have cost at least $40 million, but the fire continues to expand. China, which depends on coal for much of its heating and electricity, has hundreds of smoldering mine fires; one has been burning for 400 years. According to a recent study from the International Institute for Aerospace Survey in the Netherlands, these fires consume up to 200 million tons of coal every year and emit as much carbon dioxide as all the cars in the United States. Toxic fumes, explosive methane, and other hazardous emissions are also released from these fires.

Open-pit mines are used to extract massive beds of metal ores and other minerals. The size of modern open pits can be hard to comprehend. The Bingham Canyon mine, near Salt Lake City, Utah, is 800 m (2,640 ft) deep and nearly 4 km (2.5 mi) wide at the top. More than 5 billion tons of copper ore and waste material have been removed from the hole since 1906. A chief environmental challenge of open-pit mining is that groundwater accumulates in the pit. In metal mines, a toxic soup results. No one yet knows how to detoxify these lakes, which endanger wildlife and nearby watersheds.

Half the coal used in the United States comes from surface or strip mines (fig. 14.12). Since coal is often found in expansive, horizontal beds, the entire land surface can be stripped away to cheaply and quickly expose the coal. The overburden, or surface material, is placed back into the mine, but usually in long ridges called spoil banks. Spoil banks are very susceptible to erosion and chemical weathering. Since the spoil banks have no topsoil (the complex organic mixture that supports vegetation—see chapter 9), revegetation occurs very slowly.

The 1977 federal Surface Mining Control and Reclamation Act (SMCRA) requires better restoration of strip-mined lands,

**FIGURE 14.12** Some giant mining machines stand as tall as a 20-story building and can scoop up thousands of cubic meters of rock per hour.

**FIGURE 14.13** Mountaintop removal mining is a relatively new, and deeply controversial, method of extracting Appalachian coal.

especially where mines replaced prime farmland. Since then, the record of strip-mine reclamation has improved substantially. Complete mine restoration is expensive, often more than $10,000 per hectare. Restoration is also difficult because the developing soil is usually acidic and compacted by the heavy machinery used to reshape the land surface.

Bitter controversy has grown recently over mountaintop removal, a coal mining method mainly practiced in Appalachia. Long, sinuous ridge-tops are removed by giant, 20-story-tall shovels to expose horizontal beds of coal (fig. 14.13). Up to 215 m (700 ft) of ridge-top is pulverized and dumped into adjacent river valleys. The debris can be laden with selenium, arsenic, coal, and other toxic substances. At least 900 km (560 mi) of streams have been buried in West Virginia alone. Environmental lawyers have sued to stop the destruction of streams, arguing that it violates the Clean Water Act (chapter 24). In response, the Bush administration issued a "clarification" of the Clean Water Act, rewriting the law to allow stream filling by any sort of mining or industrial debris, rather than forbidding any stream destruction. Environmentalists charge that the new wording subverts the Clean Water Act, giving a green light to all sorts of stream destruction. West Virginians are deeply divided about mountaintop removal because they live in affected stream valleys but also depend on a coal economy.

The Mineral Policy Center in Washington, D.C., estimates that 19,000 km (12,000 mi) of rivers and streams in the United States are contaminated by mine drainage. The EPA estimates that cleaning up impaired streams, along with 550,000 abandoned mines in the U.S., may cost $70 billion. Worldwide, mine closing and rehabilitation costs are estimated in the trillions of dollars. Because of the volatile prices of metals and coal, many mining companies have gone bankrupt before restoring mine sites, leaving the public responsible for cleanup. In 2002, more than 500 leading mine executives and their critics convened at the Global Mining Initiative, a meeting in Toronto aimed at improving the sustainability of mining. Executives acknowledged that in the future they will increasingly be held liable for environmental damages, and they said they were seeking ways to improve the industry's social and environmental record. Jay Hair, secretary general of the International Council on Mining and Metals, stated that "environmental protection and social responsibility are important," and that mining companies were interested in participating in sustainable development. Mine executives also recognized that, increasingly, big cleanup bills will cut into company values and stock prices. Finding creative ways to keep mines cleaner from the start will make good economic sense, even if it's not easy to do.

## Processing ores also has negative consequences

Metals are extracted from ores by heating or with chemical solvents. Both processes release large quantities of toxic materials that can be even more environmentally hazardous than mining. **Smelting**—roasting ore to release metals—is a major source of air pollution. One of the most notorious examples of ecological devastation from smelting is a wasteland near Ducktown, Tennessee. In the mid-1800s, mining companies began excavating the rich copper deposits in the area. To extract copper from the ore, they built huge, open-air, wood fires, using timber from the surrounding forest. Dense clouds of sulfur dioxide released from sulfide ores poisoned the vegetation and acidified the soil over a 50 $mi^2$ (13,000 ha) area. Rains washed the soil off the denuded land, creating a barren moonscape.

Sulfur emissions from Ducktown smelters were reduced in 1907 after Georgia sued Tennessee over air pollution. In the 1930s the Tennessee Valley Authority (TVA) began treating the soil and replanting trees to cut down on erosion. Recently, upwards of $250,000 per year has been spent on this effort. While the trees and other plants are still spindly and feeble, more than two-thirds of the area is considered "adequately" covered with vegetation. Similarly, smelting of copper-nickel ore in Sudbury, Ontario, a century ago

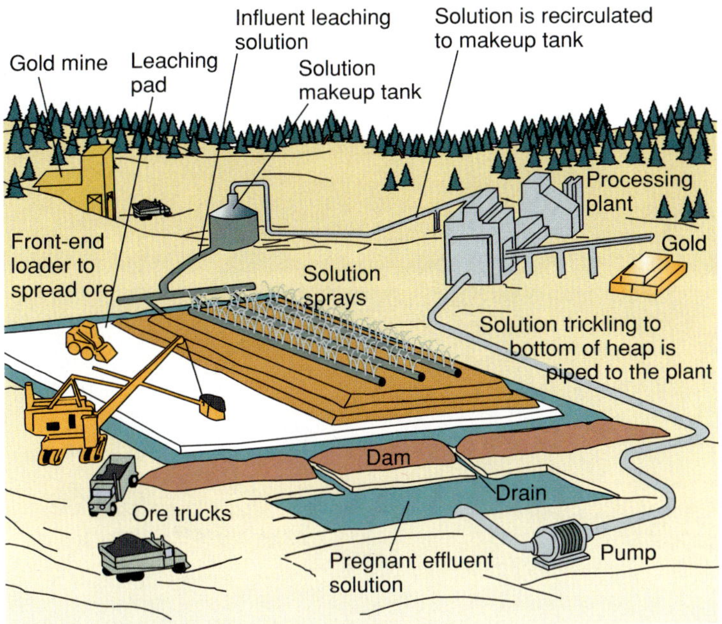


**FIGURE 14.14** In a heap-leach operation, huge piles of low-grade ore are heaped on an impervious pad and sprayed with a cyanide solution. As the leaching solution trickles through the crushed ore, it extracts gold and other precious metals. This technique is highly profitable but carries large environmental risks.

caused widespread ecological destruction that is slowly being repaired following pollution-control measures (see fig. 16.23).

As discussed in the opening case study for this chapter, chemical extraction methods have a high potential for environmental contamination. The cyanide solution sprayed on a heap-leach pile (fig. 14.14) can leak into surface or groundwater. A case in point is the Summitville Mine near Alamosa, Colorado. After extracting \$98 million in gold, the absentee owners declared bankruptcy in 1992, abandoning millions of tons of mine waste and huge leaking ponds of cyanide. The Environmental Protection Agency may spend more than \$100 million trying to clean up the mess and keep the cyanide pool from spilling into the Alamosa River.

Every year, the Blacksmith Institute compiles a list of the world's worst polluted places (table 14.3). For 2006, seven of the top ten worst places were mines and/or smelter facilities. These problems are especially disastrous in the developing world or in the former Soviet Union, where funds and political will aren't available to deal with pollution or help people suffering from terrible health effects of pollution.

## 14.5 Conserving Geological Resources

Conservation offers great potential for extending our supplies of economic minerals and reducing the effects of mining and processing. The advantages of conservation are also significant: less waste to dispose of, less land lost to mining, and less consumption of money, energy, and water resources.

**TABLE 14.3**

**The World's Worst Polluted Places**

| Place | Pollution Source |
|---|---|
| Chernobyl, Ukraine | Nuclear power plant |
| Dzerzhinsk, Russia | Weapons production |
| Haina, Dominican Republic | Lead smelter |
| Kabwe, Zambia | Lead mine and smelter |
| La Oroya, Peru | Mine and smelter |
| Lifen, China | Coal mine and power plants |
| Maiuu Suu, Kyrgyzstan | Uranium mine |
| Norlisk, Russia | Copper-nickel mine and smelter |
| Ranipet, India | Tannery |
| Rudnava Pristan/Dalnegorsk, Russia | Lead mine |

**Source:** Blacksmith Institute, 2006.

### Recycling saves energy as well as materials

Some waste products already are being exploited, especially for scarce or valuable metals. Aluminum, for instance, must be extracted from bauxite ore by electrolysis, an expensive, energy-intensive process. Recycling waste aluminum, such as beverage cans, on the other hand, consumes one-twentieth of the energy of extracting new aluminum. Today, nearly two-thirds of all aluminum beverage cans in the United States are recycled, up from only 15 percent 20 years ago. The high value of aluminum scrap (\$650 a ton versus \$60 for steel, \$200 for plastic, \$50 for glass, and \$30 for paperboard) gives consumers plenty of incentive to deliver their cans for collection. Recycling is so rapid and effective that half of all the aluminum cans now on a grocer's shelf will be made into another can within two months. The energy cost of extracting other materials is shown in table 14.4.

Platinum, the catalyst in automobile catalytic exhaust converters, is valuable enough to be regularly retrieved and recycled from used cars (fig. 14.15). Other metals commonly recycled are gold, silver, copper, lead, iron, and steel. The latter four are readily available in a pure and massive form, including copper pipes, lead batteries, and steel and iron auto parts. Gold and silver are valuable enough to warrant recovery, even through more difficult means. See chapter 21 for further discussion of this topic.

While total U.S. steel production has fallen in recent decades—largely because of inexpensive supplies from new and efficient Japanese steel mills—a new type of mill subsisting entirely on a readily available supply of scrap/waste steel and iron is a growing industry. Minimills, which remelt and reshape scrap iron and steel, are smaller and cheaper to operate than traditional integrated mills that perform every process from preparing raw ore to finishing iron and steel products. Minimills produce steel at between \$225 and \$480 per metric ton, while steel from integrated mills costs \$1,425 to \$2,250 per metric

## TABLE 14.4

### Energy Requirements in Producing Various Materials from Ore and Raw Source Materials

| Product | Energy Requirement (Mj/Kg) New | From Scrap |
|---|---|---|
| Glass | 25 | 25 |
| Steel | 50 | 26 |
| Plastics | 162 | n.a. |
| Aluminum | 250 | 8 |
| Titanium | 400 | n.a. |
| Copper | 60 | 7 |
| Paper | 24 | 15 |

**Source:** E. T. Hayes, *Implications of Materials Processing,* 1997.

ton on average. The energy cost is likewise lower in minimills: 5.3 million BTU/ton of steel compared to 16.08 million BTU/ton in integrated mill furnaces. Minimills now produce about half of all of U.S. steel production. Recycling is slowly increasing as raw materials become more scarce and wastes become more plentiful.

### New materials can replace mined resources

Mineral and metal consumption can be reduced by new materials or new technologies developed to replace traditional uses. This is a long-standing tradition; for example, bronze replaced stone technology and iron replaced bronze. More recently, the introduction of plastic pipe has decreased our consumption of copper, lead, and steel pipes. In the same way, the development of fiber-optic technology and satellite communication reduces the need for copper telephone wires.

Iron and steel have been the backbone of heavy industry, but we are now moving toward other materials. One of our primary uses for iron and steel has been machinery and vehicle parts. In automobile production, steel is being replaced by polymers (long-chain organic molecules similar to plastics), aluminum, ceramics, and new, high-technology alloys. All of these reduce vehicle weight and cost, while increasing fuel efficiency. Some of the newer alloys that combine steel with titanium, vanadium, or other metals wear much better than traditional steel. Ceramic engine parts provide heat insulation around pistons, bearings, and cylinders, keeping the rest of the engine cool and operating efficiently. Plastics and glass fiber-reinforced polymers are used in body parts and some engine components.

Electronics and communications (telephone) technology, once major consumers of copper and aluminum, now use ultra-high-purity glass cables to transmit pulses of light, instead of metal wires carrying electron pulses. Once again, this technology has been developed for its greater efficiency and lower cost, but it also affects consumption of our most basic metals.

**FIGURE 14.15** The richest metal source we have—our mountains of scrapped cars—offers a rich, inexpensive, and ecologically beneficial resource that can be "mined" for a number of metals.

## 14.6 Geological Hazards

Earthquakes, volcanic eruptions, floods, and landslides are among the geological forces that have shaped the world around us (fig. 14.16). Catastrophic events, such as the impact of a giant asteroid 65 million years ago off the coast of what is now Yucatan, or volcanic eruptions, such as those that covered 2 million km$^2$ of Siberia with basalt up to 2 km deep 250 million years ago, are thought to have triggered the mass extinctions that mark transitions between major historic eras. The asteroid impact 65 million years ago that ended the age of the dinosaurs is calculated to have created a tsunami hundreds of meters high that could have swept around the world several times before subsiding. This impact also ejected so much dust into the air that sunlight was blocked for years and a global winter decimated much of the life on the earth. A similar impact 250 million years ago off the coast of Australia is thought to be related to the end of the Permian era, when 90 to 95 percent of all marine species perished. This asteroid impact may also have caused fissures in the earth's crust that allowed the vast outpouring of lava in Siberia. Together, these cataclysmic events probably filled the atmosphere with ash and sulfuric acid aerosols that plunged the earth into a massive deep freeze.

Fortunately, such massive events are rare. Still, as the devastating 2004 Sumatran tsunami shows, geological hazards represent a

(a) A meteor impact crater in Arizona (b) Volcanic eruption (c) Glacier in Alaska

**FIGURE 14.16** Geologic events such as meteor or asteroid impacts (*a*), massive volcanic eruptions (*b*), or climate change (*c*), are thought to trigger mass extinctions that mark major eras in the earth's history.

huge threat. Diseases kill more people than any other cause. Some pandemics have been truly enormous. In the 1350s, for example, bubonic plague (the Black Death) is thought to have killed at least 135 million people in Europe and Asia. A similar pandemic in the sixth century (the Justinian plague) probably killed at least 100 million and uprooted nearly every existing major culture. Incidentally, the Justinian plague may have been triggered by an explosion of the predecessor of Indonesia's Krakatoa volcano, which exploded again in 1883. Drought-caused famines are probably the second biggest killers. A severe El Niño (chapter 15) from 1876 to 1879, for example, is reported to have caused 50 million deaths around the world. Bad weather and extremely bad governance are thought to have caused a famine in China from 1959 to 1961 that killed at least 30 million people.

Among direct natural disasters, floods take the largest number of human lives, while wind storms (hurricanes, cyclones, tornadoes) cause the greatest property damage. Table 14.5 shows the estimated loss of life in the 20 worst natural disasters in the past 250 years.

**TABLE 14.5**

**Worst Natural Disasters, 1755–2005**

| Date | Type | Deaths | Location |
|---|---|---|---|
| 1931 | Flood | 3.7 million | Yangtze River, China |
| 1959 | Flood | 2 million | Hwang Ho (Yellow) River, China |
| 1938 | Flood | 900,000 | Yangtze River, China |
| 1877 | Flood | 900,000 | Hwang Ho (Yellow) River, China |
| 1939 | Flood | 500,000 | Honan province, China |
| 1970 | Cyclone/Flood | 500,000 | Bangladesh, India |
| 2004 | Tsunami | 300,000 | Indian Ocean (mag. 9.2) |
| 1850 | Earthquake | 300,000 | Sichuan, China (mag. 8.0) |
| 1976 | Earthquake | 275,000 | Tianjin, China (mag. 8.2) |
| 1920 | Earthquake | 200,000 | Gansu, China (mag. 8.6) |
| 1927 | Earthquake | 200,000 | Xining, China (mag. 8.3) |
| 1923 | Earthquake | 140,000 | Tokyo, Japan (mag. 8.2) |
| 1935 | Flood | 142,000 | Changiyang River, China |
| 1991 | Cyclone/Flood | 138,000 | Bangladesh |
| 1948 | Earthquake | 110,000 | Turkmenistan, USSR (mag. 7.3) |
| 1883 | Tsunami | 100,000 | Java (Krakatoa volcano) |
| 1911 | Flood | 100,000 | Yangtze River, China |
| 1815 | Volcano | 92,000 | Tambora Sumbawa, Indonesia |
| 1908 | Earthquake | 83,000 | Messina, Italy (mag. 7.5) |
| 1755 | Earthquake | 77,000 | Lisbon, Portugal (mag. 8.7) |

**Source:** The Disaster Center, 2005.

As you can see, floods have been the most lethal and, because of both its geology and large population, China has suffered a majority of these disasters. The 2004 earthquake/tsunami is seventh on this list, but it was the greatest death toll of any earthquake in recorded history. The only other earthquake/tsunami on this list was the explosion of Krakatoa, at the opposite end of Sumatra from the epicenter of the 2004 quake.

### *Think About It*

What are the most dangerous geological hazards where you live? Are you aware of emergency preparedness plans? How would you evacuate your area if it becomes necessary?

## Earthquakes can be very destructive

**Earthquakes** are sudden movements in the earth's crust that occur along faults (planes of weakness) where one rock mass slides past another one. When movement along faults occurs gradually and relatively smoothly, it is called creep or seismic slip and may be undetectable to the casual observer. When friction prevents rocks from slipping easily, stress builds up until it is finally released with a sudden jerk, as was the case in the 2004 Sumatran earthquake. The point on a fault at which the first movement occurs during an earthquake is called the epicenter.

Earthquakes have always seemed mysterious, sudden, and violent, coming without warning and leaving in their wake ruined cities and dislocated landscapes (table 14.6). Cities such as Tokyo, Japan, or Mexico City, parts of which are built on soft landfill or poorly consolidated soil, usually suffer the greatest damage from earthquakes. Water-saturated soil can liquify when shaken. Buildings sometimes sink out of sight or fall down like a row of dominoes under these conditions.

**FIGURE 14.17** An earthquake near the town of Izmit in western Turkey in 1999 killed at least 50,000 people and injured an equal number. Poorly constructed buildings collapsed and trapped residents, while stronger structures nearby remained standing.

The worst death toll usually occurs in cities with poorly constructed buildings (fig. 14.17). Today contractors in earthquake zones are attempting to prevent damage and casualties by constructing buildings that can withstand tremors. The primary methods used are heavily reinforced structures, strategically placed weak spots in the building that can absorb vibration from the rest of the building, and pads or floats beneath the building on which it can shift harmlessly with ground motion. The problem of planning and building for earthquakes is at the center of a persistent dispute about nuclear waste storage at Yucca Mountain, Nevada (Exploring Science, p. 317).

As you can see in figure 14.18, the most seismically active region in the lower 48 U.S. states is along the west coast where tectonic plates are colliding. Nearly every state has had significant

### TABLE 14.6
### Worldwide Frequency of Earthquakes

| Richter Scale Magnitude* | Description | Average Number Per Year |
|---|---|---|
| 2–2.9 | Unnoticeable | 300,000 |
| 3–3.9 | Smallest felt | 49,000 |
| 4–4.9 | Minor earthquake | 6,200 |
| 5–5.9 | Damaging earthquake | 800 |
| 6–6.9 | Destructive earthquake | 120 |
| 7–7.9 | Major earthquake | 18 |
| >8 | Great earthquake | 1 or 2 |

*For every unit increase in the Richter scale, ground displacement increases by a factor of 10, while energy release increases by a factor of 30. There is no upper limit to the scale, but the largest earthquake recorded was 9.5, in Chile in 1960.

**Source:** B. Gutenberg in *Earth* by F. Press and R. Siever, 1978, W. H. Freeman & Company.

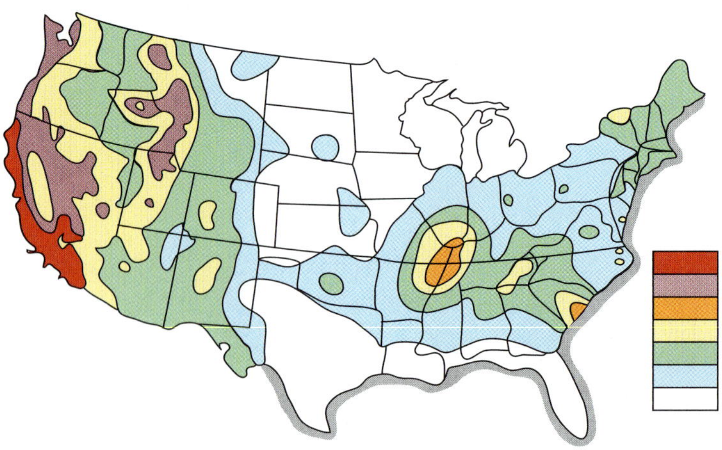

**FIGURE 14.18** A seismic map of the lower 48 states shows the risk of earthquakes. Although the highest risk is along the Pacific coast and the Rocky Mountains, you may be surprised to see that an area along the Mississippi River around New Madrid, Missouri, also has a high potential for seismic activity.

## *Exploring* SCIENCE

### Radioactive Waste Disposal at Yucca Mountain

In July 2002 the U.S. Senate voted to designate Yucca Mountain, Nevada, as the permanent resting place for 77,000 metric tons of high-level nuclear waste, most of it spent fuel from the nation's 107 nuclear power plants (see fig. 19.25). Nuclear power proponents celebrated: After twenty years of research and lobbying, the site was finally approved. With a repository on the horizon, plans for new nuclear plants could proceed, and existing plants could look forward to clearing out 50,000 tons of spent nuclear fuel currently in temporary, sometimes inadequate, storage at reactor sites. The state of Nevada, meanwhile, immediately filed a lawsuit to stop the plan. Nevada Governor Kenny Guinn charged that the Department of Energy (DOE) had lowered its scientific standards for evaluating the geological integrity of the site. Opponents charged that heavy-handed eastern politics, not sound geology, was behind the decision to put the entire nation's nuclear waste in Nevada.

The stakes at Yucca Mountain are high. The DOE and the nuclear power industry have invested more than $4 billion in research, testing, and promotion of the site. The lack of permanent storage has been an obstacle to construction of nuclear power plants since the 1980s. Further, cleanup is beginning at old military installations, such as Hanford, Washington; and Rocky Flats, Colorado. These are some of the nation's worst toxic waste sites, and permanent storage is needed for their radioactive materials.

Geology is the central problem in siting a waste repository of any kind. For high-level radioactive wastes, the DOE needed to find a

Yucca Mountain, Nevada (long ridge in center of photo), has been chosen as the United States' first permanent high-level nuclear waste storage site.
**Source:** Department of Energy.

site where a labyrinth of deep tunnels could keep extremely dangerous materials isolated and secure for 10,000 years, more than twice the length of recorded human history. (The waste will remain highly radioactive for more than 500,000 years, but the DOE considers itself unlikely to be responsible for the site for that long.) This time span requires a geologically stable area—no active faults like those in California, no volcanoes like those in Washington and Oregon. Bedrock needs to be relatively impermeable, not riddled with underground channels and sinkholes as in Florida or parts of Texas. There must be no groundwater, which would soak storage casks and mobilize radioactive materials, possibly allowing tainted water to reach the surface. This rules out the central Great Plains and most of the eastern United States, where plentiful rainfall keeps water tables high and aquifers full.

Yucca Mountain fits these requirements better than most places in the United States. But there is conflicting evidence, and conflicting interpretations of evidence, about whether even Yucca Mountain is good enough. DOE geologists insist that the only aquifers in the area are 300 m below the site; other geologists say there is evidence in the rocks that aquifers have risen in the past, suggesting that they might rise again to the level of the repository tunnels. Revelation that some of the hydrologic data were fake cast even more doubt about the safety of the site.

Critics of the site point out that the area has more than 30 known faults. Geologists disagree on how long these faults have been dormant, and on how long it will be before they shift again. In addition, there are seven dormant volcanoes in the area. While these are likely to remain dormant for 10,000 years, their activity in several hundred thousand years is hard to predict. Opponents of the site also worry about the potential dangers of shipping waste, potentially 28,000 truckloads and 10,000 rail cars over the site's 30-year lifespan. Defenders point out that high-level waste is currently stored at more than 130 temporary sites, and those sites are clearly unsafe for the long term.

Do you think Nevada should be the repository for the nation's nuclear waste? Should we build more nuclear plants even if Yucca Mountain turns out to be unacceptable, given that we will have to find something to do with existing waste? Would our economy be different today if we had waited to build nuclear power plants until a permanent storage site was available?

tremors, however. The largest North American earthquake in recorded history occurred, surprisingly, in New Madrid, Missouri, in 1811. The tremor, which had an estimated magnitude of 8.8, made bells ring as far away as Boston, and caused the Mississippi River to run backwards for several hours. The region was sparsely inhabited, so there were few deaths. If a similar quake occurred today, however, a large part of Memphis would be destroyed.

Earthquakes are almost always followed by a series of aftershocks that can continue long after the initial shock. In the case of the 2004 Sumatran quake, hundreds of aftershocks followed in succeeding months. One in April was magnitude 8.7 (about half the energy of the original 9.2 quake). It didn't cause a tsunami, but it did kill about 2,000 people on the nearby island of Nias, which is about 35 km from the epicenter of the earlier quake. No tsunami occurred—apparently because there was less displacement of the ocean floor.

Tsunamis can be generated by earthquakes, as the 2004 one was, or by landslides or underwater volcanic eruptions. The largest wave ever recorded was caused by a landslide. In July 1958, a massive rock avalanche fell into Lituya Fjord in Alaska's Glacier Bay National Park. The rockfall created a wave 525 m (over 1,700 ft) high. Several eyewitnesses lived to tell about it, including the crew of a fishing boat that was picked up from its anchorage in the fjord and thrown out into the ocean. A major concern in the Atlantic Ocean is a huge bulge forming on the Cumbre Vieja volcano on La Palma in the Canary Islands. This volcano erupts every 25 to 200 years. The last eruption was in 1971, but indications are that another eruption could be imminent. If this crater were to collapse all at once, it could dump half a trillion tons of rock into the ocean, which would send a tsunami 50 m high to the east coast of North America.

The ring of seismic activity and active volcanoes (often called the "ring of fire") around the edge of the Pacific Ocean makes it the most likely place in the world for tsunami formation. The United States already has six tsunami warning buoys in the Pacific Ocean. In January 2005, President Bush announced a $37.5 million plan to deploy 32 new Deep-ocean Assessment and Reporting of Tsunami (DART) buoys by mid-2007 in the Pacific, Atlantic, and Caribbean to protect U.S. coastal areas (fig. 14.19). Each buoy has an anchored seafloor bottom pressure recorder and a companion moored surface buoy for real-time communications. An acoustic link transmits data from the pressure recorder on the seafloor to the surface buoy. The data are then relayed by a satellite link to ground stations. The Indian Ocean has had few tsunamis in recent years, and had no such warning system in 2004. In the wake of the disaster, however, 53 nations met in Brussels and pledged to create a worldwide system, including the Indian Ocean, by 2015.

## Volcanoes eject gas and ash, as well as lava

**Volcanoes** and undersea magma vents produce much of the earth's crust. Over hundreds of millions of years, gaseous emissions from these sources formed the earth's earliest oceans and atmosphere. Many of the world's fertile soils are weathered volcanic materials. Volcanoes have also been an ever-present threat to human populations. One of the most famous historic volcanic eruptions was that of Mount Vesuvius in southern Italy, which buried the cities of Herculaneum and Pompeii in A.D. 79. The

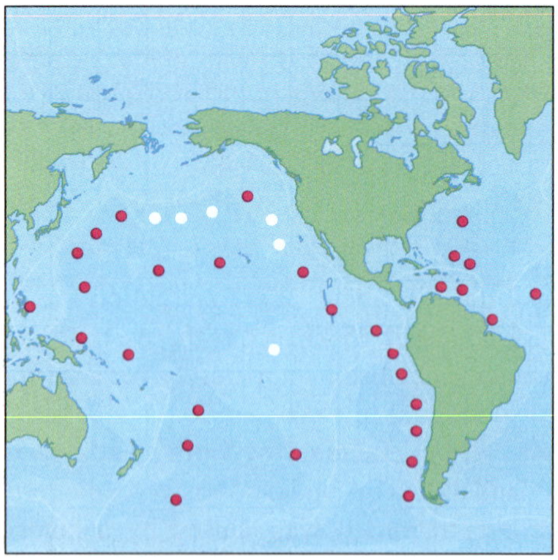


**FIGURE 14.19** Proposed location of buoys for the Deep-ocean Assessment and Reporting of Tsunami (DART) warning system maintained by the United States.

**FIGURE 14.20** Pyroclastic flows spill down the slopes of Mayon volcano in the Philippines in this September 23, 1984, image. This eruption caused no casualties because more than 73,000 people evacuated danger zones.

mountain had been giving signs of activity before it erupted, but many citizens chose to stay and take a chance on survival. On August 24, the mountain buried the two towns in ash. Thousands were killed by the dense, hot, toxic gases that accompanied the ash flowing down from the volcano.

Today, more than 500 million people live in the danger zone around volcanoes. Many assume that because the volcano hasn't erupted for years it never will again. They don't realize that the time between eruptions can be centuries, but the next big one could come any time without warning.

Nuees ardentes (French for "glowing clouds") are deadly, denser-than-air mixtures of hot gases and ash like those that inundated Pompeii and Herculaneum. Temperatures in these clouds may exceed 1,000°C, and they move at more than 100 km/hour (60 mph). Nuees ardentes destroyed the town of St. Pierre on the Caribbean island of Martinique on May 8, 1902. Mount Pelee released a cloud of nuees ardentes that rolled down through the town, killing somewhere between 25,000 and 40,000 people within a few minutes. All the town's residents died except a single prisoner being held in the town dungeon.

Mudslides are also disasters sometimes associated with volcanoes. The 1985 eruption of Nevado del Ruiz, 130 km (85 mi) northwest of Bogotá, Colombia, caused mudslides that buried most of the town of Armero and devastated the town of Chinchina. An estimated 25,000 people were killed. Heavy mudslides also accompanied the eruption of Mount St. Helens in Washington in 1980. Sediments mixed with melted snow and the waters of Spirit Lake at the mountain's base and flowed many kilometers from their source. Extensive damage was done to roads, bridges, and property, but because of sufficient advance warning, there were few casualties.

Volcanic eruptions often release large volumes of ash and dust into the air (fig. 14.20). Mount St. Helens expelled 3 $km^3$ of dust and ash, causing ash fall across much of North America. This was only a minor eruption. An eruption in a bigger class of volcanoes was that of Tambora, Indonesia, in 1815, which expelled 175 $km^2$ of dust and ash, more than 58 times that of

Mount St. Helens. These dust clouds circled the globe and reduced sunlight and air temperatures enough so that 1815 was known as the year without a summer.

It is not just a volcano's dust that blocks sunlight. Sulfur emissions from volcanic eruptions combine with rain and atmospheric moisture to produce sulfuric acid ($H_2SO_4$). The resulting droplets of $H_2SO_4$ interfere with solar radiation and can significantly cool the world climate. In 1991, Mt. Pinatubo in the Philippines emitted 20 million tons of sulfur dioxide that combined with water to form tiny droplets of sulfuric acid. This acid aerosol reached the stratosphere where it circled the globe for two years. This thin haze cooled the entire earth by 1°C and postponed global warming for several years. It also caused a 10 to 15 percent reduction in stratospheric ozone, allowing increased ultraviolet light to reach the earth's surface.

## Landslides are examples of mass wasting

Gravity constantly pulls downward on every material everywhere on earth, causing a variety of phenomena collectively termed **mass wasting** or mass movement, in which geologic materials are moved downslope from one place to another. The resulting movement is often slow and subtle, but some slope processes such as rockslides, avalanches, and land slumping can be swift, dangerous, and very obvious. Landslide is a general term for rapid downslope movement of soil or rock. In the United States alone, over $1 billion in property damage is done every year by landslides and related mass wasting.

In some areas, active steps are taken to control landslides or to limit the damage they cause. On the other hand, many human activities such as road construction, forest clearing, agricultural cultivation, and building houses on steep, unstable slopes increase both the frequency and the damage done by landslides. In some cases, people are unaware of the risks they face by locating on or under unstable hillsides. In other cases, they simply deny clear and obvious danger. Southern California, where people build expensive houses on steep hills of relatively unconsolidated soil, is often the site of large economic losses from landslides. Chapparal fires expose the soil to heavy winter rains. Resulting mudslides carry away whole neighborhoods and bury downslope areas in debris flows (fig. 14.21). Generally these processes are slow enough that few lives are lost, but the property damage can be high.

**FIGURE 14.21** Parts of an expensive house slide down the hillside in Pacific Palisades, California. Building at the edge of steep slopes made of unconsolidated sediment in an earthquake-prone region is a risky venture.

# Conclusion

We need materials from the earth to sustain our modern lifestyle, but often the methods we use to get those materials have severe environmental consequences. Still, there are ways that we can extend resources through recycling and the development of new materials and more efficient ways of using them. We can do much to repair the damage caused by resource extraction, although operations, such as that at the Beartrack Mine, will probably never be returned to their original, pristine condition. We will need to be eternally vigilant to ensure that mine pits full of toxic water or the nuclear waste disposal site at Yucca Mountain, Nevada (assuming it ever overcomes critics objections and is actually used to store nuclear waste), don't leak.

Water contamination is one of the main environmental costs of mining. Pollutants include acids, cyanide, mercury, heavy metals, and sediment. Air pollution from smelting also can cause widespread damage. We'll discuss this risk further in chapter 16.

We also should be aware of geological hazards, such as floods, earthquakes, volcanoes, and landslides. Because these hazards often occur on a geological time scale, residents who haven't experienced one of these catastrophic events assume that they never will. People move into highly risky places without considering what the consequences may be of ignoring nature. The earth may seem extremely stable to us, but it's really a highly dynamic system. We remain ignorant of these forces at our own peril.

# Reviewing Learning Outcomes

By now you should be able to explain the following points:

**14.1** Summarize the processes that shape the earth and its resources.

- Earth is a dynamic planet.
- Tectonic processes reshape continents and cause earthquakes.

**14.2** Explain how rocks and minerals are formed.

- The rock cycle creates and recycles rocks.
- Weathering and sedimentation wear down rocks.

**14.3** Think critically about economic geology and mineralogy.

- Metals are essential to our economy.
- Nonmetal minerals include gravel, clay, sand, and salts.

**14.4** Critique the environmental effects of resource extraction.

- Mining can have serious environmental impacts.
- Processing ores also has negative consequences.

**14.5** Discuss ways we could conserve geological resources.

- Recycling saves energy as well as materials.
- New materials can replace mined resources.

**14.6** Describe geological hazards.

- Earthquakes can be very destructive.
- Volcanoes eject gas and ash, as well as lava.
- Landslides are examples of mass wasting.

## Practice Quiz

1. Describe the layered structure of the earth.
2. Define *mineral* and *rock.*
3. What are tectonic plates and why are they important to us?
4. Why are there so many volcanoes, earthquakes, and tsunamis along the "ring of fire" that rims the Pacific Ocean?
5. Describe the rock cycle and name the three main rock types that it produces.
6. Figure 14.11 maps sources of some of the most important metals. What are they, and where do they come from?
7. Give some examples of nonmetal mineral resources and describe how they are used.
8. Describe some ways metals and other mineral resources can be recycled.
9. What are some environmental hazards associated with mineral extraction?
10. Describe some of the leading geological hazards and their effects.

## Critical Thinking and Discussion Questions

1. Look at the walls, floors, appliances, interior, and exterior of the building around you. How many earth materials were used in their construction?
2. Is your local bedrock igneous, metamorphic, or sedimentary? If you don't know, who might be able to tell you?
3. Suppose you live in a small, mineral-rich country, and a large, foreign mining company proposes to mine and market your minerals. How should revenue be divided between the company, your government, and citizens of the country? Who bears the greatest costs? How should displaced people be compensated? Who will make sure that compensation is fairly distributed?
4. Geological hazards affect a minority of the population who build houses on unstable hillsides, in flood-prone areas, or on faults. What should society do to ensure safety for these people and their property?
5. A persistent question in this chapter is how to reconcile our responsibility as consumers with the damage from mineral extraction and processing. How responsible are you? What are some steps you could take to reduce this damage?
6. If gold jewelry is responsible for environmental and social devastation, should we stop wearing it? Should we worry about the economy in producing areas if we stop buying gold and diamonds? What further information would you need to answer these questions?

## DATA analysis Mapping

### Volcanoes

Look at the animated map of volcanoes in the animated National Atlas. http:nationalatlas.gov/dynamic.html. (You will need Shockwave to display this movie; you can download Shockwave at the link on this page) Where are most of the volcanoes in the lower 48 states? Roll over the red dots to see photos and elevations of the volcanoes. As you move down the chain of mountains from north to south, is there a trend in elevations? Look at the Alaska map. How are the volcanoes arranged here? Based on your readings in this chapter, can you explain the pattern?

### Where Are Recent Earthquakes?

You can find out about recent earthquakes, where they were and how big they were, by looking at the USGS earthquake information page, http://earthquake.usgs.gov/. Click on the map to investigate recent events.

1. Describe where most recent earthquakes have occurred, either by dominant states or by mountain ranges/coastlines or other geographic features. Using what you know from this chapter about causes of earth movement and earthquakes, can you

explain any of the patterns you see? Find at least one concentration of earthquakes, and try to explain the geological processes that cause them. You might do this with a partner so that you can discuss likely explanations.

2. Now choose a map (conterminous U.S. Alaska, Hawaii, or Puerto Rico), and look at the magnitude of recent earthquakes: compare the size of squares to the magnitude numbers in the legend. What size class is most dominant? What proportion of earthquakes on your chosen map are in the largest size class?
3. Alaska is often a hot spot of earthquake activity. Find the Alaska map (you may have to back up a page or two), and zoom in on a concentration of earthquakes. How many earthquakes are off shore? How many are on shore?
4. Click on an earthquake in a location that is somewhat familiar to you. After you zoom in a few times, you should be redirected to some text information about that earthquake, and you may be able to find still more detailed maps. As you get closer, can you recognize features around the earthquake site? All these earthquakes occurred within the last week or less. Have you heard about these recent earthquakes in the news? Why or why not?

**For Additional Help in Studying This Chapter,** please visit our website at www.mhhe.com/cunningham10e. You will find additional practice quizzes and case studies, flashcards, regional examples, place markers for Google Earth™ mapping, and an extensive reading list, all of which will help you learn environmental science.

Planting trees is something everyone can do to reduce global warming.

CHAPTER

15

# Air, Weather, and Climate

*Climate is an angry beast and we are poking it with sticks.*

*—Wallace Broecker, Lamont-Doherty Earth Observatory—*

## LEARNING OUTCOMES

After studying this chapter, you should be able to:

**15.1** Illustrate how the atmosphere is a complex machine.

**15.2** Explain how weather happens.

**15.3** Clarify how climate can be an angry beast.

**15.4** Understand why global warming is happening.

**15.5** Evaluate how the Kyoto Protocol attempts to slow climate change.

# Case Study California Tackles Global Warming

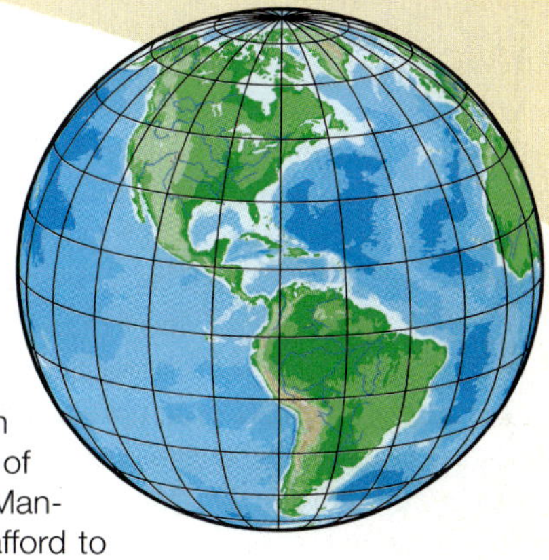

In 2006, California passed a groundbreaking law that places a cap on emissions of carbon dioxide and other global warming gases from utilities, refineries, and manufacturing plants. This law sets an ambitious goal of rolling back the state's greenhouse gas releases to 1990 levels by 2020, and to 80 percent below 1990 levels by 2050. It creates enforceable caps (maximum discharges from specific facilities) and requires regular industry emissions reports. The law specifically recommends market mechanisms for reaching these goals. That is, companies that can cut emissions below their maximum allowance can sell credits to other companies for which meeting limits is difficult or expensive. It's hoped that this will create incentives for innovation, and will produce the desired effect at the least cost and with the greatest flexibility for industry.

This law also prohibits utilities and corporations from buying power from out-of-state suppliers whose sources don't meet California's emission standards. It isn't the first time that the state has tried to do something about global warming. In 2004, California passed revolutionary legislation that required automakers to cut tailpipe emissions of carbon dioxide from cars and trucks. Industry didn't comply, so in 2006, California sued the six largest automakers, charging that their failure to control emissions is costing billions of dollars in global warming damages.

Why does all this matter? Most environmental scientists regard global warming as the most serious threat—outside of nuclear war—to our global environment. As we'll discuss in this chapter, a growing body of evidence suggests that our emissions of greenhouse gases—including carbon dioxide, methane, chlorofluorocarbons, nitrous oxide, and other chemicals—is already changing our global climate. If these releases continue to grow in the future at present rates, there could be truly disastrous consequences. We see increasing evidence of climate change all around us. Glaciers are melting, sea level is rising, seasons are changing their timing and length, and storms are growing more severe. Droughts devastate some areas, while intense rains and floods afflict other places. Nineteen of the 20 warmest years in the past 150 have occurred since 1980. Scientists warn that if trends continue as they are now, by the middle of this century there could be hundreds of millions of environmental refugees. The global economy could suffer a depression worse than that of the 1930s.

At least half the states and more than 500 American cities have taken steps to promote renewable energy and reduce greenhouse gas emissions, but California remains a leader in these efforts. California is currently the world's twelfth largest emitter of greenhouse gases and the world's eighth largest economy, while its 33 million residents make up 11 percent of the United States population. Manufacturers and retailers can't afford to ignore such a large segment of the country. Already, a number of large corporations are demanding a national policy. It's too expensive for them to have one set of rules and procedures to meet California rules (and similar limits in Europe for those doing business overseas) and another in the rest of the country.

What inspired California to take such a revolutionary step? The Bush administration has refused to place limits on greenhouse gases, claiming that it would harm the economy and cost jobs. Two recent studies, however, say just the opposite. These reports, one led by economists at the University of California, Berkeley the other by a Washington think tank that emphasizes market solutions to environmental problems, find that California's ambitious plan to limit global warming could create tens of thousands of new jobs and dramatically boost the economy. The reports didn't agree on the number of jobs added (one said 20,000 while the other estimated 83,000), or how much the economy would gain (one said \$60 billion, while the other said \$4 billion), but they agreed that the outcome would be positive.

California's leadership, as well as that of other states and cities, gives hope that we may be reaching a tipping point in public understanding of global warming. With the change in leadership in Washington in 2007, suddenly there are half a dozen bills in Congress to regulate greenhouse gases at the national level. But is it possible to do anything about such a huge problem? As you'll see in this chapter, experts argue that we already have the technology to make these changes. There isn't a single magic bullet to do the job, but a number of reasonable, even beneficial, changes can accomplish these goals. A major part of this chapter will be devoted to explaining why global warming is occurring and what we can do about it, but first, in order to understand our climate system better, we'll look at the factors that normally shape our weather and climate.

## 15.1 The Atmosphere Is a Complex Machine

We live at the bottom of a virtual ocean of air that extends upward about 500 km (300 mi). In the lowest 10 to 12 km, a layer known as the troposphere, the air moves ceaselessly, flowing and swirling, and continually redistributing heat and moisture from one part of the globe to another. The composition and behavior of the troposphere and other layers control our **weather** (daily temperature and moisture conditions in a place) and our **climate** (long-term weather patterns).

The earth's earliest atmosphere probably consisted mainly of hydrogen and helium. Over billions of years, most of that hydrogen and helium diffused into space. Volcanic emissions added carbon, nitrogen, oxygen, sulfur, and other elements to the atmosphere. Virtually all of the molecular oxygen ($O_2$) we breathe was probably produced by photosynthesis in blue-green bacteria, algae, and green plants.

Clean, dry air is mostly nitrogen and oxygen (table 15.1). Water vapor concentrations vary from near zero to 4 percent, depending on air temperature and available moisture. Minute particles and liquid droplets—collectively called **aerosols**—also are suspended in the air (fig. 15.1). Atmospheric aerosols play important roles in the earth's energy budget and in producing rain.

The atmosphere has four distinct zones of contrasting temperature, due to differences in absorption of solar energy (fig. 15.2). The layer of air immediately adjacent to the earth's surface is called the **troposphere** (*tropein* means to turn or change, in Greek). Within the troposphere, air circulates in great vertical and horizontal **convection currents,** constantly redistributing heat and moisture around the globe. The troposphere ranges in depth from about 18 km (11 mi) over the equator to about 8 km (5 mi) over the poles, where air is cold and dense. Because gravity holds most air molecules close to the earth's surface, the troposphere is much more dense than the other layers: It contains about 75 percent of the total mass of the atmosphere. Air temperature drops rapidly with increasing altitude in this layer, reaching about −60°C (−76°F) at the top of the troposphere. A sudden reversal of this temperature gradient creates a sharp boundary called the tropopause, which limits mixing between the troposphere and upper zones.

The **stratosphere** extends from the tropopause up to about 50 km (31 mi). It is vastly more dilute than the troposphere, but it has similar composition—except that it has almost no water vapor and nearly 1,000 times more **ozone** ($O_3$). This ozone absorbs some wavelengths of ultraviolet solar radiation, known as UV-B (290–330 nm, see fig. 3.10). This absorbed energy makes the atmosphere warmer toward the top of the stratosphere. Since UV radiation damages living tissues, this UV absorption in the stratosphere also protects life on the earth's surface. Recently discovered depletion of stratospheric ozone, especially over Antarctica, is allowing increased amounts of UV radiation to reach the earth's surface. If observed trends continue, this radiation could cause higher rates of skin cancer, genetic mutations, crop failures, and disruption of important biological communities, as you will see later in this chapter.

**FIGURE 15.1** The atmospheric processes that purify and redistribute water, moderate temperatures, and balance the chemical composition of the air are essential in making life possible. To a large extent, living organisms have created, and help to maintain, the atmosphere on which we all depend.

## TABLE 15.1

### Present Composition of the Lower Atmosphere*

| Gas | Symbol or Formula | Percent by Volume |
|---|---|---|
| Nitrogen | $N_2$ | 78.08 |
| Oxygen | $O_2$ | 20.94 |
| Argon | Ar | 0.934 |
| Carbon dioxide | $CO_2$ | 0.035 |
| Neon | Ne | 0.00182 |
| Helium | He | 0.00052 |
| Methane | $CH_4$ | 0.00015 |
| Krypton | Kr | 0.00011 |
| Hydrogen | $H_2$ | 0.00005 |
| Nitrous oxide | $N_2O$ | 0.00005 |
| Xenon | Xe | 0.000009 |

*Average composition of dry, clean air.

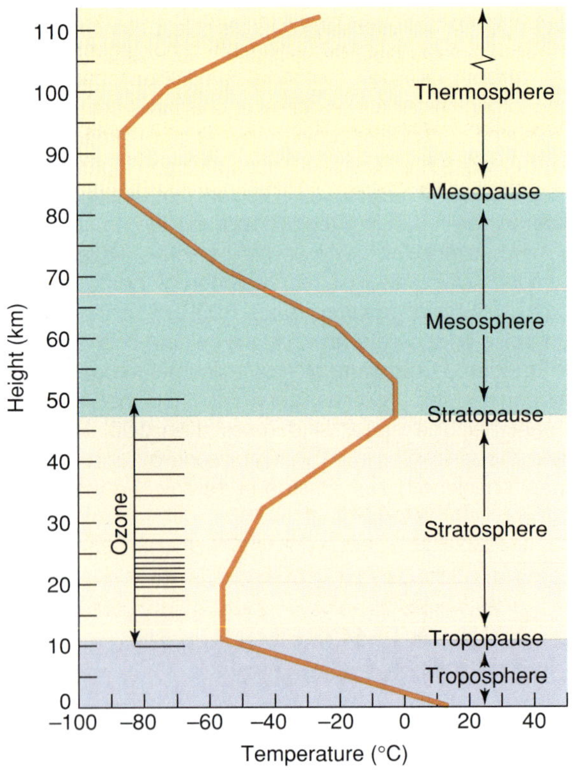


**FIGURE 15.2** Temperatures change dramatically in different layers of the atmosphere. Bars in the ozone graph represent relative concentrations of the "ozone layer" with altitude.
**Source:** Courtesy of Dr. William Culver, St. Petersburg Junior College.

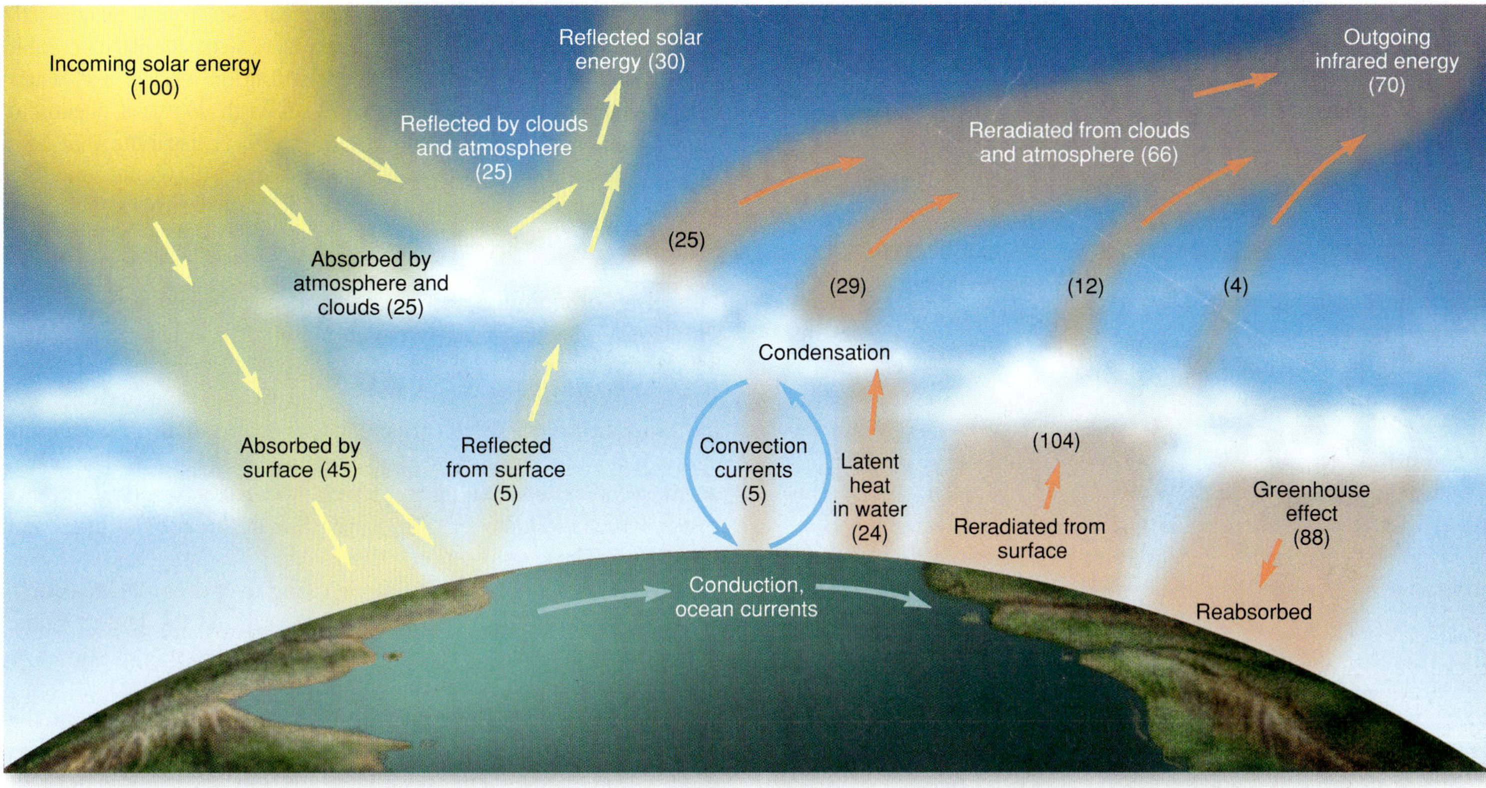


**FIGURE 15.3** Energy balance between incoming and outgoing radiation. The atmosphere absorbs or reflects about half of the solar energy reaching the earth. Most of the energy reemitted from the earth's surface is long-wave, infrared energy. Most of this infrared energy is absorbed by aerosols and gases in the atmosphere and is re-radiated toward the planet, keeping the surface much warmer than it would otherwise be. This is known as the greenhouse effect.

Unlike the troposphere, the stratosphere is relatively calm. There is so little mixing in the stratosphere that volcanic ash or human-caused contaminants can remain in suspension there for many years.

Above the stratosphere, the temperature diminishes again, creating the mesosphere, or middle layer. The thermosphere (heated layer) begins at about 80 km. This is a region of highly ionized (electrically charged) gases, heated by a steady flow of high-energy solar and cosmic radiation. In the lower part of the thermosphere, intense pulses of high-energy radiation cause electrically charged particles (ions) to glow. This phenomenon is what we know as the *aurora borealis* and *aurora australis,* or northern and southern lights.

No sharp boundary marks the end of the atmosphere. Pressure and density decrease with distance from the earth until they become indistinguishable from the near-vacuum of interstellar space.

## The sun warms our world

The sun supplies the earth with an enormous amount of energy, but that energy is not evenly distributed over the globe. Incoming solar radiation (insolation) is much stronger near the equator than at high latitudes. Of the solar energy that reaches the outer atmosphere, about one-quarter is reflected by clouds and atmospheric gases, and another quarter is absorbed by carbon dioxide, water vapor, ozone, methane, and a few other gases (fig. 15.3). This energy absorption warms the atmosphere slightly. About half of insolation reaches the earth's surface. Some of this energy is reflected by bright surfaces, such as snow, ice, and sand. The rest is absorbed by the earth's surface and by water. Surfaces that *reflect* energy have a high **albedo** (reflectivity). Most of these surfaces appear bright to us because they reflect light as well as other forms of radiative energy. Surfaces that *absorb* energy have a low albedo and generally appear dark. Black soil, pavement, dark green vegetation, and open water, for example, have low albedos (table 15.2).

**TABLE 15.2**

**Albedo (Reflectivity) of Surfaces**

| Surface | Albedo (%) |
|---|---|
| Fresh snow | 80–85 |
| Dense clouds | 70–90 |
| Water (low sun) | 50–80 |
| Sand | 20–30 |
| Water (sun overhead) | 5 |
| Forest | 5–10 |
| Black soil | 3 |
| Earth/atmosphere average | 30 |

Absorbed energy heats the absorbing surface (such as an asphalt parking lot in summer), evaporates water, or provides the energy for photosynthesis in plants. Following the second law of thermodynamics, absorbed energy is gradually reemitted as lower-quality heat energy. A brick building, for example, absorbs energy in the form of light and reemits that energy in the form of heat.

The change in energy quality is very important because the atmosphere selectively absorbs longer wavelengths. Most solar energy comes in the form of intense, high-energy light or near-infrared wavelengths. This short-wavelength energy passes relatively easily through the atmosphere to reach the earth's surface. Energy re-released from the earth's warmed surface ("terrestrial energy") is lower-intensity, longer-wavelength energy in the far-infrared part of the spectrum. Atmospheric gases, especially carbon dioxide and water vapor, absorb much of this long-wavelength energy, re-releasing it in the lower atmosphere and letting it leak out to space only slowly. This terrestrial energy provides most of the heat in the lower atmosphere. If the atmosphere were as transparent to infrared radiation as it is to visible light, the earth's average surface temperature would be about 18°C (33°F) colder than it is now.

This phenomenon is called the **greenhouse effect** because the atmosphere, loosely comparable to the glass of a greenhouse, transmits sunlight while trapping heat inside. The greenhouse effect is a natural atmospheric process that is necessary for life as we know it. However, too strong a greenhouse effect, caused by burning of fossil fuels and deforestation, may cause harmful environmental change.

The current rapid melting of Arctic sea ice is a good example both of the importance of albedo and a **positive feedback loop.** When the Arctic Ocean was entirely covered with ice, as it was most of the time previously, it acted as a giant mirror reflecting solar radiation back to space. Now that the ice is melting, open water is exposed to sunlight. The much lower albedo of water, compared to ice, means that much more solar energy is absorbed by the ocean. This leads to warmer water, which melts the ice further, leading to even more warming. This ever accelerating system could bring huge changes to our world.

## Water stores energy, and winds redistribute it

Much of the incoming solar energy is used to evaporate water. Every gram of evaporating water absorbs 580 calories of energy as it transforms from liquid to gas. Globally, water vapor contains a huge amount of stored energy, known as **latent heat**. When water vapor condenses, returning from a gas to a liquid form, the 580 calories of heat energy are released. Imagine the sun shining on the Gulf of Mexico in the winter. Warm sunshine and plenty of water allow continuous evaporation that converts an immense amount of solar (light) energy into latent heat stored in evaporated water. Now imagine a wind blowing the humid air north from the Gulf toward Canada. The air cools as it moves north (especially if it encounters cold air moving south). Cooling causes the water vapor to condense. Rain (or snow) falls as a consequence.

Note that not only water has moved from the Gulf to the Midwest: 580 calories of heat have also moved with every gram of moisture. The heat and water have moved from a place with strong incoming solar energy to a place with much less solar energy and much less water. The redistribution of heat and water around the globe are essential to life on earth. Without oceans to absorb and store heat, and wind currents to redistribute that heat in the latent energy of water vapor, the earth would undergo extreme temperature fluctuations like those of the moon, where it is 100°C (212°F) during the day and –130°C (–200°F) at night. Water performs this vital function because of its unique properties in heat absorption and energy of vaporization (chapter 3).

Uneven heating, with warm air close to the equator and colder air at high latitudes, also produces pressure differences that cause wind, rain, storms, and everything else we know as weather. As the sun warms the earth's surface, the air nearest the surface warms and expands, becoming less dense than the air above it. The warm air must then rise above the denser air. Vertical convection currents result, which circulate air from warm latitudes to cool latitudes and vice versa. These convection currents can be as small and as localized as a narrow column of hot air rising over a sun-heated rock, or they can cover huge regions of the earth. At the largest scale, the convection cells are described by a simplified model known as Hadley cells, which redistribute heat globally (fig. 15.4).

Where air rises in convection currents, air pressure at the surface is low. Where air is sinking, or subsiding, air pressure is high. On a weather map these high and low pressure centers, or rising and sinking currents of air, move across continents. In most of North America, they generally move from west to east. Rising air tends to cool with altitude, releasing latent heat that causes further rising. Very warm and humid air can rise very vigorously,

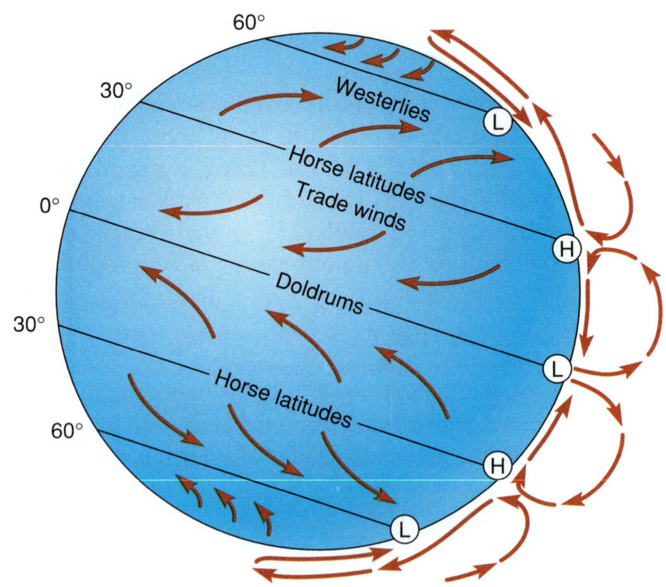


**FIGURE 15.4** General circulation patterns. The actual boundaries of the circulating convection cells vary from day to day and season to season, as do the local directions of surface winds. Surface topography also complicates circulation patterns, but within these broad regions, winds usually have a predominant and predictable direction. L and H represent low and high pressure.

especially if it is rising over a mass of very cold air. As water vapor carried aloft cools and condenses, it releases energy that fuels violent storms, which we will discuss later.

Pressure differences are an important cause of wind. There is always someplace with sinking (high pressure) air and someplace with low pressure (rising) air. Air moves from high-pressure centers toward low-pressure areas, and we call this movement wind.

## 15.2 Weather Happens

Weather is a description of the physical conditions in the atmosphere (humidity, temperature, air pressure, wind, and precipitation) over short time scales. Although most of us live in controlled environments, it's useful to know how and why weather happens.

### Why does it rain?

To understand why it rains, remember two things: Water condenses as air cools, and air cools as it rises. Any time air is rising, clouds, rain, or snow might form. Cooling occurs because of changes in pressure with altitude: Air cools as it rises (as pressure decreases); air warms as it sinks (as pressure increases). Air rises in convection currents where solar heating is intense, such as over the equator. Moving masses of air also rise over each other and cool. Air also rises when it encounters mountains. If the air is moist (if it has recently come from over an ocean or an evaporating forest region, for example), condensation and rainfall are likely as the air is lifted (fig. 15.5). Regions with intense solar

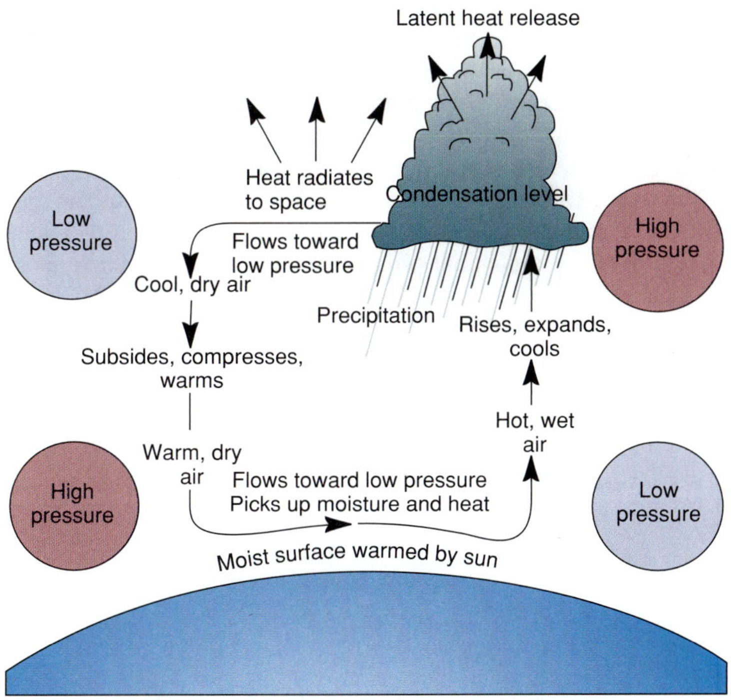


**FIGURE 15.5** Convection currents and latent energy cause atmospheric circulation and redistribute heat and water around the globe.

heating, frequent colliding air masses, or mountains tend to receive a great deal of precipitation.

Where air is sinking, on the other hand, it tends to warm because of increasing pressure. As it warms, available moisture evaporates. Rainfall occurs relatively rarely in areas of high pressure. High pressure and clear, dry conditions occur where convection currents are sinking. High pressure also occurs where air sinks after flowing over mountains. Figure 15.4 shows sinking, dry air at about 30° north and south latitudes. If you look at a world map, you will see a band of deserts at approximately these latitudes.

Another ingredient is usually necessary to initiate condensation of water vapor: condensation nuclei. Tiny particles of smoke, dust, sea salts, spores, and volcanic ash all act as condensation nuclei. These particles form a surface on which water molecules can begin to coalesce. Without them even supercooled vapor can remain in gaseous form. Even apparently clear air can contain large numbers of these particles, which are generally too small to be seen by the naked eye.

### Large-scale winds don't move in a straight line

In the Northern Hemisphere, winds generally bend clockwise (right), and in the Southern Hemisphere, they bend counterclockwise (left). This curving pattern results from the fact that the earth rotates in an eastward direction as the winds move above the surface. The apparent curvature of the winds is known as the **Coriolis effect.** On a global scale, this effect produces steady, reliable wind patterns, such as the trade winds and the midlatitude Westerlies (see fig. 15.4). Ocean currents similarly curve clockwise in the Northern Hemisphere and counterclockwise in the south. On a regional scale, the Coriolis effect produces cyclonic winds, or wind movements controlled by the earth's spin. Cyclonic winds spiral clockwise out of an area of high pressure in the Northern Hemisphere and counterclockwise into a low-pressure zone. If you look at a weather map in the newspaper, can you find this counterclockwise spiral pattern?

Why does this curving or spiraling motion occur? Imagine you were looking down on the North Pole of the rotating earth. Now imagine that the earth was a merry-go-round in a playground, with the North Pole at its center and the equator around the edge. As it spins counterclockwise (eastward), the spinning edge moves very fast (a full rotation, 39,800 km, every 24 hours for the real earth, or more than 1,600 km/hour). Near the center, though, there is very little eastward velocity. If you threw a ball from the edge toward the center, it would be traveling faster (edge speed) than the middle. It would appear, to someone standing on the merry-go-round, to curve toward the right. If you threw the ball from the center toward the edge, it would start out with no eastward velocity, but the surface below it would spin eastward, making the ball end up, to a person on the merry-go-round, west of its starting point.

Winds move above the earth's surface much as the ball does. If you were looking down at the South Pole, you would see the earth spinning clockwise, and winds—or thrown balls—would appear to bend left. Incidentally, this effect does not apply to the

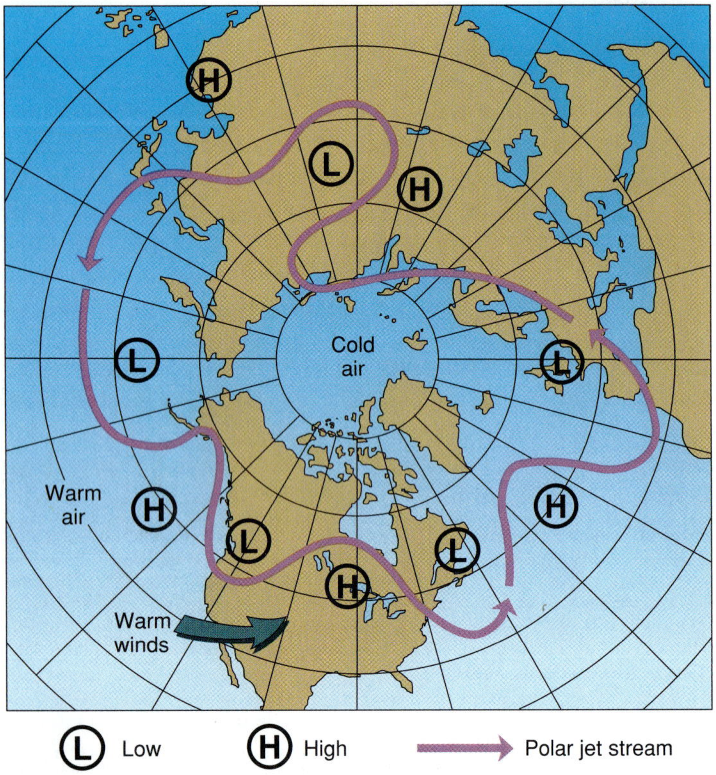


**FIGURE 15.6** A typical pattern of the arctic circumpolar vortex. This large, circulating mass of cold air sends "fingers," or lobes, across North America and Eurasia, spreading storms in their path. If the vortex becomes stalled, weather patterns stabilize, causing droughts in some areas and excess rain elsewhere.

way a sink drains. It's a myth that bathtubs and sinks spiral in opposite directions in the Northern and Southern Hemispheres. Those movements are far too small to be affected by the spinning of the earth.

At the top of the troposphere are **jet streams,** hurricane-force winds that circle the earth. These powerful winds follow an undulating path approximately where the vertical convection currents known as the Hadley and Ferrell cells meet. The approximate path of one jet stream over the Northern Hemisphere is shown in figure 15.6. Although we can't perceive jet streams on the ground, they are important to us because they greatly affect weather patterns. Sometimes jet streams dip down near the top of the world's highest mountains, exposing mountain climbers to violent, brutally cold winds.

## Ocean currents modify our weather

Warm and cold ocean currents strongly influence climate conditions on land. Surface ocean currents result from wind pushing on the ocean surface, as well as from the Coriolis effect. As surface water moves, deep water wells up to replace it, creating deeper ocean currents. Differences in water density—depending on the temperature and saltiness of the water—also drive ocean circulation. Huge cycling currents called gyres carry water north and south, redistributing heat from low latitudes to high latitudes (see fig. 17.6). The Alaska current, flowing from Alaska southward to California, keeps San Francisco cool and foggy during the summer.

The Gulf Stream, one of the best known currents, carries warm Caribbean water north past Canada's maritime provinces to northern Europe. This current is immense, some 800 times the volume of the Amazon, the world's largest river. The heat transported from the Gulf keeps Europe much warmer than it should be for its latitude. As the warm Gulf Stream passes Scandinavia and swirls around Iceland, the water cools and evaporates, becomes dense and salty, and plunges downward, creating a strong, deep, southward current. Oceanographer Wallace Broecker calls this the ocean conveyor system (see fig. 17.6).

Ocean circulation patterns were long thought to be unchanging, but now oceanographers believe that currents can shift abruptly. About 11,000 years ago, for example, as the earth was gradually warming at the end of the Pleistocene ice age, a huge body of meltwater, called Lake Agassiz, collected along the south margin of the North American ice sheet. At its peak, it contained more water than all the current freshwater lakes in the world. Drainage of this lake to the east was blocked by ice covering what is now the Great Lakes. When that ice dam suddenly gave way, it's estimated that some 163,000 $km^3$ of fresh water roared down the St. Lawrence Seaway and out into the North Atlantic, where it layered on top of the ocean and prevented the sinking of deep, cold, dense seawater. This, in turn, stopped the oceanic conveyor and plunged the whole planet back into an ice age (called the Younger Dryas after a small tundra flower that became more common in colder conditions) that lasted for another 1,300 years.

Could this happen again? Meltwater from Greenland glaciers is now flooding into the North Atlantic just where the Gulf Stream sinks and creates the deep south-flowing current. Already, evidence shows that the deep return flow has weakened by about 30 percent. Even minor changes in the strength or path of the Gulf Stream might give northern Europe a climate more like that of Siberia—not a pleasant prospect for inhabitants of the area.

***Think About It***

Find London and Stockholm on a globe. Then find cities in North America at a similar latitude. Temperatures in London and Stockholm rarely get much below freezing. How do you think their climate compares with the cities you've identified in North America? Why are they so different?

## Seasonal winds and monsoons have powerful effects

While figure 15.4 shows regular global wind patterns, large parts of the world, especially the tropics, receive seasonal winds and

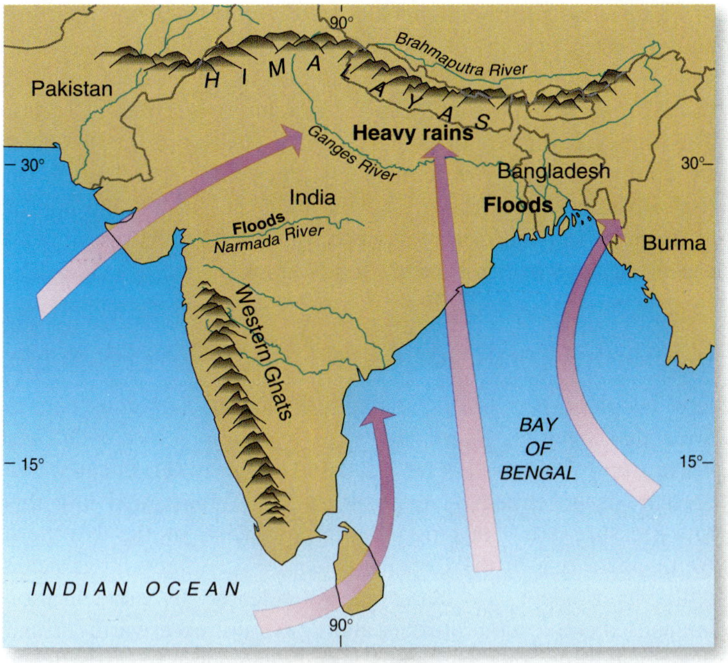


**FIGURE 15.7** Summer monsoon air flows over the Indian subcontinent. Warming air rises over the plains of central India in the summer, creating a low-pressure cell that draws in warm, wet oceanic air. As this moist air rises over the Western Ghats or the Himalayas, it cools and heavy rains result. These monsoon rains flood the great rivers, bringing water for agriculture, but also causing much suffering.

**FIGURE 15.8** Failure of monsoon rains brings drought, starvation, and death to both livestock and people in the Sahel desert margin of Africa. Although drought is a fact of life in Africa, many governments fail to plan for it, and human suffering is much worse than it needs to be.

rainy seasons that are essential for sustaining both ecosystems and human life. Sometimes these seasonal rains are extreme, causing disastrous flooding. In 2003, immense floods in China forced 100 million people from their homes. Many more millions elsewhere in Asia were similarly affected.

Sometimes the rains fail, causing crop failures and famine. The most regular seasonal winds and rains are known as **monsoons.** In India and Bangladesh, monsoon rains come when seasonal winds blow hot, humid air from the Indian Ocean (fig. 15.7). Strong convection currents lift this air, causing heavy rain across the subcontinent. When the rising air reaches the Himalayas, it rises even further, creating some of the heaviest rainfall in the world. During the 5-month rainy season of 1970, a weather station in the foothills of the Himalayas recorded 25 m (82 ft) of rain!

Tropical and subtropical regions around the world have seasonal rainy and dry seasons (see the discussion of tropical biomes, chapter 5). The main reason for this variable climate is that the region of most intense solar heating and evaporation shifts through the year. Remember that the earth's axis of rotation is at an angle. In December and January the sun is most intense just south of the equator; in June and July the sun is most intense just north of the equator. Wherever the sun shines most directly, evaporation and convection currents—and rainfall and thunderstorms—are very strong. As the earth orbits the sun, the tilt of its axis creates seasons with varying amounts of wind, rain, and heat or cold. Seasonal rains support seasonal tropical forests, and they fill some of the world's greatest rivers, including the Ganges and the Amazon. As the year shifts from summer to winter, solar heating weakens, the rainy season ends, and little rain may fall for months.

During the 1970s and 1980s, rains failed repeatedly in parts of the Sahel in northern Africa. Although rains have often been irregular in this region, increasing populations and civil disorder made these droughts result in widespread starvation and death (fig. 15.8). Climate scientists predict that droughts in some areas, and severe storms in others, will become more common with global climate change.

## Frontal systems create local weather

The boundary between two air masses of different temperature and density is called a front. When cooler air displaces warmer air, we call the moving boundary a **cold front.** Since cold air tends to be more dense than warm air, a cold front will hug the ground and push under warmer air as it advances. As warm air is forced upward, it cools adiabatically (without loss or gain of energy), and its cargo of water vapor condenses and precipitates. Upper layers of a moving cold air mass move faster than those in contact with the ground because of surface friction or drag, so the boundary profile assumes a curving, "bull-nose" appearance (fig. 15.9, *left*). Notice that the region of cloud formation and precipitation is relatively narrow. Cold fronts generate strong convective currents and often are accompanied by violent surface winds and destructive storms. An approaching cold front generates towering clouds called thunderheads that reach into the stratosphere where the jet stream pushes the cloud tops into a

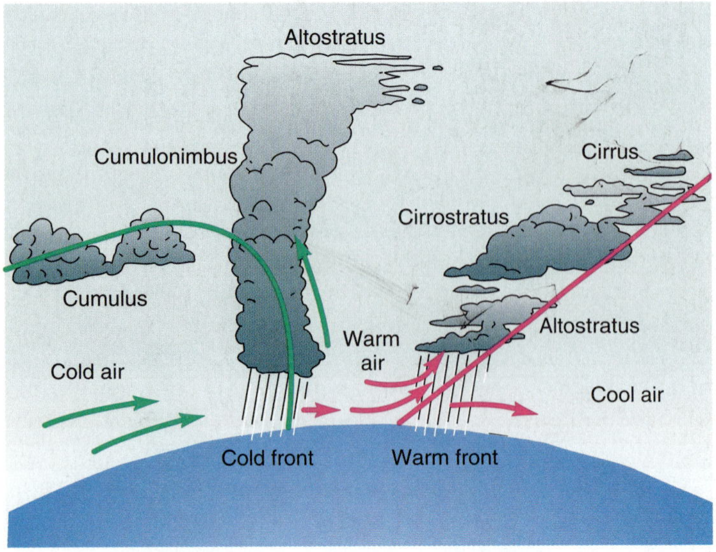


**FIGURE 15.9** A cold front assumes a bulbous, "bull-nose" appearance because ground drag retards forward movement of surface air. As warm air is lifted up over the advancing cold front, it cools, producing precipitation. When warm air advances, it slides up over cooler air in front and produces a long, wedge-shaped zone of clouds and precipitation. The high cirrus clouds that mark the advancing edge of the warm air mass may be 1,000 km and 48 hours ahead of the front at ground level.

characteristic anvil shape. The weather after the cold front passes is usually clear, dry, and invigorating.

If the advancing air mass is warmer than local air, a **warm front** results. Since warm air is less dense than cool air, an advancing warm front will slide up over cool, neighboring air parcels, creating a long, wedge-shaped profile with a broad band of clouds and precipitation (fig. 15.9, *right*). Gradual uplifting and cooling of air in the warm front avoids the violent updrafts and strong convection currents that accompany a cold front. A warm front will have many layers of clouds at different levels. The highest layers are often wispy cirrus (mare's tail) clouds that are composed mainly of ice crystals. They may extend 1,000 km (621 mi) ahead of the contact zone with the ground and appear as much as 48 hours before any precipitation. A moist warm front can bring days of drizzle and cloudy skies.

## Cyclonic storms can cause extensive damage

Few people experience a more powerful and dangerous natural force than cyclonic storms spawned by low-pressure cells over warm tropical oceans. As we discussed earlier in this chapter, low pressure is generated by rising warm air. Winds swirl into this low-pressure area, turning counterclockwise in the Northern Hemisphere due to the Coriolis effect. When rising air is laden with water vapor, the latent energy released by condensation intensifies convection currents and draws up more warm air and water vapor. As long as a temperature differential exists between air and ground and a supply of water vapor is available, the storm cell will continue to pump energy into the atmosphere.

Called **hurricanes** in the Atlantic and eastern Pacific, typhoons in the western Pacific, or cyclones in the Indian Ocean, winds near the center of these swirling air masses can reach hundreds of kilometers per hour and cause tremendous suffering and destruction. Often hundreds of kilometers across, these giant storms can generate winds as high as 320 km/hr (200 mph) and push walls of water called storm surges far inland (fig. 15.10*a*). In July 1931, torrential rains spawned by a typhoon caused massive flooding on China's Yangtze River that killed an

(a) Hurricane Floyd 1999

(b) Gulf Shores, Alabama 2005

(c) A tornado touches down

**FIGURE 15.10** (a) Hurricane Floyd was hundreds of kilometers wide as it approached Florida in 1999. Note the hole, or eye, in the center of the storm. (b) Destruction caused by Hurricane Katrina in 2005. More than 230,000 $km^2$ (90,000 $mi^2$) of coastal areas were devastated by this massive storm and many cities were almost completely demolished. (c) Tornadoes are much smaller than hurricanes, but can have stronger local winds.

estimated 3.7 million people, the largest known storm death toll in human history. A similar storm in July 1959 caused flooding of the Huang Ho (or Yellow River) that killed 2 million people.

On August 29, 2005, Hurricane Katrina slammed into the Gulf of Mexico shoreline. The category 4 storm, with 232 km/ hr winds (145 mph), pushed a storm surge up to 9 m (29 ft) high onto coastal areas and caused the greatest natural disaster in North American history. Coastal areas from Florida to Louisiana were damaged and many ocean-front cities were almost completely destroyed (fig. 15.10*b*).

**Tornadoes,** swirling funnel clouds that form over land, also are considered to be cyclonic storms, although their rotation isn't generated by Coriolis forces. While never as large or powerful as hurricanes, tornadoes can be just as destructive in the limited areas where they touch down (fig. 15.10*c*). Some meteorologists consider smaller, less destructive phenomena such as waterspouts, dust devils, and smoke funnels over forest fires to be miniature tornadoes, but others think they have very different origins.

Tornadoes are generated on the American Great Plains by giant "supercell" frontal systems where strong, dry air cold fronts from Canada collide with warm humid air moving north from the Gulf of Mexico. Greater air temperature differences cause more powerful storms, which is why most tornadoes occur in the spring, when arctic cold fronts penetrate far south over the warming plains. As warm air rises rapidly over dense, cold air, intense vertical convection currents generate towering thunderheads with anvil-shaped leading edges and domed tops up to 20,000 m (65,000 ft) high. Water vapor cools and condenses as it rises, releasing latent heat and accelerating updrafts within the supercell. Penetrating into the stratosphere, the tops of these clouds can encounter jet streams, which help create even stronger convection currents.

Sometimes a supercell isn't organized enough to spawn tornadoes, but strong downdrafts, called **downbursts,** can generate straight-line winds well over 160 km/hr (100 mph). In January 2007, straight-line winds in Europe reached hurricane strength, downing trees, grounding airplanes, derailing trains, and killing dozens of people.

The same strong updrafts that create tornadoes can also generate hail. Falling water droplets rise on powerful updrafts, freeze, fall again, and rise again. Droplets accumulate water, or ice, as they rise and fall, until they become heavy enough to resist being lofted by updrafts. If they remain frozen all the way to the ground, we call them hail. If you cut open large hail stones, you often can see accretion rings where successive layers of ice were added.

There seems to be a connection between strong monsoon winds over West Africa and the frequency of killer hurricanes in the Caribbean and along the Atlantic coast of North and Central America. In years when the Atlantic surface temperatures are high, a stronger than average monsoon trough forms over Africa. This pulls in moist maritime air that brings rain (and life) to the Sahel (the margin of the Sahara). This trough gives rise to tropical depressions (low-pressure storms) that follow one another in regular waves across the Atlantic. The weak trade winds associated with "wet" years allow these storms to organize and gain strength. Evaporation from the warm surface water provides energy.

***Think About It***

How would you describe the weather patterns where you live? Which of the factors described so far in this chapter are most important in determining your local weather?

During the 1970s and 1980s when the Sahel had devastating droughts, the weather was relatively quiet in North America. Only one killer hurricane (winds over 70 km/hr or 112 mph) reached the United States in two decades. By contrast, in 1995 when ocean surface temperatures were high and rain returned to the Sahel, seven hurricanes and six named tropical storms were spawned in the Atlantic. During the summer of 1996, satellite images of the Atlantic showed five active hurricanes roaring across the Caribbean at the same time. Are we witnessing a long-term climatic change or merely a temporary aberration?

## 15.3 Climate Can Be an Angry Beast

If weather is a description of physical conditions in the atmosphere, then climate is the *pattern* of weather in a region over long time periods. The interactions of atmospheric systems are so complex that climatic conditions are never exactly the same at any given location from one time to the next. While it is possible to discern patterns of average conditions over a season, year, decade, or century, complex fluctuations and cycles within cycles make generalizations doubtful and forecasting difficult. We always wonder whether anomalies in local weather patterns represent normal variations, a unique abnormality, or the beginnings of a shift to a new regime.

### Climates have changed dramatically throughout history

Ice cores from glaciers have revolutionized our understanding of climate history. In this research, a hollow tube is drilled down through the ice. Every 30 m, or so, the tube is pulled up and an ice cylinder is pushed out of the center (fig. 15.11). Back illumination reveals a pattern

**FIGURE 15.11** Dr. Mark Twickler, of the University of New Hampshire, holds a section of the 3,000 m Greenland ice sheet core, which records 250,000 years of climate history.

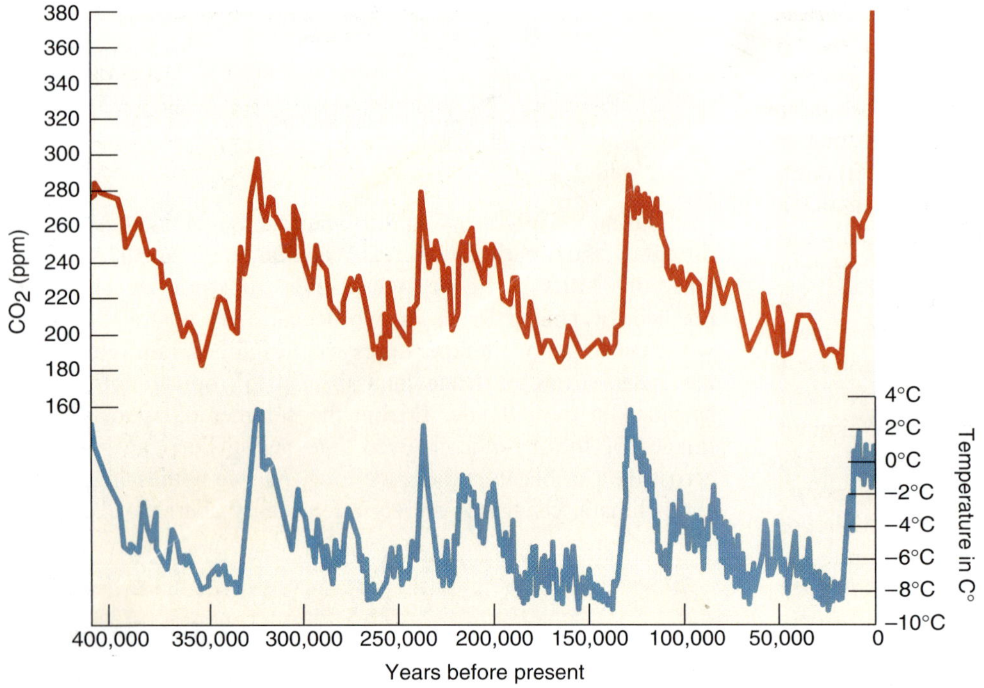


**FIGURE 15.12** Atmospheric $CO_2$ concentrations and global mean temperatures estimated from the antarctic Vostok ice core. Note the relatively rapid changes and close correlation in both climate and atmospheric chemistry.
**Source:** Data from United Nations Environment Programme.

of lighter and darker bands caused by annual snow accumulation on top of the glacier (which gives us an estimate of climate at the time). Gas bubbles trapped in the ice give us data on atmospheric composition when the snow was deposited. Ash layers and sulfate concentrations in the ice can be correlated with volcanic eruptions. The longest ice record ever collected is the Vostok ice core, which is 3,100 m long, and gives us a record of both global temperatures and atmospheric $CO_2$ over the past 420,000 years (fig. 15.12). A team of Russian scientists worked for 37 years at a site about 1,000 km from the South Pole to extract this ice core. A similar core, nearly as long as the Vostok, has been drilled from the Greenland ice sheet. Other glaciers throughout the world have also now been cored. All these ice samples show that climate has varied dramatically over time, but that there is a close correlation between atmospheric temperatures and $CO_2$ concentrations.

Major climatic changes, such as those of the Ice Ages, can have catastrophic effects on living organisms. If climatic change is gradual, species may have time to adapt or migrate to more suitable locations. Where climatic change is relatively abrupt, many organisms are unable to respond before conditions exceed their tolerance limits. Whole communities may be destroyed, and if the climatic change is widespread, many species may become extinct.

Perhaps the most well-studied example of this phenomenon is the great die-off that occurred about 65 million years ago at the end of the Cretaceous period. Most dinosaurs—along with 75 percent of all previously existing plant and animal species—became extinct, apparently as a result of sudden cooling of the earth's climate. Geologic evidence suggests that this catastrophe was not an isolated event. There appear to have been several great climatic changes, perhaps as many as a dozen, in which large numbers of species were exterminated (see table 11.3, p. 235).

## What causes catastrophic climatic swings?

There are many explanations for climatic catastrophes. One of these is that changes in solar energy associated with 11-year sunspot cycles or 22-year solar magnetic cycles might play a role. Another theory is that a regular 18.6-year cycle of shifts in the angle at which our moon orbits the earth alters tides and atmospheric circulation in a way that affects climate. A theory that has received attention in recent years suggests that orbital variations as the earth rotates around the sun might be responsible for cyclic weather changes.

**Milankovitch cycles,** named after Serbian scientist Milutin Milankovitch, who first described them in the 1920s, are periodic shifts in the earth's orbit and tilt (fig. 15.13). The earth's elliptical orbit stretches and shortens

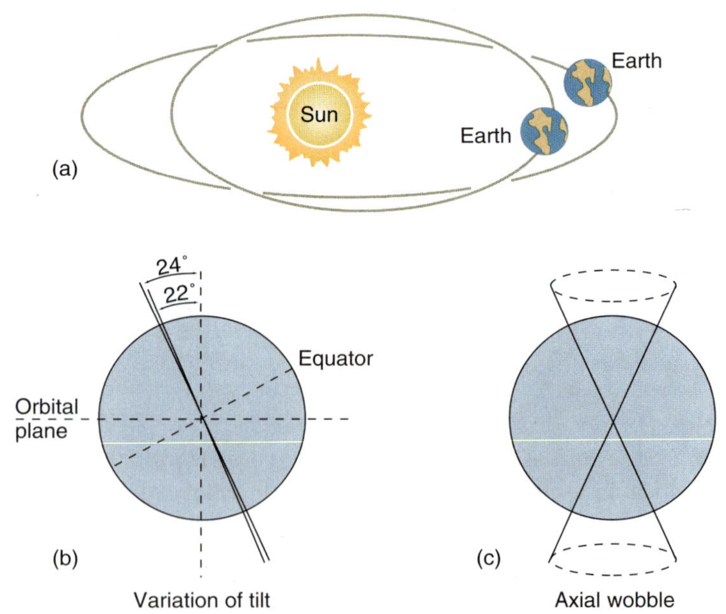


**FIGURE 15.13** Milankovitch cycles, which may affect long-term climate conditions: (a) changes in the elliptical shape of the earth's orbit, (b) shifting tilt of the axis, and (c) wobble of the earth.

in a 100,000-year cycle, while the axis of rotation changes its angle of tilt in a 40,000-year cycle. Furthermore, over a 26,000-year period, the axis wobbles like an out- of-balance spinning top. These variations change the distribution and intensity of sunlight reaching the earth's surface and, consequently, global climate. Bands of sedimentary rock laid in the oceans seem to match both these Milankovitch cycles and the periodic cold spells associated with worldwide expansion of glaciers every 100,000 years or so.

A historical precedent for global climate change that had disastrous effects on humans was the "little ice age" that began in the 1400s. Temperatures dropped so that crops failed repeatedly in parts of northern Europe that once were good farmland. Scandinavian settlements in Greenland founded during the warmer period around A.D. 1000 lost contact with Iceland and Europe as ice blocked shipping lanes. It became too cold to grow crops, and fish that once migrated along the coast stayed farther south. The settlers slowly died out, perhaps in battles with Inuit people who were driven south from the high Arctic by colder weather.

Evidence from ice cores drilled in the Greenland ice cap suggests that world climate may change much more rapidly than previously thought. During the last major interglacial period 135,000 to 115,000 years ago, it appears that temperatures flipped suddenly from warm to cold or vice versa over a period of years or decades rather than centuries.

Volcanic eruptions may have been responsible for some of these climate flips. When Mt. Toba exploded in western Sumatra about 73,000 years ago, for example, it was the largest volcanic cataclysm in the past 28 million years. It's estimated that Mt. Toba ejected at least 2,800 $km^3$ of material compared to only 1 $km^3$ emitted by Mt. St. Helens in Washington State in 1980. Mt. Toba produced enough ash to cover the entire earth an average of 10 cm deep. The signal from this eruption can be detected in the Greenland ice core. The sulfuric acid and particulate material ejected into the atmosphere from Mt. Toba are estimated to have dimmed incoming sunlight by 75 percent and to have cooled the whole planet by as much as 16°C for more than 160 years. The ensuing volcanic winter is calculated to have reduced terrestrial biomass by 90 percent, and to have created a genetic bottleneck still evident in many species, including humans.

## El Niño/Southern Oscillations are powerful cycles

What do sea surface temperatures near Indonesia, anchovy fishing off the coast of Peru, and the direction of tropical trade winds have to do with rainfall and temperatures over the United States? Strangely enough, they all may be interrelated. If the terms *El Niño, La Niña,* and *Southern Oscillation* are not yet a part of your vocabulary, perhaps they should be. They describe a connection between the ocean and atmosphere that appears to affect these factors as well as weather patterns throughout the world.

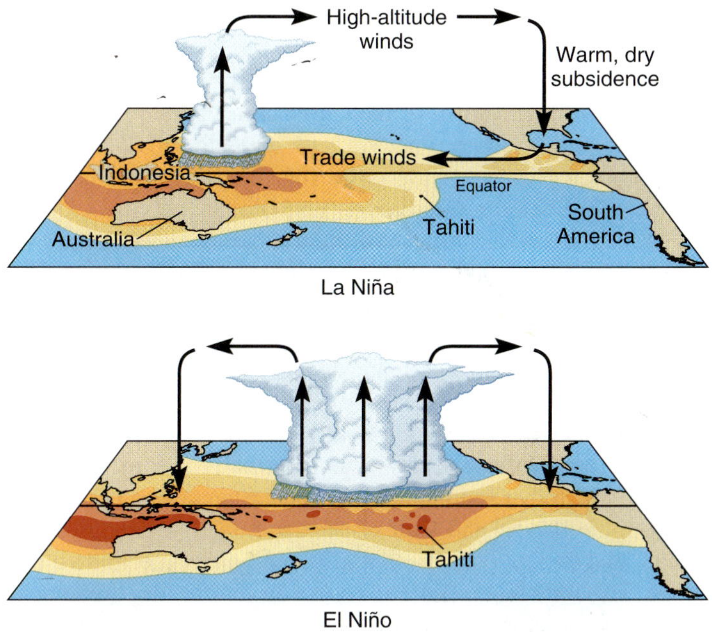


**FIGURE 15.14** The El Niño/La Niña Southern Oscillation Cycle. During El Niño years, surface trade winds that normally push warm water westward toward Indonesia weaken and allow this pool of water to flow eastward, bringing storms to the Americas.

The core of this climatic system is a huge pool of warm surface water in the Pacific Ocean that sloshes slowly back and forth between Indonesia and South America like water in a giant bathtub. Most years, this pool is held in the western Pacific by steady equatorial trade winds pushing ocean surface currents westward (fig. 15.14). These surface winds are generated by a huge low-pressure cell formed by upwelling convection currents of moist air warmed by the ocean. Towering thunderheads created by rising air bring torrential summer rains to the tropical rainforests of Northern Australia and Southeast Asia. Winds high in the troposphere carry a return flow back to the eastern Pacific where dry subsiding currents create deserts from Chile to southern California. Surface waters driven westward by the trade winds are replaced by upwelling of cold, nutrient-rich, deep waters off the west coast of South America that support dense schools of anchovies and other finfish.

Every three to five years, for reasons that we don't fully understand, the Indonesian low collapses and the mass of warm surface water surges back east across the Pacific. One theory is that the high cirrus clouds atop the cloud columns block the sun and cool the ocean surface enough to reverse trade winds and ocean surface currents so they flow eastward rather than westward. Another theory is that eastward-flowing deep currents called baroclinic waves periodically interfere with coastal upwelling, warming the sea surface off South America and eliminating the temperature gradient across the Pacific. At any rate, the shift in position of the tropical depression sets off a chain of events lasting a year or more with repercussions in weather systems across North and South America and perhaps around the world.

Peruvian fishermen were the first to notice irregular cycles of rising ocean temperatures that resulted in disappearance of the anchovy schools on which they depended. They named these events **El Niño** (Spanish for the Christ child) because they often occur around Christmas time. We have come to call the intervening years **La Niña** (or little girl). Together, this cycle is called the El Niño Southern Oscillation (ENSO).

Actually, sea surface temperature measurements in the central Pacific near Tahiti are the best measurement of this giant oscillation. In May 1998, as an unusually strong El Niño was ending, surface temperatures in this region plunged 7°C (12.5°F) in a little over a week, a dramatic event considering the enormous heat-storing capacity of ocean waters.

How does the ENSO cycle affect you? During an El Niño year, the northern jet stream—which normally is over Canada—splits and is drawn south over the United States. This pulls moist air from the Pacific and Gulf of Mexico inland, bringing intense storms and heavy rains from California across the midwestern states. The intervening La Niña years bring hot, dry weather to these same areas. An exceptionally long El Niño event from 1991 to 1995 broke a seven-year drought over the western United States and resulted in floods of the century in the Mississippi Valley. Oregon, Washington, and British Columbia, on the other hand, tend to have warm, sunny weather in El Niño years rather than their usual rain. Droughts in Australia and Indonesia during El Niño episodes cause disastrous crop failures and forest fires, including one in Borneo in 1983 that burned 3.3 million ha (8 million acres). An even stronger El Niño in 1997 spread health-threatening forest fire smoke over much of Southeast Asia.

Are ENSO events becoming stronger or more irregular because of global climate change? Studies of corals up to 130,000 years old suggest this is true. Furthermore, there are signs that warm ocean surface temperatures are spreading. In addition to the pool of warm water in the western Pacific associated with La Niña years, oceanographers recently discovered a similar warm region in the Indian Ocean. High sea surface temperatures spawn larger and more violent storms such as hurricanes and typhoons. On the other hand, increased cloud cover would raise the albedo while upwelling convection currents generated by these storms could pump heat into the stratosphere. This might have an overall cooling effect and act as a safety valve for global warming.

A longer-scale ocean/weather connection called the **Pacific Decadal Oscillation (PDO)** involves a very large pool of warm water that moves back and forth across the North Pacific every 30 years or so. From about 1977 to 1997, surface water temperatures in the middle and western part of the North Pacific Ocean were cooler than average, while waters off the western United States were warmer. During this time, salmon runs in Alaska were bountiful, while those in Washington and Oregon were greatly diminished. In 1997, however, ocean surface temperatures along the coast of western North America turned significantly cooler, perhaps marking a return to conditions that prevailed between 1947 and 1977. Under this cooler regime, Alaskan salmon runs declined while those in Washington and Oregon improved somewhat. A similar oscillation occurs in the North Atlantic. Together these decades-long climate patterns can be seen in tree-ring growth throughout North America.

## 15.4 Global Warming Is Happening

As the opening case study for this chapter shows, many scientists regard anthropogenic (human-caused) global climate change to be the most important environmental issue of our times. The possibility that humans might alter world climate is not a new idea. In 1895, Svante Arrhenius, who subsequently received a Nobel Prize for his work in chemistry, predicted that $CO_2$ released by coal burning could cause global warming. Most of Arrhenius' contemporaries dismissed his calculations as impossible, but his work seems remarkably prescient now.

### A scientific consensus is emerging

The first evidence that human activities are increasing atmospheric $CO_2$ came from an observatory on top of the Mauna Loa volcano in Hawaii. The observatory was established in 1957 as part of an International Geophysical Year, and was intended to provide data on air chemistry in a remote, pristine environment. Surprisingly, measurements showed $CO_2$ levels increasing about 0.5 percent per year, rising from 315 ppm in 1958 to 383 ppm in 2006 (fig. 15.15). This increase isn't a straight line, however. Because a majority of the world's land and vegetation are in the Northern Hemisphere, northern seasons dominate the signal. Every May, $CO_2$ levels drop slightly as plant growth on northern continents use $CO_2$ in photosynthesis. During the northern winter, levels rise again as respiration releases $CO_2$.

The **Intergovernmental Panel on Climate Change (IPCC)** brings together scientists and government representatives from 130 countries to review scientific evidence on the causes and likely effects of human-caused climate change. In 2007, the IPCC issued its fourth report. The result of 6 years of work by 2,500 scientists, the four volumes of the report represent a consensus by more than 90 percent of all the scientists working on climate

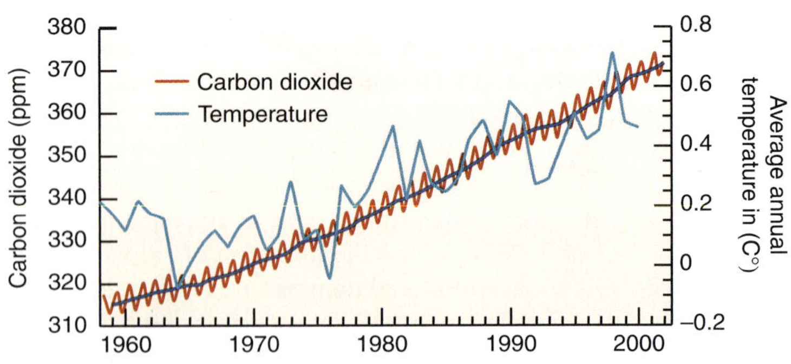


**FIGURE 15.15** Carbon dioxide concentrations on Mauna Loa (monthly and annual averages) and global annual average temperature variations from 1960 to 2000.
**Data source:** NASA Earth Observatory, 2002.

change. This report has the strongest expression of certainty to date from the climate community. It says that global warming is "very likely" caused by humans. This implies a 90 percent probability that changes we are now seeing are caused by human actions. It's reported, however, that most delegates wanted the statement to say that human causation is "virtually certain," meaning a 99 percent probability. To achieve a consensus, the lead authors accepted the less definite wording. Most scientists, however, regard the evidence as "indisputable."

The estimated range of global temperature increase by the end of this century also has been increased in this report. The third IPCC report, published in 2001, calculated that average global surface temperatures would increase 1.4–5.8°C by 2100. The fourth report gives a variety of scenarios depending on population and economic growth, energy conservation and efficiency, and adoption (or lack thereof) of greenhouse gas controls. This gives a range of increases from 1.1–6.4°C (2–11.5°F) higher than now by 2100. According to the IPCC, the "best estimate," for temperature rise is now 1.8–4.0°C (3.2–7.8°F). To put that in perspective, the change in average global temperature between now and the middle of the last glacial period is estimated to be about 5°C.

The worst outcome from global warming could mean more than 1 million human deaths and hundreds of billions of dollars in costs by 2100. For most people, the harshest effects of global warming will be from extreme weather, including droughts, floods, heat waves, and hurricanes. These extremes have already increased significantly in the past decade, and are expected to get even worse in the future.

Sea levels are projected to rise 17–57 cm (7–23 in.) by the end of this century. If recent rapid melting of polar ice sheets and Greenland glaciers continues, it could add another 15 cm (6 in.) to that total. If nothing is done to halt greenhouse emissions, complete melting of Greenland's ice sheet would raise sea level by more than 60 m (nearly 20 ft). This would flood most of Florida, a broad swath of the Gulf Coast, most of Manhattan Island, Shanghai, Hong Kong, Tokyo, Kolkata, Mumbai, and many of the other largest cities in the world.

Following the release of the IPCC report, 45 nations called for a new global environmental body to regulate greenhouse emissions, perhaps with policing powers to punish violators. Conspicuously absent from this initiative were the world's biggest emitters of these gases, including the United States, China, and India. The Bush administration praised the IPCC report but said it still opposes mandatory cuts in greenhouse gas emissions. Voluntary programs, the president maintains, will be enough.

## Greenhouse gases have many sources

Since preindustrial times atmospheric concentrations of $CO_2$, $CH_4$, and $N_2O$ have climbed by over 31 percent, 151 percent, and 17 percent, respectively. Carbon dioxide is by far the most important cause of anthropogenic climate change (fig. 15.16*a*). Burning fossil fuels, making cement, burning forests and grasslands, and other human activities release nearly 30 billion tons of $CO_2$ every year, on average, containing some 8 billion tons of carbon (fig. 15.16*b*).

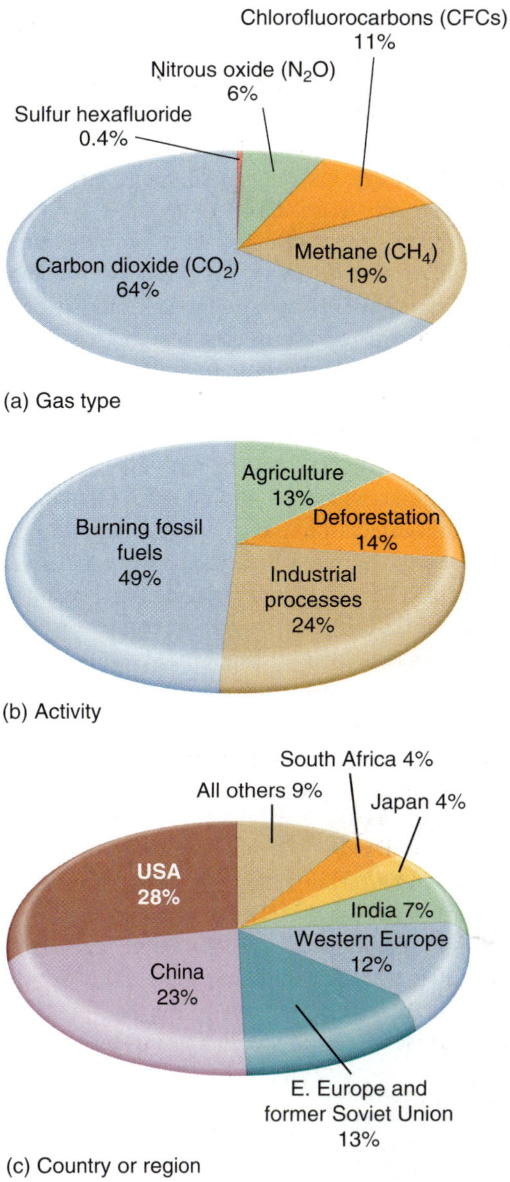


**FIGURE 15.16** Percentage contributions to global warming by (a) gas type, (b) source activity, and (c) country or region.
Data from world Resources Institute.

About 3 billion tons of this excess carbon is taken up by terrestrial ecosystems, and around 2 billion tons are absorbed by the oceans, leaving an annual atmospheric increase of some 3 billion tons per year. If current trends continue, $CO_2$ concentrations could reach about 500 ppm (approaching twice the preindustrial level of 280 ppm) by the end of the twenty-first century.

Although rarer than $CO_2$, methane absorbs 23 times as much infrared and is accumulating in the atmosphere about twice as fast as $CO_2$. Methane is released by ruminant animals, wet-rice paddies, coal mines, landfills, wetlands, and pipeline leaks. Hydroelectricity is usually promoted as a clean power source that offers an alternative to fossil fuels, but this may not always be true.

Philip Fearnside, an ecologist at Brazil's National Institute for Amazon Research, calculates that decaying vegetation in the reservoir behind the Cura-Una dam in Para Province emits so much carbon dioxide and methane every year that it causes three and a half times as much global warming as would generating the same amount of energy by burning fossil fuels. Chlorofluorocarbons (CFCs) also are powerful infrared absorbers. CFC releases in developed countries have declined since many of their uses were banned, but increasing production in developing countries, such as China and India, remains a problem. Nitrous oxide is produced by burning organic material and by soil denitrification. As fig. 15.16*a* shows, CFCs and $N_2O$ together are thought to account for only about 17 percent of human-caused global warming.

The United States, with less than 5 percent of the world's population, produces 28 percent of all anthropogenic $CO_2$ (fig. 15.16*c*). China, with 1.3 billion people, is second in total $CO_2$ emissions, but fourteenth in per capita production. Japan and western Europe, which by most measures have standards of living at least equal to the United States, produce only about half as much $CO_2$ per person. Eastern Europe and the former Soviet Union have many old, inefficient factories and coal-buring power plants that emit high $CO_2$ levels with relatively little economic return.

## Evidence of climate change is overwhelming

There is little doubt that our planet is warming. The American Geophysical Union, one of the nation's largest and most respected scientific organizations, says, "As best as can be determined, the world is now warmer than it has been at any point in the last two millennia, and, if current trends continue, by the end of the century it will likely be hotter than at any point in the last two million years." Some of the evidence for this warming includes the following:

- Over the last century the average global temperature has climbed about 0.6°C (1°F). Nineteen of the 20 warmest years in the past 150 have occurred since 1980. The hottest year since the instrumental temperature record began was 2005, and 2007 is expected to be even hotter.
- Polar regions have warmed much faster than the rest of the world. In Alaska, western Canada, and eastern Russia, average temperatures have increased as much as 4°C (7°F) over the past 50 years. Permafrost is melting; houses, roads, pipelines, sewage systems, and transmission lines are being damaged as the ground sinks beneath them. Trees are tipping over, and a beetle infestation in Alaska (made possible by warmer winters) killed 250 million spruce trees on the Kenai Peninsula. Coastal villages, such as Shishmaref, Point Hope, and Barrow, are having to relocate inland as ice breakup and increasingly severe storms erode the arctic shoreline.
- Arctic sea ice is only half as thick now as it was 30 years ago and the area covered by sea ice has decreased by more than 1 million $km^2$ (an area larger than Texas and Oklahoma combined) in just three decades. By 2040, the Arctic Ocean could be totally ice-free in the summer, if present trends continue.

**FIGURE 15.17** Diminishing arctic sea ice prevents polar bears from hunting seals, their main food source.

This is bad news for polar bears, which depend on the ice to hunt seals. An aerial survey in 2005 found bears swimming across as much as 260 km (160 mi) of open water to reach the pack ice. When the survey was repeated after a major storm, dozens of bears were missing or spotted dead in the water. In 2007 the United States considered adding polar bears to the endangered species list because of loss of arctic ice (fig. 15.17). Loss of sea ice is also devastating for Inuit people whose traditional lifestyle depends on ice for travel and hunting.

- Ice shelves on the Antarctic Peninsula are breaking up and disappearing rapidly. A recent survey found that 90 percent of the glaciers on the peninsula are now retreating an average of 50 m per year. The Greenland ice cap also is reported to be melting twice as fast as it did a few years ago. Because ice shelves are floating, they don't affect sea level when they melt. Greenland's massive ice cap, however, holds enough water to raise sea level by about 7 m (about 23 ft) if it all melts.
- Alpine glaciers everywhere are retreating rapidly (fig. 15.18). Mt. Kilimanjaro has lost 85 percent of its famous ice cap since 1915. By 2015, all the permanent ice on the mountaintop is expected to be gone. In 1972, Venezuela had six glaciers; now it has only two. When Montana's Glacier National Park was created in 1910, it held some 150 glaciers. Now, fewer than 30, greatly shrunken glaciers remain (fig. 15.18). If current trends continue, all will have melted by 2030.
- So far, the oceans have been buffering the effects of our greenhouse emissions both by absorbing $CO_2$ directly and by storing heat. Deep-diving sensors show that the oceans are absorbing 0.85 watts per $m^2$ more than is radiated back to space. This absorption slows current warming, but also means that even if we reduce our greenhouse gas emissions today, it will take centuries to dissipate that stored heat. The higher levels of $CO_2$ being absorbed are acidifying

**FIGURE 15.18** Alpine glaciers everywhere are retreating rapidly. These images show the Grinnell Glacier in 1914 and 1998. By 2030, if present melting continues, there will be no glaciers in Glacier National Park.

the oceans, and could have adverse effects on sea life. Mollusks and corals, for example, have a more difficult time making calcium carbonate shells and skeletons at the lower pH.

- Sea level has risen worldwide approximately 15–20 cm (6–8 in.) in the past century. About one-quarter of this increase is ascribed to melting glaciers; roughly half is due to thermal expansion of seawater. If all of Antarctica were to melt, the resulting rise in sea level could be several hundred meters.
- Satellite images and surface measurements show that growing seasons are now as much as three weeks longer in a band across northern Eurasia and North America than they were 30 years ago (see fig. 1.9). New plants and animals, never before seen in the far north, are now extending their territories, while native species, such as musk ox, caribou, walrus, and seal populations are declining as climate changes.
- Droughts are becoming more frequent and widespread. In Africa, for example, droughts have increased about 30 percent since 1970. In 2005, the United Nations reported that 60 million people in 36 countries needed emergency food aid. Drought was the single largest cause of famine, although wars, economics, and internal politics always exacerbate these emergencies.
- Biologists report that many animals are breeding earlier or extending their range into new territory as the climate changes. In Europe and North America, for example, 57 butterfly species have either died out at the southern end of their range, or extended the northern limits, or both. Plants, also, are moving into new territories. Given enough time and a route for migration, many species may adapt to new conditions, but we now are forcing them to move much faster than many achieved at the end of the last ice age (fig. 15.19).
- The disappearance of amphibians, such as the beautiful golden toads from the cloud forests of Costa Rica or western toads from Oregon's Cascade Range, are thought to have been caused at least in part by changing weather patterns. Emperor and Adélie penguin populations have declined by half over the past 50 years as the ice shelves on which they depend for feeding and breeding disappear.
- Coral reefs worldwide are "bleaching," losing key algae and resident organisms, as water temperatures rise above 30°C (85°F). With reefs nearly everywhere threatened by pollution, overfishing, and other stressors, scientists worry that rapid climate change could be the final blow for many species in these complex, biologically rich ecosystems.
- Storms are becoming stronger and more damaging. The 2005 Atlantic storm season was the most severe on record, with 26 named tropical storms, twice as many as the average

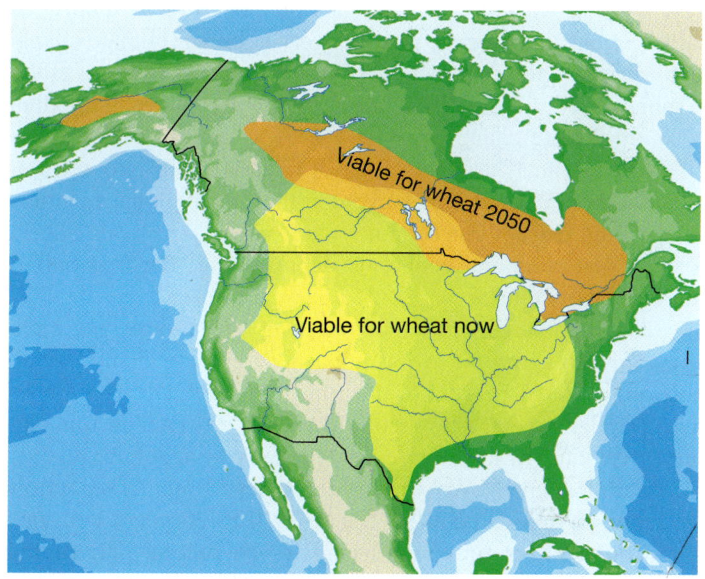


**FIGURE 15.19** Most of the central United States is suitable for growing wheat now, but if current trends continue, the climatic conditions for wheat could be in central Canada in 2050.

**FIGURE 15.20** More than 1 million people had to be evacuated from New Orleans in 2005 when a storm surge created by Hurricane Katrina breached the levees and allowed Lake Pontchartrain to flood into the city.

over the past 30 years. Some of this increased frequency could be a natural cycle, but the greater strength of these storms (an unprecedented 7 hurricanes reached level 5, the highest category) undoubtedly reflects higher sea surface temperatures. Hurricane Katrina, which demolished a 230,000 $km^2$ (90,000 $mi^2$) area of the U.S. Gulf Coast (an area about the size of Great Britain) and flooded New Orleans (fig. 15.20) caused at least $200 billion in damage, the greatest storm loss in American history.

## Global warming will be expensive

In 2006, Sir Nicholas Stern, former chief economist of the World Bank, issued a study on behalf of the British government on the costs of global climate change. It was one of the strongest worded warnings to date from a government report. He said, "Scientific evidence is now overwhelming: climate change is a serious global threat, and it demands an urgent global response." If we don't act soon, Stern estimates, immediate costs of climate change will be at least 5 percent of the global GDP each year. If a wider range of risks is taken into account, the damage could equal 20 percent of the annual global economy. That would disrupt our economy and society on a scale similar to the great wars and economic depression of the first half of the twentieth century.

The report estimates that reducing greenhouse gas emissions now to avoid the worst impacts of climate change would cost only about 1 percent of the annual global GDP. That means that a dollar invested now could save us $20 later in this century. "We can't wait the five years it took to negotiate Kyoto," Sir Nicholas says, "We simply don't have time." The actions we take—or fail to take—in the next 10 to 20 years will have a profound effect on those living in the second half of this century and in the next. Energy production, Stern suggests, will have to be at least 80 percent decarbonized by 2050 to stabilize our global climate.

As chapter 2 points out, global climate change is a moral and ethical issue as well as a practical one. Many religious leaders are joining with scientists and business leaders to campaign for measures to reduce greenhouse gas emissions. Ultimately, the people likely to suffer most from global warming are the poorest in Africa, Asia, and Latin America who have contributed least to the problem. Those of us in the richer countries will likely have resources to blunt problems caused by climate change, but residents of poorer countries will have fewer options. The Stern report says that without action, at least 200 million people could become refugees as their homes are hit by drought or floods. Furthermore, there's a question of intergenerational equity. What kind of world are we leaving to our children and grandchildren? What price will they pay if we fail to act?

The Stern review recommends four key elements for combating climate change. They are: (1) *emissions trading* to promote cost-effective emissions reductions, (2) *technology sharing* that would double research investment in clean energy technology and accelerate spread of that technology to developing countries, (3) *reduce deforestation,* which is a quick and highly cost-effective way to reduce emissions, and (4) *help poorer countries* by honoring pledges for development assistance to adapt to climate change.

The fourth IPCC report—issued in 2007—estimates the cost of preventing $CO_2$ doubling to be even less than Sir Nicholas calculates. The IPCC says it would cost only 0.12 percent of annual global GDP to reduce carbon emissions the 2 percent per year necessary to stabilize world climate. The Bush administration continues to claim this to be too expensive even though reports, such as that described in the opening case study for this chapter, predict that regulating carbon could stimulate the global economy and create millions of jobs.

A vigorous market for emissions trading already exists. In 2006, about 700 million tons of carbon equivalent credits were exchanged for some (U.S.) $3.5 billion. In 2008, trade is expected to grow to the equivalent of 4.8 billion tons of carbon reduction. Ultimately, this market may grow to $500 billion per year by 2050, and could represent an important source of development assistance for poorer countries. Market mechanisms for controlling climate change are discussed further in chapter 23.

In many cases, both organisms and human infrastructure will not be able to move or adapt quickly enough to accommodate the rates at which climate is now changing. Tropical areas will probably not get much warmer than they are now, but some middle- and high-latitude areas could experience unaccustomed heat (fig. 15.21). Of course there could also be some good results from climate change. Residents of northern Canada, Siberia, and Alaska may enjoy warmer temperatures, longer growing seasons, and longer ice-free shipping seasons from their ports. The opening of the fabled Northwest Passage through the Arctic Ocean could cut thousands of kilometers off a trip from London to Vladivostok. Stronger monsoons might

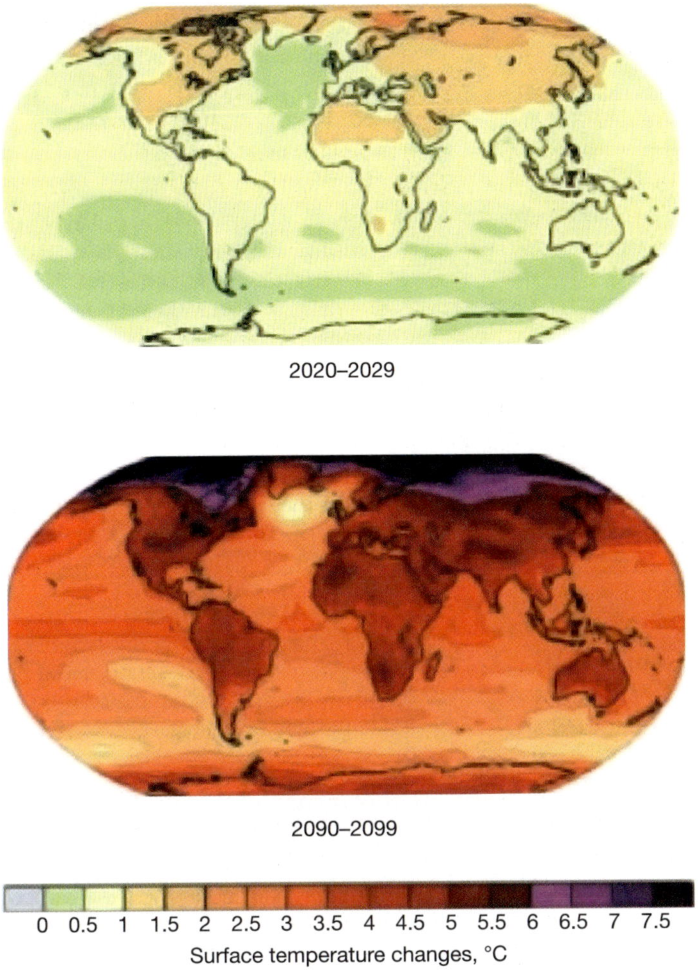


**FIGURE 15.21** Surface temperature projections from IPCC scenario B1. This storyline assumes that global population peaks in mid-century and declines thereafter. It also infers rapid introduction of new, cleaner, and more-efficient technologies, but without additional climate initiatives.
**Source:** IPCC, 2007.

bring welcome rains to Africa's Sahel, but we're already seeing extreme droughts in other parts of Africa that put millions of people at risk of food shortages.

> ***Think About It***
>
> Looking at the predictions for the greatest warming by the end of this century shown in figure 15.21, which of the signs of warming discussed in this chapter can you locate on the map?

Carbon dioxide is a fertilizer for plants, and higher $CO_2$ levels bring faster growth for many plant species. Sherwood Idso of the U.S. Department of Agriculture claims that increased $CO_2$ concentrations could trigger lush plant growth that would make crops abundant. But other researchers point out that the effects of elevated $CO_2$ in real-world situations may be very different from conditions in a greenhouse (Exploring Science, p. 340).

Enough water is stored in glacial ice caps in Greenland and Antarctica to raise sea levels around 100 m (300 ft) if they all melt. About one-third of the world's population now live in areas that would be flooded if that happens. Even the 75 cm (30 in.) sea-level rise expected by 2050 will flood much of south Florida, Bangladesh, Pakistan, and many other low-lying coastal areas. Most of the world's largest urban areas are on coastlines. Wealthy cities such as New York or London can probably afford to build dikes to keep out rising seas, but poorer cities such as Jakarta, Kolkata, or Manila might simply be abandoned as residents flee to higher ground. Several small island countries such as the Maldives, the Bahamas, Kiribati, and the Marshall Islands could become uninhabitable if sea levels rise a meter or more. The South Pacific nation of Tuvalu has already announced that it is abandoning its island homeland. All 11,000 residents will move to New Zealand, perhaps the first of many climate change refugees.

Insurance companies worry that the $2 trillion in insured property along U.S. coastlines is at risk increased from a combination of high seas and catastrophic storms. At least 87,000 homes in the United States within 150 m (500 ft) of a shoreline are in danger of coastal erosion or flooding in the next 50 years. Accountants warn that loss of land and structures to flooding and coastal erosion together with damage to fishing stocks, agriculture, and water supplies could raise worldwide insurance claims from about $50 billion, which they were in the 1970s, to more than $150 billion per year in 2010. Some of this increase in insurance claims is that more people are living in dangerous places, but extra-severe storms only exacerbate this problem.

Infectious diseases are likely to increase as the insects and rodents that carry them spread to new areas. Already we have seen diseases such as malaria, dengue fever, and West Nile virus appear in parts of North America where they had never before been reported. Coupled with the movement of hundreds of millions of environmental refugees and greater crowding as people are forced out of areas made uninhabitable by rising seas and changing climates, the spread of epidemic diseases will also increase.

An ominous consequence of arctic permafrost melting might be the release of vast stores of methane hydrate now locked in frozen ground and in sediments in the ocean floor. Together with the increased oxidation of high-latitude peat lands potentially caused by warmer and dryer conditions, release of these carbon stores could add as much $CO_2$ to the atmosphere as all the fossil fuels ever burned. We could trigger a disastrous positive feedback loop in which the effects of warming cause even more warming. On the other hand, increased ocean evaporation might intensify snowfall at high latitudes so that arctic glaciers and snow pack would increase rather than decrease. Ironically, the increased

# *Exploring* SCIENCE

## Carbon-Enrichment Studies

Among the greenhouse gases, none has been studied more than carbon dioxide ($CO_2$). As you've learned already in this chapter, $CO_2$ has been analyzed in 400,000-year-old air bubbles taken from glacial ice, measured atop Hawaii's Mauna Loa volcano since 1957, and implicated in a whole host of current climate changes. No one disputes that $CO_2$ levels in the troposphere increased from 280 to 383 ppm between 1750 and 2006. Few argue that it may exceed 500 ppm—nearly twice its preindustrial concentration—before the twenty-first century ends, but we're not sure what the effects of these high $CO_2$ levels may be on ecological processes.

**FIGURE 1** Student researchers collect samples for analysis at Cedar Creek's Free Air Carbon Enrichment study. The ring of striped poles blow $CO_2$ over the prairie plants growing in experimental plots.

Plants need $CO_2$ to carry out photosynthesis. We know that many plant species flourish in elevated $CO_2$ levels. That's why greenhouses often pipe in extra $CO_2$ to accelerate growth. Some horticulturists predict that doubling $CO_2$ atmospheric concentrations will produce such lush vegetation that the earth will return to a verdant Eden. Others contend that how plants behave in the ideal conditions of a greenhouse or growth chamber is very different from what happens in the real world. How can we know what to expect if current trends continue?

In 1997, Peter Reich and his colleagues began experiments at the University of Minnesota's Cedar Creek Natural History Area that attempt to mimic natural conditions more faithfully than is possible in a growth chamber. At Cedar Creek, the researchers erected several large rings of towers from which $CO_2$ is released (fig. 1). Sensors track wind speeds and $CO_2$ concentrations within the ring, while a computer regulates gas flows to maintain 560 ppm throughout the experimental plot. Even when winds are swirling around the research site, the computer can adjust gas flows to keep $CO_2$ levels constant. Because the $CO_2$ isn't contained, these experiments are called Free Air Carbon Enrichment (FACE). Control plots have exactly the same design as the experimental ones except that the control towers blow plain air rather than $CO_2$-enriched air. The Cedar Creek Long-Term Ecological Research is an important center for the study of the relationship between biodiversity and prairie ecosystem stability, so the FACE research also included several different levels of native prairie species richness, as well as different fertilizer treatments, in experimental plots.

Dr. Reich and colleagues discovered that the response of plants to elevated $CO_2$ and nitrogen fertilization depended on how many species were in a test plot. In a paper published in 2001 in *Nature,* they showed that the amount of plant material (biomass) produced each year rose when $CO_2$ and nitrogen were added to the test plots. But test plots containing 9 or 16 plant species benefited much more than test plots with a single plant species (fig. 2). In a later paper in the *Proceedings of the National Academy of Sciences,* Dr. Reich and his team also demonstrated that, if test plots contained ecologically different plants (e.g., a legume and a summer-blooming grass), more biomass developed than in test plots with a single one of those functional plant groups. In other words, vegetation that contained a single plant species responded positively to rising carbon dioxide, but several plant species with different ecological functions responded much better, even without fertilization. The different species and functional groups compensated for each other in the face of a varying climate and environment.

Similar research is being done on forest species and agricultural crops. Peter Curtis, an Ohio State University researcher, writing in the journal *Phytologist,* summarized the results of dozens of FACE experiments on a number of species. In grain crop plants the total seed weight—and therefore food yield—increased by up to 25 percent with a doubling of $CO_2$ concentration, but in all species except legumes the nutritional value of those plants fell. In one study of tall fescue (*Festuca alatior*), one of the most important pasture grasses in United States, the digestible protein that grazers could extract from the $CO_2$-stimulated grass actually fell below levels in existing forage, despite fertilization. So, while we may have luxuriant plant growth under elevated $CO_2$, it may be less valuable for certain uses than what we have today.

In the end, raising $CO_2$ through human activities represents a huge global experiment not unlike the FACE experiments, but with an important difference. If $CO_2$ reaches 500 ppm by 2100, we'll be stuck with those conditions—and their effects—for many years. What basic research tells us in the meantime about plant responses, including our important crop plants, will be important in understanding how to adapt to the challenges we face.

Dr. Reich says, "The idea that carbon dioxide with uniformly increase plant growth and therefore provide a large and predictable sink for carbon dioxide is naive at best. This type of model fails to consider the real complexity of the carbon dioxide fertilization effect. Species composition, community diversity, and the amount of nitrogen in the system interact to produce varying responses. In particular, reduced biodiversity and the widespread limited amount of nitrogen in the soil will reduce the $CO_2$ fertilization effect in terrestrial plant biomass."

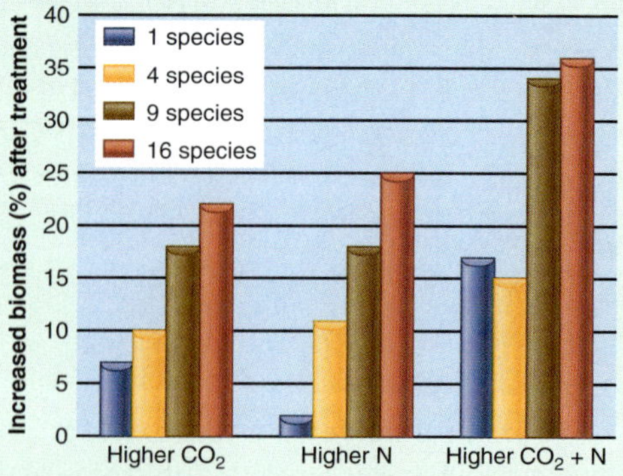


**FIGURE 2** The Cedar Creek BioCON study demonstrates strength in numbers. Single species plots did not respond as well to $CO_2$ increases as multi-species plots.

Data From Peter Reich.

albedo of colder, snow-covered surfaces might then trigger a new ice age. Clearly, we're disturbing a complex system and facing consequences that we don't fully understand.

> ***Think About It***
>
> Most of us find it difficult to contemplate global climate change. The potential effects are so disastrous and the time scale is so long that we simply put it out of our minds. How can we frame the issue in positive, practical terms that people can do something about immediately?

## 15.5 The Kyoto Protocol Attempts to Slow Climate Change

One of the centerpieces of the 1992 United Nations Earth Summit meeting in Rio de Janeiro was the Framework Convention on Climate Change, which set an objective of stabilizing greenhouse gas emissions to reduce the threats of global warming. At a follow-up conference in Kyoto, Japan, in 1997, 160 nations agreed to roll back $CO_2$, methane, and nitrous oxide emissions about 5 percent below 1990 levels by 2012. Three other greenhouse gases, hydrofluorocarbons, perfluorocarbons, and sulfur hexafluoride, would also be reduced, although from what level was not decided. Known as the **Kyoto Protocol,** this treaty sets different limits for individual nations, depending on their output before 1990. Poorer nations, such as China and India, were exempted from emission limits to allow development to increase their standard of living. Wealthy countries created the problem, the poorer nations argue, and the wealthy should deal with it.

Although the United States took a lead role in negotiating a compromise at Kyoto that other countries could accept, President George W. Bush refused to honor U.S. commitments. Claiming that reducing carbon emissions would be too costly for the U.S. economy, he said, "We're going to put the interests of our own country first and foremost." The United States, which depends solely on voluntary efforts to limit greenhouse gas emissions, had a 2 percent rise in those releases in 2007. At this rate, the U.S. will be 25 percent above 1990 emissions by 2012.

Meanwhile, 126 countries have ratified the Kyoto Protocol. If the U.S. continues to refuse to comply, it could have severe ramifications on U.S. corporations engaged in international business. Thousands of American businesses—including most of the largest ones—fall into this category. Having to modify their products and practices for overseas markets while not doing so for domestic sales would be expensive and would put them at a disadvantage with competitors not subject to these costs. Moreover, because Kyoto is based on a global cap-and-trade program, companies that get in on it early will have an advantage—they can buy cheap emissions credits before the price gets bid up.

We'll discuss market mechanisms for controlling climate change further in chapter 23. In 2007, the heads of ten of the largest business conglomerates in America joined four environmental groups to call for strong national legislation to achieve significant reductions of greenhouse gas emissions. The corporations included Alcoa, BP America, Caterpillar, Duke Energy, DuPont, FPL Group, General Electric, Lehman Brothers, Pacific Gas & Electric, and PNM Resources. The nongovernmental organizations were Environmental Defense, Pew Center, Natural Resources Defense Council, and World Resources Institute. Many companies would prefer a single national standard rather than a jumble of conflicting local and state rules. Knowing that climate controls are inevitable, they'd rather know now how they'll have to adapt rather than wait until a crisis causes us to demand sudden, radical changes.

Even if the United States joins the rest of the world in accepting the Kyoto provisions, however, many climatologists argue that much more will have to be done to combat global warming. According to some calculations, we will have to reduce $CO_2$ emissions by 60 percent or more to stabilize atmospheric concentrations. In the next section, we'll look at some of the ways we might accomplish this.

### There are many ways we can control greenhouse emissions

Is the situation hopeless? Can we do anything to reverse the path we're on? No single step can reverse our release of greenhouse gases, but many simple policy changes could help stabilize our climate. One of the most encouraging analyses in this area is that of Robert Socolow and his colleagues at Princeton University. They say we already have the scientific, technical, and industrial know-how to stabilize our $CO_2$ emissions over the next half century, while still meeting the world's energy needs. In several recent publications, this group examines a portfolio of 14 policy options for reducing our $CO_2$ output. No single step can do the entire job by itself, they conclude, but the potential benefits of the entire portfolio are large enough that not every option must be used.

To avoid a doubling of atmospheric $CO_2$ we need to reduce our annual carbon emissions by about 7 billion tons by 2054. To make this easier to understand, the Princeton group divides the "stabilization triangle" between our current path and the flat emissions that we need in order to avoid doubling $CO_2$ levels into seven "wedges" (fig. 15.22). Each wedge represents 1 GT (1 billion tons) of carbon emissions avoided in 2054, or a cumulative total of 25 billion tons over the next 50 years, compared to a "business as usual" scenario.

Because most of our $CO_2$ emissions come from fossil fuel combustion, energy conservation and a switch to renewable fuels probably are the first places we should look to reduce our climate problem. For example, 1.5 wedges (1.5 GT reduction in carbon emissions) could be accomplished by doubling our average fuel economy from the expected 30 miles per gallon in 2054 to 60 mpg, and by halving the number of miles driven each year by switching to walking, biking, or using mass transit. Simply installing the most efficient lighting and appliances available, along with improved insulation in buildings, could save another 2 GT of carbon emissions. Similarly, improving power plant efficiency and reducing the energy consumption of industrial processes could reduce carbon output one wedge (1 GT) by 2054. Capturing and storing $CO_2$

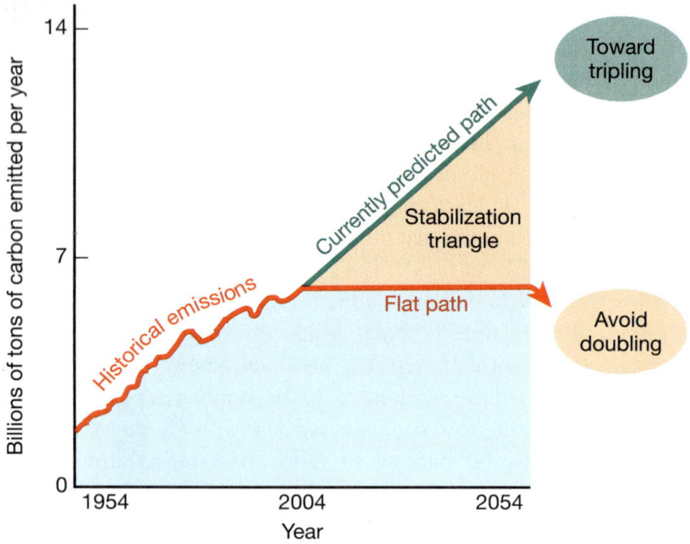


**FIGURE 15.22** What would it take to stabilize our carbon emissions? Some authors argue that we already have the science and technology to do so. All we need is the political will.
Data from Socolow, et al. Environment vol. 46, no. 10 p. 11, 2004.

released by power plants, methane wells, and other large sources could save another billion tons of carbon. Much of this $CO_2$ could be injected into oil wells to improve crude oil production. Building 2 million wind turbines (of the current one-megawatt size) or doubling our current global nuclear generating capacity would each reduce carbon emissions by about 1 GT in 50 years. This would, of course, require solving the energy storage problems with wind power as well as public worries about the safety of nuclear reactors and storing radioactive waste. Increasing production of biofuels, such as ethanol and vegetable oil, 50-fold, could replace 1 GT of carbon from fossil fuels. Halting tropical forest destruction and replanting 250 million ha of forest (or 400 million ha in temperate regions) would avoid almost 1 GT of carbon emissions, while practices such as conservation tillage and planting of cover crops could reduce carbon releases by another GT.

As you can see, altogether these policy changes add up to considerably more than the 7 GT per year reduction we need to make by 2054 to avoid doubling atmospheric $CO_2$ concentrations. At least seven other immediately available options exist for reducing greenhouse gases by the equivalent of 1 GT of carbon. If we did all of them, we could reverse the present trajectory and move toward zero greenhouse emissions.

There also are many ways individuals can contribute to this effort (What Can You Do? p. 342). Perhaps the easiest way to become carbon neutral is to simply pay someone to do it for you. Many organizations will now plant a tree for you or arrange some other carbon reduction to offset the costs of your lifestyle. British Petroleum, for example, will compensate for the $CO_2$ emitted by a year's driving of the average car for about $40. You could make a guilt-free round-trip airplane flight from New York to Los Angeles for about $6.25. Or Carbonfund.org will offset for all the greenhouse gas emissions (both direct and indirect) for the average American for $99 per year.

There are some worries about these programs, however. In some places, planting trees is inappropriate. Fast-growing trees, such as the eucalyptus or pines often used in reforestation programs, can dry streams and springs and deplete soil nutrients. Furthermore, if the trees are grown for a few years and then harvested and burned, you haven't done anything for the planet. If you're going to finance tree planting, ask where the trees will be planted and how they'll be cared for.

## *What Can You Do?*

### Reducing Carbon Dioxide Emissions

Individuals can help reduce global warming. Although some actions may cause only a small impact, collectively they add up. Many of these options save money in the long run and have other environmental benefits such as reducing air pollution and resource consumption.

1. Drive less, walk, bike, take public transportation, carpool, or buy a vehicle that gets at least 30 mpg (12.6 km/l). Average annual $CO_2$ reduction: about 20 lbs for each gallon of gasoline saved. Increasing your mileage by 10 mpg, for example, eliminates about 2,500 lbs (1.1 metric tons) of $CO_2$ per year. Becoming entirely self-propelled would save, on average, about 12,000 lbs (5.45 metric tons) per year.
2. Plant trees to shade your house during the summer, and paint your house a light color if you live in a warm climate or a dark color in a cold climate. Average annual $CO_2$ reduction: about 5,000 lbs (2.2 metric tons).
3. Insulate your house and seal all drafts. Average annual $CO_2$ reduction for the highest efficiency insulation, weatherstripping, and windows: about 5,000 lbs (2.2 metric tons).
4. Replace old appliances with new, energy-efficient models. Average annual $CO_2$ reduction for the most efficient refrigerator, for instance: about 3,000 lbs (1.4 metric tons).
5. Produce less waste. Buy minimally packaged goods and reusable products. Recycle. Average annual $CO_2$ reduction: about 1,000 lbs (0.45 metric tons) for 25 percent less garbage.
6. Turn down your thermostat in the winter and turn it up in the summer. Average annual $CO_2$ reduction: about 500 lbs (0.23 metric tons) for every 1°C (1.8°F) change.
7. Replace standard light bulbs with long-lasting compact fluorescent ones. Average annual $CO_2$ reduction: about 500 lbs (0.23 metric tons) for every bulb.
8. Wash laundry in warm or cold water, not hot. Average annual $CO_2$ reduction: about 500 lbs (0.23 metric tons) for two loads per week.
9. Set your water heater thermostat no higher than 48.9°C (120°F). Average annual $CO_2$ reduction: about 500 lbs (0.23 metric tons) for each 5°C (9°F) temperature change.
10. Buy renewable energy from your local utility if possible. Potential annual $CO_2$ reduction: about 30,000 lbs (13.5 metric tons).

## Progress is being made

Many countries are working to reduce greenhouse emissions. The United Kingdom, for example, had already rolled $CO_2$ emissions back to 1990 levels by 2000 and vowed to reduce them 60 percent by 2050. Britain already has started to substitute natural gas for coal, promote energy efficiency in homes and industry, and raise its already high gasoline tax. Plans are to "decarbonize" British society and to decouple GNP growth from $CO_2$ emissions. A revenue-neutral carbon levy is expected to lower $CO_2$ releases and trigger a transition to renewable energy over the next five decades. New Zealand Prime Minister Helen Clark has pledged that her country will be the first to be **"carbon neutral,"** that is, to reduce net greenhouse gas emissions to zero, although she didn't say when this will occur.

Germany, also, has reduced its $CO_2$ emissions at least 10 percent by switching from coal to gas and by encouraging energy efficiency throughout society. Atmospheric scientist Steve Schneider calls this a "no regrets" policy; even if we don't need to stabilize our climate, many of these steps save money, conserve resources, and have other environmental benefits. Nuclear power also is being promoted as an alternative to fossil fuels. It's true that nuclear reactions don't produce greenhouse gases, but security worries and unresolved problems of how to store wastes safely make this option unacceptable to many people.

Many people believe renewable energy sources offer the best solution to climate problems. Chapter 20 discusses options for conserving energy and switching to renewable sources, such as solar, wind, geothermal, biomass, and fuel cells. Denmark, the world's leader in wind power, now gets 20 percent of its electricity from windmills. Plans are to generate half of the nation's electricity from offshore wind farms by 2030. Even China reduced its $CO_2$ emissions 20 percent between 1997 and 2005 through greater efficiency in coal burning and industrial energy use.

In addition to reducing its output, there are options for capturing and storing $CO_2$. Dr. Klaus S. Lackner of Columbia University has proposed building synthetic "trees" to capture $CO_2$ from the air (fig. 15.23). A calcium hydroxide solution trickling across the "leaves" would absorb $CO_2$. A structure with 2,500 $m^2$ of collecting area could remove 90,000 tons of $CO_2$ per year from the air (1,000 times more than a natural tree), an amount equal to the $CO_2$ produced by 15,000 cars. About 250,000 synthetic trees worldwide would be needed to soak up the 22 billion tons of $CO_2$ produced annually by burning fossil fuels. Critics claim that synthesizing the hydroxide needed for this reaction and disposing of the vast amount of calcium carbonate it would create may take more energy than the fuel that produced the $CO_2$ released. The first of these "trees" is now being built in Arizona. Planting real trees can also be effective if they're allowed to mature to old-growth status or are made into products, such as window frames and doors, that will last for many years before they're burned or recycled. Farmland also can serve as a carbon sink if farmers change their crop mixture and practice minimum till cultivation that keeps carbon in the soil.

**FIGURE 15.23** Synthetic "trees," could capture $CO_2$ from the air. Each of these structures would be about 100 m (300 ft) tall and would have an array of metal slats 50 × 50 m across. Each of these trees could remove 90,000 tons of $CO_2$ per year from the air, an amount equal to the $CO_2$ produced by 15,000 cars. It would take a quarter of a million of them to remove all the $CO_2$ produced annually by humans.

In 1988 John H. Martin suggested that phytoplankton growth in the ocean may be limited by iron deficiency. This theory was tested in 2000, when researchers spread about 3.5 tons of dissolved iron over 75 $km^2$ of the South Pacific. Monitored by satellite, a ten-fold rise in chlorophyll concentration eventually spread over about 1,700 $km^2$ and was calculated to have pulled several thousand tons of $CO_2$ out of the air. Still, some ecologists warn about such large-scale tinkering. We don't know whether the carbon fixed by phytoplankton growth will be stored in sediments or simply eaten by predators and returned immediately to the atmosphere. It's possible that dying algae might create an anoxic zone that would devastate oceanic food webs.

Another way to store $CO_2$ is to inject it into underground strata or deep ocean waters. Since 1996, Norway's Statoil has been pumping more than 1 million metric tons of $CO_2$ per year into an aquifer 1,000 m below the seafloor at a North Sea gas well. This enhances oil recovery. It also saves money because otherwise the company would have to pay a $50 per ton carbon tax on its emissions. Around the world, deep, briny aquifers could store a century worth of $CO_2$ output at current fossil fuel consumption rates.

It might also be possible to pump liquid $CO_2$ into deep ocean trenches, where it would form lakes contained by enormous water pressures. There are worries, however, about what this might do to deep ocean fauna and what might happen if earthquakes or landslides caused a sudden release of this $CO_2$. This suggestion got a boost, however, with the recent discovery of a natural lake of liquid $CO_2$ off the coast of Japan. Microbial life around the edge of the lake was flourishing.

In Canada, an Alberta power plant is injecting $CO_2$ into a coal seam too deep to be mined. The $CO_2$ releases natural gas, which is burned to produce electricity and more $CO_2$. Proponents

**FIGURE 15.24** Burning fossil fuels produces about half our greenhouse gas emissions, and transportation accounts for about half our fossil fuel consumption. Driving less, choosing efficient vehicles, carpooling, and other conservation measures are among our most important personal choices in the effort to control global warming.

of **carbon management,** as these various projects are called, argue that it may be cheaper to clean up fossil fuel effluents than to switch to renewable energy sources.

Most attention is focused on $CO_2$ because it lasts in the atmosphere, on average, for about 120 years. Methane and other greenhouse gases are much more powerful infrared absorbers but remain in the air for a much shorter time. NASA's Dr. James Hansen made a controversial suggestion, however, that the best short-term attack on warming might come by focusing not on carbon dioxide but on methane. Reducing gas pipeline leaks would conserve this valuable resource as well as help the environment.

Methane from landfills, oil wells, and coal mines that once would have been simply vented into the air is now being collected and used to generate electricity. Rice paddies are a major methane source. Changing flooding schedules and fertilization techniques can reduce anoxia (oxygen depletion) that produces marsh gas. Finally, ruminant animals (such as cows, camels, and buffalo) create large amounts of digestive system gas. Modified diets can reduce methane releases significantly.

Soot, while not mentioned in the Kyoto Protocol, might also be important in global warming. Dark, airborne particles absorb both ultraviolet and visible light, converting them to heat energy. According to some calculations, reducing soot from diesel engines, coal-fired generators, forest fires, and wood stoves might reduce net global warming by 40 percent within three to five years. Curbing soot emissions would also have beneficial health effects.

Some individual cities and states have announced their own plans to combat global warming. Among the first of these were Toronto, Copenhagen, and Helsinki, which pledged to reduce $CO_2$ emissions 20 percent from 1990 levels by 2010. Some corporations are following suit. British Petroleum has set a goal of cutting $CO_2$ releases from all its facilities by 10 percent before 2010. Each of us can make a contribution in this effort. As Professor Socolow and his colleagues point out, simply driving less and buying high-mileage vehicles could save about 1.5 billion tons of carbon emissions by 2054 (fig. 15.24).

In the midst of all the debate about how serious the consequences of global climate change may or may not be, we need to remember that many of the proposed solutions are advantageous in their own right. Even if climate change turns out not to be as much of a threat as we think now, they have other positive benefits. Moving from fossil fuels to renewable energy sources such as solar or wind power, for example, would free us from dependence on foreign oil and improve air quality. Planting trees makes cities more pleasant places to live and provides habitat for wildlife. Making buildings more energy efficient and buying high-mileage vehicles saves money in the long run. Walking, biking, and climbing stairs are good for your health as well as reducing traffic congestion and energy consumption. Reducing waste, recycling, and other forms of sustainable living improve our environment in many ways in addition to helping fight climate change. It's important to focus on these positive effects rather than to look only at the gloom-and-doom scenarios for global climate catastrophes. As the Irish statesman and philosopher Edmund Burke said, "Nobody made a greater mistake than he who did nothing because he could do only a little."

## Conclusion

We appear to be at a tipping point in our attitudes toward climate change. Where, a few years ago, most people in America thought that this was just a wild theory of a small group of crazy scientists, we've gone to a widespread acceptance that the science of global warming is indisputable. The surprising success of former Vice President Al Gore's documentary, *An Inconvenient Truth,* has been a powerful force in this social transformation. Perhaps more important is that many people can actually see tangible proof in their own lives that something strange is happening to our climate.

But is it too late to do anything? Most scientists believe there is time, if we act quickly and effectively, to avoid the worst consequences of global climate change. It's encouraging to know that we don't need a technological miracle to save us from this impending catastrophe. We have the ability to dramatically reduce our greenhouse gas emissions right now. And it will probably save money and create jobs to do so.

It's said that Napoleon Bonaparte ordered his government to plant trees along all French roads to provide shade and fodder for local people. "But sir," the bureaucrats protested, "it will take a generation for the trees to grow." "All the more reason to start immediately," Bonaparte responded.

## Reviewing Learning Outcomes

By now you should be able to explain the following points:

**15.1** Illustrate how the atmosphere is a complex machine.

- The sun warms our world.
- Water stores energy, and winds redistribute it.

**15.2** Explain how weather happens.

- Why does it rain?
- Large-scale winds don't move in a straight line.
- Ocean currents modify our weather.
- Seasonal winds and monsoons have powerful effects.
- Frontal systems create local weather.
- Cyclonic storms can cause extensive damage.

**15.3** Clarify how climate can be an angry beast.

- Climates have changed dramatically throughout history.
- What causes catastrophic climate swings?
- El Niño/Southern Oscillations are powerful cycles.

**15.4** Understand why global warming is happening.

- A scientific consensus is emerging.
- Greenhouse gases have many sources.
- Evidence of climate change is overwhelming.
- Global warming will be expensive.

**15.5** Evaluate how the Kyoto Protocol attempts to slow climate change.

- There are many ways we can control greenhouse emissions.
- Progress is being made.

## Practice Quiz

1. What are weather and climate? How do they differ?
2. Name and describe the four layers of air in the atmosphere.
3. What is the greenhouse effect?
4. What are the ENSO cycle and the PDO?
5. What are the jet streams? How do they influence weather patterns?
6. How do tornadoes form and why are they so destructive?
7. Describe the Coriolis effect. What causes it?
8. Explain some of the observed or expected effects of climate change.
9. Summarize some actions we could take, collectively and as individuals, to combat global climate change.
10. Why has the United States not ratified the Kyoto Protocol? What effects may this have for the U.S. economy?

## Critical Thinking and Discussion Questions

1. In most places, weather changes in highly variable, seemingly chaotic ways. How would you distinguish between meaningful patterns and random variations?
2. Most scientists say global warming is real; politicians, however, dispute this claim. How can we decide what to do when the science is uncertain and the risk is great?
3. Has there been a major change recently in the weather where you live? Can you propose any reasons for such changes?
4. What forces determine the climate in your locality? Are they the same for neighboring states?
5. Which of the seven options recommended by Professor Socolow do you think has the greatest chance of success?
6. Would you favor building nuclear power plants to reduce $CO_2$ emissions? Why or why not?

## DATA analysis Understanding Methane Trends

Methane has about 23 times the climate change potential of $CO_2$. Although methane is a much more minor component of the atmosphere than $CO_2$, it has been accumulating at a faster rate. In 2007 the IPCC reported that atmospheric methane concentrations have increased about 2.5 times from preindustrial levels (measured from ice core data). The IPCC concluded it is "*very likely*" that the observed increase is do to anthropogenic activities, predominantly agriculture and fossil fuel use. Look at the graphs on this page and answer the following questions:

1. What do the curves look different on these two graphs? (Hint: look at the x axis)
2. Is one graph more accurate than the other?
3. Do these graphs give different impressions about methane trends?
4. What are the units of measurement in these two graphs?
5. Assuming that the industrial revolution started in about 1800, what was the preindustrial methane concentration?
6. In which graph is it easier to determine the 1800 concentration?
7. What was the atmospheric methane concentration in 2000?
8. How does the current world methane concentration differ from 10,000 years ago?

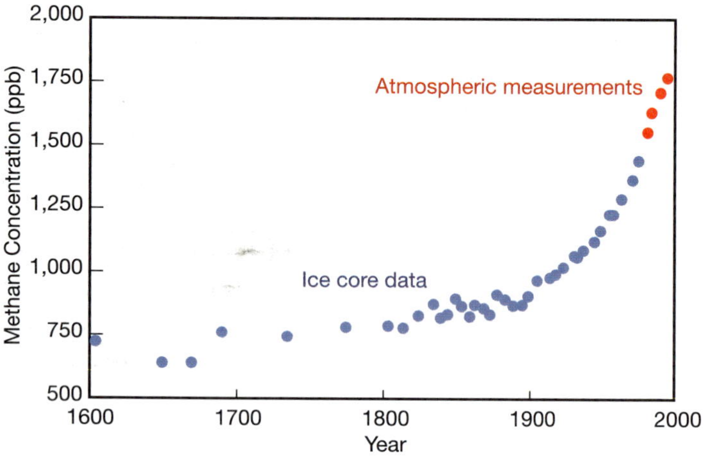


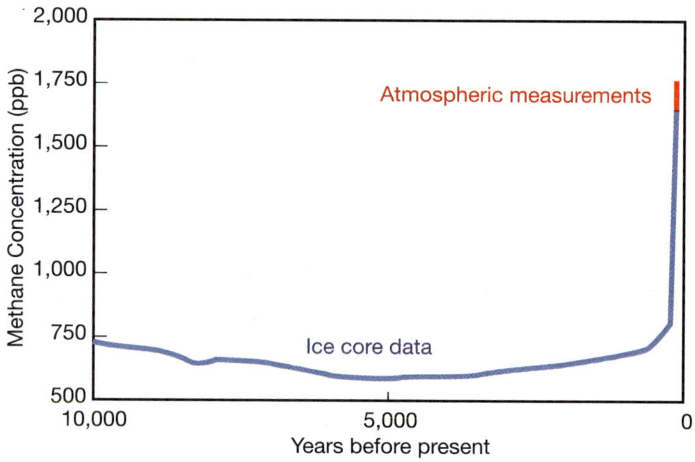


Changes in atmospheric methane measured from ice core data.

**For Additional Help in Studying This Chapter,** please visit our website at www.mhhe.com/cunningham10e. You will find additional practice quizzes and case studies, flashcards, regional examples, place markers for Google Earth™ mapping, and an extensive reading list, all of which will help you learn environmental science.

Coal-burning power plants produce about two-thirds of the sulfur oxides, one-third of all nitrogen oxides, and half of the mercury emitted in the United States each year.

CHAPTER 16

# Air Pollution

*The only thing we have to fear is fear itself.*

*—Franklin D. Roosevelt—*

## LEARNING OUTCOMES

After studying this chapter, you should be able to:

**16.1** Describe the air around us.

**16.2** Identify natural sources of air pollution.

**16.3** Discuss human-caused air pollution.

**16.4** Explain how climate topography and atmospheric processes affect air quality.

**16.5** Compare the effects of air pollution.

**16.6** Evaluate air pollution control.

**16.7** Summarize current conditions and future prospects.

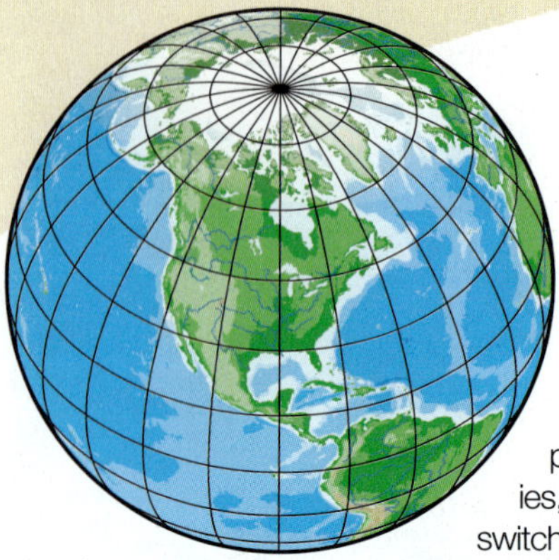

# Case Study Controlling Mercury Pollution

Often referred to as quicksilver, mercury is the only common metal that is liquid at room temperature. It's useful in a wide variety of industrial processes, and found in a host of commercial products including paints, batteries, fluorescent light bulbs, electrical switches, pesticides, skin creams, antifungal agents, and old thermometers.

Mercury also is a powerful neurotoxin that destroys the brain and central nervous system at high doses. Even tiny amounts of this toxic metal can cause neurological and developmental defects in young children. In a survey of freshwater fish from 260 lakes across the United States, the EPA found that every fish sampled had at least low levels of mercury. More than half the fish contained mercury levels unsafe for women of childbearing age, and three-quarters exceed the safe limit for young children. Fifty states have issued warnings about eating freshwater or ocean fish; mercury contamination is by far the most common reason for these advisories. In 1994, the EPA declared mercury a hazardous pollutant regulated under the Clean Air Act. Municipal and medical incinerators were ordered to reduce their mercury emissions by 90 percent. Chlorine-caustic soda plants and gold mining operations, the largest commercial mercury consumers, also agreed to reduce mercury releases by about 50 percent. Left unregulated, however, were the 1,032 coal-burning power plants, which are estimated to emit some 48 tons of mercury per year, or nearly half of total annual U.S. emissions.

Finally, in 2000, the EPA declared that mercury from power plants also is a risk to public health. Had the agency applied existing air-toxin regulations, utilities would have required installation of maximum achievable control technology, which would reduce power plant emissions by about 90 percent (similar to that achieved by incinerators) within three to five years. Rather than impose mandatory rules on industry, the EPA chose to regulate mercury through "cap and trade" market mechanisms that allow utilities to buy and sell pollution rights rather than have government specified pollution controls. This plan aims to reduce mercury releases by 70 percent by 2018, but the Congressional Research Service predicts that this reduction will be unlikely before 2030.

Critics contend that while cap and trade systems work well for some pollutants, they are an inappropriate form of regulation for a toxic substance, such as mercury. They say it will allow utilities to continue to emit mercury for years longer than necessary. It also will create mercury "hot spots," often in poor communities, where utilities find it cheaper to purchase pollution credits from distant facilities rather than clean up their own plants. Many eastern states oppose this approach because they suffer from high mercury pollution generated far away and blown in by prevailing winds (fig. 16.1).

Officially, the EPA estimates the benefits from reducing mercury emissions from power plants to be only $50 million a year, while the cost to industry would be $750 million a year. An independent study, on the other hand, concluded that mercury controls could save nearly $5 billion a year through reduced neurological and cardiac harm. Part of the discrepancy is that the EPA estimates that three-quarters of the mercury deposited in U.S. waterways comes from abroad and that 70 percent of the U.S. mercury emissions fell on other countries or the oceans. Thus, changes in mercury emissions from U.S. power plants would benefit the global environment more than U.S. residents. Part of the debate here is whether we have an ethical responsibility to the rest of the world for the pollutants we emit.

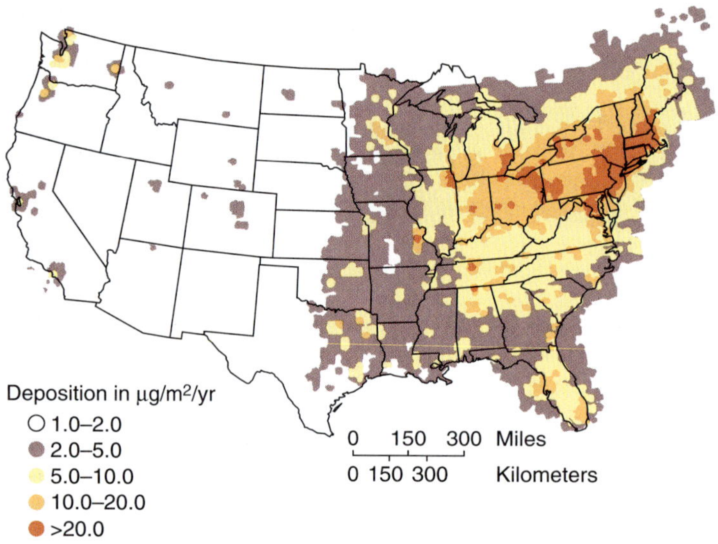


**FIGURE 16.1** Atmospheric mercury deposition in the United States. Due to prevailing westerly winds, and high levels of industrialization, eastern states have high mercury deposition.
**Source:** EPA, 1998.

Meanwhile, in the Allegheny Mountains of West Virginia, a huge coal-fired power plant is adding fuel to the mercury debate. Aptly named Mount Storm, it's a 1,600-megawatt Goliath that just a few years ago ranked second in the nation in mercury emissions. Since new pollution controls were installed to scrub sulfur and nitrogen oxides from its smokestack, however, 95 percent of its mercury also is being captured at no extra cost. This success story casts an odd light on the United States' mercury rules. If existing technology can remove so much mercury economically, why wait until 2018 to impose similarly stringent limits on other power plants? Utility executives, however, protest that technology that works for a specific plant may not be suitable everywhere.

This case study illustrates both the importance and the difficulty of regulating air pollution. Highly mobile, widely dispersed, produced by a variety of sources, and having diverse impacts, air pollutants can be challenging to regulate. Often air quality controversies—such as mercury control—pit a diffuse public interest (improving general health levels or IQ a few points) versus a very specific private interest (making utilities pay millions of dollars per year to control pollutants). In this chapter, we'll look at the major types and sources of air pollution, as well as more on the controversy about how best to ensure a healthy environment.

**FIGURE 16.2** On a smoggy day in Shanghai (*left*) visibility is less than 1 km. Twenty-four hours later, after a rainfall (*right*), the air has cleared dramatically.

## 16.1 The Air Around Us

How does the air taste, feel, smell, and look in your neighborhood? Chances are that wherever you live, the air is contaminated to some degree. Smoke, haze, dust, odors, corrosive gases, noise, and toxic compounds are present nearly everywhere, even in the most remote, pristine wilderness. Air pollution is generally the most widespread and obvious kind of environmental damage. According to the Environmental Protection Agency (EPA), some 147 million metric tons of air pollution (not counting carbon dioxide or wind-blown soil) are released into the atmosphere each year in the United States by human activities. Total worldwide emissions of these pollutants are around 2 billion metric tons per year.

Over the past 30 years, however, air quality has improved appreciably in most cities in Europe, North America, and Japan. Many young people might be surprised to learn that a generation ago most American cities were much dirtier than they are today. This is an encouraging example of improvement in environmental conditions. Our success in controlling some of the most serious air pollutants gives us hope for similar progress in other environmental problems.

While developed countries have been making progress, however, air quality in the developing world has been getting much worse. Especially in the burgeoning megacities of rapidly industrializing countries (chapter 22), air pollution often exceeds World Health Organization standards most of the time. In many Chinese cities, for example, airborne dust, smoke, and soot often are ten times higher than levels considered safe for human health (fig. 16.2). Currently, seven of the ten smoggiest cities in the world are in China.

## 16.2 Natural Sources of Air Pollution

It is difficult to give a simple, comprehensive definition of pollution. The word comes from the Latin *pollutus,* which means made foul, unclean, or dirty. Some authors limit the use of the term to damaging materials that are released into the environment by human activities. There are, however, many natural sources of air quality degradation. Volcanoes spew out ash, acid mists, hydrogen sulfide, and other toxic gases (fig. 16.3). Sea spray and decaying vegetation are major sources of reactive sulfur compounds in the air. Trees and bushes emit millions of tons of volatile organic compounds (terpenes and isoprenes), creating, for example, the blue haze that gave the Blue Ridge Mountains their name. Pollen, spores, viruses, bacteria, and other small bits of organic material in the air cause widespread suffering from allergies and airborne infections. Storms in arid regions raise dust clouds that transport millions of tons of soil and can be detected half a world away. Bacterial metabolism of decaying vegetation in swamps and of cellulose in the guts of termites and ruminant animals is responsible for as much as two-thirds of the methane (natural gas) in the air.

Does it make a difference whether smoke comes from a natural forest fire or one started by humans? In many cases, the chemical compositions of pollutants from natural and human-related sources are identical, and their effects are inseparable. Sometimes, however, materials in the atmosphere are considered innocuous at naturally occurring levels, but when humans add to these levels, overloading of natural cycles or disruption of essential processes can occur. While the natural sources of suspended

**FIGURE 16.3** Natural pollution sources, such as volcanoes, can be important health hazards.

particulate material in the air outweigh human sources at least tenfold worldwide, in many cities more than 90 percent of the airborne particulate matter is anthropogenic (human-caused).

## 16.3 Human-Caused Air Pollution

What are the major types of anthropogenic air pollutants and where do they come from? In this section, we will define some general categories and sources of air pollution.

### We Categorize pollutants according to their source

**Primary pollutants** are those released directly from the source into the air in a harmful form (fig. 16.4). **Secondary pollutants,** by contrast, are modified to a hazardous form after they enter the air or are formed by chemical reactions as components of the air mix and interact. Solar radiation often provides the energy for these reactions. Photochemical oxidants and atmospheric acids formed by these mechanisms are probably the most important secondary pollutants in terms of human health and ecosystem damage. We will discuss several important examples of such pollutants in this chapter.

**Fugitive emissions** are those that do not go through a smokestack. By far the most massive example of this category is dust from soil erosion, strip mining, rock crushing, and building construction (and destruction). In the United States, natural and

**FIGURE 16.4** Primary pollutants are released directly from a source into the air. A point source is a specific location of highly concentrated discharge, such as this smokestack.

anthropogenic sources of fugitive dust add up to some 100 million metric tons per year. The amount of $CO_2$ released by burning fossil fuels and biomass is nearly equal in mass to fugitive dust. Fugitive industrial emissions are also an important source of air pollution. Leaks around valves and pipe joints contribute as much as 90 percent of the hydrocarbons and volatile organic chemicals emitted from oil refineries and chemical plants.

The U.S. Clean Air Act of 1970 designated seven major pollutants (sulfur dioxide, carbon monoxide, particulates, hydrocarbons, nitrogen oxides, photochemical oxidants, and lead) for which maximum **ambient air** (air around us) levels are mandated. These seven **conventional** or **criteria pollutants** contribute the largest volume of air-quality degradation and also are considered the most serious threat of all air pollutants to human health and welfare. Figure 16.5 shows the major sources of the first six criteria pollutants. Table 16.1 shows an estimate of the total annual worldwide emissions of some important air pollutants. Now let's look more closely at the sources and characteristics of each of these major pollutants.

### We also categorize pollutants according to their content

Seven "conventional" pollutants were regulated by the original Clean Air Act.

#### *Sulfur Compounds*

Natural sources of sulfur in the atmosphere include evaporation of sea spray, erosion of sulfate-containing dust from arid soils, fumes from volcanoes and fumaroles, and biogenic emissions of hydrogen sulfide ($H_2S$) and organic sulfur-containing compounds, such as dimethylsulfide, methyl mercaptan, carbon disulfide, and carbonyl

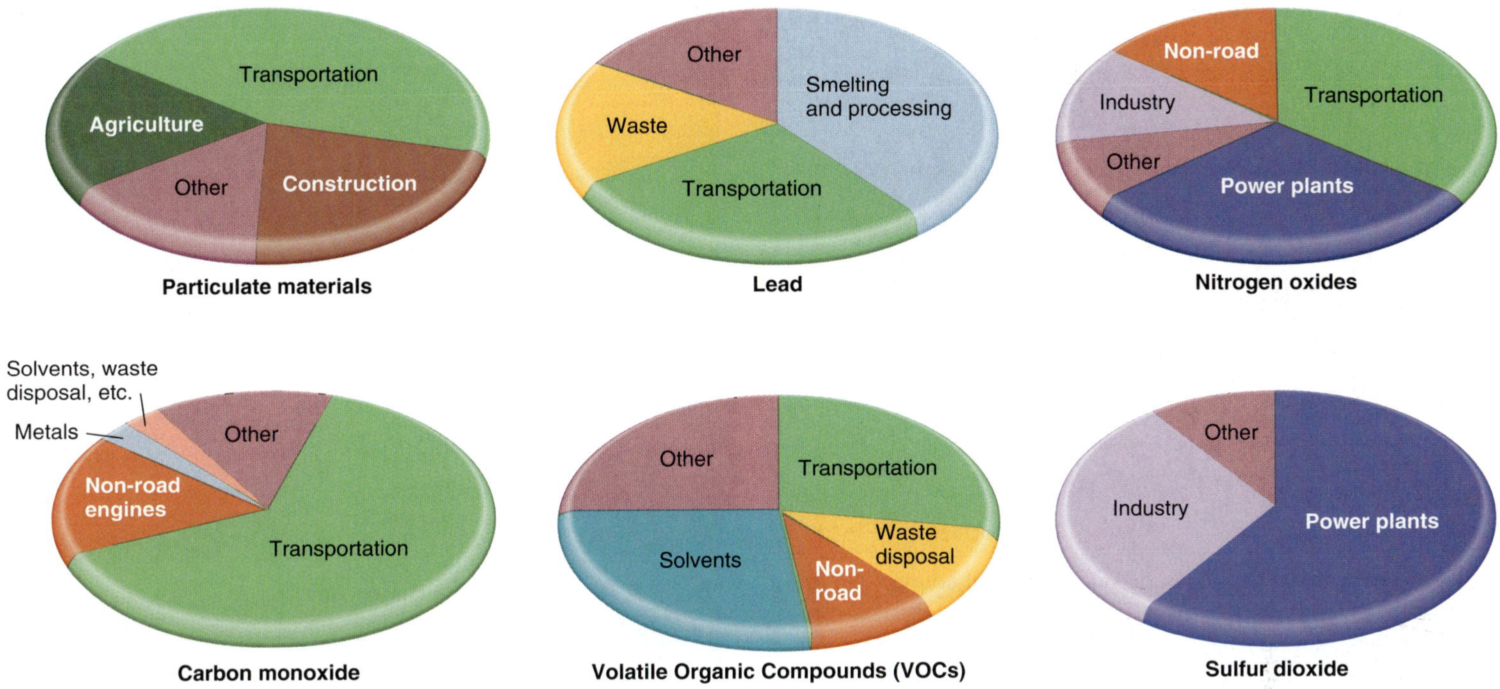


**FIGURE 16.5** Anthropogenic sources of six of the primary "criteria" air pollutants in the United States.
**Source:** UNEP, 1999.

**TABLE 16.1**

## Estimated Fluxes of Pollutants and Trace Gases to the Atmosphere

| | | Approximate Annual Flux (Millions of Metric Tons/Yr) | |
|---|---|---|---|
| **Species** | **Sources** | **Natural** | **Anthropogenic** |
| $CO_2$ (carbon dioxide) | Respiration, fossil fuel burning, land clearing, industrial processes | 370,000 | 29,600* |
| $CH_4$ (methane) | Rice paddies and wetlands, gas drilling, landfills, animals, termites | 155 | 350 |
| CO (carbon monoxide) | Incomplete combustion, $CH_4$ oxidation, biomass burning, plant metabolism | 1,580 | 930 |
| NMHC (nonmethane hydrocarbons) | Fossil fuels, industrial uses, plant isoprenes and other biogenics | 860 | 92 |
| $NO_x$ (nitrogen oxides) | Fossil fuel burning, lightning, biomass burning, soil microbes | 90 | 140 |
| $SO_x$ (sulfur oxides) | Fossil fuel burning, industry, biomass burning, volcanoes, oceans | 35 | 79 |
| SPM (suspended particulate materials) | Biomass burning, dust, sea salt, biogenic aerosols, gas-to-particle conversion | 583 | 362 |

*Only 27.3 percent of this amount—or 8 billion tons—is carbon.
**Source:** UNEP, 1999.

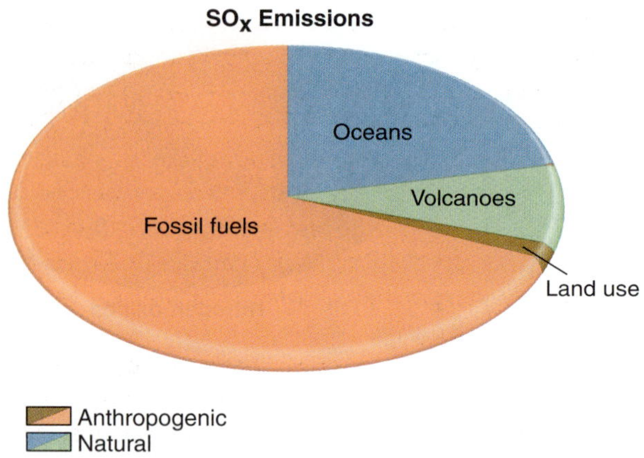


**FIGURE 16.6** Sulfur fluxes into the atmosphere.
**Source:** UNEP, 1999.

sulfide. Total yearly emissions of sulfur from all sources amount to some 114 million metric tons (fig. 16.6). Worldwide, anthropogenic sources represent about two-thirds of the total sulfur flux, but in most urban areas they contribute as much as 90 percent of the sulfur in the air. The predominant form of anthropogenic sulfur is sulfur dioxide ($SO_2$) from combustion of sulfur-containing fuel (coal and oil), purification of sour (sulfur-containing) natural gas or oil, and industrial processes, such as smelting of sulfide ores. China and the United States are the largest sources of anthropogenic sulfur, primarily from coal burning.

**Sulfur dioxide** is a colorless corrosive gas that is directly damaging to both plants and animals. Once in the atmosphere, it can be further oxidized to sulfur trioxide ($SO_3$), which reacts with water vapor or dissolves in water droplets to form sulfuric acid ($H_2SO_4$), a major component of acid rain. Very small solid particles or liquid droplets can transport the acidic sulfate ion ($SO_4^{-2}$) long distances through the air or deep into the lungs where it is very damaging. Sulfur dioxide and sulfate ions are probably second only to smoking as causes of air pollution-related health damage. Sulfate particles and droplets reduce visibility in the United States as much as 80 percent. Some of the smelliest and most obnoxious air pollutants are sulfur compounds, such as hydrogen sulfide from pig manure lagoons or mercaptans (organosulfur thiols) from papermills (fig. 16.7).

**FIGURE 16.7** The most annoying pollutants from this paper mill are pungent organosulfur thiols and sulfides. Chlorine bleaching can also produce extremely dangerous organochlorines, such as dioxins.

### *Nitrogen Compounds*

**Nitrogen oxides** are highly reactive gases formed when nitrogen in fuel or combustion air is heated to temperatures above 650°C (1,200°F) in the presence of oxygen, or when bacteria in soil or water oxidize nitrogen-containing compounds. The initial product, nitric oxide (NO), oxidizes further in the atmosphere to nitrogen dioxide ($NO_2$), a reddish brown gas that gives photochemical smog its distinctive color. Because of their interconvertibility, the general term $NO_x$ is used to describe these gases. Nitrogen oxides combine with water to make nitric acid ($HNO_3$), which is also a major component of atmospheric acidification.

The total annual emissions of reactive nitrogen compounds into the air are about 230 million metric tons worldwide (see table 16.1). Anthropogenic sources account for 60 percent of these emissions (fig. 16.8). About 95 percent of all human-caused $NO_x$ in the United States is produced by fuel combustion in transportation and electric power generation. Nitrous oxide ($N_2O$) is an intermediate in soil denitrification that absorbs ultraviolet light and plays an important role in climate modification (chapter 15). Excess nitrogen is causing fertilization and eutrophication of inland waters and coastal seas. It also may be adversely affecting terrestrial plants both by excess fertilization and by encouraging growth of weedy species that crowd out native varieties.

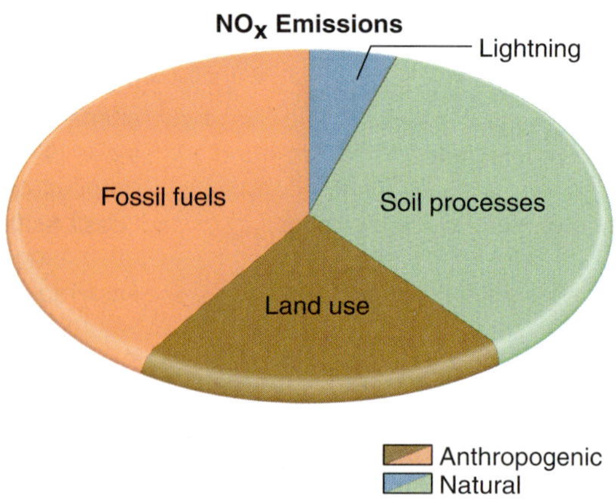


**FIGURE 16.8** Worldwide sources of reactive nitrogen gases in the atmosphere.
**Source:** UNEP, 1999.

## *Carbon Oxides*

The predominant form of carbon in the air is carbon dioxide ($CO_2$). It is usually considered nontoxic and innocuous, but increasing atmospheric levels (about 0.5 percent per year) due to human activities is now causing global climate change that may have disastrous effects on both human and natural communities. As table 16.1 shows, more than 90 percent of the $CO_2$ emitted each year is from respiration (oxidation of organic compounds by plant and animal cells). These releases are usually balanced, however, by an equal uptake by photosynthesis in green plants.

Anthropogenic (human-caused) $CO_2$ releases are difficult to quantify because they spread across global scales. The best current estimate from the Intergovernmental Panel on Climate Change (IPCC) is that between 7 and 8 billion tons of carbon (in the form of $CO_2$) are released each year by fossil fuel combustion and that another 1 to 2 billion tons are released by forest and grass fires, cement manufacturing, and other human activities. Typically, terrestrial ecosystems take up about 3 billion tons of this excess carbon every year, while oceanic processes take up another 2 billion tons. This leaves an average of at least 3 billion tons to accumulate in the atmosphere. The actual releases and uptakes vary greatly, however, from year to year. Some years almost all anthropogenic $CO_2$ is reabsorbed; in other years, almost none of it is. The ecological processes that sequester $CO_2$ depend strongly on temperature, nutrient availability, and other environmental factors.

United States negotiators at the Global Climate meetings claim that forests and soils in North America act as carbon sinks—that is, they take up more carbon than is released by other activities. Over the past decade, $CO_2$ levels in air coming ashore on the U.S. West Coast have averaged about 2 ppm higher than air leaving from the East Coast. If we assume that there isn't a major inflow of $CO_2$-depleted air entering from Canada or Mexico, this would mean that somewhere between 1.6 and 2.2 billion tons of $CO_2$ are being taken up every year than are being released in the United States. Other countries doubt these measurements, however, and refuse to give the United States credit for this large carbon sequestration.

**Carbon monoxide** (CO) is a colorless, odorless, nonirritating but highly toxic gas produced by incomplete combustion of fuel (coal, oil, charcoal, or gas), incineration of biomass or solid waste, or partially anaerobic decomposition of organic material. CO inhibits respiration in animals by binding irreversibly to hemoglobin. About 1 billion metric tons of CO are released to the atmosphere each year, half of that from human activities. In the United States, two-thirds of the CO emissions are created by internal combustion engines in transportation. Land-clearing fires and cooking fires also are major sources. About 90 percent of the CO in the air is consumed in photochemical reactions that produce ozone.

## *Particulate Material*

An **aerosol** is any system of solid particles or liquid droplets suspended in a gaseous medium. For convenience, we generally describe all atmospheric aerosols, whether solid or liquid, as **particulate material.** This includes dust, ash, soot, lint, smoke, pollen, spores, algal cells, and many other suspended materials. Anthropogenic particulate emissions amount to about 362 million metric tons per year worldwide. Wind-blown dust, volcanic ash, and other natural materials may contribute considerably more suspended particulate material.

Particulates often are the most apparent form of air pollution since they reduce visibility and leave dirty deposits on windows, painted surfaces, and textiles. Respirable particles smaller than 2.5 micrometers are among the most dangerous of this group because they can be drawn into the lungs, where they damage respiratory tissues. Asbestos fibers and cigarette smoke are among the most dangerous respirable particles in urban and indoor air because they are carcinogenic.

Diesel fumes also are highly toxic because they contain both fine particulates and chemicals such as benzene, dioxins, and mercury. The EPA has proposed new rules to require low-sulfur fuel and antipollution devices, particularly for off-road engines such as bulldozers, tractors, pumps, and generators. Epidemiologists estimate that these new standards will prevent more than 360,000 asthma attacks and 8,300 premature deaths annually. Diesel owners protest that expenses will be exorbitant, but Europe has had standards ten times more stringent than the EPA proposes for several years.

In the 1930s, America experienced terrible soil erosion known as the dust bowl. Poor agricultural practices and policies, coupled with years of drought, left soil on 5 million ha of the southern plains exposed to the wind. Billowing clouds of dust darkened the skies for days and reached as far as Washington, D.C. In 1935, at the peak of the drought, an estimated 850 million tons of topsoil blew away in "black blizzards."

Soil conservation techniques have reduced dust storms in North America, but deserts and dust storms have increased elsewhere. Soil scientists report that 3 billion tons of sand and soil blow from drylands around the world every year. Although these storms start out gritty, coarse sediments soon fall out, leaving finer silts and clays to rise up to 4,500 m and can travel thousands of kilometers before settling out. Dust from Africa's Sahara desert regularly crosses the Atlantic and raises particulate levels above federal health standards in Miami and San Juan, Puerto Rico (fig. 16.9).

There are some benefits from these storms. Research has shown that existence of the Amazon rainforest depends on mineral nutrients carried in dust from Africa. Remarkably, more than half the 50 million tons of dust transported to South America each year has been traced to the bed of the former Lake Chad in Africa (see Chapter 17). There also are adverse effects. Huge storms blow out of China's Gobi desert. Every spring, dust clouds from China shut down airports and close schools in Japan and Korea. The dust plume follows the jet stream across the Pacific to Hawaii and then to the west coast of North America, where it sometimes makes up as much as half the particulate air pollution in Seattle, Washington. Some Asian dust storms have polluted the skies as far east as Savannah, Georgia, and Portland, Maine.

As we discussed in chapter 9, as much as one-third of the earth's surface is threatened by desertification. Population growth and poverty drive people into fragile, marginal lands that blow away when rains fail. Poor farming practices expose soils to wind erosion. The result is an escalating crisis that not only threatens food production but also pollutes air around the globe. The resulting haze reduces visibility in remote locations such as California's Sequoia National Park or Big Bend National Park in Texas.

**FIGURE 16.9** A massive dust storm extends more than 1,600 km (1,000 mi) from the coast of western Sahara and Morocco. Storms such as this can easily reach the Americas, and they have been linked both to the decline of coral reefs in the Caribbean and to the frequency and intensity of hurricanes formed in the eastern Atlantic Ocean.

Human health also suffers when dust fills the air. Epidemiological studies have found that cities with chronically high levels of particulates have higher death rates, mostly from heart and lung disease. Emergency-room visits and death rates rise in days following a dust storm. Some of this health risk comes from the particles themselves, which clog tiny airways and make breathing difficult. The dust also carries pollen, bacteria, viruses, fungi, herbicides, acids, radioactive isotopes, and heavy metals between continents. As we mentioned in the opening case study for this chapter, roughly half the mercury contamination falling on North America is thought to come from Asia; much of it may arrive attached to airborne particulates. Scientists once thought that living organisms couldn't survive the trip across oceans, but thick dust helps shield cells from sunlight and desiccation.

Airborne dust is considered the primary source of allergies worldwide. Saharan dust storms are suspected of raising asthma rates in Trinidad and Barbados, where cases have increased 17-fold in 30 years. *Aspergillus sydowii,* a soil fungus from Africa, has been shown to be causing death of corals and sea fans in remote reefs in the Caribbean. Europe also receives airborne pathogens via dust storms. Outbreaks of foot-and-mouth disease in Britain have been traced to dust storms from North Africa. The recent discovery of nanobacteria—the smallest known self-replicating organisms, about 100 times smaller than regular bacteria—in dust clouds suggests an even wider role of airborne pathogens in global disease. Although the effect of these tiny cells is controversial, they have been implicated in diseases as different as heart disease, kidney stone formation, and HIV.

### *Metals and Halogens*

Many toxic metals are mined and used in manufacturing processes or occur as trace elements in fuels, especially coal. These metals are released to the air in the form of metal fumes or suspended particulates by fuel combustion, ore smelting, and disposal of wastes. Worldwide atmospheric lead emissions amount to about 2 million metric tons per year, or two-thirds of all metallic air pollution. Most of this lead is from leaded gasoline. Lead is a metabolic poison and a neurotoxin that binds to essential enzymes and cellular components and inactivates them.

Banning leaded gasoline is one of the most successful pollution-control measures in American history. Since 1986, when the ban was enforced, children's average blood lead levels have dropped 90 percent and average IQs have risen three points. Now, 50 nations have renounced leaded gasoline. The global economic benefit of this step is estimated to be more than $200 billion per year.

Mercury, described in the opening case study of this chapter, also is an extremely toxic substance. Young children and developing fetuses are particularly vulnerable. The most famous case of mercury poisoning occurred in Minamata, Japan, in the 1950s. A chemical factory dumped mercury-contaminated waste into the ocean, where it concentrated in fish. Babies whose mothers ate mercury-contaminated fish suffered profound neurological disabilities including deafness, blindness, mental retardation, and cerebral palsy. In adults, mercury poisoning can cause numbness, loss of muscle control, dementia, and death. The most dangerous form, dimethyl mercury, is so toxic that a single drop on your skin can kill you.

Mercury contamination is the most common cause of impairment of U.S. rivers and lakes. In 2007, researchers sampled more than 2,700 fish from 626 rivers and streams in 12 western states and found mercury in every one of them. Forty-five states have issued warnings about eating locally caught freshwater fish (fig. 16.10). Pregnant women and young children also are advised to limit their consumption of certain seafood, including swordfish, marlin, tuna,

**FIGURE 16.10** Mercury contamination is the most common cause of impairment of U.S. rivers and lakes. Forty-five states have issued warnings about eating locally caught freshwater fish. Long-lived, top predators are especially likely to bioaccumulate toxic concentrations of mercury.

shark, and lobster. The U.S. National Institutes of Health (NIH) estimates that one woman in 12 in the United States has more mercury in her blood than 5.8 ug/l considered safe by the EPA. Between 300,000 and 600,000 of the 4 million children born each year in the United States are exposed in the womb to mercury levels that could cause diminished intelligence or developmental problems. According to the NIH, elevated mercury levels cost the U.S. economy $8.7 billion a year in higher medical and educational costs and lost productivity in the workforce.

Mercury comes from many sources. It's released by volcanoes, weathering of rocks, and evaporation from oceans and wetlands. Eighty-five percent of anthropogenic emissions come from smokestacks, primarily power plants and municipal and medical waste incinerators. Most mercury in the air is elemental metallic vapor. In this relatively insoluble and unreactive form, it can circumnavigate the globe for up to two years. When mercury gets into wetlands, however, it can be converted by microbes into an organic form that is absorbed by other organisms and concentrated as it passes through food webs. Top predators such as game fish can accumulate dangerously high levels of mercury.

Environment Canada's models indicate that 38 percent of the mercury pollution in the Great Lakes region, home to more than 9 million Canadians, comes from U.S. sources. They also note that airborne mercury from industrial regions, including the United States, is having "a serious and disproportionate impact on Canada's Northern and Arctic communities." Meanwhile, federal, provincial, and territorial governments in Canada are working under nationwide standards to reduce up to 90 percent of mercury emissions by 2010. Some states in America have similar programs. Massachusetts, for example, which has long had a serious mercury contamination problem, is requiring all its power plants to install scrubbing equipment that will remove 95 percent of mercury from smokestacks. Some observers wonder why the federal government can't do the same.

### Think About It

Industry has a good idea of how much it will cost to install mercury scrubbers on power plants. But how much is it worth to add a point or two to a child's IQ? Would the answer be different if the child were yours?

Other toxic metals of concern are nickel, beryllium, cadmium, thallium, uranium, cesium, and plutonium. Some 780,000 tons of arsenic, a highly toxic metalloid, are released from metal smelters, coal combustion, and pesticide use each year. Halogens (fluorine, chlorine, bromine, and iodine) are highly reactive and generally toxic in their elemental form. Chlorofluorocarbons (CFCs) have been banned for most uses in industrialized countries, but about 600 million tons of these highly persistent chemical compounds are used annually worldwide in spray propellants, refrigeration compressors, and for foam blowing. They diffuse into the stratosphere where they release chlorine and fluorine atoms that destroy the ozone shield that protects the earth from ultraviolet radiation. We'll return to this topic later in this chapter.

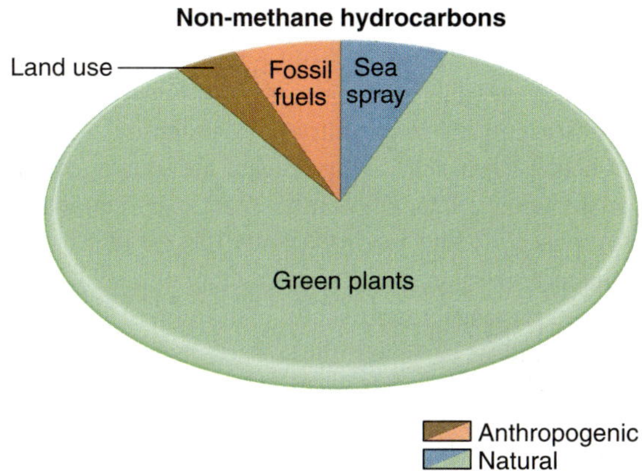


**FIGURE 16.11** Sources of non-methane hydrocarbons in the atmosphere.
**Source:** UNEP, 1999.

## *Volatile Organic Compounds*

**Volatile organic compounds (VOCs)** are organic chemicals that exist as gases in the air. Plants are the largest source of VOCs, releasing an estimated 350 million tons of isoprene ($C_5H_8$) and 450 million tons of terpenes ($C_{10}H_{15}$) each year (fig. 16.11). About 400 million tons of methane ($CH_4$) are produced by natural wetlands and rice paddies and by bacteria in the guts of termites and ruminant animals. These volatile hydrocarbons are generally oxidized to CO and $CO_2$ in the atmosphere.

In addition to these natural VOCs, a large number of other synthetic organic chemicals, such as benzene, toluene, formaldehyde, vinyl chloride, phenols, chloroform, and trichloroethylene, are released into the air by human activities. About 28 million tons of these compounds are emitted each year in the United States, mainly unburned or partially burned hydrocarbons from transportation, power plants, chemical plants, and petroleum refineries. These chemicals play an important role in the formation of photochemical oxidants.

## *Photochemical Oxidants*

**Photochemical oxidants** are products of secondary atmospheric reactions driven by solar energy (fig. 16.12). One of the most important of these reactions involves formation of singlet (atomic)

Atmospheric oxidant production:

1. NO + VOC $\longrightarrow$ $NO_2$ (nitrogen dioxide)
2. $NO_2$ + UV $\longrightarrow$ NO + O (nitric oxide + atomic oxygen)
3. O + $O_2$ $\longrightarrow$ $O_3$ (ozone)
4. $NO_2$ + VOC $\longrightarrow$ PAN, etc. (peroxyacetyl nitrate)

Net results:

NO + VOC + $O_2$ + UV $\longrightarrow$ $O_3$, PAN, and other oxidants

**FIGURE 16.12** Secondary production of urban smog oxidants by photochemical reactions in the atmosphere.

oxygen by splitting nitrogen dioxide ($NO_2$). This atomic oxygen then reacts with another molecule of $O_2$ to make **ozone** ($O_3$). Ozone formed in the stratosphere provides a valuable shield for the biosphere by absorbing incoming ultraviolet radiation. In ambient air, however, $O_3$ is a strong oxidizing reagent and damages vegetation, building materials (such as paint, rubber, and plastics), and sensitive tissues (such as eyes and lungs). Ozone has an acrid, biting odor that is a distinctive characteristic of photochemical smog. Hydrocarbons in the air contribute to accumulation of ozone by removing NO in the formation of compounds, such as peroxyacetyl nitrate (PAN), which is another damaging photochemical oxidant.

### *Air Toxins*

Although most air contaminants are regulated because of their potential adverse effects on human health or environmental quality, a special category of toxins is monitored by the U.S. EPA because they are particularly dangerous. Called **hazardous air pollutants (HAPs),** these chemicals include carcinogens, neurotoxins, mutagens, teratogens, endocrine system disrupters, and other highly toxic compounds (chapter 8). Twenty of the most "persistent bioaccumulative toxic chemicals" (see table 8.2) require special reporting and management because they remain in ecosystems for long periods of time, and accumulate in animal and human tissues. Most of these chemicals are either metal compounds, chlorinated hydrocarbons, or volatile organic compounds.

Only about 50 locations in the United States regularly measure concentrations of HAPs in ambient air. Often the best source of information about these chemicals is the **Toxic Release Inventory (TRI)** collected by the EPA as part of the community right-to-know program. Established by Congress in 1986, the TRI requires 23,000 factories, refineries, hard rock mines, power plants, and chemical manufacturers to report on toxin releases (above certain minimum amounts) and waste management methods for 667 toxic chemicals. Although this total is less than 1 percent of all chemicals registered for use, and represents a limited range of sources, the TRI is widely considered the most comprehensive source of information about toxic pollution in the United States.

In 2005, U.S. industries released 4 billion pounds (1.8 million metric tons) of toxic chemicals into the environment from 24,000 facilities and disposed of roughly four times that much through waste management and recycling methods. This represents a 52 percent reduction from releases in 1999. Of the environmental releases, 65 percent was discharged on land, 27 percent (486,000 metric tons) was released into the air, 4 percent was discharged into water, and 4 percent was injected into deep wells. Twenty chemicals accounted for 88 percent of the total releases, with metals and mining waste comprising a vast majority of that amount. Nearly 200,000 metric tons of persistent bioaccumulative chemicals are emitted annually, with mercury and lead compounds comprising 97 percent of that total.

While most HAP releases are decreasing, discharges of mercury and dioxins—both of which are bioaccumulative and toxic at extremely low levels—have increased in recent years. Dioxins are created mainly by burning plastics and medical waste containing chlorine. The EPA reports that 100 million Americans live in areas where the cancer rate from HAPs exceeds 10 in 1 million or ten times the normally accepted standard for action (fig. 16.13). Benzene, formaldehyde, acetaldehyde, and 1,3 butadiene are responsible for most of this HAP cancer risk. Furthermore, twice that many (70 percent of the U.S. population) live in areas where the non-cancer risk of death exceeds 1 in 1 million. To help residents track local air quality levels, the EPA recently estimated the concentration of HAPs in localities across the continental United States (over 60,000 census tracts). You can access this information on the Environmental Defense Fund web page at www.scorecard.org/env-releases/hap/.

## Unconventional pollutants also are important

In addition to toxic air pollutants, some other unconventional forms of air pollution deserve mention. **Aesthetic degradation** includes any undesirable changes in the physical characteristics or chemistry of the atmosphere. Noise, odors, and light pollution are examples of atmospheric degradation that may not be life-threatening but reduce the quality of our lives. This is a very subjective category. Odors and noise (such as loud music) that are offensive to some may be attractive to others. Often the most sensitive device for odor detection is the human nose. We can smell styrene, for example, at 44 parts per billion (ppb). Trained panels of odor testers often are used to evaluate air samples. Factories that emit noxious chemicals sometimes spray "odor maskants" or perfumes into smokestacks to cover up objectionable odors.

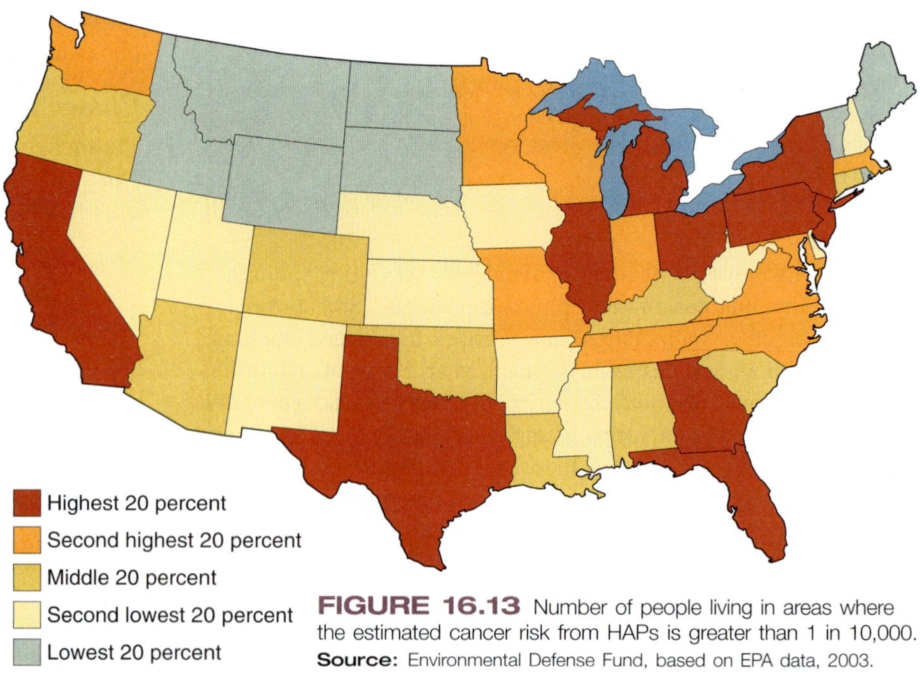


**FIGURE 16.13** Number of people living in areas where the estimated cancer risk from HAPs is greater than 1 in 10,000. **Source:** Environmental Defense Fund, based on EPA data, 2003.

In most urban areas, it is difficult or impossible to see stars in the sky at night because of dust in the air and stray light from buildings, outdoor advertising, and streetlights. This light pollution has become a serious problem for astronomers.

## Indoor air is more dangerous for most of us than outdoor air

We have spent a considerable amount of effort and money to control the major outdoor air pollutants, but we have only recently become aware of the dangers of indoor air pollutants. The EPA has found that indoor concentrations of toxic air pollutants are often higher than outdoors. Furthermore, people generally spend more time inside than out and therefore are exposed to higher doses of these pollutants.

Smoking is without doubt the most important air pollutant in the United States in terms of human health. The Surgeon General estimates that more than 400,000 people die each year in the United States from emphysema, heart attacks, strokes, lung cancer, or other diseases caused by smoking. These diseases are responsible for 20 percent of all mortality in the United States, or four times as much as infectious agents. Lung cancer has now surpassed breast cancer as the leading cause of cancer deaths for U.S. women. Advertising aimed at making smoking appear stylish and liberating has resulted in a 600 percent increase in lung cancer among women since 1950.

Total costs for early deaths and smoking-related illnesses in the United States are estimated to be $100 billion per year. Eliminating smoking probably would save more lives than any other pollution-control measure. Smoking restrictions in many places have resulted in dramatic declines in second-hand smoke exposure to nonsmokers, EPA data show. In just a decade after indoor smoking bans were passed, levels of tobacco by-products in nonsmokers' blood dropped 75 percent. With increasing restrictions on smoking in Western countries, tobacco companies are now turning their attention to developing countries. Persuading consumers (especially women, who traditionally don't smoke) that American cigarettes are modern and stylish could recruit billions of new customers—and cause hundreds of millions of cancer deaths.

In some cases, indoor air in homes has concentrations of chemicals that would be illegal outside or in the workplace. The EPA has found that concentrations of such compounds as chloroform, benzene, carbon tetrachloride, formaldehyde, and styrene can be seventy times higher in indoor air than in outdoor air. "Green design" principles can make indoor spaces both healthier and more pleasant (Exploring Science, p. 358).

In the less-developed countries of Africa, Asia, and Latin America where such organic fuels as firewood, charcoal, dried dung, and agricultural wastes make up the majority of household energy, smoky, poorly ventilated heating and cooking fires represent the greatest source of indoor air pollution (fig. 16.14). The World Health Organization (WHO) estimates that 2.5 billion people—nearly half the world's population—are adversely affected by pollution from this source. Women and small children spend long hours each day around open fires or unventilated stoves in enclosed spaces. The levels of carbon monoxide, particulates, aldehydes, and other toxic chemicals can be 100 times higher than would be legal for outdoor ambient concentrations in the United States. Designing and building cheap, efficient, nonpolluting energy sources for the developing countries would not only save shrinking forests but would make a major impact on health as well.

**FIGURE 16.14** Smoky cooking and heating fires may cause more ill health effects than any other source of indoor air pollution except tobacco smoking. Some 2.5 billion people, mainly women and children, spend hours each day in poorly ventilated kitchens and living spaces where carbon monoxide, particulates, and cancer-causing hydrocarbons often reach dangerous levels.

# 16.4 Climate, Topography, and Atmospheric Processes

Topography, climate, and physical processes in the atmosphere play an important role in transport, concentration, dispersal, and removal of many air pollutants. Wind speed, mixing between air layers, precipitation, and atmospheric chemistry all determine whether pollutants will remain in the locality where they are produced or go elsewhere. In this next section, we will survey some environmental factors that affect air pollution levels.

## Temperature inversions trap pollutants

**Temperature inversions** occur when a stable layer of warmer air overlays cooler air, reversing the normal temperature decline with increasing height and preventing convection currents from dispersing pollutants. Several mechanisms create inversions. When a cold front slides under an adjacent warmer air mass or when cool air subsides down a mountain slope to displace warmer air in the valley below, an inverted temperature gradient is established. These inversions are usually not stable, however, because winds

## *Exploring* SCIENCE — Indoor Air

How safe is the air in your home, office, or school room? As we decrease air-infiltration into buildings to conserve energy, we often trap indoor air pollutants within spaces where most of us spend the vast majority of our time. In what has come to be known as "sick building syndrome," people complain of headaches, fatigue, nausea, upper-respiratory problems, and a wide variety of allergies from workplace or home exposure to airborne toxins. While these symptoms often are vague and difficult to verify scientifically, the U.S. Environmental Protection Agency estimates that sick building syndrome may cost \$60 billion a year in medical expenses, absenteeism, and reduced productivity.

What might be making us sick? Mold spores are probably the greatest single cause of allergic reactions in indoor air. Moisture trapped in air-tight houses often accumulates in walls where molds flourish. Air ducts provide both a good environment for growth of pathogens such as Legionnaire's disease bacteria as well as a path for their dispersal. Legionnaire's pneumonia is much more prevalent than most people realize in places like California and Australia where air-conditioning is common. Uranium-bearing rocks and sediment are widespread across North America. When uranium decays, it produces carcinogenic radon gas that can seep into buildings. The EPA warns that one home in ten in the United States may exceed the recommended maximum radon concentration of 4 picocuries per liter.

In addition, we are exposed to a variety of synthetic chemicals emitted from carpets, wall coverings, building materials, and combustion gases (see fig. 8.11). You might be surprised to learn how many toxic, synthetic compounds are used to construct buildings and make furniture. Formaldehyde, for instance is a component of more than 3,000 products, including building materials such as particle board, waferboard, and urea-formaldehyde foam insulation. Vinyl chloride is used in plastic plumbing pipe, floor and wall coverings, and countertops. Volatile organic solvents make up as much as half the volume of some paint. New carpets and drapes typically contain up to two dozen chemical compounds designed to kill bacteria and molds, resist stains, bind fibers, and retain colors.

What can you do if you suspect that your living spaces are exposing you to materials that make you sick? Probably few students will be in a position anytime soon to build a new house with nontoxic materials, but there are some principles from the emerging field of "green design" that you might apply if you're house hunting, redecorating your apartment, or interviewing for a job. Low-volatile paint is now available for indoor use. Nontoxic, formaldehyde-free plywood, particle board, and insulation can be used in new construction. Nonallergenic carpets, drapes, and wall coverings are available, but some architects recommend natural wood, stone, and plaster surfaces that are easier to clean and less allergenic than any fabric.

High rates of air exchange can help rid indoor air of moisture, odors, mold spores, radon, and toxins. Does that mean energy inefficiency? Not necessarily. Air-to-air heat exchangers keep heat in during the winter and out during the summer, while still providing a healthy rate of fresh-air flow. Bathrooms and kitchens should have outdoor vents. Gas or oil furnaces should be checked for carbon monoxide production. Although many cooks prefer gas stoves because they heat quickly, they can produce toxic carbon monoxide and nitrogen oxides. Contact your city housing authority or county extension service for further tips on how to make your home, work, or study environment healthier. No matter what your situation, there are things that each of us can do to make our indoor air cleaner and safer.

---

accompanying these air exchanges tend to break up the temperature gradient fairly quickly and mix air layers.

The most stable inversion conditions are usually created by rapid nighttime cooling in a valley or basin where air movement is restricted. Los Angeles is a classic example of the conditions that create temperature inversions and photochemical smog (fig. 16.15). The city is surrounded by mountains on three sides and the climate is dry and sunny. Millions of automobiles and trucks create high pollution levels. Skies are generally clear at night, allowing rapid radiant heat loss, and the ground cools quickly. Surface air layers are cooled by conduction, while upper layers remain relatively warm. Density differences retard vertical mixing. During the night, cool, humid, onshore breezes slide in under the contaminated air, squeezing it up against the cap of warmer air above and concentrating the pollutants accumulated during the day.

Morning sunlight is absorbed by the concentrated aerosols and gaseous chemicals of the inversion layer. This complex mixture quickly cooks up a toxic brew of hazardous compounds. As the ground warms later in the day, convection currents break up the temperature gradient and pollutants are carried back down to the surface where more contaminants are added. Nitric oxide (NO) from automobile exhaust is oxidized to nitrogen dioxide. As nitrogen oxides are used up in reactions with unburned hydrocarbons, the ozone levels begin to rise. By early afternoon, an acrid brown haze fills the air, making eyes water and throats burn. In the 1970s, before pollution controls were enforced, the Los Angeles basin often would reach 0.34 ppm or more by late afternoon and the pollution index could be 300, the stage considered a health hazard.

## Cities create dust domes and heat islands

Even without mountains to block winds and stabilize air layers, many large cities create an atmospheric environment quite different from the surrounding conditions. Sparse vegetation and high levels of concrete and glass in urban areas allow rainfall to run off quickly and create high rates of heat absorption during the day and radiation at night. Tall buildings create convective updrafts that sweep pollutants into the air. Temperatures in the center of large cities are frequently 3–5°C (5–9°F) higher than the surrounding countryside. Stable air masses created by this "heat island" over the city concentrate pollutants in a "dust dome." Rural areas downwind from major industrial areas often have significantly decreased visibility and increased rainfall (due to increased condensation nuclei in the dust plume) compared to neighboring areas with cleaner air. In the late 1960s,

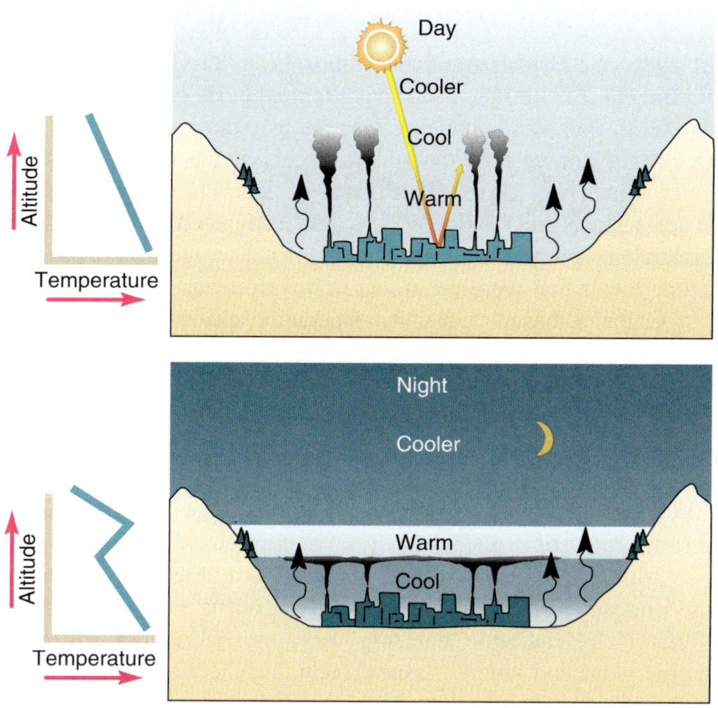


**FIGURE 16.15** Atmospheric temperature inversions occur where ground level air cools more quickly than upper levels. This temperature differential prevents mixing and traps pollutants close to the ground.

for instance, areas downwind from Chicago and St. Louis reported up to 30 percent more rainfall than upwind regions.

Aerosols and dust in urban air seem to trigger increased cloud-to-ground lightning strikes. Houston and Lake Charles, Louisiana, for instance, which have many petroleum refineries, have among the highest number of lightning strikes in the United States and twice as many as nearby areas with similar climate but cleaner air.

## Wind currents carry pollutants intercontinentally

Dust and contaminants can be carried great distances by the wind. Areas downwind from industrial complexes often suffer serious contamination, even if they have no pollution sources of their own (fig. 16.16). Pollution from the industrial belt between the Great Lakes and the Ohio River Valley, for example, regularly contaminates the Canadian Maritime Provinces, and sometimes can be traced as far as Ireland. As we've already seen in this chapter, as much as 70 percent of the mercury that falls on North America may come from abroad, much of it from Asia.

Studies of air pollutants over southern Asia reveal a 3 km thick toxic cloud of ash, acids, aerosols, dust, and photochemical reactants regularly covers the entire Indian subcontinent for much of the year. Nobel laureate Paul Crutzen estimates that up to 2 million people in India alone die each year from atmospheric pollution. Produced by forest fires, the burning of agricultural wastes, and dramatic increases in the use of fossil fuels, the Asian smog layer cuts the amount of solar energy reaching the earth's surface beneath it by up to 15 percent. Meteorologists suggest

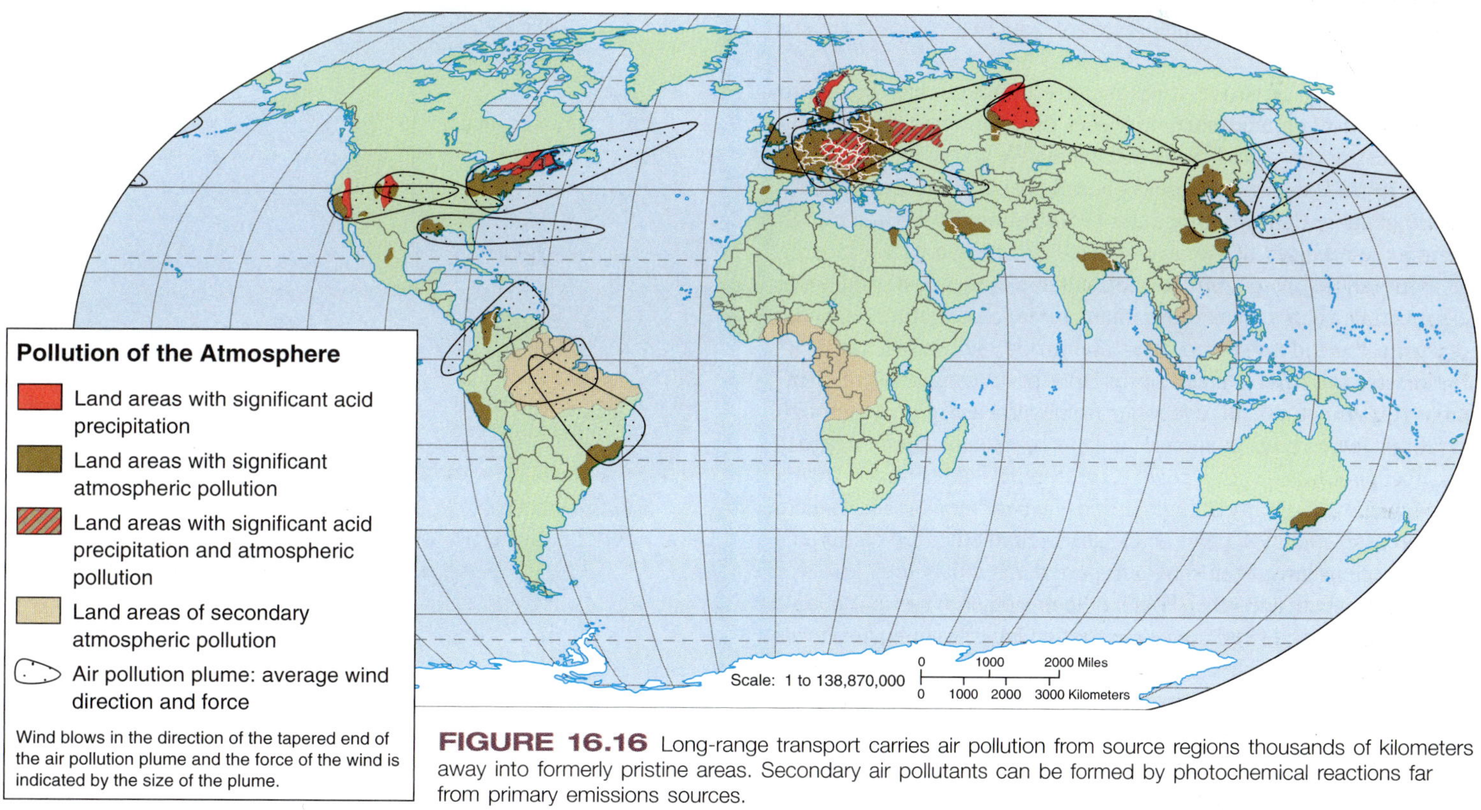


**FIGURE 16.16** Long-range transport carries air pollution from source regions thousands of kilometers away into formerly pristine areas. Secondary air pollutants can be formed by photochemical reactions far from primary emissions sources.

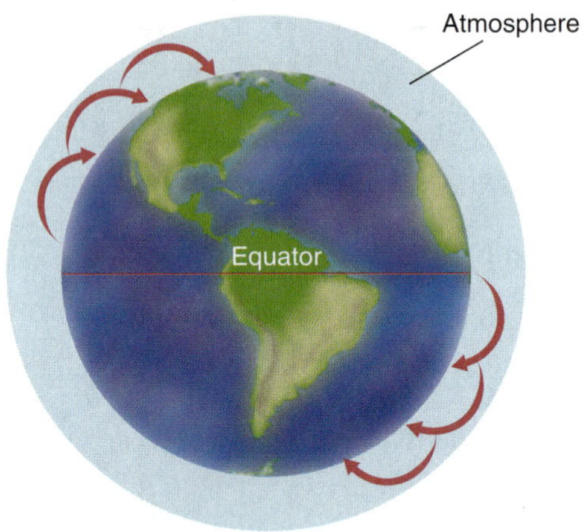


**FIGURE 16.17** Air pollutants evaporate from warmer areas and then condense and precipitate in cooler regions. Eventually, this "grasshopper" redistribution leads to accumulation in the Arctic and Antarctic.

that the cloud—80 percent of which is human-made—could disrupt monsoon weather patterns and may be disturbing rainfall and reducing rice harvests over much of South Asia. Shifting monsoon flows may also have contributed to catastrophic floods in Nepal, Bangladesh, and eastern India that killed at least 1,000 people in 2002, and left more than 25 million homeless.

When this "Asian Brown Cloud" drifts out over the Indian Ocean at the end of the monsoon season, it cools sea temperatures and may be changing El Niño/Southern Oscillation patterns in the Pacific Ocean as well (chapter 15). As UN Environment Programme executive director, Klaus Töpfer, said, "There are global implications because a pollution parcel like this, which stretches three km high, can travel half way round the globe in a week."

Increasingly sensitive monitoring equipment has begun to reveal industrial contaminants in places usually considered among the cleanest in the world. Samoa, Greenland, and even Antarctica and the North Pole, all have heavy metals, pesticides, and radioactive elements in their air. Since the 1950s, pilots flying in the high Arctic have reported dense layers of reddish-brown haze clouding the arctic atmosphere. Aerosols of sulfates, soot, dust, and toxic heavy metals such as vanadium, manganese, and lead travel to the pole from the industrialized parts of Europe and Russia.

In a process called "grasshopper" transport, or atmosphere distillation, volatile compounds evaporate from warm areas, travel through the atmosphere, then condense and precipitate in cooler regions (fig. 16.17). Over several years, contaminants accumulate in the coldest places, generally at high latitudes where they bioaccumulate in food chains. Whales, polar bears, sharks, and other top carnivores in polar regions have been shown to have dangerously high levels of pesticides, metals, and other HAPs in their bodies. The Inuit people of Broughton Island, well above the Arctic Circle, have higher levels of polychlorinated biphenyls (PCBs) in their blood than any other known population, except victims of industrial accidents. Far from any source of this industrial by-product, these people accumulate PCBs from the flesh of fish, caribou, and other animals they eat. This exacerbates the cultural crisis caused by climate change.

## Stratospheric ozone is destroyed by chlorine

In 1985, the British Antarctic Atmospheric Survey announced a startling and disturbing discovery: **Stratospheric ozone** levels over the South Pole were dropping precipitously during September and October every year as the sun reappears at the end of the long polar winter (fig. 16.18). This ozone depletion has been occurring at least since the 1960s but was not recognized because earlier researchers programmed their instruments to ignore changes in ozone levels that were presumed to be erroneous.

Chlorine-based aerosols, especially **chlorofluorocarbons (CFCs)** and other halon gases, are the principal agents of ozone depletion. Nontoxic, nonflammable, chemically inert, and cheaply produced, CFCs were extremely useful as industrial gases and in refrigerators, air conditioners, styrofoam inflation, and aerosol spray cans for many years. From the 1930s until the 1980s, CFCs were used all over the world and widely dispersed through the atmosphere.

Although ozone is a pollutant in the ambient air, ozone in the stratosphere is important because it absorbs much of the ultraviolet (UV) radiation entering the atmosphere. UV radiation harms plant and animal tissues, including the eyes and the skin. A 1 percent loss of ozone could result in about a million extra human skin cancers per year worldwide if no protective measures are taken. Excessive UV exposure could reduce agricultural production and disrupt ecosystems. Scientists worry, for example, that high UV levels in Antarctica could reduce populations of

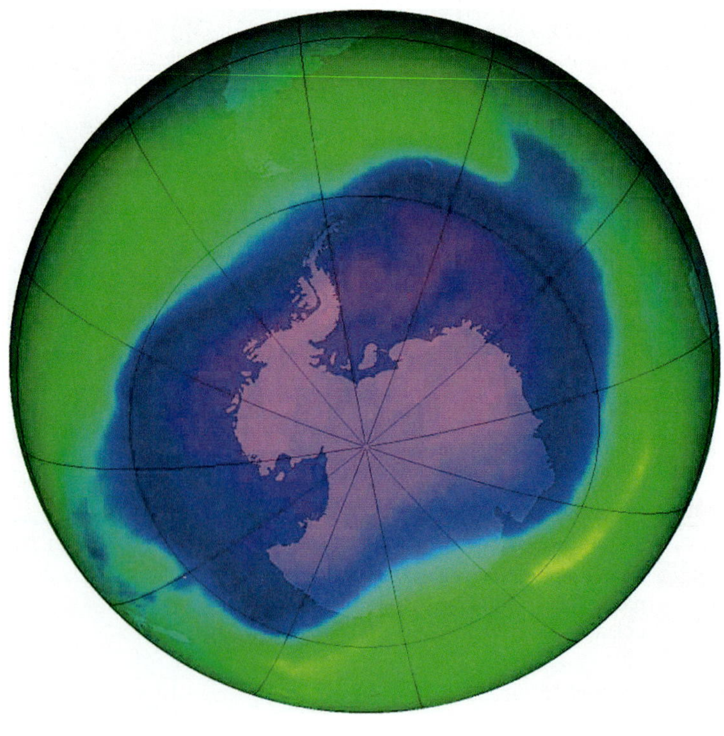

**FIGURE 16.18** In 2006, stratospheric ozone was depleted over an area (*dark, irregular circle*) that covered 29.5 million km$^2$, or more than the entire Antarctic continent. Although CFC production is declining, this was the largest area ever recorded.

**TABLE 16.2**

**Stratospheric Ozone Destruction by Chlorine Atoms and UV Radiation**

| Step | Products |
|---|---|
| 1. $CFCl_3$ (chlorofluorocarbon) + UV energy | $CFCl_2 + Cl$ |
| 2. $Cl + O_3$ | $ClO + O_2$ |
| 3. $O_2$ + UV energy | $2O$ |
| 4. $ClO + 2O$ | $O_2 + Cl$ |
| 5. Return to step 2 | |

plankton, the tiny floating organisms that form the base of a food chain that includes fish, seals, penguins, and whales in Antarctic seas.

Antarctica's exceptionally cold winter temperatures (−85–−90°C) help break down ozone. During the long, dark, winter months, strong winds known as the circumpolar vortex isolate Antarctic air and allow stratospheric temperatures to drop low enough to create ice crystals at high altitudes—something that rarely happens elsewhere in the world. Ozone and chlorine-containing molecules are absorbed on the surfaces of these ice particles. When the sun returns in the spring, it provides energy to liberate chlorine ions, which readily bond with ozone, breaking it down to molecular oxygen (table 16.2). It is only during the Antarctic spring (September through December) that conditions are ideal for rapid ozone destruction. During that season, temperatures are still cold enough for high-altitude ice crystals, but the sun gradually becomes strong enough to drive photochemical reactions.

As the Antarctic summer arrives, temperatures moderate somewhat, the circumpolar vortex breaks down, and air from warmer latitudes mixes with Antarctic air, replenishing ozone concentrations in the ozone hole. Slight decreases worldwide result from this mixing, however. Ozone re-forms naturally, but not nearly as fast as it is destroyed. Since the chlorine atoms are not themselves consumed in reactions with ozone, they continue to destroy ozone for years, until they finally precipitate or are washed out of the air. Almost every year since it was discovered, the Antarctic ozone hole has grown. In 2006 the region of ozone depletion covered 29.5 million $km^2$ (larger than North America).

Although not as pronounced, about 10 percent of all stratospheric ozone worldwide has been destroyed in recent years, and levels over the Arctic have averaged 40 percent below normal. Ozone depletion has been observed over the North Pole as well, although it is not as concentrated as that in the south.

## The Montreal Protocol is a resounding success

The discovery of stratospheric ozone losses brought about a remarkably quick international response. In 1987 an international meeting in Montreal, Canada, produced the Montreal Protocol, the first of several major international agreements on phasing out most use of CFCs by 2000. As evidence accumulated, showing that losses were larger and more widespread than previously thought, the deadline for the elimination of all CFCs (halons, carbon tetrachloride, and methyl chloroform) was moved up to 1996, and a $500 million fund was established to assist poorer countries in switching to non-CFC technologies. Fortunately, alternatives to CFCs for most uses already exist. The first substitutes are hydrochlorofluorocarbons (HCFCs), which release much less chlorine per molecule. Eventually, scientists hope to develop halogen-free molecules that work just as well and are no more expensive than CFCs.

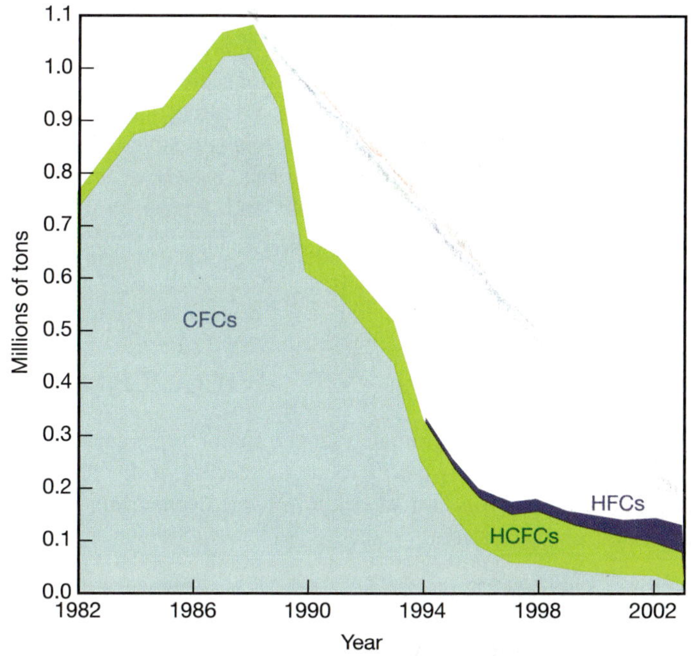


**FIGURE 16.19** The Montreal Protocol has been remarkably successful in eliminating CFC production. The remaining HFC and HCFC use in primarily in developing countries, such as China and India.

The Montreal Protocol is often cited as the most effective international environmental agreement ever established. Global CFC production has been cut by more than 95 percent since 1988 (fig. 16.19). Some of that has been replaced by hydrochloroflourocarbons (HCFCs), which release chlorine, but not as much as CFCs. The amount of chlorine entering the atmosphere already has begun to decrease, suggesting that stratospheric $O_3$ levels should be back to normal by about 2049. You might wonder, then, why the 2006 $O_3$ hole was the largest ever. The answer is global warming (see chapter 15). Greenhouse gases are warming the troposphere, which is causing the stratosphere to cool. This increases ice crystal formation over the Antarctic, and results in more $O_3$ depletion.

There's another interesting connection to climate change. Under the Montreal Protocol, China, India, Korea, and Argentina were allowed to continue to produce 72,000 tons (combined) of CFCs per year until 2010. Most of the funds appropriated through the Montreal Protocol are going to these countries to help them phase out CFC production and to destroy their existing stocks. Because CFCs are potent greenhouse gases, this phase-out also makes these countries eligible for credits in the climate trading market. In 2006, nearly two-thirds of the greenhouse gas emissions credits traded internationally were for HFC-23 elimination, and almost half of all payments went to China. Some critics think this is double-dipping, but if it eliminates a dangerous risk to all of us, isn't it worth it?

In 1995, chemists Sherwood Rowland, Mario Molina, and Paul Crutzen shared the Nobel Prize for their work on atmospheric chemistry and stratospheric ozone. This was the first Nobel Prize for an environmental issue.

## 16.5 Effects of Air Pollution

So far we have looked primarily at the major types and sources of air pollutants. Now we will focus more closely on the effects of those pollutants on human health, physical materials, ecosystems, and global climate.

### Polluted air is dangerous

The World Health Organization estimates that some 5 to 6 million people die prematurely every year from illnesses related to air pollution. Heart attacks, respiratory diseases, and lung cancer all are significantly higher in people who breathe dirty air, compared to matching groups in cleaner environments. Residents of the most polluted cities in the United States, for example, are 15 to 17 percent more likely to die of these illnesses than those in cities with the cleanest air. This can mean as much as a 5- to 10-year decrease in life expectancy if you live in the worst parts of Los Angeles or Baltimore, compared to a place with clean air. Of course your likelihood of suffering ill health from air pollutants depends on the intensity and duration of exposure as well as your age and prior health status. You are much more likely to be at risk if you are very young, very old, or already suffering from some respiratory or cardiovascular disease. Some people are super-sensitive because of genetics or prior exposure. And those doing vigorous physical work or exercise are more likely to succumb than more sedentary folks.

Conditions are often much worse in other countries than Canada or the United States. The United Nations estimates that at least 1.3 billion people around the world live in areas where outdoor air is dangerously polluted. In Madrid, Spain, smog is estimated to shave one-half year off the life of each resident. This adds up to more than 50,000 years lost annually for the whole city. In China, city dwellers are four to six times more likely than country folk to die of lung cancer. As mentioned earlier, the greatest air quality problem is often in poorly ventilated homes in poorer countries where smoky fires are used for cooking and heating. Billions of women and children spend hours each day in these unhealthy conditions. The World Health Organization estimates that 2 million children under age 5 die each year from acute respiratory diseases exacerbated by air pollution.

In industrialized countries, one of the biggest health threats from air pollution is from soot or fine particulate material. We once thought that particles smaller than 10 μm (10 millionths of a meter) were too small to be trapped in the lungs. Now we know that small particles (less than 2.5 μm diameter) called PM2.5 are an even greater risk than larger ones. They have been linked with heart attacks, asthma, bronchitis, lung cancer, immune suppression, and abnormal fetal development, among other health problems. Fine particulates have many sources. Until recently, power plants were the largest source, but with recent clean air rules, they will be required to install filters and precipitators to remove at least 70 percent of their particulate emissions.

**FIGURE 16.20** Soot and fine particulate material from diesel engines, wood stoves, power plants, and other combustion sources have been linked to asthma, heart attacks, and a variety of other diseases.

Diesel engines have long been a major source of both soot and $SO_2$ in the United States (fig. 16.20). Under a new rule announced in 2006, new engines in trucks and buses, in combination with low-sulfur diesel fuel that is now required nationwide, will reduce particulate emissions by up to 98 percent when the rule is fully implemented in 2012. These standards will also be applied to off-road vehicles, such as tractors, bulldozers, locomotives, and barges, whose engines previously emitted more soot than all the nation's cars, trucks, and buses together. The sulfur content of diesel fuel is now 500 parts per million (ppm) compared to an average of 3,400 ppm before the regulations were imposed. By 2012, only 15 ppm of sulfur will be allowed in diesel fuel. Europe has had low-sulfur fuel and clean diesel engines since the early 1990s, but in spite of this, diesel buses and trucks are banned from the centers of many European cities.

In some rural areas, smoke from wood stoves or burning crops remain an important soot source. Resort towns, such as Telluride and Aspen, Colorado, are beginning to limit or ban wood stoves and open fireplaces because of the pollution they produce. The U.S. EPA estimates that at least 160 million Americans—more than half the population—live in areas with unhealthy concentrations of fine particulate matter. More than 450 counties in 32 states are considered in nonattainment of clean air rules. The EPA reports that PM2.5 levels have decreased about 30 percent over the past 25 years, but health officials argue that the remaining pollution should also be cleaned up.

## How does pollution harm us?

The most common route of exposure to air pollutants is by inhalation, but direct absorption through the skin or contamination of food and water also are important pathways. Because they are strong oxidizing agents, sulfates, $SO_2$, $NO_x$, and $O_3$ act as irritants that damage delicate tissues in the eyes and respiratory passages. Fine particulates penetrate deep into the lungs and are irritants in their own right, as well as carrying metals and other HAPs on their surfaces. Inflammatory responses set in motion by these irritants impair lung function and trigger cardiovascular problems as the heart tries to compensate for lack of oxygen by pumping faster and harder. If the irritation is really severe, so much fluid seeps into lungs through damaged tissues that the victim actually drowns.

Carbon monoxide binds to hemoglobin and decreases the ability of red blood cells to carry oxygen. Asphyxiants such as this cause headaches, dizziness, heart stress, and can even be lethal if concentrations are high enough. Lead also binds to hemoglobin and reduces oxygen-carrying capacity at high levels. At lower levels, lead causes long-term damage to critical neurons in the brain that results in mental and physical impairment and developmental retardation.

Some important chronic health effects of air pollutants include bronchitis and emphysema.

**Bronchitis** is a persistent inflammation of bronchi and bronchioles (large and small airways in the lung) that causes mucus buildup, a painful cough, and involuntary muscle spasms that constrict airways. Severe bronchitis can lead to emphysema, an irreversible **chronic obstructive lung disease** in which airways become permanently constricted and alveoli are damaged or even destroyed. Stagnant air trapped in blocked airways swells the tiny air sacs in the lung (alveoli), blocking blood circulation. As cells die from lack of oxygen and nutrients, the walls of the alveoli break down, creating large empty spaces incapable of gas exchange (fig. 16.21). Thickened walls of the bronchioles lose elasticity and breathing becomes more difficult. Victims of emphysema make a characteristic whistling sound when they breathe. Often they need supplementary oxygen to make up for reduced respiratory capacity.

Irritants in the air are so widespread that about half of all lungs examined at autopsy in the United States have some degree of alveolar deterioration. The Office of Technology Assessment (OTA) estimates that 250,000 people suffer from pollution-related bronchitis and emphysema in the United States, and some 50,000 excess deaths each year are attributable to complications of these diseases, which are probably second only to heart attack as a cause of death.

Smoking is undoubtedly the largest cause of obstructive lung disease and preventable death in the world. The World Health Organization says that tobacco kills some 3 million people each year. This makes it rank with AIDS as one of the world's leading killers. Because of cardiovascular stress caused by carbon monoxide in smoke and chronic bronchitis and emphysema, about twice as many people die of heart failure as die from lung cancer associated with smoking.

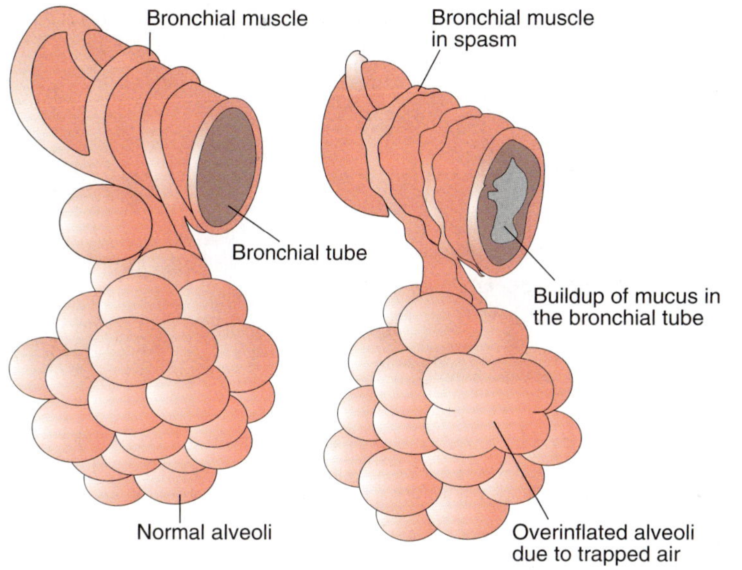


**FIGURE 16.21** Bronchitis and emphysema can result in constriction of airways and permanent damage to tiny, sensitive air sacs called alveoli, where oxygen diffuses into blood vessels.

## Plants are susceptible to pollution damage

In the early days of industrialization, fumes from furnaces, smelters, refineries, and chemical plants often destroyed vegetation and created desolate, barren landscapes around mining and manufacturing centers. The copper-nickel smelter at Sudbury, Ontario, is a spectacular and notorious example of air pollution effects on vegetation and ecosystems. In 1886, the corporate ancestors of the International Nickel Company (INCO) began open-bed roasting of sulfide ores at Sudbury. Sulfur dioxide and sulfuric acid released by this process caused massive destruction of the plant community within about 30 km of the smelter. Rains washed away the exposed soil, leaving a barren moonscape of blackened bedrock (fig. 16.22). Super-tall, 400 m smokestacks were installed in the 1950s and sulfur scrubbers were added 20 years later. Emissions were reduced by 90 percent and the surrounding ecosystem is beginning to recover (fig. 16.23). Similar destruction occurred at many other sites during the nineteenth century. Copperhill, Tennessee; Butte, Montana; and the Ruhr Valley in Germany are some well-known examples, but these areas also are showing signs of recovery since corrective measures were taken.

There are two probable ways that air pollutants damage plants. They can be directly toxic, damaging sensitive cell membranes much as irritants do in human lungs. Within a few days of exposure to toxic levels of oxidants, mottling (discoloration) occurs in leaves due to chlorosis (bleaching of chlorophyll), and then necrotic (dead) spots develop (fig. 16.24). If injury is severe, the whole plant may be killed. Sometimes these symptoms are so distinctive that positive identification of the source of damage is possible. Often, however, the symptoms are vague and difficult to separate from diseases or insect damage.

**FIGURE 16.22** In 1975, acid precipitation from the copper-nickel smelters (tall stacks in background) had killed all the vegetation and charred the pink granite bedrock black for a large area around Sudbury, Ontario.

**FIGURE 16.23** By 2005, a scrubby forest was growing again around Sudbury, but the rock surfaces remain burned black.

Certain combinations of environmental factors have **synergistic effects** in which the injury caused by exposure to two factors together is more than the sum of exposure to each factor individually. For instance, when white pine seedlings are exposed to subthreshold concentrations of ozone and sulfur dioxide individually, no visible injury occurs. If the same concentrations of pollutants are given together, however, visible damage occurs. In alfalfa, however, $SO_2$ and $O_3$ together cause less damage than either one alone. These complex interactions point out the unpredictability of future effects of pollutants. Outcomes might be either more or less severe than previous experience indicates.

Pollutant levels too low to produce visible symptoms of damage may still have important effects. Field studies using open-top chambers (fig. 16.25) and charcoal-filtered air show that yields in some sensitive crops, such as soybeans, may be reduced as much as 50 percent by currently existing levels of oxidants in ambient air. Some plant pathologists suggest that ozone and photochemical oxidants are responsible for as much as 90 percent of agricultural, ornamental, and forest losses from air pollution. The total costs of this damage may be as much as $10 billion per year in North America alone.

## Acid deposition has many negative effects

Most people in the United States became aware of problems associated with **acid precipitation** (the deposition of wet acidic solutions or dry acidic particles from the air) within the last decade or so, but English scientist Robert Angus Smith coined the term "acid rain" in his studies of air chemistry in Manchester, England, in the 1850s. By the 1940s, it was known that pollutants, including atmospheric acids, could be transported long distances by wind currents. This was thought to be only an academic curiosity until it was shown that precipitation of these acids can have far-reaching ecological effects.

We describe acidity in terms of pH (chapter 3). Values below 7 are acidic, while those above 7 are alkaline. Normal, unpolluted rain generally has a pH of about 5.6 due to carbonic acid created by $CO_2$ in air. Volcanic emissions, biological decomposition, and chlorine and sulfates from ocean spray can drop the pH of rain well below 5.6, while alkaline dust can raise it above 7. In industrialized areas, anthropogenic acids in the air usually far outweigh those from natural sources. Acid rain is only one form in which acid deposition occurs. Fog, snow, mist, and dew also trap and deposit atmospheric contaminants. Furthermore, fallout of dry sulfate, nitrate, and chloride particles can account for as much as half of the acidic deposition in some areas.

### *Aquatic Effects*

It has been known for about 30 years that acids—principally $H_2SO_4$ and $HNO_3$—generated by industrial and automobile emissions in northwestern Europe are carried by prevailing winds to Scandinavia where they are deposited in rain, snow, and dry precipitation. The thin, acidic soils and oligotrophic lakes and streams in the mountains of southern Norway and Sweden have been severely affected by this acid deposition. Some 18,000 lakes

**FIGURE 16.24** Soybean leaves exposed to 0.8 parts per million sulfur dioxide for 24 hours show extensive chlorosis (chlorophyll destruction) in white areas between leaf veins.

**FIGURE 16.25** An open-top chamber tests air pollution effects on plants under normal conditions for rain, sun, field soil, and pest exposure.

in Sweden are now so acidic that they will no longer support game fish or other sensitive aquatic organisms.

Generally, reproduction is the most sensitive stage in fish life cycles. Eggs and fry of many species are killed when the pH drops to about 5.0. This level of acidification also can disrupt the food chain by killing aquatic plants, insects, and invertebrates on which fish depend for food. At pH levels below 5.0, adult fish die as well. Trout, salmon, and other game fish are usually the most sensitive. Carp, gar, suckers, and other less desirable fish are more resistant.

In the early 1970s, evidence began to accumulate suggesting that air pollutants are acidifying many lakes in North America. Studies in the Adirondack Mountains of New York revealed that about half of the high-altitude lakes (above 1,000 m or 3,300 ft) are acidified and have no fish. Areas showing lake damage correlate closely with average pH levels in precipitation (fig. 16.26). Some 48,000 lakes in Ontario are endangered and nearly all of Quebec's surface waters, including about 1 million lakes, are believed to be highly sensitive to acid deposition.

Sulfates account for about two-thirds of the acid deposition in eastern North America and most of Europe, while nitrates contribute most of the remaining one-third. In urban areas, where transportation is the major source of pollution, nitric acid is equal to or slightly greater than sulfuric acids in the air. A vigorous program of pollution control has been undertaken by both Canada and the United States, and $SO_2$ and $NO_x$ emissions have decreased dramatically over the past three decades over much of North America.

### Forest Damage

In the early 1980s, disturbing reports appeared of rapid forest declines in both Europe and North America. One of the earliest was a detailed ecosystem inventory on Camel's Hump Mountain in Vermont. A 1980 survey showed that seedling production, tree density, and viability of spruce-fir forests at high elevations

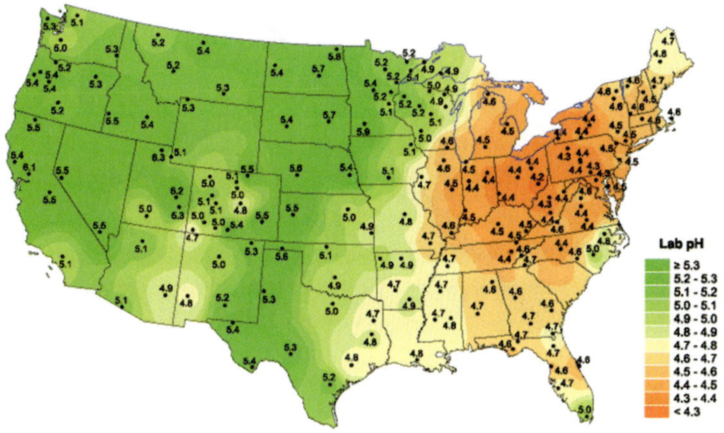


**FIGURE 16.26** Acid precipitation over the United States.
**Source:** National Atmospheric Deposition Program/National Trends Network, 2000. http://nadp.sws.uiuc.edu.

had declined about 50 percent in 15 years. A similar situation was found on Mount Mitchell in North Carolina where almost all red spruce and Fraser fir above 2,000 m (6,000 ft) are in a severe decline. Nearly all the trees are losing needles and about half of them are dead (fig. 16.27). The stress of acid rain and fog, other air pollutants, and attacks by an invasive insect called the woody aldegid are killing the trees.

Many European countries reported catastrophic forest destruction in the 1980s. It still isn't clear what caused this injury. In the longest-running forest-ecosystem monitoring record in North America, researchers at the Hubbard Brook Experimental Forest in New Hampshire have shown that forest soils have become depleted of natural buffering reserves of basic cations such as calcium and magnesium through years of exposure to acid rain. Replacement of these cations by hydrogen and aluminum ions seems to be one of the main causes of plant mortality.

**FIGURE 16.27** A Fraser fir forest on Mount Mitchell, North Carolina, killed by acid rain, insect pests, and other stressors.

**FIGURE 16.28** Atmospheric acids, especially sulfuric and nitric acids, have almost completely eaten away the face of this medieval statue. Each year, the total losses from air pollution damage to buildings and materials amounts to billions of dollars.

### *Buildings and Monuments*

In cities throughout the world, some of the oldest and most glorious buildings and works of art are being destroyed by air pollution. Smoke and soot coat buildings, paintings, and textiles. Limestone and marble are destroyed by atmospheric acids at an alarming rate. The Parthenon in Athens, the Taj Mahal in Agra, the Colosseum in Rome, frescoes and statues in Florence, medieval cathedrals in Europe (fig. 16.28). and the Lincoln Memorial and Washington Monument in Washington, D.C., are slowly dissolving and flaking away because of acidic fumes in the air. Medieval stained glass windows in Cologne's gothic cathedral are so porous from etching by atmospheric acids that pigments disappear and the glass literally crumbles away. Restoration costs for this one building alone are estimated at three to 1.5 billion Euros (U.S. $1.8 billion).

On a more mundane level, air pollution also damages ordinary buildings and structures. Corroding steel in reinforced concrete weakens buildings, roads, and bridges. Paint and rubber deteriorate due to oxidization. Limestone, marble, and some kinds of sandstone flake and crumble. The Council on Environmental Quality estimates that U.S. economic losses from architectural damage caused by air pollution amount to about $4.8 billion in direct costs and $5.2 billion in property value losses each year.

## Smog and haze reduce visibility

We have realized only recently that pollution affects rural areas as well as cities. Even supposedly pristine places like our national parks are suffering from air pollution. Grand Canyon National Park, where maximum visibility used to be 300 km, is now so smoggy on some winter days that visitors can't see the opposite rim only 20 km across the canyon. Mining operations, smelters, and power plants (some of which were moved to the desert to improve air quality in cities like Los Angeles) are the main culprits. Similarly, the vistas from Shenandoah National Park just outside Washington, D.C., are so hazy that summer visibility is often less than 1.6 km because of smog drifting in from nearby urban areas.

Historical records show that over the past four or five decades human-caused air pollution has spread over much of the United States. Researchers report that a gigantic "haze blob" as much as 3,000 km across covers much of the eastern United States in the summer, cutting visibility as much as 80 percent. Smog and haze are so prevalent that it's hard for people to believe that the air once was clear. Studies indicate, however, that if all human-made sources of air pollution were shut down, the air would clear up in a few days and there would be about 150 km visibility nearly everywhere rather than the 15 km to which we have become accustomed.

# 16.6 Air Pollution Control

"Dilution is the solution to pollution" was one of the early approaches to air pollution control. Tall smokestacks were built to send emissions far from the source, where they became unidentifiable and largely untraceable. But dispersed and diluted pollutants are now the source of some of our most serious pollution problems. We are finding that there is no "away" to which we can throw our waste products. While most of the discussion in this section focuses on industrial solutions, each of us can make important personal contributions to this effort (What Can You Do? p. 366).

## *What Can You Do?*

### Saving Energy and Reducing Pollution

- Conserve energy: carpool, bike, walk, use public transport, buy compact fluorescent bulbs, and energy-efficient appliances (see chapter 20 for other suggestions).
- Don't use polluting two-cycle gasoline engines if cleaner four-cycle models are available for lawn mowers, boat motors, etc.
- Buy refrigerators and air conditioners designed for CFC alternatives. If you have old appliances or other CFC sources, dispose of them responsibly.
- Plant a tree and care for it (every year).
- Write to your Congressional representatives and support a transition to an energy-efficient economy.
- If green-pricing options are available in your area, buy renewable energy.
- If your home has a fireplace, install a high-efficiency, clean-burning, two-stage insert that conserves energy and reduces pollution up to 90 percent.
- Have your car tuned every 10,000 miles (16,000 km) and make sure that its antismog equipment is working properly. Turn off your engine when waiting longer than one minute. Start trips a little earlier and drive slower—it not only saves fuel but it's safer, too.
- Use latex-based, low-volatile paint rather than oil-based (alkyd) paint.
- Avoid spray can products. Light charcoal fires with electric starters rather than petroleum products.
- Don't top off your fuel tank when you buy gasoline; stop when the automatic mechanism turns off the pump. Don't dump gasoline or used oil on the ground or down the drain.
- Buy clothes that can be washed rather than dry-cleaned.

## The most effective strategy for controlling pollution is to minimize production

Since most air pollution in the developed world is associated with transportation and energy production, the most effective strategy would be conservation: Reducing electricity consumption, insulating homes and offices, and developing better public transportation could all greatly reduce air pollution in the United States, Canada, and Europe. Alternative energy sources, such as wind and solar power, produce energy with little or no pollution, and these and other technologies are becoming economically competitive (chapter 20). In addition to conservation, pollution can be controlled by technological innovation.

*Particulate removal* involves filtering air emissions. Filters trap particulates in a mesh of cotton cloth, spun glass fibers, or asbestos-cellulose. Industrial air filters are generally giant bags 10 to 15 m long and 2 to 3 m wide. Effluent gas is blown through the bag, much like the bag on a vacuum cleaner. Every few days or weeks, the bags are opened to remove the dust cake. Electrostatic precipitators are the most common particulate controls in power plants. Ash particles pick up an electrostatic surface charge as they pass between large electrodes in the effluent stream (fig. 16.29). Charged particles then collect on an oppositely charged collecting plate. These precipitators consume a large amount of electricity, but maintenance is relatively simple, and collection efficiency can be as high as 99 percent. The ash collected by both of these techniques is a solid waste (often hazardous due to the heavy metals and other trace components of coal or other ash source) and must be buried in landfills or other solid-waste disposal sites.

*Sulfur removal* is important because sulfur oxides are among the most damaging of all air pollutants in terms of human health and ecosystem viability. Switching from soft coal with a high sulfur content to low-sulfur coal is the surest way to reduce sulfur emissions. High-sulfur coal is frequently politically or economically expedient, however. In the United States, Appalachia, a region of chronic economic depression, produces most high-sulfur coal. In China, much domestic coal is rich in sulfur. Switching to cleaner oil or gas would eliminate metal effluents as well as sulfur. Cleaning fuels is an alternative to switching. Coal can be crushed, washed, and gasified to remove sulfur and metals before combustion. This improves heat content and firing properties, but may replace air pollution with solid-waste and water pollution problems; furthermore, these steps are expensive.

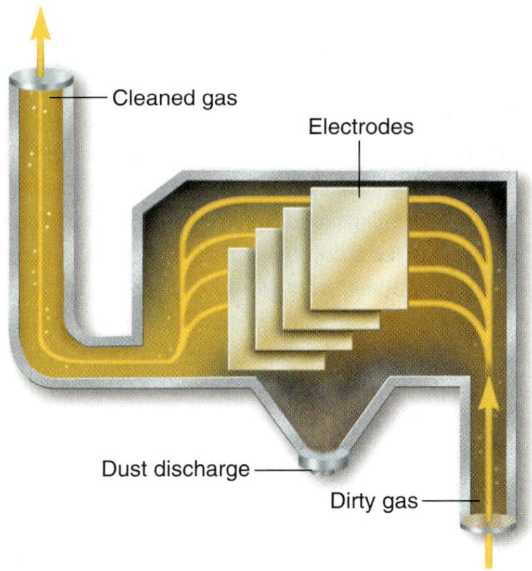


**FIGURE 16.29** An electrostatic precipitator traps particulate material on electrically charged plates as effluent makes its way to the smokestack.

Sulfur can also be removed to yield a usable product instead of simply a waste disposal problem. Elemental sulfur, sulfuric acid, or ammonium sulfate can all be produced using catalytic converters to oxidize or reduce sulfur. Markets have to be reasonably close and fly ash contamination must be reduced as much as possible for this procedure to be economically feasible.

*Nitrogen oxides* ($NO_x$) can be reduced in both internal combustion engines and industrial boilers by as much as 50 percent by carefully controlling the flow of air and fuel. Staged burners, for example, control burning temperatures and oxygen flow to prevent formation of $NO_x$. The catalytic converter on your car uses platinum-palladium and rhodium catalysts to remove up to 90 percent of $NO_x$, hydrocarbons, and carbon monoxide at the same time.

*Hydrocarbon controls* mainly involve complete combustion or controlling evaporation. Hydrocarbons and volatile organic compounds are produced by incomplete combustion of fuels or by solvent evaporation from chemical factories, paints, dry cleaning, plastic manufacturing, printing, and other industrial processes. Closed systems that prevent escape of fugitive gases can reduce many of these emissions. In automobiles, for instance, positive crankcase ventilation (PCV) systems collect oil that escapes from around the pistons and unburned fuel and channels them back to the engine for combustion. Controls on fugitive losses from industrial valves, pipes, and storage tanks can have a significant impact on air quality. Afterburners are often the best method for destroying volatile organic chemicals in industrial exhaust stacks.

## Fuel switching and fuel cleaning also are effective

Switching from soft coal with a high sulfur content to low-sulfur coal can greatly reduce sulfur emissions. This may eliminate jobs, however, in such areas as Appalachia that are already economically depressed. Changing to another fuel, such as natural gas or nuclear energy, can eliminate all sulfur emissions as well as those of particulates and heavy metals. Natural gas is more expensive and more difficult to ship and store than coal, however, and many people prefer the sure dangers of coal pollution to the uncertain dangers of nuclear power (chapter 19). Alternative energy sources, such as wind and solar power, are preferable to either fossil fuel or nuclear power, and are becoming economically competitive (chapter 20) in many areas. In the interim, coal can be crushed, washed, and gassified to remove sulfur and metals before combustion. This improves heat content and firing properties but may replace air pollution with solid waste and water pollution problems.

Ironically, reducing air pollution could increase global warming. The amount of sunlight reaching the earth's surface is reported to have been declining over the past few decades, especially over large cities. This "global dimming" has been ascribed to greater reflection of sunlight by atmospheric aerosols such as sulfate droplets and fine particulates. If they are removed, dimming may reverse and the earth may warm more.

## Clean air legislation is controversial

Throughout history, countless ordinances have prohibited emission of objectionable smoke, odors, and noise. Air pollution traditionally has been treated as a local problem, however. The Clean Air Act of 1963 was the first national legislation in the United States aimed at air pollution control. The act provided federal grants to states to combat pollution but was careful to preserve states' rights to set and enforce air quality regulations. It soon became obvious that some pollution problems cannot be solved on a local basis.

In 1970, an extensive set of amendments essentially rewrote the Clean Air Act. These amendments identified the "criteria" pollutants discussed earlier in this chapter, and established primary and secondary standards for ambient air quality. Primary standards (table 16.3) are intended to protect human health, while secondary standards are set to protect materials, crops, climate, visibility, and personal comfort.

Since 1970 the Clean Air Act has been modified, updated, and amended many times. The most significant amendments were in the 1990 update. Amendments have involved acrimonious debate, with bills sometimes languishing in Congress from one session to the next because of disputes over burdens of responsibility and cost and definitions of risk. A 2002 report concluded that simply by enforcing existing clean air legislation, the United States could save at least another 6,000 lives per year and prevent 140,000 asthma attacks.

**FIGURE 16.30** Should old power plants be required to install costly pollution-control equipment? This is the critical issue in the "new source review" under the Clean Air Act.

**TABLE 16.3**

**National Ambient Air Quality Standards (NAAQS)**

| Pollutant | Primary (Health-Based) Averaging Time | Standard Concentration |
|---|---|---|
| TSP[a] | Annual geometric mean[b] | 50 μg/m$^3$ |
| | 24 hours | 150 μg/m$^3$ |
| $SO_2$ | Annual arithmetic mean[c] | 80 μg/m$^3$ (0.03 ppm) |
| | 24 hours | 120 μg/m$^3$ (0.14 ppm) |
| CO | 8 hours | 10 mg/m$^3$ (9 ppm) |
| | 1 hour | 40 mg/m$^3$ (35 ppm) |
| $NO_2$ | Annual arithmetic mean | 80 μg/m$^3$ (0.05 ppm) |
| $O_3$ | Daily max 8 hour avg. | 157 μg/m$^3$ (0.08 ppm) |
| Lead | Maximum quarterly avg. | 1.5 μg/m$^3$ |

[a]Total suspended particulate material.

[b]The geometric mean is obtained by taking the nth root of the product of n numbers. This tends to reduce the impact of a few very large numbers in a set.

[c]An arithmetic mean is the average determined by dividing the sum of a group of data points by the number of points.

Throughout its history the Clean Air Act has been controversial. Victims of air pollution demand more protection; industry and special interest groups complain that controls are too expensive.

One of the most contested aspects of the act is the "new source review," which was established in 1977. This provision was originally adopted because industry argued that it would be intolerably expensive to install new pollution-control equipment on old power plants and factories that were about to close down anyway. Congress agreed to "grandfather" or exempt existing equipment from new pollution limits with the stipulation that when they were upgraded or replaced, more stringent rules would apply (fig. 16.30). The result was that owners kept old facilities operating precisely because they were exempted from pollution control. In fact, corporations poured millions into aging power plants and factories, expanding their capacity rather than build new ones. Thirty years later, most of those grandfathered plants are still going strong, and continue to be among the biggest contributors to smog and acid rain.

The Clinton administration attempted to force utilities to install modern pollution control on old power plants when they replaced or repaired equipment. President Bush, however, said that determining which facilities are new, and which are not, represented a cumbersome and unreasonable imposition on industries. The EPA subsequently announced it would abandon new source reviews, depending instead on voluntary emissions controls and a trading program for air pollution allowances.

Environmental groups generally agree that cap-and-trade (which sets maximum amounts for pollutants, and then lets facilities facing costly cleanup bills to pay others with lower costs to reduce emissions on their behalf) has worked well for sulfur dioxide. When trading began in 1990, economists estimated that eliminating 10 million tons of sulfur dioxide would cost $15 billion per year. Left to find the most economical ways to reduce emissions, however, utilities have been able to reach clean air goals for one-tenth that price. A serious shortcoming of this approach is that while trading has resulted in overall pollution reduction, some local "hot spots" remain where owners have found it cheaper to pay someone else to reduce pollution than to do it themselves. Knowing that the average person is enjoying cleaner air isn't much comfort if you're living in one of the persistently dirty areas.

Many environmentalists argue that carbon dioxide should be classified as a pollutant because of its role in global warming. Some also complain that market mechanisms allow industry to postpone installing pollution controls and forces residents of many states to continue to breathe dirty air for far longer than is necessary. And industry contends that "command and control" mechanisms aren't effective because they don't provide an incentive to continue to search for new, more efficient means of pollution control. What do you think? Which of these regulatory approaches would you favor? Does the kind of pollutant or its effects influence your answer?

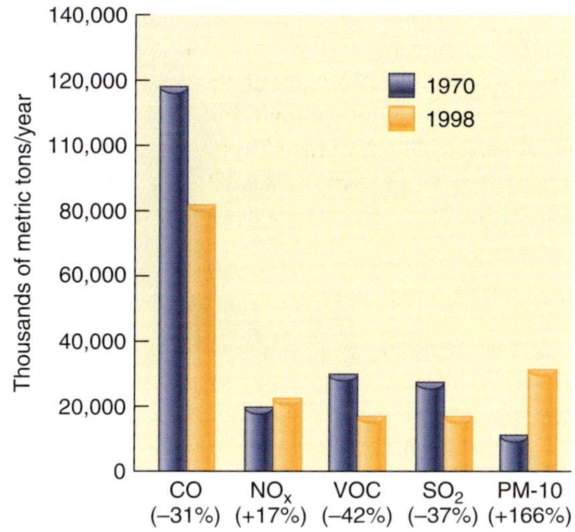


**FIGURE 16.31** Air pollution trends in the United States, 1970 to 1998. Although population and economic activity increased during this period, emissions of all "criteria" air pollutants, except for nitrogen oxides and particulate matter, decreased significantly.
**Source:** Environmental Protection Agency, 2002.

Particulate matter (mostly dust and soot) is produced by agriculture, fuel combustion, metal smelting, concrete manufacturing, and other activities. Industrial cities, such as Baltimore, Maryland, and Baton Rouge, Louisiana, also have continuing problems. Eighty-five other urban areas are still considered nonattainment regions. In spite of these local failures, however, 80 percent of the United States now meets the National Ambient Air Quality Standards. This improvement in air quality is perhaps the greatest environmental success story in our history.

## 16.7 Current Conditions and Future Prospects

Although the United States has not yet achieved the Clean Air Act goals in many parts of the country, air quality has improved dramatically in the last decade in terms of the major large-volume pollutants. For 23 of the largest U.S. cities, the number of days each year in which air quality reached the hazardous level is down 93 percent from a decade ago. Of 97 metropolitan areas that failed to meet clean air standards in the 1980s, 41 are now in compliance. For many cities, this is the first time they met air quality goals in 20 years.

There have been some notable successes and some failures. The EPA estimates that between 1970 and 1998, lead fell 98 percent, $SO_2$ declined 35 percent, and CO shrank 32 percent (fig. 16.31). Filters, scrubbers, and precipitators on power plants and other large stationary sources are responsible for most of the particulate and $SO_2$ reductions. Catalytic converters on automobiles are responsible for most of the CO and $O_3$ reductions.

The only conventional "criteria" pollutants that have not dropped significantly are particulates and $NO_x$. Because automobiles are the main source of $NO_x$, cities, such as Nashville, Tennessee, and Atlanta, Georgia, where pollution comes largely from traffic, still have serious air quality problems. Rigorous pollution controls are having a positive effect on Southern California air quality. Los Angeles, which had the dirtiest air in the nation for decades, wasn't even in the top 20 polluted cities in 2005.

### Air pollution remains a problem in many places

The outlook is not so encouraging in other parts of the world. The major metropolitan areas of many developing countries are growing at explosive rates to incredible sizes (chapter 22), and environmental quality is abysmal in many of them. Mexico City remains notorious for bad air. Pollution levels exceed WHO health standards 350 days per year, and more than half of all city children have lead levels in their blood high enough to lower intelligence and retard development. Mexico City's 131,000 industries and 2.5 million vehicles spew out more than 5,500 tons of air pollutants daily. Santiago, Chile, averages 299 days per year on which suspended particulates exceed WHO standards of 90 $mg/m^3$.

While China is making efforts to control air and water pollution (see chapter 1), many of China's 400,000 factories have no air pollution controls. Experts estimate that home coal burners and factories emit 10 million tons of soot and 15 million tons of sulfur dioxide annually and that emissions have increased rapidly over the past 20 years. Seven of the ten cities in the world with the worst air quality are in China. Sheyang, an industrial city in northern China, is thought to have the world's worst continuing particulate

**FIGURE 16.32** Air quality in Delhi, India, has improved dramatically since buses, auto-rickshaws, and taxis were required to switch from liquid fuels to compressed natural gas. This is one of the most encouraging success stories in controlling pollution in the developing world.

**FIGURE 16.33** Cubatao, Brazil, was once considered one of the most polluted cities in the world. Better environmental regulations and enforcement along with massive investments in pollution-control equipment have improved air quality significantly.

problem, with peak winter concentrations over 700 mg/m$^3$ (nine times U.S. maximum standards). Airborne particulates in Sheyang exceed WHO standards on 347 days per year. It's estimated that air pollution is responsible for 400,000 premature deaths every year in China. Beijing, Xi'an, and Guangzhou also have severe air pollution problems. The high incidence of cancer in Shanghai is thought to be linked to air pollution (see fig. 16.2).

Every year, the Blacksmith Institute compiles a list of the world's worst polluted places (see table 14.3). For 2006, air pollution was the main problem in nine of the top ten worst places, and all but two of those were mines and/or smelter complexes. These problems are especially disastrous in the developing world and the former Soviet Union, where funds and political will aren't available to deal with pollution or help people suffering from terrible health effects of pollution. You can learn more about these places at www.blacksmithinstitute.org.

Norilsk, Russia (one Blacksmith's pick) is a notorious example of toxic air pollution. Founded in 1935 as a slave labor camp, this Siberian city is considered one of the most polluted places on earth. Norilsk houses the world's largest nickel mine and heavy metals smelting complex, which discharge over 4 million tons of cadmium, copper, lead, nickel, arsenic, selenium, and zinc into the air every year. The snow turns black as quickly as it falls, the air tastes of sulfur, and the average life expectancy for factory workers is ten years below the Russian average (which already is lowest of any industrialized country). Difficult pregnancies and premature births are much more common in Norilsk than elsewhere in Russia. Children living near the nickel plant are ill twice as much as Russia's average, and birth defects are reported to affect as much as 10 percent of the population. Why do people stay in such a place? Many were attracted by high wages and hardship pay, and now that they're sick, they can't afford to move.

## There are signs of hope

Not all is pessimistic, however. There have been some spectacular successes in air pollution control. Sweden and West Germany (countries affected by forest losses due to acid precipitation) cut their sulfur emissions by two-thirds between 1970 and 1985. Austria and Switzerland have gone even further, regulating even motorcycle emissions. The Global Environmental Monitoring System (GEMS) reports declines in particulate levels in 26 of 37 cities worldwide. Sulfur dioxide and sulfate particles, which cause acid rain and respiratory disease, have declined in 20 of these cities.

Even poor countries can control air pollution. Delhi, India, for example was once considered one of the world's ten most polluted cities. Visibility often was less than 1 km on smoggy days. Health experts warned that breathing Delhi's air was equivalent to smoking two packs of cigarettes per day. Pollution levels exceeded World Health Organization standards by nearly five times. Respiratory diseases were widespread, and the cancer rate was significantly higher than surrounding rural areas. The biggest problem was vehicle emissions, which contributed about 70 percent of air pollutants (industrial emissions made up 20 percent, while burning of garbage and firewood made up most of the rest).

In the 1990s, catalytic converters were required for automobiles, and unleaded gasoline and low-sulfur diesel fuel were introduced. In 2000, more than private automobiles were required to meet European standards, and in 2002, more than 80,000 buses, auto-rickshaws, and taxis were required to switch from liquid fuels to compressed natural gas (fig. 16.32). Sulfur dioxide and carbon monoxide levels have dropped 80 percent and 70 percent, respectively, since 1997. Particulate emissions dropped by about 50 percent. Residents report that the air is dramatically clearer and more healthy. Unfortunately, rising prosperity, driven by globalization of information management, has doubled the number of vehicles on the roads, threatening this progress. Still, the gains made in New Delhi are encouraging for people everywhere.

Twenty years ago, Cubatao, Brazil, was described as the "Valley of Death," one of the most dangerously polluted places in the world. A steel plant, a huge oil refinery, and fertilizer and chemical factories churned out thousands of tons of air pollutants every year that were trapped between onshore winds and the uplifted plateau on which São Paulo sits (fig. 16.33). Trees died on the surrounding hills. Birth defects and respiratory diseases were alarmingly high. Since then, however, the citizens of Cubatao have made remarkable progress in cleaning up their environment. The end of military rule and restoration of democracy allowed residents to publicize their complaints. The environment became an important political issue. The state of São Paulo invested about $100 million and the private sector spent twice as much to clean up most pollution sources in the valley. Particulate pollution was reduced 75 percent, ammonia emissions were reduced 97 percent, hydrocarbons that cause ozone and smog were cut 86 percent, and sulfur dioxide production fell 84 percent. Fish are returning to the rivers, and forests are regrowing on the mountains. Progress is possible! We hope that similar success stories will be obtainable elsewhere.

## Conclusion

Air pollution is often the most obvious and widespread type of pollution. It can spread from a single source over the entire earth. No matter where you live, from the most remote island in the Pacific, to the highest peak in the Himalayas, to the frigid ice cap over the North Pole, there are traces of human–made contaminants, remnants of the 2 billion metric tons of pollutants released into the air worldwide every year by human activities.

There are many adverse effects of air pollution, from destroying the protective ozone layer in the stratosphere, poisoning whole forests with acid rain, and corroding building materials, to causing respiratory diseases, birth defects, heart attacks, or cancer in individual humans. We have made encouraging progress in controlling air pollution in many places. Many students aren't aware of how much worse air quality was in the industrial centers of North America and Europe a century or two ago than they are now. Cities such as London, Pittsburg, Chicago, Baltimore, and New York had air quality as bad or worse than most megacities of the developing world now. The progress in reducing air pollution in these cities gives us hope that residents can do so elsewhere as well.

The success of the Montreal Protocol in eliminating CFCs is a landmark in international cooperation on an environmental problem. While the stratospheric ozone hole continues to grow because of global warming effects and the residual chlorine in the air released decades ago, we expect the ozone depletion to end in about 50 years. This is one of the few global environmental threats that has had such a rapid and successful resolution. Let's hope that others will follow.

Progress in reducing local pollution in developing countries, such as Brazil and India, also is encouraging. Problems that once seemed overwhelming can be overcome. In some cases, it requires lifestyle changes or different ways of doing things to bring about progress, but as the Chinese philosopher Lao Tsu wrote, "A journey of a thousand miles must begin with a single step."

## Reviewing Learning Outcomes

By now you should be able to explain the following points:

**16.1** Describe the air around us.

**16.2** Identify natural sources of air pollution.

**16.3** Discuss human-caused air pollution.

- We categorize pollutants according to their source.
- We also categorize pollutants according to their content.
- Unconventional pollutants also are important.
- Indoor air is more dangerous for most of us than outdoor air.

**16.4** Explain how climate topography and atmospheric processes affect air quality.

- Temperature inversions trap pollutants.
- Cities create dust domes and heat islands.
- Wind currents carry pollutants intercontinentally.
- Stratospheric ozone is destroyed by chlorine.
- The Montreal Protocol is a resounding success.

**16.5** Compare the effects of air pollution.

- Polluted air is dangerous.
- How does pollution harm us?.
- Plants are susceptible to pollution damage.
- Acid deposition has many negative effects.
- Smog and haze reduce visibility.

**16.6** Evaluate air pollution control.

- The most effective strategy for controlling pollution is to minimize production.
- Fuel switching and fuel cleaning also are effective.
- Clean air legislation is controversial.

**16.7** Summarize current conditions and future prospects.

- Air pollution remains a problem in many places.
- There are signs of hope.

## Practice Quiz

1. Define *primary* and *secondary air pollutants.*
2. What are the seven "criteria" pollutants in the original Clean Air Act? Why were they chosen? How many more hazardous air toxins have been added?
3. What pollutants in indoor air may be hazardous to your health? What is the greatest indoor air problem globally?
4. What is acid deposition? What causes it?

5. What is an atmospheric inversion and how does it trap air pollutants?
6. What is the difference between ambient and stratospheric ozone? What is destroying stratospheric ozone?
7. What is long-range air pollution transport? Give two examples.
8. What is "new source review," and why is it controversial?
9. Which of the conventional pollutants has decreased most in the recent past and which has decreased least?
10. Give one example of current air quality problems and one success in controlling pollution in a developing country.

## Critical Thinking and Discussion Questions

1. What might be done to improve indoor air quality? Should the government mandate such changes? What values or worldviews are represented by different sides of this debate?
2. Debate the following proposition: Our air pollution blows onto someone else; therefore, installing pollution controls will not bring any direct economic benefit to those of us who have to pay for them.
3. Utility managers once claimed that it would cost $1,000 per fish to control acid precipitation in the Adirondack lakes and that it would be cheaper to buy fish for anglers than to put scrubbers on power plants. Suppose that is true. Does it justify continuing pollution?
4. Developing nations claim that richer countries created global warming and stratospheric ozone depletion, and therefore should bear responsibility for fixing these problems. How would you respond?
5. If there are thresholds for pollution effects, is it reasonable or wise to depend on environmental processes to disperse, assimilate, or inactivate waste products?
6. How would you choose between government "command and control" regulations versus market-based trading programs for air pollution control? Are there situations where one approach would work better than the other?

## DATA analysis Graphing Air Pollution Control

Reduction of acid-forming air pollutants in Europe is an inspiring success story. The first evidence of ecological damage from acid rain came from disappearance of fish from Scandinavian lakes and rivers in the 1960s. By the 1970s, evidence of air pollution damage to forests in northern and central Europe alarmed many people. International agreements, reached since the mid-1980s have been highly successful in reducing emissions of $SO_2$ and $NO_x$ as well as photochemical oxidants, such as $O_3$. The graph on this page shows reductions in $SO_2$ emissions in Europe between 1990 and 2002. The light blue area shows actual $SO_2$ emissions. Blue represents changes due to increased nuclear and renewable energy. Orange shows reductions due to energy conservation. Green shows improvement from switching to low-sulfur fuels. Purple shows declines due to increased abatement measures (flue gas scrubbers). The upper boundary of each area indicates what emissions would have been without pollution control.

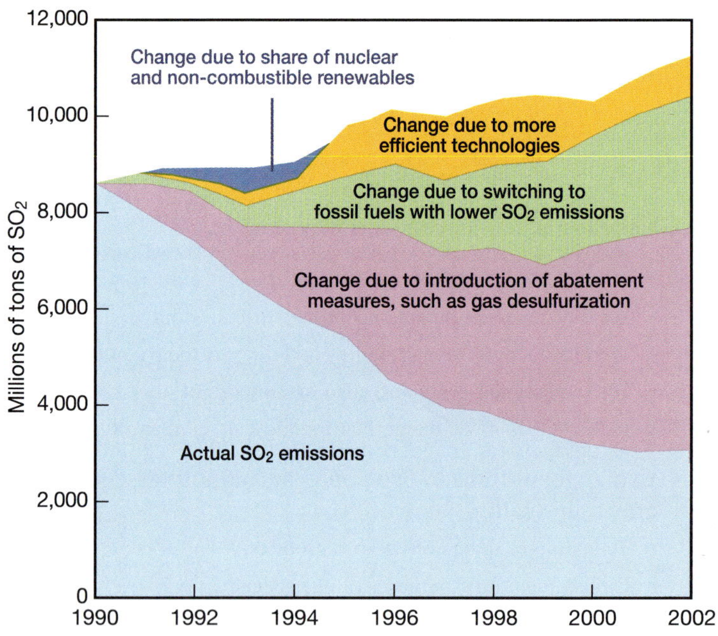


Sulfur dioxide emission reductions in Europe, 1990–2002.

1. How much have actual $SO_2$ emissions declined since 1990?
2. How much lower were $SO_2$ emissions in 2002 than they would have been without pollution control (either in percentage or actual amount)?
3. What percentage of this reduction was due to abatement measures, such as flue gas scrubbers?
4. What percent was gained by switching to low-sulfur fuels?
5. How much did energy conservation contribute?
6. What happened to nuclear power?

**For Additional Help in Studying This Chapter,** please visit our website at www.mhhe.com/cunningham10e. You will find additional practice quizzes and case studies, flashcards, regional examples, place markers for Google Earth™ mapping, and an extensive reading list, all of which will help you learn environmental science.

Water is a precious and beautiful resource.

# CHAPTER 17

# Water Use and Management

*Of all the natural resources available to human beings, water is the most essential for virtually every human activity.*

*—Anna Kajumulo Tibaijuka, UN Undersecretary General—*

## LEARNING OUTCOMES

After studying this chapter, you should be able to:

**17.1** Summarize why water is a precious resource and why shortages occur.

**17.2** Compare major water compartments.

**17.3** Summarize water availability and use.

**17.4** Investigate freshwater shortages.

**17.5** Illustrate the benefits and problems of dams and diversions.

**17.6** Understand how we might increase water supplies.

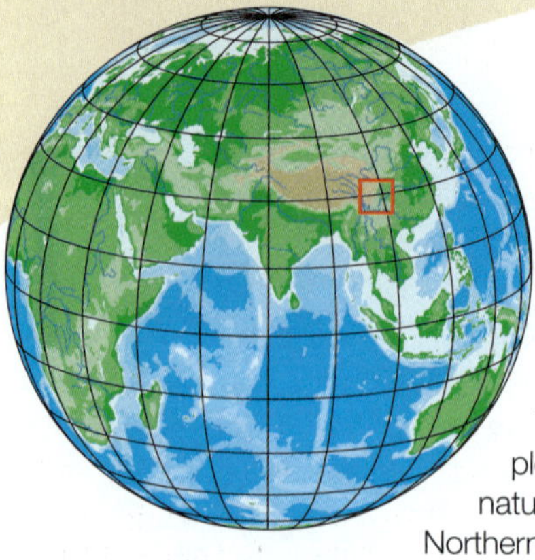

# Case Study China's South-to-North Water Diversion

Water is inequitably distributed in China. In the south, torrential monsoon rains cause terrible floods. A 1931 flood on the Yangtze displaced 56 million people and killed 3.7 million (the worst natural disaster in recorded history). Northern and western China, on the other hand, are too dry, and getting drier. At least 200 million Chinese live in areas without sufficient fresh water. The government has warned that unless new water sources are found soon, many of those people (including the capital Beijing, with 14 million residents) will have to be moved. But where could they go? Southern China has water, but doesn't need more people.

The solution, according to the government, is to transfer some of the extra water from south to north (fig. 17.1). A gargantuan project is now underway to do just that. Work has begun to build three major canals to carry water from the Yangtze River to northern China. Ultimately, it's planned to move 45 billion m$^3$ per year (more than twice the flow of the Colorado River through the U.S. Grand Canyon) up to 1,600 km (1,000 mi) north. The initial cost estimate of this scheme is about 400 billion yuan (roughly U.S.$62 billion), but it could easily be twice that much.

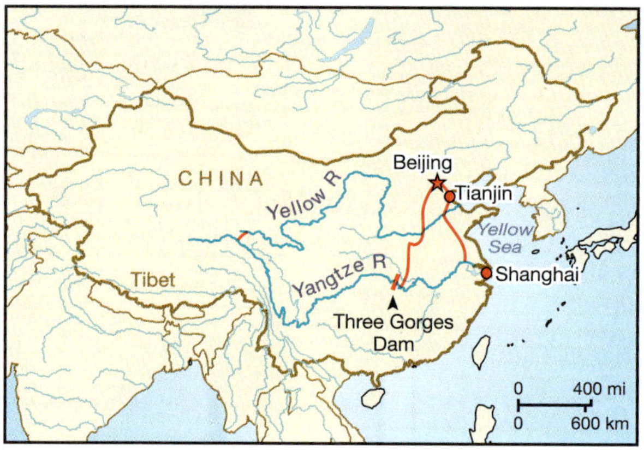


**FIGURE 17.1** With the Yellow River nearly depleted by overuse, northern China now plans canals (*red*) to deliver Yangtze water to Beijing.

The eastern route (fig. 17.2) uses the Grand Canal, built by Zhou and Sui emperors 1,500 years ago across the coastal plain between Shanghai and Beijing. This project is already operational. It's relatively easy to pump water through the existing waterways, but they're so polluted by sewage and industrial waste that northern cities—even though they're desperately dry—are reluctant to accept this water.

The central route will draw water from the reservoir behind the nearly completed Three Gorges Dam on the Yangtze. Part of the motivation for building this controversial dam and flooding the historic Three Gorges (fig. 17.3) was to provide energy and raise the river level for the South-to-North project. This middle canal will cross several major mountain ranges and dozens of rivers, including the Han and the Yellow River. Currently, work is in progress on raising the Danjiangkou Dam and enlarging its reservoir as part of this route. This will displace some 200,000 people, but planners say it's worthwhile to benefit a thousand times as many. It's hoped this segment will be finished by 2020.

The western route is the most difficult and expensive. It would tunnel through rugged mountains, across aqueducts, and over deep canyons for more than 250 km (160 mi), from the upper Yangtze to the Yellow River, where they both spill off the Tibetan Plateau. This phase won't be finished until at least 2050. If global warming melts all Tibet's glaciers, however, it may not be feasible anyway.

Planners have waited a lifetime to see this project move forward. Revolutionary leader Mao Zedong proposed it 50 years ago. Environmental scientists worry, however, that drawing down the Yangtze will worsen pollution problems (already exacerbated by the Three Gorges Dam), dry up downstream wetlands, and possibly even alter ocean circulation and climate along China's eastern coast. Although southern China has too much water during the rainy season, even there cities face water shortages because of rapidly growing populations and severe pollution problems. At least half of all major Chinese rivers are too polluted for human consumption. Drawing water away from the rivers on which millions rely only makes pollution problems worse.

**FIGURE 17.2** The 1,500-year-old Grand Canal is being cleaned and remodeled to carry water from Shanghai to Tianjin as part of the most ambitious water diversion plan in human history.

**FIGURE 17.3** The famous Three Gorges region of the Yangtze River is being flooded in part so water can be pumped 1,600 km north to Beijing.

# Case Study *continued*

China isn't unique in its water woes. As you'll read in this chapter, water shortages increasingly threaten economies, societies, and the environment in many places. Growing populations, changing climate, and demands on agriculture and industry stress supplies. Conflicts have already arisen within and between nations. Other countries have dreamed about—and some have started—massive water relocation schemes similar to China's. How to conserve and allocate water resources is an important theme in environmental science. In this chapter, we'll survey water supplies, water budgets, and strategies for using water more efficiently.

## 17.1 Water Resources

Water is a marvelous substance—flowing, rippling, swirling around obstacles in its path, seeping, dripping, trickling, constantly moving from sea to land and back again. Water can be clear, crystalline, icy green in a mountain stream, or black and opaque in a cypress swamp. Water bugs skitter across the surface of a quiet lake; a stream cascades down a stairstep ledge of rock; waves roll endlessly up a sand beach, crash in a welter of foam, and recede. Rain falls in a gentle mist, refreshing plants and animals. A violent thunderstorm floods a meadow, washing away stream banks. Water is a most beautiful and precious resource.

Water is also a great source of conflict. Some 2 billion people, a third of the world's population, live in countries with insufficient fresh water. Some experts estimate this number could double in 25 years. To understand this resource, let's first ask, where does our water come from, and why is it so unevenly distributed?

### The hydrologic cycle distributes water in our environment

The hydrologic cycle (water cycle) describes the circulation of water as it evaporates from land, water, and organisms; enters the atmosphere; condenses and is precipitated to the earth's surfaces; and moves underground by infiltration or overland by runoff into rivers, lakes, and seas (see fig. 3.19). This cycle supplies fresh water to the landmasses, maintains a habitable climate, and moderates world temperatures. Movement of water back to the sea in rivers and glaciers is a major geological force that shapes the land and redistributes material. Plants play an important role in the hydrologic cycle, absorbing groundwater and pumping it into the atmosphere by **transpiration** (transport plus evaporation). In tropical forests, as much as 75 percent of annual precipitation is returned to the atmosphere by plants. Much of this moisture falls again, keeping the rainforest humid.

Solar energy drives the hydrologic cycle by evaporating surface water. **Evaporation** is the process in which a liquid is changed to vapor (gas phase) at temperatures well below its boiling point. Water also can move between solid and gaseous states without ever becoming liquid in a process called **sublimation.** On bright, cold, windy winter days, when the air is very dry, snowbanks disappear by sublimation, even though the temperature never gets above freezing. This is the same process that causes "freezer burn" of frozen foods.

In both evaporation and sublimation, molecules of water vapor enter the atmosphere, leaving behind salts and other contaminants and thus creating purified fresh water. This is essentially distillation on a grand scale.

The amount of water vapor in the air is called humidity. Warm air can hold more water than cold air. When a volume of air contains as much water vapor as it can at a given temperature, we say that it has reached its **saturation point. Relative humidity** is the amount of water vapor in the air expressed as a percentage of the maximum amount (saturation point) that could be held at that particular temperature.

When the saturation concentration is exceeded, water molecules begin to aggregate in the process of **condensation.** If the temperature at which this occurs is above 0°C, tiny liquid droplets result. If the temperature is below freezing, ice forms. For a given amount of water vapor, the temperature at which condensation occurs is the **dew point.** Tiny particles, called **condensation nuclei,** float in the air and facilitate this process. Smoke, dust, sea salts, spores, and volcanic ash all provide such particles. Even apparently clear air can contain large numbers of these particles, which are generally too small to be seen by the naked eye. Sea salt is an excellent source of such nuclei, and heavy, low clouds frequently form in the humid air over the ocean.

A cloud, then, is an accumulation of condensed water vapor in droplets or ice crystals. Normally, cloud particles are small enough to remain suspended in the air, but when cloud droplets and ice crystals become large enough, gravity overcomes uplifting air currents, and precipitation occurs.

### Water supplies are unevenly distributed

China isn't alone in suffering from droughts in some places and excess water elsewhere. Rain falls unevenly over the planet (fig. 17.4). Some places get almost no precipitation, while others receive heavy rain almost daily. At Iquique, in Chile's Atacama Desert, no rain has fallen in recorded history. At the other end of the scale, Cherrapunji, in northeastern India, received nearly 23 m (897 in.) of rain in 1861.

Three principal factors control these global water deficits and surpluses. First, global atmospheric circulation creates regions of persistent high air pressure and low rainfall about 20° to 40° north and south of the equator (chapter 15). These same circulation patterns produce frequent rainfall near the equator and between about 40° and 60° north and south latitude. Second, proximity to water sources influences precipitation. Where prevailing winds come over oceans, they bring moisture to land. Areas far from oceans—in a windward direction—are usually relatively dry.

A third factor in water distribution is topography. Mountains act as both cloud formers and rain catchers. As air sweeps up the windward side of a mountain, air pressure decreases and air cools. As the air cools, it reaches the saturation point, and moisture condenses as either rain or snow. Thus the windward side of a

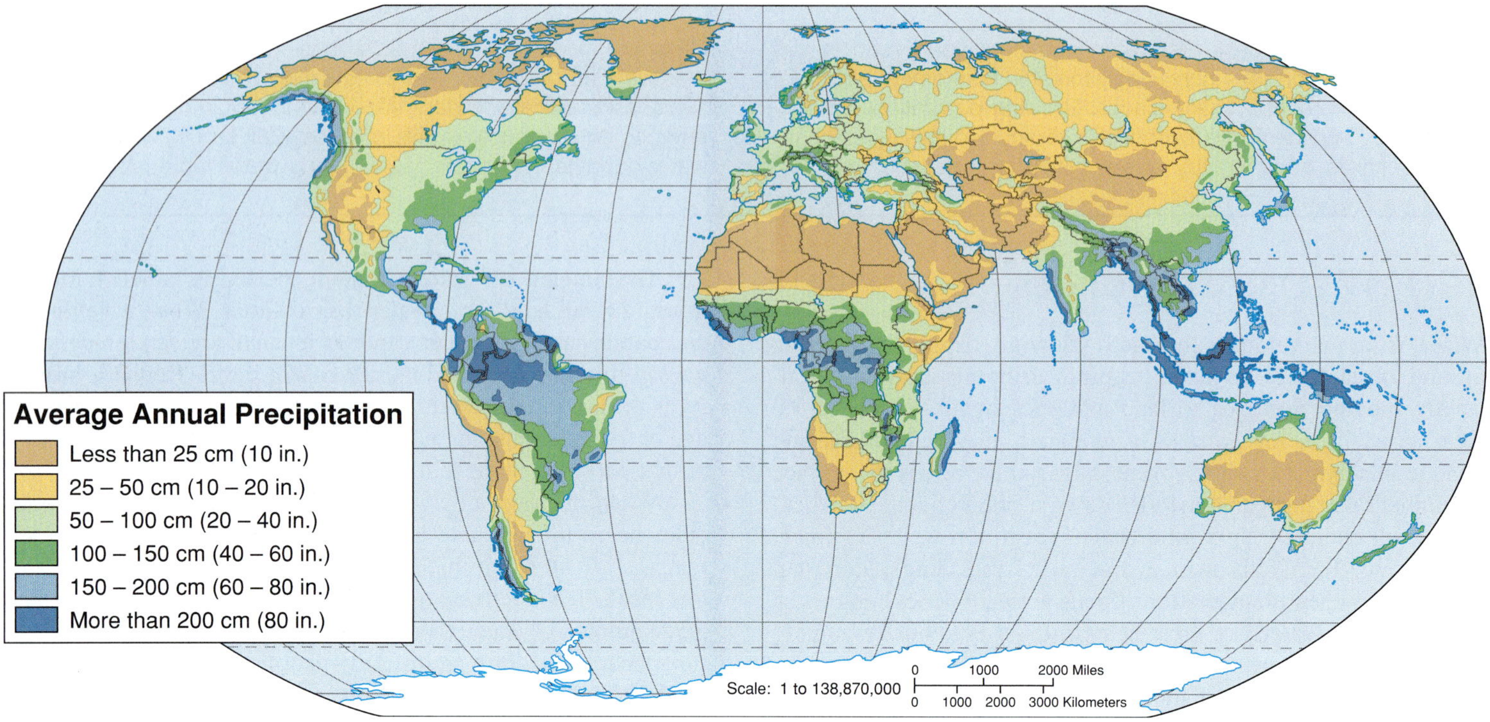


**FIGURE 17.4** Average annual precipitation. Note wet areas that support tropical rainforests occur along the equator, while the major world deserts occur in zones of dry, descending air between 20° and 40° north and south.

mountain range, as in the Pacific Northwest, is usually wet much of the year. Precipitation leaves the air drier than it was on its way up the mountain. As the air passes the mountaintop and descends the other side, air pressure rises, and the already-dry air warms, increasing its ability to hold moisture. Descending, warming air rarely produces any rain or snow. Places in the **rain shadow,** the dry, leeward side of a mountain range, receive little precipitation. A striking example of the rain shadow effect is that of Mount Waialeale, on the island of Kauai, Hawaii (fig. 17.5). The windward side of the island receives nearly 12 m of rain per year, while the leeward side, just a few kilometers away, receives just 46 cm.

Usually a combination of factors affects precipitation. In Cherrapunji, India, atmospheric circulation sweeps moisture from the warm Indian Ocean toward the high ridges of the Himalayas. Iquique, Chile, lies in the rain shadow of the Andes and in a high-pressure desert zone. Prevailing winds are from the east, so even though Iquique lies near the ocean, it is far from the winds' moisture source—the Atlantic. In the American Southwest, Australia, and the Sahara, high-pressure atmospheric conditions tend to keep the air and land dry. The global map of precipitation represents a complex combination of these forces of atmospheric circulation, prevailing winds, and topography.

Human activity also explains some regions of water deficit. As noted earlier, plant transpiration recycles moisture and produces rain. When forests are cleared, falling rain quickly enters streams and returns to the ocean. In Greece, Lebanon, parts of Africa, the Caribbean, South Asia, and elsewhere, desert-like conditions have developed since the original forests were destroyed.

***Think About It***

We have noted three important natural causes of water surpluses and deficits. Which of these might be important where you live? Does water availability affect your life style? Should it?

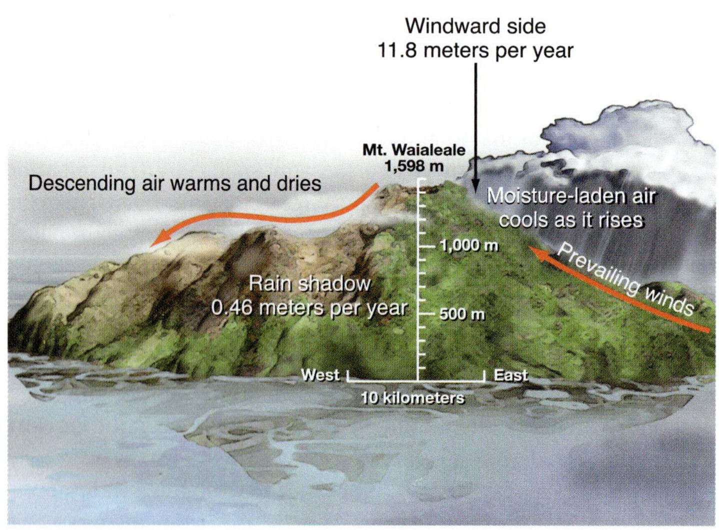


**FIGURE 17.5** Rainfall on the east side of Mount Waialeale in Hawaii is more than 20 times as much as on the west side. Prevailing trade winds bring moisture-laden sea air onshore. The air cools as it rises up the flanks of the mountain and the water it carries precipitates as rain—11.8 m (38 ft) per year!

**TABLE 17.1**

**Earth's Water Compartments—Estimated Volume of Water in Storage, Percent of Total, and Average Residence Time**

| | Volume (Thousands of $km^3$) | % Total Water | Average Residence Time |
|---|---|---|---|
| Total | 1,403,377 | 100 | 2,800 years |
| Ocean | 1,370,000 | 97.6 | 3,000 years to 30,000 years* |
| Ice and snow | 29,000 | 2.07 | 1 to 16,000 years* |
| Groundwater down to 1 km | 4,000 | 0.28 | From days to thousands of years* |
| Lakes and reservoirs | 125 | 0.009 | 1 to 100 years* |
| Saline lakes | 104 | 0.007 | 10 to 1,000 years* |
| Soil moisture | 65 | 0.005 | 2 weeks to a year |
| Biological moisture in plants and animals | 65 | 0.005 | 1 week |
| Atmosphere | 13 | 0.001 | 8 to 10 days |
| Swamps and marshes | 3.6 | 0.003 | From months to years |
| Rivers and streams | 1.7 | 0.0001 | 10 to 30 days |

*Depends on depth and other factors.

**Source:** Data from U.S. Geological Survey.

## 17.2 Major Water Compartments

The distribution of water often is described in terms of interacting compartments in which water resides for short or long times. Table 17.1 shows the major water compartments in the world. Because these compartments are very large, we use special units, such as acre-feet, to describe their volume (table 17.2).

### Oceans hold 97 percent of all water on earth

Together, the oceans contain more than 97 percent of all the *liquid* water in the world. (The water of crystallization in rocks is far larger than the amount of liquid water.) Oceans are too salty for most human uses, but they contain 90 percent of the world's living biomass. While the ocean basins really form a continuous reservoir, shallows and narrows between them reduce water exchange, so they have different compositions, climatic effects, and even different surface elevations.

**TABLE 17.2**

**Some Units of Water Measurement**

One cubic kilometer ($km^3$) equals 1 billion cubic meters ($m^3$), 1 trillion liters, or 264 billion gallons.

One acre-foot is the amount of water required to cover an acre of ground 1 foot deep. This is equivalent to 325,851 gallons, or 1.2 million liters, or 1,234 $m^3$, about the amount consumed annually by a family of four in the United States.

One cubic foot per second of river flow equals 28.3 liters per second or 449 gallons per minute.

See table at end of back for conversion factors.

Oceans play a crucial role in moderating the earth's temperature (fig. 17.6). Vast river-like currents transport warm water from the equator to higher latitudes, and cold water flows from the poles to the tropics (fig. 17.7). The Gulf Stream, which flows northeast from the coast of North America toward northern Europe, flows at a steady rate of 10–12 km per hour (6–7.5 mph) and carries more than 100 times more water than all rivers on earth put together.

In tropical seas, surface waters are warmed by the sun, diluted by rainwater and runoff from the land, and aerated by wave action. In higher latitudes, surface waters are cold and much more dense. This dense water subsides or sinks to the bottom of deep ocean basins and flows toward the equator. Warm surface water of the tropics stratifies or floats on top of this cold, dense water as currents carry warm water to high latitudes. Sharp boundaries form between different water densities, different salinities, and different temperatures, retarding mixing between these layers.

While parts of the hydrologic cycle occur on a time scale of hours or days, other parts take centuries. The average **residence time** of water in the ocean (the length of time that an individual molecule spends circulating in the ocean before it evaporates and starts through the hydrologic cycle again) is about 3,000 years. In the deepest ocean trenches, movement is almost nonexistent and water may remain undisturbed for tens of thousands of years.

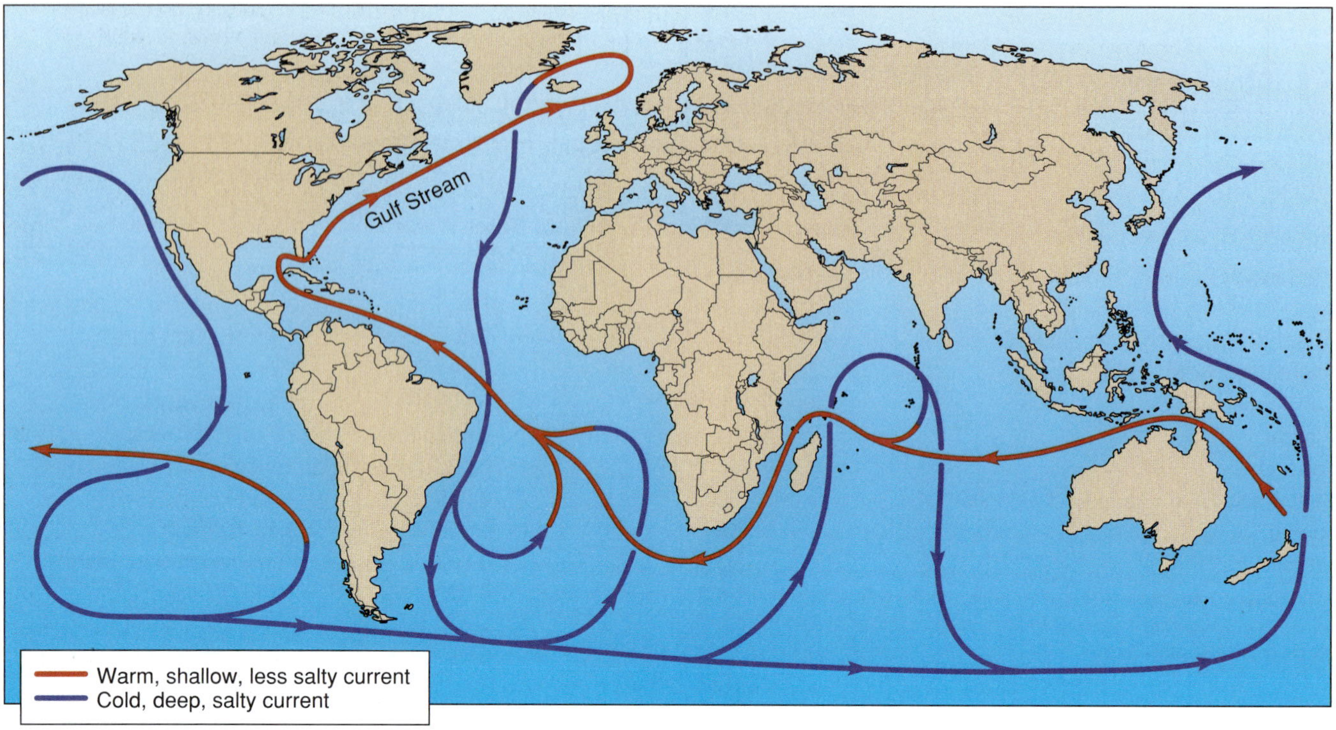


**FIGURE 17.6** Ocean currents act as a global conveyor system, redistributing warm and cold water around the globe. These currents moderate our climate. For example, the Gulf Stream keeps northern Europe much warmer than northern Canada.

## Glaciers, ice, and snow contain most surface fresh water

Of the 2.4 percent of all water that is fresh, nearly 90 percent is tied up in glaciers, ice caps, and snowfields (fig. 17.8). Glaciers are really rivers of ice flowing downhill very slowly (fig. 17.9). They now occur only at high altitudes or high latitudes, but as recently as 18,000 years ago about one-third of the continental landmass was covered by glacial ice sheets. Most of this ice has now melted and the largest remnant is in Antarctica. As much as 2 km (1.25 mi) thick, the Antarctic glaciers cover all but the highest mountain peaks and contain nearly 85 percent of all ice in the world.

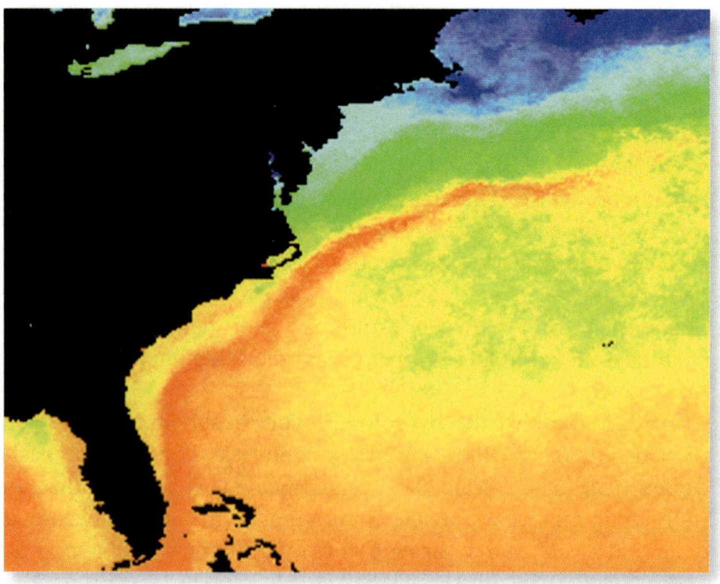

**FIGURE 17.7** Ocean currents, such as the warm Gulf Stream, redistribute heat as they flow around the globe. Here, orange and yellow indicate warm water temperatures (25–30°C); blue and green are cold (0–5°C).

A smaller ice sheet on Greenland, together with floating sea ice around the North Pole, makes up another 10 percent of the world's frozen water reservoirs. Mountain snow pack and ice constitute the remaining 5 percent.

## Groundwater stores large resources

After glaciers, the next largest reservoir of fresh water is held in the ground as **groundwater.** Precipitation that does not evaporate back into the air or run off over the surface percolates through the soil and into fractures and spaces of permeable rocks in a process called **infiltration** (fig. 17.10). Upper soil layers that hold both air and water make up the **zone of aeration.** Moisture for plant growth comes primarily from these layers. Depending on rainfall amount, soil type, and surface topography, the zone of aeration may be very shallow or quite deep. Lower soil layers where all spaces are filled with water make up the **zone of saturation.** The top of this zone is the **water table.** The water table is not flat, but undulates according to the surface topography and subsurface structure. Water tables also rise and fall seasonally, depending on precipitation and infiltration rates.

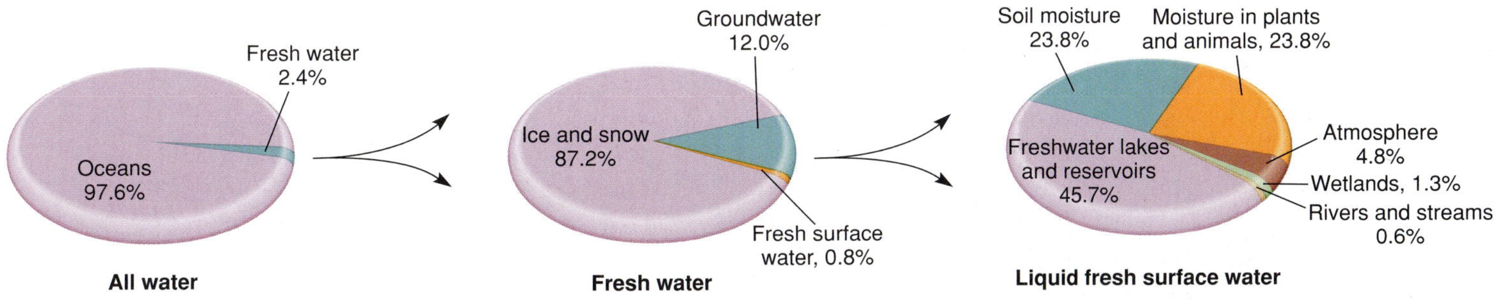


**FIGURE 17.8** Less than 1 percent of fresh water, and less than 0.02 percent of all water, is fresh, liquid surface water on which terrestrial life depends.
**Source:** U.S. Geological Survey.

Porous layers of sand, gravel, or rock lying below the water table are called **aquifers.** Aquifers are always underlain by relatively impermeable layers of rock or clay that keep water from seeping out at the bottom (fig. 17.11).

Folding and tilting of the earth's crust by geologic processes can create shapes that generate water pressure in confined aquifers (those trapped between two impervious, confining rock layers). When a pressurized aquifer intersects the surface, or if it is penetrated by a pipe or conduit, an **artesian** well or spring results from which water gushes without being pumped.

Areas in which infiltration of water into an aquifer occurs are called **recharge zones.** The rate at which most aquifers are refilled is very slow, however, and groundwater presently is being removed faster than it can be replenished in many areas. Urbanization, road building, and other development often block recharge zones and prevent replenishment of important aquifers. Contamination of surface water in recharge zones and seepage of pollutants into abandoned wells have polluted aquifers in many places, making them unfit for most uses (chapter 18). Many cities protect aquifer recharge zones from pollution or development, both as a way to drain off rainwater and as a way to replenish the aquifer with pure water.

Some aquifers contain very large volumes of water. The groundwater within 1 km of the surface in the United States is more than 30 times the volume of all the freshwater lakes, rivers, and reservoirs on the surface. While water can flow through limestone caverns in underground rivers, most movement in aquifers is a dispersed and almost imperceptible trickle through tiny fractures and spaces. Depending on geology, it can take anywhere from a few hours to several years for contaminants to move a few hundred meters through an aquifer.

## Rivers, lakes, and wetlands cycle quickly

Precipitation that does not evaporate or infiltrate into the ground runs off over the surface, drawn by the force of gravity back toward the sea. Rivulets accumulate to form streams, and streams join to form rivers. Although the total amount of water contained

**FIGURE 17.9** Glaciers are rivers of ice sliding very slowly downhill. Together, polar ice sheets and alpine glaciers contain more than three times as much fresh water as all the lakes, ponds, streams, and rivers in the world. The dark streaks on the surface of this Alaskan glacier are dirt and rocks marking the edges of tributary glaciers that have combined to make this huge flow.

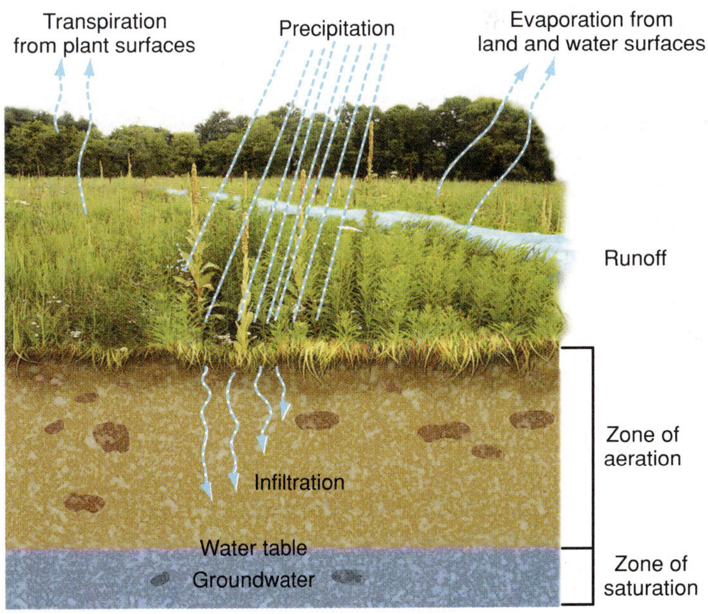


**FIGURE 17.10** Precipitation that does not evaporate or run off over the surface percolates through the soil in a process called infiltration. The upper layers of soil hold droplets of moisture between air-filled spaces. Lower layers, where all spaces are filled with water, make up the zone of saturation, or groundwater.

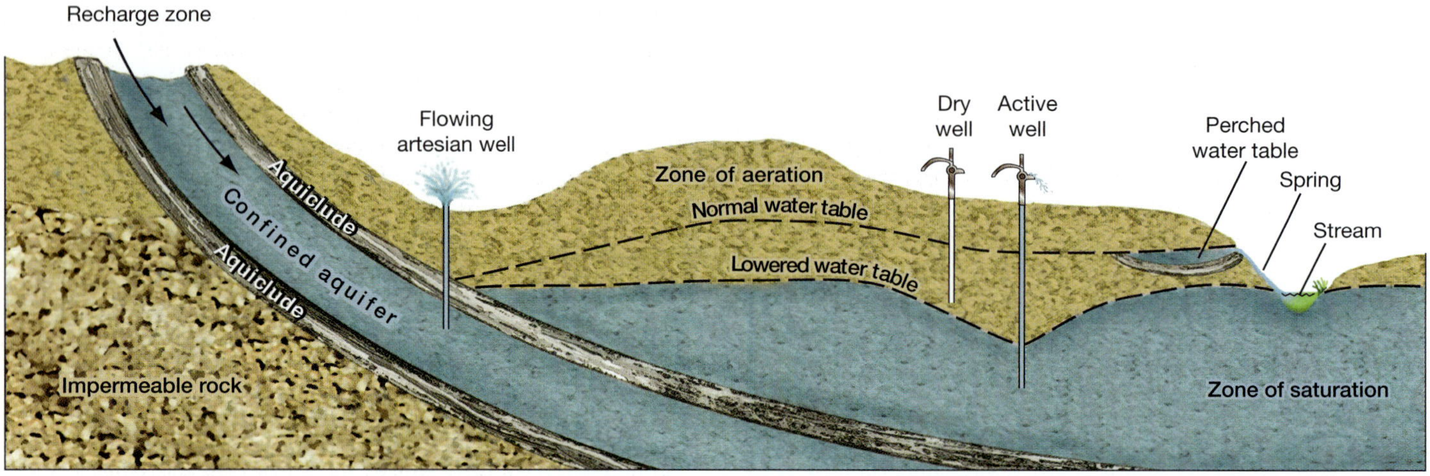


**FIGURE 17.11** An aquifer is a porous or cracked layer of rock. Impervious rock layers (aquicludes) keep water within a confined aquifer. Pressure from uphill makes an artesian well flow freely. Pumping can create a cone of depression, which leaves shallower wells dry.

at any one time in rivers and streams is small compared to the other water reservoirs of the world (see table 17.1), these surface waters are vitally important to humans and most other organisms. Most rivers, if they were not constantly replenished by precipitation, meltwater from snow and ice, or seepage from groundwater, would begin to diminish in a few weeks.

We measure the size of a river in terms of its **discharge,** the amount of water that passes a fixed point in a given amount of time. This is usually expressed as liters or cubic feet of water per second. The 16 largest rivers in the world carry nearly half of all surface runoff on earth. The Amazon is by far the largest river in the world (table 17.3), carrying roughly ten times the volume of the Mississippi. Several Amazonian tributaries such as the Maderia, Rio Negro, and Ucayali would be among the world's top rivers in their own right.

Ponds are generally considered to be small temporary or permanent bodies of water shallow enough for rooted plants to grow over most of the bottom. Lakes are inland depressions that hold standing fresh water year-round. Maximum lake depths range from a few meters to over 1,600 m (1 mi) in Lake Baikal in Siberia. Surface areas vary in size from less than one-half hectare (one acre) to large inland seas, such as Lake Superior or the Caspian Sea, covering hundreds of thousands of square kilometers. Both ponds and lakes are relatively temporary features on the landscape because they eventually fill with silt or are emptied by cutting of an outlet stream through the barrier that creates them.

While lakes contain nearly 100 times as much water as all rivers and streams combined, they are still a minor component of total world water supply. Their water is much more accessible than groundwater or glaciers, however, and they are important in many ways for humans and other organisms.

Wetlands play a vital and often unappreciated role in the hydrologic cycle. Their lush plant growth stabilizes soil and holds back surface runoff, allowing time for infiltration into aquifers and producing even, year-long stream flow. In the United States, about 20 percent of the 1 billion ha of land area was once wetland. In the past 200 years, more than one-half of those wetlands have been drained, filled, or degraded. Agricultural drainage accounts for the bulk of the losses.

When wetlands are disturbed, their natural water-absorbing capacity is reduced and surface waters run off quickly, resulting in floods and erosion during the rainy season and dry, or nearly dry, stream beds the rest of the year. This has a disastrous effect on biological diversity and productivity, as well as on human affairs.

**TABLE 17.3**

**Major Rivers of the World**

| River | Countries in River Basin | Average Annual Discharge at ($m^3$/sec) |
|---|---|---|
| Amazon | Brazil, Peru | 175,000 |
| Orinoco | Venezula, Colombia | 45,300 |
| Congo | Congo | 39,200 |
| Yangtze | Tibet, China | 28,000 |
| Bramaputra | Tibet, India, Bangladesh | 19,000 |
| Mississippi | United States | 18,400 |
| Mekong | China, Laos, Burma, Thailand, Cambodia, Vietnam | 18,300 |
| Parana | Paraguay, Argentina | 18,000 |
| Yenisey | Russia | 17,200 |
| Lena | Russia | 16,000 |

1 $m^3$ = 264 gallons.
**Source:** World Resources Institute.

## The atmosphere is among the smallest of compartments

The atmosphere is among the smallest of the major water reservoirs of the earth in terms of water volume, containing less than 0.001 percent of the total water supply. It also has the most rapid turnover rate. An individual water molecule resides in the atmosphere for about ten days, on average. While water vapor makes up only a small amount (4 percent maximum at normal temperatures) of the total volume of the air, movement of water through the atmosphere provides the mechanism for distributing fresh water over the landmasses and replenishing terrestrial reservoirs.

***Think About It***

Locate the ten rivers in table 17.3 on the physiographic map in the back of your book. Also, check their approximate locations in figure 17.4. How many of these rivers are tropical? In rainy regions? In populous regions? How might some of these rivers affect their surrounding environment or populations?

# 17.3 Water Availability and Use

Clean, fresh water is essential for nearly every human endeavor. Perhaps more than any other environmental factor, the availability of water determines the location and activities of humans on earth (fig. 17.12). **Renewable water supplies** are made up, in general, of surface runoff plus the infiltration into accessible freshwater aquifers. About two-thirds of the water carried in rivers and streams every year occurs in seasonal floods that are too large or violent to be stored or trapped effectively for human uses. Stable runoff is the dependable, renewable, year-round supply of surface water. Much of this occurs, however, in sparsely inhabited regions or where technology, finances, or other factors make it difficult to use it productively. Still, the readily accessible, renewable water supplies are very large, amounting to some 1,500 $km^3$ (about 400,000 gal) per person per year worldwide.

**FIGURE 17.12** Water has always been the key to survival. Who has access to this precious resource and who doesn't has long been a source of tension and conflict.

When discussing water use, you will encounter several different units of volume (table 17.2). Becoming familiar with these terms will help you evaluate the statistics that you read.

## Many people lack access to clean water

As you can see in figure 17.4. South America, West Central Africa, and South and Southeast Asia all have areas of very high rainfall. Brazil and the Democratic Republic of Congo, because they have high precipitation levels and large land areas, are among the most water-rich countries on earth. Canada and Russia, which are both very large, also have large annual water supplies. The highest per capita water supplies generally occur in countries with wet climates and low population densities. Iceland, for example, has about 160 million gallons per person per year. In contrast, Bahrain, where temperatures are extremely high and rain almost never falls, has essentially no natural fresh water. Almost all of Bahrain's water comes from imports and desalinized seawater. Egypt, in spite of the fact that the Nile River flows through it, has only about 11,000 gallons of water annually per capita, or about 15,000 times less than Iceland.

Periodic droughts create severe regional water shortages. Droughts are most common and often most severe in semiarid zones where moisture availability is the critical factor in determining plant and animal distribution. Undisturbed ecosystems often survive extended droughts with little damage, but introduction of domestic animals and agriculture disrupts native vegetation and undermines natural adaptations to low moisture levels.

Droughts are often cyclic, and land-use practices exacerbate their effects. In the United States, the cycle of drought seems to be about 30 years. There were severe dry years in the 1870s, 1900s, 1930s, 1950s, and 1970s. The worst of these in economic and social terms were the 1930s. Poor soil conservation practices and a series of dry years in the Great Plains combined to create the "dust bowl." Wind stripped topsoil from millions of hectares of land, and billowing dust clouds turned day into night. Thousands of families were forced to leave farms and migrate to cities.

Much of the western United States continues to be plagued by drought and overexploitation of limited water supplies (fig.17.13). The El Niño, Southern Oscillation (ENSO) system plays an important role in droughts in North America and elsewhere. There now is a great worry that global warming (see chapter 15) will bring about major climatic changes and make droughts both more frequent and more severe than in the past in some places.

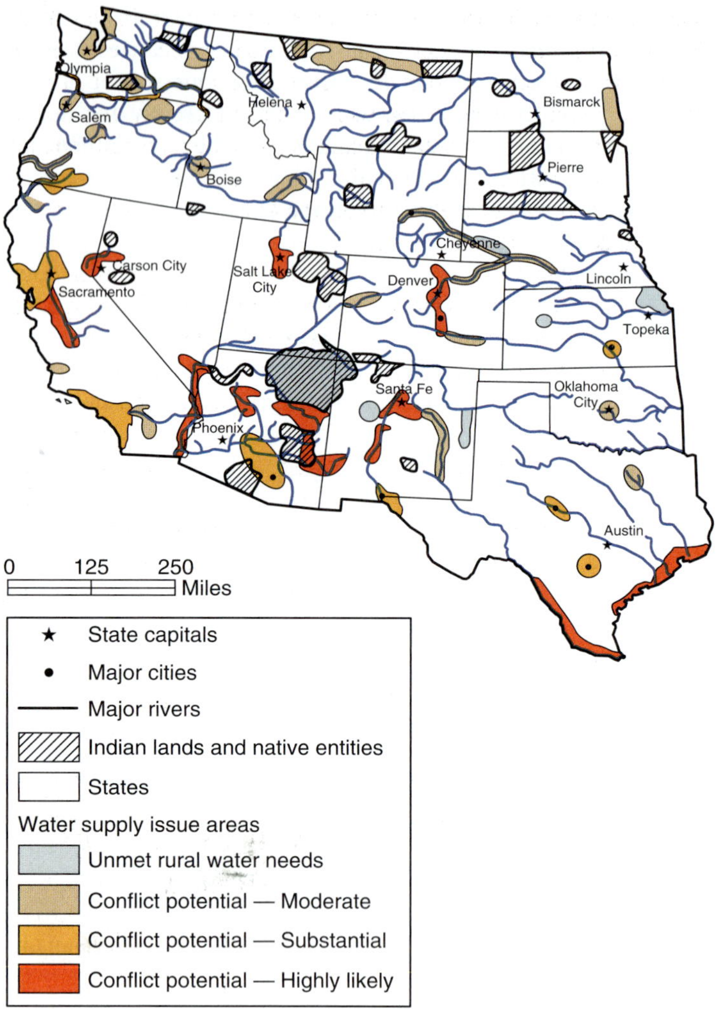


**FIGURE 17.13** Rapidly growing populations in arid regions are straining available water supplies. By 2025, the Department of the Interior warns, shortages could cause conflicts in many areas.
**Source:** Data from U.S. Department of Interior.

## Water consumption is less than withdrawal

Most water we use eventually returns to rivers and streams. Therefore, it is important to distinguish between withdrawal and consumption. **Withdrawal** is the total amount of water taken from a lake, river, or aquifer for any purpose. Much of this water is employed in nondestructive ways and is returned to circulation in a form that can be used again. **Consumption** is the fraction of withdrawn water that is lost in transmission, evaporation, absorption, chemical transformation, or otherwise made unavailable for other purposes as a result of human use. Note that much water that is withdrawn but not consumed may be **degraded**—polluted or heated so that it is unsuitable for other uses.

Many societies have always treated water as if there is an inexhaustible supply. It has been cheaper and more convenient for most people to dump all used water and get a new supply than to determine what is contaminated and what is not. The natural cleansing and renewing functions of the hydrologic cycle do replace the water we need if natural systems are not overloaded or damaged. Water is a renewable resource, but renewal takes time. The rate at which many of us are using water now may make it necessary to conscientiously protect, conserve, and replenish our water supply.

## Water use is increasing

Human water use has been increasing about twice as fast as population growth over the past century (fig. 17.14). Water use is stabilizing in industrialized countries, but demand will increase in developing countries where supplies are available. The average amount of water withdrawn worldwide is about 646 $m^3$ (170,544 gal) per person per year. This overall average hides great discrepancies in the proportion of annual runoff withdrawn in different areas. Some countries with a plentiful water supply withdraw a very small percentage of the water available to them. Canada, Brazil, and the Congo, for instance, withdraw less than 1 percent of their annual renewable supply.

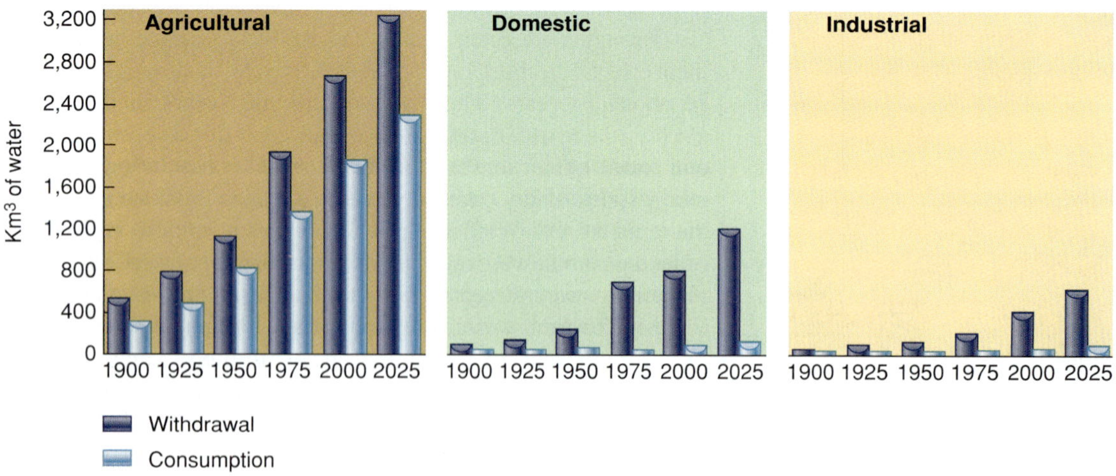


**FIGURE 17.14** Growth of water withdrawal and consumption, by sector, with projected levels to 2025.
**Source:** UNEP, 2002.

By contrast, in countries such as Libya and Israel, where water is one of the most crucial environmental resources, groundwater and surface water withdrawal together amount to more than 100 percent of their renewable supply. They are essentially "mining" water—extracting groundwater faster than it is being replenished. Obviously, this is not sustainable in the long run.

The total annual renewable water supply in the United States amounts to an average of about 9,000 $m^3$ (nearly 2.4 million gal) per person per year. We now withdraw about one-fifth of that amount, or some 5,000 l (1,300 gal) per person per day, including industrial and agricultural water. By comparison, the average water use in Haiti is less than 30 l (8 gal) per person per day.

## Agriculture is the greatest water consumer worldwide

We can divide water use into three major sectors: agricultural, domestic, and industrial. Of these, agriculture accounts for by far the greatest use and consumption. Worldwide, crop irrigation is responsible for two-thirds of water withdrawal and 85 percent of consumption. Evaporation and seepage from unlined irrigation canals are the principal consumptive water losses. Agricultural water use varies greatly, of course. Over 90 percent of water used in India is agricultural; in Kuwait, where water is especially precious, only 4 percent is used for crops. In the United States, which has both a large industrial sector and a highly urbanized population, about half of all water withdrawal, and about 80 percent of consumption, is agricultural.

A tragic case of water overconsumption is the Aral Sea, which lies in Kazakhstan and Uzbekistan (see map at the end of this book). Once the fourth largest inland water body in the world, this giant saline lake lost 75 percent of its surface area and 80 percent of its volume between 1975 and 2004 (fig. 17.15) when, under the former Soviet Union, 90 percent of the natural flow of the Amu Dar'ya and Syr Dar'ya Rivers was diverted to irrigate rice and cotton. Towns that once were prosperous fish processing and shipping ports now lie 100 km from the lake shore.

Vozrojdenie Island, which was used for biological weapons productions in the Soviet era, has become connected to the mainland. The salt concentration in the remaining water doubled, and fishing, which once produced 20,000 tons per year, ceased completely. Today, more than 200,000 tons of salt, sand, and toxic chemicals are blown every day from the dried lake bottom. This polluted cloud is destroying pastures, poisoning farm fields, and damaging the health of residents who remain in the area.

As water levels dropped, the lake split into two lobes. The "Small Aral" in Kazakhstan is now being reclaimed. Some of the river flow has been restored (mainly because Soviet-style rice and cotton farming have been abandoned), and a dam has been built to separate this small lobe from the larger one in Uzbekistan. Water levels in the small, northern lake have risen more than 8 m and surface area has expanded by 30 percent. With cleaner water pouring into the Small Aral, native fish are being reintroduced, and it's hoped that commercial fishing might one day be resumed. The fate of the larger lake remains clouded. There may never be enough water to refill it, and if there were, the toxins left in the lake bed could make it unusable anyway.

An similar catastrophe has befallen Lake Chad in northern Africa. Sixty thousand years ago, during the last ice age, this area was a verdant savanna sprinkled with freshwater lakes and occupied by crocodiles, hippopotamuses, elephants, and gazelles. At that time, Lake Chad was about the present size of the Caspian Sea (400,000 km$^2$). Climate change has turned the Sahara into a desert, and by the mid-1960s, Lake Chad had shrunk to 25,000 km$^2$ (as large as the United States' Lake Erie). With a maximum depth of 7 m, the lake is highly sensitive to climate, and it expands and contracts dramatically. Persistent drought coupled with increased demand by massive irrigation projects in the

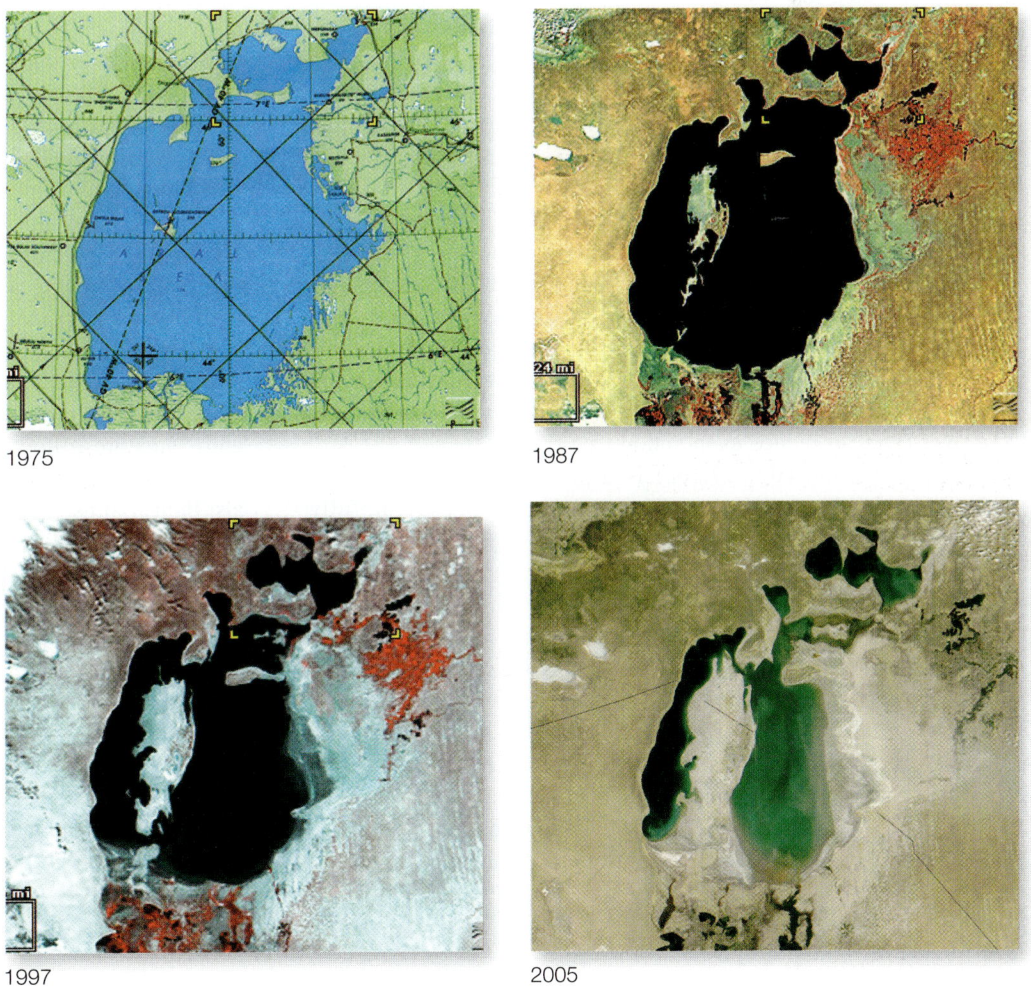


**FIGURE 17.15** For 30 years, rivers feeding the Aral Sea have been diverted to irrigate cotton and rice fields. The Aral Sea has lost more than 80 percent of its water. The "Small Aral" (*upper right lobe*) has separated from the main lake, and is now being refilled.

**FIGURE 17.16** Agricultural irrigation consumes more water than any other use. Methods vary from flood and furrow (a), which use extravagant amounts of water but also flush salts from soils, to sprinklers (b), to highly efficient drip systems (c).

1970s and 1980s has reduced Lake Chad to less than 1,000 $km^2$. The silty sand left on the dry lake bed is whipped aloft by strong winds funneled between adjacent mountain ranges. In the winter, the former lake bed, known as the Bodélé Depression, produces an average of 700,000 tons of dust every day. About 40 million tons of this dust are transported annually from Africa to South America, where it is thought to be the main source of mineral nutrients for the Amazon rainforest (chapter 16).

Irrigation can be very inefficient. Traditionally, the main method has been flood or furrow irrigation, in which water- floods a field (fig. 17.16*a*). As much as half of this water can be lost directly through evaporation. Much of the rest runs off before it is used by plants. In arid lands, flood irrigation is needed to help remove toxic salts from soil, but these salts contaminate streams, lakes, and wetlands downstream. Repeated flood irrigation also waterlogs the soil, reducing crop growth. Sprinkler systems can also be inefficient (fig. 17.16*b*). Water spraying high in the air quickly evaporates, rather than watering crops. In recent years, growing pressure on water resources has led to more efficient sprinkler systems that hang low over crops to reduce evaporation (see fig. 9.22).

Drip irrigation (fig. 17.16*c*) is a promising technology for reducing irrigation water use. These systems release carefully regulated amounts of water just above plant roots, so that nearly all water is used by plants. Only about 1 percent of the world's croplands currently use these systems, however.

Irrigation infrastructure, such as dams, canals, pumps, and reservoirs, is expensive. Irrigation is also the economic foundation of many regions. In the United States, the federal government has taken responsibility for providing irrigation for nearly a century. The argument for doing so is that irrigated agriculture is a public good that cannot be provided by individual farmers. A consequence of this policy has frequently been heavily subsidized crops whose costs, in water and in dollars, far outweigh their value.

## Domestic and industrial water use are greatest in wealthy countries

Worldwide, domestic water use accounts for about one-fifth of water withdrawals. Because little of this water evaporates or seeps into the ground, consumptive water use is slight, about 10 percent on average. Where sewage treatment is unavailable, water can be badly degraded by urban uses, however. In wealthy countries, each person uses about 500 to 800 l per day (180,000 to 280,000 l per year), far more than in developing countries (30 to 150 l per day). In North America, the largest single user of domestic water is toilet flushing (fig. 17.17). On average, each person in the United States uses about 50,000 l (13,000 gal) of drinking-quality water annually to flush toilets. Bathing accounts for nearly a third of water use, followed by laundry and washing. In western cities such as Palm Desert and Phoenix, lawn watering is also a major water user.

Urban and domestic water use have grown approximately in proportion with urban populations, about 50 percent between 1960 and 2000. Although individual water use seems slight on the scale of world water withdrawals, the cumulative effect of inefficient appliances, long showers, liberal lawn-watering, and other uses is enormous. California has established increasingly stringent standards for washing machines, toilets, and other appliances, in order to reduce urban water demands. Many other cities and states are following this lead to reduce domestic water use.

Industry accounts for 20 percent of global freshwater withdrawals. Industrial use rates range from 70 percent in industrialized parts of Europe to less than 5 percent in countries with little industry.

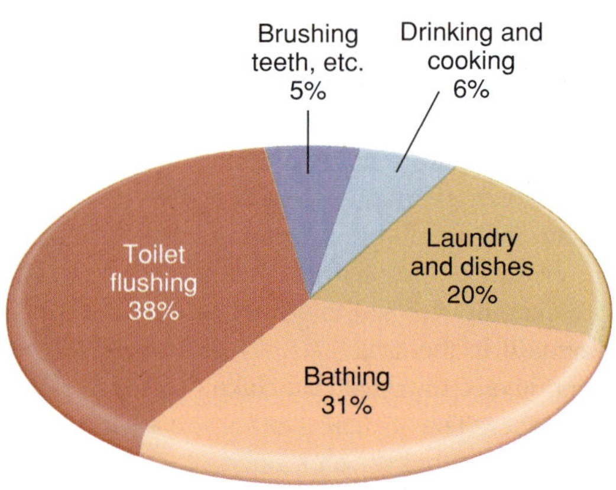


**FIGURE 17.17** Typical household water use in the United States. Where could you save the most water?
**Source:** EPA, 2004.

Power production, including hydropower, nuclear, and thermoelectric power, make up 50 to 70 percent of industrial uses, and industrial processes make up the remainder. As with domestic water, little of this water is made unavailable after use, but it is often degraded by defouling agents, chlorine, or heat when it is released to the environment. The greatest industrial producer of degraded water is mining. Ores must be washed and treated with chemicals such as mercury and cyanide (chapter 14). As much as 80 percent of water used in mining and processing is released with only minimal treatment. In developed countries, industries have greatly improved their performance in recent decades, however. Water withdrawal and consumption have both fallen relative to industrial production.

**FIGURE 17.18** Village water supplies in Ghana.

## 17.4 Freshwater Shortages

As the case study for this chapter shows, parts of China are desperately dry. The Gobi Desert is expanding rapidly, and its leading edge is now only 100 km (60 mi) from Beijing. Of 600 major Chinese cities, more than two-thirds have water shortages, and 100 of those cities have severe water scarcity. Pollution exacerbates this situation. Nearly half the population is reported to drink unsafe water.

In addition to the South-to-North water diversion project, the Chinese government has spent nearly 900 billion yuan (about U.S.$125 billion) over the past five years on sewage treatment, desalinization plants, and distribution systems. It's reported that 91 percent of all industrial waste water is now treated, although illegal dumping of toxic chemicals is much too common—particularly from factories shielded by local or regional governments. So far, only 39 percent of domestic sewage is treated in China, but that's probably ten times the amount treated a few decades ago. In 2006, 20 million rural Chinese got safe drinking water. The government has promised to provide clean water to everyone in the country by 2020.

### Many countries experience water scarcity and stress

The United Nations estimates that at least a billion people worldwide lack access to safe drinking water, and 2.6 billion don't have adequate sanitation. Depending on what happens to our global climate and human population growth (see the Data Analysis box at the end of this chapter) those numbers could be much higher in 50 years than they are now.

At least 45 countries—most of them in Africa or the Middle East—are considered to have serious water stress. They can't provide the 1,000 $m^3$ of water per person annually to meet the essential needs of their citizens. As is the case in some parts of China, the problem often isn't so much the total water supply as it is lack of access to clean water. In Mali, for example, 88 percent of the population lacks clean water; in Ethiopia it is 94 percent. Rural people often have less access to safe water than do city dwellers. In 33 of the most water-stressed countries, 60 percent of urban residents can get clean water as opposed to only 20 percent of those in the country.

Many developing countries have adequate water, but lack delivery systems. More than two-thirds of the world's households have to fetch water from outside the home. Women and children do most of this work, which takes several hours per day of heavy carrying. Improved public systems bring many benefits to poor families. In Mozambique, for example, the World Bank reports that the average time women spent carrying water decreased from 2 hours per day to only 25 minutes when village wells were installed. The time saved could be used to garden, tend livestock, trade in the market, care for children, or even rest!

Clean water is available in most countries—for those who can pay for it. Water from vendors (fig. 17.18) often is the only source for many in the crowded slums and shantytowns around the major cities of the developing world. Although the quality is often questionable, this water generally costs about ten times more than a piped city supply. Naturally, sanitation levels decline when water is so expensive. A typical poor family in Lima, Peru, for instance, uses one-sixth as much water as a middle-class American family but pays three times as much for it. Following government recommendations that all water be boiled to prevent cholera would take up to one-third of the total income for such a poor family.

### Would you fight for water?

Many environmental scientists have warned that water shortages could lead to wars between nations. *Fortune* magazine has written that "water will be to the 21st century what oil was to the 20th." With one-third of all humans living in areas with water stress now, the situation could become much worse as population grows and climate change dries up some areas and brings more severe storms to others. Already we've seen skirmishes—if not outright warfare—over insufficient water. In Kenya, for instance, nomadic tribes have fought over dwindling water and grazing. An underlying cause of the genocide now occurring in the Darfur region of Sudan is water scarcity. When rain was plentiful, Arab pastoralists and African farmers coexisted peacefully. Drought—perhaps caused by global warming—has upset that truce. The hundreds of thousands who have fled to Chad could be considered climate refugees as well as war victims.

Although they haven't usually risen to the level of war, there have been at least 37 military confrontations in the past 50 years in which water has been at least one of the motivating factors. Thirty of those conflicts have been between Israel and its neighbors. India, Pakistan, and Bangladesh also have confronted each other over water rights, and Turkey and Iraq threatened to send their armies to protect access to the water in the Tigris and Euphrates Rivers. Water can even be used as a weapon. As chapter 13 reports, Saddam Hussein cut off water flow into the massive Iraq marshes as a way of punishing his enemies among the Marsh Arabs. Drying of the marshes drove 140,000 people from their homes and destroyed a unique way of life. It also caused severe ecological damage to what is regarded by many as the original Garden of Eden.

Public anger over privatization of the public water supply in Bolivia sparked a revolution that overthrew the government in 2000. Water sales are already a $400-billion-a-year business. Multinational corporations are moving to take control of water systems in many countries. Who owns water and how much they are able to charge for it could become the question of the century. Investors are now betting on scarce water resources by buying future water rights. One Canadian water company, Global Water Corporation, puts it best: "Water has moved from being an endless commodity that may be taken for granted to a rationed necessity that may be taken by force."

Freshwater shortages may become much worse in many places because of global climate change. In 2007, the Intergovernmental Panel on Climate Change (IPCC) issued its fourth report on climate change. Figure 17.19 shows a summary of predictions from several models on likely changes in global precipitation for the period 2090–2099 compared to 1980–1999. White areas are where less than two-thirds of the models agree on likely outcomes; stippled areas are where more than 90 percent of the models agree. How does this map compare to figure 17.4? Which areas do you think are most likely to suffer from water shortages by the end of this century? If you lived in one of those areas, would you fight for water?

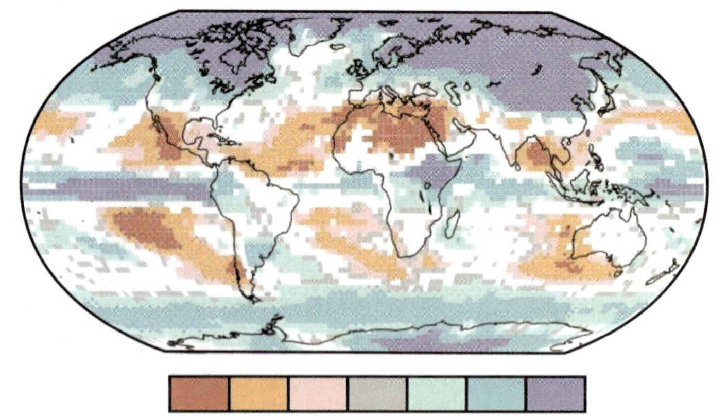


**FIGURE 17.19** Relative changes in precipitation (in percentage) for the period 2090–2099 compared to 1980–1999, predicted by the Intergovernmental Panel on Climate Change.
**Source:** IPCC, 2007.

## 17.5 Dams and Diversions

One way to make more water available is to store runoff in lakes or reservoirs and to ship it to places that need it. People have been moving water around for thousands of years, although none of these projects has ever been as big as the Chinese South-to-North plan. Some of the great civilizations (Mesopotamia, Egypt, China, Harrapa, and the Inca, for example) were based on large-scale water systems that controlled floods and brought irrigation water to farm fields. In fact, some historians argue that organizing people to build and operate these systems was the catalyst for emergence of civilization. Some of these systems are still in use. Roman aqueducts built 2,000 years ago are still in use. Those early water engineers probably never dreamed of moving water on the scale now being proposed.

According to the World Dam Commission, there were only about 250 high dams (more than 15 m tall) in the world before 1900. In the twentieth century, however, at least 45,000 dams were built, about half of them in China. Other countries with many dams include Turkey, Japan, Iran, India, Russia, Brazil, Canada, and the United States. The total cost of this building boom is estimated to have been $2 trillion. At least one-third aren't justified on economic grounds, and less than half have planned for social or environmental impacts.

The Army Corps of Engineers and the Bureau of Reclamation are the primary agencies responsible for building federal dams and water diversion projects in the United States. Building dams provided cheap, renewable power, created jobs for workers, stimulated regional economic development, stored water to reduce flooding, and allowed farming on lands that would otherwise be too dry (fig. 17.20). But not everyone agrees that these dams are an unmitigated benefit. Their storage reservoirs drown free-flowing rivers and often submerge towns and valuable riparian farmlands. They block fish migrations and change aquatic habitats essential for endemic species. In this section, we'll look at some of the advantages and disadvantages of dams, as well as what can be done to mitigate their effects.

### Dam failure can be disastrous

When a dam collapses, it can send a wall of water roaring down the river valley. One of the most infamous catastrophes in American history was the Johnstown flood of 1889. The city was built in the Little Conemaugh River valley just east of Pittsburg, Pennsylvania. Twenty kilometers upstream, and 150 m above Johnstown was a 5 km long lake created by an old, poorly maintained private dam. On May 31, 1889, the dam failed, sending 20 million tons of water crashing down the narrow valley. At times the wall of flood water grew to 60 ft in height. Moving at 60 km/hr, it swept away everything in its path, including much of Johnstown. More than 2,200 people died.

A much larger disaster occurred in China in 1975. Heavy monsoon rains caused 62 modern dams in China's Henan Province to fall like dominoes. At least 230,000 people were killed directly or died in subsequent famine and epidemics.

**FIGURE 17.20** Hoover Dam powers Las Vegas, Nevada. Lake Mead, behind the dam, loses about 1.3 billion $m^3$ per year to evaporation.

**FIGURE 17.21** A Yangtze River town. This photo was taken just before the three Gorges Dam was finished and the lower, older section of the town was flooded. Most residents moved up to the large, new apartment buildings on the hills above the reservoir.

If the Three Gorges Dam on the Yangtze were to fail similarly, it could cause a flood of biblical proportions. More than 100 million people live downstream. The dam is built on an active seismic fault. If it were ruptured by an earthquake or landslide upstream, it could send a wall of water 200 m high racing downstream. What would be your chances of survival if you were in its path? More than 3,200 Chinese dams have failed since 1949.

The Rogun Dam on the Vakhsh River in Tajikistan also is regarded as highly risky. Construction started in 1987, but was abandoned in 1993 after the collapse of the Soviet Union. In 2006, the Tajik government decided to build the dam by itself. Designed to be 335 m high (1,100 ft), it would be the tallest in the world. This dam also is being built in an active seismic zone at the edge of a tectonic plate. If the reservoir behind the dam is ever fully filled, calculations are that its weight will likely trigger earthquakes. The dam, built of rock and earth fill, probably will be highly vulnerable to seismic action.

## Dams often displace people and damage ecosystems

The 600 km long reservoir created by China's Three Gorges Dam flooded 1,500 towns and displaced 1.4 million people. Not all those people were unhappy about being resettled. Many merely moved uphill to newly constructed towns that were much better than the ancient river towns where they formerly lived (fig. 17.21). But many farmers, who were promised equal land to that of what they lost, have been disappointed by what they were offered.

Similarly, a series of dams on India's Narmada River has been the focus of decades of protest. Many of the 1 million villagers and tribal people being displaced by this project have engaged in mass resistance and civil disobedience when police try to remove them forcibly. Some have vowed to drown rather than leave their homes. They regard the river as sacred, and don't believe it should be shackled with dams. Furthermore, they don't trust government promises of resettlement.

Canada, also, has been the site of decades of protest by First Nations people over flooding of ancestral lands for hydroelectric projects. The James Bay project, built by Hydro-Quebec between 1971 and 2004, diverted three major rivers flowing west into Hudson Bay and created huge lakes that flooded more than 10,000 $km^2$ (4,000 $mi^2$) of forest and tundra, to generate 26,000 megawatts of electrical power. In 1984, shortly after Phase I of this project was completed, 10,000 caribou drowned trying to follow their usual migration route across the newly flooded land. The loss of traditional hunting and fishing sites has been culturally devastating for native Cree people. In addition, mercury, leeched out of rocks in recently submerged land, has entered the food chain, and many residents of the area suffer from mercury poisoning.

In a similar, but less well-known case, Manitoba Hydro has diverted most water in the Churchill River on the west shore of Hudson Bay into the Nelson River to generate hydroelectricity. More than 125,000 $km^2$ (50,000 $mi^2$) of tundra and boreal forest have been flooded. This area was traditional hunting land of First Nations people. Much of this land is underlain by permafrost, which is slowly melted by the impounded water, causing the lake shore to continually cave in and the broad, shallow lakes to expand ever further. Trees that fall into the lakes when the shore collapses become a navigational hazard for native people, who depend on boats for travel. As is the case in Quebec, most of the electricity generated by these dams is exported to the United States, where it's marketed as renewable energy. It's true that the hydropower is renewable, but the land that's being destroyed isn't.

## Dams kill fish

Dams are especially lethal for migratory fish, such as salmon. Adult fish are blocked from migrating to upstream spawning areas. And juvenile fish die if they go through hydroelectric turbines. The slack water in reservoirs behind dams is also a serious problem. Juvenile salmon evolved to ride the surge of spring runoff downstream to the ocean in two or three weeks. Reservoirs slow this journey to as much as three months, throwing off the time-sensitive physiological changes that allow the fish to survive in salt water when they reach the ocean. Reservoirs expose young salmon to predators, and warm water in reservoirs increases disease in both young and older fish.

Some dams have fish ladders—a cascading series of pools and troughs—that allow fish to bypass the dam. Another option is to move both adults and juveniles by barge. This can result in the strange prospect of barges of wheat moving downstream while passing barges of fish moving the opposite direction. Both these options are expensive and only partially effective in restoring blocked salmon runs.

The tide may be turning against dams. In 1998, the Army Corps announced that it would no longer be building large dams and diversion projects. In the few remaining sites where dams might be built, public opposition is so great that getting approval for projects is unlikely. Instead, the new focus may be on removing existing dams and restoring natural habitats. Former Interior Secretary Bruce Babbitt said, "Of the 75,000 large dams in the United States, most were built a long time ago and are now obsolete, expensive, and unsafe. They were built with no consideration of the environmental costs. As operating licenses come up for renewal, removal and restoration to original stream flows will be one of the options." (See What Do You Think? p. 389).

An example of this option is the removal of four aging dams on California's Klamath River. For years, Indian tribes, commercial fishermen, and conservation groups have urged the government to remove these dams because of their adverse impacts on salmon. PacifiCorp, which owns and operates the dams, offered to trap fish and truck them around the dams, but in a move that surprised everyone, the National Marine Fisheries Service ordered that fish ladders, that would cost about $300 million, be built at the dams. Because removing the dams would cost about $100 million less than modifying them, it seems possible that they may soon be demolished.

## Sedimentation limits reservoir life

Rivers with high sediment loads can fill reservoirs quickly (fig. 17.22). In 1957 the Chinese government began building the Sanmenxia Dam on the Huang He (Yellow River) in Shaanxi Province. From the beginning, engineers warned that the river carried so much sediment that the reservoir would have a very limited useful life. Dissent was crushed, however, and by 1960, the dam began filling the river valley and inundating fertile riparian fields that once had been part of China's traditional granaries.

Within two years, sediment accumulation behind the dam had become a serious problem. It blocked the confluence of the Wei and Yellow Rivers and backed up the Wei so it threatened to flood the historic city of Xi'an. By 1962, the reservoir was almost completely filled with sediment and hydropower production dropped by 80 percent. The increased elevation of the riverbed raised the underground water table and caused salinization of wells and farm fields. By 1991, the riverbed was 4.6 m above the surrounding landscape. The river is only kept in check by earthen dams that frequently fail and flood the surrounding countryside. By the time the project was complete, more than 400,000 people had been relocated, far more than planners expected.

**FIGURE 17.22** This dam is now useless because its reservoir has filled with silt and sediment.

Problems are similar, although not so severe, in some American rivers. As the muddy Colorado River slows behind the Glen Canyon and Boulder Dams, it drops its load of suspended sand and silt. More than 10 million metric tons of sediment collect every year behind these dams. Imagine a line of 20,000 dump trucks backed up to Lake Mead and Lake Powell every day, dumping dirt into the water. Within about a century, these reservoirs could be full of mud and useless for either water storage or hydroelectric generation. Elimination of normal spring floods—and the sediment they would usually drop to replenish beaches—has changed the riverside environment in the Grand Canyon. Invasive species crowd out native riparian plants. Beaches that campers use have disappeared. Boulders dumped in the canyon by side streams fill the riverbed. On several occasions, dam managers have released large surges of water from the Glen Canyon Dam to try to replicate normal spring floods. The results have been gratifying, but they don't last long. The canyon needs regular floods to maintain its character.

These lakes also lose huge amounts of water through evaporation in the dry desert air and seepage into porous rock. Together, these two lakes lose more than 2 billion $m^3$ of water every year. During a severe drought between 1995 and 2004, Lake Powell lost more than 60 percent of its volume and the lake surface dropped more than 50 m (150 ft), leaving a giant bathtub ring of precipitated salts on canyon walls (fig. 17.23). At its lowest point, the reservoir was only a few meters above the minimum level for hydroelectric generation. River runners and canyon aficionados hoped this would convince Congress to breach the dam and let the river run free, but better rains and snows in 2006 began to refill the lake.

The accumulating sediments that clog reservoirs also represents a loss of valuable nutrients. The Aswan High Dam in Egypt, for example, was built to supply irrigation water to make

## What Do You Think?

### Should We Remove Dams?

The first active hydroelectric dam in the United States to be breached against the wishes of its owners was the 162-year-old Edwards Dam, on the Kennebec River in Augusta, Maine. For years, the U.S. Fish and Wildlife Service advocated removal of this dam, which blocked migration of the endangered Atlantic salmon. After the dam was destroyed in 1999, anglers reported seeing salmon, striped bass, shad, alewive, and sturgeon upstream of the dam. This was the first of at least 140 dams removed over the next five years.

A much larger project is the removal of the Elwah and Glines Dams in Olympic National Park in Washington. Before these dams were built a century ago to provide power to lumber and paper mills in the town of Port Angeles, the Elwah River was one of the most productive salmon rivers in the world. Fifty kilogram (110 lb) King salmon once migrated upstream to spawn. Destruction of these dams is scheduled for 2008. Simply breaching the dams won't restore the salmon, however. Deep sediment beds deposited behind the dams will have to be removed or stabilized to restore the streambed (chapter 13). The costs for dam removal and stream restoration are expected to be at least \$200 million.

Another challenging dam removal project is now underway in Montana. The Milltown Dam on the Clark Fork River just east of Missoula is a toxic nightmare. For more than a century, mining, milling, and smelting upstream at Anaconda and Butte (see fig 13.34) dumped millions of tons of waste into the river. In 2004, the Environmental Protection Agency declared the river from the Milltown Dam upstream for 200 km (120 mi) the largest Superfund site in the United States.

The greatest concentration of contaminated sediment is behind the dam, which is now old and unstable. The high levels of arsenic, copper, lead, zinc, nickel, and other toxins that would be released if the dam fails threaten water supplies of Missoula and the water quality in the Columbia, one of the most important rivers in the American west. But you can't just blow out the dam and hope for the best. In 2006, work started to build a bypass channel around the dam and to remove 2.6 million m$^3$ of toxic sediment. When this is completed, restoration work will reconstruct the streambed and revegetate the surrounding land.

A more controversial topic is the proposed removal of four high dams on the Snake River. These dams were authorized by Congress in 1945 to generate electricity and make Lewistown, Idaho—750 km inland—a seaport. At the time, Idaho produced half of the chinook salmon in the Columbia watershed. Salmon and steelhead trout runs almost immediately plummeted when the dams closed off the river. In 1991, Snake River sockeye became the first of 13 salmon and steelhead stocks to be declared endangered species. The government was forced to propose a plan for species recovery.

Over the next 15 years, federal judges have repeatedly rejected proposed plans said to be inadequate for salmon protection. The major issue is what to do about the dams. On one side, irrigators and businesses that benefit from cheap water, power, and shipping maintain that the dams are indispensable. On the other side are Native American tribes, conservation groups, and commercial and sport fishermen. They point out that the dams produce a very small fraction of the electricity consumed in the Pacific Northwest, benefit very few farmers, and that other shipping options exist. The gains from tourist revenue and commercial salmon fishing could more than offset the potential losses from removing the dams.

Pressure for salmon recovery intensified in 2006, when only three sockeye—all hatchery fish—made the 1,600 km journey from the Pacific to the headwaters of the Salmon River in central Idaho. U.S. District Judge James Redden, who has rejected federal plans for salmon recovery three times, warned recalcitrant agencies that time has run out. Either they submit a viable restoration plan, or he will order breaching of the four Snake River dams.

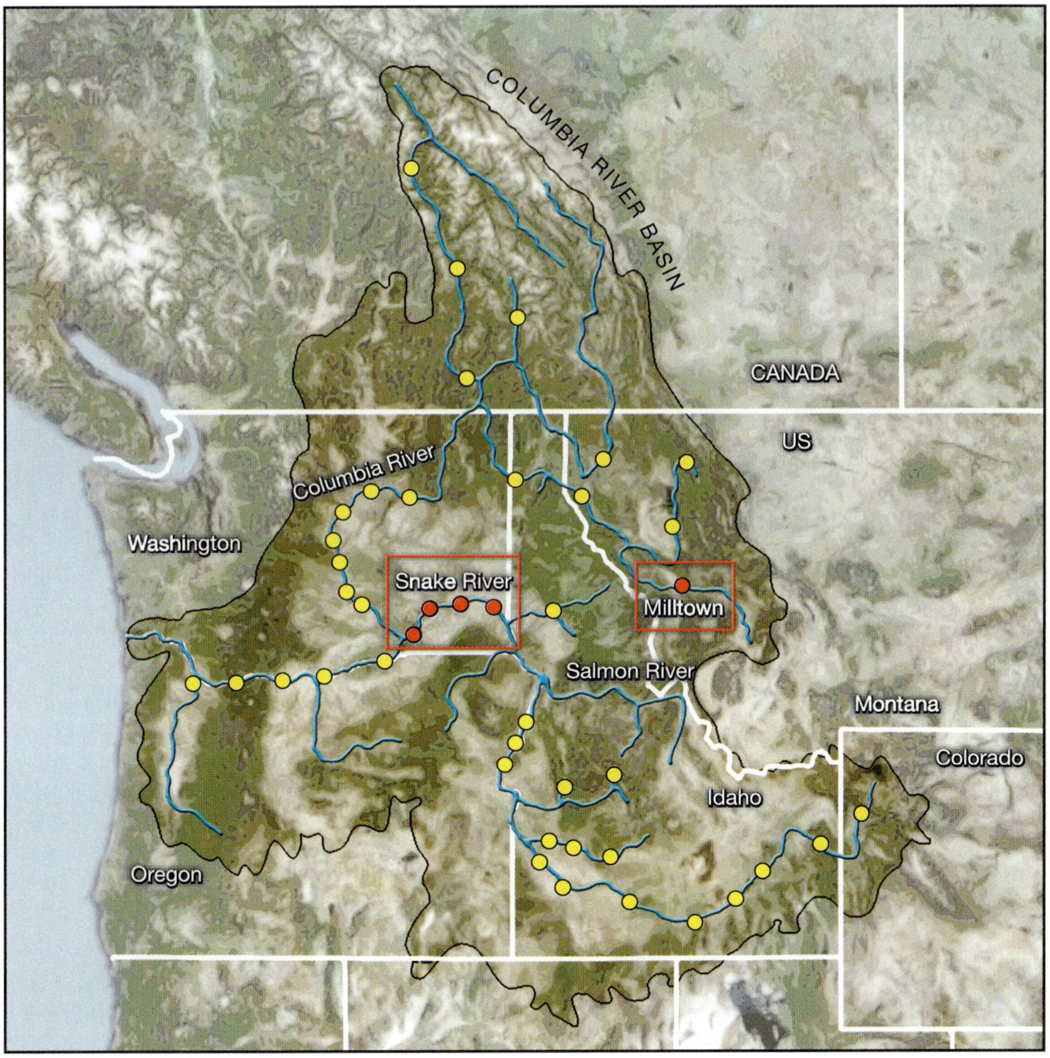

What do you think? How would you weigh the different ecological, economic, cultural, and health impacts of outdated dams? Is it simply an economic calculation of relative costs and benefits, or should ethical considerations be included? How would you compare the value of endangered salmon survival with the ability of farmers to grow food? These questions don't have simple answers. As you read subsequent chapters on water pollution, public policy, and environmental economics, keep in mind the practical questions about removing dams.

agriculture more productive. Although thousands of hectares are being irrigated, the water available is only about half that anticipated because of evaporation in Lake Nasser behind the dam, and seepage losses in unlined canals that deliver the water. Controlling the annual floods of the Nile also has stopped the deposition of nutrient-rich mud on which farmers depended for fertilizing their fields. Commercial fertilizer used to replace naturally provided nutrients costs more than $100 million annually. Furthermore, the nutrients carried by the river once supported a rich fishery in the Mediterranean that was a valuable food source for Egypt. After the dam was built, sardine fishing declined 97 percent. To make matters worse, snails living in shallow irrigation canals has led to an epidemic of schistosomiasis, a debilitating disease caused by parasitic flatworms, for which the snails are an alternate host. In some areas, 80 percent of the residents are infected.

## Diversion projects sometimes dry up rivers

While China's South-to-North project probably won't deplete the Yangtze, in many places diversion projects take so much water out of smaller rivers that they are completely dried up for much of the year. China's Huang He (Yellow River), for example, loses so much water to evaporation and diversions that it's reduced to a muddy trickle in much of its lower section, and is completely dry 226 days per year. Similarly, the Colorado River, which once carved the United States' Grand Canyon, is so salty and depleted by the time it reaches the Mexican border that it's useless for agriculture. Most of the year, no water at all reaches the Sea of Cortez.

Environmental scientists point out the ecological disruption caused by overdraft of water from streams. Obviously, fish and other aquatic organisms suffer when the medium in which they live disappears. Anglers, whitewater boaters and others who enjoy the beauty of free-flowing rivers mourn the loss of canyons drowned in reservoirs, and streams dried up by diversion projects (fig. 17.24). Many battles have been fought over the relative value of natural aquatic ecosystems versus the economic gains from appropriating the water for agriculture, industry, or domestic use. Management agencies charged with regulating resources as a public good have to find a balance between competing demands.

California's Owens Valley and Mono Lake, just east of Yosemite National Park, are historic examples of the conflict and controversy over water diversion. In 1905, Frederic Eaton, mayor of Los Angeles, and his friend, William Mulholland, the superintendent of the L.A. Department of Water and Power, surreptitiously bought farms in the Owens Valley. Eventually, Los Angeles became the largest landowner in the valley, controlling more than 90 percent of the water rights. The city built a huge aqueduct to transport the water about 400 km over

**FIGURE 17.23** Lake Powell, on the Colorado River, loses more than 1 billion m$^3$ of water to evaporation and seepage every year. During a severe drought between 1995 and 2004, the lake lost more than 60 percent of its volume and its surface dropped more than 50 m (150 ft).

**FIGURE 17.24** The recreational and aesthetic values of free-flowing wild rivers and wilderness lakes may be their greatest assets. Competition between *in situ* values and extractive uses can lead to bitter fights and difficult decisions.

**FIGURE 17.25** Mono Lake in eastern California. Diversion of tributary rivers to provide water for Los Angeles has shrunk the lake's surface area by one-third, threatening migratory bird flocks that feed here. These formations were created underwater where calcium-rich springs entered the brine-laden lake.

the mountains. The movie *Chinatown* is a fictional account of this project. So much water was diverted that the Owens River was completely dry for most of its 100 km course, and Owens Lake, which was fed by the river, disappeared.

In 1941, Los Angeles needed even more water and began tapping streams that fed Mono Lake just to the north of the Owens Valley. Deprived of its freshwater sources, the volume of the lake was cut in half. The lake surface fell by about 20 m, exposing strange tufa towers created where mineral-rich underwater springs had once entered the salty lake (fig. 17.25). Salinity of the lake water doubled, killing the prolific brine shrimp that nourished huge flocks of migratory waterfowl that once stopped here to feed.

After years of legal wrangling, the California Water Resources Board ruled in 1994 that Los Angeles must allow some water to replenish Mono Lake. The recovery plan set 2014 as the target date to return the lake to its 1964 level. So far, the surface of the lake has risen about 3 m. Annual turnover of the lake, which brought nutrients to the surface where the brine shrimp could access them, has resumed again. The ecology of tributary streams is also recovering, although it will take several decades for complete revival. Similarly, in 2006, Los Angeles began to allow a small amount of water to flow, once again, down the Owens River.

In another inspiring river restoration story, the Deschutes River, a tributary of the Columbia, also is being revived. Much of the river flow was diverted a century ago to irrigate farms in central Oregon. Native American tribes on the Warm Springs Reservation, downstream from this diversion, sued over the destruction of their fishing rights. As part of the settlement, irrigation districts upstream are implementing conservation measures that will return some water to the river. By lining canals and switching from flood irrigation to more efficient sprinkler systems, they can save water but get the same crop yields. Both native people and farmers, seeing the disastrous confrontation over water rights in the Klamath River just to their south, decided it's better to work cooperatively rather than risk solutions that no one likes.

## Groundwater is depleted when withdrawals exceed recharge

Groundwater is the source of nearly 40 percent of the fresh water for agricultural and domestic use in the United States. Nearly half of all Americans and about 95 percent of the rural population depend on groundwater for drinking and other domestic purposes. Overuse of these supplies causes several kinds of problems, including drying of wells, natural springs, and disappearance of surface water features such as wetlands, rivers, and lakes.

In many areas of the United States, groundwater is being withdrawn from aquifers faster than natural recharge can replace it. On a local level, this causes a cone of depression in the water table, as is shown in figure 17.11. A heavily pumped well can lower the local water table so that shallower wells go dry. On a broader scale, heavy pumping can deplete a whole aquifer. The Ogallala Aquifer, for example, underlies eight states in the arid high plains between Texas and North Dakota (fig. 17.26). As deep as 400 m (1,200 ft) in its center, this porous bed of sand, gravel, and sandstone once held more water than all the freshwater lakes, streams, and rivers on earth. Excessive pumping for

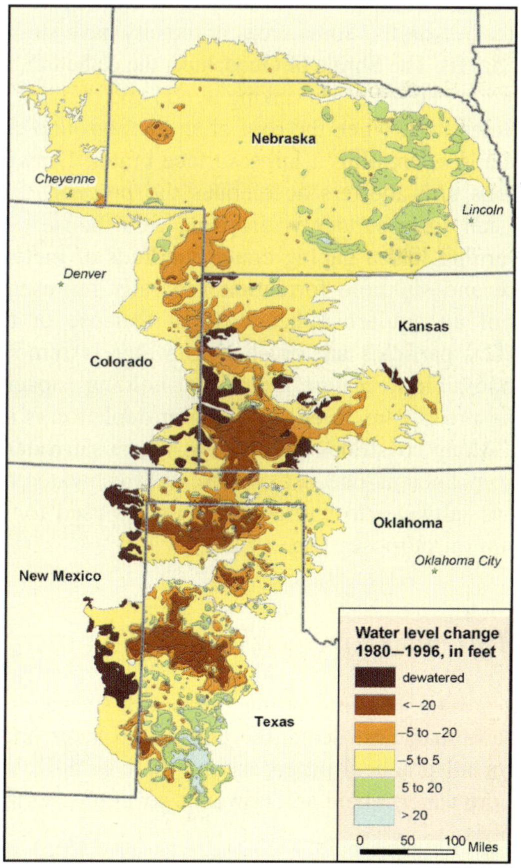


**FIGURE 17.26** The Ogallala/High Plains regional aquifer supports a multimillion-dollar agricultural economy, but withdrawal far exceeds recharge. Some countries are down to less than 3 m of saturated thickness.

irrigation and other uses has removed so much water that wells have dried up in many places, and farms, ranches, even whole towns are being abandoned.

Many aquifers have slow recharge rates, so it will take thousands of years to refill them once they are emptied. Much of the groundwater we now are using probably was left there by the glaciers thousands of years ago. It is fossil water, in a sense. It will never be replaced in our lifetimes, and is, essentially, a nonrenewable resource. Covering aquifer recharge zones with urban development or diverting runoff that once replenished reservoirs ensures that they will not refill.

Withdrawal of large amounts of groundwater causes porous formations to collapse, resulting in **subsidence** or settling of the surface above. The U.S. Geological Survey estimates that the San Joaquin Valley in California, for example, has sunk more than 10 m in the last 50 years because of excessive groundwater pumping. Around the world, many cities are experiencing subsidence. Many are coastal cities, built on river deltas or other unconsolidated sediments. Flooding is frequently a problem as these coastal areas sink below sea level (chapter 13). Some inland areas also are affected by severe subsidence. Mexico City is one of the worst examples. Built on an old lake bed, it has probably been sinking since Aztec times. In recent years, however, rapid population growth and urbanization (chapter 22) have caused groundwater overdrafts. Some areas of the city have sunk as much as 8.5 m (25.5 ft). The Shrine of Guadalupe, the cathedral, and many other historic monuments are sinking at odd and perilous angles.

**Sinkholes** form when the roof of an underground channel or cavern collapses, creating a large surface crater. Drawing water from caverns and aquifers accelerates the process of collapse. Sinkholes can form suddenly, dropping cars, houses, and trees without warning into a gaping crater hundreds of meters across. Subsidence and sinkhole formation generally represent permanent loss of an aquifer. When caverns collapse or the pores between rock particles are crushed as water is removed, it is usually impossible to restore their water-holding capacity.

A widespread consequence of aquifer depletion is **saltwater intrusion.** Along coastlines and in areas where saltwater deposits are left from ancient oceans, overuse of freshwater reservoirs often allows saltwater to intrude into aquifers used for domestic and agricultural purposes (fig. 17.27).

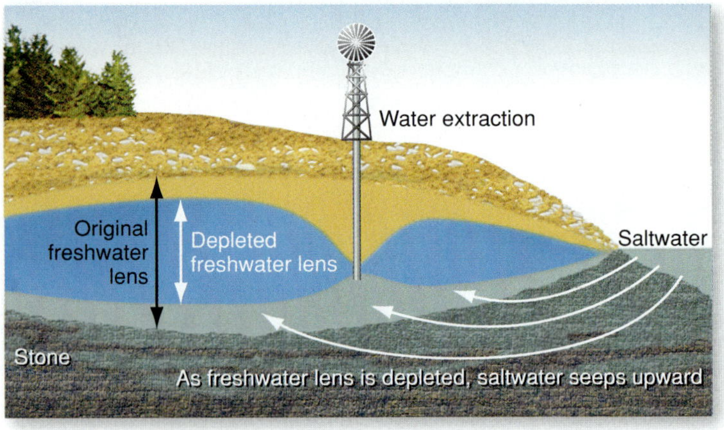


**FIGURE 17.27** Saltwater intrusion into a coastal aquifer as the result of groundwater depletion. Many coastal regions of the United States are losing freshwater sources due to saltwater intrusion.

## 17.6 Increasing Water Supplies

Where do present and impending freshwater shortages leave us now? On a human time scale, the amount of water on the earth is fixed, for all practical purposes, and there is little we can do to make more water. There are, however, several ways to increase local supplies.

In the dry prairie states of the 1800s and early 1900s, desperate farmers paid self-proclaimed "rainmakers" in efforts to save their withering crops. Centuries earlier, Native Americans danced and prayed to rain gods. We still pursue ways to make rain. Seeding clouds with dry ice or potassium iodide particles has been tested for many years with mixed results. Recently, researchers have been having more success using hygroscopic salts that seem to significantly increase rainfall amounts. This technique is being tested in Mexico, South Africa, and the western United States. There is a concern, however, that rain induced to fall in one area decreases the precipitation somewhere else. Furthermore, there are worries about possible contamination from the salts used to seed clouds.

### Desalination provides expensive water

A technology that might have great potential for increasing freshwater supplies is **desalination** of ocean water or brackish saline lakes and lagoons. The most common methods of desalination are distillation (evaporation and recondensation) or reverse osmosis (forcing water under pressure through a semipermeable membrane whose tiny pores allow water to pass but exclude most salts and minerals). In 2007, the global capacity was about 40 million $m^3$ per day, less than 0.2 percent of all freshwater withdrawals worldwide. This is expected to grow to about 100 million $m^3$ per day by 2015. Middle Eastern oil-rich states produce about 60 percent of desalinated water. Saudi Arabia is the largest single producer, at about 34 percent of world total. The United States is second, at 20 percent. Although desalination is still three to four times more expensive than most other sources of fresh water, it provides a welcome water supply in such places as Oman and Bahrain where there is no other access to fresh water. If a cheap, inexhaustible source of energy were available, however, the oceans could supply all the water we would ever need.

### Domestic conservation can save water

We could probably save as much as half of the water we now use for domestic purposes without great sacrifice or serious changes in our lifestyles. Simple steps, such as taking shorter showers, stopping leaks, and washing cars, dishes, and clothes as efficiently as possible, can go a long way toward forestalling the water shortages that many authorities predict. Isn't it better to adapt to more conservative uses now when we have a choice than to be forced to do it by scarcity in the future?

**FIGURE 17.28** By using native plants in a natural setting, residents of Phoenix save water and fit into the surrounding landscape.

The use of conserving appliances, such as low-volume shower heads and efficient dishwashers and washing machines, can reduce water consumption greatly (What Can You Do? p. 394). If you live in an arid part of the country, you might consider whether you really need a lush green lawn that requires constant watering, feeding, and care. Planting native ground cover in a "natural lawn" or developing a rock garden or landscape in harmony with the surrounding ecosystem can be both ecologically sound and aesthetically pleasing (fig. 17.28). There are about 30 million ha (75 million acres) of cultivated lawns, golf courses, and parks in the United States. They receive more water, fertilizer, and pesticides per hectare than any other kind of land.

Our largest domestic water use is toilet flushing (see fig. 17.17). There are now several types of waterless or low-volume toilets. Waterless composting systems can digest both human and kitchen wastes by aerobic bacterial action, producing a rich, nonoffensive compost that can be used as garden fertilizer. There are also low-volume toilets that use recirculating oil or aqueous chemicals to carry wastes to a holding tank, from which they are periodically taken to a treatment plant. Anaerobic digesters use bacterial or chemical processes to produce usable methane gas from domestic wastes. These systems provide valuable energy and save water but are more difficult to operate than conventional toilets. Few cities are ready to mandate waterless toilets, but a number of cities (including Los Angeles, California; Orlando, Florida; Austin, Texas; and Phoenix, Arizona) have ordered that water-saving toilets, showers, and faucets be installed in all new buildings. The motivation was twofold: to relieve overburdened sewer systems and to conserve water.

Significant amounts of water can be reclaimed and recycled. In California, water recovered from treated sewage constitutes the fastest growing water supply, growing about 30 percent per year. Despite public squeamishness, purified sewage effluent is being used for everything from agricultural irrigation to flushing toilets (fig. 17.29). In a statewide first, San Diego is currently piping water from the local sewage plant directly into a drinking-water reservoir. Residents of Singapore and Queensland, Australia, also are now drinking purified sewage effluent. "Don't rule out desalination because it's expensive, or recycling because it sounds yucky," says Morris Iemma, premier of New South Wales. "We're not getting rain; we have no choice."

**FIGURE 17.29** Recycled water is being used in California and Arizona for everything from agriculture, to landscaping, to industry. Some cities even use treated sewage effluent for human drinking-water supplies.

## Recycling can reduce consumption

In many developing countries as much as 70 percent of all the agricultural water used is lost to leaks in irrigation canals, application to areas where plants don't grow, runoff, and evaporation. Better farming techniques, such as leaving crop residue on fields and ground cover on drainage ways, intercropping, use of mulches, and low-volume irrigation, could reduce these water losses dramatically.

Nearly half of all industrial water use is for cooling of electric power plants and other industrial facilities. Some of this water use could be avoided by installing dry cooling systems similar to the radiator of your car. In many cases, cooling water could be reused for irrigation or other purposes in which water does not have to be drinking quality. The waste heat carried by this water could be a valuable resource if techniques were developed for using it.

## Prices and policies have often discouraged conservation

Through most of U.S. history, water policies have generally worked against conservation. In the well-watered eastern United States, water policy was based on riparian usufructuary (use) rights—those who lived along a river bank had the right to use as much water as they liked as long as they didn't interfere with its quality or availability to neighbors downstream. It was assumed that the supply would always be endless and that water had no value until it was used. In the drier western regions where water often is a limiting

## What Can You Do?

### Saving Water and Preventing Pollution

Each of us can conserve much of the water we use and avoid water pollution in many simple ways.

- Don't flush every time you use the toilet. Take shorter showers; don't wash your car so often.
- Don't let the faucet run while washing hands, dishes, food, or brushing your teeth. Draw a basin of water for washing and another for rinsing dishes. Don't run the dishwasher when half full.
- Dispose of used motor oil, household hazardous waste, batteries, etc., responsibly. Don't dump anything down a storm sewer that you wouldn't want to drink.
- Avoid using toxic or hazardous chemicals for simple cleaning or plumbing jobs. A plunger or plumber's snake will often unclog a drain just as well as caustic acids or lye. Hot water and soap will clean brushes more safely than organic solvents.
- If you have a lawn, use water sparingly. Water your grass and garden at night, not in the middle of the day. Consider planting native plants, low-maintenance ground cover, a rock garden, or some other xeriphytic landscaping.
- Use water-conserving appliances: low-flow showers, low-flush toilets, and aerated faucets.
- Use recycled (gray) water for lawns, house plants, car washing.
- Check your toilet for leaks. A leaky toilet can waste 50 gallons per day. Add a few drops of dark food coloring to the tank and wait 15 minutes. If the tank is leaking, the water in the bowl will change color.

resource, water law is based primarily on the Spanish system of prior appropriation rights, or "first in time are first in right." Even if the prior appropriators are downstream, they can legally block upstream users from taking or using water flowing over their property. But the appropriated water had to be put to "beneficial" use by being consumed. This creates a policy of "use it or lose it." Water left in a stream, even if essential for recreation, aesthetic enjoyment, or to sustain ecological communities, is not being appropriated or put to "beneficial" (that is, economic) use. Under this system, water rights can be bought and sold, but water owners frequently are reluctant to conserve water for fear of losing their rights.

In most federal "reclamation" projects, customers were charged only for the immediate costs of water delivery. The costs of building dams and distribution systems was subsidized, and the potential value of competing uses was routinely ignored. Farmers in California's Central Valley, for instance, for many years paid only about one-tenth of what it cost the government to supply water to them. This didn't encourage conservation. Subsidies created by underpriced water amounted to as much as $500,000 per farm per year in some areas.

Growing recognition that water is a precious and finite resource has changed policies and encouraged conservation across the United States. Despite a growing population, the United States is now saving some 144 million liters (38 million gal) per day—or enough water to fill Lake Erie in a decade—compared to per capita consumption rates of 20 years ago. With 37 million more people in the United Sates now than there were in 1980, we get by with 10 percent less water. New requirements for water-efficient fixtures and low-flush toilets in many cities help to conserve water on the home front. More efficient irrigation methods on farms also are a major reason for the downward trend.

Charging a higher proportion of real costs to users of public water projects has helped encourage conservation, and so have water marketing policies that allow prospective users to bid on water rights. Both the United States and Australia have had effective water pricing and allocation policies that encourage the most socially beneficial uses and discourage wasteful water uses. Market mechanisms for water allotment can be sensitive, however, in developing countries where farmers and low-income urban residents could be outbid for irreplaceable water supplies.

It will be important, as water markets develop, to be sure that environmental, recreational, and wildlife values are not sacrificed to the lure of high-bidding industrial and domestic uses. Given prices based on real costs of using water and reasonable investments in public water supplies, pollution control, and sanitation, the World Bank estimates that everyone in the world could have an adequate supply of clean water by the year 2030 (fig. 17.30). We will discuss the causes, effects, and solutions for water pollution in chapter 18.

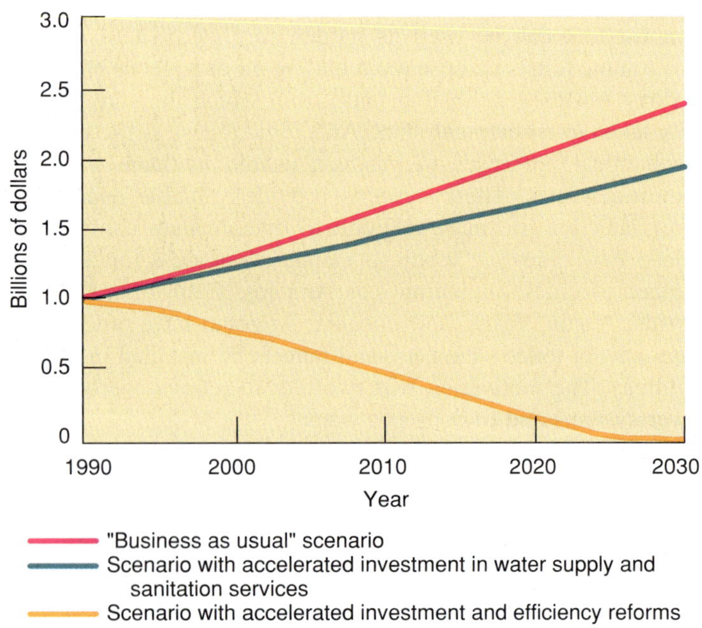


**FIGURE 17.30** Three scenarios for government investments on clean water and sanitation services, 1990 to 2030.

**Source:** World Bank estimates based on research paper by Dennis Anderson and William Cavendish, "Efficiency and Substitution in Pollution Abatement: Simulation Studies in Three Sectors."

## Conclusion

Water is a precious resource. As human populations grow and climate change affects rainfall patterns, water is likely to become even more scarce in the future. Already, about 2 billion people live in water-stressed countries (where there are inadequate supplies to meet all demands), and at least half those people don't have access to clean drinking water. Depending on population growth rates and climate change, it's possible that by 2050 there could be 7 billion people (about 60 percent of the world population) living in areas with water stress or scarcity. Conflicts over water rights are becoming more common between groups within countries and between neighboring countries that share water resources. This is made more likely by the fact that most major rivers cross two or more countries before reaching the sea. Many experts agree with *Fortune* magazine that "water will be to the 21st century what oil was to the 20th."

There are many ways to make more water available. Huge diversion projects, such as the Chinese South-to-North diversion, are already underway. Building dams and shipping water between watersheds, however, can have severe ecological and social effects. Perhaps a better way is to practice conservation and water recycling. These efforts, also, are underway in many places, and show great promise for meeting our needs for this irreplaceable resource. There are things you can do as an individual to save water and prevent pollution. Even if you don't have water shortages now where you live, it may be wise to learn how to live in a water-limited world.

## Reviewing Learning Outcomes

By now you should be able to explain the following points:

**17.1** Summarize why water is a precious resource and why shortages occur.

- The hydrologic cycle distributes water in our environment.
- Water supplies are unevenly distributed.

**17.2** Compare major water compartments.

- Oceans hold 97 percent of all water on earth.
- Glaciers, ice, and snow contain most surface fresh water.
- Groundwater stores large resources.
- Rivers, lakes, and wetlands cycle quickly.
- The atmosphere is among the smallest of compartments.

**17.3** Summarize water availability and use.

- Many people lack access to clean water.
- Water consumption is less than withdrawal.
- Water use is increasing.
- Agriculture is the greatest water consumer worldwide.
- Domestic and industrial water use are greatest in wealthy countries.

**17.4** Investigate freshwater shortages.

- Many countries experience water scarcity and stress.
- Would you fight for water?

**17.5** Illustrate the benefits and problems of dams and diversions.

- Dam failure can be disastrous.
- Dams often displace people and damage ecosystems.
- Dams kill fish.
- Sedimentation limits reservoir life.
- Diversion projects sometimes dry up rivers.
- Groundwater is depleted when withdrawals exceed recharge.

**17.6** Understand how we might increase water supplies.

- Desalination provides expensive water.
- Domestic conservation can save water.
- Recycling can reduce consumption.
- Prices and policies have often discouraged conservation.

## Practice Quiz

1. What is the difference between withdrawal, consumption, and degradation of water?
2. Explain how water can enter and leave an aquifer (see fig. 17.11).
3. Describe the changes in water withdrawal and consumption by sector shown in figure 17.14.
4. Describe some problems associated with dam building and water diversion projects.
5. Describe the path a molecule of water might follow through the hydrologic cycle from the ocean to land and back again.
6. Where are the five largest rivers in the world (table 17.3)?
7. How do mountains affect rainfall distribution? Does this affect your part of the country?
8. Identify and explain three consequences of overpumping aquifers.
9. How much water is fresh (as opposed to saline) and where is it?
10. Explain how saltwater intrusion happens (fig. 17.27).

## Critical Thinking and Discussion Questions

1. What changes might occur in the hydrologic cycle if our climate were to warm or cool significantly?
2. Why does it take so long for the deep ocean waters to circulate through the hydrologic cycle? What happens to substances that contaminate deep ocean water or deep aquifers in the ground?
3. Are there ways you could use less water in your own personal life? What obstacles prevent you from taking these steps?
4. Should we use up underground water supplies now or save them for some future time?
5. How should we compare the values of free-flowing rivers and natural ecosystems with the benefits of flood control, water diversion projects, hydroelectric power, and dammed reservoirs?
6. Would it be feasible to change from flush toilets and using water as a medium for waste disposal to some other system? What might be the best way to accomplish this?

## DATA analysis Graphing Global Water Stress and Scarcity

According to the United Nations, **water stress** is when annual water supplies drop below 1,700 $m^3$ per person. **Water scarcity** is defined as annual water supplies below 1,000 $m^3$ per person. More than 2.8 billion people in 48 countries will face either water stress or scarcity conditions by 2025. Of these countries, 40 are expected to be in West Asia or Africa. By 2050, far more people could be facing water shortages, depending both on population projections and scenarios for water supplies based on global warming and consumption patterns. The graph in this box shows an estimate for water stress and scarcity in 1995 together with three possible scenarios (high, medium, and low population projections) for 2050. You'll remember from chapter 7 that according to the 2004 UN population revision, the low projection for 2050 is about 7.6 billion, the medium projection is 8.9 billion, and the high projection is 10.6 billion.

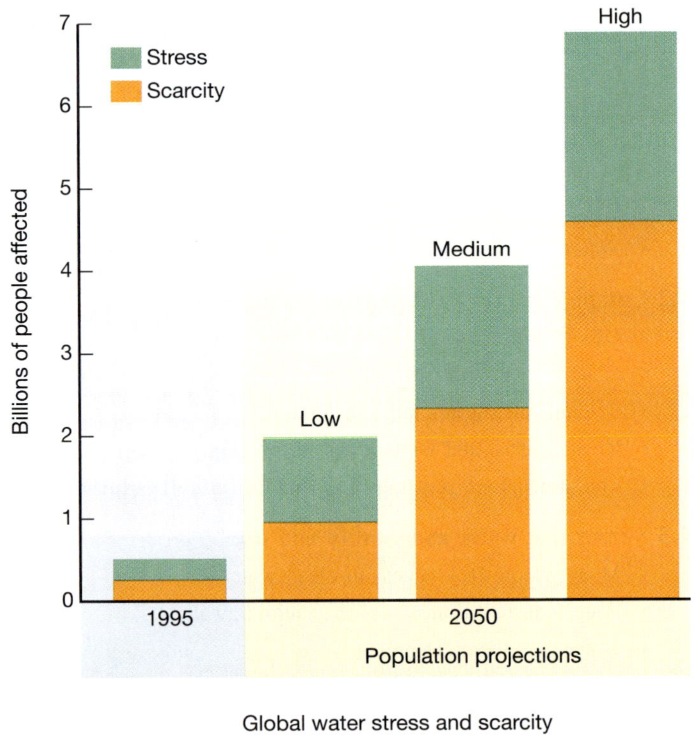


Global water stress and scarcity.

1. What are the combined numbers of people could experience water stress and scarcity under the low, medium, and high scenarios in 2050?
2. What proportion (percentage) of 7.6 billion, 8.9 billion, and 10.6 billion would this be?
3. How does the percentage of the population in these two categories vary in the three estimates?
4. Why is the proportion of people in the scarce category so much larger in the high projection?
5. How many liters are in 1,000 $m^3$? How many gallons?
6. How does 1,000 $m^3$ compare to the annual consumption by the average family of four in the United States? (Hint: Look at table 17.2 and the table of units of measurement conversions at the end of this book).
7. Why isn't the United States (as a whole) considered to be water stressed?

**For Additional Help in Studying This Chapter,** please visit our website at www.mhhe.com/cunningham10e. You will find additional practice quizzes and case studies, flashcards, regional examples, place markers for Google Earth™ mapping, and an extensive reading list, all of which will help you learn environmental science.

Arcata, California, built an artificial marsh as a low-cost, ecologically based treatment system for sewage effluent.

CHAPTER 18

# Water Pollution

*Water, water everywhere; nor any drop to drink.*

*—Samuel Taylor Coleridge—*

## LEARNING OUTCOMES

After studying this chapter, you should be able to:

**18.1** Define water pollution.

**18.2** Describe the types and effects of water pollutants.

**18.3** Investigate water quality today.

**18.4** Explain water pollution control.

**18.5** Summarize water legislation.

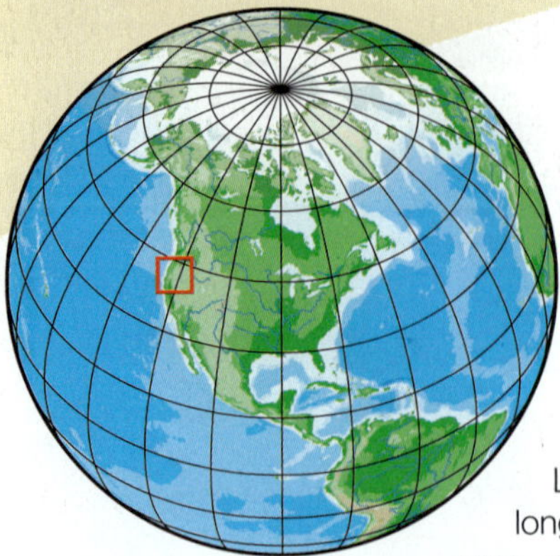

# Case Study A Natural System for Wastewater Treatment

Arcata, a small town in northwestern California's redwood country, sits at the north end of Humbolt Bay, about 450 km (280 mi) north of San Francisco. Logging, fishing, and farming, long the economic engines of the area, are now giving way to tourism, education, and diversified industries. Many of the 17,000 residents of Arcata choose to live there because of the outstanding environmental quality and recreational opportunities of the area. Until the early 1970s, however, Arcata's sewage system discharged unchlorinated sewage effluent directly into the nearly enclosed bay. Algal blooms discolored the water, while fishing and swimming appeal declined.

In 1974 California enacted a policy prohibiting discharge of untreated wastewater into bays and estuaries. State planners proposed a large and expensive regional sewage treatment plant for the entire Humbolt Bay area. The plan was for large interceptor sewers to encircle the bay with a major underwater pipeline crossing the main navigation channel. Effluent from the proposed plant was to be released offshore into an area of shifting sea bottom in heavy winter storms. The cost and disruption required by this project prompted city officials to look for alternatives.

Ecologists from Humbolt State University pointed out that natural processes could be harnessed to solve Arcata's wastewater problems at a fraction of the cost and disturbance of a conventional treatment plant. After several years of study and experimentation, the city received approval to build a constructed wetland for wastewater treatment. This solved two problems at the same time. Arcata's waterfront was blighted by an abandoned lumbermill pond, channelized sloughs, marginal pasture land, and an abandoned sanitary landfill. Building a new wetland on this degraded area would beautify it while also solving the sewage treatment problem.

Today, Arcata's waterfront has been transformed into 40 ha (about 100 acres) of freshwater and saltwater marshes, brackish ponds, tidal sloughs, and estuaries (fig. 18.1). This diverse habitat supports a wide variety of plants and animals, and is now a wildlife sanctuary and a major tourist attraction. As a home or rest stop for over 200 bird species, the wetland is regarded as one of the best birding sites along the Pacific North Coast. In 1987, the city was awarded a prize by the Ford Foundation for innovative local government projects that included $100,000 to build an interpretive center in the marsh, The wetland is now an outstanding place for outdoor education, scientific study, and recreation.

Arcata's constructed wetland is also an unqualified success in wastewater treatment. After primary clarification to remove grit and sediment, wastewater passes through several oxidation ponds that remove about half the organic material (measured by biological oxygen demand) and suspended solids. Sludge captured by the clarifiers goes to digesters that generate methane gas, which is burned to provide heat that speeds the digestion process. Effluent from the oxidation ponds passes through treatment marshes, where sunlight and oxygen kill pathogens while aquatic plants and animals remove remaining organic material along with nitrogen, phosphorus, and other plant nutrients. The effluent is treated with chlorine gas to kill any residual pathogens. Finally, the treated water trickles through several large enhancement marshes that allow chlorine to evaporate and complete pollutant removal.

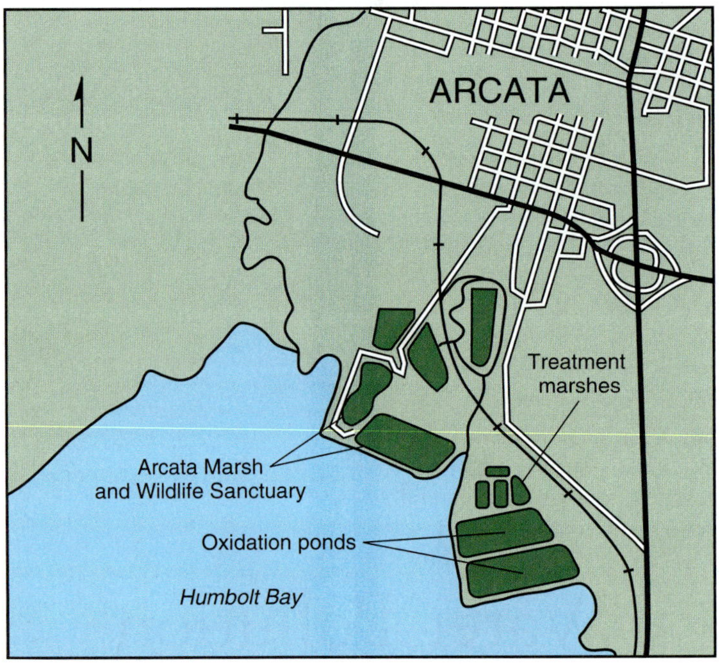


**FIGURE 18.1** Arcata's constructed wetlands remove pathogens, pollutants, and nutrients from municipal wastewater using native biological communities and ecological processes in a setting that also serves as a wildlife sanctuary and tourist attraction.

After 20 years of successful operation, Arcata's constructed wetland has inspired many other communities to find ecological solutions to their problems. Arcata Bay now produces more than half the oysters grown in California. The city also operates a wastewater aquaculture project where salmon, trout, and other fish species are raised in a mixture of wastewater effluent and seawater. Economically, the project is a success, providing excellent water quality treatment far below the cost of conventional systems. And the innovative approach of working with nature rather than against it has made Arcata an important ecotourist destination. Around the world, hundreds of communities have followed Arcata's model and now use constructed wetlands to eliminate up to 98 percent of the pollutants from their wastewater.

In this chapter, we'll look at both the causes and effects of water pollution as well as our options for controlling or treating water contaminants.

## 18.1 Water Pollution

Most students today are too young to appreciate that water in most industrialized countries was once far more polluted and dangerous than it is now. Thirty years ago, Lake Erie was on the brink of ecological collapse. The Cuyahoga River, choked with oil and floating debris, burned regularly. Factories and cities routinely dumped untreated chemicals, metals, oil, solvents, and sewage into rivers and lakes. Toxic solvents and organic chemicals were commonly dumped or buried in the ground, poisoning groundwater that we're now paying billions to clean up. In 1972, President Nixon signed the Clean Water Act, which has been called the United States' most successful and popular environmental legislation. This act established a goal that all the nation's waters should be "fishable and swimmable." While this goal is far from being achieved, the Clean Water Act remains popular because it protects public health (thus saving taxpayer dollars), as well as reducing environmental damage. In addition, water has an aesthetic appeal: The view of a clean lake, river, or seashore makes people happy, and water provides for recreation, so many people feel their quality of life has improved as water quality has been restored.

Clean water is a national, as well as global, priority. Recent polls have found repeatedly that 90 percent of Americans believe we should invest more in clean water and 70 percent would support establishing a trust fund to help communities repair water facilities.

We still have a long way to go in improving water quality. While pollution from factory pipes has been vastly reduced in the past 30 years, erosion from farm fields, construction sites, and streets has, in many areas, gotten worse since 1972. Airborne mercury, sulfur, and other substances are increasingly contaminating lakes and wetlands. Concentrated livestock production and agricultural runoff, threaten underground water as well as surface water systems. Increasing industrialization in developing countries has led to widespread water pollution in impoverished regions with little environmental regulation.

### Water pollution is anything that degrades water quality

Any physical, biological, or chemical change in water quality that adversely affects living organisms or makes water unsuitable for desired uses can be considered pollution. There are natural sources of water contamination, such as poison springs, oil seeps, and sedimentation from erosion, but in this chapter we will focus primarily on human-caused changes that affect water quality or usability.

Pollution-control standards and regulations usually distinguish between point and nonpoint pollution sources. Factories, power plants, sewage treatment plants, underground coal mines, and oil wells are classified as **point sources** because they discharge pollution from specific locations, such as drain pipes, ditches, or sewer outfalls (fig. 18.2). These sources are discrete and identifiable, so they are relatively easy to monitor and regulate. It is generally possible to divert effluent from the waste streams of these sources and treat it before it enters the environment.

**FIGURE 18.2** Sewer outfalls, industrial effluent pipes, acid draining out of abandoned mines, and other point sources of pollution are generally easy to recognize.

In contrast, **nonpoint sources** of water pollution are scattered or diffuse, having no specific location where they discharge into a particular body of water. Nonpoint sources include runoff from farm fields and feedlots (fig. 18.3), golf courses, lawns and gardens, construction sites, logging areas, roads, streets, and parking lots. Whereas point sources may be fairly uniform and predictable throughout the year, nonpoint sources are often highly episodic. The first heavy rainfall after a dry period may flush high concentrations of gasoline, lead, oil, and rubber residues off city streets, for instance, while subsequent runoff may have lower levels of these pollutants. Spring snowmelt carries high levels of atmospheric acid deposition into streams and lakes in some areas. The irregular timing of these

**FIGURE 18.3** This bucolic scene looks peaceful and idyllic, but allowing cows to trample stream banks is a major cause of bank erosion and water pollution. Nonpoint sources such as this have become the leading unresolved cause of stream and lake pollution in the United States.

events, as well as their multiple sources and scattered location, makes them much more difficult to monitor, regulate, and treat than point sources.

Perhaps the ultimate in diffuse, nonpoint pollution is **atmospheric deposition** of contaminants carried by air currents and precipitated into watersheds or directly onto surface waters as rain, snow, or dry particles. The Great Lakes, for example, have been found to be accumulating industrial chemicals such as PCBs and dioxins, as well as agricultural toxins such as the insecticide toxaphene that cannot be accounted for by local sources alone. The nearest sources for many of these chemicals are sometimes thousands of kilometers away (chapter 16).

Amounts of these pollutants can be quite large. It is estimated that there are 600,000 kg of the herbicide atrazine in the Great Lakes, most of which is thought to have been deposited from the atmosphere. Concentration of persistent chemicals up the food chain can produce high levels in top predators. Several studies have indicated health problems among people who regularly eat fish from the Great Lakes.

Ironically, lakes can be pollution sources as well as recipients. In the past 12 years, about 26,000 metric tons of PCBs have "disappeared" from Lake Superior. Apparently, these compounds evaporate from the lake surface and are carried by air currents to other areas where they are redeposited.

## 18.2 Types and Effects of Water Pollutants

Although the types, sources, and effects of water pollutants are often interrelated, it is convenient to divide them into major categories for discussion (table 18.1). Let's look more closely at some of the important sources and effects of each type of pollutant.

### Infectious agents remain an important threat to human health

The most serious water pollutants in terms of human health worldwide are pathogenic organisms (chapter 8). Among the most important waterborne diseases are typhoid, cholera, bacterial and amoebic dysentery, enteritis, polio, infectious hepatitis, and schistosomiasis. Malaria, yellow fever, and filariasis are transmitted by insects that have aquatic larvae. Altogether, at least 25 million deaths each year are blamed on these water-related diseases. Nearly two-thirds of the mortalities of children under 5 years old are associated with waterborne diseases.

The main source of these pathogens is from untreated or improperly treated human wastes. Animal wastes from feedlots or fields near waterways and food processing factories with inadequate waste treatment facilities also are sources of disease-causing organisms.

In developed countries, sewage treatment plants and other pollution-control techniques have reduced or eliminated most of the worst sources of pathogens in inland surface waters. Furthermore, drinking water is generally disinfected by chlorination so epidemics of waterborne diseases are rare in these countries. The United Nations estimates that 90 percent of the people in developed countries have adequate (safe) sewage disposal, and 95 percent have clean drinking water.

The situation is quite different in less-developed countries (fig. 18.4). The United Nations estimates that at least 2.5 billion people in these countries lack adequate sanitation, and that about half these people also lack access to clean drinking water. Conditions are especially bad in remote, rural areas where sewage treatment is usually primitive or nonexistent, and purified water is either unavailable or too expensive to obtain (fig. 18.5). The World Health Organization estimates

**TABLE 18.1**

**Major Categories of Water Pollutants**

| Category | Examples | Sources |
|---|---|---|
| **A. Causes Health Problems** | | |
| 1. Infectious agents | Bacteria, viruses, parasites | Human and animal excreta |
| 2. Organic chemicals | Pesticides, plastics, detergents, oil, and gasoline | Industrial, household, and farm use |
| 3. Inorganic chemicals | Acids, caustics, salts, metals | Industrial effluents, household cleansers, surface runoff |
| 4. Radioactive materials | Uranium, thorium, cesium, iodine, radon | Mining and processing of ores, power plants, weapons production, natural sources |
| **B. Causes Ecosystem Disruption** | | |
| 1. Sediment | Soil, silt | Land erosion |
| 2. Plant nutrients | Nitrates, phosphates, ammonium | Agricultural and urban fertilizers, sewage, manure |
| 3. Oxygen-demanding wastes | Animal manure and plant residues | Sewage, agricultural runoff, paper mills, food processing |
| 4. Thermal | Heat | Power plants, industrial cooling |

**FIGURE 18.4** New Delhi residents wash food in polluted water. At least a billion people lack access to clean water.

**FIGURE 18.6** Our national goal of making all surface waters in the United States "fishable and swimmable" has not been fully met, but scenes like this have been reduced by pollution-control efforts.

that 80 percent of all sickness and disease in less-developed countries can be attributed to waterborne infectious agents and inadequate sanitation.

If everyone had pure water and satisfactory sanitation, the World Bank estimates that 200 million fewer episodes of diarrheal illness would occur each year, and 2 million childhood deaths would be avoided. Furthermore, 450 million people would be spared debilitating roundworm or fluke infections. Surely these are goals worth pursuing.

Detecting specific pathogens in water is difficult, time-consuming, and costly; thus, water quality control personnel usually analyze water for the presence of **coliform bacteria,** any of the many types that live in the colon or intestines of humans and other animals. The most common of these is *Eschericha coli* (or *E. coli*). Many strains of bacteria are normal symbionts in mammals, but some, such as *Shigella, Salmonella,* or *Lysteria* can cause fatal diseases. It is usually assumed that if any coliform bacteria are present in a water sample, infectious pathogens are present also.

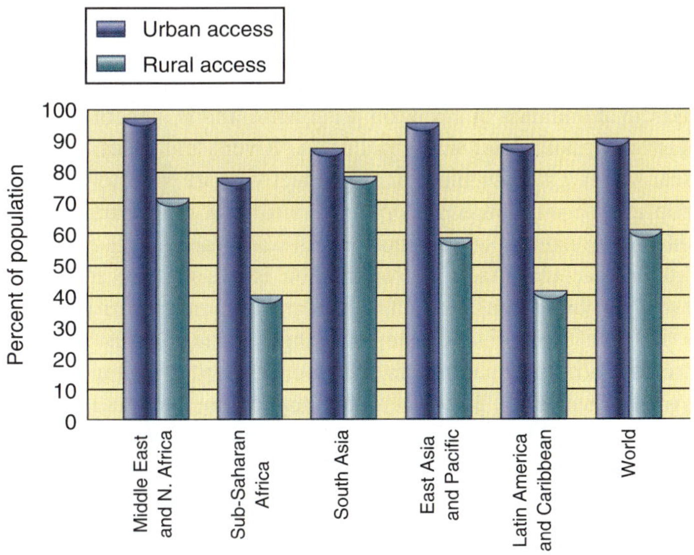


**FIGURE 18.5** Proportion of people in developing regions with access to safe drinking water.
**Source:** UNESCO, 2002.

To test for coliform bacteria, a water sample (or a filter through which a measured water sample has passed) is placed in a dish containing a nutrient medium that supports bacterial growth. After 24 hours in an incubator, living cells will have produced small colonies. If *any* colonies are found in drinking water samples, the U.S. Environmental Protection Agency considers the water unsafe and requiring disinfection. The EPA-recommended maximum coliform count for swimming water is 200 colonies per 100 ml, but some cities and states allow higher levels. If the limit is exceeded, the contaminated pool, river, or lake usually is closed to swimming (fig. 18.6).

## Bacteria are detected by measuring oxygen levels

The amount of oxygen dissolved in water is a good indicator of water quality and of the kinds of life it will support. Water with an oxygen content above 6 parts per million (ppm) will support game fish and other desirable forms of aquatic life. Water with less than 2 ppm oxygen will support mainly worms, bacteria, fungi, and other detritus feeders and decomposers. Oxygen is added to water by diffusion from the air, especially when turbulence and mixing rates are high, and by photosynthesis of green plants, algae, and cyanobacteria. Oxygen is removed from water by respiration and chemical processes that consume oxygen.

Organic waste such as sewage, paper pulp, or food waste is rich in nutrients, especially nitrogen and phosphorus. These nutrients stimulate the growth of oxygen-demanding decomposing

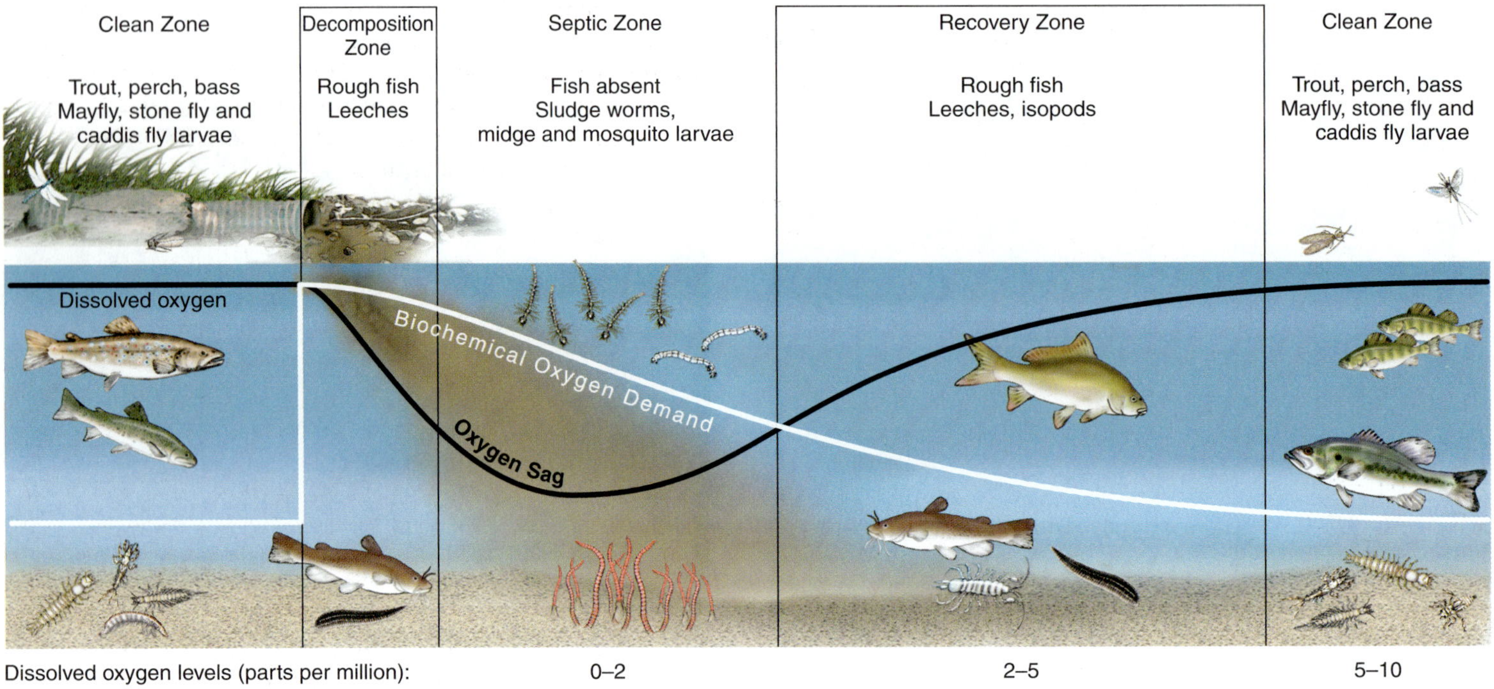


**FIGURE 18.7** Oxygen sag downstream of an organic source. A great deal of time and distance may be required for the stream and its inhabitants to recover.

bacteria. **Biochemical oxygen demand (BOD)** is thus a useful test for the presence of organic waste in water. Usually, BOD tests involve incubating a water sample for five days, then comparing oxygen levels in the water before and after incubation. An alternative method, called the chemical oxygen demand (COD), uses a strong oxidizing agent (dichromate ion in 50 percent sulfuric acid) to completely break down all organic matter in a water sample. This method is much faster than the BOD test, but it records inactive organic matter as well as bacteria, so it is less useful. A third method of assaying pollution levels is to measure **dissolved oxygen (DO) content** directly, using an oxygen electrode. The DO content of water depends on factors other than pollution (for example, temperature and aeration), so it is best for indicating the health of the aquatic system.

The effects of oxygen-demanding wastes on rivers depends to a great extent on the volume, flow, and temperature of the river water. Aeration occurs readily in a turbulent, rapidly flowing river, which is, therefore, often able to recover quickly from oxygen-depleting processes. Downstream from a point source, such as a municipal sewage plant discharge, a characteristic decline and restoration of water quality can be detected either by measuring dissolved oxygen content or by observing the flora and fauna that live in successive sections of the river.

The oxygen decline downstream is called the **oxygen sag** (fig. 18.7). Upstream from the pollution source, oxygen levels support normal populations of clean-water organisms. Immediately below the source of pollution, oxygen levels begin to fall as decomposers metabolize waste materials. Rough fish, such as carp, bullheads, and gar, are able to survive in this oxygen-poor environment where they eat both decomposer organisms and the waste itself. Further downstream, the water may become so oxygen-depleted that only the most resistant microorganisms and invertebrates can survive. Eventually, most of the nutrients are used up, decomposer populations are smaller, and the water becomes oxygenated once again. Depending on the volumes and flow rates of the effluent plume and the river receiving it, normal communities may not appear for several miles downstream.

## Nutrient enrichment leads to cultural eutrophication

Water clarity (transparency) is affected by sediments, chemicals, and the abundance of plankton organisms, and is a useful measure of water quality and water pollution. Rivers and lakes that have clear water and low biological productivity are said to be **oligotrophic** (*oligo* = little + *trophic* = nutrition). By contrast, **eutrophic** (*eu* + *trophic* = truly nourished) waters are rich in organisms and organic materials. Eutrophication is an increase in nutrient levels and biological productivity. Some amount of eutrophication is a normal part of successional changes in most lakes. Tributary streams bring in sediments and nutrients that stimulate plant growth. Over time, ponds or lakes may fill in, eventually becoming marshes. The rate of eutrophication and succession depends on water chemistry and depth, volume of inflow, mineral content of the surrounding watershed, and the biota of the lake itself.

As with BOD, eutrophication often results from nutrient enrichment sewage, fertilizer runoff, even decomposing leaves in street gutters can produce a human-caused increase in biological productivity called **cultural eutrophication.** Cultural eutrophication can also result from higher temperatures, more sunlight

**FIGURE 18.8** Eutrophic lake. Nutrients from agriculture and domestic sources have stimulated growth of algae and aquatic plants. This reduces water quality, alters species composition, and lowers the lake's recreational and aesthetic values.

reaching the water surface, or a number of other changes. Increased productivity in an aquatic system sometimes can be beneficial. Fish and other desirable species may grow faster, providing a welcome food source.

Often, however, eutrophication has undesirable results. Elevated phosphorus and nitrogen levels stimulate "blooms" of algae or thick growths of aquatic plants (fig. 18.8). Bacterial populations also increase, fed by larger amounts of organic matter. The water often becomes cloudy or turbid and has unpleasant tastes and odors. In extreme cases, plants and algae die and decomposers deplete oxygen in the water. Collapse of the aquatic ecosystem can result.

## Eutrophication can cause toxic tides and "dead zones"

According to the Bible, the first plague to afflict the Egyptians when they wouldn't free Moses and the Israelites was that the water in the Nile turned into blood. All the fish died and the people were unable to drink the water, a terrible calamity in a desert country. Some modern scientists believe this may be the first recorded history of a **red tide** or a bloom of deadly aquatic microorganisms. Red tides—and other colors, depending on the species involved—have become increasingly common in slow-moving rivers, brackish lagoons, estuaries, and bays, as well as nearshore ocean waters where nutrients and wastes wash down our rivers.

Eutrophication in marine ecosystems occurs in nearshore waters and partially enclosed bays or estuaries. Some areas such as the Gulf of Mexico, the Caspian Sea, the Baltic, and Bohai Bay in the Yellow Sea tend to be in especially critical condition. During the tourist season, the coastal population of the Mediterranean, for example, swells to 200 million people. Eighty-five percent of the effluents from large cities go untreated into the sea. Beach pollution, fish kills, and contaminated shellfish result. Extensive "dead zones" often form where rivers dump nutrients into estuaries and shallow seas. The second largest in the world occurs during summer months in the Gulf of Mexico at the mouth of the Mississippi River. (Exploring Science, p. 404). Studies indicate that as human populations, cities, and agriculture expand, these hypoxic zones will be come increasingly common.

It appears that fish and other marine species die in these hypoxic zones not only because oxygen is depleted, but also because of high concentrations of harmful organisms including toxic algae, pathogenic fungi, parasitic protists, and other predators. One of the most notorious and controversial of these is *Pfiesteria piscicida,* a dinoflagellate (a single-celled organism that swims with two whip-like flagella). Researchers at North Carolina State University reported in 1997 that they found this new species in Pamlico Sound, where thousands of fish died during a toxic tide. Initial reports described this organism as having a complex life cycle including dormant cysts, free-floating amoebae, and swimming flagellates. *Pfiesteria* cells were reported to secrete nerve-damaging toxins, allowing amoeba forms to attack wounded fish. *Pfiesteria* toxins have also been blamed for human health problems including nerve damage. Other studies, however, have raised doubts over the complexity of *Pfiesteria's* life cycle and its role in both fish mortality and human illnesses. In 2007, however, researchers succeeded in identifying the *Pfiesteria* toxin. It's an organic molecule containing a copper atom linked to two thiol (sulfur-containing) ligands. This toxin creates free radicals that destroy cells and tissues. As the compound generates free radicals, it decomposes. The fact that it remains active for only a few days is part of the reason it has been so elusive.

## Inorganic pollutants include metals, salts, acids, and bases

Some toxic inorganic chemicals are released from rocks by weathering, are carried by runoff into lakes or rivers, or percolate into groundwater aquifers. This pattern is part of natural mineral cycles (chapter 3). Humans often accelerate the transfer rates in these cycles thousands of times above natural background levels through the mining, processing, using, and discarding of minerals.

In many areas, toxic, inorganic chemicals introduced into water as a result of human activities have become the most serious form of water pollution. Among the chemicals of greatest concern are heavy metals, such as mercury, lead, tin, and cadmium. Supertoxic elements, such as selenium and arsenic, also have reached hazardous levels in some waters. Other inorganic materials, such as acids, salts, nitrates, and chlorine, that normally are not toxic at low concentrations may become concentrated enough to lower water quality or adversely affect biological communities.

### *Metals*

Many metals, such as mercury, lead, cadmium, tin, and nickel, are highly toxic in minute concentrations. Because metals are highly persistent, they can accumulate in food webs and have a cumulative effect in top predators—including humans.

## *Exploring* SCIENCE

### Studying the Dead Zone

In the 1980s shrimp boat crews noticed that certain locations off the Gulf Coast of Louisiana were emptied of all aquatic life. Since the region supports shrimp, fish, and oyster fisheries worth \$250 to \$450 million per year, these "dead zones" were important to the economy as well as to the Gulf's ecological systems. In 1985, Nancy Rabelais, a scientist working with Louisiana Universities Marine Consortium, began mapping areas of low oxygen concentrations in the Gulf waters. Her results, published in 1991, showed that vast areas, just above the floor of the Gulf, had oxygen concentration less than 2 parts per million (ppm), a level that eliminated all animal life except primitive worms. Healthy aquatic systems usually have about 10 ppm dissolved oxygen. What caused this hypoxic (oxygen-starved) area to develop?

Rabelais and her team tracked the phenomenon for several years, and it became clear that the dead zone was growing larger over time, that poor shrimp harvests coincided with years when the zone was large, and that the size of the dead zone, which ranges from 5,000 to 20,000 km$^2$ (about the size of New Jersey), depended on rainfall and runoff rates from the Mississippi River. Excessive nutrients, mainly nitrogen, from farms and cities far upstream on the Mississippi River, were the suspected culprit.

How did Rabelais and her team know that nutrients were the problem? They noticed that each year, 7–10 days after large spring rains in the agricultural parts of the upper Mississippi watershed, oxygen concentrations in the Gulf drop from 5 ppm to below 2 ppm. These rains are known to wash soil, organic debris, and last year's nitrogen-rich fertilizers from farm fields. The scientists also knew that saltwater ecosystems normally have little available nitrogen, a key nutrient for algae and plant growth. Pulses of agricultural runoff were followed by a profuse growth of algae and phytoplankton (tiny floating plants). Such a burst of biological activity produces an excess of dead plant cells and fecal matter that drifts to the seafloor. Shrimp, clams, oysters, and other filter feeders normally consume this debris, but they can't keep up with the sudden flood of material. Instead, decomposing bacteria in the sediment break down the debris, and they consume most of the available dissolved oxygen as well. Putrefying sediments also produce hydrogen sulfide, which further poisons the water near the seafloor.

In well-mixed water bodies, as in the open ocean, oxygen from upper layers of water is frequently mixed into lower water layers. Warm, protected water bodies are often stratified, however, as abundant sunlight keeps the upper layers warmer, and less dense, than lower layers. Denser lower layers cannot mix with upper layers unless strong currents or winds stir the water.

Many enclosed coastal waters, including Chesapeake Bay, Long Island Sound, the Mediterranean Sea, and the Black Sea, tend to be stratified and suffer hypoxic conditions that destroy bottom and near-bottom communities. There are about 200 dead zones around the world, and the number has doubled each decade since dead zones were first observed in the 1970s. The Gulf of Mexico is second in size behind a 100,000 km$^2$ dead zone in the Baltic Sea.

Can dead zones recover? Yes. Water is a forgiving medium, and organisms use nitrogen quickly. In 1996 in the Black Sea region, farmers in collapsing communist economies cut their nitrogen applications by half out of economic necessity; the Black Sea dead zone disappeared, while farmers saw no drop in their crop yields. In the Mississippi watershed, farmers can afford abundant fertilizer, and they fear they can't afford to risk underfertilizing. Because of the great geographic distance between the farm states and the Gulf, Midwestern states have been slow to develop an interest in the dead zone. At the same time, concentrated feedlot production of beef and pork is rapidly increasing, and feedlot runoff is the fastest growing, and least regulated, source of nutrient enrichment in rivers.

In 2001, federal, state, and tribal governments forged an agreement to cut nitrogen inputs by 30 percent and reduce the size of the dead zone to 5,000 km$^2$. This agreement represented astonishingly quick research and political response to scientific results, but it doesn't appear to be enough. Computer models suggest that it would take a 40–45 percent reduction in nitrogen to achieve the 5,000 km$^2$ goal.

Human activities have increased the flow of nitrogen reaching U.S. coastal waters by four to eight times since the 1950s. Phosphorus, another key nutrient, has tripled. This case study shows how water pollution can connect far-distant places, such as Midwestern farmers and Louisiana shrimpers.

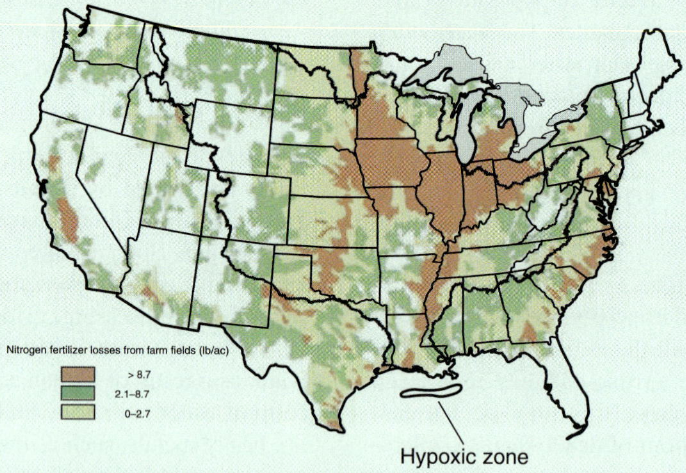


The Mississippi River drains 40 percent of the conterminous United States, including the most heavily farmed states. Nitrogen fertilizer produces a summer "dead zone" in the Gulf of Mexico.

Currently, the most widespread toxic metal contamination problem in North America is mercury released from coal-burning power plants. As chapter 16 mentions, an EPA survey of 2,500 fish from 260 lakes across the United States found at least low levels of mercury in every fish sampled. More than half the fish contained mercury levels unsafe for women of childbearing age, and three-quarters exceed the safe limit for young children.

Fifty states have issued warnings about eating freshwater or ocean fish; mercury contamination is by far the most common reason for these advisories. Top marine predators, such as shark, swordfish, marlin, king mackerel, and blue-fin tuna, should be avoided completely. You should check local advisories about the safety of fish caught in your local lakes, rivers, and coastal areas. If no advice is available, eat no more than one meal of such fish per week. Public health officials estimate that 600,000 American children now have mercury levels in their bodies high enough to cause mental and developmental problems, while one woman in six in the United States has blood-mercury concentrations that would endanger a fetus.

Mine drainage and leaching of mining wastes are serious sources of metal pollution in water. A survey of water quality in eastern Tennessee—where there has been a great deal of surface mining—found that 43 percent of all surface streams and lakes and more than half of all groundwater used for drinking supplies were contaminated by acids and metals from mine drainage. In some cases, metal levels were 200 times higher than what is considered safe for drinking water.

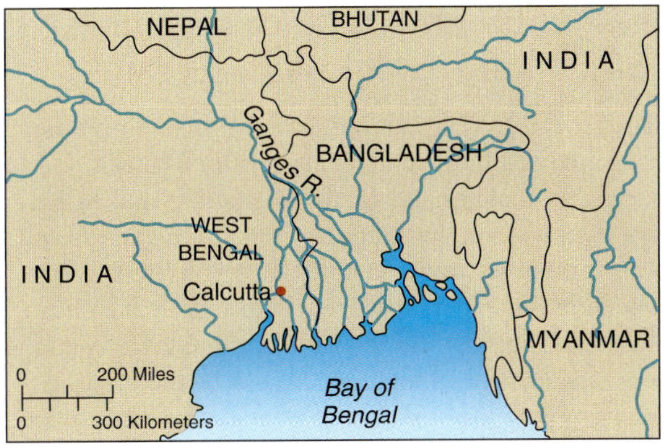


**FIGURE 18.9** West Bengal and adjoining areas of Bangladesh have hundreds of millions of people who may be exposed to dangerous arsenic levels in well water.

### *Nonmetallic Salts*

Some soils contain high concentrations of soluble salts, including toxic selenium and arsenic. You have probably heard of poison springs and seeps in the desert, where percolating groundwater brings these materials to the surface. Irrigation and drainage of desert soils can mobilize these materials on a larger scale and result in serious pollution problems, as in Kesterson Marsh in California, where selenium poisoning killed thousands of migratory birds in the 1980s.

Salts, such as sodium chloride (table salt), that are nontoxic at low concentrations also can be mobilized by irrigation and concentrated by evaporation, reaching levels that are toxic for many plants and animals. Globally, 20 percent of the world's irrigated farmland is estimated to be affected by salinization, and half that land has enough salt buildup to decrease yields significantly. In the northern United States, millions of tons of sodium chloride and calcium chloride are used every year to melt road ice. Leaching of these road salts into surface waters is having adverse effects on some aquatic ecosystems.

Perhaps the largest human population threatened by naturally occurring arsenic in groundwater is in West Bengal, India, and adjacent areas of Bangladesh (fig. 18.9). Arsenic occurs naturally in the sediments that make up the Ganges River delta. Rapid population growth, industrialization, and intensification of irrigated agriculture have consumed or polluted limited surface water supplies. In an effort to provide clean drinking water for local residents, thousands of deep tube wells were sunk in the 1960s throughout the area. Much of this humanitarian effort was financed by loans from the World Bank.

By the 1980s, health workers became aware of widespread signs of chronic arsenic poisoning among Bengali villagers. Symptoms include watery and inflamed eyes, gastrointestinal cramps, gradual loss of strength, scaly skin and skin tumors, anemia, confusion, and, eventually, death. Some villages have had wells for centuries without a problem; why is arsenic poisoning appearing now? One theory is that excessive withdrawals now lower the water table during the dry season, exposing arsenic-bearing minerals to oxidation, which converts normally insoluble salts to soluble oxides. When aquifers refill during the next monsoon season, dissolved arsenic can be pumped out. Health workers estimate that the total number of potential victims in India and Bangladesh may exceed 200 million people.

### *Acids and Bases*

Acids are released as by-products of industrial processes, such as leather tanning, metal smelting and plating, petroleum distillation, and organic chemical synthesis. Coal mining is an especially important source of acid water pollution. Sulfur compounds in coal react with oxygen and water to make sulfuric acid. Thousands of kilometers of streams in the United States have been acidified by acid mine drainage, some so severely that they are essentially lifeless.

Coal and oil combustion also leads to formation of atmospheric sulfuric and nitric acids (chapter 16), which are disseminated by long-range transport processes and deposited via precipitation (acidic rain, snow, fog, or dry deposition) in surface waters. Where soils are rich in such alkaline material as limestone, these atmospheric acids have little effect because they are neutralized. In high mountain areas or recently glaciated regions where crystalline bedrock is close to the surface and lakes are oligotrophic, however, there is little buffering capacity (ability to neutralize acids) and aquatic ecosystems can be severely disrupted. These effects were first recognized in the mountains of northern England and Scandinavia about 30 years ago.

Aquatic damage due to acid precipitation has been reported in about 200 lakes in the Adirondack Mountains of New York State and in several thousand lakes in eastern Quebec, Canada. Game fish, amphibians, and sensitive aquatic insects are generally the first to be killed by increased acid levels in the water. If acidification is severe enough, aquatic life is limited to a few resistant species of mosses and fungi. Increased acidity may result in leaching of toxic metals, especially aluminum, from soil and rocks, making water unfit for drinking or irrigation, as well.

## Organic pollutants include pesticides and other industrial substances

Thousands of different natural and synthetic organic chemicals are used in the chemical industry to make pesticides, plastics, pharmaceuticals, pigments, and other products that we use in everyday life. Many of these chemicals are highly toxic (chapter 8). Exposure to very low concentrations (perhaps even parts per quadrillion in the case of dioxins) can cause birth defects, genetic disorders, and cancer. Some can persist in the environment because they are resistant to degradation and toxic to organisms that ingest them. Contamination of surface waters and groundwater by these chemicals is a serious threat to human health.

The two most important sources of toxic organic chemicals in water are improper disposal of industrial and household wastes and runoff of pesticides from farm fields, forests, roadsides, golf courses, and other places where they are used in large quantities. The U.S. EPA estimates that about 500,000 metric tons of pesticides are used in the United States each year. Much of this material washes into the nearest waterway, where it passes through ecosystems and may accumulate in high levels in nontarget organisms. The bioaccumulation of DDT in aquatic ecosystems was one of the first of these pathways to be understood. Dioxins, and other chlorinated hydrocarbons (hydrocarbon molecules that contain chlorine atoms) have been shown to accumulate to dangerous levels in the fat of salmon, fish-eating birds, and humans and to cause health problems similar to those resulting from toxic metal compounds (fig. 18.10).

As chapter 8 reports, atrazine, the most widely used herbicide in America, has been shown to disrupt normal sexual development in frogs at concentrations as low as 0.1 ppb. This level is found regularly wherever farming occurs. Could this be a problem for us as well?

**FIGURE 18.10** The deformed beak of this young robin is thought to be due to dioxins, DDT, and other toxins in its mother's diet.

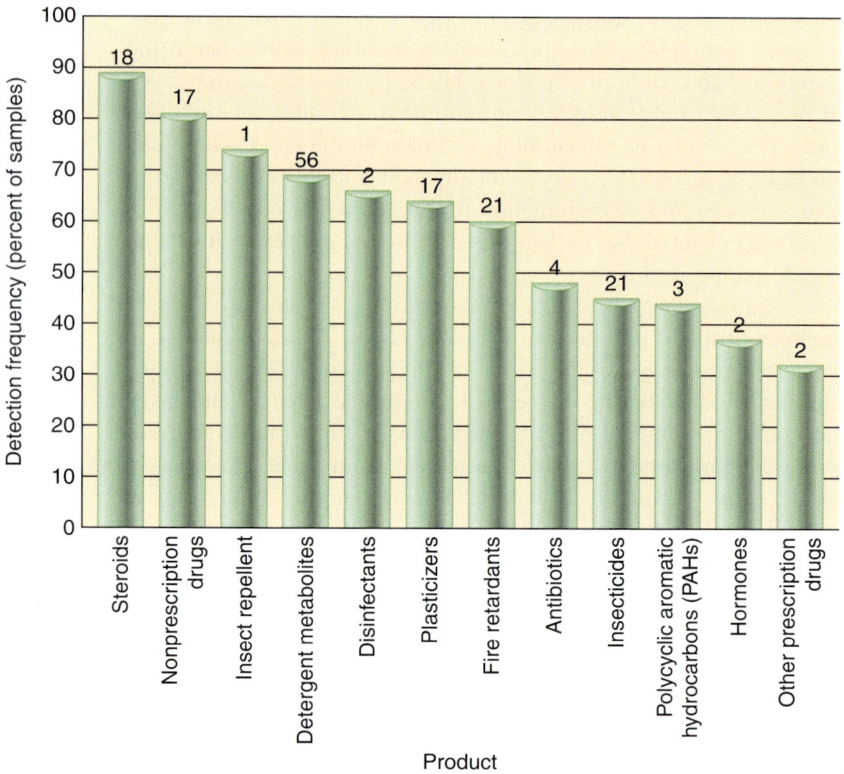


**FIGURE 18.11** Detection frequency of organic, wastewater contaminants in a recent USGS survey. Maximum concentrations in water samples are shown above the bars in micrograms per liter. Dominant substances included DEET insect repellent, caffeine, and triclosan, which comes from antibacterial soaps.

Hundreds of millions of tons of hazardous organic wastes are thought to be stored in dumps, landfills, lagoons, and underground tanks in the United States (chapter 21). Many, perhaps most, of these sites have leaked toxic chemicals into surface waters or groundwater or both. The EPA estimates that about 26,000 hazardous waste sites will require cleanup because they pose an imminent threat to public health, mostly through water pollution.

Countless additional organic compounds enter our water unmonitored. In 2002, the USGS released the first-ever study of pharmaceuticals and hormones in streams. Scientists sampled 130 streams, looking for 95 contaminants, including antibiotics, natural and synthetic hormones, detergents, plasticizers, insecticides, and fire retardants (fig. 18.11). All these substances were found, usually in low concentrations. One stream had 38 of the compounds tested. Drinking water standards exist for only 14 of the 95 substances. A similar study found the same substances in groundwater, which is much harder to clean than surface waters. What are the effects of these widely used chemicals, on our environment or on people consuming the water? Nobody knows. This study is a first step toward filling huge gaps in our knowledge about their distribution, though.

**FIGURE 18.12** A plume of sediment and industrial waste flows from this drainage canal into Lake Erie.

## Sediment also degrades water quality

Rivers have always carried sediment to the oceans, but erosion rates in many areas have been greatly accelerated by human activities. Some rivers carry astounding loads of sediment. Erosion and runoff from croplands contribute about 25 billion metric tons of soil, sediment, and suspended solids to world surface waters each year. Forests, grazing lands, urban construction sites, and other sources of erosion and runoff add at least 50 billion additional tons. This sediment fills lakes and reservoirs, obstructs shipping channels, clogs hydroelectric turbines, and makes purification of drinking water more costly. Sediments smother gravel beds in which insects take refuge and fish lay their eggs. Sunlight is blocked so that plants cannot carry out photosynthesis, and oxygen levels decline. Murky, cloudy water also is less attractive for swimming, boating, fishing, and other recreational uses (fig. 18.12).

Sediment also can be beneficial. Mud carried by rivers nourishes floodplain farm fields. Sediment deposited in the ocean at river mouths creates valuable deltas and islands. The Ganges River, for instance, builds up islands in the Bay of Bengal that are eagerly colonized by land-hungry people of Bangladesh. Louisiana's coastal wetlands require constant additions of sediment from the muddy Mississippi to counteract coastal erosion. These wetlands are now disappearing at a disastrous rate: Levees now channel the river and its load straight out to the Gulf of Mexico, where sediments are dumped beyond the continental shelf.

## Thermal pollution is dangerous for organisms

Raising or lowering water temperatures from normal levels can adversely affect water quality and aquatic life. Water temperatures are usually much more stable than air temperatures, so aquatic organisms tend to be poorly adapted to rapid temperature changes. Lowering the temperature of tropical oceans by even one degree can be lethal to some corals and other reef species. Raising water temperatures can have similar devastating effects on sensitive organisms. Oxygen solubility in water decreases as temperatures increase, so species requiring high oxygen levels are adversely affected by warming water.

Humans cause thermal pollution by altering vegetation cover and runoff patterns, as well as by discharging heated water directly into rivers and lakes.

The cheapest way to remove heat from an industrial facility is to draw cool water from an ocean, river, lake, or aquifer, run it through a heat-exchanger to extract excess heat, and then dump the heated water back into the original source. A **thermal plume** of heated water is often discharged into rivers and lakes, where raised temperatures can disrupt many processes in natural ecosystems and drive out sensitive organisms. Nearly half the water we withdraw is used for industrial cooling. Electric power plants, metal smelters, petroleum refineries, paper mills, food-processing factories, and chemical manufacturing plants all use and release large amounts of cooling water.

To minimize thermal pollution, power plants frequently are required to construct artificial cooling ponds or cooling towers in which heat is released into the atmosphere and water is cooled before being released into natural water bodies.

Some species find thermal pollution attractive. Warm water plumes from power plants often attract fish, birds, and marine mammals that find food and refuge there, especially in cold weather. This artificial environment can be a fatal trap, however. Organisms dependent on the warmth may die if they leave the plume or if the flow of warm water is interrupted by a plant shutdown. Endangered manatees in Florida, for example, are attracted to the abundant food and warm water in power plant thermal plumes and are enticed into spending the winter much farther north than they normally would. On several occasions, a midwinter power plant breakdown has exposed a dozen or more of these rare animals to a sudden thermal shock that they could not survive.

# 18.3 Water Quality Today

Surface-water pollution is often both highly visible and one of the most common threats to environmental quality. In more developed countries, reducing water pollution has been a high priority over the past few decades. Billions of dollars have been spent on control programs and considerable progress has been made. Still much remains to be done. In developed countries, poor water quality often remains a serious problem. In this section, we will look at progress as well as continuing obstacles in this important area.

## The Clean Water Act protects our water

Like most developed countries, the United States and Canada have made encouraging progress in protecting and restoring water quality in rivers and lakes over the past 40 years. In 1948, only about one-third of Americans were served by municipal sewage systems, and most of those systems discharged sewage without any treatment or with only primary treatment (the bigger lumps of waste are removed). Most people depended on cesspools and septic systems to dispose of domestic wastes.

The 1972 Clean Water Act established a National Pollution Discharge Elimination System (NPDES), which requires an easily revoked permit for any industry, municipality or other entity dumping wastes in surface waters. The permit requires disclosure of what is being dumped and gives regulators valuable data and evidence for litigation. As a consequence, only about 10 percent of our water pollution now comes from industrial or municipal point sources. One of the biggest improvements has been in sewage treatment.

Since the Clean Water Act was passed in 1972, the United States has spent more than $180 billion in public funds and perhaps ten times as much in private investments on water pollution control. Most of that effort has been aimed at point sources, especially to build or upgrade thousands of municipal sewage treatment plants. As a result, nearly everyone in urban areas is now served by municipal sewage systems and no major city discharges raw sewage into a river or lake except as overflow during heavy rainstorms.

This campaign has led to significant improvements in surface-water quality in many places. Fish and aquatic insects have returned to waters that formerly were depleted of life-giving oxygen. Swimming and other water-contact sports are again permitted in rivers, lakes, and at ocean beaches that once were closed by health officials.

The Clean Water Act goal of making all U.S. surface waters "fishable and swimmable" has not been fully met, but in 1999 the EPA reported that 91.4 percent of all monitored river miles and 87.5 percent of all assessed lake acres are suitable for their designated uses. This sounds good, but you have to remember that not all water bodies are monitored. Furthermore, the designated goal for some rivers and lakes is merely to be "boatable." Water quality doesn't have to be very high to be able to put a boat in it. Even in "fishable" rivers and lakes, there isn't a guarantee that you can catch anything other than rough fish like carp or bullheads, nor can you be sure that what you catch is safe to eat. Even with billions of dollars of investment in sewage treatment plants, elimination of much of the industrial dumping and other gross sources of pollutants, and a general improvement in water quality, the EPA reports that 21,000 water bodies still do not meet their designated uses. According to the EPA, an overwhelming majority of the American people—almost 218 million—live within 16 km (10 mi) of an impaired water body.

In 1998, a new regulatory approach to water quality assurance was instituted by the EPA. Rather than issue standards on a river by river approach or factory by factory permit discharge, the focus is being changed to watershed-level monitoring and protection. Some 4,000 watersheds are monitored for water quality (fig. 18.13). You can find information about your watershed at www.epa.gov/owow/tmdl/. The intention of this program is to give the public more and better information about the health of their watersheds. In addition, states will have greater flexibility as they identify impaired water bodies and set priorities, and new tools will be used to achieve goals. States are required to identify waters not meeting water quality goals and to develop **total maximum daily loads (TMDL)** for each pollutant and each listed water body. A TMDL is the amount of a particular pollutant that a water body can receive from both point and nonpoint sources. It considers seasonal variation and includes a margin of safety.

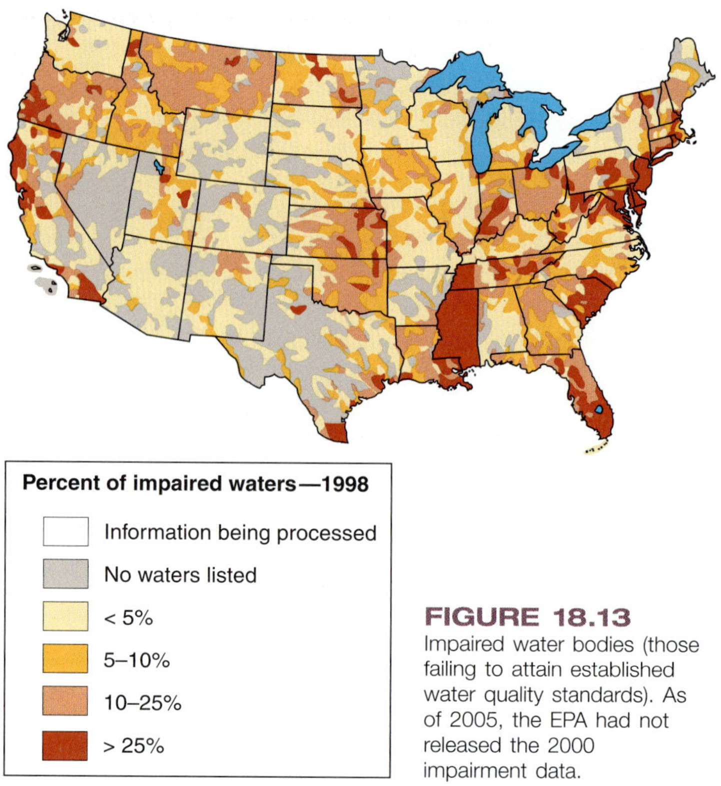


**FIGURE 18.13** Impaired water bodies (those failing to attain established water quality standards). As of 2005, the EPA had not released the 2000 impairment data.

By 1999, all 56 states and territories had submitted TMDL lists, and the EPA had approved most of them. Of the 3.5 million mi (5.6 million km) of rivers monitored, only 300,000 mi (480,000 km) fail to meet their clean water goals. Similarly, of 40 million lake acres (99 million ha), only 12.5 percent (in about 20,000 lakes) failed to meet their goal. To give states more flexibility in planning, the EPA has proposed new rules that include allowances for reasonably foreseeable increases in pollutant loadings to encourage "Smart Growth." In the future, TMDLs also will include load allocations from all nonpoint sources, including air deposition and natural background levels.

An encouraging example of improved water quality is seen in Lake Erie. Although widely regarded as "dead" in the 1960s, the lake today is promoted as the "walleye capital of the world." Bacteria counts and algae blooms have decreased more than 90 percent since 1962. Water that once was murky brown is now clear. Interestingly, part of the improved water quality is due to immense numbers of invasive zebra mussels, which filter the lake water very efficiently. Swimming is now officially safe along 96 percent of the lake's shoreline. Nearly 40,000 nesting pairs of double-crested cormorants nest in the Great Lakes region, up from only about 100 in the 1970s. Anglers now complain that the cormorants eat too many fish. In 1998 wildlife agents found 800 cormorants shot to death in a rookery on Galloo Island at the east end of Lake Ontario.

Canada's 1970 Water Act has produced comparable results. Seventy percent of all Canadians in towns over 1,000 population are now served by some form of municipal sewage treatment. In Ontario, the vast majority of those systems include tertiary treatment. After ten years of controls, phosphorus levels in the Bay of Quinte in the northeast corner of Lake Ontario have dropped nearly by half, and algal blooms that once turned waters green are less frequent and less intense than they once were. Elimination of mercury discharges from a pulp and paper mill on the Wabigoon-English River system in western Ontario has resulted in a dramatic decrease in mercury contamination. Twenty years ago this mercury contamination was causing developmental retardation in local residents. Extensive flooding associated with hydropower projects has raised mercury levels in fish to dangerous levels elsewhere, however.

## Water quality problems remain

The greatest impediments to achieving national goals in water quality in both the United States and Canada are sediment, nutrients, and pathogens, especially from nonpoint discharges of pollutants. These sources are harder to identify and to reduce or treat than are specific point sources. About three-fourths of the water pollution in the United States comes from soil erosion, fallout of air pollutants, and surface runoff from urban areas, farm fields, and feedlots. In the United States, as much as 25 percent of the 46,800,000 metric tons (52 million tons) of fertilizer spread on farmland each year is carried away by runoff (fig. 18.14).

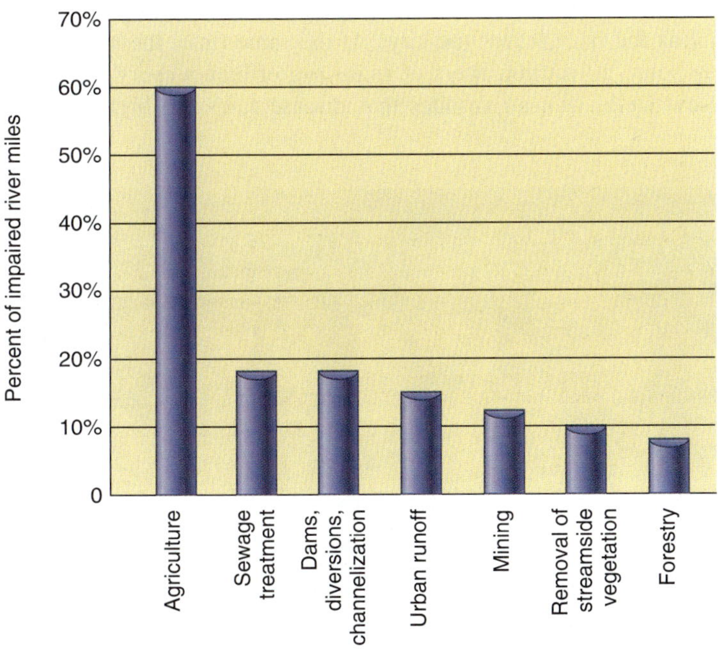


**FIGURE 18.14** Percentage of impaired river miles in the United States by source of damage. Totals add up to more than 100 percent because one river can be affected by many sources.
**Source:** USDA and Natural Resources Conservation Service.

Cattle in feedlots produce some 129,600,000 metric tons (144 million tons) of manure each year, and the runoff from these sites is rich in viruses, bacteria, nitrates, phosphates, and other contaminants. A single cow produces about 30 kg (66 lb) of manure per day, or about as much as that produced by ten people. Some feedlots have 100,000 animals with little provision for capturing or treating runoff water. Imagine drawing your drinking water downstream from such a facility. Pets also can be a problem. It is estimated that the wastes from about a half million dogs in New York City are disposed of primarily through storm sewers, and therefore do not go through sewage treatment.

Loading of both nitrates and phosphates in surface water have decreased from point sources but have increased about fourfold since 1972 from nonpoint sources. Fossil fuel combustion has become a major source of nitrates, sulfates, arsenic, cadmium, mercury, and other toxic pollutants that find their way into water. Carried to remote areas by atmospheric transport, these combustion products now are found nearly everywhere in the world. Toxic organic compounds, such as DDT, PCBs, and dioxins, also are transported long distances by wind currents.

## Developing countries often have serious water pollution

Japan, Australia, and most of western Europe also have improved surface-water quality in recent years. Sewage treatment in the wealthier countries of Europe generally equals or surpasses that in the United States. Sweden, for instance, serves 98 percent of its population with at least secondary sewage treatment (compared with 70 percent in the United States), and the other 2 percent have primary treatment. Poorer countries have much less to spend on sanitation. Spain serves only 18 percent of its population with even primary sewage treatment. In Ireland, it is only 11 percent, and in Greece, less than 1 percent of the people have even primary treatment. Most of the sewage, both domestic and industrial, is dumped directly into the ocean.

The fall of the "iron curtain" in 1989 revealed appalling environmental conditions in much of the former Soviet Union and its satellite states in eastern and central Europe. The countries closest geographically and socially to western Europe, the Czech Republic, Hungary, East Germany, and Poland, have made massive investments and encouraging progress toward cleaning up environmental problems. Parts of Russia itself, however, along with former socialist states in the Balkans and Central Asia, remain some of the most polluted places on earth. In Russia, for example, only about half the tap water is fit to drink. In cities like St. Petersburg, even boiling and filtering isn't enough to make municipal water safe. As we saw in chapter 17, at least 200 million Chinese live in areas without sufficient fresh water. Sadly, pollution makes much of the limited water unusable (fig. 18.15). It's estimated that 70 percent of China's surface water is unsafe for human consumption, and that the water in half the country's major rivers is so contaminated that it's unsuited for any use, even agriculture. The situation in Shanxi Province exemplifies the problems of water pollution in China. An industrial powerhouse,

**FIGURE 18.15** Half the water in China's major rivers is too polluted to be suitable for any human use. Although the government has spent billions of yuan in recent years, dumping of industrial and domestic waste continues at dangerous levels.

in the north-central part of the country, Shanxi has about one-third of China's known coal resources and currently produces about two-thirds of the country's energy. In addition to power plants, major industries include steel mills, tar factories, and chemical plants.

Economic growth has been pursued in recent decades at the expense of environmental quality. According to the Chinese Environmental Protection Agency, the country's ten worst polluted cities are all in Shanxi. Factories have been allowed to exceed pollution discharges with impunity. For example, 3 million tons of wastewater is produced every day in the province with two-thirds of it discharged directly into local rivers without any treatment. Locals complain that the rivers, which once were clean and fresh, now run black with industrial waste. Among the 26 rivers in the province, 80 percent were rated Grade V (unfit for any human use) or higher in 2006. More than half the wells in Shanxi are reported to have dangerously high arsenic levels. Many of the 85,000 reported public protests in China in 2006 involved complaints about air and water pollution.

There are also some encouraging pollution-control stories. In 1997, Minamata Bay in Japan, long synonymous with mercury poisoning, was declared officially clean again. Another important success is found in Europe, where one of its most important rivers has been cleaned up significantly through international cooperation. The Rhine, which starts in the rugged Swiss Alps and winds 1,320 km through five countries before emptying through a Dutch delta into the North Sea, has long been a major commercial artery into the heart of Europe. More than 50 million people live in its catchment basin and nearly 20 million get their drinking water from the river or its tributaries. By the 1970s, the Rhine had become so polluted that dozens of fish species disappeared and swimming was discouraged along most of its length.

Efforts to clean up this historic and economically important waterway began in the 1950s, but a disastrous fire at a chemical warehouse near Basel, Switzerland, in 1986 provided the impetus for major changes. Through a long and sometimes painful series of international conventions and compromises, land-use practices, waste disposal, urban runoff, and industrial dumping have been changed and water quality has significantly improved. Oxygen concentrations have gone up fivefold since 1970 (from less than 2 mg/l to nearly 10 mg/l or about 90 percent of saturation) in long stretches of the river. Chemical oxygen demand has fallen fivefold during this same period, and organochlorine levels have decreased as much as tenfold. Many species of fish and aquatic invertebrates have returned to the river. In 1992, for the first time in decades, mature salmon were caught in the Rhine.

The less-developed countries of South America, Africa, and Asia have even worse water quality than do the poorer countries of Europe. Sewage treatment is usually either totally lacking or woefully inadequate. In some urban areas, 95 percent of all sewage is discharged untreated into rivers, lakes, or the ocean. Low technological capabilities and little money for pollution control are made even worse by burgeoning populations, rapid urbanization, and the shift of much heavy industry (especially the dirtier ones) from developed countries where pollution laws are strict to less-developed countries where regulations are more lenient.

Appalling environmental conditions often result from these combined factors (fig. 18.16). Two-thirds of India's surface waters are contaminated sufficiently to be considered dangerous to human health. The Yamuna River in New Delhi has 7,500 coliform bacteria per 100 ml (37 times the level considered safe for swimming in the United States) *before* entering the city. The coliform count increases to an incredible 24 *million* cells per 100 ml as the river leaves the city! At the same time, the river picks up some 20 million liters of industrial effluents every day from New Delhi. It's no wonder that disease rates are high and life

**FIGURE 18.16** Ditches in this Haitian slum serve as open sewers into which all manner of refuse and waste are dumped. The health risks of living under these conditions are severe.

expectancy is low in this area. Only 1 percent of India's towns and cities have any sewage treatment, and only eight cities have anything beyond primary treatment.

In Malaysia, 42 of 50 major rivers are reported to be "ecological disasters." Residues from palm oil and rubber manufacturing, along with heavy erosion from logging of tropical rainforests, have destroyed all higher forms of life in most of these rivers. In the Philippines, domestic sewage makes up 60 to 70 percent of the total volume of Manila's Pasig River. Thousands of people use the river not only for bathing and washing clothes but also as their source of drinking and cooking water. China treats only 2 percent of its sewage. Of 78 monitored rivers in China, 54 are reported to be seriously polluted. Of 44 major cities in China, 41 use "contaminated" water supplies, and few do more than rudimentary treatment before it is delivered to the public.

## Groundwater is hard to monitor and clean

About half the people in the United States, including 95 percent of those in rural areas, depend on underground aquifers for their drinking water. This vital resource is threatened in many areas by overuse and pollution and by a wide variety of industrial, agricultural, and domestic contaminants. For decades it was widely assumed that groundwater was impervious to pollution because soil would bind chemicals and cleanse water as it percolated through. Springwater or artesian well water was considered to be the definitive standard of water purity, but that is no longer true in many areas.

One of the serious sources of groundwater pollution throughout the United States is MTBE (methyl tertiary butyl ether), a suspected carcinogen. MTBE is a gasoline additive that has been used since the 1970s to reduce the amount of carbon monoxide and unburned hydrocarbons in vehicle exhaust. By the time the health dangers of MTBE were confirmed in the late 1990s, aquifers across the country had been contaminated—mainly from leaking underground storage tanks at gas stations. About 250,000 of these tanks are leaking MTBE into groundwater nationwide. In one U.S. Geological Survey (USGS) study, 27 percent of shallow urban wells tested contained MTBE. The additive is being phased out, but plumes of tainted water will continue to move through aquifers for decades to come. (Surface waters have also been contaminated, especially by two-stroke engines, such as those on personal watercraft.)

Treating MTBE-laced aquifers is expensive but not impossible. Douglas MacKay of the University of Waterloo in Ontario suggests that if oxygen could be pumped into aquifers, then naturally occurring bacteria could metabolize (digest) the compound. It could take decades or even centuries for natural bacteria to eliminate MTBE from a water supply, however. Water can also be pumped out of aquifers, reducing the flow and spread of contamination. Thus far, little funding has been invested in finding cost-effective remedies, however.

The U.S. EPA estimates that every day some 4.5 trillion l (1.2 trillion gal) of contaminated water seep into the ground in the United States from septic tanks, cesspools, municipal and industrial landfills and waste disposal sites, surface impoundments, agricultural fields, forests, and wells (fig. 18.17). The most toxic of these are probably waste disposal sites. Agricultural chemicals and wastes are responsible for the largest total volume of pollutants and area affected. Because deep underground aquifers often have residence times of thousands of years, many contaminants are extremely stable once underground. It is possible,

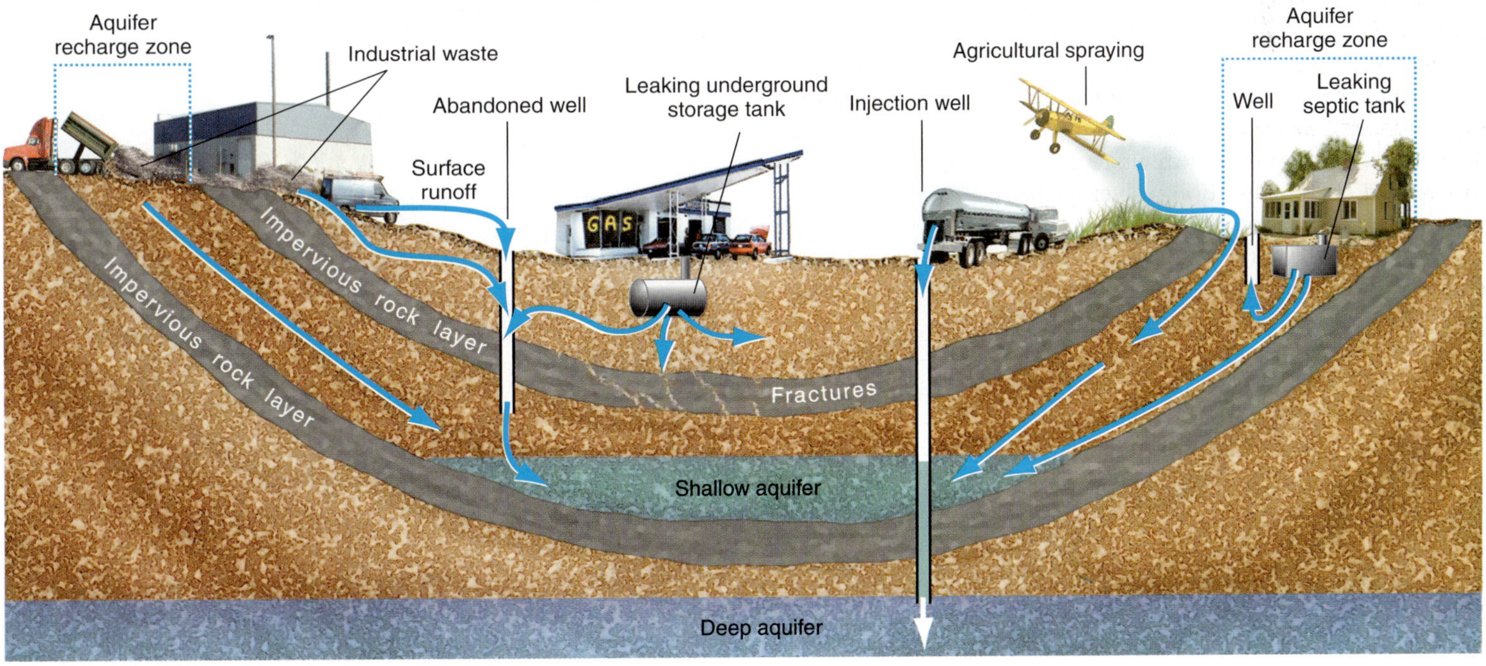


**FIGURE 18.17** Sources of groundwater pollution. Septic systems, landfills, and industrial activities on aquifer recharge zones leach contaminants into aquifers. Wells provide a direct route for injection of pollutants into aquifers.

but expensive, to pump water out of aquifers, clean it, and then pump it back.

In farm country, especially in the Midwest's corn belt, fertilizers and pesticides commonly contaminate aquifers and wells. Herbicides such as atrazine and alachlor are widely used on corn and soybeans and show up in about half of all wells in Iowa, for example. Nitrates from fertilizers often exceed safety standards in rural drinking water. These high nitrate levels are dangerous to infants (nitrate combines with hemoglobin in the blood and results in "blue-baby" syndrome). They also are transformed into cancer-causing nitrosamines in the human gut. In Florida, 1,000 drinking water wells were shut down by state authorities because of excessive levels of toxic chemicals, mostly ethylene dibromide (EDB), a pesticide used to kill nematodes (roundworms) that damage plant roots.

Although most of the leaky, single-walled underground storage tanks once common at filling stations and factories have now been removed and replaced by more modern ones, a great deal of soil in American cities remains contaminated by previous careless storage and disposal of petroleum products. Considering that a single gallon (3.8 l) of gasoline can make a million gallons of water undrinkable, soil contamination remains a serious problem.

In addition to groundwater pollution problems, contaminated surface waters and inadequate treatment make drinking water unsafe in many areas. A 1996 survey concluded that nearly 20,000 public drinking water systems in the United States expose consumers to contaminants such as lead, pesticides, and pathogens at levels that violate EPA rules (fig. 18.18). A vast majority of these systems are small, serving fewer than 3,000 people, but altogether some 50 million people are sometimes at risk. Problems often occur because small systems can't afford modern purification and distribution equipment, regular testing, and trained operators to bring water quality up to acceptable standards.

Every year epidemiologists estimate that around 1.5 million Americans fall ill from infections caused by fecal contamination. In 1993, for instance, a pathogen called cryptosporidium got into the Milwaukee public water system, making 400,000 people sick and killing at least 100 people. The total costs of these diseases amount to billions of dollars per year. Preventive measures such as protecting water sources and aquifer recharge zones, providing basic treatment for all systems, installing modern technology and distribution networks, consolidating small systems, and strengthening the Clean Water Act and the Safe Drinking Water Act would cost far less. Unfortunately, in the present climate of budget-cutting and anti-regulation, these steps seem unlikely.

## There are few controls on ocean pollution

Coastal zones, especially bays, estuaries, shoals, and reefs near large cities or the mouths of major rivers, often are overwhelmed by human-caused contamination. Suffocating and sometimes poisonous blooms of algae regularly deplete ocean waters of oxygen and kill enormous numbers of fish and other marine life. High levels of toxic chemicals, heavy metals, disease-causing organisms, oil, sediment, and plastic refuse are adversely affecting some of the most attractive and productive ocean regions. The potential losses caused by this pollution amount to billions of dollars each year.

Discarded plastic flotsam and jetsam are lightweight and nonbiodegradable. They are carried thousands of miles on ocean currents and last for years (fig. 18.19). Even the most remote beaches of distant islands are likely to have bits of polystyrene foam containers or polyethylene packing material that were discarded half a world away. It has been estimated that some 6 million metric tons of plastic bottles, packaging material, and other litter are tossed from ships every year into the ocean where they ensnare

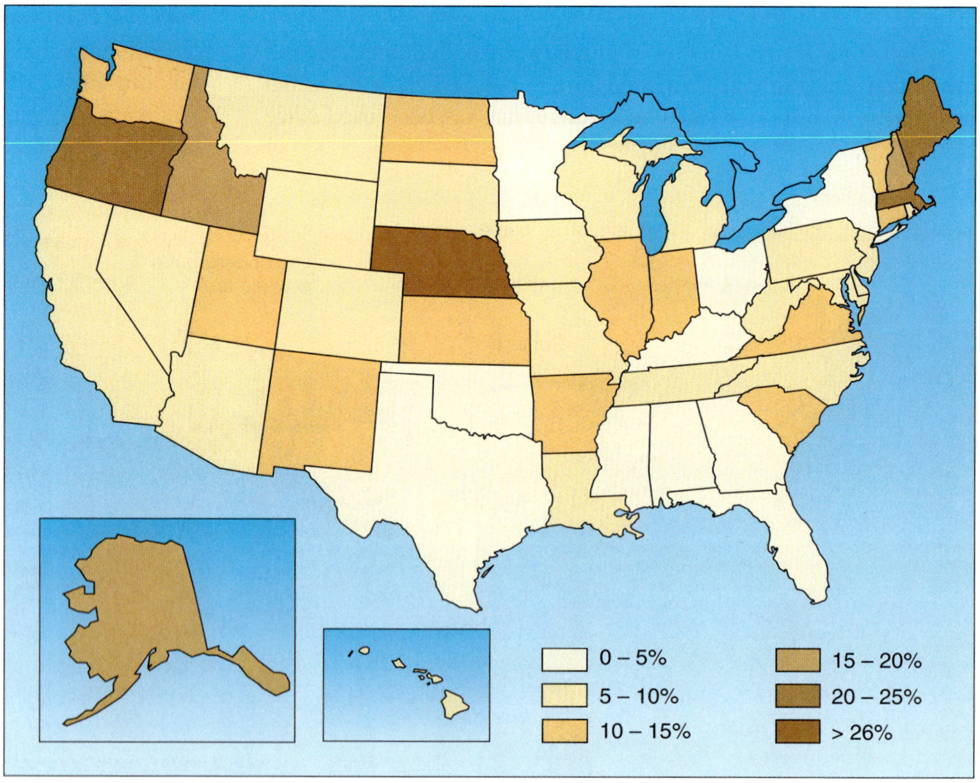


**FIGURE 18.18** Percentage of drinking-water systems within states with violations of EPA health standards in 2000.
**Source:** EPA Safe Drinking Water Information System, 2001.

**FIGURE 18.19** Beach pollution, including garbage, sewage, and contaminated runoff, is a growing problem associated with ocean pollution.

and choke seabirds, mammals (fig. 18.20), and even fish. Sixteen states now require that six-pack yokes be made of biodegradable or photodegradable plastic, limiting their longevity as potential killers. In one day, volunteers in Texas gathered more than 300 tons of plastic refuse from Gulf Coast beaches.

**FIGURE 18.20** A deadly necklace. Marine biologists estimate that castoff nets, plastic beverage yokes, and other packing residue kill hundreds of thousands of birds, mammals, and fish each year.

Few coastlines in the world remain uncontaminated by oil or oil products. Oceanographers estimate that somewhere between 3 million and 6 million metric tons of oil are discharged into the world's oceans each year from both land- and sea-based operations. About half of this amount is due to maritime transport. Most oil spills result not from catastrophic, headliner accidents, but from routine open-sea bilge pumping and tank cleaning. These procedures are illegal but are easily carried out once ships are beyond sight of land. Much of the rest comes from land-based municipal and industrial runoff or from atmospheric deposition of residues from refining and combustion of fuels.

The transport of huge quantities of oil creates opportunities for major oil spills through a combination of human and natural hazards. Military conflict in the Middle East and oil drilling in risky locations, such as the notoriously rough North Sea and the Arctic Ocean, make it likely that more oil spills will occur. Plans to drill for oil along the seismically active California and Alaska coasts have been controversial because of the damage that oil spills could cause to these biologically rich coastal ecosystems.

## 18.4 Water Pollution Control

Appropriate land-use practices and careful disposal of industrial, domestic, and agricultural wastes are essential for control of water pollution.

### Source reduction is often the cheapest way to reduce pollution

The cheapest and most effective way to reduce pollution is usually to avoid producing it or releasing it to the environment in the first place. Elimination of lead from gasoline has resulted in a widespread and significant decrease in the amount of lead in surface waters in the United States. Studies have shown that as much as 90 percent less road deicing salt can be used in many areas without significantly affecting the safety of winter roads. Careful handling of oil and petroleum products can greatly reduce the amount of water pollution caused by these materials. Although we still have problems with persistent chlorinated hydrocarbons spread widely in the environment, the banning of DDT and PCBs in the 1970s has resulted in significant reductions in levels in wildlife.

Modifying agricultural practices in headwater streams in the Chesapeake Bay watershed and the Catskill Mountains of New York have had positive and cost-effective impacts on downstream water quality (What Do You Think? pg. 414).

Industry can reduce pollution by recycling or reclaiming materials that otherwise might be discarded in the waste stream. Both of these approaches usually have economic as well as environmental benefits. It turns out that a variety of valuable metals can be recovered from industrial wastes and reused or sold for other purposes. The company benefits by having a product to sell, and the municipal sewage treatment plant benefits by not having to deal with highly toxic materials mixed in with millions of gallons of other types of wastes.

## What Do You Think?

### Watershed Protection in the Catskills

New York City has long been proud of its excellent municipal drinking water. Drawn from the rugged Catskill Mountains 100 km (60 mi) north of the city, stored in hard-rock reservoirs, and transported through underground tunnels, the city water is outstanding for so large an urban area. Yielding 450,000 $m^3$ (1.2 billion gal) per day, and serving more than 9 million people, this is the largest surface-water storage and supply complex in the world (see figure). As the metropolitan agglomeration has expanded, however, people have moved into the area around the Catskill Forest Preserve, and water quality is not as high as it was a century ago.

When the 1986 U.S. Safe Drinking Water Act mandated filtration of all public surface-water systems, the city was faced with building an $8 billion water treatment plant that would cost up to $500 million per year to operate. In 1989, however, the EPA ruled that the city could avoid filtration if it could meet certain minimum standards for microbial contaminants such as bacteria, viruses, and protozoan parasites. In an attempt to avoid the enormous cost of filtration, the city proposed land-use regulations for the five counties (Green, Ulster, Sullivan, Schoharie, and Delaware) in the Catskill/Delaware watershed from which it draws most of its water.

With a population of 50,000 people, the private land within the 520 $km^2$ (200 $mi^2$) watershed is mostly devoted to forestry, small farms, housing, and recreational activities. Among the changes the city called for was elimination of storm water runoff from barnyards, feedlots, or grazing areas into watersheds. In addition, farmers would be required to reduce erosion and surface runoff from crop fields and logging operations. Property owners objected strenuously to what they regarded as onerous burdens that would cost enough to put many of them out of business. They also bristled at having the huge megalopolis impose rules on them. It looked like a long and bitter battle would be fought through the courts and the state legislature.

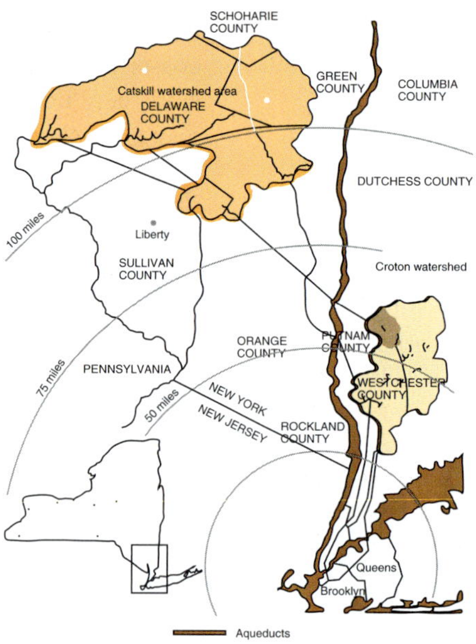

Investment in pollution prevention efforts in the Catskills has saved New York City billions of dollars in water filtration costs.

To avoid confrontation, a joint urban/rural task force was set up to see if a compromise could be reached, and to propose alternative solutions to protect both the water supply and the long-term viability of agriculture in the region. The task force agreed that agriculture is the "preferred land use" on private land, and that agriculture has "significant present and future environmental benefits." In addition, the task force proposed a voluntary, locally developed and administered program of "whole farm planning and best management approaches" very similar to ecosystem-based, adaptive management (chapter 9).

This grass-roots program, financed mainly by the city, but administered by local farmers themselves, attempts to educate landowners, and provides alternative marketing opportunities that help protect the watershed. Economic incentives are offered to encourage farmers and foresters to protect the water supply. Collecting feedlot and barnyard runoff in infiltration ponds together with solid conservation practices such as terracing, contour plowing, strip farming, leaving crop residue on fields, ground cover on waterways, and cultivation of perennial crops such as orchards and sugarbush have significantly improved watershed water quality. As of 1999, about 400 farmers—close to the 85 percent participation goal—have signed up for the program. The cost, so far, to the city has been about $50 million—or less than 1 percent of constructing a treatment plant.

Although landowners often object to any restrictions on development, many in the Catskills have found that land-use rules also protect rural lifestyles. Protection of the forests and waters has also helped the area retain in recreational economy and regional identity. Watershed management saved New York billions of dollars; it can also save traditional land uses and livelihoods.

What do you think? Are land-use restrictions a reasonable approach for saving on water treatment? How much should cities pay for watershed protection?

## Controlling nonpoint sources requires land management

Among the greatest remaining challenges in water pollution control are diffuse, nonpoint pollution sources. Unlike point sources, such as sewer outfalls or industrial discharge pipes, which represent both specific locations and relatively continuous emissions, nonpoint sources have many origins and numerous routes by which contaminants enter ground and surface waters. It is difficult to identify—let alone monitor and control—all these sources and routes. Some main causes of nonpoint pollution are:

- *Agriculture:* The EPA estimates that 60 percent of all impaired or threatened surface waters are affected by sediment from eroded fields and overgrazed pastures; fertilizers, pesticides, and nutrients from croplands; and animal wastes from feedlots.
- *Urban runoff:* Pollutants carried by runoff from streets, parking lots, and industrial sites contain salts, oily residues, rubber, metals, and many industrial toxins. Yards, golf courses, parklands, and urban gardens often are treated with far more fertilizers and pesticides per unit area than farmlands. Excess chemicals are carried by storm runoff into waterways.
- *Construction sites:* New buildings and land development projects such as highway construction affect relatively small areas but produce vast amounts of sediment, typically ten to twenty times as much per unit area as farming (fig. 18.21).
- *Land disposal:* When done carefully, land disposal of certain kinds of industrial waste, sewage sludge, and biodegradable garbage can be a good way to dispose of unwanted materials. Some poorly run land disposal sites, abandoned dumps, and leaking septic systems, however, contaminate local waters.

**FIGURE 18.21** Erosion on construction sites produces a great deal of sediment and is a major cause of nonpoint water pollution. Builders are generally required to install barriers to contain sediments, but these measures are often ineffectual.

Generally, soil conservation methods (chapter 9) also help protect water quality. Applying precisely determined amounts of fertilizer, irrigation water, and pesticides saves money and reduces contaminants entering the water. Preserving wetlands that act as natural processing facilities for removing sediment and contaminants helps protect surface and groundwaters.

In urban areas, reducing materials carried away by storm runoff is helpful. Citizens can be encouraged to recycle waste oil and to minimize use of fertilizers and pesticides. Regular street sweeping greatly reduces contaminants. Runoff can be diverted away from streams and lakes. Many cities are separating storm sewers and municipal sewage lines to avoid overflow during storms.

A good example of watershed management is seen in Chesapeake Bay, the United States' largest estuary. Once fabled for its abundant oysters, crabs, shad, striped bass, and other valuable fisheries, the Bay had deteriorated seriously by the early 1970s. Citizens' groups, local communities, state legislatures, and the federal government together established an innovative pollution-control program that made the bay the first estuary in America targeted for protection and restoration.

Among the principal objectives of this plan is reducing nutrient loading through land-use regulations in the six watershed states to control agricultural and urban runoff. Pollution prevention measures such as banning phosphate detergents also are important, as are upgrading wastewater treatment plants and improving compliance with discharge and filling permits. Efforts are underway to replant thousands of hectares of seagrasses and to restore wetlands that filter out pollutants. Since the 1980s, annual phosphorous discharges into the bay dropped 40 percent. Nitrogen levels, however, have remained constant or have even risen in some tributaries. Although progress has been made, the goals of reducing both nitrogen and phosphate levels by 40 percent and restoring viable fish and shellfish populations are still decades away. Still, as former EPA Administrator Carol Browner says, it demonstrates the "power of cooperation" in environmental protection.

## Human waste disposal occurs naturally when concentrations are low

As we have already seen, human and animal wastes usually create the most serious health-related water pollution problems. More than 500 types of disease-causing (pathogenic) bacteria, viruses, and parasites can travel from human or animal excrement through water. In this section, we will look at how to prevent the spread of these diseases.

### *Natural Processes*

In the poorer countries of the world, most rural people simply go out into the fields and forests to relieve themselves as they have always done. Where population densities are low, natural processes eliminate wastes quickly, making this a feasible method of sanitation. The high population densities of cities make this practice unworkable, however. Even major cities of many less-developed countries are often littered with human waste which has been left for rains to wash away or for pigs, dogs, flies, beetles, or other scavengers to consume. This is a major cause of disease, as well as being extremely unpleasant. Studies have shown that a significant portion of the airborne dust in Mexico City is actually dried, pulverized human feces.

Where intensive agriculture is practiced—especially in wet rice paddy farming in Asia—it has long been customary to collect "night soil" (human and animal waste) to be spread on the fields as fertilizer. This waste is a valuable source of plant nutrients, but it is also a source of disease-causing pathogens in the food supply. It is the main reason that travelers in less-developed countries must be careful to surface sterilize or cook any fruits and vegetables they eat. Collecting night soil for use on farm fields was common in Europe and America until about 100 years ago when the association between pathogens and disease was recognized.

Until about 50 years ago, most rural American families and quite a few residents of towns and small cities depended on a pit toilet or "outhouse" for waste disposal. Untreated wastes tended to seep into the ground, however, and pathogens sometimes contaminated drinking water supplies. The development of septic tanks and properly constructed drain fields represented a considerable improvement in public health (fig. 18.22). In a typical septic system, wastewater is first drained into a septic tank. Grease and oils rise to the top and solids settle to the bottom, where they are subject to bacterial decomposition. The clarified effluent from the septic tank is channeled out through a drainfield of small perforated pipes embedded in gravel just below the surface of the soil. The rate of aeration is high in this drainfield so that pathogens (most of which are anaerobic) will be killed, and soil microorganisms can metabolize any nutrients carried by the water. Excess water percolates up through the gravel and evaporates. Periodically, the solids in the septic tank are pumped out into a tank truck and taken to a treatment plant for disposal.

Where land is available and population densities are not too high, this can be an effective method of waste disposal. It is widely used in rural areas, but aging, leaky septic systems can be a huge cumulative problem. As chapter 13 points out, the Chesapeake Bay

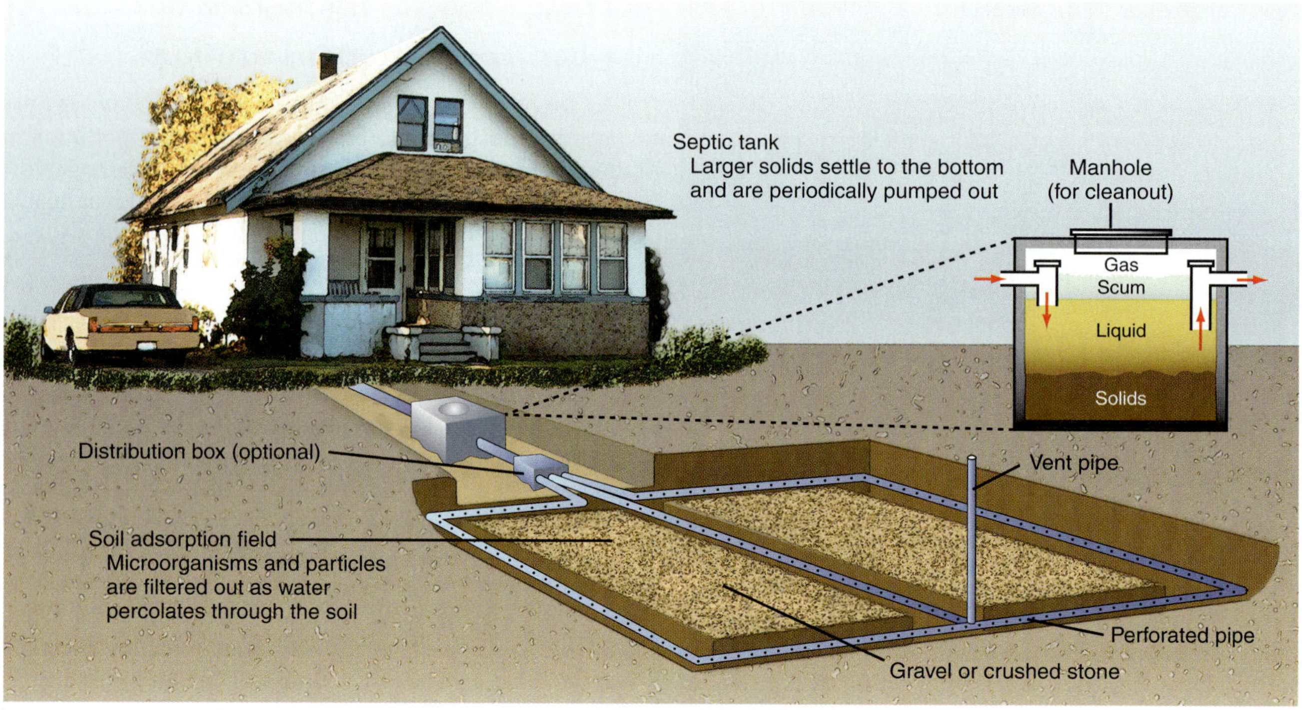


**FIGURE 18.22** A domestic septic tank and drain field system for sewage and wastewater disposal. To work properly, a septic tank must have healthy microorganisms, which digest toilet paper and feces. For this reason, antimicrobial cleaners and chlorine bleach should never be allowed down the drain.

watershed has 420,000 individual septic systems, which constitute a major source of nutrients. Maryland, alone, plans to spend $7.5 million annually to upgrade failing septic systems.

### *Municipal Sewage Treatment*

Over the past 100 years, sanitary engineers have developed ingenious and effective municipal wastewater treatment systems to protect human health, ecosystem stability, and water quality. This topic is an important part of pollution control, and is a central focus of every municipal government; therefore, let's look more closely at how a typical municipal sewage treatment facility works.

**Primary treatment** is the first step in municipal waste treatment. It physically separates large solids from the waste stream. As raw sewage enters the treatment plant, it passes through a metal grating that removes large debris (fig. 18.23*a*). A moving screen then filters out smaller items. Brief residence in a grit tank allows sand and gravel to settle. The waste stream then moves to the primary sedimentation tank where about half the suspended, organic solids settle to the bottom as sludge. Many pathogens remain in the effluent and it is not yet safe to discharge into waterways or onto the ground.

**Secondary treatment** consists of biological degradation of the dissolved organic compounds. The effluent from primary treatment flows into a trickling filter bed, an aeration tank, or a sewage lagoon. The trickling filter is simply a bed of stones or corrugated plastic sheets through which water drips from a system of perforated pipes or a sweeping overhead sprayer. Bacteria and other microorganisms in the bed catch organic material as it trickles past and aerobically decompose it.

Aeration tank digestion is also called the activated sludge process. Effluent from primary treatment is pumped into the tank and mixed with a bacteria-rich slurry (fig. 18.23*b*). Air pumped through the mixture encourages bacterial growth and decomposition of the organic material. Water flows from the top of the tank and sludge is removed from the bottom. Some of the sludge is used as an inoculum for incoming primary effluent. The remainder would be valuable fertilizer if it were not contaminated by metals, toxic chemicals, and pathogenic organisms. The toxic content of most sewer sludge necessitates disposal by burial in a landfill or incineration. Sludge disposal is a major cost in most municipal sewer budgets (fig. 18.24). In some communities this is accomplished by land farming, composting, or anaerobic digestion, but these methods don't inactivate metals and some other toxic materials.

Where space is available for sewage lagoons, the exposure to sunlight, algae, aquatic organisms, and air does the same job more slowly but with less energy costs. Effluent from secondary treatment processes is usually disinfected with chlorine, UV light, or ozone to kill harmful bacteria before it is released to a nearby waterway.

**Tertiary treatment** removes plant nutrients, especially nitrates and phosphates, from the secondary effluent. Although wastewater is usually free of pathogens and organic material after secondary treatment, it still contains high levels of inorganic nutrients, such as nitrates and phosphates. When discharged into

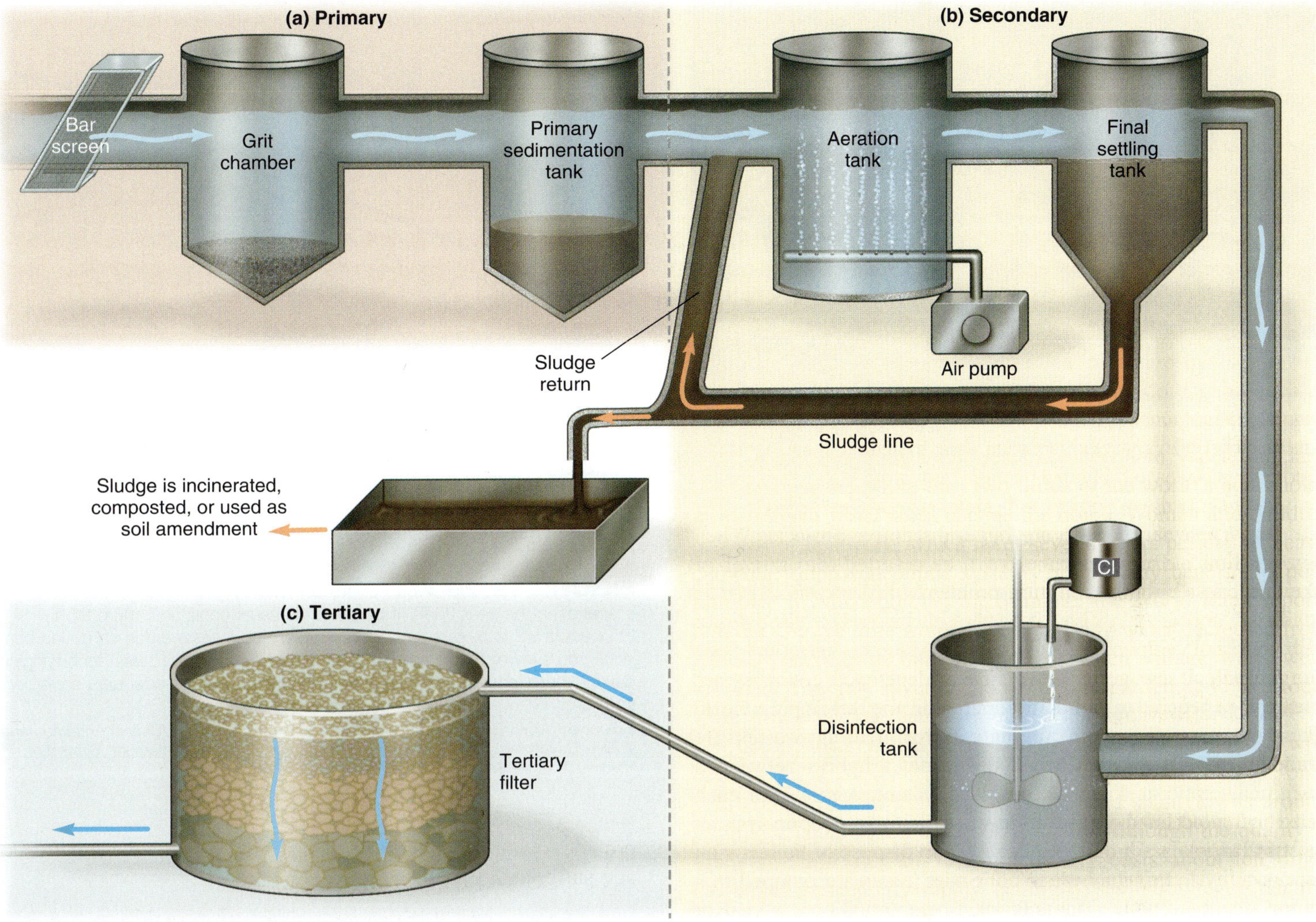


**FIGURE 18.23** (a) Primary sewage treatment removes only solids and suspended sediment. (b) Secondary treatment, through aeration of activated sludge (or biosolids), followed by sludge removal and chlorination of effluent, kills pathogens and removes most organic material. (c) During tertiary treatment, passage through a trickling bed evaporator and/or a lagoon or marsh further removes inorganic nutrients, oxidizes any remaining organics, and reduces effluent volume.

surface waters, these nutrients stimulate algal blooms and eutrophication. To preserve water quality, these nutrients also must be removed. Passage through a wetland or lagoon can accomplish this. Alternatively, chemicals often are used to bind and precipitate nutrients (see fig. 18.23*c*).

In many American cities, sanitary sewers are connected to storm sewers, which carry runoff from streets and parking lots. Storm sewers are routed to the treatment plant rather than discharged into surface waters because runoff from streets, yards, and industrial sites generally contains a variety of refuse, fertilizers, pesticides, oils, rubber, tars, lead (from gasoline), and other undesirable chemicals. During dry weather, this plan works well. Heavy storms often overload the system, however, causing bypass dumping of large volumes of raw sewage and toxic surface runoff directly into receiving waters. To prevent this overflow, cities are spending hundreds of millions of dollars to separate storm

**FIGURE 18.24** "Well, if *you* can't use it, do you know anyone who *can* use 3,000 tons of sludge every day?"

**FIGURE 18.25** In India, a poplar plantation thrives on raw sewage water piped directly from nearby homes. Innovative solutions like this can make use of nutrients that would pollute water systems.

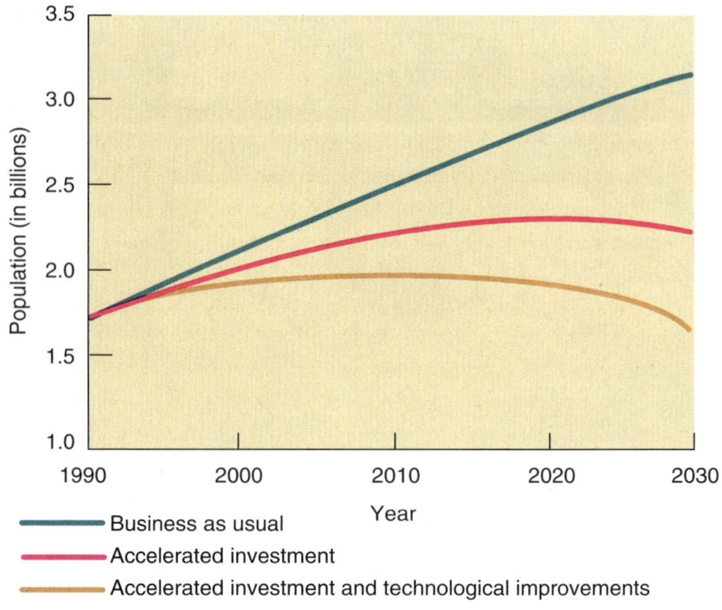


**FIGURE 18.26** World population without adequate sanitation—three scenarios in the year 2030. If business as usual continues, more than 3 billion people will lack safe sanitation. Accelerated investment in sanitation services could lower this number. Higher investment, coupled with technological development, could keep the number of people without adequate sanitation from growing even though the total population increases.

**Source:** World Bank estimates based on research paper by Dennis Anderson and William Cavendish, "Efficiency and Substitution in Pollution Abatement: Simulation Studies in Three Sectors."

and sanitary sewers. These are huge, disruptive projects. When they are finished, surface runoff will be diverted into a river or lake and cause another pollution problem.

### *Low-Cost Waste Treatment*

The municipal sewage systems used in developed countries are often too expensive to build and operate in the developing world where low-cost, low-tech alternatives for treating wastes are needed. One option is **effluent sewerage,** a hybrid between a traditional septic tank and a full sewer system. A tank near each dwelling collects and digests solid waste just like a septic system. Rather than using a drainfield, however, to dispose of liquids—an impossibility in crowded urban areas—effluents are pumped to a central treatment plant. The tank must be emptied once a year or so, but because only liquids are treated by the central facility, pipes, pumps, and treatment beds can be downsized and the whole system is much cheaper to build and run than a conventional operation.

Another alternative is to use natural or artificial wetlands to dispose of wastes. Constructed wetlands can cut secondary treatment costs to one-third of mechanical treatment costs, or less. As the opening case study for this chapter shows, this can be a critical savings for small municipalities.

Constructed wetland waste treatment systems are now operating in many American cities and many developing countries. Effluent from these operations can be used to irrigate crops or raise fish for human consumption if care is taken to first destroy pathogens (fig. 18.25). Usually 20 to 30 days of exposure to sun, air, and aquatic plants is enough to make the water safe. These systems make an important contribution to human food supplies. A 2,500 ha (6,000-acre) waste-fed aquaculture facility in Calcutta, for example, supplies about 7,000 metric tons of fish annually to local markets. The World Bank estimates that some 3 billion people will be without sanitation services by the middle of the next century under a business-as-usual scenario (fig. 18.26). With investments in innovative programs, however, sanitation could be provided to about half those people and a great deal of misery and suffering could be avoided.

## Water remediation may involve containment, extraction, or phytoremediation

Remediation means finding remedies for problems. Just as there are many sources for water contamination, there are many ways to clean it up. New developments in environmental engineering are providing promising solutions to many water pollution problems.

Containment methods confine or restrain dirty water or liquid wastes *in situ* (in place) or cap the surface with an impermeable layer to divert surface water or groundwater away from the site and to prevent further pollution. Where pollutants are buried too deeply to be contained mechanically, materials sometimes can be injected to precipitate, immobilize, chelate, or solidify them. Bentonite slurries, for instance, can effectively stabilize liquids in porous substrates. Similarly, straw or other absorbent material is spread on surface spills to soak up contaminants.

Extraction techniques pump out polluted water so it can be treated. Many pollutants can be destroyed or detoxified by chemical reactions that oxidize, reduce, neutralize, hydrolyze, precipitate, or otherwise change their chemical composition.

Where chemical techniques are ineffective, physical methods may work. Solvents and other volatile organic compounds, for instance, can be stripped from solution by aeration and then burned in an incinerator. Some contaminants can be removed by semipermeable membranes or resin filter beds that bind selectively to specific materials. Some of the same techniques used to stabilize liquids *in situ* can also be used *in vitro* (in a reaction vessel). Metals, for instance, can be chelated or precipitated in insoluble, inactive forms.

Often, living organisms can be used effectively and inexpensively to clean contaminated water. We call this bioremediation (chapter 21). Restored wetlands, for instance, along stream banks or lake margins can be very effective in filtering out sediment and removing pollutants. They generally cost far less than mechanical water treatment facilities and provide wildlife habitat as well.

Lowly duckweed (*Lemna* sp.), the green scum you often see covering the surface of eutrophic ponds, grows fast and can remove large amounts of organic nutrients from water. Under optimal conditions, a few square centimeters of these tiny plants can grow to cover nearly a hectare (about 2.5 acres) in four months. Large duckweed lagoons are being used as inexpensive, low-tech sewage treatment plants in developing countries. Where conventional wastewater purification typically costs \$300 to \$600 per person served, a duckweed system can cost one-tenth as much. The duckweed can be harvested and used as feed, fuel, or fertilizer. Up to 35 percent of its dry mass is protein—about twice as much as alfalfa, a popular animal feed.

Where space for open lagoons is unavailable, bioremediation can be carried out in reaction vessels. This has the advantage of controlling conditions more precisely and doesn't release organisms into the environment. Some of the most complex, holistic systems for water purification are designed by Ocean Arks International (OAI) in Falmouth, Massachusetts. Their "living machines" combine living organisms—chosen to perform specific functions—in contained environments. In a typical living machine, water flows through a series of containers, each with a distinct ecological community designed for a particular function. Wastes generated by the inhabitants of one vessel become the food for inhabitants of another. Sunlight provides the primary source of energy.

OAI has created or is in the process of building water treatment plants in a dozen states and foreign countries. Designs range from remediating toxic wastes from Superfund sites to simply treating domestic wastes. Starting with microorganisms in aerobic and anaerobic environments where different kinds of wastes are metabolized or broken down, water moves through a series of containers containing hundreds of different kinds of plants and animals, including algae, rooted aquatic plants, clams, snails, and fish, each chosen to provide a particular service. Technically, the finished water is drinkable, although few people feel comfortable doing so. More often, the final effluent is used to flush toilets or for irrigation. Called ecological engineering, this novel approach can save resources and money as well as clean up our environment and serve as a valuable educational tool (fig. 18.27).

**FIGURE 18.27** In-house wastewater treatment in Oberlin College's Environmental Studies building. Constructed wetlands outside, and tanks inside, allow water plants to filter water and use nutrients.

## 18.5 Water Legislation

Water pollution control has been among the most broadly popular and effective of all environmental legislation in the United States. It has not been without controversy, however. In this section, we will look at some of the major issues concerning water quality laws and their provisions (table 18.2).

### The Clean Water Act was ambitious, bipartisan, and largely successful

Passage of the Clean Water Act of 1972 was a bold, bipartisan step determined to "restore and maintain the chemical, physical, and biological integrity of the Nation's waters" that made clean water a national priority. Along with the Endangered Species Act and the Clean Air Act, this is one of the most significant and effective pieces of environmental legislation ever passed by the U.S. Congress. It also is an immense and complex law, with more than 500 sections regulating everything from urban runoff, industrial discharges, and municipal sewage treatment to land-use practices and wetland drainage.

The ambitious goal of the Clean Water Act was to return all U.S. surface waters to "fishable and swimmable" conditions. For specific "point" sources of pollution such as industrial discharge pipes or sewage outfalls, the act requires discharge permits and **best practicable control technology (BPT).** It sets national

**TABLE 18.2**

## Some Important U.S. and International Water Quality Legislation

1. *Federal Water Pollution Control Act* (1972). Established uniform nationwide controls for each category of major polluting industries.
2. *Marine Protection Research and Sanctuaries Act* (1972). Regulates ocean dumping and established sanctuaries for protection of endangered marine species.
3. *Ports and Waterways Safety Act* (1972). Regulates oil transport and the operation of oil handling facilities.
4. *Safe Drinking Water Act* (1974). Requires minimum safety standards for every community water supply. Among the contaminants regulated are bacteria, nitrates, arsenic, barium, cadmium, chromium, fluoride, lead, mercury, silver, pesticides; radioactivity and turbidity also regulated. This act also contains provisions to protect groundwater aquifers.
5. *Resource Conservation and Recovery Act* (RCRA) (1976). Regulates the storage, shipping, processing, and disposal of hazardous wastes and sets limits on the sewering of toxic chemicals.
6. *Toxic Substances Control Act* (TOSCA) (1976). Categorizes toxic and hazardous substances, establishes a research program, and regulates the use and disposal of poisonous chemicals.
7. *Comprehensive Environmental Response, Compensation, and Liability Act* (CERCLA) (1980) and *Superfund Amendments and Reauthorization Act* (SARA) (1984). Provide for sealing, excavation, or remediation of toxic and hazardous waste dumps.
8. *Clean Water Act* (1985) (amending the 1972 Water Pollution Control Act). Sets as a national goal the attainment of "fishable and swimmable" quality for all surface waters in the United States.
9. *London Dumping Convention* (1990). Calls for an end to all ocean dumping of industrial wastes, tank washing effluents, and plastic trash. The United States is a signatory to this international convention.

goals of **best available, economically achievable technology (BAT),** for toxic substances and zero discharge for 126 priority toxic pollutants. As we discussed earlier in this chapter, these regulations have had a positive effect on water quality. While not yet swimmable or fishable everywhere, surface-water quality in the United States has significantly improved on average over the past quarter century. Perhaps the most important result of the act has been investment of $54 billion in federal funds and more than $128 billion in state and local funds for municipal sewage treatment facilities.

Not everyone, however, is completely happy with the Clean Water Act. Industries, state and local governments, farmers, land developers, and others who have been forced to change their operations or spend money on water protection often feel imposed upon. One of the most controversial provisions of the act has been Section 404, which regulates draining or filling of wetlands. Although the original bill only mentions wetlands briefly, this section has evolved through judicial interpretation and regulatory policy to become one of the principal federal tools for wetland protection. Many people applaud the protection granted to these ecologically important areas that were being filled in or drained at a rate of about half a million hectares per year before the passage of the Clean Water Act. Farmers, land developers, and others who are prevented from converting wetlands to other uses often are outraged by what they consider "taking" of private lands.

Another sore point for opponents of the Clean Water Act are what are called "unfunded mandates," or requirements for state or local governments to spend money that is not repaid by Congress. You will notice that the $128 billion already spent by cities to install sewage treatment and stormwater diversion to meet federal standards far exceeds the $54 billion in congressional assistance for these projects. Estimates are that local units of government could be required to spend another $130 billion to finish the job without any further federal funding. Small cities that couldn't afford or chose not to participate in earlier water quality programs,

## What Can You Do?

### Steps You Can Take to Improve Water Quality

Individual actions have important effects on water quality. Here are some steps you can take to make a difference.

- Compost your yard waste and pet waste. Nutrients from decayed leaves, grass, and waste are a major urban water pollutant. Many communities have public compost sites available.
- Don't fertilize your lawn or apply lawn chemicals. Untreated grass can be just as healthy, and it won't poison your pets or children.
- Make sure your car doesn't leak fluids, oil, or solvents on streets and parking lots, from which contaminants wash straight into rivers and lakes. Recycle motor oil at a gas station or oil change shop.
- Don't buy lawn mowers, personal watercraft, or other vehicles with two-cycle engines, which release abundant fuel and oil into air and water. Instead, buy more efficient, four-stroke engines.
- Visit your local sewage treatment plant. Often public tours are available or group tours can be arranged, and these sites can be fascinating.
- Keep informed about water policy debates at local and federal levels. Policies change often, and public input is important.

in which the federal government paid up to 90 percent of the costs, are especially hard hit by requirements that they upgrade municipal sewer and water systems. They now are faced with carrying out those same projects entirely on their own funds.

## Clean water reauthorization remains contentious

Opponents of federal regulation have tried repeatedly to weaken or eliminate the Clean Water Act. They regard restriction of their "right" to dump toxic chemicals and waste into wetlands and waterways to be an undue loss of freedom. They resent being forced to clean up municipal water supplies, and call for cost/benefit analysis that places greater weight on economic interests in all environmental planning. Most of all, they view any limitation on use of private property to be a "taking" for which they should be fully compensated.

Even those who support the Clean Water Act in principle would like to see it changed and strengthened. Among these proposals are a shift from "end-of-the-pipe" focus on removing specific pollutants from effluents to more attention to changing industrial processes so toxic substances won't be produced in the first place. Another important issue is nonpoint pollution from agricultural runoff and urban areas, which has become the largest source of surface-water degradation in the United States. Regulating these sources remains a difficult problem.

Environmentalists also would like to see stricter enforcement of existing regulations, mandatory minimum penalties for violations, more effective community right-to-know provisions, and increased powers for citizen lawsuits against polluters. Studies have found that, in practice, polluters are given infrequent and light fines for polluting. Under the current law, using data that polluters themselves are required to submit, groups such as the Natural Resources Defense Council and the Citizens for a Better Environment have won million-dollar settlements in civil lawsuits (the proceeds generally are applied to clean-up projects) and some transgressors have even been sent to jail. Not surprisingly, environmentalists want these powers expanded, while polluters find them very disagreeable.

## Other important legislation also protects water quality

In addition to the Clean Water Act, several other laws help to regulate water quality in the United States and abroad. Among these is the Safe Drinking Water Act, which regulates water

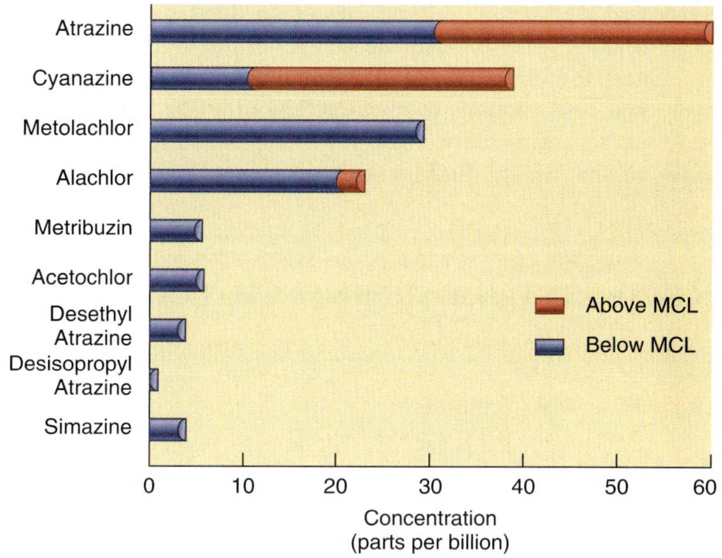


**FIGURE 18.28** Nine herbicide active ingredients and metabolites found in drinking water samples in Fort Wayne, Indiana, in spring of 1995, compared to federal Maximum Concentration Levels (MCL). The use of these chemicals, and their presence in water sources, has increased in subsequent years.

**Source:** B. Cohen and E. Bondoc, "Weed Killers by the Glass," Environmental Working Group, 1995.

quality in commercial and municipal systems. Critics complain that standards and enforcement policies are too lax, especially for rural water districts and small towns. Some researchers report pesticides, herbicides, and lead in drinking water at levels they say should be of concern (fig. 18.28). Atrazine, for instance, was detected in 96 percent of all surface-water samples in one study of 374 communities across 12 states. Remember, however, that simply detecting a toxic compound is not the same as showing dangerous levels.

The Superfund program for remediation of toxic waste sites was created in 1980 by the Comprehensive Environmental Response, Compensation, and Liability Act (CERCLA) and was amended by the Superfund Amendments and Reauthorization Act (SARA) of 1984. This program is designed to provide immediate response to emergency situations and to provide permanent remedies for abandoned or inactive sites. These programs provide many jobs for environmental science majors in monitoring and removal of toxic wastes and landscape restoration. A variety of methods have been developed for remediation of problem sites.

# Conclusion

Forty years ago, rivers in the United States were so polluted that some caught fire while others ran red, black, orange, or other unnatural colors with toxic industrial wastes. Many cities still dumped raw sewage into local rivers and lakes, so that warnings had to be posted to avoid any bodily contact. We've made huge progress since that time. Not all rivers and lakes are "fishable or swimmable," but federal, state, and local pollution controls have greatly improved our water quality in most places.

In rapidly developing countries, such as China and India, water pollution remains a serious threat to human health and ecosystem well-being. Billions of people don't have access to clean drinking water or adequate sanitation. It will take a massive investment to correct this growing problem. But there are relatively low-cost solutions to many pollution issues. The example of Arcata's constructed wetland for ecological sewage treatment shows us that we can find low-tech, inexpensive ways to reduce pollution. Living machines for water treatment in individual buildings or communities also offer hope for better ways to treat our wastes. Perhaps you can use the information you've learned by studying environmental science to make constructive suggestions for your own community.

## Reviewing Learning Outcomes

By now you should be able to explain the following points:

**18.1** Define water pollution.

- Water pollution is anything that degrades water quality.

**18.2** Describe the types and effects of water pollutants.

- Infectious agents remain an important threat to human health.
- Bacteria are detected by measuring oxygen levels.
- Nutrient enrichment leads to cultural eutrophication.
- Eutrophication can cause toxic tides and "dead zones."
- Inorganic pollutants include metals, salts, acids, and bases.
- Organic pollutants include pesticides and other industrial substances.
- Sediment also degrades water quality.
- Thermal pollution is dangerous for organisms.

**18.3** Investigate water quality today.

- The Clean Water Act protects our water.
- Water quality problems remain.
- Developing countries often have serious water pollution.
- Groundwater is hard to monitor and clean.
- There are few controls on ocean pollution.

**18.4** Explain water pollution control.

- Source reduction is often the cheapest way to reduce pollution.
- Controlling nonpoint sources requires land management.
- Human waste disposal occurs naturally when concentrations are low.
- Water remediation may involve containment, extraction, or phytoremediation.

**18.5** Summarize water legislation.

- The Clean Water Act was ambitious, bipartisan, and largely successful.
- Clean water reauthorization remains contentious.
- Other important legislation also protects water quality.

## Practice Quiz

1. Define *water pollution.*
2. List eight major categories of water pollutants and give an example for each category.
3. Describe eight major sources of water pollution in the United States. What pollution problems are associated with each source?
4. What are red tides, and why are they dangerous?
5. What is eutrophication? What causes it?
6. What is an oxygen sag? How much dissolved oxygen, in ppm, is present at each stage?
7. What are the origins and effects of siltation?
8. Describe primary, secondary, and tertiary processes for sewage treatment. What is the quality of the effluent from each of these processes?
9. Why do combined storm and sanitary sewers cause water quality problems? Why does separating them also cause problems?
10. What pollutants are regulated by the Clean Water Act? What goals does this act set for abatement technology?
11. What is MTBE? Why is it so widespread and hard to control?
12. Describe remediation techniques and how they work.

## Critical Thinking and Discussion Questions

1. Cost is the greatest obstacle to improving water quality. How would you decide how much of the cost of pollution control should go to private companies, government, or individuals?
2. How would you define *adequate* sanitation? Think of some situations in which people might have different definitions for this term.

3. What sorts of information would you need to make a judgment about whether water quality in your area is getting better or worse? How would you weigh different sources, types, and effects of water pollution?
4. Imagine yourself in a developing country with a severe shortage of clean water. What would you miss most if your water supply were suddenly cut by 90 percent?
5. Proponents of deep well injection of hazardous wastes argue that it will probably never be economically feasible to pump water out of aquifers more than 1 kilometer below the surface. Therefore, they say, we might as well use those aquifers for hazardous waste storage. Do you agree? Why or why not?
6. Arsenic contamination in Bangladesh results from geological conditions, World Bank and U.S. aid, poverty, government failures, and other causes. Who do you think is responsible for finding a solution? Why? Would you answer differently if you were a poor villager in Bangladesh?

## DATA analysis Examining Pollution Sources

Understanding the origins of pollution is the first step toward considering policies for reducing it. The chapter you have just read includes several graphs displaying pollution sources. The following questions ask you to think more about these sources:

1. In figure 18.14 (p. 409), which sources do you think are mainly point sources? Which are mainly nonpoint sources? How might you consider urban runoff to be either point or nonpoint? (If you need ideas to start, see figure 18.17, p. 411)
2. Explain the units on the Y-axis in figure 18.14. Does the graph show you the amount of actual pollution, by weight or volume, from different sources? Why might it be useful to show the extent of impacts?
3. Based on your knowledge of water contaminants, describe the kinds of pollutants you would expect from each of the sources shown:

   Agriculture ______________________________

   Sewage treatment ______________________________

   Dams, diversions, channelization ______________________________

   Urban runoff ______________________________

   Mining ______________________________

   Removal of streamside vegetation ______________________________

   Forestry ______________________________
4. How would you design a sampling strategy to assess water pollution on your school campus?
5. Figure 18.11 (p. 406) shows some of the new developments in water pollution assessment. Conventional treatment systems were not designed to remove thousands of newly invented chemical compounds, or increasingly widespread compounds, including those shown in the figure. Explain the units used on the Y-axis. Also explain why the numbers (maximum concentrations found) are shown.
6. How many of the pollutants shown do *you* use? Don't forget to include caffeine (which is classed as a nonprescription drug) and antibacterial soaps (disinfectants). Steroids include cholesterol, which occurs naturally in foods.
7. Try to think of additional substances that you might contribute to wastewater.
8. This graph results from a reconnaissance study done by the U.S. Geological Survey. The researchers wanted to assess whether a list of 95 contaminants could be detected at all in public waterways. If you wanted to design a study like this, what sorts of sites would you select for sampling? How might your results and your conclusions differ if you did a random sample?

**For Additional Help in Studying This Chapter,** please visit our website at www.mhhe.com/cunningham10e. You will find additional practice quizzes and case studies, flashcards, regional examples, place markers for Google Earth™ mapping, and an extensive reading list, all of which will help you learn environmental science.

Fishing once provided a livelihood for most residents of this small Norwegian village, but heavily-subsidized deep-sea trawlers and floating fish processing factories from far-distant countries have now depleted the resource. The economic choices we make often have unintended consequences.

CHAPTER 23

# Ecological Economics

*Unleashing the energy and creativity in each human being is the answer to poverty.*

*—Muhammad Yunus—*

## LEARNING OUTCOMES

After studying this chapter, you should be able to:

**23.1** Analyze economic worldviews.

**23.2** Scrutinize population, technology, and scarcity.

**23.3** Investigate natural resource accounting.

**23.4** Summarize how market mechanisms can reduce pollution.

**23.5** Study trade, development, and jobs.

**23.6** Evaluate green business.

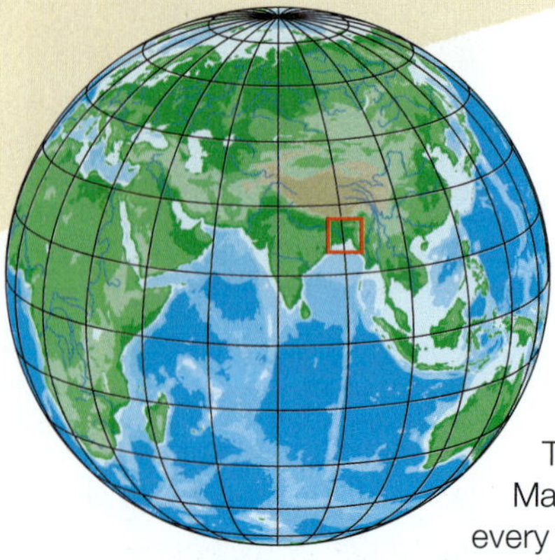

# Case Study Loans That Change Lives

Ni Made is a young mother of two children who lives in a small Indonesian village. Her husband is a day laborer who makes only a few dollars per day—when he can find work. To supplement their income, Made goes to the village market every morning to sell a drink she makes out of boiled pandamus leaves, coconut milk, and pink tapioca (fig. 23.1). A small loan would allow her to rent a covered stall during the rainy season and to offer other foods for sale. The extra money she could make could change her life. But traditional banks consider Made too risky to lend to, and the amounts she needs too small to bother with.

Around the world, billions of poor people find themselves in the same position as Made; they're eager to work to build a better life for themselves and their families, but lack resources to succeed. Now, however, a financial revolution is sweeping around the world. Small loans are becoming available to the poorest of the poor. This new approach was invented by Dr. Muhammad Yunus, professor of rural economics at Chittagong University in Bangladesh. Talking to a woman who wove bamboo mats in a village near his university, Dr. Yunus learned that she had to borrow the few taka she needed each day to buy bamboo and twine. The interest rate charged by the village moneylenders consumed nearly all her profits. Always living on the edge, this woman, and many others like her, couldn't climb out of poverty.

To break this predatory cycle, Dr. Yunus gave the woman and several of her neighbors small loans totaling about 1,000 taka (about $20). To his surprise, the money was paid back quickly and in full. So he offered similar amounts to other villagers with similar results. In 1983, Dr. Yunus started the Grameen (village) Bank to show that "given the support of financial capital, however small, the poor are fully capable of improving their lives." His experiment has been tremendously successful. By 2007, the Grameen Bank had almost 7 million customers, 97 percent of them women. It had loaned more than $6 billion with 98 percent repayment, nearly twice the collection rate of commercial Bangladesh banks.

The Grameen Bank provides credit to poor people in rural Bangladesh without the need for collateral. It depends, instead, on mutual trust, accountability, participation, and creativity of the borrowers themselves. Microcredit is now being offered by hundreds of organizations in 43 other countries. Institutions from the World Bank to religious charities make small loans to worthy entrepreneurs. Wouldn't you like to be part of this movement? Well, now you can. You don't have to own a bank to help someone in need.

A brilliant way to connect entrepreneurs in developing countries with lenders in wealthy countries is offered by Kiva, a San Francisco-based technology start-up. The idea for Kiva, which means unity or cooperation in Swahili, came from Matt and Jessica Flannery. Jessica had worked in East Africa with the Village Enterprise Fund, a California nonprofit that provides training, capital, and mentoring to small businesses in developing countries. Jessica and Matt wanted to help some of the people she had met, but they weren't wealthy enough to get into microfinancing on their own. Joining with four other young people with technology experience, they created Kiva, which uses the power of the Internet to help the poor.

Kiva partners with about a dozen development nonprofits with staff in developing countries. The partners identify hardworking entrepreneurs who deserve help. They then post a photo and brief introduction to each one on the Kiva web page. You can browse the collection to find someone whose story touches you. The minimum loan is generally $25. Your loan is bundled with that of others until it reaches the amount needed by the borrower. You make your loan using your credit card (through PayPal, so it's safe and easy). The loan is generally repaid within 12 to 18 months (although without interest). At that point, you can either withdraw the money, or use it to make another loan.

The in-country staff keeps track of the people you're supporting and monitors their progress, so you can be confident that your money will be well used. In less than a year, Kiva has raised more than $2 million and helped thousands of people like Made change their lives.

**FIGURE 23.1** A small amount of seed money would allow this young mother to expand her business and help provide for her family.

# Case Study *continued*

Loan requests often are on their web page for only a few hours before being filled. Wouldn't you like to take part in this innovative person-to-person human development project? Check out Kiva.org.

In this chapter, we'll look further at both microlending and conventional financing for human development. We'll also look at the role of natural resources in national economies, and how ecological economics is bringing ecological insights into economic analysis. We'll examine cost-benefit analysis as well as other measures of human well-being and genuine progress. Finally, we'll look at how market mechanisms can help us solve environmental problems, and how businesses can contribute to sustainability.

## 23.1 Economic Worldviews

Economy is the management of resources to meet our needs in the most efficient manner possible. It typically deals with choices and alternatives. Because we don't have unlimited ability to produce, distribute, or consume goods and services, economists ask, "Should we make bread or bullets? What are the trade-offs between them, and which would result in the greatest benefits?" (fig. 23.2). Interestingly, ecology and economy are derived from the same root words and concerns. *Oikos* (ecos) is the Greek word for household. Economics is the *nomos,* or counting, of the household resources. Ecology is the *logos,* or logic of how the household works. In both disciplines, the household is expanded to include the whole world. As you will read in this chapter, economics provides important tools for understanding and managing resources. Economics has also evolved as our understanding of resources has changed.

**FIGURE 23.2** Bread or bullets? What are the costs and benefits of each? And what are the trade-offs between them?

### Can development be sustainable?

By now it is clear that security and living standards for the world's poorest people are inextricably linked to environmental protection. One of the most important questions in environmental science is how we can continue improvements in human welfare within the limits of the earth's natural resources. *Development* means improving people's lives. *Sustainability* means living on the earth's renewable resources without damaging the ecological processes that support us all (table 23.1). **Sustainable development** is an effort to marry these two ideas. A popular definition describes this goal as "meeting the needs of the present without compromising the ability of future generations to meet their own needs."

But is this possible? As you've learned elsewhere in this book, many people argue that our present population and economic levels are exhausting the world's resources. There's no way, they insist, that more people can live at a higher standard without irreversibly degrading our environment. Others claim that there's enough for everyone if we just share equitably and live modestly. Let's look a little deeper into this important debate.

### Our definitions of resources shape how we use them

**Capital** is any form of wealth available for use in the production of more wealth. Economists distinguish between *natural capital* (goods and services provided by nature), *human* or *cultural capital* (knowledge, experience, and human enterprise), and *manufactured* or *built capital* (tools, infrastructure, and technology). Social scientists would add *social capital*

**TABLE 23.1**

**Goals for Sustainable Natural Resource Use**

- Harvest rates for renewable resources (those like organisms that regrow or those like fresh water that are replenished by natural processes) should not exceed regeneration rates.
- Waste emissions should not exceed the ability of nature to assimilate or recycle those wastes.
- Nonrenewable resources (such as minerals) may be exploited by humans, but only at rates equal to the creation of renewable substitutes.

**FIGURE 23.3** Nonrenewable resources, such as the oil from this forest of derricks in Huntington Beach, California, are irreplaceable. Once they're exhausted (as this oil field was half a century ago) they will never be restored on a human time scale.

**FIGURE 23.4** Nature provides essential ecological services, such as the biological productivity, water storage and purification, and biodiversity protection in this freshwater marsh and its surrounding forest. Ironically, while biological resources are infinitely renewable, if they're damaged by our actions they may be lost forever.

(shared values, trust, cooperative spirit, and community organization) to this list. A **resource** is anything with potential use in creating wealth or giving satisfaction. Natural resources can be either renewable or nonrenewable, as well as tangible or intangible. Although we generally define these terms from a human perspective, many resources we use are important to other species as well.

In general, **nonrenewable resources** are the earth's geological endowment: the minerals, fossil fuels, and other materials present in fixed amounts in the environment (fig. 23.3). Although many of these resources are renewed or recycled by geological or ecological processes, the time scales to do so are so long by human standards that the resource will be gone once present supplies are exhausted. Predictions abound that we are in imminent danger of running out of one or another of these exhaustible resources. The actual available supplies of many commodities—such as metals—can be effectively extended, however, by more efficient use, recycling, substitution of one material for another, or better extraction from dilute or remote sources.

**Renewable resources** are things that can be replenished or replaced. They include sunlight—our ultimate energy source—and the biological organisms and biogeochemical cycles that provide essential ecological services (fig. 23.4 and table 23.2).

Because biological organisms and ecological processes are self-renewing, we often can harvest surplus organisms or take advantage of ecological services without diminishing future availability, if we do so carefully. Unfortunately, our stewardship of these resources often is less than ideal. Even once vast biological populations such as passenger pigeons, American bison (buffalo), and Atlantic cod, for instance, were exhausted by overharvesting in only a few years. Similarly, we are now upsetting climatic systems with potentially disastrous results (see chapter 15). Mismanagement of renewable resources often makes them more ephemeral and limited than fixed geological resources.

Abstract or **intangible resources** include open space, beauty, serenity, wisdom, diversity, and satisfaction (fig. 23.5). Paradoxically, these resources can be both infinite *and* exhaustible. There is no upper limit to the amount of beauty, knowledge, or compassion that can exist in the world, yet they can be easily destroyed. A single piece of trash can ruin a beautiful vista, or a single cruel remark can spoil an otherwise perfect day. On the other hand, unlike tangible resources that usually are reduced by use or sharing, intangible resources often are increased by use and multiplied by being shared. Nonmaterial assets can be important economically. Information management and tourism—both based on intangible resources—have become two of the largest and most powerful industries in the world.

**TABLE 23.2**

## Important Ecological Services

1. *Regulate* global energy balance; chemical composition of the atmosphere and oceans; local and global climate; water catchment and groundwater recharge; production, storage, and recycling of organic and inorganic materials; maintenance of biological diversity.
2. *Provide* space and suitable substrates for human habitation, crop cultivation, energy conversion, recreation, and nature protection.
3. *Produce* oxygen, fresh water, food, medicine, fuel, fodder, fertilizer, building materials, and industrial inputs.
4. *Supply* aesthetic, spiritual, historic, cultural, artistic, scientific, and educational opportunities and information.

**Source:** R. S. de Groot, *Investing in Natural Capital*, 1994.

**FIGURE 23.5** Scenic beauty, solitude, and relatively untouched nature in this Colorado wilderness are treasured by many people, yet hard to evaluate in economic terms.

**FIGURE 23.6** Informal markets such as this one in Bali, Indonesia, may be the purest example of willing sellers and buyers setting prices based on supply and demand.

## Classical economics examines supply and demand

**Classical economics** originally was a branch of moral philosophy concerned with how individual interest and values intersect with larger social goals. The founder of modern Western economics, Adam Smith (1723–1790), was an ethicist concerned with individual freedom of choice. Smith's landmark book *Inquiry into the Nature and Causes of the Wealth of Nations,* published in 1776, argued that

> Every individual endeavors to employ his capital so that its produce may be of the greatest value. He generally neither intends to promote the public interest, nor knows how much he is promoting it. He intends only his own security, only his own gain. And he is in this led by an *invisible hand* to promote an end which was no part of his intention. By pursuing his own interests he frequently promotes that of society more effectually than when he really intends to.

This statement often is taken as justification for the capitalist system, in which market competition between willing sellers and buyers is believed to bring about the greatest efficiency of resource use, the optimum balance between price and quality, and to be critical for preserving individual liberty (fig. 23.6). As we will discuss later in this chapter, however, *laissez faire* market systems rarely incorporate factors such as social or environmental costs.

David Ricardo (1772–1823), an important contemporary of Smith, introduced a better understanding of the relation between supply and demand in economics. **Demand** is the amount of a product or service that consumers are willing and able to buy at various possible prices, assuming they are free to express their preferences. **Supply** is the quantity of that product being offered for sale at various prices, other things being equal. Classical economics proposes that there is a direct, inverse relationship between supply and demand (fig. 23.7). As prices rise, the supply increases and demand falls. The reverse holds true as price decreases. The difference between the cost of production and the price buyers are willing to pay, Ricardo called "rent." Today we call it profit.

In a free market of independent and intelligent buyers and sellers, supply and demand should come into a **market equilibrium,** represented in figure 23.7 by the intersection of the two curves. In real life, prices are not determined strictly by total supply and demand as much as what economists called **marginal costs and benefits.** Sellers ask themselves, "What would it cost to produce one more unit of this product or service? Suppose I

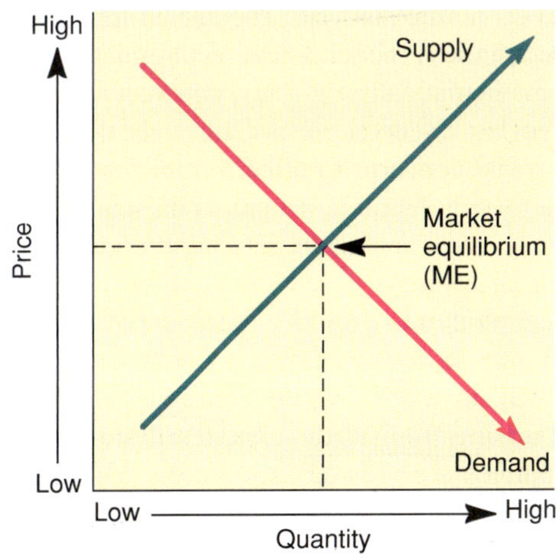


**FIGURE 23.7** Classic supply/demand curves. When price is low, supply is low and demand is high. As prices rise, supply increases but demand falls. Market equilibrium is the price at which supply and demand are equal.

add one more worker or buy an extra supply of raw materials, how much profit could I make?" Buyers ask themselves similar questions, "How much would I benefit and what would it cost if I bought one more widget?" If both buyer and seller find the marginal costs and benefits attractive, a sale is made.

There are exceptions, however, to this theory of supply and demand. Consumers will buy some things regardless of cost. Raising the price of cigarettes, for instance, doesn't necessarily reduce demand. We call this price inelasticity. Other items have **price elasticity;** that is, they follow supply/demand curves exactly. When price goes up, demand falls and vice versa.

John Stuart Mill (1806–1873), another important classical philosopher/economist, believed that perpetual growth in material well-being is neither possible nor desirable. Economies naturally mature to a steady state, he believed, leaving people free to pursue nonmaterialistic goals. He didn't regard this equilibrium state to be necessarily one of stagnation or poverty. In *Principles of Political Economy,* he wrote, "It is scarcely necessary to remark that a stationary condition of capital and population implies no stationary state of human improvement. There would be as much scope as ever for all kinds of mental culture, and moral and social progress; as much room for improving the art of living, and much more likelihood of its being improved when minds cease to be engrossed by the art of getting on." This view has much in common with the concepts of a steady-state economy and sustainable development that we will examine later in this chapter.

## Neoclassical economics emphasizes growth

Toward the end of the nineteenth century, the field of economics divided into two broad camps. **Political economy** continued the tradition of moral philosophy and concerned itself with social structures, value systems, and relationships among the classes. This group included reformers such as Karl Marx and E. F. Schumacher, along with many socialists, anarchists, and utopians. The other camp, called **neoclassical economics,** adapted principles of modern science to economic analysis. The late Milton Friedman was a leader in free-market, neoclassical economics. This approach strives to be mathematically rigorous, noncontextual, abstract, and predictive. Neoclassical economists claim to be objective and value free, leaving social concerns to other disciplines. Like their classical predecessors, they retain an emphasis on scarcity and the interaction of supply and demand in determining prices and resource allocation (fig. 23.8).

Continued economic growth is considered to be both necessary and desirable in the neoclassical worldview. Growth is seen as the only way to maintain full employment and avoid class conflict arising from inequitable distribution of wealth. Natural resources are viewed as merely a factor of production rather than a critical supply of materials, services, and waste sinks by neoclassical economics. Because factors of production are thought to be interchangeable and substitutable, materials and services provided by the environment are not considered indispensable. As one resource becomes scarce, neoclassical economists believe, substitutes will be found.

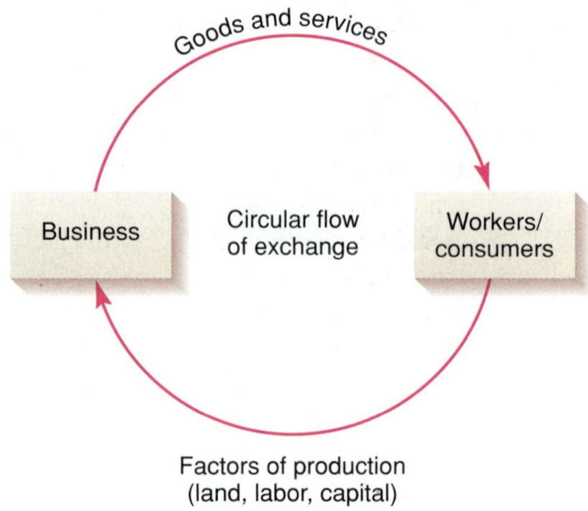


**FIGURE 23.8** The neoclassical model of the economy focuses on the flow of goods, services, and factors of production (land, labor, capital) between business and individual workers and consumers. The social and environmental consequences of these relationships are irrelevant in this view.

## Ecological economics incorporates principles of ecology

Classical and neoclassical economics have usually focused on human resources, such as buildings, roads, or labor. Natural systems are essential for economic productivity; rivers absorb wastewater, bacteria decompose waste, and winds carry away smoke that would otherwise debilitate workers, but classical and neoclassical economics treat these natural services as external to the costs of production. For example, energy producers have calculated profits against the price of coal, the costs of labor, investments in new buildings, and furnaces, but they have not accounted for the climate's absorption of carbon dioxide, sulfur dioxide, or mercury, nor have they accounted for the public health costs associated with polluted air. These have been **external costs,** or costs outside the accounting system. **Internal costs,** in contrast, are expenses considered to be the normal cost of doing business.

Often resources are externalized (treated as external costs) because they're free. Air or sunlight, for example, are provided without charge by nature and, thus, are usually excluded from accounting systems. *Natural resource economics* assigns a value to some of nature's resources, such as clean water, forests, or even biodiversity. Like neoclassical economics, however, natural resource economics assumes that natural services are abundant, and thus cheap, while manufactured capital is limited, and thus expensive (fig. 23.9).

**Ecological economics** focuses on the value of natural services and tries to include those services into price calculations. More than natural resource economics, ecological economics assumes that natural resources are limited and valuable, while manufactured capital is abundant (fig. 23.10).

This is because ecological economics focuses explicitly on ecological concepts such as systems (chapter 2), thermodynamics, and material cycles (chapter 3). In a system, all components are interdependent. Disrupting one component (such as climate

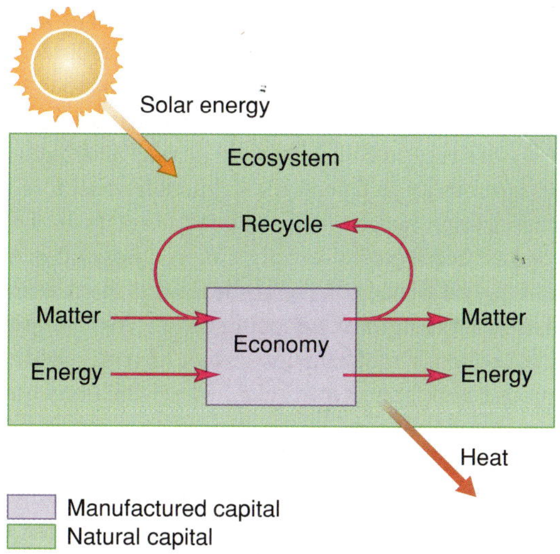


**FIGURE 23.9** In natural resource economics, the economy is seen as an open subset of the ecosystem. Matter and energy are consumed in the form of raw resources and rejected as waste. Material is recycled by ecological processes, while heat is ejected back to space. Manufactured or "built" capital is regarded as scarce—and therefore valuable—while natural capital is regarded as plentiful and, therefore, cheap.

conditions) risks destabilizing other components (such as agricultural production) in unpredictable and possibly catastrophic ways. Thermodynamics and material cycles teach that energy and materials cycle through systems, constantly being reused; one organism's waste is another's nutrient or energy source. This perspective allows us to see environmental resources, and other species, as limited in supply, valuable, and often fragile.

Systems analysis also raises the concern of carrying capacities for human populations and questions the idea of unlimited economic growth. Many ecological economists, such as Herman

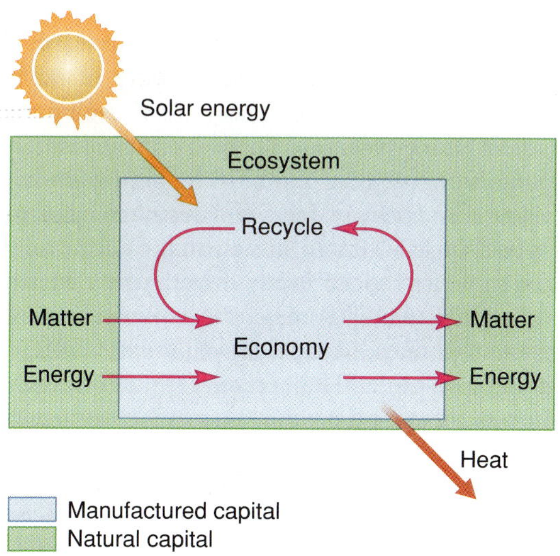


**FIGURE 23.10** An ecological economic view reverses ideas of scarcity. Because material recycling is considered to be integral to the economy, manufactured or built capital is seen to be large compared to limited supplies of natural capital.

Daly of the University of Maryland, promote a **steady-state economy,** characterized by low human birth rates and death rates, the use of renewable energy sources, material recycling, and an emphasis on durability, efficiency, and stability. "Throughput," the volume of materials and energy consumed and of waste produced, should be minimized. This model contrasts sharply with the neoclassical emphasis on growth based on ever-increasing consumption and waste production. The steady-state idea reflects the notion of an ecosystem in equilibrium, or a population below its carrying capacity, where overall conditions remain stable and catastrophic conditions are unlikely to develop.

Ecological economics tries to make producers account for social costs, as well as environmental costs. For example, if a power plant releases soot from its smokestack, populations downwind suffer from air pollution. Suppose the population living downwind incurs health expenses, lost work days, and illness as a result. Is it the power plant's responsibility to pay for these health expenses and lost wages? Traditional economics has said no, that society should bear responsibility. Ecological economists say yes, these costs should be considered part of the cost of power production, and the added expenses should be included in the price of electricity. In general, the cost of cleaning up a power plant usually is lower than the cost of health care and lost productivity in a population. But calculating who should pay, and how much, requires that accountants internalize these external costs.

In economic terms, the extra costs of illness and lost work days, as well as other costs associated with pollution, are market inefficiencies; they represent inefficient overall use of resources (money, time, energy, materials) because of incomplete accounting of costs and benefits.

Ecological economics also questions the basic economic assumption that all goods can be compared according to their monetary value, so that all goods, including bread and bullets, or guns and butter, are essentially interchangeable and can be compared on the same monetary scale. Ecological economists propose that some aspects of nature are irreplaceable and essential, such as beauty, space, wild animals, or quiet. If natural amenities are essential, then it follows that human population growth and resource consumption must have limits.

Ecological economists have developed methods to measure well-being, such as the Genuine Progress Index, discussed later in this chapter. These measurements allow us to assess growth in terms beyond simply the amount of money changing hands.

***Think About It***

What do you suppose are the internal costs (those calculated into the price) when you buy a gallon of gas? What might be some external costs? How many of those costs are geographically expressed where you live? Are there any costs that might be paid for through your taxes? Are there any that are not paid for by your taxes? (*Hint:* Think of the conditions necessary to get oil safely delivered to your gas station. How are those conditions created or maintained?)

**FIGURE 23.11** Adding more cattle to the Brazilian Cerrado (savanna) increases profits for individual ranchers, but is bad for biodiversity and environmental quality.

### Communal property resources are a classic problem in ecological economics

In 1968, biologist Garret Hardin wrote a widely quoted article entitled **"The Tragedy of the Commons"** in which he argued that any commonly held resource inevitably is degraded or destroyed because the narrow self-interests of individuals tend to outweigh public interests. Hardin offered as a metaphor the common woodlands and pastures held by most colonial New England villages. In deciding how many cattle to put on the commons, Hardin explained, each villager would attempt to maximize his or her own personal gain. Adding one more cow to the commons could mean a substantially increased income for an individual farmer. The damage done by overgrazing, however, would be shared among all the farmers (fig. 23.11). Hardin concluded that the only solution would be either to give coercive power to the government or to privatize the resource.

Hardin intended this dilemma, known in economics as the "free-rider" problem, to warn about human overpopulation and resource availability. Other authors have used his metaphor to explain such diverse problems as African famines, air pollution, fisheries declines, and urban crime. What Hardin was really describing, however, was an **open access system** in which there are no rules to manage resource use. In fact, many communal resources have been successfully managed for centuries by cooperative arrangements among users. Some examples include Native American management of wild rice beds and hunting grounds; Swiss village-owned mountain forests and pastures; Maine lobster fisheries; communal irrigation systems in Spain, Bali, and Laos; and nearshore fisheries almost everywhere in the world. A large body of literature in economics and social sciences describes how these cooperative systems work. Among the features shared by **communal resource management systems** are: (1) community members have lived on the land or used the resource for a long time and anticipate that their children and grandchildren will as well, thus giving them a strong interest in sustaining the resource and maintaining bonds with their neighbors; (2) the resource has clearly defined boundaries; (3) the community group size is known and enforced; (4) the resource is relatively scarce and highly variable so that the community is forced to be interdependent; (5) management strategies appropriate for local conditions have evolved over time and are collectively enforced; that is, those affected by the rules have a say in them; (6) the resource and its use are actively monitored, discouraging anyone from cheating or taking too much; (7) conflict resolution mechanisms reduce discord; and (8) incentives encourage compliance with rules, while sanctions for noncompliance keep community members in line.

Rather than being the only workable solution to problems in common pool resources, privatization and increasing external controls often prove to be disastrous. In places where small villages have owned and operated local jointly held forests and fishing grounds for generations, nationalization and commodification of resources generally have led to rapid destruction of both society and ecosystems. Where communal systems once enforced restraint over harvesting, privatization encouraged narrow self-interest and allowed outsiders to take advantage of the weakest members of the community.

A tragic example is the forced privatization of Indian reservations in the United States. Failing to recognize or value local knowledge and forcing local people to participate in a market economy allowed outsiders to disenfranchise native people and resulted in disastrous resource exploitation. Learning to distinguish between open access systems and communal property regimes is important in understanding how best to manage natural resources.

## 23.2 Population, Technology, and Scarcity

Are we about to run out of essential natural resources? It stands to reason that if we consume a fixed supply of nonrenewable resources at a constant rate, eventually we'll use up all the economically recoverable reserves. There are many warnings in the environmental literature that our extravagant depletion of nonrenewable resources sooner or later will result in catastrophe. The dismal prospect of Malthusian diminishing returns and a life of misery, starvation, and social decay inspire many environmentalists to call for an immediate change to voluntary simplicity and lower consumption rates. Models for exploitation rates of nonrenewable resources—called Hubbert curves, after Stanley Hubbert who developed them—often closely match historic experience for natural resource depletion (fig. 23.12).

Many economists, however, contend that neither supply/demand relationships nor economically recoverable reserves are rigidly fixed. Human ingenuity and enterprise often allow us to respond to scarcity in ways that postpone or alleviate the dire effects predicted by modern Jeremiahs. In the next section we will look at some of the arguments for and against limits to growth of the global economy.

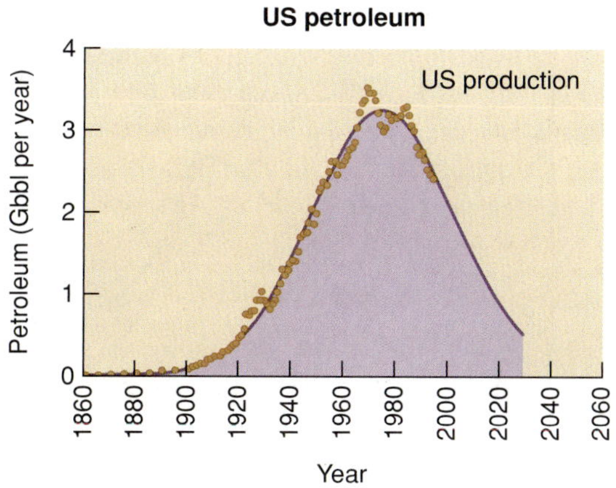


**FIGURE 23.12** United States petroleum production, 1860 to 2000. Dots indicate actual production. The bell-shaped curve is a theoretical Hubbert curve for a nonrenewable resource. The shaded area under the curve, representing 220 Gbbl (Gbbl = Gigabarrels or billions of standard 42-gallon barrels), is an estimate of the total economically recoverable resource.

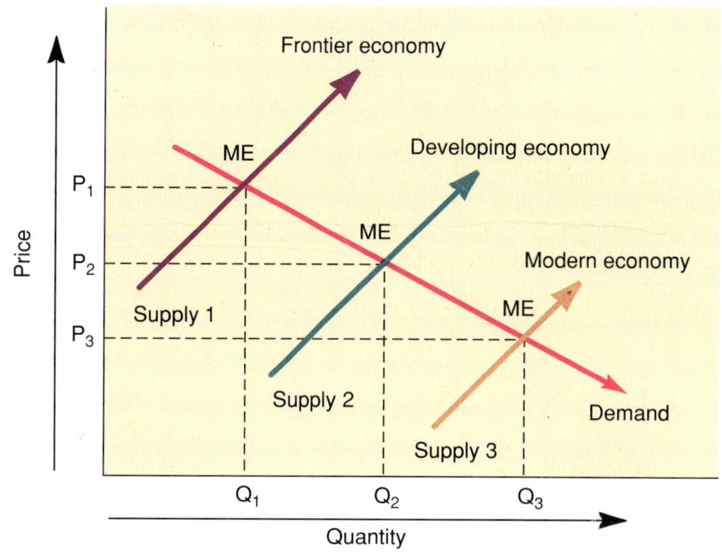


**FIGURE 23.13** Supply and demand curves at three different stages of economic development. At each stage there is a market equilibrium point at which supply and demand are in balance. As the economy becomes more efficient, the equilibrium shifts so there is a larger quantity available at a lower price than before. (P = price, Q = quantity, ME = market equilibrium)

## Scarcity can lead to innovation

In a pioneer or frontier economy, methods for harvesting resources and turning them into useful goods and services tend to be inefficient and wasteful. This may not matter, however, if the supply of resources exceeds the demand for them. The loggers, for example, who swarmed across the Great Lakes Forest at the beginning of the twentieth century, wasted a vast amount of wood. Their inefficiency didn't seem important, however, because the supply of trees appeared infinite, while labor, capital, and means for getting lumber to markets was scarce. As markets and societies develop, however, better technology and more efficient systems allow people to create the same amount of goods and services using far fewer resources. We now produce hundreds of times the crop yield with less labor from the same land that pioneers farmed.

This increasing technological efficiency can dramatically shift supply and demand relationships (fig. 23.13). As technology makes goods and services cheaper to produce, the quantity available at a given price can increase greatly. The market equilibrium or the point at which supply and demand equilibrate will shift to lower prices and higher quantities as a market matures.

Scarcity can actually serve as a catalyst for innovation and change (fig. 23.14). As materials become more expensive and difficult to obtain, it becomes cost-effective to discover new supplies or to use available ones more efficiently. The net effect is as if a new supply of resources has been created or discovered. Several important factors play a role in this cycle of technological development:

- Technical inventions can increase efficiency of extraction, processing, use, and recovery of materials.
- Substitution of new materials or commodities for scarce ones can extend existing supplies or create new ones. For instance, substitution of aluminum for copper, concrete for structural steel, grain for meat, and synthetic fibers for natural ones all remove certain limits to growth.
- Trade makes remote supplies of resources available and may also bring unintended benefits in information exchange and cultural awakening.
- Discovery of new reserves through better exploration techniques, more investment, and looking in new areas becomes rewarding as supplies become limited and prices rise.
- Recycling becomes feasible and accepted as resources become more valuable. Recycling now provides about 37 percent of the iron and lead, 20 percent of the copper, 10 percent of the aluminum, and 60 percent of the antimony that is consumed each year in the United States.

## Carrying capacity is not necessarily fixed

Despite repeated warnings that rapidly growing populations and increasing affluence are bound to exhaust natural resources and result in rapid price increases, technological developments of the sort described earlier have resulted in price decreases for most raw materials over the last hundred years. Consider copper for example. Twenty years ago worries about impending shortages led the United States to buy copper and store it in strategic stockpiles. Estimates of the amount of this important metal needed for electric motors, telephone lines, transoceanic cables, and other uses essential for industrialized society far exceeded known reserves. It looked as if severe shortages and astronomical price increases were inevitable. But then aluminum power lines, satellites, fiber optics, integrated circuits, microwave transmission, and other inventions greatly diminished the need for copper. Although prices

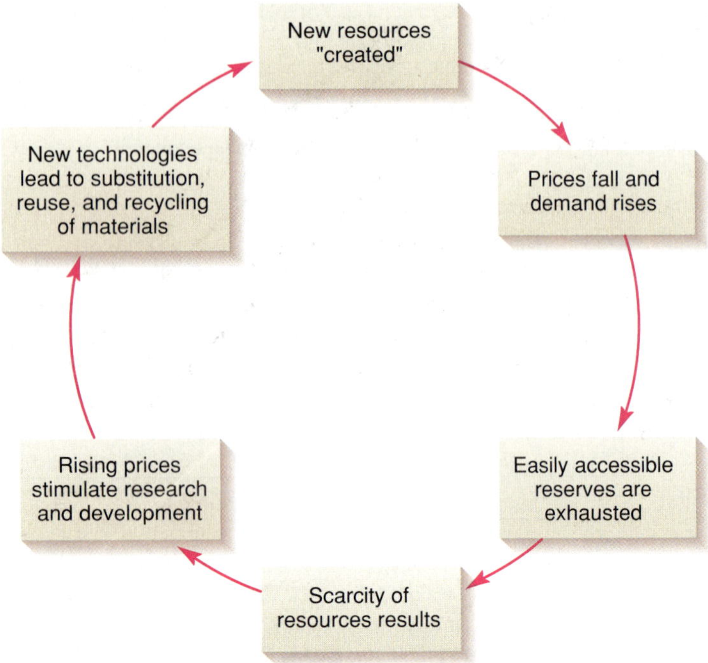


**FIGURE 23.14** Scarcity/development cycle. Paradoxically, resource use and depletion of reserves can stimulate research and development, the substitution of new materials, and the effective creation of new resources.

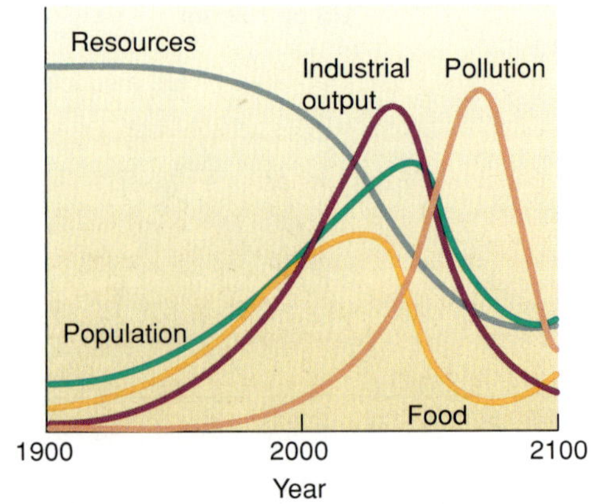


**FIGURE 23.15** A run of one of the world models in *Limits to Growth.* This model assumes business-as-usual for as long as possible until Malthusian limits cause industrial society to crash. Notice that pollution continues to increase well after industrial output, food supplies, and population have all plummeted.

are highly variable because of world politics and trade policies, the general trend for most materials has been downward in this century. It is as if the carrying capacity of our natural resources—at least in terms of copper—has been increased.

Economists generally believe that this pattern of substitutability and technological development is likely to continue into the future. Ecologists generally disagree. There are bound to be limits, they argue, to how many people our environment can support. An interesting example of this debate occurred in 1980. Ecologist Paul Ehrlich bet economist Julian Simon that increasing human populations and growing levels of material consumption would inevitably lead to price increases for natural resources. They chose a package of five metals—chrome, copper, nickel, tin, and tungsten—priced at the time at $1,000. If, in ten years, the combined price (corrected for inflation) was higher than $1,000, Simon would pay the difference. If the combined price had fallen, Ehrlich would pay. In 1990 Ehrlich sent Simon a check for $576.07; the price for these five metals had fallen 47.6 percent.

Does this prove that resource abundance will continue indefinitely? Hardly. Ehrlich claims that the timing and set of commodities chosen simply were the wrong ones. The fact that we haven't yet run out of raw materials doesn't mean that it will never happen. Many ecological economists now believe that some nonmarket resources such as ecological processes may be more irreplaceable than tangible commodities like metals. What do you think? Are we approaching the limits of nature to support more humans and more consumption? Which resources, if any, do you think are most likely to be limiting in the future?

## Economic models compare growth scenarios

In the early 1970s, an influential study of resource limitations was funded by the Club of Rome, an organization of wealthy business owners and influential politicians. The study was carried out by a team of scientists from the Massachusetts Institute of Technology headed by the late Donnela Meadows. The results of this study were published in the 1972 book *Limits to Growth.* A complex computer model of the world economy was used to examine various scenarios of different resource depletion rates, growing population, pollution, and industrial output. Given the Malthusian assumptions built into this model, catastrophic social and environmental collapse seemed inescapable.

Figure 23.15 shows one example of the world model. Food supplies and industrial output rise as population grows and resources are consumed. Once past the carrying capacity of the environment, however, a crash occurs as population, food production, and industrial output all decline precipitously. Pollution continues to grow as society decays and people die, but, eventually, it also falls. Notice the similarity between this set of curves and the "boom and bust" population cycles described in chapter 6.

Many economists criticized these results because they discount technological development and factors that might mitigate the effects of scarcity. In 1992, the Meadows group published updated computer models in *Beyond the Limits* that include technological progress, pollution abatement, population stabilization, and new public policies that work for a sustainable future. If we adopt these changes sooner rather than later, the computer shows an outcome like that in figure 23.16, in which all factors stabilize sometime in this century at an improved standard of living for everyone. Of course neither of these computer models shows what will happen, only what some possible outcomes *might* be, depending on the choices we make.

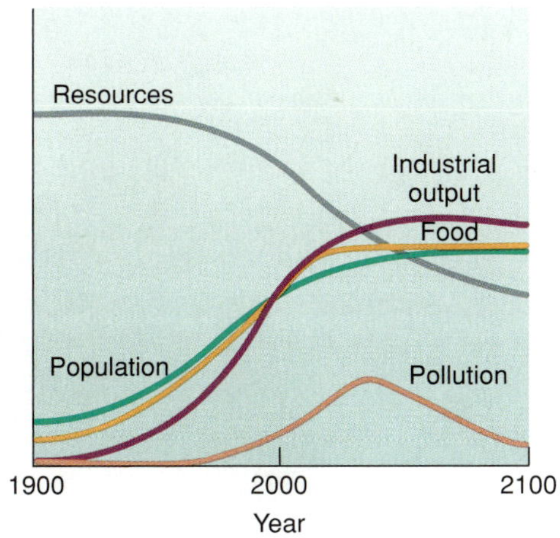


**FIGURE 23.16** A run of the world model from *Beyond the Limits.* This model assumes that population and consumption are curbed, new technologies are introduced, and sustainable environmental policies are embraced immediately, rather than after resources are exhausted.

### Why not conserve resources?

Even if large supplies of resources are available or the technological advances to mitigate scarcity exist, wouldn't it be better to reduce our use of natural resources so they will last as long as possible? Will anything be lost if we're frugal now and leave more to be used by future generations? Many economists would argue that resources are merely a means to an end rather than an end in themselves. They have to be used to have value.

If you bury your money in a jar in the backyard, it will last a long time but may not be worth much when you dig it up. If you invest it productively, you may have much more in the future than you do now. Furthermore, a window of opportunity for investment may be open now but not later. Suppose that 300 years ago our ancestors had decided that the Industrial Revolution should not be allowed to continue because it required too much resource consumption. There certainly would be more easily accessible resources available now, but would we be better off than we are? What do you think? Where is the proper balance between using our resources now or saving them for the future?

## 23.3 Natural Resource Accounting

How can we determine the value of environmental goods and services? Some of the most crucial natural resources are not represented by monetary prices in the marketplace. Certain resource allocation decisions are political or social. Others are simply ignored. Groundwater, sunlight, clean air, biological diversity, and other assets that we all share in common often are treated as public goods that anyone can use freely. Until they are transformed by human activities, natural resources are regarded as having little value. Ecological economics calls for recognition of the real value of those resources in calculating economic progress. In this section we'll look at some suggestions for how we could do this.

### Gross national product is our dominant growth measure

The most common way to measure a nation's output is **gross national product (GNP).** GNP can be calculated in two ways. One is the money flow from households to businesses in the form of goods and services purchased (see fig. 23.8). The other is to add up all the costs of production in the form of wages, rent, interest, taxes, and profit. In either case, a subtraction is made for capital depreciation, the wear and tear on machines, vehicles, and buildings used in production. Some economists prefer **gross domestic product (GDP),** which includes only the economic activity within national boundaries. Thus the vehicles made and sold by Ford in Europe don't count in GDP.

Both GNP and GDP are criticized as a measure of real progress or well-being because it doesn't attempt to distinguish between economic activities that are beneficial or harmful. A huge oil spill that pollutes beaches and kills wildlife, for example, shows up as a positive addition to GNP because of the business generated by cleanup efforts.

Ecological economists also criticize GNP because it doesn't account for natural resources used up or ecosystems damaged by economic activities. Robert Repeto of the World Resources Institute estimates that soil erosion in Indonesia, for instance, reduces the value of crop production about 40 percent per year. If natural capital were taken into account, Indonesian GNP would be reduced by at least 20 percent annually.

Similarly, Costa Rica experienced impressive increases in timber, beef, and banana production between 1970 and 1990. But decreased natural capital during this period represented by soil erosion, forest destruction, biodiversity losses, and accelerated water runoff add up to at least $4 billion or about 25 percent of annual GNP.

### Alternate measures account for well-being

A number of systems have been proposed as alternatives to GNP that reflect genuine progress and social welfare. In their 1989 book, Herman Daly and John Cobb proposed a **genuine progress index (GPI)** that takes into account real per capita income, quality of life, distributional equity, natural resource depletion, environmental damage, and the value of unpaid labor. They point out that while per capita GNP in the United States nearly doubled between 1970 and 2000, per capita GPI increased only 4 percent (fig. 23.17). Some social service organizations would add to this index the costs of social breakdown and crime, which would decrease real progress even further over this time span.

A newer measure is the **Environmental Performance Index (EPI)** created by researchers at Yale and Columbia Universities to evaluate national sustainability and progress toward achievement of the United Nations Millennium Development Goals. The EPI is based on sixteen indicators tracked in six categories: environmental health, air quality, water resources, productive natural resources, biodiversity and habitat, and sustainable energy. The top-ranked countries—New Zealand, Sweden, Finland, the Czech

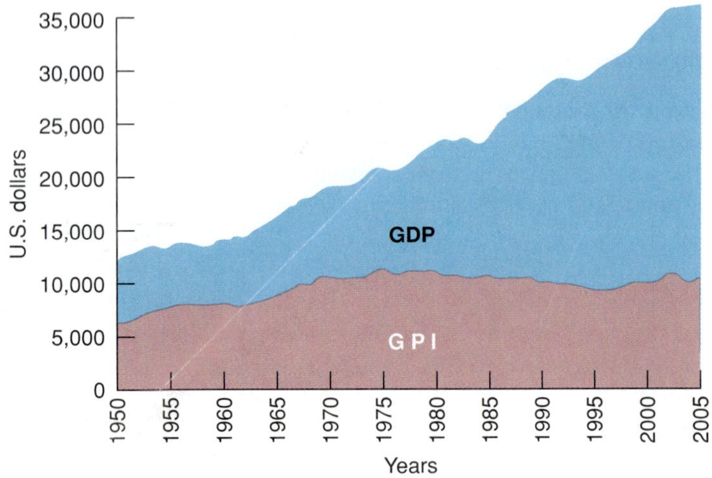


**FIGURE 23.17** Although per capita GDP in the United States nearly doubled between 1970 and 2000 in inflation-adjusted dollars, a genuine progress index that takes into account natural resource depletion, environmental damage, and options for future generations hardly increased at all.
**Source:** Data from *Redefining Progress,* 2006.

**FIGURE 23.18** Raising agricultural productivity and rural incomes are high priorities of the UN Millennium Development Goals.

Republic, and the United Kingdom—all commit significant resources and effort to environmental protection. In 2006, the United States ranked 28th in the EPI, or lower than Malaysia, Costa Rica, Columbia, and Chile, all of which have between 6 and 15 times lower GDP than the U.S. See Data Analysis (p. 539) for a graph of human development index (HDI) versus EPI.

The United Nations Development Program (UNDP) uses a benchmark called the human development index (HDI) to track social progress. HDI incorporates life expectancy, educational attainment, and standard of living as critical measures of development. Gender issues are accounted for in the gender development index (GDI), which is simply HDI adjusted or discounted for inequality or achievement between men and women.

In its annual Human Development Report, the UNDP compares country-by-country progress. As you might expect, the highest development levels are generally found in North America, Europe, and Japan. In 2006, Norway ranked first in the world in both HDI and GDI. The United States ranked eighth while Canada was sixth. The 25 countries with the lowest HDI in 2006 were all in Africa. Haiti ranks the lowest in the Western Hemisphere.

Although poverty remains widespread in many places, encouraging news also can be found in development statistics. Poverty has fallen more in the past 50 years, the UNDP reports, than in the previous 500 years. Child death rates in developing countries as a whole have been more than halved. Average life expectancy has increased by 30 percent while malnutrition rates have declined by almost a third. The proportion of children who lack primary school has fallen from more than half to less than a quarter. And the share of rural families without access to safe water has fallen from nine-tenths to about one-quarter.

Some of the greatest progress has been made in Asia. China and a dozen other countries with populations that add up to more than 1.6 billion, have decreased the proportion of their people living below the poverty line by half. Still, in the 1990s the number of people with incomes less than $1 per day increased by almost 100 million to 1.3 billion—and the number appears to be growing in every region except Southeast Asia and the Pacific. Even in industrial countries, more than 100 million people live below the poverty line and 37 million are chronically unemployed.

Economic growth can be a powerful means of reducing poverty, but its benefits are not automatic. The GNP of Honduras, for instance, grew 2 percent per year in the 1980s and yet poverty doubled. To combat poverty, the UNDP calls for "pro-poor growth" designed to spread benefits to everyone. Specifically, some key elements of this policy would be to: (1) create jobs that pay a living wage, (2) lessen inequality, (3) encourage small-scale agriculture, microenterprises, and the informal sector, (4) foster technological progress, (5) reverse environmental decline in marginal regions, (6) speed demographic transitions, and (7) provide education for all. Since about three-quarters of the world's poorest people live in rural areas, raising agricultural productivity and incomes is a high priority for these actions (fig. 23.18).

## New approaches incorporate nonmarket values

New tools and new approaches are needed to represent nature in national accounting and human development. Among the natural resource characteristics that ecological economists suggest be taken into account include:

- Use values: the price we pay to use or consume a resource
- Option value: preserving options for the future
- Existence value: those things we like to know still exist even though we may never use or even see them
- Aesthetic values: aspects we appreciate for their beauty
- Cultural values: factors important for cultural identity
- Scientific and educational values: information or experience-rich aspects of nature

How can we measure these values of natural resources and ecological services when they are not represented in market systems? Ecological economists often have to resort to "shadow pricing" or other indirect valuation methods for natural resources. For instance, what is the worth of a day of canoeing on a wild river? We might measure opportunity costs such as how much we pay to get to the river or to rent a canoe. The direct out-of-pocket costs might represent only a small portion, however, of what it is really worth to participants. Another approach is contingent valuation in which potential resource users are asked, "How much would you be willing to pay for this experience?" or "What price would you be willing to accept to sell your access or forego this opportunity?" These approaches are controversial because people may report what they think they ought to pay rather than what they would really pay for these activities.

Several ecological economists have attempted to put a price on the goods and services provided by natural ecosystems. Although many ecological processes have no direct market value, we can estimate replacement costs, contingent values, shadow prices, and other methods of indirect assessment to determine a rough value. For instance, we now dispose of much of our wastes by letting nature detoxify them. How much would it cost if we had to do this ourselves? Ecological economists look at everything from recreational beaches to forest lumber to hidden services such as the ocean's regulation of atmospheric carbon dioxide to pollination of crops by insects.

The estimated annual value of all ecological goods and services provided by nature range from $16 trillion to $54 trillion, with a median worth of $33 trillion, or about three-fourths the combined annual GNPs of all countries in the world (table 23.3). These estimates are probably understated because they omit ecosystem services from several biomes, such as deserts and tundra, that are poorly understood in terms of their economic contributions. The most valuable ecosystems in terms of biological processes are wetlands and coastal estuaries because of their high level of biodiversity and their central role in many biogeochemical cycles.

In 2003, economists from Cambridge University (U.K.) estimated that protecting a series of nature reserves representing samples of all major biomes would cost (U.S.) $45 billion per year, but would preserve ecological services worth between $4.4 trillion to $5.2 trillion annually.

**TABLE 23.3**

**Estimated Annual Value of Ecological Services**

| Ecosystem Services | Value (Trillion $U.S.) |
|---|---|
| Soil formation | 17.1 |
| Recreation | 3.0 |
| Nutrient cycling | 2.3 |
| Water regulation and supply | 2.3 |
| Climate regulation (temperature and precipitation) | 1.8 |
| Habitat | 1.4 |
| Flood and storm protection | 1.1 |
| Food and raw materials production | 0.8 |
| Genetic resources | 0.8 |
| Atmospheric gas balance | 0.7 |
| Pollination | 0.4 |
| All other services | 1.6 |
| **Total value of ecosystem services** | **33.3** |

**Source:** Adapted from R. Costanza et al., "The Value of the World's Ecosystem Services and Natural Capital," *Nature*, Vol. 387 (1997).

## Cost-benefit analysis aims to optimize resource use

One way to evaluate public projects is to analyze the costs and benefits they generate in a **cost-benefit analysis (CBA).** This process attempts to assign values to resources as well as to social and environmental effects of carrying out or not carrying out a given undertaking. It tries to find the optimal efficiency point at which the marginal cost of pollution control equals the marginal benefits (fig. 23.19).

CBA is one of the main conceptual frameworks of resource economics and is used by decision makers around the world as a way of justifying the building of dams, roads, and airports, as well as in considering what to do about biodiversity loss, air pollution, and global climate change. Deeply entrenched in bureaucratic practice and administrative culture, this technique has become much more widespread in American public affairs since the Reagan administration's executive orders in the 1980s calling for the application of CBA to all regulatory decisions and legislative proposals. Many conservatives see CBA as a way of eliminating what they consider to be unnecessary and burdensome requirements to protect clean air, clear water, human health, or biodiversity. They would like to add a requirement that all regulations be shown to be cost-effective.

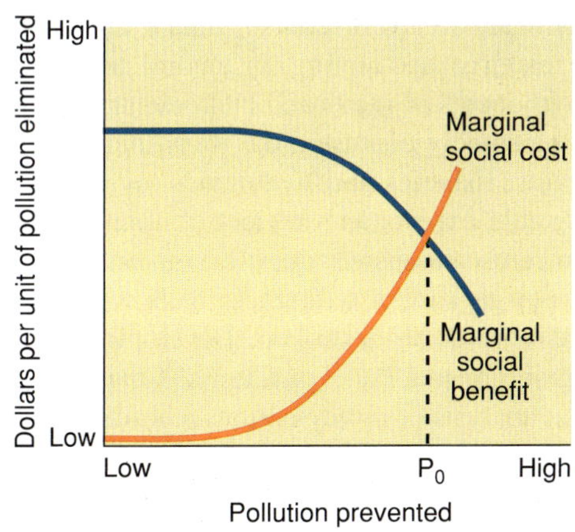


**FIGURE 23.19** To achieve maximum economic efficiency, regulations should require pollution prevention up to the optimum point ($P_o$) at which the costs of eliminating pollution just equal the social benefits of doing so.

**FIGURE 23.20** What is the value of solitude or beauty? How would you assign costs and benefits to a scene such as this?

The first step in CBA is to identify who or what might be affected by a particular plan. What are the potential outcomes and results? What alternative actions might be considered? After identifying and quantifying all the contingent factors, an attempt is made to assign monetary costs and benefits to each component. Usually the direct expenses of a project are easy to ascertain. How much will you have to pay for land, materials, and labor? The monetary worth of lost opportunities—to swim or fish in a river, or to see birds in a forest—on the other hand, are much harder to appraise. How would you put a price on good health or a long life? It's also important to ask who will bear the costs and who will reap the benefits of any proposal. Are there consequences that cannot be given a monetary price? What is a bug or a bird worth, for instance, or the opportunity for solitude or inspiration (fig. 23.20)? Eventually, the decision maker compares all the costs and benefits to see whether the project is justified or whether some alternative action might bring more benefits at less cost.

At the same time that some environmentalists are using CBA as a way of obtaining recognition for environmental issues within the terms of mainstream discourse, others are challenging this method of resource accounting as amoral and deeply flawed. Grassroots opponents of roads and hydroelectric dams around the world have repeatedly contested the ways that CBA values land, forests, streams, fisheries and livelihoods, as well as its reliance on unaccountable experts and neglect of equity issues. Ordinary people often refuse to answer questions about how much money they would pay to save a wilderness or how much they would accept to allow it to be destroyed. Developing world delegates to the Intergovernmental Panel on Climate Change angrily rejected a cost-benefit analysis of policy options regarding global warming that assigned a higher value to lives of people in industrialized countries compared to those in less-developed nations.

A study by the Economic Policy Institute of Washington, D.C., found that costs for complying with environmental regulations are almost always less than industry and even governments estimate they will be. For example, electric utilities in the United States claimed that it would cost $4 to $5 billion to meet the 1990 Clean Air Act. But by 1996, utilities were actually saving $150 million per year. Similarly, when CFCs were banned, automobile manufacturers protested it would add $1,200 to the cost of each new car. The actual cost was about $40.

Some other criticisms of CBA include its absence of standards, inadequate attention to alternatives, and the placing of monetary values on intangible and diffuse or distant costs and benefits. Who judges how costs and benefits will be estimated? How can we compare things as different as the economic gain from cheap power with loss of biodiversity or the beauty of a free-flowing river? Critics claim that placing monetary values on everything could lead to a belief that only money and profits count and that any behavior is acceptable as long as you can pay for it. Sometimes speculative or even hypothetical results are given specific numerical values in CBA and then treated as if they are hard facts. Risk-assessment techniques (see chapter 8) may be more appropriate for comparing uncertainties.

## 23.4 Market Mechanisms Can Reduce Pollution

We are becoming increasingly aware that our environment and economy are mutually interconnected. Natural resources and ecological services are essential for a healthy economy, and a vigorous economy can provide the means to solve environmental problems. In this section, we'll explore some of these links.

### Using market forces

As you've learned from previous chapters in this book, most scientists regard global climate change as the most serious environmental problem we face currently. In 2006, the business world got a harsh warning about this problem from British economist, Sir Nicolas Stern. Commissioned by the British treasury department to assess the threat of global warming, Sir Nicolas, who formerly was chief economist at the World Bank, issued a 700-page study that concluded that if we don't act to control greenhouse gases, the damage caused by climate change could be equivalent to losing as much as 20 percent of the global GDP every year. This could have an impact on our lives and environment greater than the worldwide depression or the great wars of the twentieth century.

We have many options for combating climate change, but economists believe that **market forces** can reduce pollution more efficiently than rigid rules and regulations. Assessing a tax, for example, on each ton of carbon emitted could have the desired effect of reducing greenhouse gases and controlling climate change, but could still allow industry to search for the most cost-effective ways to achieve these goals. It also creates a continuing incentive to search for better ways to reduce emissions. The more you reduce your discharges, the more you save.

The cost of climate control won't be cheap. Sir Nicolas calculates that it will take about $500 billion per year (1 percent of global GDP) to avoid the worst impacts of climate change if we

**FIGURE 23.21** Markets for low-carbon energy could be worth $500 billion per year by 2050, and could create millions of high-paying jobs.

act now. Half a trillion dollars is a lot of money, but it's a bargain compared to his estimates of $10 trillion in annual losses and costs of climate change in 50 years if we don't do something. And the longer we wait, the more expensive carbon reduction and adaptation are going to be.

On the other hand, reducing greenhouse gas emissions and adapting to climate change will create significant business opportunities, as new markets are created in low-carbon energy technologies and services (fig. 23.21). These markets could create millions of jobs and be worth hundreds of billions of dollars every year. Already, Europe has more than 5 million jobs in renewable energy, and the annual savings from solar, wind, and hydro power are saving the European Union about $10 billion per year in avoided oil and natural gas imports. Being leaders in the fields of renewable energy and carbon reduction gives pioneering countries a tremendous business advantage in the global marketplace. Markets for low-carbon energy could be worth $500 billion per year by 2050, according to the Stern report.

## Is emissions trading the answer?

They Kyoto Protocol, which was negotiated in 1997, and has been ratified by every industrialized nation in the world except the United States and Australia, sets up a mechanism called **emissions trading** to control greenhouse gases. This is also called a **cap-and-trade** approach. The first step is to mandate upper limits (the cap), on how much each country, sector, or specific industry is allowed to emit. Companies that can cut pollution by more than they're required can sell the **credit** to other companies that have more difficulty meeting their mandated levels. Suppose you've just built a state-of-the-art integrated gasification combined cycle (IGCC) power plant (chapter 19) that allows you to capture and store $CO_2$ for about $20 per ton. You're allowed to emit a thousand tons of $CO_2$ per day, but you could easily cut your releases below that amount.

Suppose, further, that your neighboring utility has a dirty, old coal-fired power plant for which it would cost $60 per ton to keep $CO_2$ out of the smokestack. You might strike a deal with your neighbor. For $40 per ton, you'll reduce your emissions enough to meet his mandated cap. You make $20 per ton, and your neighbor saves $20 per ton. Both benefit. But if your neighbor can find an even cheaper way to **offset** his carbon emissions, he's free to do so. This creates an incentive to continually search for ever more cost-effective ways to reduce emissions.

Here's another example of carbon trading that might have even greater application to your life. Suppose you drive an old car that doesn't get very good mileage. You may feel guilty about the $CO_2$ you're emitting, but perhaps you can't afford to buy a new, more efficient vehicle. There are several organizations that will now sell an offset to you. For about $20 per ton (or about $100 per year) for the average American car, they'll plant trees, build a windmill, or provide solar lights to a village in a developing country to compensate for your emissions. You can take pride in being **carbon-neutral** at a far lower price than buying a new automobile.

## Sulfur trading offers a good model

The 1990 U.S. Clean Air Act created one of the first market-based systems for reducing air pollution. It mandated a decrease in acid-rain-causing sulfur dioxide ($SO_2$) from power plants and other industrial facilities. A $SO_2$ targeted reduction was set at 10 million tons per year, leaving it to industry to find the most efficient way to do this. The government expected that meeting this goal would cost companies up to $15 billion per year, but the actual cost has been less than one-tenth of that. Prices on the sulfur exchange have varied from $60 to $800 per ton depending on the availability and price of new technology, but most observers agree that the market has found much more cost-effective ways to achieve the desired goal than rigid rules would have created.

This program is regarded as a shining example of the benefits of market-based approaches. There are complaints, however, that while nationwide emissions have come down, "hot spots" remain where local utilities have paid for credits rather than install pollution abatement equipment. If you're living in one of these hot spots and continuing to breathe polluted air, it's not much comfort to know that nationwide average air quality has improved. Currently, credits and allowances of more than 30 different air pollutants are traded in international markets.

## Carbon trading is already at work

Climate change is revolutionizing global economics. In 2006, approximately (U.S.) $28 billion worth of climate credits, equivalent to 1 billion tons of $CO_2$, traded hands on international markets. It's expected that this market will expand by nearly five-fold in 2008. By far the most active market currently is the Amsterdam-based European Climate Exchange. The United States has a climate market in Chicago, but at this point, participation is only voluntary because the U.S. doesn't have mandatory emissions limits, and carbon credits are selling for only about one-tenth the price they are in Europe.

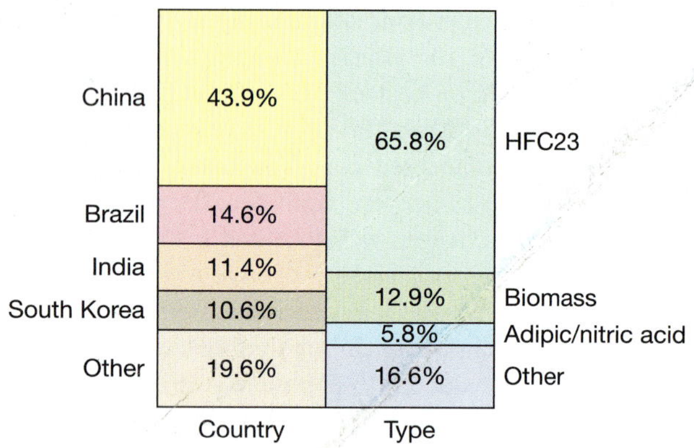


**FIGURE 23.22** Worldwide emissions reductions payments by country and type. Currently, four countries are collecting 80 percent of all proceeds from emissions trading, and two-thirds of those payments are going for relatively cheap HFC23 incineration. Is this fair?
**Source:** United Nations, 2007.

In 2006, more than 80 percent of the international emissions payments went to just four countries, and nearly two-thirds of those payments were for reductions of the refrigerant HFC-23 (fig. 23.22). Most entrepreneurs are uninterested in deals less than about $250,000. Smaller projects just aren't worth the time and expense of setting them up. In one of the biggest deals so far, a consortium of British bankers signed a contract to finance an incinerator on a large chemical factory in Quzhou, in China's Zhejiang Province. The incinerator will destroy hydrofluorocarbon (HFC-23) that previously had been vented into the air. This has a double benefit: HFC-23 destroys stratospheric ozone, and it also is a potent greenhouse gas (approximately 11,700 times as powerful as $CO_2$). The $500 million deal will remove the climate-changing equivalent of the $CO_2$ emitted by 1 million typical American cars each driven 20,000 km per year. But the incinerator will cost only $5 million—a windfall profit to be split between the bankers and the factory owners.

There's a paradox in this deal. HFC production in China and India is soaring because a growing middle class fuels a demand for refrigerators and air conditioners. The huge payments flowing into these countries under the Kyoto Protocol are helping their economies grow and increasing middle-class affluence, and thus creating more demand for refrigerators and air conditioners. Furthermore, air conditioners using this refrigerant are much less energy efficient than newer models, so their increasing numbers are driving the demand for electricity, which currently is mostly provided by coal-fired power plants.

Critics of our current emissions markets point out that this mechanism was originally intended to encourage the spread of renewable energy and nonpolluting technology to developing countries in places such as sub-Saharan Africa. It was envisioned as a way to spread solar panels, windmills, tree farms, and other technologies that would provide climate control and also speed development of the poorest people. Instead, marketing emission credits, so far, is benefiting primarily bankers, consultants, and factory owners and is leading to short-term fixes rather than fundamental, long-term solutions.

## 23.5 Trade, Development, and Jobs

Trade can be a powerful tool in making resources available and raising standards of living. Think of the things you now enjoy that might not be available if you had to live exclusively on the resources available in your immediate neighborhood. Too often, the poorest, least powerful people suffer in this global marketplace. To balance out these inequities, nations can deliberately invest in economic development projects. In this section, we'll look at some aspects of trade, development, business, and jobs that have impacts on our environment and welfare.

### International trade brings benefits but also intensifies inequities

The banking and trading systems that regulate credit, currency exchange, shipping rates, and commodity prices were set up by the richer and more powerful nations in their own self-interest. The General Agreement on Tariffs and Trade (GATT) and World Trade Organization (WTO) agreements, for example, negotiated primarily between the largest industrial nations, regulate 90 percent of all international trade.

These systems tend to keep the less-developed countries in a perpetual role of resource suppliers to the more-developed countries. The producers of raw materials, such as mineral ores or agricultural products, get very little of the income generated from international trade (fig. 23.23).

Policies of the WTO and the IMF have provoked criticism and resistance in many countries. As a prerequisite for international development loans, the IMF frequently requires debtor nations to adopt harsh "structural adjustment" plans that slash welfare programs and impose cruel hardships on poor people. The WTO has issued numerous rulings that favor international

**FIGURE 23.23** What do we really pay for when we purchase a dollar's worth of coffee?

trade over pollution prevention or protection of endangered species. Trade conventions such as the North American Free Trade Agreement (NAFTA) have been accused of encouraging a "race to the bottom" in which companies can play one country against another and move across borders to find the most lax labor and environmental protection standards.

### Aid often doesn't help the people who need it

No single institution has more influence on financing and policies of developing countries than the World Bank. Of some $25 billion loaned each year for development projects by international agencies, about two-thirds comes from the World Bank. Founded in 1945 to fund reconstruction of Europe and Japan, the World Bank shifted its emphasis to aid developing countries in the 1950s. Many of its projects have had adverse environmental and social effects, however. Its loans often go to corrupt governments and fund ventures such as nuclear power plants, huge dams, and giant water diversion schemes. Former U.S. Treasury Secretary Paul O'Neill said that these loans have driven poor countries "into a ditch" by loading them with unpayable debt. He said that funds should not be loans, but rather grants to fight poverty.

### Microlending helps the poorest of the poor

As the opening case study for this chapter shows, tiny loans can change the lives of the poorest of the poor. The Grameen Bank, founded by Dr. Muhammad Yunus, has assisted millions of people—most of them low-status women who have no other way to borrow money at reasonable interest rates. This model is now being used by hundreds of other development agencies around the world (fig. 23.24). Even in the United States, organizations assist microenterprises with loans, grants, and training. The Women's Self-Employment Project in Chicago, for instance, teaches job skills to single mothers in housing projects. Similarly, "tribal circle" banks on Native American reservations successfully finance microscale economic development ventures.

One of the most important innovations of the Grameen Bank is that borrowers take out loans in small groups. Everyone in the group is responsible for each other's performance. The group not only guarantees loan repayment, it helps businesses succeed by offering support, encouragement, and advice. Where banks depend on the threat of foreclosure and a low credit rating to ensure debt repayment, the Grameen Bank has something at least as powerful for poor villagers—the threat of letting down your neighbors and relatives. Becoming a member of a Grameen group also requires participation in a savings program that fosters self-reliance and fiscal management.

Microlending is built on a series of small loans. You start with a very small amount: the equivalent of only a few dollars. When that loan is repaid, you can take out another one. This creates an ongoing relationship between the borrowers and lenders. Success builds on success. And as each loan is granted, it allows the lender to learn about other aspects of the borrower's life, and to help in ways other than merely providing money. The process of running

**FIGURE 23.24** With a loan of only a few dollars, this Chinese coal deliverer could buy his own cart and more than double his daily income. If you could make a tiny loan that would change his life, wouldn't you do it?

a successful business and repaying each installment of the loan on time transforms many individuals. Women, who previously had little power or self-esteem, are empowered with a sense of pride and accomplishment. Dr. Yunus discovered that money going to families through women helped the families much more than the same amount of money in men's salaries.

The most recent venture for the Grameen Bank is providing mobile phone service to rural villages. Supplying mobile phones to poor women not only allows them to communicate, it provides another business opportunity. They rent out their phone to neighbors, giving the owner additional income, and linking the whole village to the outside world. Suddenly, people who had no access to communication can talk with their relatives, order supplies from the city, check on prices at the regional market, and decide when and where to sell their goods and services. This is a great example of "bottom-up development." Founded in 1996, Grameen Phone now has 2.5 million subscribers and is Bangladesh's largest mobile phone company.

In 2006, Dr. Yunus received the Nobel Peace Prize for his groundbreaking work in helping the poor. He has formed a political party called "Nagorik Shakti," (Citizen Power), and is running for national office. Changing the lives of the poor continues to be his highest priority.

## 23.6 Green Business

During the first Industrial Revolution 200 years ago, raw materials such as lumber, minerals, and clean water seemed inexhaustible, while nature was regarded as something to be tamed and civilized. The "solution to pollution is dilution," was a common assumption in disposal of unwanted wastes. Today business leaders are increasingly discovering that they can save money by greening up their business practices. Fuel efficiency saves money and cuts greenhouse gases. Conversion to less-hazardous chemicals can cut disposal costs, health costs, and production costs. In addition, these companies win public praise and new customers

## What Do You Think?

### Eco-efficient Business Practices

In 1994, in response to customers' concerns about health problems caused by chemical fumes from new carpeting, wall coverings, and other building materials, Ray Anderson, founder and CEO of Interface, Inc., a billion-dollar-a-year interior furnishing company, decided to review company environmental policy. What he found was that the company really didn't have an environmental vision other than to obey all relevant laws and comply with regulations. He also learned that carpeting—of which Interface was the world's third-largest manufacturer—is one of the highest volume and longest lasting components in landfills. A typical carpet is made of nylon embedded in fiberglass and polyvinyl chloride. After a useful life of about 12 years, most carpeting is ripped up and discarded. Every year, more than 770 million $m^2$ (920 million $yd^2$) of carpet weighing 1.6 billion kg (3.5 billion lbs) ends up in U.S. landfills. The only recycling that most manufacturers do is to shave off some of the nylon for remanufacture. Everything else is buried in the ground where it will last at least 20,000 years.

At about the same time that Interface was undergoing its environmental audit, Anderson was given a copy of Paul Hawken's book *The Ecology of Commerce.* Reading it, he said, was like "a spear through the chest." He vowed to turn his company around, to make its goal sustainability instead of simply maximizing profits. Rather than sell materials, Interface would focus on selling service. The key is what Anderson calls an "evergreen lease." First of all, the carpet is designed to be completely recyclable. Where most flooring companies merely sell carpet, Interface offers to lease carpets to customers. As carpet tiles wear out, old ones are removed and replaced as part of the lease. The customer pays no installation or removal charges, only a monthly fee for constantly fresh-looking and functional carpeting. Everything in old carpet is used to make new product. Only after many reincarnations as carpet, are materials finally sent to the landfill.

Dramatic changes have been made at Interface's 26 factories. Toxic air emissions have been nearly eliminated by changing manufacturing processes and substituting nontoxic materials for more dangerous ones. Solar power and methane from a landfill are replacing fossil fuel use. Interface may be the first carbon-neutral manufacturing company in America. Less waste is produced as more material is recycled and products are designed for eco-efficiency. The total savings from pollution prevention and recycling in 2007 was $150 million.

Not only has Interface continued to be an industry leader, it was named one of the "100 Best Companies to Work For in America" by *Fortune* magazine. Ray Anderson has become a popular speaker on the topic of eco-efficiency and clean production. He cochairs the President's Council on Sustainable Development, was named Entrepreneur of the Year by Ernest & Young, and was the Georgia Conservancy's Conservationist of the Year in 1998. Anderson's book, *Mid-Course Correction: Toward a Sustainable Enterprise,* published in 1999 by Chelsea Green, has won critical acclaim.

Ray Anderson.
*Courtesy Ray Anderson, Interface, Inc.*

Transforming an industry as large as interior furnishing has not been an overnight success. "Like aircraft carriers," Anderson says, "big businesses don't turn on a dime." Still, he has shown that the principles of sustainability and financial success can coexist and can lead to a new prosperity that includes both environmental and human dividends. His motto, that we should "put back more than we take and do good to the Earth, not just no harm," has become a vision for a new industrial revolution that now is reaching many companies beyond his own.

#### *Ethical Considerations*

What responsibilities do businesses have to protect the environment or save resources beyond the legal liabilities spelled out in the law? None whatever, according to conservative economist Milton Friedman. In fact, Friedman argues, it would be unethical for corporate leaders to consider anything other than maximizing profits. To spend time or resources doing anything other than making profits and increasing the value of the company is a betrayal of their duty. What do you think? Should social justice, sustainability, or environmental protection be issues of concern to corporations?

---

by demonstrating an interest in our shared environment. By conserving resources, they also help ensure the long-term survival at their own corporations.

Known by a variety of names, including eco-efficiency, clean production, pollution prevention, industrial ecology, natural capitalism, restorative technology, the natural step, environmentally preferable products, design for the environment, and the next industrial revolution, this movement has had some remarkable successes and presents an encouraging pathway for how we might achieve both environmental protection and social welfare. Some of the leaders in this new approach to business include Paul Hawken, William McDonough, Ray Anderson, Amory Lovins, David Crockett, and John and Nancy Todd.

Operating in a socially responsible manner consistent with the principles of sustainable development and environmental protection, they have shown, can be good for employee morale, public relations, and the bottom line simultaneously. Environmentally conscious or "green" companies such as the Body Shop, Patagonia, Aveda, Malden Mills, Johnson and Johnson, and Interface, Inc. (What Do You Think? p. 534) consistently earn high marks from community and environmental groups. Conserving resources, reducing pollution, and treating employees and customers fairly may cost a little more initially, but can save money and build a loyal following in the long run.

## New business models follow concepts of ecology

Paul Hawken's 1993 book, *The Ecology of Commerce,* was a seminal influence in convincing many people to reexamine the role of business and economics in environmental and social welfare. Basing his model for a new industrial revolution on the principles of ecology, Hawken points out that almost nothing is discarded or unused in nature. The wastes from one organism become the food of another. Industrial processes, he argues, should be designed on a similar principle (table 23.4). Rather than a linear pattern in which we try to maximize the throughput of material and minimize labor, products and processes should be designed to

- be energy efficient;
- use renewable materials;
- be durable and reusable or easily dismantled for repair and remanufacture, nonpolluting throughout their entire life cycle;
- provide meaningful and sustainable livelihoods for as many people as possible;
- protect biological and social diversity;
- use minimum and appropriate packaging made of reusable or recyclable materials.

We can do all this and at the same time increase profits, reduce taxes, shrink government, increase social spending, and restore our environment, Hawken claims. Recently, Hawken has served as chairperson for The Natural Step in America, a movement started in Sweden by Dr. K. H. Robert, a physician concerned about the increase in environmentally related cancers. Through a consensus process, a group of 50 leading scientists endorsed a description of the living systems on which our economy and lives depend. More than 60 major European corporations and 55 municipalities have incorporated sustainability principles (table 23.5) into their operations.

Another approach to corporate responsibility is called the **triple bottom line.** Rather than reporting only net profits as a measure of success, ethically sensitive corporations include environmental effects and social justice programs as indications of genuine progress.

Corporations committed to eco-efficiency and clean production include such big names as Monsanto, 3M, DuPont, Duracell, and Johnson and Johnson. Following the famous three Rs—reduce, reuse, recycle—these firms have saved money and gotten welcome publicity. Savings can be substantial. Slashing energy use and redesigning production to use less raw material and to produce less waste is reported to have saved DuPont $3 billion over the past decade, while also reducing greenhouse emissions 72 percent.

**TABLE 23.4**

**Goals for an Eco-Efficient Economy**

- Introduce no hazardous materials into the air, water, or soil.
- Measure prosperity by how much natural capital we can accrue in productive ways.
- Measure productivity by how many people are gainfully and meaningfully employed.
- Measure progress by how many buildings have no smokestacks or dangerous effluents.
- Make the thousands of complex governmental rules unnecessary that now regulate toxic or hazardous materials.
- Produce nothing that will require constant vigilance from future generations.
- Celebrate the abundance of biological and cultural diversity.
- Live on renewable solar income rather than fossil fuels.

**TABLE 23.5**

**The Natural Step: System Conditions for Sustainability**

1. Minerals and metals from the earth's crust must not systematically increase in nature.
2. Materials produced by human society must not systematically increase in nature.
3. The physical basis for biological productivity must not be systematically diminished.
4. The use of resources must be efficient and just with respect to meeting human needs.

***Think About It***

Most designs for environmental efficiency involve relatively simple rethinking of production or materials. Many of us might be able to save money, time, or other resources in our own lives just by thinking ahead. Think about your own daily life: Could you use new strategies to reduce consumption or waste in recreational activities, cooking, or shopping? In transportation? In housing choices?

## Efficiency starts with design of products and processes

Our current manufacturing system often is incredibly wasteful. On average, for every truckload of products delivered in the United States, 32 truckloads of waste are produced along the way. The automobile is a typical example. Industrial ecologist, Amory Lovins, calculates that for every 100 gallons (380 l) of gasoline burned in your car engine, only one percent (1 gal or 3.8 l)

**TABLE 23.6**

**McDonough Design Principles**

Inspired by the way living systems actually work, Bill McDonough offers three simple principles for redesigning processes and products:

1. *Waste equals food.* This principle encourages elimination of the concept of waste in industrial design. Every process should be designed so that the products themselves, as well as leftover chemicals, materials, and effluents, can become "food" for other processes.
2. *Rely on current solar income.* This principle has two benefits: First, it diminishes, and may eventually eliminate, our reliance on hydrocarbon fuels. Second, it means designing systems that sip energy rather than gulping it down.
3. *Respect diversity.* Evaluate every design for its impact on plant, animal, and human life. What effects do products and processes have on identity, independence, and integrity of humans and natural systems? Every project should respect the regional, cultural, and material uniqueness of its particular place.

actually moves passengers. All the rest is used to move the vehicle itself. The wastes produced—carbon dioxide, nitrogen oxides, unburned hydrocarbons, rubber dust, heat—are spread through the environment where they pollute air, water, and soil.

Architect William McDonough urges us to rethink design approaches (table 23.6). In the first place, he says, we should question whether the product is really needed. Could we provide the same service in a more eco-efficient manner? According to McDonough, products should be divided into three categories:

1. *Consumables* are products like food, natural fabrics, or paper that can harmlessly go back to the soil as compost.
2. *Service products* are durables such as cars, TVs, and refrigerators. These products should be leased to the customer to provide their intended service, but would always belong to the manufacturer. Eventually they would be returned to the maker, who would be responsible for recycling or remanufacturing the product. Knowing that they will have to dismantle the product at the end of its life will encourage manufacturers to design for easy disassembly and repair.
3. *Unmarketables* are compounds like radioactive isotopes, persistent toxins, and bioaccumulative chemicals. Ideally, no one would make or use these products. But because eliminating their use will take time, McDonough suggests that in the mean time these materials should belong to the manufacturer and be molecularly tagged with the maker's mark. If they are discovered to be discarded illegally, the manufacturer would be held liable.

Following these principles, McDonough Bungart Design Chemistry has created nontoxic, easily recyclable materials to use in buildings and for consumer goods. Among some important and innovative "green office" projects designed by the McDonough and Partners architectural firm are the Environmental Defense Fund headquarters in New York City, the Environmental Studies Center at Oberlin College in Ohio (see fig. 20.10), the European Headquarters for Nike in Hilversum, the Netherlands, and the Gap Corporate Offices in San Bruno, California (fig. 23.25). Intended to promote employee well-being and productivity as well as eco-efficiency, the Gap building has high ceilings, abundant skylights, windows that open, a full-service fitness center (including pool), and a landscaped atrium for each office bay that brings the outside in. The roof is covered with native grasses. Warm interior tones and natural wood surfaces (all wood used in the building was harvested by certified sustainable methods) give a friendly feeling. Paints, adhesives, and floor coverings are low toxicity and the building is one-third more energy efficient than strict California laws require. A pleasant place to work, the offices help recruit top employees and improve both effectiveness and retention. As for the bottom line, Gap, Inc. estimates that the increased energy and operational efficiency will have a four- to eight-year payback.

## Green consumerism gives the public a voice

Consumer choice can play an important role in persuading businesses to produce eco-friendly goods and services (What Can You Do? p. 537). Increasing interest in environmental and social sustainability has caused an explosive growth of green products. The National Green Pages published by Co-Op America currently lists more than 2,000 green companies. You can find ecotravel agencies, telephone companies that donate profits to environmental groups, entrepreneurs selling organic foods, shade-grown coffee, straw-bale houses, geodesic-dome kits, paint thinner made from orange peels, sandals made from recycled auto tires, earthworms for composting, and a plethora of hemp products

**FIGURE 23.25** The award-winning Gap, Inc. corporate offices in San Bruno, California, demonstrate some of the best features of environmental design. A roof covered with native grasses provides insulation and reduces runoff. Natural lighting, an open design, and careful relation to its surroundings all make this a good place to work.

including burgers, ale, clothing, shoes, rugs, balm, shampoo, and insect repellent. Although these eco-entrepreneurs represent a tiny sliver of the $7 trillion per year U.S. economy, they often serve as pioneers in developing new technologies and offering innovative services.

In some industries eco-entrepreneurs have found profitable niches (organic, naturally colored clothing, for example) within a larger market. In other cases, once a consumer demand has built up, major companies add green products or services to their inventory. Natural foods, for instance, have grown from the domain of a few funky, local co-ops to a $7 billion market segment. Most supermarket chains now carry some organic food choices. Similarly, natural-care health and beauty products are now more than 10 percent of a $33 billion industry. By supporting these products, you can ensure that they will continue to be available and, perhaps, even help expand their penetration into the market.

## Environmental protection creates jobs

For years business leaders and politicians have portrayed environmental protection and jobs as mutually exclusive. Pollution control, protection of natural areas and endangered species, limits on use of nonrenewable resources, they claim, will strangle the economy and throw people out of work. Ecological economists dispute this claim, however. Their studies show that only 0.1 percent of all large-scale layoffs in the United States in recent years were due to government regulations (fig. 23.26). Environmental protection, they argue, is not only necessary for a healthy economic system, it actually creates jobs and stimulates business.

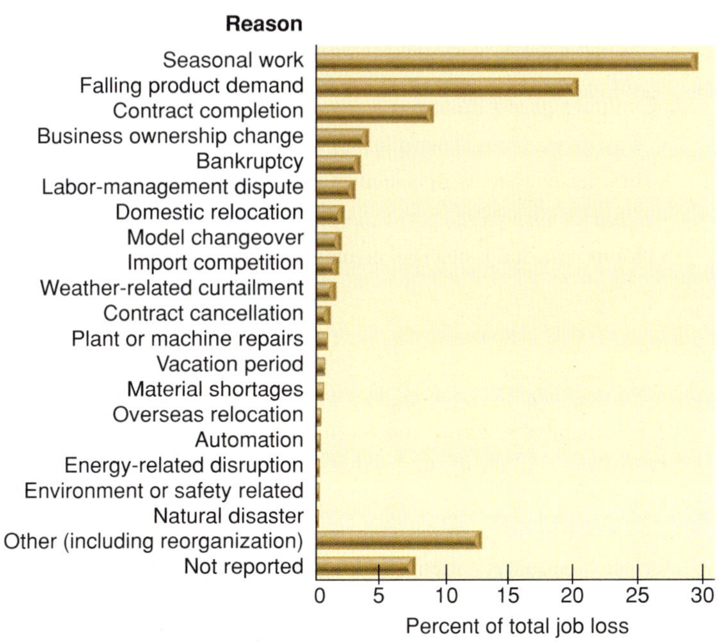


**FIGURE 23.26** Although opponents of environmental regulation often claim that protecting the environment costs jobs, studies by economist E. S. Goodstein show that only 0.1 percent of all large-scale layoffs in the United States were the result of environmental laws.
**Source:** E. S. Goodstein, Economic Policy Institute, Washington, D.C.

### What Can You Do?

### Personally Responsible Consumerism

There are many things that each of us can do to lower our ecological impacts and support green businesses through responsible consumerism and ecological economics.

- Practice living simply. Ask yourself if you really need more material goods to make your life happy and fulfilled.
- Rent, borrow, or barter when you can. Do you really need to have your own personal supply of tools, machines, and other equipment that you use, at most, once per year?
- Recycle or reuse building materials: doors, windows, cabinets, appliances. Shop at salvage yards, thrift stores, yard sales, or other sources of used clothes, dishes, appliances, etc.
- Consult the National Green Pages from Co-Op America for a list of eco-friendly businesses. Write one letter each month to a company from which you buy goods or services and ask them what they are doing about environmental protection and human rights.
- Buy green products. Look for efficient, high quality materials that will last and that are produced in the most environmentally friendly manner possible. Subscribe to clean energy programs if they are available in your area. Contact your local utility and ask that they provide this option if they don't now.
- Buy products in bulk if you can or look for the least amount of packaging. Choose locally grown or locally made products made under humane conditions by workers who receive a fair wage.
- Think about the total life-cycle costs of the things you buy, including environmental impacts, service charges, energy use, and disposal costs as well as initial purchase price.
- Stop junk mail. Demand that your name be removed from mass-mailing lists.
- Invest in socially and environmentally responsible mutual funds or green businesses when you have money for investment.

Recycling, for instance, makes more new jobs than extracting virgin raw materials. This doesn't necessarily mean that recycled goods are more expensive than those from virgin resources. We're simply substituting labor in the recycling center for energy and huge machines used to extract new materials in remote places.

Japan, already a leader in efficiency and environmental technology, has recognized the multibillion dollar economic potential of green business. The Japanese government is investing (U.S.) $4 billion per year on research and development that targets seven areas, ranging from utilitarian projects such as biodegradable plastics and heat-pump refrigerants to exotic schemes such as carbon-dioxide-fixing algae and hydrogen-producing microbes.

A World Wildlife Fund study predicts that over the next decade, energy conservation, renewable energy sources, and other environmental protection programs could result in 2 million new jobs in the United States if new technologies are embraced and promoted. The net economic impact could be as much as $10 billion per year.

## Conclusion

At the 1972 Stockholm Conference on the Human Environment, Indira Gandhi, then prime minister of India, stated that "Poverty is the greatest polluter of them all." What she meant was that the world's poorest people are too often both the victims and agents of environmental degradation. They are forced to meet short-term survival needs at the cost of long-term sustainability. But "charity is not an answer to poverty," Dr. Muhammad Yunus says, "It only helps poverty to continue. It creates dependency and takes away individual's initiative . . . Poverty isn't created by the poor, it's created by the institutions and policies that surround them . . . All we need to do is to make appropriate changes in the institutions and policies, and/or create new ones." The microcredit revolution he started may be the key for breaking the cycle of poverty and changing the lives of the poor.

Emissions trading could also be a way to aid poor countries. It could encourage the spread of renewable energy and nonpolluting technology to sub-Saharan Africa, Asia, and Latin America. Many of the countries where biodiversity is highest, and programs to stop deforestation, replant trees, and install solar, wind, and biomass generators can do the most good, are also poor countries most in need of development aid. Incorporating ecological knowledge into our economic policies could help us value nature as well as assist in meeting the UN Millennium goals of eradicating extreme poverty and hunger while also ensuring environmental sustainability.

## Reviewing Learning Outcomes

By now you should be able to explain the following points:

**23.1** Analyze economic worldviews.

- Can development be sustainable?
- Our definitions of resources shape how we use them.
- Classical economics examines supply and demand.
- Neoclassical economics emphasizes growth.
- Ecological economics incorporates principles of ecology.
- Communal property resources are a classic problem in ecological economics.

**23.2** Scrutinize population, technology, and scarcity.

- Scarcity can lead to innovation.
- Carrying capacity is not necessarily fixed.
- Economic models compare growth scenarios.
- Why not conserve resources?

**23.3** Investigate natural resource accounting.

- Gross national product is our dominant growth measure.
- Alternate measures account for well-being.
- New approaches incorporate nonmarket values.
- Cost-benefit analysis aims to optimize resource use.

**23.4** Summarize how market mechanisms can reduce pollution.

- Using market forces.
- Is emissions trading the answer?
- Sulfur trading offers a good model.
- Carbon trading is already at work.

**23.5** Study trade, development, and jobs.

- International trade brings benefits but also intensifies inequities.
- Aid often doesn't help the people who need it.
- Microlending helps the poorest of the poor.

**23.6** Evaluate green business.

- New business models follow concepts of ecology.
- Efficiency starts with design of products and processes.
- Green consumerism gives the public a voice.
- Environmental protection creates jobs.

## Practice Quiz

1. Define *economics* and distinguish between classical, neoclassical, and ecological economics.
2. Define *resources* and give some examples of renewable, nonrenewable, and intangible resources.
3. List three economic categories of resources and describe the differences among them.
4. Describe the relationship between supply and demand.
5. Identify some important ecological services on which our economy depends.
6. Describe how cost-benefit ratios are determined and how they are used in natural resource management.
7. Explain how scarcity and technological progress can extend resource availability and extend the carrying capacity of the environment.
8. Describe how GNP is calculated and explain why this may fail to adequately measure human welfare and environmental quality. Discuss some alternative measures of national progress.
9. What is microlending, and what are its benefits?
10. List some of the characteristics of an eco-efficient economic system.

## Critical Thinking and Discussion Questions

1. When the ecologist warns that we are using up irreplaceable natural resources and the economist rejoins that ingenuity and enterprise will find substitutes for most resources, what underlying premises and definitions shape their arguments?
2. How can intangible resources be infinite and exhaustible at the same time? Isn't this a contradiction in terms? Can you find other similar paradoxes in this chapter?
3. What would be the effect on the developing countries of the world if we were to change to a steady-state economic system? How could we achieve a just distribution of resource benefits while still protecting environmental quality and future resource use?
4. Resource use policies bring up questions of intergenerational justice. Suppose you were asked: "What has posterity ever done for me?" How would you answer?
5. If you were doing a cost-benefit study, how would you assign a value to the opportunity for good health or the existence of rare and endangered species in faraway places? Is there a danger or cost in simply saying some things are immeasurable and priceless and therefore off limits to discussion?
6. If natural capitalism or eco-efficiency has been so good for some entrepreneurs, why haven't all businesses moved in this direction?

## DATA analysis Evaluating Human Development

The human development index (HDI) is a measure created by the United Nations Development Programme to track social progress. HDI incorporates life expectancy, adult literacy, children's education, and standard of living indicators to measure human development. The 2006 report draws on statistics from 175 countries. While there has been encouraging progress in most world regions, the index shows that widening inequality is taking a toll on global human development.

The graph shows trends in the HDI by world region. Study this graph carefully, and answer the following questions: (*Hint:* you may have to use a search engine to find some answers.)

1. Which region has the highest HDI rating?
2. What does OECD stand for?
3. Which region has made the greatest progress over the past 30 years, and how much has its HDI increased?
4. Which region has shown the least progress in human development?
5. What historic events could explain the reduction in Europe and the CIS between 1990 and 1995?
6. How much lower is the HDI ranking of sub-Saharan Africa from the OECD?

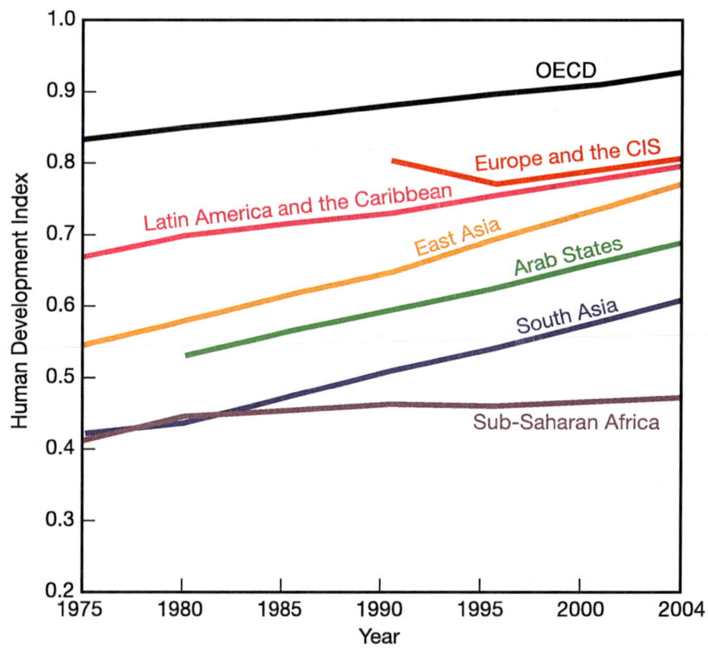


Trends in human development, 1975–2004.
**Source:** United Nations Development Programme, 2006.

**For Additional Help in Studying This Chapter,** please visit our website at www.mhhe.com/cunningham10e. You will find additional practice quizzes and case studies, flashcards, regional examples, place markers for Google Earth™ mapping, and an extensive reading list, all of which will help you learn environmental science.

# UNIT 1

## Environmental Philosophy

*Environmental debates are rooted in questions of values—what is right? what is just?—and inevitably political in nature. It is worth stressing that people who consider themselves to be environmentalists can be found on both sides of most of the issues in this book. They differ in what they see as their own self-interest and even in what they see as humanity's long-term interest.*

*Understanding the general issues raised in this section is useful preparation for examining the more specific controversies that follow in later sections.*

- Is the Precautionary Principle a Sound Approach to Risk Analysis?
- Is Sustainable Development Compatible with Human Welfare?
- Should a Price Be Put on the Goods and Services Provided by the World's Ecosystems?

# ISSUE 1

## Is the Precautionary Principle a Sound Approach to Risk Analysis?

**YES: Nancy Myers**, from "The Rise of the Precautionary Principle: A Social Movement Gathers Strength," *Multinational Monitor* (September 2004)

**NO: Bernard D. Goldstein**, from "The Precautionary Principle: Is It a Threat to Toxicological Science?" *International Journal of Toxicology* (January/February 2006)

### ISSUE SUMMARY

**YES:** Nancy Myers, communications director for the Science and Environmental Health Network, argues that because the precautionary principle "makes sense of uncertainty," it has gained broad international recognition as being crucial to environmental policy.

**NO:** Bernard D. Goldstein, professor of Environmental and Occupational Health at the University of Pittsburgh, argues that although the precautionary principle is potentially valuable, it poses a risk that scientific (particularly toxicological) risk assessment will be displaced to the detriment of public health, social justice, and the field of toxicology itself.

The traditional approach to environmental problems has been reactive. That is, first the problem becomes apparent—wildlife or people sicken and die, or drinking water or air tastes foul. Then researchers seek the cause of the problem, and regulators seek to eliminate or reduce that cause. The burden is on society to demonstrate that harm is being done and that a particular cause is to blame.

An alternative approach is to presume that *all* human activities—construction projects, new chemicals, new technologies, etc.—have the potential to cause environmental harm. Therefore, those responsible for these activities should prove in advance that they will not do harm and should take suitable steps to prevent any harm from happening. A middle ground is occupied by the "precautionary principle," which has played an increasingly important part in environmental law ever since it first appeared in Germany in the mid-1960s. On the international scene, it has been applied to climate change, hazardous waste management, ozone depletion, biodiversity, and fisheries management. In 1992 the

Rio Declaration on Environment and Development, listing it as Principle 15, codified it thus:

> In order to protect the environment, the precautionary approach shall be widely applied by States according to their capabilities. When there are threats of serious or irreversible damage, lack of full scientific certainty shall not be used as a reason for postponing cost-effective measures to prevent environmental degradation.

Other versions of the principle also exist, but all agree that when there is reason to think—not absolute proof—that some human activity is or might be harming the environment, precautions should be taken. Furthermore, the burden of proof should be on those responsible for the activity, not on those who may be harmed. This has come to be broadly accepted as a basic tenet of ecologically or environmentally sustainable development. See Paul L. Stein, "Are Decision-Makers Too Cautious With the Precautionary Principle?" *Environmental and Planning Law Journal* (February 2000), and Marco Martuzzi and Roberto Bertollini, "The Precautionary Principle, Science and Human Health Protection," *International Journal of Occupational Medicine and Environmental Health* (January 2004).

The precautionary principle also contributes to thinking in the areas of risk assessment and risk management in general. Human activities—the manufacture of chemicals and other products; the use of pesticides, drugs, and fossil fuels; the construction of airports and shopping malls; and even agriculture—can damage health and the environment. Some people insist that action need not be taken against any particular activity until and unless there is solid, scientific proof that it is doing harm, and even then risks must be weighed against each other. Others insist that mere suspicion should be grounds enough for action.

Since solid, scientific proof can be very difficult to obtain, the question of just how much proof is needed to justify action is vital. Not surprisingly, if action threatens an industry, that industry's advocates will argue against taking precautions, generally saying that more proof is needed. A good example can be found in Stuart Pape, "Watch Out for the Precautionary Principle," *Prepared Foods* (October 1999): "In recent months, U.S. food manufacturers have experienced a rude introduction to the 'Precautionary Principle.' . . . European regulators have begun to adopt extreme definitions of the Principle in order to protect domestic industries and place severe restrictions on the use of both old and new materials without justifying their action upon sound science."

In the following selections, Nancy Myers, communications director for the Science and Environmental Health Network, argues that the precautionary principle has gained broad international recognition as crucial to environmental policy for good reason. "The essence of the Precautionary Principle is that when lives and the future of the planet are at stake, people must act on . . . clues and prevent as much harm as possible, despite imperfect knowledge and even ignorance." Professor Bernard D. Goldstein argues that although the precautionary principle is potentially valuable, it poses a risk that scientific (particularly toxicological) risk assessment will be displaced to the detriment of public health, social justice, and the field of toxicology itself.

## YES

**Nancy Myers**

# The Rise of the Precautionary Principle: A Social Movement Gathers Strength

Ed Soph is a jazz musician and professor at the University of North Texas in Denton, a growing town of about 100,000 just outside Dallas, Texas. In 1997, Ed and his wife Carol founded Citizens for Healthy Growth, a Denton group concerned about the environment and future of their town. The Sophs and their colleagues—the group now numbers about 400—are among the innovative pioneers who are implementing the Precautionary Principle in the United States.

The Sophs first came across the Precautionary Principle in 1998, in the early days of the group's campaign to prevent a local copper wire manufacturer, United Copper Industries, from obtaining an air permit that would have allowed lead emissions. Ed remembers the discovery of the Wingspread Statement on the Precautionary Principle—a 1998 environmental health declaration holding that "When an activity raises threats of harm to human health or the environment, precautionary measures should be taken even if some cause and effect relationships are not fully established scientifically"—as "truly a life-changing experience." Using the Precautionary Principle as a guide, the citizens refused to be drawn into debates on what levels of lead, a known toxicant, might constitute a danger to people's health. Instead, they pointed out that a safer process was available and insisted that the wise course was not to issue the permit. The citizens prevailed.

The principle helped again in 2001, when a citizen learned that the pesticides 2,4-D, simazine, Dicamba and MCPP were being sprayed in the city parks. "The question was, given the 'suspected' dangers of these chemicals, should the city regard those suspicions as a reassurance of the chemicals' safety or as a warning of their potential dangers?" Ed recalls. "Should the city act out of ignorance or out of common sense and precaution?"

Soph learned that the Greater Los Angeles School District had written the Precautionary Principle into its policy on pesticide use and had turned to Integrated Pest Management (IPM), a system aimed at controlling pests without the use of toxic chemicals. The Denton group decided to advocate for a similar policy. They persuaded the city's park district to form a focus group of park users

From *Multinational Monitor*, September 2004, pp. 9–15. 

and organic gardening experts. The city stopped spraying the four problem chemicals and initiated a pilot IPM program.

The campaign brought an unexpected economic bonus to the city. In the course of their research, parks department staff discovered that corn gluten was a good turf builder and natural broadleaf herbicide. But the nearest supplier of corn gluten was in the Midwest, and that meant high shipping costs for the city. Meanwhile, a corn processing facility in Denton was throwing away the corn gluten it produced as a byproduct. The parks department made the link, and everyone was pleased. The local corn company was happy to add a new product line; the city was happy about the expanded local business and the lower price for a local product; and the environmental group chalked up another success.

The citizens of Denton, Texas, did not stop there. They began an effort to improve the community's air pollution standards. They got arsenic-treated wood products removed from school playgrounds and parks and replaced with nontoxic facilities. "The Precautionary Principle helped us define the problems and find the solutions," Ed says.

But, as he wrote in an editorial for the local paper, "The piecemeal approach is slow, costly and often more concerned with mitigation than prevention." Taking a cue from Precautionary Principle pioneers in San Francisco, they also began lobbying for a comprehensive new environmental code for the community, based on the Precautionary Principle.

In June 2003, San Francisco's board of supervisors had become the first government in the United States to embrace the Precautionary Principle. A new environmental code drafted by the city's environment commission put the Precautionary Principle at the top, as Article One. Step one in implementing the code was a new set of guidelines for city purchasing, pointing the way toward "environmentally preferable" purchases by careful analysis and choice of the best alternatives. The White Paper accompanying the ordinance pointed out that most of the city's progressive environmental policies were already in line with the Precautionary Principle, and that the new code provided unity and focus to the policies rather than a radically new direction.

That focus is important; too often, environmental matters seem like a long, miscellaneous and confusing list of problems and solutions.

Likewise in Denton, the Precautionary Principle has not been a magic wand for transforming policy, but it has put backbone into efforts to enact truly protective and far-sighted environmental policies. Ed Soph points out that, in his community as in others, growth had often been dictated by special interests in the name of economic development, and the environment got short shrift.

"Environmental protection and pollution prevention in our city have been a matter, not of proactive policy, but of reaction to federal and state mandates, to the threat of citizens' lawsuits, and to civic embarrassment. Little thought is given to future environmental impacts," he told the city council when he argued for a new environmental code.

He added, "The toxic chemical pollution emitted by area industries has been ignored or accepted for all the ill-informed or selfish reasons that we are too familiar with. The Precautionary Principle dispels that ignorance and empowers concerned citizens with the means to ensure a healthier future."

The Precautionary Principle has leavened the discussion of environmental and human health policy on many fronts—in international treaty negotiations and global trade forums, in city resolutions and national policies, among conservationists and toxicologists, and even in corporate decision making.

Two treaties negotiated in 2000 incorporated the principle for the first time as an enforceable measure. The Cartagena Protocol on Biosafety allows countries to invoke the Precautionary Principle in decisions on admitting imports of genetically modified organisms. It became operative in June 2003. The Stockholm Convention on Persistent Organic Pollutants prescribes the Precautionary Principle as a standard for adding chemicals to the original list of 12 that are banned by the treaty. This treaty went into force in February 2004.

## Making Sense of Uncertainty

Understanding the need for the Precautionary Principle requires some scientific sophistication. Ecologists say that changes in ecological systems may be incremental and gradual, or surprisingly large and sudden. When change is large enough to cause a system to cross a threshold, it creates a new dynamic equilibrium that has its own stability and does not change back easily. These new interactions become the norm and create new realities.

Something of this new reality is evident in recently observed changes in patterns of human disease:

- Chronic diseases and conditions affect more than 100 million men, women, and children in the United States—more than a third of the population. Cancer, asthma, Alzheimer's disease, autism, birth defects, developmental disabilities, diabetes, endometriosis, infertility, multiple sclerosis and Parkinson's disease are becoming increasingly common.
- Nearly 12 million children in the United States (17 percent) suffer from one or more developmental disabilities. Learning disabilities alone affect at least 5 to 10 percent of children in public schools, and these numbers are increasing. Attention deficit hyperactivity disorder conservatively affects 3 to 6 percent of all school children. The incidence of autism appears to be increasing.
- Asthma prevalence has doubled in the last 20 years.
- Incidence of certain types of cancer has increased. The age-adjusted incidence of melanoma, non-Hodgkins lymphoma, and cancers of the prostate, liver, testis, thyroid, kidney, breast, brain, esophagus and bladder has risen over the past 25 years. Breast cancer, for example, now strikes more women worldwide than any other type of cancer, with rates increasing 50 percent during the past half century. In the 1940s, the lifetime risk of breast cancer was one in 22. Today's risk is one in eight and rising.
- In the United States, the incidence of some birth defects, including male genital disorders, some forms of congenital heart disease and obstructive disorders of the urinary tract, is increasing. Sperm density is declining in some parts of the United States and elsewhere in the world.

These changes in human health are well documented. But proving direct links with environmental causative factors is more complicated.

Here is how the scientific reasoning might go: Smoking and diet explain few of the health trends listed above. Genetic factors explain up to half the population variance for several of these conditions—but far less for the majority of them—and in any case do not explain the changes in disease incidence rates. This suggests that other environmental factors play a role. Emerging science suggests this as well. In laboratory animals, wildlife and humans, considerable evidence documents a link between environmental contamination and malignancies, birth defects, reproductive disorders, impaired behavior and immune system dysfunction. Scientists' growing understanding of how biological systems develop and function leads to similar conclusions.

But serious, evident effects such as these can seldom be linked decisively to a single cause. Scientific standards of certainty (or "proof") about cause and effect are high. These standards may never be satisfied when many different factors are working together, producing many different results. Sometimes the period of time between particular causes and particular results is so long, with so many intervening factors, that it is impossible to make a definitive link. Sometimes the timing of exposure is crucial—a trace of the wrong chemical at the wrong time in pregnancy, for example, may trigger problems in the child's brain or endocrine system, but the child's mother might never know she was exposed.

In the real world, there is no way of knowing for sure how much healthier people might be if they did not live in the modern chemical stew, because the chemicals are everywhere—in babies' first bowel movement, in the blood of U.S. teenagers and in the breastmilk of Inuit mothers. No unexposed "control" population exists. But clearly, significant numbers of birth defects, cancers and learning disabilities are preventable.

Scientific uncertainty is a fact of life even when it comes to the most obvious environmental problems, such as the disappearance of species, and the most potentially devastating trends, such as climate change. Scientists seldom know for sure what will happen until it happens, and seldom have all the answers about causes until well after the fact, if ever. Nevertheless, scientific knowledge, as incomplete as it may be, provides important clues to all of these conditions and what to do about them.

The essence of the Precautionary Principle is that when lives and the future of the planet are at stake, people must act on these clues and prevent as much harm as possible, despite imperfect knowledge and even ignorance.

## Environmental Failures

A premise of Precautionary Principle advocates is that environmental policies to date have largely not met this challenge. Part of the explanation for why they have not is that the dimensions of the emerging problems are only now becoming apparent. The limits of the earth's assimilative capacity are much clearer now than they were when the first modern environmental legislation was enacted 30 years ago.

Another part of the explanation is that, although some environmental policies are preventive, most have focused on cleaning up messes after the fact—what environmentalists call "end of pipe" solutions. Scrubbers on power plant stacks, catalytic converters on tailpipes, recycling and super-sized funds dedicated to detoxifying the worst dumps have not been enough. The Precautionary Principle holds that earlier, more comprehensive and preventive approaches are necessary. Nor is it enough to address problems only after they have become so obvious that they cannot be ignored—often, literally waiting for the dead bodies to appear or for coastlines to disappear under rising tides.

The third factor in the failure of environmental policies is political, say Precautionary Principle proponents. After responding to the initial burst of concern for the environment, the U.S. regulatory system and others like it were subverted by commercial interests, with the encouragement of political leaders and, increasingly, the complicity of the court system. Environmental laws have been subjected to an onslaught of challenges since the 1980s; many have been modified or gutted, and all are enforced by regulators who have been chastened by increasing challenges to their authority by industry and the courts.

The courts, and now increasingly international trade organizations and agreements like the World Trade Organization (WTO) and the North American Free Trade Agreement (NAFTA), have institutionalized an anti-precautionary approach to environmental controls. They have demanded the kinds of proof and certainty of harms and efficacy of regulation that science often cannot provide.

## False Certainties

Ironically, one tool that has proved highly effective in the battle against environmental regulations was one that was meant to strengthen the enforcement of such laws: quantitative risk assessment. Risk assessment was developed in the 1970s and 1980s as a systematic way to evaluate the degree and likelihood of harmful side effects from products and technologies. With precise, quantitative risk assessments in hand, regulators could more convincingly demonstrate the need for action. Risk assessments would stand up in court. Risk assessments could "prove" that a product was dangerous, would cause a certain number of deaths per million, and should be taken off the market.

Or not. Quantitative risk assessment, which became standard practice in the United States in the mid-1980s and was institutionalized in the global trade agreements of the 1990s, turned out to be most useful in "proving" that a product or technology was not inordinately dangerous. More precisely, risk assessments presented sets of numbers that purported to state definitively how much harm might occur. The next question for policymakers then became: How much harm is acceptable? Quantitative risk assessment not only provided the answers; it dictated the questions.

As quantitative risk assessment became the norm, commercial and industrial interests were increasingly able to insist that harm must be proven "scientifically"—in the form of a quantitative risk assessment demonstrating harm in excess of acceptable limits—before action was taken to stop a process or product. These exercises were often linked with cost-benefit assessments that heavily weighted the immediate monetary costs of regulations and gave little, if any, weight to costs to the environment or future generations.

Although risk assessments tried to account for uncertainties, those projections were necessarily subject to assumptions and simplifications. Quantitative risk assessments usually addressed a limited number of potential harms, often missing social, cultural or broader environmental factors. These risk assessments have consumed enormous resources in strapped regulatory agencies and have slowed the regulatory process. They have diverted attention from questions that could be answered: Do better alternatives exist? Can harm be prevented?

The slow pace of regulation, the insistence on "scientific certainty," and the weighting toward immediate monetary costs often give the benefit of doubt to products and technologies, even when harmful side effects are suspected. One result is that neither international environmental agreements nor national regulatory systems have kept up with the increasing pace and cumulative effects of environmental damage.

A report by the European Environment Agency in 2001 tallied the great costs to society of some of the most egregious failures to heed early warnings of harm. Radiation, ozone depletion, asbestos, Mad Cow disease and other case studies show a familiar pattern: "Misplaced 'certainty' about the absence of harm played a key role in delaying preventive actions," the authors conclude.

They add, "The costs of preventive actions are usually tangible, clearly allocated and often short term, whereas the costs of failing to act are less tangible, less clearly distributed and usually longer term, posing particular problems of governance. Weighing up the overall pros and cons of action, or inaction, is therefore very difficult, involving ethical as well as economic considerations."

## The Precautionary Approach

As environmentalists looked at looming problems such as global warming, they were appalled at the inadequacy of policies based on quantitative risk assessment. Although evidence was piling up rapidly that human activities were having an unprecedented effect on global climate, for example, it was difficult to say when the threshold of scientific certainty would be crossed. Good science demanded caution about drawing hard and fast conclusions. Yet, the longer humanity waited to take action, the harder it would be to reverse any effect. Perhaps it was already too late. Moreover, action would have to take the form of widespread changes not only in human behavior but also in technological development. The massive shift away from fossil fuels that might yet mitigate the effects of global warming would require rethinking the way humans produce and use energy. Nothing in the risk-assessment-based approach to policy prepared society to do that.

The global meetings called to address the coming calamity were not helping much. Politicians fiddled with blame and with protecting national economic interests while the globe heated up. Hard-won and heavily compromised agreements such as the 1997 Kyoto agreement on climate change were quickly mired in national politics, especially in the United States, the heaviest fossil-fuel user of all.

In the United States and around the globe, a different kind of struggle had been going on for decades: the fight for attention to industrial pollution in communities. From childhood lead poisoning in the 1930s to Love Canal in the 1970s, communities had always faced an uphill battle in proving that pollution and toxic products were making them sick. Risk assessments often made the case that particular hazardous waste dumps were safe, or that a single polluting industry could not possibly have caused the rash of illnesses a community claimed. But these risk assessments missed the obvious fact that many communities suffered multiple environmental assaults, compounded by other effects of poverty. A landmark 1987 report by the United Church of Christ coined the term "environmental racism" and confirmed that the worst environmental abuses were visited on communities of color. This growing awareness generated the international environmental justice movement.

In early 1998, a small conference at Wingspread, the Johnson Foundation's conference center in Racine, Wisconsin, addressed these dilemmas head-on. Participants groped for a better approach to protecting the environment and human health. At that time, the Precautionary Principle, which had been named in Germany in the 1970s, was an emerging precept of international law. It had begun to appear in international environmental agreements, gaining reference in a series of protocols, starting in 1984, to reduce pollution in the North Sea; the 1987 Ozone Layer Protocol; and the Second World Climate Conference in 1990.

At the Rio Earth Summit in 1992, precaution was enshrined as Principle 15 in the Rio Declaration on Environment and Development: "In order to protect the environment, the precautionary approach shall be widely applied by states according to their capabilities. Where there are threats of serious or irreversible damage, lack of full scientific certainty shall not be used as a reason for postponing cost-effective measures to prevent environmental degradation."

In the decade after Rio, the Precautionary Principle began to appear in national constitutions and environmental policies worldwide and was occasionally invoked in legal battles. For example:

- The Maastricht Treaty of 1994, establishing the European Union, named the Precautionary Principle as a guide to EU environment and health policy.
- The Precautionary Principle was the basis for arguments in a 1995 International Court of Justice case on French nuclear testing. Judges cited the "consensus flowing from Rio" and the fact that the Precautionary Principle was "gaining increasing support as part of the international law of the environment."
- At the World Trade Organization in the mid-1990s, the European Union invoked the Precautionary Principle in a case involving a ban on imports of hormone-fed beef.

## THE U.S. CHAMBER OF COMMERCE: POLICY BRIEF AGAINST THE PRECAUTIONARY PRINCIPLE

### Objective

To ensure that regulatory decisions are based on scientifically sound and technically rigorous risk assessments, and to oppose the adoption of the Precautionary Principle as the basis for regulation.

### Summary of the Issue

The U.S. Chamber supports a science-based approach to risk management; where risk is assessed based on scientifically sound and technically rigorous analysis. Under this approach, regulatory actions are justified where there are legitimate, scientifically ascertainable risks to human health, safety, or the environment. That is, the greater the risk, the greater the degree of regulatory scrutiny. This standard has served the nation well, and has led to astounding breakthroughs in the fields of science, health care, medicine, biotechnology, agriculture, and many other fields. There is, however, a relatively new theory, known as the Precautionary Principle that is gaining popularity among environmentalists and other groups. The Precautionary Principle says that when the risks of a particular activity are unclear or unknown, assume the worst and avoid the activity. It is essentially a policy of risk avoidance.

The regulatory implications of the Precautionary Principle are huge. For instance, the Precautionary Principle holds that since the existence and extent of global warming and climate change are not known one should assume the worst and immediately restrict the use of carbon-based fuels. However the nature and extent of key environmental, health, and safety concerns requires careful scientific and technical analysis. That is why the U.S. Chamber has long supported the use of sound science, cost-benefit analysis, risk assessment and a full understanding of uncertainty when assessing a particular regulatory issue.

The Precautionary Principle has been explicitly incorporated into various laws and regulations in the European Union and various international bodies. In the United States, radical environmentalists are pushing for its adoption as a basis for regulating biotechnology, food and drug safety, environmental protection, and pesticide use.

### U.S. Chamber Strategy

- Support a science-based approach to risk management, where risk is assessed based on scientifically sound and technically rigorous standards.
- Oppose the domestic and international adoption of the Precautionary Principle as a basis for regulatory decision making.
- Educate consumers, businesses, and federal policymakers about the implications of the Precautionary Principle.

The Wingspread participants believed the Precautionary Principle was not just another weak and limited fix for environmental problems. They believed it could bring far-reaching changes to the way those policies were formed and implemented. But action to prevent harm in the face of scientific uncertainty alone did not translate into sound policies protective of the environment and human health. Other norms would have to be honored simultaneously and as an integral part of a precautionary decision-making process. Several other principles had often been linked with the Precautionary Principle in various statements of the principle or in connection with precautionary policies operating in Northern European countries. The statement released at the end of the meeting, the Wingspread Statement on the Precautionary Principle, was the first to put four of these primary elements on the same page—acting upon early evidence of harm, shifting the burden of proof, exercising democracy and transparency, and assessing alternatives. These standards form the basis of what has come to be known as the overarching or comprehensive Precautionary Principle or approach:

> When an activity raises threats of harm to human health or the environment, precautionary measures should be taken even if some cause and effect relationships are not fully established scientifically.
>
> In this context the proponent of an activity, rather than the public, should bear the burden of proof.
>
> The process of applying the Precautionary Principle must be open, informed and democratic and must include potentially affected parties. It must also involve an examination of the full range of alternatives, including no action.

The conference generated widespread enthusiasm for the principle among U.S. environmentalists and academics as well as among some policymakers. That was complemented by continuing and growing support for the principle among Europeans as well as ready adoption of the concept in much of the developing world. And in the years following Wingspread, the Precautionary Principle has gained new international status.

**Bernard D. Goldstein**

**NO**

# The Precautionary Principle: Is It a Threat to Toxicological Science?

The central figure in developing the *1970 US Clean Air Act,* Senator Edmund Muskie, rather plaintively asked that the scientific information underlying this act be obtained from a one-handed scientist. He was tired of hearing scientists saying "on the one hand . . . , yet on the other hand. . . ." The Senator was expressing the desire of lawmakers for indisputable scientific facts on which to base regulatory actions. Unfortunately, this is not the way science works, particularly in situations where there is uncertainty concerning the questions being asked, or the terms being used.

Lack of clarity in definition or consistency in use is a particular problem when considering the Precautionary Principle. My discussion of the impact of the Precautionary Principle will inevitably fall into the "on the one hand . . . , yet on the other hand . . ." category that made Senator Muskie so unhappy. I will argue that the precautionary principle is both good public health and that it is causing public health problems; that it will both increase and decrease the importance of toxicological science in decision making; and that the European Community (EC) is both using the Precautionary Principle in some innovative ways to protect public health and the environment, and that the EC is intentionally abusing the precautionary principle by using it as a means to erect artificial trade barriers not warranted by toxicological science and risk analysis.

## Definitions of the Precautionary Principle

The precautionary principle was developed primarily in Europe, beginning in the 1980s in Germany. One of the first times it received major prominence was in 1992 at the United Nations Conference on Environment and Development in Rio de Janeiro. The primary goal of this meeting was to establish an environmental agenda that would guide nations toward sustainable development. At this meeting the precautionary principle was adopted in the following form:

From *International Journal of Toxicology,* vol. 25, no. 1, January/February 2006, pp. 3–7. 

> "Nations shall use the precautionary approach to protect the environment. Where there are threats of serious or irreversible damage, scientific uncertainty shall not be used to postpone cost-effective measures to prevent environmental degradation."

It is informative to compare this definition with the more recent Wingspread Statement:

> "When an activity raises threat [of harm] to human health or the environment, precautionary measures should be taken even if some cause and effect relationships are not fully established scientifically."

Wingspread is notable for a number of differences from the earlier Rio declaration. The Rio declaration includes a statement that the threat of harm should be serious and irreversible, but no threat of harm is specified in Wingspread; the Rio declaration states that the action should be cost-effective, but cost-effectiveness is not part of theWingspread statement; the whole thrust of the Wingspread statement is positive in that measures should be taken, whereas in Rio it is negative in that the lack of uncertainty should not postpone taking measures; and finally the Rio statement focuses solely on the environment whereas the Wingspread statement goes beyond the environment to the broad field of public health.

The problem of appropriately defining the precautionary principle is not simply academic. Like sustainable development, as a concept the precautionary principle is supportable by everyone. But the precautionary principle also has been adopted in various national and international treaties and by governmental entities in a way that provides legal teeth to what is otherwise a nebulous policy concept. The Wingspread approach is beginning to take hold in the United States where the Precautionary Principle has been approved by the San Francisco and Berkeley City Councils and is under consideration by other local authorities. Where laws and actions are at stake, definitions become very important. The European approach, which often infuriates its trading partners including the United States, is to state that the precautionary principle is enshrined within European policy—but to never define it. Rather, a policy or action is judged as to whether it will result in a precautionary approach. This has led to what Carruth and I call the Cynical American Definition:

> "The Precautionary Principle is a nebulous doctrine invented by Europeans as a means to erect a trade barrier against any item produced more efficiently elsewhere."

There are many different actions that can be taken under the heading of the precautionary principle, or the precautionary approach as it is often described. It is useful to consider such actions under the public health and prevention nomenclature of primary and secondary prevention. Simply put, primary prevention is an approach that focuses on there never being a problem in the first place; e.g., an individual does not start smoking or, even better, there are no cigarettes being produced. Secondary prevention depends upon the early detection of a problem, such as detecting high blood pressure before someone has a stroke or is otherwise symptomatic. Much of what is done as a result of environmental risk assessment and risk management is secondary prevention, particularly when it is related to the standard approaches to dealing with known toxic chemicals. Many risk assessors would argue that the precautionary principle is already part of risk assessment. Among the potentially precautionary standard risk assessment practices are 10-fold "safety" factors; the 95% upper confidence limit; and the use of exposure models containing prudent default assumptions. The primary preventive approaches of the precautionary principle go well beyond more stringent risk assessment. Its components include taking preventive action in the face of uncertainty; shifting the burden of proof to the proponents of an activity such as industry of government; exploring a wide range of alternatives; and increasing public involvement in decision making.

## The Precautionary Principle and World Trade

The implications of how to define the precautionary principle have nowhere been more evident than for some of the trade issues to which it is being applied. In order to keep a level playing field, various world trade agreements depend upon harmonizing risk assessments so that a nation cannot just arbitrarily decide that another nation's product is unhealthy. Unfortunately, there are a variety of examples in which it appears that the European community has used the precautionary principle as a means to exclude items from elsewhere in the world. Before describing these as inappropriate uses of toxicological science, let me first emphasize that we in the United States have much to learn from the EC as it wrestles with core issues of democracy that impact on the control of environmental hazards. The current debates about the governance relationships among citizens, states, and the multistate European confederation are unparalleled in the US since the Federalist Papers; and are of profound global importance.

As I do not want current broader EU-US issues to cloud the discussion, I will start with an issue that does not involve the United States. An example of action based upon the precautionary principle that fits under the heading of secondary prevention is the adoption by the EC of a very stringent standard for aflatoxin. This was done in essence by adding an additional safety factor for the genotoxicity of aflatoxins. For all nondairy products designated for human consumption, the EC standard is one-fifth that of the United States and for milk it is one-tenth of that of the United States. This has resulted in the exclusion of an estimated $700 million yearly of trade from Sub-Saharan Africa to the advantage of European growers. These Sub-Saharan nations are among the poorest in the world, with relatively fixed trade routes following old colonial patterns. The difference in risk is less than one cancer per year in Europe. The Joint FAO/WHO Expert Committee on Food Additives (JECFA), the expert body of the Food and Agricultural Organization and the World Health Organization, at the request of the Codex Alimentarius Commission, reviewed the difference between the US and European standard and found that it was of no health significance. There is also concern that the EC standard is halting the usual harvesting of Brazil nuts in the Amazon area with the result that the trees are being cut down rather than providing resources for indigenous groups.

An example of a trade barrier related to primary prevention aspects of the precautionary principle is that of the EU's ban on beef from cattle that have previously been treated with estrogenic growth agents. Canada and the US brought this issue to the World Trade Organization (WTO) who decided in their favor. The EU argued that despite the lack of recognition of risk by JECFA, or even its own scientific bodies, the precautionary principle permitted them to ban beef from hormone-treated animals as a risk could not be ruled out, and that a formal risk assessment was not needed. Among the arguments advanced by Canada and the US were that the EC was inconsistent in not having limits on natural estrogens, and was inconsistent in permitting the use in swine of antibiotics whose residues in pork were carcinogenic in laboratory animals. The WTO ruling basically said that the precautionary principle was not yet generally accepted as a legal doctrine capable of supplanting risk assessment. Much of the subsequent EU activity in justifying the precautionary principle can be seen as a means of reversing the WTO opinion. The use of the precautionary principle as a trade barrier will likely be central to the forthcoming WTO hearings on the EU's ban of genetically modified foods.

The use of a primary preventive precautionary approach also frequently occurs in US law. An interesting example demonstrating a switch to primary precaution is the hazardous air pollutant (HAP) provisions of the *1990 Clean Air Act* amendments. Prior to 1990, to regulate a compound as a HAP, the Environmental Protection Agency (EPA) had to go through an involved and time-consuming process. It began by listing a compound as one reasonably anticipated to cause adverse health effects (and is not regulated by setting an ambient standard). After suitable hearings, the EPA would then use risk-based and other considerations to decide which sources of the listed compound to regulate. The burden of proof was on the government to show the potential for adverse consequences. Relatively few compounds had been regulated under this approach, primarily those believed to be human carcinogens. In 1990, faced with impatience at this slow approach, and the recognition that common air contaminants responsible for many tons of emissions were uncontrolled, Congress dramatically changed the HAP regulatory process. Congress listed over 180 compounds to be regulated for which the only rule-making procedure was the onerous one of attempting to remove them from the list by demonstrating safety, i.e., the burden of proof was switched to the industry. Further, Congress specified that emission sources should be required to use the maximum available control technology (MACT), reasoning that all sources should do as well as the best (defined operationally as the upper 12th percentile). Risk considerations are taken into account only secondarily. If, after MACT, there is still a one in a million risk to the maximally exposed individual (MEI), additional steps are to be taken. These steps are currently under consideration and likely will go through court challenges in the next few years. The use of the MEI as the target in a public health bill is problematic—it does not matter whether the source is in the middle of a populous major metropolitan area or upwind of the Mojave Desert.

If a similar switch in the burden of proof, and away from risk based approaches, was under consideration today, the debate would be phrased in terms of the precautionary principle. As it is now 15 years from the passage of the *1990 Clean Air Act* amendments, the change in the HAP provisions allows an opportunity to consider the effectiveness of the precautionary principle in action. Carruth and I have pointed out potential public health shortfalls to the precautionary approach taken in the HAP amendments. From the vantage point of a toxicologist interested in understanding the mechanism of chemical effects, the most important is that once a compound is listed, why would there be any interest in studying it further? In fact, there does appear to be a decline in EPA's research interest in HAPs in recent years. Another issue is that the compounds on the HAP list have varying degrees of toxicity, e.g., benzene and toluene. But if they are both on the list and subject to the same controls, what is the incentive to use the least toxic agent? Also, nonlisted compounds are likely to be in the large universe of compounds for which there is less toxicological information and even more likelihood for unwanted surprise. Yet the HAP amendments may well push toward use of nonlisted compounds as it allows the industry source to avoid regulation. There is also a concern that the MACT standards, which have taken years and much debate to establish, will inhibit investment in even better technology—for example, air pollution control technology that might result from an advance in materials science. All of these concerns can be counteracted by provisions enacted into the 1990 HAP amendments—but whether they will be is uncertain. Accordingly, despite the *1990 Clean Air Act* HAP amendments likely leading to a decrease in tonnage of air pollutants emitted, it is still questionable whether there will be any significant lessening of toxicity directly caused by HAPs.

## Registration Evaluation and Authorization of Chemicals (Reach)

Currently under consideration by the EU is a new far-reaching act that is firmly based upon the precautionary principle (Registration, Evaluation, and Authorisation of Chemicals 2005). It is a work in progress, with amendments being acted upon that may change the details. REACH does have provisions for prioritization of chemicals based upon exposure, or based upon hazard—but not based upon risk. In many ways, REACH is the antithesis of the *US Toxic Substances Control Act* through its requirement of a substantial amount of safety data for all chemicals, new or existing in commerce. Through this requirement it presumably will counteract the major problems we have had in the US with existing chemicals, such as the counterproductive move toward using MTBE and other oxygenated fuels without adequate toxicological testing. Also, through its insistence on the testing of all chemicals, REACH presumably will avoid the potential problem posed by unlisted chemicals in the HAP provisions. Like the *1990 US Clean Air Act* HAP amendments, REACH is justified by its proponents in part on the basis of frustration with the slowness of existing regulatory approaches. This expectation of a speedier regulatory response has not been borne out for the US HAP amendments—15 years after the passage of the act much remains unfinished, including the inevitable court cases.

Alan Boobis has pointed out that there are far fewer toxicologists in Europe than in the US, and has questioned whether there are sufficient toxicologists to accomplish the task. Accordingly, REACH may greatly increase the demand for toxicologists; although in my view, much of these funds would be better spent on developing the scientific basis for newer equivalents of the Ames test that would be effective in primary prevention. A similar criticism can be made of the current approach to testing high production volume chemicals.

## Toxicology as Primary Prevention

Toxicology should be given more recognition as a primary precautionary approach. An excellent example is the Ames test, which was developed based upon scientific advances in understanding basic bacterial genetics; the role of mutagenesis in carcinogenesis; and the importance of liver metabolism in activating indirect carcinogens. To develop new consumer and industrial products, the chemical industry routinely uses the Ames test, and similar tests, to weed out potentially harmful chemicals from their development processes. The focusing of new chemical development and marketing on less harmful products is an excellent example of a primary preventive precautionary approach achieved through advances in toxicological sciences.

Unfortunately, the contributions of toxicology to primary precautionary approaches through safety assessment are not well understood by many who advocate the precautionary principle. Some are distrustful of scientists and are opposed to what they see as a technocracy that is controlling their lives. This is not surprising, in view of the frequency with which industry or government has used scientific uncertainty as a basis to stall needed actions. The justifiable concern is that without a precautionary approach, industry will operate under the CATNIP principle—Cheapest Available Technology Not Involving Prosecution.

## Communication with Those Who Advocate the Precautionary Principle

Potentially problematic is the lack of communication between those active in the field of toxicology or environmental risk assessment and those most strongly advocating the precautionary principle. Of note is that of the 34 signers of the Wingspread Statement on the Precautionary Principle described above, none appear to have been members of the Society of Toxicology or of the Society for Risk Analysis. This lack of communication among different disciplines concerned with environmental protection may well limit understanding of the role of toxicological science in preventing disease. As this could be of grave consequence to our field, our reaching out to those who are advocates of the precautionary principle is particularly important. Let me emphasize that many, and perhaps most, active supporters of the precautionary principle are well aware of the role of science in preventing disease, and that there is a growing literature on the subject of science and the precautionary principle.

## Conclusion

Among the major external threats to the field of toxicology are animal rights activists who would rather put humans and animal pets at risk rather than subject any laboratory animal to study; so called ethicists who seem opposed to any controlled experimental exposures of human subjects to chemical agents, even at levels at which we are exposed in our homes and in gasoline stations; the whole field of toxic torts; and, I would add, the precautionary principle. Each of these threats has their potentially valuable side in promoting the use of toxicology or developing new approaches—such as in vitro techniques that can replace laboratory animals. Similarly, although the precautionary principle can be a threat to the field of toxicology, it also challenges us to think carefully about the goals of toxicology and of the importance of broadly communicating about what we do and why we do it.

# POSTSCRIPT

## Is the Precautionary Principle a Sound Approach to Risk Analysis?

Ronald Bailey, in "Precautionary Tale," *Reason* (April 1999), defines the precautionary principle as "precaution in the face of any actions that may affect people or the environment, no matter what science is able—or unable—to say about that action." "No matter what science says" is not quite the same thing as "lack of full scientific certainty." Indeed, Bailey turns the precautionary principle into a straw man and thereby endangers whatever points he makes that are worth considering. One of those points is that widespread use of the precautionary principle would hamstring the development of the Third World. Roger Scruton, in "The Cult of Precaution," *National Interest* (Summer 2004), calls the Precautionary Principle "a meaningless nostrum" that is used to avoid risk and says it "clearly presents an obstacle to innovation and experiment," which are essential. Bernard D. Goldstein and Russellyn S. Carruth remind us in "Implications of the Precautionary Principle: Is It a Threat to Science?" *International Journal of Occupational Medicine and Environmental Health* (January 2004), that there is no substitute for proper assessment of risk. Jonathan Adler, in "The Precautionary Principle's Challenge to Progress," in Ronald Bailey, ed., *Global Warming and Other Eco-Myths* (Prima, 2002), argues that because the precautionary principle does not adequately balance risks and benefits, "the world would be safer without it." Peter M. Wiedemann and Holger Schutz, "The Precautionary Principle and Risk Perception: Experimental Studies in the EMF Area," *Environmental Health Perspectives* (April 2005), report that "precautionary measures may trigger concerns, amplify . . . risk perceptions, and lower trust in public health protection." Cass R. Sunstein, *Laws of Fear: Beyond the Precautionary Principle* (Cambridge, 2005), criticizes the precautionary principle in part because, he says, people overreact to tiny risks. John D. Graham, dean of the Frederick S. Pardee RAND Graduate School, argues in "The Perils of the Precautionary Principle: Lessons from the American and European Experience," Heritage Lecture #818 (delivered January 15, 2004, at the Heritage Foundation, Washington, D.C.), that the precautionary principle is so subjective that it permits "precaution without principle" and threatens innovation and public and environmental health. It must therefore be used cautiously.

The 1992 Rio Declaration emphasized that the precautionary principle should be "applied by States according to their capabilities" and that it should be applied in a cost-effective way. These provisions would seem to preclude the draconian interpretations that most alarm the critics. Yet, say David Kriebel et al., in "The Precautionary Principle in Environmental Science," *Environmental Health Perspectives* (September 2001), "environmental scientists should be aware of the policy uses of their work and of their social responsibility to do science that protects human health and the environment." Businesses are also conflicted, writes Arnold Brown in "Suitable Precautions," *Across the Board* (January/February 2002), because the precautionary principle tends to slow decision making, but he maintains that "we will all have to learn and practice anticipation." Professor Goldstein mentions the European Union's REACH (Registration Evaluation and Authorization of Chemicals) program. It took effect on June 1, 2007. See http://ec.europa.eu/environment/chemicals/reach/reach_intro.htm for details.

# ISSUE 2

## Is Sustainable Development Compatible with Human Welfare?

**YES: Jeremy Rifkin**, from "The European Dream: Building Sustainable Development in a Globally Connected World," *E Magazine* (March/April 2005)

**NO: Ronald Bailey**, from "Wilting Greens," *Reason* (December 2002)

**ISSUE SUMMARY**

**YES:** Jeremy Rifkin, president of the Foundation on Economic Trends, argues that Europeans pride themselves on their quality of life, and their emphasis on sustainable development promises to maintain that quality of life into the future.

**NO:** Environmental journalist Ronald Bailey states that sustainable development results in economic stagnation and threatens both the environment and the world's poor.

**O**ver the last 30 years, many people have expressed concerns that humanity cannot continue indefinitely to increase population, industrial development, and consumption. The trends and their impacts on the environment are amply described in numerous books, including historian J. R. McNeill's *Something New Under the Sun: An Environmental History of the Twentieth-Century World* (W. W. Norton, 2000).

"Can we keep it up?" is the basic question behind the issue of sustainability. In the 1960s and 1970s, this was expressed as the "Spaceship Earth" metaphor, which said that we have limited supplies of energy, resources, and room and that we must limit population growth and industrial activity, conserve, and recycle in order to avoid crucial shortages. "Sustainability" entered the global debate in the early 1980s, when the United Nations secretary general asked Gro Harlem Brundtland, a former prime minister and minister of environment in Norway, to organize and chair the World Commission on Environment and Development and produce a "global agenda for change." The resulting report, *Our Common Future* (Oxford University Press, 1987), defined *sustainable development* as "development that meets the needs of the present without compromising the ability of future generations to meet their own needs." It recognized that limits on population size and resource use cannot be known precisely; that problems may arise not suddenly but rather gradually, marked by rising costs; and that limits may be redefined by changes in technology. The report also recognized that limits exist and must be taken into account when governments, corporations, and individuals plan for the future.

The Brundtland report led to the UN Conference on Environment and Development held in Rio de Janeiro in 1992. The Rio conference set sustainability firmly on the global agenda and made it an essential part of efforts to deal with global environmental issues and promote equitable economic development. In brief, sustainability means such things as cutting forests no faster than they can grow back, using groundwater no faster than it is recharged by precipitation, stressing renewable energy sources rather than exhaustible fossil fuels, and farming in such a way that soil fertility does not decline. In addition, economics must be revamped to take into account environmental costs as well as capital, labor, raw materials, and energy costs. Many add that the distribution of the Earth's wealth must be made more equitable as well.

Given continuing growth in population and demand for resources, sustainable development is clearly a difficult proposition. Some think it can be done, but others think that for sustainability to work, either population or resource demand must be reduced. Not surprisingly, many people see sustainable development as in conflict with business and industrial activities, private property rights, and such human freedoms as the freedoms to have many children, to accumulate wealth, and to use the environment as one wishes. Economics professor Jacqueline R. Kasun, in "Doomsday Every Day: Sustainable Economics, Sustainable Tyranny," *The Independent Review* (Summer 1999), goes so far as to argue that sustainable development will require sacrificing human freedom, dignity, and material welfare on a road to tyranny. Lester R. Brown, "Picking up the Tab," *USA Today Magazine* (May 2007), suggests that because governments currently spend some $700 billion a year subsidizing environmentally destructive activities such as automobile driving, one essential step toward sustainability is to end the subsidies.

In the following selections, Jeremy Rifkin, president of the Foundation on Economic Trends, argues that Europeans pride themselves on a quality of life that in some ways exceeds that of Americans, and their emphasis on inclusivity, diversity, sustainable development, social rights, and individual human rights promises to maintain that quality of life into the future. Ronald Bailey argues that preserving the environment, eradicating poverty, and limiting economic growth are incompatible goals. Indeed, vigorous economic growth provides wealth for all and leads to environmental protection.

YES

Jeremy Rifkin

# The European Dream: Building Sustainable Development in a Globally Connected World

A growing number of Americans are beginning to wonder why Europe has leaped ahead of the U.S. to become the most environmentally advanced political space in the world today. To understand why Europe has left America behind in the race to create a sustainable society, we need to look at the very different dreams that characterize the American and European frame of mind.

Ask Americans what they most admire about the U.S.A. and they will likely cite the individual opportunity to get ahead—at least until recently. The American Dream is based on a simple but compelling covenant: Anyone, regardless of the station to which they are born, can leverage a good public education, determination and hard work to become a success in life. We can go from "rags to riches."

Ask Europeans what they most admire about Europe and they will invariably say "the quality of life." Eight out of 10 Europeans say they are happy with their lives and when asked what they believe to be the most important legacy of the 20th century, 58 percent of Europeans picked their quality of life, putting it second only to freedom in a list of 11 legacies.

While the American Dream emphasizes individual success, the European Dream emphasizes collective well-being. The reason for this lies in the divergent histories of the two continents. America's founders came over from Europe 200 years ago in the waning days of the Protestant Reformation and the early days of the European Enlightenment. They took these two streams of European thought, froze them in time, and kept them alive in their purest form until today. Americans are the most devoutly Christian and Protestant people in the industrial world and the fiercest champions of the capitalist marketplace and the nation-state.

Both the Protestant Reformation and the Enlightenment emphasized the central role of the individual in history. John Calvin exhorted the faithful that every person stands alone with his or her God. Adam Smith, in turn, argued that all individuals pursue their own self interest in the marketplace. This individualist strain fit the American context far better than it did the European setting. In a wide-open frontier, every new immigrant did indeed stand alone and had to secure his or her survival with little or no social supports.

Americans, even today, are taught by his or her parents that to be free they must learn to be self-sufficient and independent, and that they cannot depend on others.

Europeans, however, never fully bought the idea of the individual alone in the universe. Europe was already densely populated and without a frontier by the late 18th century. Walled cities and tightly packed human settlement demanded a more communal way of life. While Americans defined freedom in terms of individual autonomy and mobility, Europeans defined freedom by their communal relationships.

In America there was enough cheap and free land and resources so that newcomers could become rich. In Europe, well-defined class boundaries—a remnant of the feudal aristocracy—made it far more difficult for an individual born in a lesser station of life to rise to the top and become wealthy. So while Americans preferred to pursue happiness individually, Europeans pursued happiness collectively by emphasizing the quality of life of the community. Today, Americans devote less than 11 percent of their Gross Domestic Product (GDP) to social benefits, compared to 26 percent in Europe.

## Doing It Better

So, what does Europe do better than America? It works hard to create a remarkably high quality of life for all of its people. The European Dream focuses on inclusivity, diversity, sustainable development, social rights and universal human rights. And it works. While Americans are 28 percent wealthier per capita than Europeans, in many ways, Europeans experience a higher quality of life, clear evidence that, in the long run, cooperation rather than competition is sometimes a surer path to happiness.

Europe and the U.S. have nearly opposite approaches to the question of environmental stewardship. At the heart of the difference is the way Americans and Europeans perceive risk. We Americans take pride in being a risk-taking people. We come from immigrant stock, people who risked their lives to journey to the new world and start over, often with only a few coins in their pockets and a dream of a better life. When Europeans and others are asked what they most admire about Americans, our risk-taking, "can-do" attitude generally tops the list. Where others see difficulties and obstacles, Americans see opportunities.

Our optimism is deeply entwined with our faith in science and technology. It has been said that Americans are a nation of tinkerers. When I was growing up, the engineer was held in as high esteem as the cowboy, admired for his efforts to improve the lot of society and contribute to the progress and welfare of civilization.

On the other side of the water, the sensibilities are different. It's not that Europeans aren't inventive. One could make the case that over the course of history Europe has produced most of the great scientific insights and not a few of the major inventions. But with their longer histories, Europeans are far more mindful of the dark side of science and technology.

## Saying No to GE Foods

In recent years, the European Union (EU) has turned upside down the standard operating procedure for introducing new technologies and products into the marketplace and society, much to the consternation of the United States. The turnaround started with the controversy over genetically engineered (GE) foods and the introduction of genetically modified organisms (GMOs). The U.S. government gave the green light to the widespread introduction of GE foods in the mid 1990s, and by the end of the decade more than half of America's agricultural land was given over to GE crops. No new laws were enacted to govern the potential harmful environmental and health impacts. Instead, existing statutes were invoked, and no special handling or labeling of the products was required.

In Europe, massive opposition to GMOs erupted across the continent. Farmers, environmentalists and consumer organizations staged protests and political parties and governments voiced concern. A defacto moratorium on the planting of GE crops and sale of GE food products was put into effect. Meanwhile, the major food processors, distributors and retailers pledged not to sell any products containing GE traits.

The EU embarked on a lengthy review process to assess the environmental and health risks of introducing GE food products. In the end, it established tough new protections designed to mitigate the potential harm. The measures included procedures to segregate and track GE grain and food products from the fields to the retail stores to ensure against contamination; labeling of GMOs at every stage of the food process to ensure transparency; and independent testing as well as more rigorous testing requirements by the companies producing GE seeds and other GMOs.

The EU is forging ahead on a wide regulatory front, changing the very conditions and terms by which new scientific and technological pursuits and products are introduced into the marketplace and the environment. Its bold initiatives put the EU far ahead of the rest of the world. Behind all of its newfound regulatory zeal is the looming question of how best to model global risks and create a sustainable and transparent approach to economic development.

## Ensuring Safety

In May of 2003, the EU proposed sweeping new regulatory controls on chemicals to mitigate toxic impacts on the environment and human and animal health. The proposed new law would require new companies to register and test for the safety of more than 30,000 chemicals at an estimated cost to the producers of nearly eight billion Euros. Under existing rules, 99 percent of the total volume of chemicals sold in Europe have not passed through any environmental and health testing and review process. In the past, there was no way to even know what kind of chemicals were being used by industry, making it nearly impossible to track potential health risks. The new regulations will change all of that. The "REACH" system—which stands for Registration, Evaluation and Authorization of Chemicals—requires the companies to conduct safety and environmental tests to prove that the products they are producing are safe. If they can't, the products will be banned from the market.

The new procedures represent an about face to the way the chemical industry is regulated in the U.S. In America, new chemicals are generally assessed to be safe and the burden is primarily put on the consumer, the public or the government to prove that they cause harm. The EU has reversed the burden of proof. Former EU Environmental Commissioner Margot Wallstrom makes the point: "No longer do public authorities need to prove they [the products] are dangerous. The onus is now on industry to prove that the products are safe."

Making companies prove that their chemical products are safe before they are sold is a revolutionary change. It's very difficult to conceive of the U.S. entertaining the kind of risk prevention regulatory regime that the EU has rolled out. In a country where corporate lobbyists spend millions of dollars influencing congressional legislation, the chances of ever having a similar regulatory regime to the one being implemented in Europe would be nigh on impossible.

## BUILDING THE HYDROGEN ECONOMY

At the very top of the list of environmental priorities for the EU is the plan to become a fully integrated renewable-based hydrogen economy by mid-century. The EU has led the world in championing the Kyoto Protocol on Climate Change, and to ensure compliance it has made a commitment to produce 22 percent of its electricity and 12 percent of all of its energy using renewable sources by 2010. Although a number of member states are lagging behind on meeting their renewable energy targets, the very fact that the EU has set benchmarks puts it far ahead of the U.S. in making the shift from fossil fuels to renewable energy sources. The Bush administration has consistently fought Congressional attempts to establish similar benchmarks for ushering in a U.S.-based renewable energy regime.

In June of 2003, EU President Romano Prodi said, "It is our declared goal of achieving a step-by-step shift toward a fully integrated hydrogen economy, based on renewable energy sources, by the middle of the century." He added that creating this economy would be the next critical step in integrating Europe after the introduction of the Euro.

The European hydrogen game plan is being implemented with a sense of history in mind. Great Britain became the world's leading power in the 19th century because it was the first country to harness its vast coal reserves with steam power. The U.S., in turn, became the world's preeminent power in the 20th century because it was the first country to harness its vast oil reserves with the internal-combustion engine. The multiplier effects of both energy revolutions were extraordinary. The EU is determined to lead the world into the third great energy revolution of the modern era.—*J.R.*

GMOs and chemical products represent just part of the new "risk prevention" agenda taking shape in Brussels. In early 2003, the EU adopted a new rule prohibiting electronics manufacturers from selling products in the EU that contain mercury, lead and other heavy metals. Another new regulation requires the manufacturers of all consumer electronics and household appliances to cover the costs for recycling their products. American companies complain that compliance with the new regulations will cost them hundreds of millions of dollars a year.

All of these strict new rules governing risk prevention would come as a shock to Americans who believe that the U.S. has the most vigilant regulatory oversight regime in the world for governing risks to the environment and public health. Although that was the case 30 years ago, it no longer is today.

The EU is the first governing institution in history to emphasize human responsibilities to the global environment as a centerpiece of its political vision. Europe's new sensitivity to global risks has led it to champion the Kyoto Protocol on climate change, the Biodiversity Treaty, the Chemical Weapons Convention and many others. The U.S. government has refused, to date, to ratify any of the above agreements.

## A New Era

In Europe, intellectuals are increasingly debating the question of the great shift from a risk-taking age to a risk-prevention era. That debate is virtually non-existent in the U.S., where risk-taking is seen as a virtue. The new European intellectuals argue that vulnerability is the underbelly of risks. A sense of vulnerability can motivate people to band together in common cause. The EU stands as a testimonial to collective political engagement arising from a sense of risk and shared vulnerability.

What's changed qualitatively in the last half century since the dropping of the atomic bombs on Hiroshima and Nagasaki is that risks are now global in scale, open ended in duration, incalculable in their consequences and not compensational. Their impact is universal, which means that no one can escape their potential effects. Risks have now become truly democratized, making everyone vulnerable. When everyone is vulnerable, then traditional notions of calculating and pooling risks become virtually meaningless. This is what European academics call a risk society.

Americans aren't there yet. While some academics speak to global risks and vulnerabilities and a significant minority of Americans express their concerns about global risks, from climate change to loss of biodiversity, the sense of utter vulnerability just isn't as strong on this side of the Atlantic. Europeans say we have blinders on. In reality, it's more nuanced than that. Call it delusional, but the sense of personal empowerment is so firmly embedded in the American mind, that even when pitted against growing evidence of potentially overwhelming global threats, most Americans shrug such notions off as overly pessimistic and defeatist. "Individuals can move mountains." Most Americans believe that. Fewer Europeans do.

The EU has already institutionalized a litmus test that cuts to the core of the differences between America and Europe. It's called "the precautionary principle" and it has become the centerpiece of EU regulatory policy governing science and technology in a globalizing world.

## The Precautionary Principle

In November 2002, the EU adopted a new policy on the use of the precautionary principle to regulate science and new products derived from technology innovations. According to the EU, reviews occur in "cases where scientific evidence is insufficient, inconclusive or uncertain and preliminary scientific evaluation indicates that there are reasonable grounds for concern that the potentially dangerous effects on the environment, human, animal or plant health may be

---

### THE TRANSITION TO ORGANIC AGRICULTURE

Europe is taking the lead in the shift to sustainable farming practices and organic food production. While the organic food sector is soaring in the U.S.—it represents the fastest-growing segment of the food industry—the government has done little to encourage it. Although the U.S. Department of Agriculture fields a small organic food research program, it amounts to only $3 million, less than .004 percent of its $74 billion budget. While American consumers are increasing their purchases of organic food, less than 0.3 percent of total U.S. farmland is currently in organic production.

By contrast, many of the EU member states have made the transition to organic agriculture a critical component of their economic development plans and have even set benchmarks. Germany, which has often been the leader in setting new environmental goals for the continent, has announced its intention to bring 20 percent of its agricultural output into organic production by the year 2020. (Organic agricultural output is now 3.2 percent of all farm output in Germany.)

The Netherlands, Sweden, Great Britain, Finland, Norway, Germany, Switzerland, Denmark, France and Austria also have national programs to promote the transition to organic food production. Denmark and Sweden enjoy the highest consumption of organic vegetables in Europe and both countries project that their domestic markets for organic food will soon reach or exceed 10 percent of domestic consumption.

Sweden has set a goal of having 20 percent of its total cultivated farm area in organic production by 2005. Italy already has 7.2 percent of its farmland under organic production while Denmark is close behind with seven percent.

Great Britain doubled its organic food production in 2002 and now boasts the second-highest sales of organic food in Europe, after Germany. According to a recent survey, nearly 80 percent of British households buy organic food. By comparison, only 33 percent of American consumers buy any organic food.—J.R.

---

inconsistent with the high level of protection chosen by the EU." The key term is "uncertain." When there is sufficient evidence to suggest a potential negative impact, but not enough to know for sure, the precautionary principle allows regulatory authorities to err on the side of safety. They can suspend the activity altogether, modify it, employ alternative scenarios, monitor the activity or create experimental protocols to better understand its effects.

The precautionary principle allows governments to respond with a lower threshold of scientific certainty than in the past. "Scientific certainty" has been tempered by the notion of "reasonable grounds for concern." The precautionary principle gives authorities the flexibility to respond to events in real time, either before or while they are unfolding.

Advocates of the precautionary principle cite the introduction of halocarbons and the tear in the ozone hole in the Earth's upper atmosphere, the outbreak of mad cow disease in cattle, growing antibiotic resistant strains of bacteria caused by the over-administering of antibiotics to farm animals and the widespread deaths caused by asbestos, benzene and polychlorinated biphenyls (PCBs).

The precautionary principle has been finding its way into international treaties and covenants. It was first recognized in 1982 when the United Nations General Assembly incorporated it into the World Charter for Nature. The precautionary principle was subsequently included in the Rio Declaration on Environment and Development in 1992, the Framework Convention on Climate Change in 1992, the Treaty on EU (Maastricht Treaty) in 1992, the Cartagena Protocol on Biosafety in 2000 and the Stockholm Convention on Persistent Organic Pollutants (POPs) in 2001.

## Valuing Nature

Americans, by and large, view nature as a treasure trove of useful resources waiting to be harnessed for productive ends. While Europeans share America's utilitarian perspective, they also have a love for the intrinsic value of nature. One can see it in Europeans' regard for the countryside and their determination to maintain natural landscapes, even if it means providing government assistance in the way of special subsidies, or foregoing commercial development. Nature figures prominently in Europeans' dream of a quality of life. Europeans spend far more time visiting the countryside on weekends and during their vacations than Americans.

The balancing of urban and rural time is less of a priority for most Americans, many of whom are just as likely to spend their weekends at a shopping mall, while their European peers are hiking along country trails. Anyone who spends significant time among Europeans knows that they have a great affinity for rural getaways. Almost everyone I know in Europe—among the professional and business class—has some small second home in the country somewhere—a dacha usually belonging to the family for generations. While working people may not be as fortunate, on any given weekend they can be seen exiting the cities en masse, motoring their way into the nearest rural enclave or country village for a respite from urban pressures.

The strongly held values about rural life and nature is one reason why Europe has been able to support green parties across the continent, with substantial representation in national parliaments as well as in the European Parliament. By contrast, not a single legislator at the federal level in the U.S. is a member of the Green Party.

There is another dimension to the European psyche that makes Europeans supportive of the precautionary principle—their sense of "connectedness."

Because we Americans place such a high premium on autonomy, we are far less likely to see the deep connectedness of things. We tend to see the world in terms of containers, each isolated from the whole and capable of standing alone. We like everything around us to be neatly bundled, autonomous, and self contained. The new view of science that is emerging in the wake of globalization is quite different. Nature is viewed as a myriad of symbiotic relationships, all embedded in a larger whole, of which they are an integral part. In this new vision of nature, nothing is autonomous, everything is connected.

By championing a host of global environmental treaties and accords taking the precautionary approach to regulation, the EU has shown a willingness to act on its commitment to sustainable development and global environmental stewardship. The fact that its commitments in most areas remain weak and are often vacillating is duly noted. But, at least Europe has established a new agenda for conducting science and technology that, if followed, could begin to wean the world from the old ways and toward a second Enlightenment.

**Ronald Bailey** **NO**

# Wilting Greens

It's clear that we've suffered a number of major defeats," declared Andrew Hewett, executive director of Oxfam Community Aid, at the conclusion of the World Summit on Sustainable Development, held in Johannesburg, South Africa, in September. Greenpeace climate director Steve Sawyer complained, "What we've come up with is absolute zero, absolutely nothing." The head of an alliance of European green groups proclaimed, "We barely kept our heads above water."

It wasn't supposed to be this way. Environmental activists hoped the summit would set the international agenda for sweeping environmental reform over the next 15 years. Indeed, they hoped to do nothing less than revolutionize how the world's economy operates. Such fundamental change was necessary, said the summiteers, because a profligate humanity consumes too much, breeds too much, and pollutes too much, setting the stage for a global ecological catastrophe.

But the greens' disappointment was inevitable because their major goals—preserving the environment, eradicating poverty, and limiting economic growth—are incompatible. Economic growth is a prerequisite for lessening poverty, and it's also the best way to improve the environment. Poor people cannot afford to worry much about improving outdoor air quality, let alone afford to pay for it. Rather than face that reality, environmentalists increasingly invoke "sustainable development." The most common definition of the phrase comes from the 1987 United Nations report *Our Common Future:* development that "meets the needs of the present without compromising the ability of future generations to meet their own needs."

For radical greens, sustainable development means economic stagnation. The Earth Island Institute's Gar Smith told Cybercast News, "I have seen villages in Africa . . . that were disrupted and destroyed by the introduction of electricity." Apparently, the natives no longer sang community songs or sewed together in the evenings. "I don't think a lot of electricity is a good thing," Smith added. "It is the fuel that powers a lot of multinational imagery." He doesn't want poor Africans and Asians "corrupted" by ads for Toyota and McDonald's, or by Jackie Chan movies.

Indian environmentalist Sunita Narain decried the "pernicious introduction of the flush toilet" during a recent PBS/BBC television debate hosted by Bill Moyers. Luckily, most other summiteers disagreed with Narain's curious disdain for sanitation. One of the few firm goals set at the confab was that adequate sanitation should be supplied by 2015 to half of the 2.2 billion people now lacking it.

Sustainable development boils down to the old-fashioned "limits to growth" model popularized in the 1970s. Hence Daniel Mittler of Friends of the Earth International moaned that "the summit failed to set the necessary economic and ecological limits to globalization." The *Jo'burg Memo*, issued by the radical green Heinrich Böll Foundation before the summit, summed it up this way: "Poverty alleviation cannot be separated from wealth alleviation."

The greens are right about one thing: The extent of global poverty is stark. Some 1.1 billion people lack safe drinking water, 2.2 billion are without adequate sanitation, 2.5 billion have no access to modern energy services, 11 million children under the age of 5 die each year in developing countries from preventable diseases, and 800 million people are still malnourished, despite a global abundance of food. Poverty eradication is clearly crucial to preventing environmental degradation, too, since there is nothing more environmentally destructive than a hungry human.

Most summit participants from the developing world understood this. They may be egalitarian, but unlike their Western counterparts they do not aim to make everyone equally poor. Instead, they want the good things that people living in industrialized societies enjoy.

That explains why the largest demonstration during the summit, consisting of more than 10,000 poor and landless people, featured virtually no banners or chants about conventional environmentalist issues such as climate change, population control, renewable resources, or biodiversity. Instead, the issues were land reform, job creation, and privatization.

The anti-globalization stance of rich activists widens this rift. Environmentalists claim trade harms the environment and further impoverishes people in the developing world. They were outraged by the dominance of trade issues at the summit.

"The leaders of the world have proved that they work as employees for the transnational corporations," asserted Friends of the Earth Chairman Ricardo Navarro. Indian eco-feminist Vandana Shiva added, "This summit has become a trade summit, it has become a trade show." Yet the U.N.'s own data underscore how trade helps the developing world. As fact sheets issued by the U.N. put it, "During the 1990s the economies of developing countries that were integrated into the world economy grew more than twice as fast as the rich countries. The 'non-globalizers' grew only half as fast and continue to lag further behind."

By invoking a zero sum version of sustainable development, environmentalists not only put themselves at odds with the developing world; they ignore the way in which economic growth helps protect the environment. The real commons from which we all draw is the growing pool of scientific, technological, and institutional concepts, and the capital they create. Past generations have left us far more than they took, and the result has been an explosion in human well-being, longer life spans, less disease, more and cheaper food, and expanding political freedom.

Such progress is accompanied by environmental improvement. Wealthier is healthier for both people and the environment. As societies become richer and more technologically adept, their air and water become cleaner, they set aside more land for nature, their forests expand, they use less land for agriculture, and more people cherish wild species. All indications suggest that the 21st century will be the century of ecological restoration, as humanity uses physical resources ever more efficiently, disturbing the natural world less and less.

In their quest to impose a reactionary vision of sustainable development, the disappointed global greens will turn next to the World Trade Organization, the body that oversees international trade rules. During the summit, the WTO emerged as the greens' bête noire. As Friends of the Earth International's Daniel Mittler carped, "Instead of using the [summit] to respond to global concerns over deregulation and liberalization, governments are pushing the World Trade Organization's agenda." "See you in Cancun!" promised Greenpeace's Steve Sawyer, referring to the location of the next WTO ministerial meeting in September 2003. That confab will build on the WTO's Doha Trade Round, launched last year, which is aimed at reducing the barriers to trade for the world's least developed countries.

The WTO may achieve worthy goals that eluded the Johannesburg summit, such as eliminating economically and ecologically ruinous farm and energy subsidies and opening developed country markets to the products of developing nations. Free marketeers and greens might even form an alliance on those issues.

But environmentalists want to use the WTO to implement their sustainable development agenda: global renewable energy targets, regulation based on the precautionary principle, a "sustainable consumption and production project," a worldwide eco-labeling scheme. According to Greenpeace's Sawyer, nearly everyone at the Johannesburg summit agreed "there is something wrong with unbridled neoliberal capitalism."

Let's hope the greens fail at the WTO just as they did at the U.N. summit. Their sustainable development agenda, supposedly aimed at improving environmental health, instead will harm the natural world, along with the economic prospects of the world's poorest people. The conflicting goals on display at the summit show that at least some of the world's poor are wise to that fact.

# POSTSCRIPT

## Is Sustainable Development Compatible with Human Welfare?

The first of the Rio Declaration's 22 principles states, "Human beings are at the centre of concerns for sustainable development. They are entitled to a healthy and productive life in harmony with nature." Any solution to the sustainability problem therefore should not infringe human welfare. This makes any solution that involves limiting or reducing human population or blocking improvements in standard of living very difficult to sell. Yet solutions may be possible. David Malin Roodman suggests in *The Natural Wealth of Nations: Harnessing the Market for the Environment* (W. W. Norton, 1998) that taxing polluting activities instead of profit or income would stimulate corporations and individuals to reduce such activities or to discover nonpolluting alternatives. In "Building a Sustainable Society," *State of the World 1999* (W. W. Norton, 1999), he adds recommendations for citizen participation in decision making, education efforts, and global cooperation, without which we are heading for "a world order [that] almost no one wants." (He is referring to a future of environmental crises, not the "new world order" feared by many conservatives, in which national policies are dictated by international [UN] regulators.) Roodman's recommendations may actually be on the way to reality. Arun Agrawal and Maria Carmen Lemos, "A Greener Revolution in the Making? Environmental Governance in the 21st Century," *Environment* (June 2007), argue that budget cuts and globalization are eroding the power of the state in favor of international "hybrid" arrangements that stress public-private partnerships, markets, and community and local participation.

Julie Davidson, in "Sustainable Development: Business as Usual or a New Way of Living?" *Environmental Ethics* (Spring 2000), notes that efforts to achieve sustainability cannot by themselves save the world. But such efforts may give us time to achieve new and more suitable values. It is thus heartening to see that the UN World Summit on Sustainable Development was held in Johannesburg, South Africa, in August 2002. Its aim was to strengthen partnerships between governments, business, nongovernmental organizations, and other stakeholders and to seek to eradicate poverty and make more equal the distribution of the benefits of globalization. See Gary Gardner, "The Challenge for Johannesburg: Creating a More Secure World," *State of the World 2002* (W. W. Norton, 2002), and the United Nations Environmental Programme's Global Environmental Outlook 3 (Earthscan, 2002), prepared as a "global state of the environment report" in preparation for the Johannesburg Summit.

The World Council of Churches brought to the Johannesburg Summit an emphasis on social justice. Martin Robra, in "Justice—The Heart of Sustainability," *Ecumenical Review* (July 2002), writes that the dominant stress on economic growth "has served, first and foremost, the interests of the powerful economic players. It has further marginalized the poor sectors of society, simultaneously undermining their basic security in terms of access to land, water, food, employment, and other basic services and a healthy environment."

Is social justice or equity worth this emphasis? Or is sustainability more a matter of population control, of shielding the natural environment from human impacts, or of economics? A. J. McMichael, C. D. Butler, and Carl Folke, in "New Visions for Addressing Sustainability," *Science* (December 12, 2003), argue that it is wrong to separate—as did the Johannesburg Summit—achieving sustainability from other goals such as reducing fertility and poverty and improving social equity, living conditions, and health. They observe that human population and lifestyle affect ecosystems, ecosystem health affects human health, human health affects population and lifestyle. "A more integrated . . . approach to sustainability is urgently needed," they say, calling for more collaboration among researchers and other fields. Yet there remains reason to focus on single threats. In February 2007, Sigma Xi and the United Nations Foundation released the Scientific Expert Group Report on Climate Change and Sustainable Development, *Confronting Climate Change: Avoiding the Unmanageable and Managing the Unavoidable* (http://www.sigmaxi.org/about/news/UNSEGReport.shtml) (Executive Summary, *American Scientist,* February 2007). Among its many points is that climate change from global warming is a huge threat to sustainability. Even in spite of feasible attempts at mitigation and adaptation, there is a serious risk of "intolerable impacts on human well-being."

Anthony R. Leiserowitz, Robert W. Kates, and Thomas M. Parris, "Do Global Attitudes and Behaviors Support Sustainable Development?" *Environment* (November 2005), find that though the world's people appear to support the component concepts of sustainable development, there is a mismatch between that support and their behavior. In the long term, they say, what is needed is a "shift from materialist to post-materialist values, from anthropocentric to ecological worldview, and a redefinition of the good life." Unfortunately, that shift remains in the future. P. Aarne Vesilind, Lauren Heine, and Jamie Hendry, "The Moral Challenge of Green Technology," *TRAMES: A Journal of the Humanities & Social Sciences* (No.1, 2006), "conclude that the unregulated free market system is incompatible with our search for sustainability. Experience has shown that if green technology threatens profits, green technology loses and profitability wins."

# ISSUE 3

## Should a Price Be Put on the Goods and Services Provided by the World's Ecosystems?

**YES: John E. Losey and Mace Vaughan**, from "The Economic Value of Ecological Services Provided by Insects," *BioScience* (April 2006)

**NO: Marino Gatto and Giulio A. De Leo**, from "Pricing Biodiversity and Ecosystem Services: The Never-Ending Story," *BioScience* (April 2000)


**ISSUE SUMMARY**

**YES:** John E. Losey and Mace Vaughan argue that even conservative estimates of the value of the services provided by wild insects are enough to justify increased conservation efforts. They say that "everyone would benefit from the facilitation of the vital services these insects provide."

**NO:** Professors of applied ecology Marino Gatto and Giulio A. De Leo contend that the pricing approach to valuing nature's services is misleading because it falsely implies that only economic values matter.


Human activities frequently involve trading a swamp or forest or mountainside for a parking lot or housing development or farm. People generally agree that these developments are worthwhile projects, for they have obvious benefits. But are there costs as well? Construction costs, labor costs, and material costs can easily be calculated, but what about the swamp? The forest? The species living there?

How much is a species worth? One approach to answering this question is to ask people how much they would be willing to pay to keep a species alive. If the question is asked when there are a million species in existence, few people will likely be willing to pay much. But if the species is the last one remaining, they might be willing to pay a great deal. Most people would agree that both answers fail to get at the true value of a species, for nature is not expressible solely in terms of cash values. Yet some way must be found to weigh the effects of human activities on nature against the benefits gained from those activities. If it is not, we will continue to degrade the world's ecosystems and threaten our own continued well-being.

Traditional economics views nature as a "free good." That is, forests generate oxygen and wood, clouds bring rain, and the sun provides warmth, all without charge to the humans who benefit. At the same time, nature has provided ways for people to dispose of wastes—such as dumping raw sewage into rivers or emitting smoke into the air—without paying for the privilege. This "free" waste disposal has turned out to have hidden costs in the form of the health effects of pollution (among other things), but it has been up to individuals and governments to bear the costs associated with those effects. The costs are real, but in general, they have not been borne by the businesses and other organizations that produced them. They have thus come to be known as "external" costs.

Environmental economists have recognized the problem of external costs, and government regulators have devised a number of ways to make those who are responsible accept the bill, such as instituting requirements for pollution control and fining those who exceed permitted emissions. Yet some would say that this approach does not help enough.

The *ecosystem services* approach recognizes that undisturbed ecosystems do many things that benefit us. A forest, for instance, slows the movement of rain and snowmelt into streams and rivers; if the forest is removed, floods may follow (a connection that recently forced China to deemphasize forest exploitation). Swamps filter the water that seeps through them. Food chains cycle nutrients necessary for the production of wood and fish and other harvests. Bees pollinate crops and make food production possible. These services are valuable—even essential—to us, and anything that interferes with them must be seen as imposing costs just as significant as the illnesses associated with pollution.

How can those costs be assessed? In 1997 Robert Costanza and his colleagues published an influential paper entitled "The Value of the World's Ecosystem Services and Natural Capital" in the May 15 issue of the journal *Nature*. In it, the authors listed a variety of ecosystem services and attempted to estimate what it would cost to replace those services if they were somehow lost. The total bill for the entire biosphere came to $33 trillion (the middle of a $16–54 trillion range), compared to a global gross national product of $25 trillion. Costanza et al. stated that this was surely an underestimate. Janet N. Abramovitz, "Putting a Value on Natures 'Free' Services," *WorldWatch* (January/February 1998), argues that nature's services are responsible for the vast bulk of the value in the world's economy and that attaching economic value to those services may encourage their protection.

In the following selections, John E. Losey, associate professor of entomology at Cornell University, and Mace Vaughan, Conservation Director at the Xerces Society for Invertebrate Conservation, argue that even conservative estimates of the value of the services provided by wild insects are enough to justify increased conservation efforts. Ecologists Marino Gatto and Giulio A. De Leo argue that the pricing approach to valuing nature's services is misleading because it ignores equally important "nonmarket" values.

YES

John E. Losey and Mace Vaughan

# The Economic Value of Ecological Services Provided by Insects

Natural systems provide ecological services on which humans depend. Countless organisms are involved in these complex interactions that put food on our tables and remove our waste. Although human life could not persist without these services, it is difficult to assign them even an approximate economic value, which can lead to their conservation being assigned a lower priority for funding or action than other needs for which values (economic or otherwise) are more readily calculated. Estimating even a minimum value for a subset of the services that functioning ecosystems provide may help establish a higher priority for their conservation.

In this article we focus on the vital ecological services provided by insects. Several authors have reviewed the economic value of ecological services in general, but none of these reviews focused specifically on insects. Insects comprise the most diverse and successful group of multicellular organisms on the planet, and they contribute significantly to vital ecological functions such as pollination, pest control, decomposition, and maintenance of wildlife species . . . Our twofold goal is to provide well-documented, conservative estimates for the value of these services and to establish a transparent, quantitative framework that will allow the recalculation of the estimates as new data become available. We also should clarify that by "value" we mean documented financial transactions—mostly the purchase of goods or services—that rely on these insect-mediated services.

We restrict our focus to services provided by "wild" and primarily by native insects; we do not include services from domesticated species (e.g., pollination from domesticated honey bees) or pest control from mass-reared insect biological-control agents (e.g., *Trichogramma* wasps). We also exclude the value of commercially produced insect-derived products, such as honey, wax, silk, or shellac, and any value derived from the capture and consumption of insects themselves. The main reasons for these exclusions are that domesticated insects that provide services or products have been covered in many other forums, and they generally do not require the active conservation that we believe is warranted by those undomesticated insects that provide services. Furthermore, in the case of products or food derived directly from wild insects, we simply do not have data to report and therefore wish to maintain a focus on ecological services.

From *BioScience,* by John E. Losey and Mace Vaughan, vol. 56, no. 4, April 2006, pp. 311–316, 318–322. 

The four insect services for which we provide value estimates were chosen not because of their importance, but because of the availability of data and an algorithm for their calculation. Three of these services (dung burial, pest control, and pollination) support the production of a commodity that has a quantifiable, published value. To be consistent in our analysis for all three of these commodities, we calculated an estimate for the amount of each commodity that depends on each service or on the amount saved in related expenses (e.g., the cost of fertilizer in our analysis of dung burial). We did not perform an in-depth analysis of how service-dependent changes in the quantity or quality of each commodity may have affected its per-unit price.

One way of looking at the economic implications of the removal of a service was provided by Southwick and Southwick, whose study involved crop pollination by honey bees. Because per-unit cost theoretically increases as supplies decrease, thus mitigating monetary losses, the costs of the service removal in the Southwick and Southwick study were lower than those calculated using our approach. However, all reported values are still within an order of magnitude of each other and, although our approach may not reflect what a consumer would pay for a commodity when these ecological services are *not* being performed, our calculations do provide a measure of the value of these crops at current estimated levels of service.

In the case of insect support of wildlife nutrition, we use a different approach to estimate costs. Instead of basing calculations on the money paid to producers for raw commodities, we use census data to find out how US consumers spent their money. By looking at the consumer end of this system, we immediately see an order-of-magnitude increase in the value reported. We believe it is important for this difference to be understood up front, because it both significantly affects our reported results and provides at least a hint of what happens when raw commodities are converted into value-added products. For example, consumers will spend potentially an order of magnitude more on jellies, pasta sauce, or hamburgers than the price paid to producers for blueberries, tomatoes, or beef.

Using the methods we describe in detail in the following sections, we estimate the annual value of four ecological services provided by primarily native insects in the United States to be more than $57 billion ($0.38 billion for dung burial, $3.07 billion for pollination, $4.49 billion for pest control of native herbivores, and $49.96 billion for recreation). We consider this estimate very conservative. If data were available to support more accurate estimates of the true value of these services (e.g., inclusion of value-added products and wages paid to those who produce such products) or to allow estimation of the value of other services, the results of our calculations would be much higher. In addition to the role of insects in the systems we analyze here, other potentially important services that insects provide could not be quantified, including suppression of weeds and exotic herbivorous species, facilitation of dead plant and animal decomposition, and improvement of the soil. Calculating the value of any of these services could add billions of dollars to our overall estimate. Nevertheless, we hope that even this minimum estimate for a subset of services provided by insects will allow these animals to be more correctly factored into land management and legislative decisions. In the following sections, we present a detailed description of how we calculated these estimates and discuss the implications of our results.

## Dung Burial

Confining large mammals in small areas creates challenging waste-management problems. Cattle production in the United States provides a particularly pertinent example, because nearly 100 million head of cattle are in production, and each animal can produce over 9000 kilograms (kg), or about 21 cubic meters, of solid waste per year. Fortunately, insects—especially beetles in the family Scarabaeidae—are very efficient at decomposing this waste. In doing so, they enhance forage palatability, recycle nitrogen, and reduce pest habitat, resulting in significant economic value for the cattle industry.

Dung beetles process a substantial amount of the cattle dung accumulated annually in the United States. . . .

The importance of this service is illustrated by the success of dung beetles introduced into Australia to deal with the dung of nonnative cattle brought to that continent in 1788. Before the introduction of dung beetle species that were adapted to feed on cattle dung, Australia had no insect fauna to process cattle feces. Consequently, rangeland across the country was fouled by slowly decomposing dung. In addition, this dung provided fodder for pest species. Recent research in western Australia has revealed that populations of the pestiferous bush fly (*Musca vetustissima*) have been reduced by 80% following dung beetle introductions. . . .

Using data from Anderson and colleagues, we calculate that the average persistence—or time until complete decomposition—of an untreated dung pat on rangeland in California is 22.74 ± 0.64 months, while the average persistence of a pat treated with insecticides is 28.14 ± 0.71 months. This indicates that dung beetle activity results in a 19% decrease in the amount of time the average pat of dung makes forage unpalatable, which translates into substantial monetary savings. Note that, for the sake of this analysis, we must assume that the 19% decrease applies broadly across the United States, even though the rate of dung burial by beetles probably varies greatly depending upon the location.

**Forage fouling.** Fincher estimated a potential value for enhanced palatability based on the concept that cattle will not consume plant material that is fouled with dung. If dung beetles were totally absent, forage fouling by dung would cause estimated annual losses of 7.63 kg of beef per head of cattle ($L_{nb}$). This level of loss is in comparison with the theoretical zero loss of production if no forage were ever fouled by dung. Fortunately, the cattle industry is not saddled with the full force of this potential loss because range fouling is reduced by the current action of dung beetles.

If we assume that the 19% decrease in dung persistence translates into a 19% decrease in lost beef, then, for cattle whose dung is processed by dung beetles, the per-animal loss would be 6.18 kg ($L_b$) each year as a result of forage fouling. This assumption seems justified, since for each increment of time a given patch of forage remains fouled, it also remains unavailable for grazing. By applying these estimated losses to the 32 million head that are untreated and on pasture or rangeland, we estimate that in the absence of dung beetles, beef losses due to forage fouling would be 244 million kg of beef per year ($C_p \times L_{nb}$), whereas losses at current levels of dung beetle function would be 198 million kg ($C_p \times L_b$). With an average price over 34 years (1970–2003, corrected for inflation) of live beef cattle at $2.65 per kg ($V_c$), losses would be $647 million ($V_c \times [C_p \times L_{nb}]$) in the absence of dung beetles and $525 million ($V_c \times [C_p \times L_b]$) in the presence of dung beetles. Subtracting the estimated value at current levels of dung beetle activity from the theoretical value if no dung beetles were active, we estimate the value of the reduced forage fouling ($V_{rf}$) to be approximately $122 million.

**Nitrogen volatilization.** Another important service provided by dung beetles is promoting decomposition of dung into labile forms of nitrogen that can be assimilated by plants and thus function as fertilizer when the dung is buried. In the absence of dung beetles, cattle feces that remain on the pasture surface until they are dry lose a large proportion of their inorganic nitrogen to the atmosphere. Experiments in South Africa and the United States have shown that approximately 2% of cattle dung is composed of nitrogen, and that 80% of this nitrogen is lost if the pats dry in the sun before they are buried.

Using Gillard's estimate of 27 kg of nitrogen produced annually per animal and assuming that 80% of this nitrogen is lost in the absence of dung beetle activity, we estimate that 21.6 kg would be lost per animal each year if dung beetles were not functioning ($L_{nb}$). On the basis of our interpretation of decomposition rates, we assume that these losses will be reduced 19% by the current level of dung beetle activity, compared with the estimate for no beetle activity. Thus, we estimate a loss of 17.5 kg per year ($L_b$) at current activity levels. Multiplying these per-animal values by the total number of cattle whose dung can potentially be buried by dung beetles ($C_p$, or 32 million), 691 million kg of nitrogen would be lost annually in the United States in the absence of dung beetle activity, compared with the 560 million kg lost at current levels of activity. With nitrogen valued at $0.44 per kg ($V_n$), we estimate the value of nitrogen lost in the absence of dung beetles to be $304 million and the value of nitrogen lost at current levels of dung beetle activity to be $246 million. Subtracting the estimated value at current levels of dung beetle activity from the theoretical value if no dung beetles were active, the value of the reduction in nitrogen loss is approximately $58 million.

**Parasites.** Many cattle parasites and pest flies require a moist environment such as dung to complete their development. Burying dung and removing this habitat can reduce the density of these pests. From field observations that reflected current levels of removal, Fincher estimated the annual losses due to mortality, morbidity, and medication of beef cattle, dairy cattle, and other livestock with internal parasites. To estimate the value of dung burial for reducing these losses, we will use only the losses associated with beef cattle,

## FORMULAS USED TO ESTIMATE INSECT SERVICES

**Formula used to estimate the number of cattle in the United States whose dung can be processed by dung beetles:**

$C_p = (C_t \times P_r) \times P_{nt}$,

where

$C_p$ = head of cattle producing dung that can be processed by dung beetles,

$C_t$ = total head of cattle produced annually in the United States,

$P_r$ = the proportion of cattle that are raised on range or pasture, and

$P_{nt}$ = the proportion of cattle not treated with avermectins.

**Formula used to estimate the value of beef saved because of reduced range fouling resulting from dung burial by dung beetles:**

$V_{rf} = [V_c \times (C_p \times L_{nb})] - [V_c \times (C_p \times L_b)]$,

where

$V_{rf}$ = value of reduced forage fouling,

$V_c$ = value of cattle (per kilogram),

$C_p$ = head of cattle producing dung that can be processed by dung beetles,

$L_{nb}$ = losses (per animal) with no dung beetle activity, and

$L_b$ = losses (per animal) at current levels of dung beetle activity.

**Formula used to estimate the value of native insects for suppressing populations of potentially pestiferous native herbivorous insects:**

$V_{ni} = (NC_{ni} - CC_{ni}) \times P_i$,

where

$V_{ni}$ = the value of suppression of native insect pests by other insects,

$NC_{ni}$ = the cost of damage from native insect pests with no natural control,

$CC_{ni}$ = the cost of damage from native insect pests at current levels of natural control, and

$P_i$ = the proportion of herbivorous insects controlled primarily by other insects.

because we do not have a good estimate for the proportion of dairy cattle or other livestock that live on open pasture or rangeland. Fincher reported that beef cattle ranchers lost $428 million annually because of parasites and pests. Corrected for inflation, this is equal to $912 million in 2003 dollars. Given that 85% of beef cattle are on range or pasture and 44% of these cattle are not treated with insecticides, we calculate that 37% of the beef cattle in the United States have fewer parasites because of the facilitation of dung decomposition by dung beetles.

We go on to assume that cattle whose dung is processed by dung beetles suffer 19% fewer losses because of parasites, on the basis of our previous calculation that dung beetles accelerate decomposition by 19%. We also assume that cattle on rangeland, pasture, and feedlots all face the same level of loss from parasites in the absence of dung beetles. Following this logic, we estimate that damage from parasites is only 93% (100% – [37% × 19%]) of what it would be if dung beetles were not providing this service. In the absence of dung beetle activity, estimated losses would be $981 million instead of the current $912 million, and thus this service saves the cattle industry an estimated $70 million per year.

**Pest flies.** Using a similar algorithm, we can calculate a value for the reduction in losses due to pest flies. Fincher estimated that losses due to horn flies and face flies cost ranchers $365 million and $150 million, respectively, for a total of $515 million. Corrected for inflation, this is the equivalent of $1.7 billion in 2003. Using the calculation described above for parasites, we assume that, as a result of the processing of dung by insects, damage from parasites is only 93% of what it would have been if the service were not being provided. We estimate that losses in the absence of dung beetle activity would be $1.83 billion instead of the current $1.7 billion, and thus this service is saving the cattle industry an estimated $130 million per year.

Adding the individual values of increased forage, nitrogen recycling, and reduced parasite and fly densities due to dung processing by beetles, we arrive at a combined annual total of $380 million. This is certainly an underestimate, since these same services are being provided to an unknown proportion of pasture-raised dairy cows, horses, sheep, goats, and pigs. Furthermore, what is said for dung recycling can also be said for burying beetles and flies that decompose carcasses. While the density of carcasses is much lower than the density of dung pats, their removal is important in rangeland, natural areas, and other public areas for returning nutrients to the soil, reducing potential spread of diseases, and increasing site utility.

## Pollination by Native Insects

Pollination, especially crop pollination, is perhaps the best-known ecosystem service performed by insects. McGregor estimates that 15% to 30% of the US diet is a result, either directly or indirectly, of animal-mediated pollination. These products include many fruits, nuts, vegetables, and oils, as well as meat and dairy products produced by animals raised on insect-pollinated forage. While this estimate is probably high, it presents one of the best published measures of pollinator-dependant food in the US diet.

Here we attempt to calculate an estimate of the value of crops produced as a result of pollination by wild (i.e., unmanaged) native insects. The US government keeps records of the production of crops and, because of their value, their insect pollinators have been given some attention, especially pollination by managed insects such as the European honey bee (*Apis mellifera L.*). From these studies and personal accounts of crop scientists and entomologists, several authors make generalizations about the proportion of pollination attributed to various insect groups, mostly honey bees. These generalizations are essentially educated guesses of the percentage of necessary pollination provided by insects, and as such, they are likely to be inaccurate. The proportions that could be attributed to native, as opposed to managed, pollinators will vary widely for each crop, depending on geographic location, availability of natural habitat, and use of pesticides. In addition, cultivars of the same species can have drastically different dependencies on insect pollinators, further complicating any calculation of the value of pollinator insects.

To conduct a truly accurate economic analysis of the role of native insects in crop pollination, we would need a much better accounting of current levels of pollination by different species of managed bees (e.g., honey bee [*A. mellifera*], alfalfa leaf-cutter bee [*Megachile rotundata*], blue orchard bee [*Osmia lignaria*], alkali bee [*Nomia melanderi*]), and wild bees (e.g., bumble bees [*Bombus* spp.], southeastern blueberry bee [*Habropoda laboriosa*], squash bee [*Peponapis pruinosa*]) in crop pollination. . . . Although we still lack much of this information, the estimate we provide here for the value of crops produced as a result of wild native bee–mediated pollination is informative.

Several scientists have estimated the value of insect-pollinated crops that are dependent on honey bees, or the financial loss to society that could be expected if managed honey bees were removed from cropping systems. These authors make a variety of assumptions and take different approaches to calculating a value for honey bees. For example, Southwick and Southwick take into account the reduced crop output stemming from a lack of managed honey bees, adjusting their figures for the changes in value of each commodity as demand increases because of reduced supply. They also present a range of possible values based on assumptions of the pollination redundancy of managed honey bees and other bee pollinators, including feral honey bees and other native and normative bees. Taking all of this into account, they give a range of $1.6 billion ($2.1 billion when adjusted for inflation to represent 2003 dollars) to $5.2 billion ($6.8 billion in 2003 dollars) for the value of honey-bee pollinators. The lower estimate included effective pollination by other bees, making the managed honey bees redundant in some localities and thereby reducing their absolute value. On the high end, Southwick and Southwick estimate that honey bees are worth $5.2 billion if few or no other bees visit insect-pollinated crops.

Robinson and colleagues and Morse and Calderone take a simpler approach, summing the value of each commodity that they estimate is dependant on honey-bee pollinators. From this they generate a portion of the overall value of each crop that they attribute to pollination by honey bees and report values of $8.3 billion and $14.6 billion ($12.3 billion and $16.4 billion, respectively, when adjusted for inflation to represent 2003 dollars). This approach is more consistent with our other calculations of the value of ecosystem services, and so we choose to use it here to calculate the value of crop production that relies on native insect pollinators.

When we sum the average value of pollinator-dependent commodities reported in Morse and Calderone, we find that native pollinators—almost exclusively bees—may be responsible for almost $3.07 billion of fruits and vegetables produced in the United States. Here we must incorrectly assume that the proportion of honey bees to native species is constant in all settings. In some systems, such as agriculturally diverse, organic farms with nearby pockets of natural or seminatural habitat, native bees may be able to provide all of the pollination needs for certain crops. For example, Morse and Calderone assume that 90% of the insect pollinators of watermelon are honey bees. While this is probably true in most farms, some organic growers can rely on native bees for 100% of their melon pollination.

## Pest Control

The best estimate available suggests that insect pests and their control measures cost the US economy billions of dollars every year, but this is only a fraction of the costs that would accrue if beneficial insects such as predators and parasitoids, among other forces, did not keep most pests below economically damaging levels. We calculate the value (*V*) of these natural forces by first estimating the cost of damage caused by insect pests at current levels of control (*CC*) and then subtracting this value from the estimated higher cost that would be caused by the greater damage from these insect pests if no controls were functioning (*NC*). Finally, we calculate a value for the specific action of insect natural enemies by multiplying the value of these natural forces by an estimate of the proportion ($P_i$) of pests that are controlled by beneficial insects as opposed to other mechanisms (e.g., pathogens or climate).

Because of data limitations, we restrict our estimate to the value derived from the suppression of insect pests that attack crop plants. Beneficial insects certainly suppress populations of both weeds and insects that attack humans and livestock, but the data were not available to calculate the value of these services. As with the rest of our analysis, we also limit our calculations to pest and beneficial insects native to the United States.

Our first step was to calculate the cost of damage due to insect pests at current levels of control from natural enemies. Drawing on previously published estimates, Yudelman and colleagues presented monetary values for total production of eight major crops and for the losses to these crops attributable to insects. Using these values, we calculated a ratio of insect loss to actual yield that allowed estimation of losses due to insects for any period for which yield values have been published. Assuming $50.5 billion for total production and $7.5 billion for losses due to insects in North America from 1988 through 1990, we calculated a ratio of 0.1485. . . .

Calkins found that only 35% of the exotic pests in the United States are pests in their home range. Extending this finding, we assume that the same relationship holds true in the United States, and thus only 35% of potential insect pest species that are native to the United States reach damaging levels. In other words, we assume that 65% of the potential damage from native pest species is being suppressed, and that 65% of the potential financial cost of this damage is being saved. We make this assumption based on (a) the abundant evidence of a strong correlation between pest density and the magnitude of loss due to pest damage, and (b) the lack of evidence of a correlation between the destructiveness of a pest and the probability that it will be suppressed.

To clarify, the pool of potential pest species—from which we assume 35% actually reach pest levels—is significantly smaller than the 90,000 described insect species in the United States, because many of the described species are not herbivores, and many of those that are herbivores do not feed on cultivated plants. Only 6000 (7%) of the described species in the United States and Canada cause any damage. For our estimate, we assume that these 6000 species, although they make up only 7% of the total species, account for 35% of the species that would be pests if they were not controlled. Following this logic, we assume that the pool of potential pests would be about 17,000 species, 11,000 of which (65%) are being kept below damage levels by biological or climatic controls.

These native species are estimated to comprise 39% of all pest species in the United States. Since native pests vary greatly in the amount of damage they cause, and include some of the most damaging pests in the United States (e.g., corn rootworm, Colorado potato beetle, and potato leafhopper), we assume that they are responsible for 39% of the cost of damage from all pests in the United States. Hence, we estimate that the cost associated with native pest species at current levels of suppression by natural enemies is 39% of $18.77 billion, or $7.32 billion. We designate this value current control by native insects ($CC_{ni}$).

On the basis of these assumptions, we estimate that the $7.32 billion lost annually to native insect pests ($CC_{ni}$) is 35% of what would be lost if natural controls were not functioning. If no natural forces were functioning to control native insect pests, we estimate that they would cause $20.92 billion in damage in the United States each year ($NC_{ni}$). By subtraction, the value of pest control by our native ecosystems is approximately $13.60 billion.

However, not all of this value for natural control of insect pests is attributable to beneficial insects. Some pest suppression comes from other causes, such as pathogens, climatic conditions, and host-plant resistance. One review of the factors responsible for suppression of 68 herbivore species reported that insects (e.g., predators and parasitoids) were primarily responsible for natural control in 33% of cultivated systems. On the basis of these findings, we estimate that insects are responsible for control of 33% of pests that are suppressed by natural controls, while pathogens or bottom-up forces control the rest. Using this average, we estimate the value of natural control attributable to insects to be $4.5 billion annually (33% of $13.6 billion).

## Recreation and Commercial Fisheries

US citizens spend over $60 billion a year on hunting, fishing, and observing wildlife. Insects are a critical food source for much of this wildlife, including many birds, fish, and small mammals. Using 1996 US census data on the spending habits of Americans, adjusted for inflation to 2003 dollars, we estimated the amount of money spent on recreational activities that is dependent on services provided by insects. In this case, the predominant service is concentrating and moving nutrients through the food web.

**Small game hunting.** Since most large game are either obligate herbivores or omnivores that are not substantially dependent on insects as a source of nutrition, we restrict our estimate of the value of insects for hunting to small game species. In 1996, expenditures for small game hunting totaled $2.5 billion ($2.9 billion in 2003 dollars). To calculate the proportion of this expenditure that is dependent on insects, we use the proportion of days spent hunting for each insectivorous small game species and the dependence of these birds on insects for food.

On the basis of published reports that most galliform chicks rely on insects as a source of protein and that many cannot even digest plant material, we assume that quail, grouse, and pheasant could not survive without insects as a nutritional resource. Therefore, multiplying the proportion of hunting days spent on each of these small game birds (0.15, 0.13, and 0.23, respectively, for a total of 0.51) by the total value for small game ($2.9 billion), we estimate that insects are required for $1.48 billion in expenditures.

**Migratory bird hunting.** Insectivory in migratory birds—primarily waterfowl such as ducks and geese in the order Anseriformes—is not as predominant as in the primarily terrestrial galliform birds discussed above. According to Ehrlich and colleagues, 19 (43%) of the 44 species in this order are primarily insectivorous. Multiplying the total money spent on migratory bird hunting ($1.3 billion) by the 43% of species that are primarily insectivorous, we estimate the value of insects as food for hunted migratory birds at $0.56 billion in hunter expenditures.

**Sport and commercial fishing.** The census also provides values for sport or recreational fishing. Since most recreational fishing is in fresh water and a majority of freshwater sport fish are insectivorous, we assume that the entire value of recreational fishing ($27.9 billion) is dependent on insects. In contrast to recreational fishing, the target of most commercial fishing is saltwater fish. There are very few marine insect species, but many fish that are caught in marine systems spend part of their life cycle in fresh water, and insects are often critical sources of nutrition during these periods. Commercial fishing is not covered by the census, but data are available on the number and value of fish landed annually in the United States by commercial operations. Twenty-five of these fish species are primarily insectivorous during at least one life stage. Summing their individual values, we estimate the total value of insects for commercial fishing to be approximately $225 million. Insectivorous fish account for more than 15% of the overall value of commercial fish.

**Wildlife observation (bird watching).** The 1996 census reports that Americans spent $33.8 billion on wildlife observation. The census also asked respondents to note which types of wildlife they were watching (e.g., birds, mammals, reptiles, amphibians, insects). Because respondents were allowed to choose more than one category of wildlife, it was impossible to separate out observed groups of organisms that were dependent on insects from those that were not. Bird watching is the most inclusive category, with 96% of respondents indicating that they included birds in their observations. Thus, we assume that 96% of the budget for wildlife observation stems directly from birds, many of which are at least partly dependent on insects as a source of nutrition. Thus, we assume that bird watching accounts for 96% of $33.8 billion spent, or $32.4 billion a year, providing a conservative starting point for calculating the dependency of wildlife observation expenditures on insects.

Our next step is to estimate what proportion of this figure for bird observation was dependent on and attributable to insects. Using data from Ehrlich and colleagues, we calculate that 61% of the bird species known to breed in the United States are primarily insectivorous, and another 28% are at least partially insectivorous. To be conservative, we consider only bird species that are primarily insectivorous. This probably underestimates the importance of insectivory for birds, since many passerine and galliform birds that are listed as partially insectivorous could not survive without the vital protein that insects provide young chicks. This estimate is conservative also because it is based on bird species numbers rather than population numbers, and the passerines, which are overwhelmingly insectivorous, have relatively high population densities. Taking these factors into account, we estimate that insects are responsible for $19.8 billion, which is 61% of the $32.4 billion spent on bird observation annually in the United States.

## Discussion

We estimate the value of those insect services we address to be almost $60 billion a year in the United States, which is only a fraction of the value for all the services insects provide. The implication of this estimate is that an annual investment of tens of billions of dollars would be justified to maintain these service-providing insects, were they threatened. And indeed, these beneficial insects are under ever increasing threat from a combination of forces, including habitat destruction, invasion of foreign species, and overuse of toxic chemicals.

Fortunately, no evidence suggests a short-term drastic decline in the insects that provide these services. What the evidence does indicate, however, is a steady decline in these beneficial insects, associated with an overall decline in biodiversity, accompanied by localized, severe declines in environments heavily degraded by human impacts. New evidence indicates that in some situations, the most important species for providing ecosystem services are lost first. The overall, gradual decline in species, coupled with nonlinear changes in service levels, makes it difficult to pinpoint an optimal level of annual investment to conserve beneficial insects and maintain the services they provide. . . .

[E]ven though we provide an estimate of the total value of certain insect services, the complications of redundancy and nonlinearity make it impossible to quantitatively gauge the level of resources that are justified for efforts aimed at conserving the services that insects provide. However, our findings lead us to espouse three qualitative guidelines. First, cost-free or relatively inexpensive measures are almost certainly justified to maintain and increase current service levels. Examples include volunteer construction of nest boxes for wild pollinators and the inclusion of a diverse variety of native plant species in plantings for bank or soil stabilization and site restoration. Second, actions or investments that are estimated to have an economic return at or slightly below the break-even point, such as the use of less toxic pesticides, are probably justified because of their nontarget benefits. Third, actions that lead to substantial decreases in biodiversity should be avoided because of the high probability of a major disruption in essential services.

Finally, although we cannot provide a quantitative formula to determine the optimal level of investment in the conservation of beneficial insects that provide essential services, we do feel justified, on the basis of our estimates, in making some specific recommendations. First, we recommend that conservation funding allocated via Farm Bill programs—such as the Conservation Security Program, Conservation Reserve Program, Wetlands Reserve Program, and Environmental Quality Incentives Program—pay specific attention to insects and the role they play in ecosystems. In particular, funding to provide habitat for beneficial insects such as predators, parasitoids, and pollinators in natural, seminatural, unproductive, or fallow areas in agricultural landscapes not only provides direct benefits to growers but, by focusing on the ecological needs of insects, results in habitat that supports a great diversity of wildlife.

Second, we recommend that ecosystem services performed by insects be taken into account in land-management decisions. Specifically, maintaining ecosystem services should be a goal of land management. With this goal in mind, specific practices such as grazing, burning, and pesticide use should be tailored to protect insect biodiversity. . . .

We believe it is imperative that some federal and local funds be directed toward the study of these beneficial insects and the vital services they provide so that conservation efforts can be optimally allocated, either through the agricultural programs listed above or through other means.

These steps are just a beginning. With greater attention, research, and conservation, the valuable services that insects provide can not only be sustained but increase in capacity. As a result, growers will be able to practice a more sustainable form of agriculture while spending less on managing pest insects or acquiring managed pollinators; ranchers will get more productivity out of their land; and wildlife lovers will find that the birds and fish they hunt occur in greater abundance than in the past few decades. In less direct but no less important ways, everyone would benefit from the facilitation of the vital services that insects provide. Judging from our estimate of the value of these four services, increased investment in the conservation of these services is justified.

Marino Gatto and
Giulio A. De Leo

# NO

# Pricing Biodiversity and Ecosystem Services: The Never-Ending Story

In 1844, the French engineer Jules Juvénal Dupuit introduced cost–benefit analysis to evaluate investment projects. . . . The application of cost–benefit analysis to ecological issues fell out of favor three decades ago, and it was gradually replaced by multicriteria analysis in the decision-making process for projects that have an impact on the environment. Although multicriteria analysis is currently used for environmental impact assessments [EIA] in many nations, [recently] the concept of cost–benefit analysis has again become fashionable, along with the various pricing techniques associated with it, such as contingent valuation methods, hedonic prices, and costs of replacement of ecological services. . . . Economists have generated a wealth of virtuosic variations on the theme of assessing the societal value of biodiversity, but most of these techniques are invariably based on price—that is, on a single scale of values, that of goods currently traded on world markets.

Perhaps the most famous recent study on the issue of pricing biodiversity and ecological services is that by Costanza et al., who argued that if the importance of nature's free benefits could be adequately quantified in economic terms, then policy decisions would better reflect the value of ecosystem services and natural capital. Drawing on earlier studies aimed at estimating the value of a wide variety of ecosystem goods and services, Costanza et al. estimated the current economic value of the entire biosphere at $16–54 trillion per year, with an average value of approximately $33 trillion per year. By contrast, the gross national product of the United States totals approximately $18 trillion per year. The paper, as its authors intended, stimulated much discussion, media attention, and debate. A special issue of *Ecological Economics* (April 1998) was devoted to commentaries on the paper, which, with few exceptions, were laudatory. Some economists have questioned the actual numbers, but many scientists have praised the attempt to value biodiversity and ecosystem functions.

Although Costanza et al. acknowledged that their estimates were crude and imperfect, they also pointed the way to improved assessments. In particular, they noted the need to develop comprehensive ecological economic models that could adequately incorporate the complex interdependencies between

From *BioScience,* vol. 50, no. 4, April 2000, by Marino Gatto and Giulio A. De Leo, pp. 347–354.

ecosystems and economic systems, as well as the complex individual dynamics of both types of systems. Despite the authors' caveats and the fact that many economists have been circumspect in applying their own tools to decisions regarding natural systems, the monetary approach is perceived by scientists, policymakers, and the general public as extremely appealing; a number of biologists are also of the opinion that attaching economic values to ecological services is of paramount importance for preserving the biosphere and for effective decision-making in all cases where the environment is concerned.

In this article, we espouse a contrary view, stressing that, for most of the values that humans attach to biodiversity and ecosystem services, the pricing approach is inadequate—if not misleading and obsolete—because it implies erroneously that complex decisions with important environmental impacts can be based on a single scale of values. We contend that the use of cost–benefit analysis as the exclusive tool for decision-making about environmental policy represents a setback relative to the existing legislation of the United States, Canada, the European Union, and Australia on environmental impact assessment, which explicitly incorporates multiple criteria (technical, economic, environmental, and social) in the process of evaluating different alternatives. We show that there are sound methodologies, mainly developed in business and administration schools by regional economists and by urban planners, that can assist decision-makers in evaluating projects and drafting policies while accounting for the nonmarket values of environmental services.

## The Limitations of Cost–Benefit Analysis and Contingent Valuation Methods

Historically, the first important implementation of cost–benefit analysis at the political level came in 1936, with passage of the US Flood Control Act. This legislation stated that a public project can be given a green light if the benefits, to whomsoever they accrue, are in excess of estimated costs. This concept implies that all benefits and costs are to be considered, not just actual cash flows from and to government coffers. However, public agencies (e.g., the US Army Corps of Engineers) quickly ran into a problem: They were not able to give a monetary value to many environmental effects, even those that were predictable in quantitative terms. For instance, engineers could calculate the reduction of downstream water flow resulting from construction of a dam, and biologists could predict the river species most likely to become extinct as a consequence of this flow reduction. However, public agencies were not able to calculate the cost of each lost species. Therefore, many ingenious techniques for the monetary valuation of environmental goods and services have been devised since the 1940s. These techniques fall into four basic categories.

- **Conventional market approaches.** These approaches, such as the replacement cost technique, use market prices for the environmental service that is affected. For example, degradation of vegetation in developing countries leads to a decrease in available fuelwood. Consequently, animal dung has to be used as a fuel instead of a fertilizer, and farmers must therefore replace dung with chemical fertilizers. By computing the cost of these chemical fertilizers, a monetary value for the degradation of vegetation can then be calculated.

- **Household production functions.** These approaches, such as the travel cost method, use expenditures on commodities that are substitutes or complements for the environmental service that is affected. The travel cost method was first proposed in 1947 by the economist Harold Hotelling, who, in a letter to the director of the US National Park Service, suggested that the actual traveling costs incurred by visitors could be used to develop a measure of the recreation value of the sites visited.
- **Hedonic pricing.** This form of pricing occurs when a price is imputed for an environmental good by examining the effect that its presence has on a relevant market-priced good. For instance, the cost of air and noise pollution is reflected in the price of plots of land that are characterized by different levels of pollution, because people are willing to pay more to build their houses in places with good air quality and little noise. . . .
- **Experimental methods.** These methods include contingent valuation methods, which were devised by the resource economist Siegfried V. Ciriacy-Wantrup. Contingent valuation methods require that individuals express their preferences for some environmental resources by answering questions about hypothetical choices. In particular, respondents to a contingent valuation methods questionnaire will be asked how much they would be willing to pay to ensure a welfare gain from a change in the provision of a nonmarket environmental commodity, or how much they would be willing to accept in compensation to endure a welfare loss from a reduced provision of the commodity.

Among these pricing techniques, the contingent valuation methods approach is the only one that is capable of providing an estimate of existence values, in which biologists have a special interest. Existence value was first defined by Krutilla as the value that individuals may attach to the mere knowledge that rare and diverse species, unique natural environments, or other "goods" exist, even if these individuals do not contemplate ever making active use of or benefiting in a more direct way from them. The name "contingent valuation" comes from the fact that the procedure is contingent on a constructed or simulated market, in which people are asked to manifest, through questionnaires and interviews, their demand function for a certain environmental good (i.e., the price they would pay for one extra unit of the good versus the availability of the good). . . .

The limits of cost–benefit analysis were discussed in the 1960s, after more than two decades of experimentation. In particular, many authors pointed out that cost–benefit analysis encouraged policymakers to focus on things that can be measured and quantified, especially in cash terms, and to disregard problems that are too large to be assessed easily. Therefore, the associated price might not reflect the "true" value of social equity, environmental services, natural capital, or human health. In particular, economists themselves recognize that the increasingly popular contingent valuation methods are undermined by several conceptual problems, such as free-riding, overbidding, and preference reversal.

When it comes to monetary valuation of the goods and services provided by natural ecosystems and landscapes specifically, a number of additional problems undermine the effectiveness of pricing techniques and cost–benefit analysis. These problems include the very definition of "existence" value, the dependence of pricing techniques on the composition of the reference group, and the significance of the simulated market used in contingent valuation.

**The definition of "existence" value** A classic example of contingent valuation methods is to ask for the amount of money individuals are willing to pay to ensure the continued existence of a species such as the blue whale. However, the existence value of whales does not take into account potential indirect services and benefits provided by these mammals. It is just the value of the existence of whales for humans, that is, the satisfaction that the existence of blue whales provides to people who want them to continue to exist. Therefore, there is a real risk that species with very low or no aesthetic appeal or whose biological role has not been properly advertised will be given a low value, even if they play a fundamental ecological function. Without adequate information, most people do not understand the extent, importance, and gravity of most environmental problems. As a consequence, people may react emotionally and either underestimate or overestimate risks and effects.

Therefore, it is not surprising that five of the seven guidelines issued by the National Oceanic and Atmospheric Administration [NOAA] about how to conduct contingent valuation discuss how to properly inform and question respondents to produce reliable estimates (e.g., in-person interviews are preferred to telephone surveys to elicit values). Of course, acquisition of reliable and complete information is always possible in theory, but in practice strict adherence to NOAA guidelines makes contingent valuation methods expensive and time consuming.

**Difficulties with the reference group for pricing** Pricing techniques such as contingent valuation methods provide information about individual willingness to pay or willingness to accept, which must be summed up in the final balance of cost–benefit analysis. Therefore, the outcome of cost–benefit analysis depends strongly on the group of people that is taken as a reference for valuation—particularly on their income. Van der Straaten noted that the Exxon *Valdez* oil spill in 1989 provides a good example of this dependence. The population of the United States was used as a reference group to calculate the damage to the existence value of the affected species and ecosystems using contingent valuation methods. Exxon was ultimately ordered to pay $5 billion to compensate the people of Alaska for their losses. This huge figure was a consequence of the high income of the US population. If the same accident had occurred in Siberia, where salaries are lower, the outcome would certainly have been different.

This example shows that contingent valuation methods simply provide information about the preferences of a particular group of people but do not necessarily reflect the ecological importance of ecosystem goods and services. Moreover, the outcome of cost–benefit analysis depends on which individual willingness to pay or willingness to accept are included in the cost–benefit analysis. If the quality of the Mississippi River is at issue, should the analysis be restricted to US citizens living close to the river, or should the willingness to pay of Californians and New Yorkers be included too? According to Krutilla's definition of existence value, for many environmental goods and ecological services that may ultimately affect ecosystem integrity at the global level, the preferences of the entire human population should potentially be considered in the analysis. Because practical reasons obviously preclude doing so, contingent valuation methods will inevitably only provide information about the preferences of specific groups of people. For many of the ecological services that may be considered the heritage of humanity, contingent valuation methods analyses performed locally in a particular economic situation should be extrapolated only with great caution to other areas. The process of placing a monetary value on biodiversity and ecosystem functioning through nonuser willingness to pay is performed in the same way as for user willingness to pay, but the identification of people who do not use an environmental good directly and still have a legitimate interest in its preservation is problematic.

**Significance of the simulated market** Contingent valuation methods are contingent on a market that is constructed or simulated, not real. It is difficult to believe in the efficiency of what Adam Smith called the "invisible hand" of the market for a process that is the artificial production of economic advisors and does not possess the dynamic feedback that characterizes real competitive markets. Is it even possible to simulate a market where units of biodiversity are bought and sold? As Friend stated, "these contingency evaluation methods (CVM) tend to create an illusion of choice based on psychology (willingness) and ideology (the need to pay) which is supposed, somewhat mysteriously, to reflect an equilibrium between the consumer demand for and producer supply of environmental goods and services."

Many additional criticisms of pricing ecological services are more familiar to biologists. For many ecological services, there is simply no possibility of technological substitution. Moreover, the precise contribution of many species is not known, and it may not be known until the species is close to extinction. . . . In addition, specific ecosystem services, as evaluated by Costanza et al., should not be separated from one another and valued individually because the importance of any piece of biodiversity cannot be determined without considering the value of biodiversity in the aggregate. And finally, the use of marginal value theory may be invalidated by the erratic and catastrophic behavior of many ecological systems, resulting in potentially detrimental effects on the health of humans, the productivity of renewable resources, and the vitality and stability of societies themselves.

Despite the efforts of many economists, we believe that some goods and services, especially those related to ecosystems, cannot reasonably be given a monetary value, although they are of great value to humans. Economists coined the term "intangibles" to define these goods. Cost–benefit analysis cannot easily deal with intangibles. As Nijkamp wrote, more than 20 years ago, "the only reasonable way to take account of intangibles in the traditional cost–benefit analysis seems to be the use of a balance with a debit and a credit side in which all intangible project effects (both positive and negative) are represented in their own (qualitative or quantitative) dimensions" as secondary information. In other words, the result of cost–benefit analysis is primarily a single number, the net monetary benefit that comprises all the effects that can be sensibly converted into monetary returns and costs.

## Commensurability of Different Objectives and Multicriteria Analysis

Cost–benefit analysis includes intangibles in the decision-making process only as ancillary information, with the main focus being on those effects that can be converted to monetary value. This approach is not a balanced solution to the problem of making political decisions that are acceptable to a wide number of social groups with a range of legitimate interests. . . .

However, even if the attempt to put a price on everything is abandoned, it is not necessary to give up the attempt to reconcile economic issues with social and environmental ones. Social scientists long ago developed multicriteria techniques to reach a decision in the face of multiple different and structurally incommensurable goals. The most important concept in multicriteria analysis was actually conceived by an Italian economist, Vilfredo Pareto, at the end of the nineteenth century. It is best explained by a simple example. Suppose that a natural area hosting several rare species is a target for the development of a mining activity. Alternative mining projects can have different effects in terms of profits from mining (measured in dollars) and in terms of sustained biodiversity (measured in suitable units, for instance, through the Shannon index). Profit from mining can be corrected using welfare economics to include those environmental and social effects that can be priced (e.g., the benefit of providing jobs to otherwise unemployed people, the cost of treating lung disease of miners, and the cost of the loss of the tourists who used to visit the natural area). . . .

The methods of multicriteria analysis are intended to assist the decision-maker in choosing among . . . alternatives . . . (a task that is particularly difficult when there are several incommensurable objectives, not just two). Nevertheless, the initial step of determining [these] alternatives is of enormous importance, for three reasons. First, [doing so] makes perfect sense even if there is no way of pricing a certain environmental good because each objective can be expressed in its own proper units without reduction to a common scale. Second, the determination of all the feasible alternatives . . . requires the joint effort of a multidisciplinary team that includes, for example, economists, engineers, and biologists and that must predict the effects of alternative decisions on all of the different environmental and social components to which humans are sensitive and which, therefore, deserve consideration. Third, the determination of [feasible alternatives] allows the objective elimination of inadequate alternatives because [they are] independent of the subjective perception of welfare . . . [and] in essence describe the tradeoff between the various incommensurable objectives when every effort is made to achieve the best results in all respects; the attention of the authority that must make the final decision is thus directed toward genuine potential solutions because nonoptimal decisions have already been discarded.

It should be noted that a cost–benefit analysis does not elicit tradeoffs between incommensurable goods because it also gives a green light to projects . . . , provided that the benefits that can be converted into a monetary scale exceed the costs. . . . Cost–benefit analysis, however, is not useful for eliciting the tradeoffs between two incommensurable goods, neither of which is monetary. For instance, there might be a conflict between the goals of preserving wildlife within a populated area and minimizing the risk that wild animals are vectors of dangerous diseases. A multicriteria analysis can describe this tradeoff, whereas a cost–benefit analysis cannot.

Another philosophical point concerning the issue of commensurability is the question of implicit pricing. Economists often argue that to make a decision is to put an implicit price on such intangibles as human life or aesthetics and, therefore, to reduce their value to a common scale (as pointed out also by Costanza et al.). . . .

## Environmental Impact Assessment and Multiattribute Decision-Making

Because of the flaws of cost–benefit analysis, many countries have taken a different approach to decision-making through the use of environmental impact assessment legislation (e.g., the United States in 1970, with the signing of the National Environmental Policy Act, NEPA; France in 1976, with the act 76/629; the European Union in 1985, with the directive 85/337). Environmental impact assessment procedures, if properly carried out, represent a wiser approach than setting an a priori value of biodiversity and ecosystem services because these procedures explicitly recognize that each situation, and every regulatory decision, responds to different ethical, economic, political, historical, and other conditions and that the final decision must be reached by giving appropriate consideration to several different objectives. As Canter noted, all projects, plans, and policies that are expected to have a significant environmental impact would ideally be subject to environmental impact assessment.

The breadth of goals embraced by environmental impact assessment is much wider than that of cost–benefit analysis. Environmental impact assessment provides a conceptual framework and formal procedures for comparing different alternatives to a proposed project (including the possibilities of not development a site, employing different management rules, or using mitigation measures); for fostering interdisciplinary team formation to investigate all possible environmental, social, and economic consequences of a proposed activity; for enhancing administrative review procedures and coordination among the agencies involved in the process; for producing the necessary documentation to enhance transparency in the decision-making process and the possibility of reviewing all the objective and subjective steps that resulted in a given conclusion; for encouraging broad public participation and the input of different interest groups; and for including monitoring and feedback procedures. Classical multiattribute analysis can be used to rank different alternatives. . . . Ranking usually requires the use of value functions to transform environmental and other indicators (e.g., biological oxygen demand or animal density) to levels of satisfaction on a normalized scale, and the weighting of factors to combine value functions and to rank the alternatives. These weights explicitly reflect the relative importance of the different environmental, social, and economic compartments and indicators.

A wide range of software packages for decision support can assist experts in organizing the collected information; in documenting the various phases of EIA; in guiding the assignment of importance weights; in scaling, rating, and ranking alternatives; and in conducting sensitivity analysis for the overall decision-making process. This last step, of testing the robustness and consistency of multiattribute analysis results, is especially important because it shows how sensitive the final ranking is to small or large changes in the set of weights and value functions, which often reflect different and subjective perspectives. It is important to stress that, although the majority of environmental impact assessments have been conducted on specific projects, such as road construction or the location of chemical plants, there is no conceptual barrier to extending the procedure to evaluation of plans, programs, policies, and regulations. In fact, according to NEPA, the procedure is mandatory for any federal action with an important impact on the environment. The extension of environmental impact assessment to a level higher than a single project is termed "strategic environmental assessment" and has received considerable attention.

## Conclusions

An impressive literature is available on environmental impact assessment and multiattribute analysis that documents the experience gained through 30 years of study and application. Nevertheless, these studies seem to be confined to the area of urban planning and are almost completely ignored by present-day economists as well as by many ecologists. Somewhere between the assignment of a zero value to biodiversity (the old-fashioned but still used practice, in which environmental impacts are viewed as externalities to be discarded from the balance sheet) and the assignment of an infinite value (as advocated by some radical environmentalists), lie more sensible methods to assign value to biodiversity than the price tag techniques suggested by the new wave of environmental economists. Rather than collapsing every measure of social and environmental value onto a monetary axis, environmental impact assessment and multiattribute analysis allow for explicit consideration of intangible nonmonetary values along with classical economic assessment, which, of course, remains important. It is, in fact, possible to assess ecosystem values and the ecological impact of human activity without using prices.

Concepts such as Odum's eMergy [the available energy of one kind previously required to be used up directly and indirectly to make the product or service] and Rees' ecological footprint [the area of land and water required to support a defined economy or population at a specified standard of living], although perceived by some as naive, may aid both ecologists and economists in addressing this important need.

To summarize our viewpoint, economists should recognize that cost–benefit analysis is only part of the decision-making process and that it lies at the same level as other considerations. Ecologists should accept that monetary valuation of biodiversity and ecosystem services is possible (and even helpful) for part of its value, typically its use value. We contend that the realistic substitute for markets, when they fail, is a transparent decision-making process, not old-style cost–benefit analysis. The idea that, if one could get the price right, the best and most effective decisions at both the individual and public levels would automatically follow is, for many scientists, a sort of Panglossian obsession. In reality, there is no simple solution to complex problems. We fear that putting an a priori monetary value on biodiversity and ecosystem services will prevent humans from valuing the environment other than as a commodity to be exploited, thus reinvigoraing the old economic paradigm that assumes a perfect substitution between natural and human-made capital. As Rees wrote, "for all its theoretical attractiveness, ascribing money values to nature's services is only a partial solution to the present dilemma and, if relied on exclusively, may actually be counterproductive."

# POSTSCRIPT

## Should a Price Be Put on the Goods and Services Provided by the World's Ecosystems?

In "Can We Put a Price on Nature's Services?" *Report From the Institute for Philosophy and Public Policy* (Summer 1997), Mark Sagoff objects that trying to attach a price to ecosystem services is futile because it legitimizes the accepted cost-benefit approach and thereby undermines efforts to protect the environment from exploitation. The March 1998 issue of *Environment* contains environmental economics professor David Pearce's detailed critique of the 1997 Costanza et al. study. Pearce objects chiefly to the methodology, not the overall goal of attaching economic value to ecosystem services. Costanza et al. reply to Pearce's objections in the same issue. Pearce and Edward B. Barbier have published *Blueprint for a Sustainable Economy* (Earthscan, 2000), in which they discuss how governments worldwide are now applying economics to environmental policy.

Despite the controversy over the worth of assigning economic values to various aspects of nature, researchers continue the effort. Gretchen C. Daily et al., in "The Value of Nature and the Nature of Value," *Science* (July 21, 2000), discuss valuation as an essential step in all decision making and argue that efforts "to capture the value of ecosystem assets . . . can lead to profoundly favorable effects." Daily and Katherine Ellison continue the theme in *The New Economy of Nature: The Quest to Make Conservation Profitable* (Island Press, 2002). In "What Price Biodiversity?" *Ecos* (January 2000), Steve Davidson describes an ambitious program funded by the Commonwealth Scientific and Industrial Research Organization (CSIRO) and the Myer Foundation that is aimed at developing principles and methods for objectively valuing "ecosystem services—the conditions and processes by which natural ecosystems sustain and fulfil human life—and which we too often take for granted. These include such services as flood and erosion control, purification of air and water, pest control, nutrient cycling, climate regulation, pollination, and waste disposal." Jim Morrison, "How Much Is Clean Water Worth?" *National Wildlife* (February/March 2005), argues that ecosystem services such as cleaning water, controlling floods, and pollinating crops have sufficient economic value to make it profitable to spend millions of dollars to protect natural systems.

Stephen Farber, et al., "Linking Ecology and Economics for Ecosystem Management," *Bioscience* (February 2006), find "the valuation of ecosystem services . . . necessary for the accurate assessment of the trade-offs involved in different management options." Ecosystem valuation is currently being used to justify restoration efforts, "linking the science to human welfare," as shown (for example) in Chungfu Tong, et al., "Ecosystem Service Values and Restoration in the Urban Sanyang Wetland of Wenzhou, China," *Ecological Engineering* (March 2007).

# Internet References . . .

## ECOLEX: A Gateway to Environmental Law

This site, sponsored by the United Nations and the World Conservation Union, is a comprehensive resource for environmental treaties, national legislation, and court decisions.

**http://www.ecolex.org/index.php**

## Environmental Defense

Environment Defense (once The Environmental Defense Fund) is dedicated to "protecting the environmental rights of all people, including future generations." Guided by science, Environmental Defense evaluates environmental problems and works "to create solutions that win lasting economic and social support because they are nonpartisan, cost-efficient, and fair."

**http://www.environmentaldefense.org/home.cfm**

## Office of Environmental Justice

The U.S. Environmental Protection Agency (EPA) pursues environmental justice under the Office of Enforcement and Compliance Assurance as part of its "firm commitment to the issue of environmental justice and its integration into all programs, policies, and activities, consistent with existing environmental laws and their implementing regulations."

**http://www.epa.gov/compliance/environmentaljustice/index.html**

## The Heritage Foundation

The Heritage Foundation is a think-tank whose mission is to formulate and promote conservative public policies on many issues, including environmental ones. It bases its work on the principles of free enterprise, limited government, individual freedom, and traditional American values.

**http://www.heritage.org/**

## SourceWatch

SourceWatch is a collaborative project of the Center for Media and Democracy. Its primary focus is on documenting the interconnections and agendas of public relations firms, think tanks, industry-funded organizations and industry-friendly experts that work to influence public opinion and public policy on behalf of corporations, governments, and special interests.

**http://www.sourcewatch.org**

## The National Endangered Species Act Reform Coalition

The National Endangered Species Act Reform Coalition is a broad based coalition of roughly 150 member organizations, representing millions of individuals across the United States, that is dedicated to improving and updating the Endangered Species Act. Our membership includes rural irrigators, municipalities, farmers, electric utilities, and many other individuals and organizations that are directly affected by the ESA.

**http://www.nesarc.org/**

# ISSUE 14

## Is Genetic Engineering the Answer to Hunger?

**YES: Gerald D. Coleman**, from "Is Genetic Engineering the Answer to Hunger?" *America* (February 21, 2005)

**NO: Sean McDonagh,** from "Genetic Engineering Is Not the Answer," *America* (May 2, 2005)

### ISSUE SUMMARY

**YES:** Gerald D. Coleman argues that genetically engineered crops are useful, healthful, and nonharmful, and although caution may be justified, such crops can help satisfy the moral obligation to feed the hungry.

**NO:** Sean McDonagh argues that those who wish to feed the hungry would do better to address land reform, social inequality, lack of credit, and other social issues.

In the early 1970s scientists first discovered that it was technically possible to move genes—the biological material that determines a living organism's physical traits—from one organism to another and thus (in principle) to give bacteria, plants, and animals new features. Most researchers in molecular genetics were excited by the potentialities that suddenly seemed within their reach. However, a few researchers—as well as many people outside the field—were disturbed by the idea; they thought that genetic mix-and-match games might spawn new diseases, weeds, and pests. Researchers in support of genetic experimentation responded by declaring a moratorium on their own work until suitable safeguards (in the form of government regulations) could be devised.

A 1987 National Academy of Sciences report said that genetic engineering posed no unique hazards. And, despite continuing controversy, by 1989 the technology had developed tremendously: researchers could obtain patents for mice with artificially added genes ("transgenic" mice); firefly genes had been added to tobacco plants to make them glow (faintly) in the dark; and growth hormone produced by genetically engineered bacteria was being used to grow low-fat pork and increase milk production in cows. The growing biotechnology industry promised more productive crops that made their own fertilizer and pesticide. Proponents argued that genetic engineering was in no significant way different from traditional selective breeding. Critics argued that genetic engineering was unnatural and violated the rights of both plants and animals to their "species integrity"; that expensive, high-tech, tinkered animals gave the competitive advantage to big agricultural corporations and drove small farmers out of business; and that putting human genes into animals, plants, or bacteria was downright offensive. Trey Popp, "God and the New Foodstuffs," *Science & Spirit* (March/April 2006), discusses objections to genetically modified rice containing human genes. For a summary of events related to the development of agricultural biotechnology, see "Biotechnology Timeline: Chronology of Key Events," *International Debates* (March 2006).

In 1992 the U.S. Office of Science and Technology issued guidelines to bar regulations that are based on the assumption that genetically engineered crops pose greater risks than similar crops produced by traditional breeding methods. The result was the rapid commercial introduction of crops that were genetically engineered to make the bacterial insecticide Bt and to resist herbicides and disease, among other things. Rice has now been engineered for herbicide and stress resistance, attempts have been made to improve nutrient content and yield, and "Transgenic rice can serve as a biofactory for the production of molecules of pharmaceutical and industrial utility. The drive to apply transgenic rice for public good as well as commercial gains has fueled research to an all time high"; see Hitesh Kathuria, et al., "Advances in Transgenic Rice Biotechnology," *Critical Reviews in Plant Sciences* (2007). In 2006, some 70 engineered crop varieties were grown on over 102 million hectares (252 million acres) in 22 countries (see GMO-Compass.org, http://www.gmo-compass.org/eng/agri_biotechnology/gmo_planting/). Commercial GM crops are chiefly soybeans, corn, cotton, and canola. Sales of genetically engineered crop products are expected to reach $210 billion by 2014.

Skepticism about the benefits remains, but in 2000, the national academies of science of the United States, the United Kingdom, China, Brazil, India, Mexico, and the third world recognized that though the use of genetically modified crops has some worrisome potentials that deserve further research and continuing caution, those crops hold the potential to feed the world during the twenty-first century while also protecting the environment (Royal Society of London, et al., "Transgenic Plants and World Agriculture," July 2000; http://books.nap.edu/catalog.php?record_id=9889). Genetic engineering is also poised to improve turfgrass and help produce biofuels (see Derek Burke, "Biofuels—Is There a Role for GM?" *Biologist*, February 2007).

The following selections illustrate the different current perspectives from within the Roman Catholic Church. Gerald D. Coleman argues that genetically engineered crops are useful, healthful, and nonharmful, and though caution may be justified, such crops can help satisfy the moral obligation to feed the hungry. Sean McDonagh, a priest who writes frequently on environmental matters, argues that ethical and environmental problems connected to the implementation of genetically engineered crops stand in the way of their promise. Those who wish to feed the hungry would do better to address land reform, social inequality, lack of credit, and other social issues.

**YES** ↩ **Gerald D. Coleman**

# Is Genetic Engineering the Answer to Hunger?

**B**oth the developed and developing worlds are facing a critical moral choice in the controversial issue of genetically modified food, also known as genetically modified organisms and genetically engineered crops. Critics of these modifications speak dismissively of biotech foods and genetic pollution. On the other hand, proponents like Nina Federoff and Nancy Marie Brown, authors of *Mendel in the Kitchen: A Scientist's View of Genetically Modified Foods* (2004), promote genetically modified organisms (GMs or GMOs) as "the miracle of seed science and fertilizers."

To mark the 20th anniversary of U.S. diplomatic relations with the Holy See, the U.S. Embassy to the Holy See, in cooperation with the Pontifical Academy of Sciences, hosted a conference last fall at Rome's Gregorian University on "Feeding a Hungry World: The Moral Imperative of Biotechnology." Archbishop Renato Martino, who heads the Pontifical Council for Justice and Peace and has been a strong and outspoken proponent of GMOs, told Vatican Radio: "The problem of hunger involves the conscience of every man. For this reason the Catholic Church follows with special interest and solicitude every development in science to help the solution of a plight that affects . . . humanity."

Americans have grown accustomed, perhaps unwittingly, to GMO products. In the United States, for example, 68 percent of the soybeans, 70 percent of the cotton crop, 26 percent of corn and 55 percent of canola are genetically engineered. GMOs represent an estimated 60 percent of all American processed foods. A recent study by the National Center for Food and Agriculture found that farmers in the United States investing in biotech products harvested 5.3 billion additional pounds of crops and realized $22 billion in increased income. Most of the world's beer and cheese is made with GMOs, as are hundreds of medications. In an article published last October, James Nicholson, then U.S. Ambassador to the Vatican and an aggressive promoter of U.S. policy in Vatican circles, wrote that "millions of Americans, Canadians, Australians, Argentines and other people have been eating genetically modified food for nearly a decade—without one proven case of an illness, allergic reaction or even the hiccups. . . . Mankind has been genetically altering food throughout

Gerald D. Coleman, "Is Genetic Engineering the Answer to Hunger?" *America*, vol. 192, Feb. 21, 2005, p. 14. 

human history." And according to its supporters, biotechnology helps the environment by reducing the use of pesticides and tilling.

The World Health Association recently reported that more than 3.7 billion people around the world are now malnourished, the largest number in history. To this, opponents of GMOs reply that the "real problems" causing hunger, especially in the developing world, are poverty, lack of education and training, unequal land distribution and lack of access to markets. The moral point they advance is that distribution, not production, is the key to solving hunger.

Another significant moral issue relates to "intellectual property policies" and the interest of companies in licensing potentially valuable discoveries. The Rev. Giulio Albanese, head of the Missionary News Agency, insists that unless the problem of intellectual property is resolved in favor of the poor, it represents a "provocation" to developing countries: "The concern of many in the missionary world over the property rights to GM seeds . . . cannot but accentuate the dependence of the poor nations on the rich ones." In response to this concern, a proposal was made recently (reported in *Science* magazine on March 19) that research universities cooperate to seek open licensing provisions that would allow them to share their intellectual property through a "developing-country license." Universities would still retain rights for research and education and maintain negotiating power with the biotechnology and pharmaceutical industries. Catholic social ethics would support this type of proposal, since it places the good of people over amassing profit.

Three moral paths suggest themselves:

1. *Favor the use of GMOs.* Nobel Prize winner Norman Borlaug, who developed the Green Revolution wheat and rice strains, recently wrote: "Biotechnology absolutely should be part of Africa's agricultural reform. African leaders would be making a grievous error if they turn their back on it." Proponents at the Rome conference agreed, arguing that the use of GMOs decreases pesticide use, creates more nutrient-filled crops that require less water and have greater drought resistance, produces more food at a lower cost and uses less land. One small-scale South African farmer concluded, "We need this technology. We don't want always to be fed food aid. We want access to this technology so that one day we can also become commercial farmers."

   This position concludes that the use of GMOs amounts to a moral obligation.
2. *Condemn the use of GMOs.* Many Catholic bishops take an opposing stance. Perhaps the clearest statement comes from the National Conference of Bishops of Brazil and their Pastoral Land Commission. Their argument is threefold: the use of GMOs involves potential risks to human health; a small group of large corporations will be the greatest beneficiaries, with grave damage to the family farmers; and the environment will be gravely damaged.

   The bishops of Botswana, South Africa and Swaziland agree: "We do not believe that agro-companies or gene technologies will help our farmers to produce the food that is needed in the 21st century." Roland Lesseps and Peter Henriot, two Jesuits working in

Zambia who are experts on agriculture in the developing world, state their opposition on principle: "Nature is not just useful to us as humans, but is valued and loved in itself, for itself, by God in Christ. . . . The right to use other creatures does not give us the right to abuse them."

In a similar but distinct criticism, the executive director of the U.S. National Catholic Rural Conference, David Andrews, C.S.C., feels that "the Pontifical Academy of Sciences has allowed itself to be subordinated to the U.S. government's insistent advocacy of biotechnology and the companies which market it." Sean McDonagh states: "With patents [on genetically engineered food], farmers will never own their own food. . . ." He believes that "corporate greed" is at the heart of the GMO controversy. Biowatch's Elfrieda Pschorn-Strauss agrees: "With GM crops, small-scale farmers will become completely reliant on and controlled by big foreign companies for their food supply."

This position concludes that the use of GMOs is morally irresponsible.

3. *Approach the use of GMOs with caution.* Two years ago Pope John Paul II declared that GMO agriculture could not be judged solely on the basis of "short-term economic interests," but needed to be subject to "a rigorous scientific and ethical process of verification." This cautionary stance has been adopted by the Catholic Bishops Conference of the Philippines in urging its government to postpone authorization of GMO corn until comprehensive studies have been made: "We have to be careful because, once it is there, how can we remedy its consequences?"

In 2003 the Rural Life Committee of the North Dakota Conference of Churches also called for "rigorous examination" to understand fully the outcomes of the use of GMOs. This document endorses the "Precautionary Principle" formulated in 1992 by the United Nations Conference on Environment and Development in order to avoid "potential harm and unforeseen and unintended consequences."

This view mandates restraint and places the fundamental burden on demonstrating safety. The arguments are based on three areas of concern: the impact on the natural environment, the size of the benefit to the small farmer if the owners and distributors are giant companies like Bristol-Myers and Monsanto and the long-term effects of GMOs on human and animal health and nutrition.

This position concludes that the use of GMOs should be approached with caution.

While the "Precautionary Principle" seems prudent, there is simultaneously a strong moral argument that a war on hunger is a grave, universal need. Last year, 10 million people died of starvation. Every 3.6 seconds someone dies from hunger—24,000 people each day. Half of sub-Saharan Africans are malnourished, and this number is expected to increase to 70 percent by 2010. It was a moral disgrace that in 2002 African governments gave in to GMO opponents and returned to the World Food Program tons of GMO corn simply because it was produced in the U.S. by biotechnology.

The Roman conference gives solid reasons that GMOs are useful, healthful and nonharmful. After all, organisms have been exchanging genetic information for centuries. The tomato, corn and potato would not exist today if human engineering had not transferred genes between species.

The *Catechism of the Catholic Church* teaches that we have a duty to "make accessible to each what is needed to lead a truly human life." The very first example given is food. In *Populorum Progressio* (1967), *Sollicitude Rei Socialis* (1987) and *Centesimus Annus* (1991), Paul VI and later John Paul II forcefully insisted that rich countries have an obligation to help the poor, just as global economic interdependence places us on a moral obligation to be in solidarity with poor nations. Likewise, The Challenge of Faithful Citizenship, published by the U.S. bishops in 2004, argues that the church's preferential option for the poor entails "a moral responsibility to commit ourselves to the common good at all levels."

At the same time, it is critical that farmers in developing countries not become dependent on GMO seeds patented by a small number of companies. Intellectual knowledge must be considered the common patrimony of the entire human family. As the U.S. bishops have stated, "Both public and private entities have an obligation to use their property, including intellectual and scientific property, to promote the good of all people" (*For I Was Hungry and You Gave Me Food, 2003*).

The Catholic Church sees deep sacramental significance in wheat and bread, and insists on the absolute imperative to feed and care for the poor of the world. A vital way to promote and ensure the dignity of every human being is to enable them to have their daily bread.

**Sean McDonagh**

**NO**

# Genetic Engineering Is Not the Answer

In 1992 the then-chief executive of Monsanto, Robert Shapiro, told the *Harvard Business Review* that genetically modified crops will be necessary to feed a growing world population. He predicted that if population levels were to rise to 10 billion, humanity would face two options: either open up new land for cultivation or increase crop yields. Since the first choice was not feasible, because we were already cultivating marginal land and in the process creating unprecedented levels of soil erosion, we would have to choose genetic engineering. This option, Shapiro argued, was merely a further improvement on the agricultural technologies that gave rise to the Green Revolution that saved Asia from food shortages in the 1960s and 1970s.

Genetically engineered crops might seem an ideal solution. Yet both current data and past examples show problems and provoke doubts as to their necessity.

## The Green Revolution

The Green Revolution involved the production of hybrid seeds by crossing two genetically distant parents, which produced an offspring plant that gave increased yield. Critics of genetic engineering question the accepted wisdom that its impact has been entirely positive. Hybrid seeds are expensive and heavily reliant on fertilizers and pesticides. And because they lose their vigor after the first planting, the farmer must purchase new seeds for each successive planting.

In his book *Geopolitics and the Green Revolution*, John H. Perkins describes the environmentally destructive and socially unjust aspects of the Green Revolution. One of its most important negative effects, he says, is that it has contributed to the loss of three-quarters of the genetic diversity of major food crops and that the rate of erosion continues at close to 2 percent per year. The fundamental importance of genetic diversity is illustrated by the fact that when a virulent fungus began to destroy wheat fields in the United States and Canada in 1950, plant breeders staved off disaster by cross-breeding five Mexican wheat varieties with 12 imported ones. In the process they created

Sean McDonagh, "Genetic Engineering Is Not the Answer." *America*, vol. 192, May 2, 2005, p. 8.

a new strain that was able to resist so-called "stem rust." The loss of these varieties would have been a catastrophe for wheat production globally.

## The Terminator Gene

The development by a Monsanto-owned company of what is benignly called a Technology Protection System—a more apt name is terminator technology—is another reason for asserting that the feed-the-world argument is completely spurious. Genetically engineered seeds that contain the terminator gene self-destruct after the first crop. Once again, this forces farmers to return to the seed companies at the beginning of each planting season. If this technology becomes widely used, it will harm the two billion subsistence farmers who live mainly in the poor countries of the world. Sharing seeds among farmers has been at the very heart of subsistence farming since the domestication of staple food crops 11,000 years ago. The terminator technology will lock farmers into a regime of buying genetically engineered seeds that are herbicide tolerant and insect resistant, tethering them to the chemical treadmill.

On an ethical level, a technology that, according to Professor Richard Lewontin of Harvard University, "introduces a 'killer' transgene that prevents the germ of the harvested grain from developing" must be considered grossly immoral. It is a sin against the poor, against previous generations who freely shared their knowledge of plant life with us, against nature itself and finally against the God of all creativity. To set out deliberately to create seeds that self-destruct is an abomination no civilized society should tolerate. Furthermore, there is danger that the terminator genes could spread to neighboring crops and to the wild and weedy relatives of the plant that has been engineered to commit suicide. This would jeopardize the food security of many poor people.

The current situation promoting genetically modified organisms also means supporting the patenting of living organisms—both crops and animals. I find it difficult to understand the support that Cardinal Renato Martino, prefect of the Pontifical Council for Justice and Peace, seems to be giving to genetically modified organisms, given the Catholic Church's strong pro-life position. In my book *Patenting Life? Stop!* I argue that "patenting life is a fundamental attack on the understanding of life as interconnected, mutually dependent and a gift of God to be shared with everyone. Patenting opts for an atomized, isolated understanding of life." The Indian scientist and activist Dr. Vandana Shiva believes that patented crops will lead to food dictatorship by a handful of northern transnational corporations. This would certainly be a recipe for hunger and starvation—in conflict with Catholic social teaching on food and agriculture.

## No Higher Yield, No Reduction in Chemicals

Early in 2003 a researcher at the Institute of Development Studies at Sussex University in England published an analysis of the GMO crops that biotech companies are developing for Africa. Among the plants studied were cotton, maize and the sweet potato. The GMO research on the sweet potato is now approaching its 12th year and has involved the work of 19 scientists; to date it has cost $6 million. Results indicated that yield has increased by 18 percent. On the other hand, conventional sweet potato breeding, working with a small budget, has produced a virus-resistant variety with a 100 percent yield increase.

Claims that GMOs lead to fewer chemicals in agriculture are also being challenged. A comprehensive study using U.S. government data on the use of chemicals on genetically engineered crops was carried out by Charles Benbrook, head of the Northwest Science and Environmental Policy Center in Sandpoint, Idaho. He found that when GMOs were first introduced, they needed 25 percent fewer chemicals for the first three years. But in 2001, 5 percent more chemicals were sprayed compared with conventional crop varieties. Dr. Benbrook stated: "The proponents of biotechnology claim GMO varieties substantially reduce pesticide use. While true in the first few years of widespread planting, it is not the case now. There's now clear evidence that the average pound of herbicide applied per acre planted to herbicide-tolerant varieties have increased compared to the first few years."

## Toward a Solution

Hunger and famine around the world have more to do with the absence of land reform, social inequality, bias against women farmers and the scarcity of cheap credit and basic agricultural tools than with lack of agribusiness superseeds. This fact was recognized by those who attended the World Food Summit in Rome in November 1996. People are hungry because they do not have access to food production processes or the money to buy food. Brazil, for example, is the third largest exporter of food in the world, yet one-fifth of its population, over 30 million people, do not have enough food to eat. Clearly hunger there is not due to lack of food but to the unequal distribution of wealth and the fact that a huge number of people are landless.

Do the proponents of genetically engineered food think that agribusiness companies will distribute such food free to the hungry poor who have no money? There was food in Ireland during the famine in the 1840s, for example, but those who were starving had no access to it or money to buy it.

As a Columban missionary in the Philippines, I saw something similar during the drought caused by El Niño in 1983. There was a severe food shortage among the tribal people in the highlands of Mindanao. The drought destroyed their cereal crops, and they could no longer harvest food in the tropical forest because it had been cleared during the previous decades. Even during the height of the drought, an agribusiness corporation was exporting tropical fruit from the lowlands. There was also sufficient rice and corn in the lowlands, but the tribal people did not have the money to buy it. Had it not been for food aid from nongovernmental organizations, many of the tribal people would have starved.

In 1990 the World Food Program at Brown University calculated that if the global food harvests over the previous few years were distributed equitably among all the people of the world, it could provide a vegetarian diet for over

6 billion people. In contrast, a meat-rich diet, favored by affluent countries and currently available to the global elite, could feed only 2.6 billion people. Human society is going to be faced with the option of getting protein from plants or from animals. If we opt for animal protein, the consequence will be a much less equitable world, with increasing levels of human misery.

Those who wish to banish hunger should address the social and economic inequalities that create poverty and not claim that a magic-bullet technology will solve all the problems.

# POSTSCRIPT

## Is Genetic Engineering the Answer to Hunger?

Kathleen Hart, in *Eating in the Dark* (Pantheon, 2002), expresses horror at the fact that "Frankenfood" is not labeled and that U.S. consumers are not as alarmed by genetically modified foods as European consumers are. The worries—and the scientific evidence to support them—are summarized by Kathryn Brown, in "Seeds of Concern," and Karen Hopkin, in "The Risks on the Table," *Scientific American* (April 2001). In the same issue, Sasha Nemecek poses the question "Does the World Need GM Foods?" to two prominent figures in the debate: Robert B. Horsch, vice president of the Monsanto Corporation and recipient of the 1998 National Medal of Technology for his work on modifying plant genes, says yes; Margaret Mellon, of the Union of Concerned Scientists, says no, adding that much more work needs to be done with regard to safety. The May 2002 U.S. General Accounting Office Report to Congressional Requesters, "Genetically Modified Foods: Experts View Regimen of Safety Tests as Adequate, but FDA's Evaluation Process Could Be Enhanced," urges more attention to verifying safety testing performed by biotechnology companies. Carl F. Jordan, in "Genetic Engineering, the Farm Crisis, and World Hunger," *Bioscience* (June 2002), says that a major problem is already apparent in the way agricultural biotechnology is widening the gap between the rich and the poor.

Charles Mann, in "Biotech Goes Wild," *Technology Review* (July/August 1999), discusses the continuing "lack of a rigorous regulatory framework to sort out the risks inherent in agricultural biotech." Margaret Kriz, in "Global Food Fight," *National Journal* (March 4, 2000), describes the January 2000 Montreal meeting, in which representatives of 130 countries reached "an agreement that requires biotechnology companies to ask permission before importing genetically altered seeds [and] forces food companies to clearly identify all commodity shipments that may contain genetically altered grain." Also see "Environmental Effects of Transgenic Plants: The Scope and Adequacy of Regulation," a report of the National Research Council's Committee on Environmental Impacts Associated With Commercialization of Transgenic Crops (National Academy Press, 2002).

Gregory Conko and C. S. Prakash, in "The Attack on Plant Biotechnology," in Ronald Bailey, ed., *Global Warming and Other Eco-Myths: How the Environmental Movement Uses False Science to Scare Us to Death* (Prima, 2002), say that genetically engineered crops have successfully increased yields and decreased pesticide usage, have not had notable bad environmental side effects, and will be essential for feeding the world's growing population. Jerry D. Glover, Cindy M. Cox, and John P. Reganold, "Future Farming: A Return to Roots?" *Scientific*

*American* (August 2007), argue that current work aimed at using biotechnology to turn present annual crops such as corn, wheat, and rice into perennial crops will boost the food supply while greatly reducing soil erosion and water pollution. Lee Silver, "Why GM Is Good for Us," *Newsweek (Atlantic Edition)* (March 20, 2006), reports that genetically modified foods may actually have fewer allergy-related problems than organic foods. On the other hand, Jeffrey M. Smith, "Frankenstein Peas," *The Ecologist* (April 2006), reports on a study that genetically modified peas fed to mice caused serious immune system problems.

The UN Food and Agriculture Organization's 2004 annual report, *The State of Food and Agriculture 2003-2004*, FAO Agriculture Series No. 35 (Rome, 2004), maintains that the biggest problem with genetically engineered crops is that the technology has focused so far on crops of interest to large commercial firms. GM crops have not spread fast enough to small farmers, although where they have been introduced into developing countries, they have yielded economic gains and reduced the use of toxic chemicals. The report concludes that there have been no adverse health or environmental consequences so far. Continued safety will require more research and governmental regulation and monitoring. Jerry Cayford notes in "Breeding Sanity into the GM Food Debate," *Issues in Science and Technology* (Winter 2004) that the issue is one of social justice as much as it is one of science. Who will control the world's food supply? Which philosophy—democratic competition or technocratic monopoly—will prevail? Theodoros H. Varzakas, et al., "The Politics and Science Behind GMO Acceptance," *Critical Reviews in Food Science & Nutrition* (June 2007), notes that in Europe a number of food scares (including Mad Cow disease) have alarmed consumers, and GMOs "despite the intense reactions from Non Governmental Organizations and consumer organizations have entered our lives with inadequate legislative measures to protect consumers from their consumption."

# ISSUE 15

## Is a Large-Scale Shift to Organic Farming the Best Way to Increase World Food Supply?

**YES: Brian Halweil,** from "Can Organic Farming Feed Us All?" *World Watch* (May/June 2006)

**NO: John J. Miller,** from "The Organic Myth: A Food Movement Makes a Pest of Itself," *National Review* (February 9, 2004)

### ISSUE SUMMARY

**YES:** Brian Halweil, senior researcher at the Worldwatch Institute, argues that organic agriculture is potentially so productive that it could sustainably increase world food supply, although the future may be more likely to see a mix of organic and nonorganic techniques.

**NO:** John J. Miller argues that organic farming is not productive enough to feed today's population, much less larger future populations, it is prone to dangerous biological contamination, and it is not sustainable.

There was a time when all farming was organic. Fertilizer was compost and manure. Fields were periodically left fallow (unfarmed) to recover soil moisture and nutrients. Crops were rotated to prevent nutrient exhaustion. Pesticides were nonexistent. And farmers were at the mercy of periodic droughts (despite irrigation) and insect infestations.

As population grew, so did the demand for food. In Europe and America, the concomitant demand for fertilizer led in the nineteenth century to a booming trade in guano mined from Caribbean and Pacific islands where deposits of seabird dung could be a hundred and fifty feet thick. When the guano deposits were exhausted, there was an agricultural crisis that was relieved only by the invention of synthetic nitrogen-containing fertilizers early in the twentieth century. See Jimmy Skaggs, *The Great Guano Rush: Entrepreneurs and American Overseas Expansion* (St. Martin's, 1994), and G. J. Leigh, *The World's Greatest Fix: A History of Nitrogen and Agriculture* (Oxford, 2004). Unfortunately, synthetic fertilizers do not maintain the soil's content of organic matter (humus). This deficit can be amended by tilling in sewage sludge, but the public is not usually very receptive to the idea, partly because of the "yuck factor," but also because sewage sludge may contain human pathogens and chemical contaminants.

Synthetic pesticides, beginning with DDT, came into use in the 1940s. Their history is nicely outlined in Keith S. Delaplane, *Pesticide Usage in the United States: History, Benefits, Risks, and Trends* (University of Georgia Extension Bulletin 1121) (http://pubs.caes.uga.edu/caespubs/pubcd/B1121.htm). When they turned out to have problems—target species quickly became resistant, and when the chemicals reached human beings and wildlife on food and in water, they proved to be toxic—some people sought alternatives. These alternatives to synthetic fertilizers and pesticides (among other things) are what is usually meant by "organic farming." Proponents of organic farming have called it holistic, biodynamic, ecological, and natural and claimed a number of advantages for its practice. They say it both preserves the health of the soil and provides healthier food for people. They also argue that it should be used more, even to the point of replacing chemical-based "industrial" agriculture. Because proponents of chemicals hold that fertilizers and pesticides are essential to produce food in the quantities that a world population of over six billion people requires, and to hold food prices down to affordable levels, one strand of debate has been over whether organic agriculture can do the job.

In the following selections, Worldwatch researcher Brian Halweil argues that organic agriculture is potentially so productive that it could sustainably increase world food supply beyond current levels while strengthening the economic position of small farmers. However, the realities of the marketplace are such that many farmers are likely to use a mixture of organic and nonorganic techniques known as "low-input" farming. John J. Miller argues that organic farming is not productive enough to feed today's population, much less larger future populations, it is prone to dangerous biological contamination, and it is not sustainable. "Wishful thinking is at the heart of the organic-food movement."

## YES

**Brian Halweil**

# Can Organic Farming Feed Us All?

The only people who think organic farming can feed the world are delusional hippies, hysterical moms, and self-righteous organic farmers. Right?

Actually, no. A fair number of agribusiness executives, agricultural and ecological scientists, and international agriculture experts believe that a large-scale shift to organic farming would not only *increase* the world's food supply, but might be the only way to eradicate hunger.

This probably comes as a surprise. After all, organic farmers scorn the pesticides, synthetic fertilizers, and other tools that have become synonymous with high-yield agriculture. Instead, organic farmers depend on raising animals for manure, growing beans, clover, or other nitrogen-fixing legumes, or making compost and other sources of fertilizer that cannot be manufactured in a chemical plant but are instead grown—which consumes land, water, and other resources. (In contrast, producing synthetic fertilizers consumes massive amounts of petroleum.) Since organic farmers can't use synthetic pesticides, one can imagine that their fields suffer from a scourge of crop-munching bugs, fruitrotting blights, and plant-choking weeds. And because organic farmers depend on rotating crops to help control pest problems, the same field won't grow corn or wheat or some other staple as often.

As a result, the argument goes, a world dependent on organic farming would have to farm more land than it does today—even if it meant less pollution, fewer abused farm animals, and fewer carcinogenic residues on our vegetables. "We aren't going to feed 6 billion people with organic fertilizer," said Nobel Prize-winning plant breeder Norman Borlaug at a 2002 conference. "If we tried to do it, we would level most of our forest and many of those lands would be productive only for a short period of time." Cambridge chemist John Emsley put it more bluntly: "The greatest catastrophe that the human race could face this century is not global warming but a global conversion to 'organic firming'—an estimated 2 billion people would perish."

In recent years, organic farming has attracted new scrutiny, not just from critics who fear that a large-scale shift in its direction would cause billions to starve, but also from farmers and development agencies who actually suspect that such a shift could *better* satisfy hungry populations. Unfortunately, no one had ever systematically analyzed whether in fact a widespread shift to organic farming would run up against a shortage of nutrients and a lack of yields—until recently. The results are striking.

## High-Tech, Low-Impact

There are actually myriad studies from around the world showing that organic farms can produce about as much, and in some settings much more, than conventional farms. Where there is a yield gap, it tends to be widest in wealthy nations, where farmers use copious amounts of synthetic fertilizers and pesticides in a perennial attempt to maximize yields. It is true that farmers converting to organic production often encounter lower yields in the first few years, as the soil and surrounding biodiversity recover from years of assault with chemicals. And it may take several seasons for farmers to refine the new approach.

But the long-standing argument that organic farming would yield just one-third or one-half of conventional farming was based on biased assumptions and lack of data. For example, the often-cited statistic that switching to organic farming in the United States would only yield one-quarter of the food currently produced there is based on a U.S. Department of Agriculture study showing that all the manure in the United States could only meet one-quarter of the nation's fertilizer needs—even though organic farmers depend on much more than just manure.

More up-to-date research refutes these arguments. For example, a recent study by scientists at the Research Institute for Organic Agriculture in Switzerland showed that organic farms were only 20 percent less productive than conventional plots over a 21-year period. Looking at more than 200 studies in North America and Europe, Per Pinstrup Andersen (a Cornell professor and winner of the World Food Prize) and colleagues recently concluded that organic yields were about 80 percent of conventional yields. And many studies show an even narrower gap. Reviewing 154 growing seasons' worth of data on various crops grown on rain-fed and irrigated land in the United States, University of California–Davis agricultural scientist Bill Liebhardt found that organic corn yields were 94 percent of conventional yields, organic wheat yields were 97 percent, and organic soybean yields were 94 percent. Organic tomatoes showed no yield difference.

More importantly, in the world's poorer nations where most of the world's hungry live, the yield gaps completely disappear. University of Essex researchers Jules Pretty and Rachel Hine looked at over 200 agricultural projects in the developing world that converted to organic and ecological approaches, and found that for all the projects—involving 9 million farms on nearly 30 million hectares—yields increased an average of 93 percent. A seven-year study from Maikaal District in central India involving 1,000 farmers cultivating 3,200 hectares found that average yields for cotton, wheat, chili, and soy were as much as 20 percent higher on the organic farms than on nearby conventionally managed ones. Farmers and agricultural scientists attributed the higher yields in this dry region to the emphasis on cover crops, compost, manure, and other practices that increased organic matter (which helps retain water) in the soils. A study from Kenya found that while organic farmers in "high-potential areas" (those with above-average rainfall and high soil quality) had lower maize yields than nonorganic farmers, organic farmers in areas with poorer resource endowments consistently outyielded conventional growers.

(In both regions, organic farmers had higher net profits, return on capital, and return on labor.)

Contrary to critics who jibe that it's going back to farming like our grandfathers did or that most of Africa already farms organically and it can't do the job, organic farming is a sophisticated combination of old wisdom and modern ecological innovations that help harness the yield-boosting effects of nutrient cycles, beneficial insects, and crop synergies. It's heavily dependent on technology—just not the technology that comes out of a chemical plant.

## High-Calorie Farms

So could we make do without the chemical plants? Inspired by a field trip to a nearby organic farm where the farmer reported that he raised an amazing 27 tons of vegetables on six-tenths of a hectare in a relatively short growing season, a team of scientists from the University of Michigan tried to estimate how much food could be raised following a global shift to organic farming. The team combed through the literature for any and all studies comparing crop yields on organic farms with those on nonorganic farms. Based on 293 examples, they came up with a global dataset of yield ratios for the world's major crops for the developed and the developing world. As expected, organic farming yielded less than conventional farming in the developed world for most food categories, while studies from the developing world showed organic farming boosting yields. The team then ran two models. The first was conservative in the sense that it applied the yield ratio for the developed world to the entire planet, i.e., they assumed that every farm regardless of location would get only the lower developed-country yields. The second applied the yield ratio for the developed world to wealthy nations and the yield ratio for the developing world to those countries.

"We were all surprised by what we found," said Catherine Badgley, a Michigan paleoecologist who was one of the lead researchers. The first model yielded 2,641 kilocalories ("calories") per person per day, just under the world's current production of 2,786 calories but significantly higher than the average caloric requirement for a healthy person of between 2,200 and 2,500. The second model yielded 4,381 calories per person per day, 75 percent greater than current availability—and a quantity that could theoretically sustain a much larger human population than is currently supported on the world's farmland. (It also laid to rest another concern about organic agriculture; see sidebar . . . .)

The team's interest in this subject was partly inspired by the concern that a large-scale shift to organic farming would require clearing additional wild areas to compensate for lower yields—an obvious worry for scientists like Badgley, who studies present and past biodiversity. The only problem with the argument, she said, is that much of the world's biodiversity exists in close proximity to farmland, and that's not likely to change anytime soon. "If we simply try to maintain biodiversity in islands around the world, we will lose most of it," she said. "It's very important to make areas between those islands friendly to biodiversity. The idea of those areas being pesticide-drenched fields is just going to be a disaster for biodiversity, especially in the

## ENOUGH NITROGEN TO GO AROUND?

In addition to looking at raw yields, the University of Michigan scientists also examined the common concern that there aren't enough available sources of non-synthetic nitrogen—compost, manure, and plant residues—in the world to support large-scale organic farming. For instance, in his book *Enriching the Earth: Fritz Haber, Carl Bosch, and the Transformation of World Food Production,* Vaclav Smil argues that roughly two-thirds of the world's food harvest depends on the Harber-Bosch process, the technique developed in the early 20th century to synthesize ammonia fertilizer from fossil fuels. (Smil admits that he largely ignored the contribution of nitrogen-fixing crops and assumed that some of them, like soybeans, are net users of nitrogen, although he himself points out that on average half of all the fertilizer applied globally is wasted and not taken up by plants.) Most critics of organic farming as a means to feed the world focus on how much manure—and how much related pastureland and how many head of livestock—would be needed to fertilize the world's organic farms. "The issue of nitrogen is different in different regions," says Don Lotter, an agricultural consultant who has published widely on organic farming and nutrient requirements. "But lots more nitrogen comes in as green manure than animal manure."

Looking at 77 studies from the temperate areas and tropics, the Michigan team found that greater use of nitrogen-fixing crops in the world's major agricultural regions could result in 58 million metric tons more nitrogen than the amount of synthetic nitrogen currently used every year. Research at the Rodale Institute in Pennsylvania showed that red clover used as a winter cover in an oat/wheat–corn–soy rotation, with no additional fertilizer inputs, achieved yields comparable to those in conventional control fields. Even in arid and semi-arid tropical regions like East Africa, where water availability is limited between periods of crop production, drought-resistant green manures such as pigeon peas or groundnuts could be used to fix nitrogen. In Washington state, organic wheat growers have matched their non-organic neighbor's wheat yields using the same field pea rotation for nitrogen. In Kenya, farmers using leguminous tree crops have doubled or tripled corn yields as well as suppressing certain stubborn weeds and generating additional animal fodder.

The Michigan results imply that no additional land area is required to obtain enough biologically available nitrogen, even without including the potential for intercropping (several crops grown in the same field at the same time), rotation of livestock with annual crops, and inoculation of soil with *Azobacter, Azospirillum,* and other free-living nitrogen-fixing bacteria.

tropics. The world would be able to sustain high levels of biodiversity much better if we could change agriculture on a large scale."

Badgley's team went out of the way to make its assumerions as conservative as possible: most of the studies they used looked at the yields of a single crop, even though many organic farms grow more than one crop in a field at the same time, yielding more total food even if the yield of any given crop may be lower. Skeptics may doubt the team's conclusions—as ecologists, they are likely to be sympathetic to organic farming—but a second recent study of the potential of a global shift to organic farming, led by Niels Halberg of the Danish Institute of Agricultural Sciences, came to very similar conclusions, even though the authors were economists, agronomists, and international development experts.

Like the Michigan team, Halberg's group made an assumption about the differences in yields with organic farming for a range of crops and then plugged those numbers into a model developed by the World Bank's International Food Policy Research Institute (IFPRI). This model is considered the definitive algorithm for predicting food output, farm income, and the number of hungry people throughout the world. Given the growing interest in organic farming among consumers, government officials, and agricultural scientists, the researchers wanted to assess whether a large-scale conversion to organic farming in Europe and North America (the world's primary food exporting regions) would reduce yields, increase world food prices, or worsen hunger in poorer nations that depend on imports, particularly those people living in the Third World's swelling megacities. Although the group found that total food production declined in Europe and North America, the model didn't show a substantial impact on world food prices. And because the model assumed, like the Michigan study, that organic farming would boost yields in Africa, Asia, and Latin America, the most optimistic scenario even had hunger-plagued sub-Saharan Africa exporting food surpluses.

"Modern non-certified organic farming is a potentially sustainable approach to agricultural development in areas with low yields due to poor access to inputs or low yield potential because it involves lower economic risk than comparative interventions based on purchased inputs and may increase farm level resilience against climatic fluctuations," Halberg's team concluded. In other words, studies from the field show that the yield increases from shifting to organic farming are highest and most consistent in exactly those poor, dry, remote areas where hunger is most severe. "Organic agriculture could be an important part of increased food security in sub-Saharan Africa," says Halberg.

That is, if other problems can be overcome. "A lot of research is to try to kill prejudices," Halberg says—like the notion that organic farming is only a luxury, and one that poorer nations cannot afford. "I'd like to kill this once and for all. The two sides are simply too far from each other and they ignore the realities of the global food system." Even if a shift toward organic farming boosted yields in hungry African and Asian nations, the model found that nearly a billion people remained hungry, because any surpluses were simply exported to areas that could best afford it.

## Wrong Question?

These conclusions about yields won't come as a surprise to many organic farmers. They have seen with their own eyes and felt with their own hands how productive they can be. But some supporters of organic farming shy away from even asking whether it can feed the world, simply because they don't think it's the most useful question. There is good reason to believe that a global conversion to organic farming would not proceed as seamlessly as plugging some yield ratios into a spreadsheet.

To begin with, organic farming isn't as easy as farming with chemicals. Instead of choosing a pesticide to prevent a pest outbreak, for example, a particular organic farmer might consider altering his crop rotation, planting a crop that will repel the pest or one that will attract its predators—decisions that require some experimentation and long-term planning. Moreover, the IFPRI study suggested that a large-scale conversion to organic farming might require that most dairy and beef production eventually "be better integrated in cereal and other cash crop rotations" to optimize use of the manure. Bringing cows back to one or two farms to build up soil fertility may seem like a no-brainer, but doing it wholesale would be a challenge—and dumping ammonia on depleted soils still makes for a quicker fix.

Again, these are just theoretical assumptions, since a global shift to organic farming could take decades. But farmers are ingenious and industrious people and they tend to cope with whatever problems are at hand. Eliminate nitrogen fertilizer and many farmers will probably graze cows on their fields to compensate. Eliminate fungicides and farmers will look for fungus-resistant crop varieties. As more and more farmers begin to farm organically, everyone will get better at it. Agricultural research centers, universities, and agriculture ministries will throw their resources into this type of farming—in sharp contrast to their current neglect of organic agriculture, which partly stems from the assumption that organic farmers will never play a major role in the global food supply.

So the problems of adopting organic techniques do not seen insurmountable. But those problems may not deserve most of our attention; even if a mass conversion over, say, the next two decades, dramatically increased food production, there's little guarantee it would eradicate hunger. The global food system can be a complex and unpredictable beast. It's hard to anticipate how China's rise as a major importer of soybeans for its feedlots, for instance, might affect food supplies elsewhere. (It's likely to drive up food prices.) Or how elimination of agricultural subsidies in wealthy nations might affect poorer countries. (It's likely to boost farm incomes and reduce hunger.) And would less meat eating around the world free up food for the hungry? (It would, but could the hungry afford it?) In other words, "Can organic farming feed the world?" is probably not even the right question, since feeding the world depends more on politics and economics than any technological innovations.

"'Can organic farming feed the world' is indeed a bogus question," says Gene Kahn, a long-time organic farmer who founded Cascadian Farms organic foods and is now vice president of sustainable development for General Mills. "The real question is, can we feed the world? Period. Can we fix the disparities in human nutrition?" Kahn notes that the marginal difference in today's organic yields and the yields of conventional agriculture wouldn't matter if food surpluses were redistributed.

But organic farming will yield other benefits that are too numerous to name. Studies have shown, for example, that the "external" costs of organic farming—erosion, chemical pollution to drinking water, death of birds and other wildlife—are just one-third those of conventional farming. Surveys from every continent show that organic farms support many more species of birds, wild plants, insects, and other wildlife than conventional farms. And tests by several governments have shown that organic foods carry just a tiny fraction of the pesticide residues of the nonorganic alternatives, while completely banning growth hormones, antibiotics, and many additives allowed in many conventional foods. There is even some evidence that crops grown organically have considerably higher levels of health-promoting antioxidants.

There are social benefits as well. Because organic farming doesn't depend on expensive inputs, it might help shift the balance towards smaller farmers in hungry nations. A 2002 report from the UN Food and Agriculture Organization noted that "organic systems can double or triple the productivity of traditional systems" in developing nations but suggested that yield comparisons offer a "limited, narrow, and often misleading picture" since farmers in these countries often adopt organic farming techniques to save water, save money, and reduce the variability of yields in extreme conditions. A more recent study by the International Fund for Agricultural Development found that the higher labor requirements often mean that "organic agriculture can prove particularly effective in bringing redistribution of resources in areas where the labour force is underemployed. This can help contribute to rural stability."

## Middle Earth

These benefits will come even without a complete conversion to a sort of organic Utopia. In fact, some experts think that a more hopeful, and reasonable, way forward is a sort of middle ground, where more and more farmers adopt the principles of organic farming even if they don't follow the approach religiously. In this scenario, both poor farmers and the environment come out way ahead. "Organic agriculture is not going to do the trick," says Roland Bunch, an agricultural extensionist who has worked for decades in Africa and the Americas and is now with COSECHA (Association of Consultants for a Sustainable, Ecological, and People-Centered Agriculture) in Honduras. Bunch knows first-hand that organic agriculture can produce more than conventional farming among poorer farmers. But he also knows that these farmers cannot get the premium prices paid for organic produce elsewhere, and that they are often unable, and unwilling, to shoulder some of the costs and risks associated with going completely organic.

Instead, Bunch points to "a middle path," of eco-agriculture, or low-input agriculture that uses many of the principles of organic farming and depends on just a small fraction of the chemicals. "These systems can immediately produce two or three times what smallholder farmers are presently producing" Bunch says. "And furthermore, it is attractive to smallholder farmers because it is less costly per unit produced." In addition to the immediate gains in food production, Bunch suggests that the benefits for the environment of this middle path will be far greater than going "totally organic," because "something like five to ten times as many smallholder farmers will adopt it

## FOOD VERSUS FUEL

Sometimes, when humans try to solve one problem, they end up creating another. The global food supply is already under serious strain: more than 800 million people go hungry every day, the world's population continues to expand, and a growing number of people in the developing world are changing to a more Western, meat-intensive diet that requires more grain and water per calorie than traditional diets do. Now comes another potential stressor: concern about climate change means that more nations are interested in converting crops into biofuels as an alternative to fossil fuels. But could this transition remove land from food production and further intensify problems of world hunger?

For several reasons, some analysts say no, at least not in the near future. First, they emphasize that nearly 40 percent of global cereal crops are fed to livestock, not humans, and that global prices of grains and oil seeds do not always affect the cost of food for the hungry, who generally cannot participate in formal markets anyway.

Second, at least to date, hunger has been due primarily to inadequate income and distribution rather than absolute food scarcity. In this regard, a biofuels economy may actually help to reduce hunger and poverty. A recent UN Food and Agriculture Organization report argued that increased use of biofuels could diversify agricultural and forestry activities, attract investment in new small and medium-sized enterprises, and increase investment in agricultural production, thereby increasing the incomes of the world's poorest people.

Third, biofuel refineries in the future will depend less on food crops and increasingly on organic wastes and residues. Producing biofuels from corn stalks, rice hulls, sawdust, or waste paper is unlikely to affect food production directly. And there are drought-resistant grasses, fast-growing trees, and other energy crops that will grow on marginal lands unsuitable for raising food.

Nonetheless, with growing human appetites for both food and fuel, biofuels' long-run potential may be limited by the priority given to food production if bioenergy systems are not harmonized with food systems. The most optimistic assessments of the long-term potential of biofuels have assumed that agricultural yields will continue to improve and that world population growth and food consumption will stabilize. But the assumption about population may prove to be wrong. And yields, organic or otherwise, may not improve enough if agriculture in the future is threatened by declining water tables or poor soil maintenance.

per unit of extension and training expense, because it behooves them economically. They aren't taking food out of their kids' mouths. If five farmers eliminate half their use of chemicals, the effect on the environment will be two and one-half times as great as if one farmer goes totally organic."

And farmers who focus on building their soils, increasing biodiversity, or bringing livestock into their rotation aren't precluded from occasionally turning to biotech crops or synthetic nitrogen or any other yield-enhancing innovations in the future, particularly in places where the soils are heavily depleted. "In the end, if we do things right, we'll build a lot of organic into conventional systems," says Don Lotter, the agricultural consultant. Like Bunch, Lotter notes that such an "integrated" approach often out-performs both a strictly organic and chemical-intensive approach in terms of yield, economics, and environmental benefits. Still, Lotter's not sure we'll get there tomorrow, since the world's farming is hardly pointed in the organic direction—which could be the real problem for the world's poor and hungry. "There is such a huge area in sub-Saharan Africa and South America where the Green Revolution has never made an impact and it's unlikely that it will for the next generation of poor farmers," argues Niels Halberg, the Danish scientist who lead the IFPR] study. "It seems that agro-ecological measures for some of these areas have a beneficial impact on yields and food insecurity. So why not seriously try it out?"

**John J. Miller** **NO**

# The Organic Myth: A Food Movement Makes a Pest of Itself

Somewhere in the cornfields of Britain, a hungry insect settled on a tall green stalk and decided to have a feast. It chewed into a single kernel of corn, filled its little belly, and buzzed off—leaving behind a tiny hole that was big enough to invite a slow decay. The agent of the decomposition was a fungus known to biologists as *Fusarium*. Farmers have a much simpler name for it: corn ear rot.

As the mold spread inside the corn, it left behind a cancer causing residue called fumonisin. This sequence repeated itself thousands and thousands of times until the infested corn was harvested and sold last year as Fresh and Wild Organic Maize Meal, Infinity Foods Organic Maize Meal, and several other products.

Consuming trace amounts of fumonisin is harmless, but large doses can be deadly. Last fall, the United Kingdom's Food Standards Agency detected alarming concentrations of the toxin in all six brands of organic corn meal subjected to testing—for a failure rate of 100 percent. The average level of contamination was almost 20 times higher than the safety threshold Europeans have set for fumonisin. The tainted products were immediately recalled from the food chain. In contrast, inspectors determined that 20 of the 24 non-organic corn meal products they examined were unquestionably safe to eat.

Despite this, millions of people continue to assume that organic foods are healthier than non-organic ones, presumably because they grow in pristine settings free from icky chemicals and creepy biotechnology. This has given birth to an energetic political movement. In 2002, activists in Oregon sponsored a ballot initiative that essentially would have required the state to slap biohazard labels on anything that wasn't produced in ways deemed fit by anti-biotech agitators. Voters rejected it, but the cause continues to percolate. Hawaiian legislators are giving serious thought to banning biotech crop tests in their state. In March, California's Mendocino County may outlaw biotech plantings altogether.

Beneath it all lurks the belief that organic food is somehow better for us. In one poll, two-thirds of Americans said that organic food is healthier. But they're wrong: It's no more nutritious than food fueled by industrial fertilizers, sprayed with synthetic pesticides, and genetically altered in science labs. And the problem isn't limited to the fungal infections that recently cursed organic corn meal in Britain; bacteria are a major source of disease in organic food as well. To complicate matters further, organic farming is incredibly inefficient. If its appeal ever grew beyond the boutique, it would pose serious threats to the environment. Consumers who go shopping for products emblazoned with the USDA's "organic" seal of approval aren't really helping themselves or the planet—and they're arguably hurting both.

## No Fear

Here's the good news: At no point in human history has food been safer than it is today, despite occasional furors like the recent one over an isolated case of mad-cow disease here in the U.S. People still get sick from food—each year, about 76 million Americans pick up at least a mild illness from what they put in their mouths—but modern agricultural methods have sanitized our fare to the point where we may eat without fear. This is true for all food, organic or otherwise.

And that raises a semantic question: What is it about organic food that makes it "organic"? The food we think of as nonorganic isn't really *in*organic, as if it were composed of rocks and minerals. In truth, everything we eat is organic—it's just not "organic" the way the organic-food movement has come to define the word.

About a decade ago, the federal government decided to wade into this semantic swamp. There was no compelling reason for this, but Congress nonetheless called for the invention of a National Organic Rule. It became official in 2002. Organic food, said the bureaucrats, is produced without synthetic fertilizers, conventional pesticides, growth hormones, genetic engineering, or germ-killing radiation. There are also varying levels of organic-ness: Special labels are available for products that are "made with organic ingredients" (which means the food is 70 percent organic), "organic" (which means 95 percent organic), and "100 percent organic." It's not at all clear what consumers are supposed to do with this information. As the Department of Agriculture explains on its website, the "USDA makes no claims that organically produced food is safer or more nutritious than conventionally produced food."

It doesn't because it can't: There's no scientific evidence whatsoever showing that organic food is healthier. So why bother with a National Organic Rule? When the thing was in development, the Clinton administration's secretary of agriculture, Dan Glickman, offered an answer: "The organic label is a marketing tool. It is not a statement about food safety." In other words, those USDA labels are intended to give people warm fuzzies for buying pricey food.

And herein lies one of the dirty secrets of organic farming: It's big business. Although the organic movement has humble origins, today most of its food isn't produced on family farms in quaint villages or even on hippie communes in Vermont. Instead, the industry has come to be dominated by large corporations that are normally the dreaded bogeymen in the minds of many organic consumers. A single company currently controls about 70 percent of the market in organic milk. California grows about $400 million per year in organic produce—and about half of it comes from just five farms. The membership list of the

Organic Trade Association includes the biggest names in agribusiness, such as Archer Daniels Midland, Gerber, and Heinz. Even Nike is a member. When its capitalist slavedrivers aren't exploiting child labor in Third World sweatshops (as they do in the fevered imaginations of campus protesters), they're promoting Nike Organics, a clothing line made from organic cotton.

## The Yum Factor

There are, in fact, good reasons to eat organic food. Often it's yummier—though this has nothing to do with the fact that it's "organic." If an organic tomato tastes better than a non-organic one, the reason is usually that it has been grown locally, where it has ripened on the vine and taken only a day or two to get from the picking to the selling. Large-scale farming operations that ship fruits and vegetables across the country or the world can't compete with this kind of homegrown quality, even though they do make it possible for people in Minnesota to avoid scurvy by eating oranges in February. Conventional produce is also a good bargain because organic foods can be expensive—the profit margins are quite high, relative to the rest of the food industry.

Unfortunately, money isn't always the sole cost. Although the overwhelming majority of organic foods are safe to eat, they aren't totally risk-free. Think of it this way: Organic foods may be fresh, but they're also fresh from the manure fields.

Organic farmers aren't allowed to enrich their soils the way most non-organic farmers do, which is with nitrogen fertilizers produced through an industrial process. In their place, many farmers rely on composted manure. When they spread the stuff in their fields, they create luscious breeding grounds for all kinds of nasty microbes. Take the dreaded *E. coli,* which is capable of killing people who ingest it. A study by the Center for Global Food Issues found that although organic foods make up about 1 percent of America's diet, they also account for about 8 percent of confirmed *E. coli* cases. Organic food products also suffer from more than eight times as many recalls as conventional ones.

Some of this problem would go away if organic farmers used synthetic sprays—but this, too, is off limits. Conventional wisdom says that we should avoid food that's been drenched in herbicides, pesticides, and fungicides. Half a century ago, there was some truth in this: Sprays were primitive and left behind chemical deposits that often survived all the way to the dinner table. Today's sprays, however, are largely biodegradable. They do their job in the field and quickly break down into harmless molecules. What's more, advances in biotechnology have reduced the need to spray. About one-third of America's corn crop is now genetically modified. This corn includes a special gene that produces a natural toxin that's safe for every living creature to eat except caterpillars with alkaline guts, such as the European corn borer, a moth larva that can ravage whole harvests. This kind of biotech innovation has helped farmers reduce their reliance on pesticides by about 50 million pounds per year.

Organic farmers, of course, don't benefit from any of this. But they do have some recourse against the bugs, weeds, and fungi that can devastate a crop: They spray their plants with "natural" pesticides. These are less effective than synthetic ones and they're certainly no safer. In rat tests, rotenone—an insecticide extracted from the roots of tropical plants—has been shown to cause the symptoms of Parkinson's disease. The Environmental Protection Agency has described pyrethrum, another natural bug killer, as a human carcinogen. Everything is lethal in massive quantities, of course, and it takes huge doses of pyrethrum to pose a health hazard. Still, the typical organic farmer has to douse his crops with it as many as seven times to have the same effect as one or two applications of a synthetic compound based on the same ingredients. Then there's one of the natural fungicides preferred by organic coffee growers in Guatemala: fermented urine. Think about that the next time you're tempted to order the "special brew" at your local organic java hut.

## St. Anthony's Fire, etc.

Fungicides are worth taking seriously—and not just because they might have prevented Britain's corn meal problem a few months ago. Before the advent of modern farming, when all agriculture was necessarily "organic," food-borne fungi were a major problem. One of the worst kinds was ergot, which affects rye, wheat, and other grains. During the Middle Ages, ergot poisoning caused St. Anthony's Fire, a painful contraction of blood vessels that often led to gangrene in limb extremities and sometimes death. Hallucinations also were common, as ergot contains lysergic acid, which is the crucial component of LSD. Historians of medieval Europe have documented several episodes of towns eating large batches of ergot-polluted food and falling into mass hysteria. There is some circumstantial evidence suggesting that ergot was behind the madness of the Salem witches: The warm and damp weather just prior to the infamous events of 1692 would have been ideal for an outbreak. Today, however, chemical sprays have virtually eradicated this affliction.

The very worst thing about organic farming requires the use of a word that doomsaying environmentalists have practically trademarked: It's not *sustainable*. Few activities are as wasteful as organic farming. Its yields are about half of what conventional farmers expect at harvest time. Norman Borlaug, who won the Nobel Peace Prize in 1970 for his agricultural innovations, has said, "You couldn't feed more than 4 billion people" on an all-organic diet.

If organic-food consumers think they're making a political statement when they eat, they're correct: They're declaring themselves to be not only friends of population control, but also enemies of environmental conservation. About half the world's land area that isn't covered with ice or sand is devoted to food production. Modern farming techniques have enabled this limited supply to produce increasing quantities of food. Yields have fattened so much in the last few decades that people refer to this phenomenon as the "Green Revolution," a term that has nothing to do with enviro-greenies and everything to do with improvements in breeding, fertilization, and irrigation. Yet even greater challenges lie ahead, because demographers predict that world population will rise to 9 billion by 2050. "The key is to produce more food," says Alex Avery of CGFI. "Growing more per acre leaves more land for nature." The alternative is to chop down rainforests so that we may dine on organic soybeans.

There's one more important reason that organics can't feed the world: There just isn't enough cow poop to go around. For fun, pretend that U.N. secretary-general Kofi Annan chowed on some ergot rye, decreed that all of humanity must eat nothing but organic food, and that all of humanity responded by saying, "What the heck, we'll give it a try." Forget about the population boom ahead. The immediate problem would be generating enough manure to fertilize all the brand-new, low-yield organic crop fields. There are a little more than a billion cattle in the world today, and each bovine needs between 3 and 30 acres to support it. Conservative estimates say it would take around 7 or 8 billion cattle to produce sufficient heaps of manure to sustain our all-organic diets. The United States alone would need about a billion head (or rear, to be precise). The country would be made up of nothing but cities and manure fields—and the experiment would give a whole new meaning to the term "fruited plains."

This is the sort of future the organic-food movement envisions—and its most fanatical advocates aren't planning to win any arguments on the merits or any consumers on the quality of organic food. In December, when a single U.S. animal was diagnosed with mad-cow disease, nobody was more pleased than Ronnie Cummins of the Organic Consumers Association, who has openly hoped for a public scare that would spark a "crisis of confidence" in American food. No such thing happened, but Cummins should be careful about what he wishes for: Germany's first case of mad-cow disease surfaced at a slaughterhouse that specializes in organic beef.

But then wishful thinking is at the heart of the organic-food movement. Its whole market rationale depends on the misperception that organic foods are somehow healthier for both consumers and Mother Earth. Just remember: Nature's Valley can't be found on any map. It's a state of mind.

# POSTSCRIPT

## Is a Large-Scale Shift to Organic Farming the Best Way to Increase World Food Supply?

Is organic food better or safer for the consumer than non-organic food? Faidon Magkos, et al., "Organic Food: Buying More Safety or Just Peace of Mind? A Critical Review of the Literature," *Critical Reviews in Food Science & Nutrition* (January 2006), report that the quality and safety of organic food are largely a matter of perception. "Relevant scientific evidence . . . is scarce, while anecdotal reports abound." Pesticide and herbicide residues may be lower, but even on non-organic food they are low. Environmental contaminants are likely to affect both organic and non-organic foods. "'Organic' does not automatically equal 'safe.'"

Is organic farming better for the environment? Soil fertility is in decline in many parts of the world; see Alfred E. Herteminck, "Assessing Soil Fertility Decline in the Tropics Using Soil Chemical Data," *Advances in Agronomy* (2006). In Africa, the situation is extraordinarily serious. According to the International Center for Soil Fertility and Agricultural Development (http://www.ifdc.org/New_Design/Whats_New/africasfailingagriculture033006.pdf), "About 75 percent of the farmland in sub-Saharan Africa is plagued by severe degradation, losing basic soil nutrients needed to grow the crops that feed Africa, according to a new report . . . on the precipitous decline in African soil health from 1980 to 2004. Africa's crisis in food production and battle with hunger are largely rooted in this 'soil health crisis.'" Proponents of organic farming argue that it is essential to relieving the crisis, but one study of changes in soil fertility, as indicated by crop yield, earthworm numbers, and soil properties, after converting from conventional to organic practices, found that different soils responded differently, with some improving and some not; see Anne Kjersti Bakken, et al., "Soil Fertility in Three Cropping Systems after Conversion from Conventional to Organic Farming, "*Acta Agriculturae Scandinavica: Section B, Soil & Plant Science* (June 2006). Richard Wood, et al., "A Comparative Study of Some Environmental Impacts of Conventional and Organic Farming in Australia," *Agricultural Systems* (September 2006), find in a comparison of organic and conventional farms "that direct energy use, energy related emissions, and greenhouse gas emissions are higher for the" former.

According to Catherine M. Cooney, "Sustainable Agriculture Delivers the Crops," *Environmental Science & Technology* (February 15, 2006), "Sustainable agriculture, such as crop rotation, organic farming, and genetically modified seeds, increased crop yields by an average of 79 percent" while also improving the lives of farmers in developing countries. Cong Tu, et al., "Responses of Soil

Microbial Biomass and N Availability to Transition Strategies from Conventional to Organic Farming Systems," *Agriculture, Ecosystems & Environment* (April 2006), note that a serious barrier to changing from conventional to organic farming, despite soil improvements, is an initial reduction in yield and increase in pests. David Pimentel, et al., "Environmental, Energetic, and Economic Comparisons of Organic and Conventional Farming Systems," *Bioscience* (July 2005), find that organic farming uses less energy, improves soil, and has yields comparable to those of conventional farming but crops probably cannot be grown as often (because of fallowing), which then reduces long-term yields. However, "because organic foods frequently bring higher prices in the marketplace, the net economic return per [hectare] is often equal to or higher than that of conventionally produced crops." This is one of the factors that prompts Craig J. Pearson to argue in favor of shifting "from conventional open or leaky systems to more closed, regenerative systems" in "Regenerative, Semiclosed Systems: A Priority for Twenty-First-Century Agriculture," *Bioscience* (May 2007).

Better economic return means that organic farming is currently good for the organic farmer. However, the advantage would disappear if the world converted to organic farming. Initial declines in yield mean the conversion would be difficult, but the longer we wait and the more population grows, the more difficult it will be. Whether the conversion is essential depends on the availability of alternative solutions to the problem, and it is worth noting that high energy prices make chemical fertilizers increasingly expensive. Sadly, Stacey Irwin, "Battle High Fertilizer Costs," *Farm Industry News* (January 2006), does not mention the possibility of using organic methods.

# Internet References . . .

## The Pesticide Action Network

The Pesticide Action Network North America (PANNA) challenges the global proliferation of pesticides.

**http://www.panna.org/**

## e.hormone

This site provides accurate, timely information about environmental hormones and their impacts.

**http://e.hormone.tulane.edu/**

## The Silicon Valley Toxics Coalition (SVTC)

SVTC was formed to engage in research, advocacy, and organizing associated with environmental and human problems caused by the rapid growth of the high-tech electronics industry.

**http://svtc.etoxics.org/site/PageServer**

## Superfund

The U.S. Environmental Protection Agency provides a great deal of information on the Superfund program, including material on environmental justice.

**http://www.epa.gov/superfund/**

## The La Hague Nuclear Reprocessing Plant

The AREVA NC La Hague site, located on the western tip of the Cotentin Peninsula in Normandy, reprocesses spent power reactor fuel to recycle reusable energy materials—uranium and plutonium—and to condition the waste into suitable final form.

**http://www.cogemalahague.com/servlet/ContentServer?pagename=cogema_en/Page/page_html_libre_full_template&c=Page&cid=1039482706828**

## Carbon Capture and Sequestration Technologies at MIT

This site describes research into technologies to capture, utilize, and store $CO_2$ from large stationary sources such as power plants. It also provides an overview of the technology and links to many other sites.

**http://sequestration.mit.edu/index.html**

## The Rewilding Institute

The mission of the Rewilding Institute is to "develop and promote the ideas and strategies to advance continental-scale conservation in North America.

**http://rewilding.org/**

## The Wildlands Project

The Wildlands project endeavors to address the extinction crisis.

**http://www.twp.org**

# UNIT 5

## Toxic Chemicals

*A great many of today's environmental issues have to do with industrial development, which expanded greatly during the twentieth century. Just since World War II, many thousands of synthetic chemicals—pesticides, plastics, and antibiotics—have flooded the environment. We have become dependent on the production and use of energy, particularly in the form of fossil fuels. We have discovered that industrial processes generate huge amounts of waste, much of it toxic. Air and water pollution have become global problems. And we have discovered that our actions may change the world for generations to come.*

- Should DDT Be Banned Worldwide?
- Do Environmental Hormone Mimics Pose a Potentially Serious Health Threat?
- Is the Superfund Program Successfully Protecting Human Health from Hazardous Materials?
- Should the United States Reprocess Spent Nuclear Fuel?
- Is Carbon Capture Technology Ready to Limit Carbon Emissions?
- Should North America's Landscape Be Restored to Its Pre-Human State?

# ISSUE 16

## Should DDT Be Banned Worldwide?

**YES: Anne Platt McGinn,** from "Malaria, Mosquitoes, and DDT," *World Watch* (May/June 2002)

**NO: Donald R. Roberts,** from Statement before the U.S. Senate Committee on Environment & Public Works, Hearing on the Role of Science in Environmental Policy-Making (September 28, 2005)

**ISSUE SUMMARY**

**YES:** Anne Platt McGinn, a senior researcher at the Worldwatch Institute, argues that although DDT is still used to fight malaria, there are other, more effective and less environmentally harmful methods. She maintains that DDT should be banned or reserved for emergency use.

**NO:** Donald R. Roberts argues that the scientific evidence regarding the environmental hazards of DDT has been seriously misrepresented by anti-pesticide activists. The hazards of malaria are much greater and, properly used, DDT can prevent them and save lives.

DDT is a crucial element in the story of environmentalism. The chemical was first synthesized in 1874. Swiss entomologist Paul Mueller was the first to notice that DDT has insecticidal properties, which, it was quickly realized, implied that the chemical could save human lives. It had long been known that more soldiers died during wars because of disease than because of enemy fire. During World War I, for example, some 5 million lives were lost to typhus, a disease carried by body lice. DDT was first deployed during World War II to halt a typhus epidemic in Naples, Italy. It was a dramatic success, and DDT was soon used routinely as a dust for soldiers and civilians. During and after the war, DDT was also deployed successfully against the mosquitoes that carry malaria and other diseases. In the United States cases of malaria fell from 120,000 in 1934 to 72 in 1960, and cases of yellow fever dropped from 100,000 in 1878 to none. In 1948 Mueller received the Nobel Prize for medicine and physiology because DDT had saved so many civilian lives.

DDT was by no means the first pesticide. But its predecessors—arsenic, strychnine, cyanide, copper sulfate, and nicotine—were all markedly toxic to humans. DDT was not only more effective as an insecticide, it was also less hazardous to users. It is therefore not surprising that DDT was seen as a beneficial substance. It was soon applied routinely to agricultural crops and used to control mosquito populations in American suburbs. However, insects quickly became resistant to the insecticide. (In any population of insects, some will be more resistant than others; when the insecticide kills the more vulnerable members of the population, the resistant ones are left to breed and multiply. This is an example of natural selection.) In *Silent Spring* (Houghton Mifflin, 1962), marine scientist Rachel Carson demonstrated that DDT was concentrated in the food chain and affected the reproduction of predators such as hawks and eagles. In 1972 the U.S. Environmental Protection Agency banned almost all uses of DDT (it could still be used to protect public health). Other developed countries soon banned it as well, but developing nations, especially those in the tropics, saw it as an essential tool for fighting diseases such as malaria. Roger Bate, director of Africa Fighting Malaria, argues in "A Case of the DDTs," *National Review* (May 14, 2001) that DDT remains the cheapest and most effective way to combat malaria and that it should remain available for use.

It soon became apparent that DDT is by no means the only pesticide or organic toxin with environmental effects. As a result, on May 24, 2001, the United States joined 90 other nations in signing the Stockholm Convention on Persistent Organic Pollutants (POPs). This treaty aims to eliminate from use the entire class of chemicals to which DDT belongs, beginning with the "dirty dozen," pesticides DDT, aldrin, dieldrin, endrin, chlordane, heptachlor, mirex, and toxaphene, and the industrial chemicals polychlorinated biphenyls (PCBs), hexachlorobenzene (HCB), dioxins, and furans. Since then, 59 countries, not including the United States and the European Union (EU), have formally ratified the treaty. It took effect in May 2004. Fiona Proffitt, in "U.N. Convention Targets Dirty Dozen Chemicals," *Science* (May 21, 2004), notes, "About 25 countries will be allowed to continue using DDT against malaria-spreading mosquitoes until a viable alternative is found."

In the following selection, Anne Platt McGinn, granting that malaria remains a serious problem in the developing nations of the tropics, especially Africa, contends that although DDT is still used to fight malaria in these nations, it is far less effective than it used to be. She argues that the environmental effects are also serious concerns and that DDT should be banned or reserved for emergency use. In the second selection, Professor Donald R. Roberts argues that the scientific evidence regarding the environmental hazards of DDT has been seriously misrepresented by anti-pesticide activists. The hazards of malaria are much greater and, properly used, DDT can prevent them and save lives. Efforts to prevent the use of DDT have produced a "global humanitarian disaster."

**YES**

**Anne Platt McGinn**

# Malaria, Mosquitoes, and DDT

This year, like every other year within the past couple of decades, uncountable trillions of mosquitoes will inject malaria parasites into human blood streams billions of times. Some 300 to 500 million full-blown cases of malaria will result, and between 1 and 3 million people will die, most of them pregnant women and children. That's the official figure, anyway, but it's likely to be a substantial underestimate, since most malaria deaths are not formally registered, and many are likely to have escaped the estimators. Very roughly, the malaria death toll rivals that of AIDS, which now kills about 3 million people annually.

But unlike AIDS, malaria is a low-priority killer. Despite the deaths, and the fact that roughly 2.5 billion people (40 percent of the world's population) are at risk of contracting the disease, malaria is a relatively low public health priority on the international scene. Malaria rarely makes the news. And international funding for malaria research currently comes to a mere $150 million annually. Just by way of comparison, that's only about 5 percent of the $2.8 billion that the U.S. government alone is considering for AIDS research in fiscal year 2003.

The low priority assigned to malaria would be at least easier to understand, though no less mistaken, if the threat were static. Unfortunately it is not. It is true that the geographic range of the disease has contracted substantially since the mid-20th century, but over the past couple of decades, malaria has been gathering strength. Virtually all areas where the disease is endemic have seen drug-resistant strains of the parasites emerge—a development that is almost certainly boosting death rates. In countries as various as Armenia, Afghanistan, and Sierra Leone, the lack or deterioration of basic infrastructure has created a wealth of new breeding sites for the mosquitoes that spread the disease. The rapidly expanding slums of many tropical cities also lack such infrastructure; poor sanitation and crowding have primed these places as well for outbreaks—even though malaria has up to now been regarded as predominantly a rural disease.

What has current policy to offer in the face of these threats? The medical arsenal is limited; there are only about a dozen antimalarial drugs commonly in use, and there is significant malaria resistance to most of them. In the absence of a reliable way to kill the parasites, policy has tended to focus on killing the mosquitoes that bear them. And that has led to an abundant use of synthetic pesticides, including one of the oldest and most dangerous: dichlorodiphenyl trichloroethane, or DDT.

DDT is no longer used or manufactured in most of the world, but because it does not break down readily, it is still one of the most commonly detected pesticides in the milk of nursing mothers. DDT is also one of the "dirty dozen" chemicals included in the 2001 Stockholm Convention on Persistent Organic Pollutants [POPs]. The signatories to the "POPs Treaty" essentially agreed to ban all uses of DDT except as a last resort against disease-bearing mosquitoes. Unfortunately, however, DDT is still a routine option in 19 countries, most of them in Africa. (Only 11 of these countries have thus far signed the treaty.) Among the signatory countries, 31—slightly fewer than one-third—have given notice that they are reserving the right to use DDT against malaria. On the face of it, such use may seem unavoidable, but there are good reasons for thinking that progress against the disease is compatible with *reductions* in DDT use.

Malaria is caused by four protozoan parasite species in the genus *Plasmodium*. These parasites are spread exclusively by certain mosquitoes in the genus *Anopheles*. An infection begins when a parasite-laden female mosquito settles onto someone's skin and pierces a capillary to take her blood meal. The parasite, in a form called the *sporozoite,* moves with the mosquito's saliva into the human bloodstream. About 10 percent of the mosquito's lode of sporozoites is likely to be injected during a meal, leaving plenty for the next bite. Unless the victim has some immunity to malaria—normally as a result of previous exposure—most sporozoites are likely to evade the body's immune system and make their way to the liver, a process that takes less than an hour. There they invade the liver cells and multiply asexually for about two weeks. By this time, the original several dozen sporozoites have become millions of *merozoites*—the form the parasite takes when it emerges from the liver and moves back into the blood to invade the body's red blood cells. Within the red blood cells, the merozoites go through another cycle of asexual reproduction, after which the cells burst and release millions of additional merozoites, which invade yet more red blood cells. The high fever and chills associated with malaria are the result of this stage, which tends to occur in pulses. If enough red blood cells are destroyed in one of these pulses, the result is convulsions, difficulty in breathing, coma, and death.

As the parasite multiplies inside the red blood cells, it produces not just more merozoites, but also *gametocytes,* which are capable of sexual reproduction. This occurs when the parasite moves back into the mosquitoes; even as they inject sporozoites, biting mosquitoes may ingest gametocytes if they are feeding on a person who is already infected. The gametocytes reproduce in the insect's gut and the resulting eggs move into the gut cells. Eventually, more sporozoites emerge from the gut and penetrate the mosquito's salivary glands, where they await a chance to enter another human bloodstream, to begin the cycle again.

Of the roughly 380 mosquito species in the genus *Anopheles,* about 60 are able to transmit malaria to people. These malaria vectors are widespread throughout the tropics and warm temperate zones, and they are very efficient at spreading the disease. Malaria is highly contagious, as is apparent from a measurement that epidemiologists call the "basic reproduction number," or BRN. The BRN indicates, on average, how many new cases a single infected person is likely to cause. For example, among the nonvectored diseases (those in which the pathogen travels directly from person to person without an intermediary like a mosquito), measles is one of the most contagious. The BRN for measles is 12 to 14, meaning that someone with measles is likely to infect 12 to 14 other people. (Luckily, there's an inherent limit in this process: as a pathogen spreads through any particular area, it will encounter fewer and fewer susceptible people who aren't already sick, and the outbreak will eventually subside.) HIV/AIDS is on the other end of the scale: it's deadly, but it burns through a population slowly. Its BRN is just above 1, the minimum necessary for the pathogen's survival. With malaria, the BRN varies considerably, depending on such factors as which mosquito species are present in an area and what the temperatures are. (Warmer is worse, since the parasites mature more quickly.) But malaria can have a BRN in excess of 100: over an adult life that may last about a week, a single, malaria-laden mosquito could conceivably infect more than 100 people.

## Seven Years, Seven Months

"Malaria" comes from the Italian "mal'aria." For centuries, European physicians had attributed the disease to "bad air." Apart from a tradition of associating bad air with swamps—a useful prejudice, given the amount of mosquito habitat in swamps—early medicine was largely ineffective against the disease. It wasn't until 1897 that the British physician Ronald Ross proved that mosquitoes carry malaria.

The practical implications of Ross's discovery did not go unnoticed. For example, the U.S. administration of Theodore Roosevelt recognized malaria and yellow fever (another mosquito-vectored disease) as perhaps the most serious obstacles to the construction of the Panama Canal. This was hardly a surprising conclusion, since the earlier and unsuccessful French attempt to build the canal—an effort that predated Ross's discovery—is thought to have lost between 10,000 and 20,000 workers to disease. So the American workers draped their water supplies and living quarters with mosquito netting, attempted to fill in or drain swamps, installed sewers, poured oil into standing water, and conducted mosquito-swatting campaigns. And it worked: the incidence of malaria declined. In 1906, 80 percent of the workers had the disease; by 1913, a year before the Canal was completed, only 7 percent did. Malaria could be suppressed, it seemed, with a great deal of mosquito netting, and by eliminating as much mosquito habitat as possible. But the labor involved in that effort could be enormous.

That is why DDT proved so appealing. In 1939, the Swiss chemist Paul Müller discovered that this chemical was a potent pesticide. DDT was first used during World War II, as a delousing agent. Later on, areas in southern Europe, North Africa, and Asia were fogged with DDT, to clear malaria-laden mosquitoes from the paths of invading Allied troops. DDT was cheap and it seemed to be harmless to anything other than insects. It was also long-lasting: most other insecticides lost their potency in a few days, but in the early years of its use, the effects of a single dose of DDT could last for up to six months. In 1948, Müller won a Nobel Prize for his work and DDT was hailed as a chemical miracle.

A decade later, DDT had inspired another kind of war—a general assault on malaria. The "Global Malaria Eradication Program," launched in 1955, became one of the first major undertakings of the newly created World Health Organization [WHO]. Some 65 nations enlisted in the cause. Funding for DDT factories was donated to poor countries and production of the insecticide climbed.

The malaria eradication strategy was not to kill every single mosquito, but to suppress their populations and shorten the lifespans of any survivors, so that the parasite would not have time to develop within them. If the mosquitoes could be kept down long enough, the parasites would eventually disappear from the human population. In any particular area, the process was expected to take three years—time enough for all infected people either to recover or die. After that, a resurgence of mosquitoes would be merely an annoyance, rather than a threat. And initially, the strategy seemed to be working. It proved especially effective on islands—relatively small areas insulated from reinfestation. Taiwan, Jamaica, and Sardinia were soon declared malaria-free and have remained so to this day. By 1961, arguably the year at which the program had peak momentum, malaria had been eliminated or dramatically reduced in 37 countries.

One year later, Rachel Carson published *Silent Spring,* her landmark study of the ecological damage caused by the widespread use of DDT and other pesticides. Like other organochlorine pesticides, DDT bioaccumulates. It's fat soluble, so when an animal ingests it—by browsing contaminated vegetation, for example—the chemical tends to concentrate in its fat, instead of being excreted. When another animal eats that animal, it is likely to absorb the prey's burden of DDT. This process leads to an increasing concentration of DDT in the higher links of the food chain. And since DDT has a high chronic toxicity—that is, long-term exposure is likely to cause various physiological abnormalities—this bioaccumulation has profound implications for both ecological and human health.

With the miseries of malaria in full view, the managers of the eradication campaign didn't worry much about the toxicity of DDT, but they were greatly concerned about another aspect of the pesticide's effects: resistance. Continual exposure to an insecticide tends to "breed" insect populations that are at least partially immune to the poison. Resistance to DDT had been reported as early as 1946. The campaign managers knew that in mosquitoes, regular exposure to DDT tended to produce widespread resistance in four to seven years. Since it took three years to clear malaria from a human population, that didn't leave a lot of leeway for the eradication effort. As it turned out, the logistics simply couldn't be made to work in large, heavily infested areas with high human populations, poor housing and roads, and generally minimal infrastructure. In 1969, the campaign was abandoned. Today, DDT resistance is widespread in *Anopheles,* as is resistance to many more recent pesticides.

Undoubtedly, the campaign saved millions of lives, and it did clear malaria from some areas. But its broadest legacy has been of much more dubious value. It engendered the idea of DDT as a first resort against mosquitoes and it established the unstable dynamic of DDT resistance in *Anopheles* populations. In mosquitoes, the genetic mechanism that confers resistance to DDT does not usually come at any great competitive "cost"—that is, when no DDT is being sprayed, the resistant mosquitoes may do just about as well as nonresistant mosquitoes. So once a population acquires resistance, the trait is not likely to disappear even if DDT isn't used for years. If DDT is reapplied to such a population, widespread resistance will reappear very rapidly. The rule of thumb among entomologists is that you may get seven years of resistance-free use the first time around, but you only get about seven months the second time. Even that limited respite, however, is enough to make the chemical an attractive option as an emergency measure—or to keep it in the arsenals of bureaucracies committed to its use.

## Malaria Taxes

In December 2000, the POPs Treaty negotiators convened in Johannesburg, South Africa, even though, by an unfortunate coincidence, South Africa had suffered a potentially embarrassing setback earlier that year in its own POPs policies. In 1996, South Africa had switched its mosquito control programs from DDT to a less persistent group of pesticides known as pyrethroids. The move seemed solid and supportable at the time, since years of DDT use had greatly reduced *Anopheles* populations and largely eliminated one of the most troublesome local vectors, the appropriately named *A. funestus* ("funestus" means deadly). South Africa seemed to have beaten the DDT habit: the chemical had been used to achieve a worthwhile objective; it had then been discarded. And the plan worked—until a year before the POPs summit, when malaria infections rose to 61,000 cases, a level not seen in decades. *A. funestus* reappeared as well, in KwaZulu-Natal, and in a form resistant to pyrethroids. In early 2000, DDT was reintroduced, in an indoor spraying program. (This is now a standard way of using DDT for mosquito control; the pesticide is usually applied only to walls, where mosquitoes alight to rest.) By the middle of the year, the number of infections had dropped by half.

Initially, the spraying program was criticized, but what reasonable alternative was there? This is said to be the African predicament, and yet the South African situation is hardly representative of sub-Saharan Africa as a whole.

Malaria is considered endemic in 105 countries throughout the tropics and warm temperate zones, but by far the worst region for the disease is sub-Saharan Africa. The deadliest of the four parasite species, *Plasmodium falciparum,* is widespread throughout this region, as is one of the world's most effective malaria vectors, *Anopheles gambiae.* Nearly half the population of sub-Saharan Africa is at risk of infection, and in much of eastern and central Africa, and pockets of west Africa, it would be difficult to find anyone who has not been exposed to the parasites. Some 90 percent of the world's malaria infections and deaths occur in sub-Saharan Africa, and the disease now accounts for 30 percent of African childhood mortality. It is true that malaria is a grave problem in many parts of the world, but the African experience is misery on a very different order of magnitude. The average Tanzanian suffers more infective bites each *night* than the average Thai or Vietnamese does in a year.

As a broad social burden, malaria is thought to cost Africa between $3 billion and $12 billion annually. According to one economic analysis, if the disease had been eradicated in 1965, Africa's GDP would now be 35 percent higher than it currently is. Africa was also the gaping hole in the global eradication program: the WHO planners thought there was little they could do on the continent and limited efforts to Ethiopia, Zimbabwe, and South Africa, where eradication was thought to be feasible.

But even though the campaign largely passed Africa by, DDT has not. Many African countries have used DDT for mosquito control in indoor spraying programs, but the primary use of DDT on the continent has been as an agricultural insecticide. Consequently, in parts of west Africa especially, DDT resistance is now widespread in *A. gambiae.* But even if *A. gambiae* were not resistant, a full-bore campaign to suppress it would probably accomplish little, because this mosquito is so efficient at transmitting malaria. Unlike most *Anopheles* species, *A. gambiae* specializes in human blood, so even a small population would keep the disease in circulation. One way to get a sense for this problem is to consider the "transmission index"—the threshold number of mosquito bites necessary to perpetuate the disease. In Africa, the index overall is 1 bite per person per month. That's all that's necessary to keep malaria in circulation. In India, by comparison, the TI is 10 bites per person per month.

And yet Africa is not a lost cause—it's simply that the key to progress does not lie in the general suppression of mosquito populations. Instead of spraying, the most promising African programs rely primarily on "bednets"—mosquito netting that is treated with an insecticide, usually a pyrethroid, and that is suspended over a person's bed. Bednets can't eliminate malaria, but they can "deflect" much of the burden. Because *Anopheles* species generally feed in the evening and at night, a bednet can radically reduce the number of infective bites a person receives. Such a person would probably still be infected from time to time, but would usually be able to lead a normal life.

In effect, therefore, bednets can substantially reduce the disease. Trials in the use of bednets for children have shown a decline in malaria-induced mortality by 25 to 40 percent. Infection levels and the incidence of severe anemia also declined. In Kenya, a recent study has shown that pregnant women who use bednets tend to give birth to healthier babies. In parts of Chad, Mali, Burkina Faso, and Senegal, bednets are becoming standard household items. In the tiny west African nation of The Gambia, somewhere between 50 and 80 percent of the population has bednets.

Bednets are hardly a panacea. They have to be used properly and retreated with insecticide occasionally. And there is still the problem of insecticide resistance, although the nets themselves are hardly likely to be the main cause of it. (Pyrethroids are used extensively in agriculture as well.) Nevertheless, bednets can help transform malaria from a chronic disaster to a manageable public health problem—something a healthcare system can cope with.

So it's unfortunate that in much of central and southern Africa, the nets are a rarity. It's even more unfortunate that, in 28 African countries, they're taxed or subject to import tariffs. Most of the people in these countries would have trouble paying for a net even without the tax. This problem was addressed in the May 2000 "Abuja Declaration," a summit agreement on infectious diseases signed by 44 African countries. The Declaration included a pledge to do away with "malaria taxes." At last count, 13 countries have actually acted on the pledge, although in some cases only by reducing rather than eliminating the taxes. Since the Declaration was signed, an estimated 2 to 5 million Africans have died from malaria.

This failure to follow through with the Abuja Declaration casts the interest in DDT in a rather poor light. Of the 31 POPs treaty signatories that have reserved the right to use DDT, 21 are in Africa. Of those 21, 10 are apparently still taxing or imposing tariffs on bednets. (Among the African countries that have *not* signed the POPs treaty, some are almost certainly both using DDT and taxing bednets, but the exact number is difficult to ascertain because the status of DDT use is not always clear.) It is true that a case can be made for the use of DDT in situations like the one in South Africa in 1999—an infrequent flare-up in a context that lends itself to control. But the routine use of DDT against malaria is an exercise in toxic futility, especially when it's pursued at the expense of a superior and far more benign technology.

## Learning to Live with the Mosquitoes

A group of French researchers recently announced some very encouraging results for a new anti-malarial drug known as G25. The drug was given to infected aotus monkeys, and it appears to have cleared the parasites from their systems. Although extensive testing will be necessary before it is known whether the drug can be safely given to people, these results have raised the hope of a cure for the disease.

Of course, it would be wonderful if G25, or some other new drug, lives up to that promise. But even in the absence of a cure, there are opportunities for progress that may one day make the current incidence of malaria look like some dark age horror. Many of these opportunities have been incorporated into an initiative that began in 1998, called the Roll Back Malaria (RBM) campaign, a collaborative effort between WHO, the World Bank, UNICEF, and the UNDP [United Nations Development Programme]. In contrast to the earlier WHO eradication program, RBM grew out of joint efforts between WHO and various African governments specifically to address African malaria. RBM focuses on household- and community-level intervention and it emphasizes apparently modest changes that could yield major progress. Below are four "operating principles" that are, in one way or another, implicit in RBM or likely to reinforce its progress.

1. Do away with all taxes and tariffs on bednets, on pesticides intended for treating bednets, and on antimalarial drugs. Failure to act on this front

certainly undercuts claims for the necessity of DDT; it may also undercut claims for antimalaria foreign aid.

2. Emphasize appropriate technologies. Where, for example, the need for mud to replaster walls is creating lots of pothole sized cavities near houses—cavities that fill with water and then with mosquito larvae—it makes more sense to help people improve their housing maintenance than it does to set up a program for squirting pesticide into every pothole. To be "appropriate," a technology has to be both affordable and culturally acceptable. Improving home maintenance should pass this test; so should bednets. And of course there are many other possibilities. In Kenya, for example, a research institution called the International Center for Insect Physiology and Ecology has identified at least a dozen native east African plants that repel *Anopheles gambiae* in lab tests. Some of these plants could be important additions to household gardens.

3. Use existing networks whenever possible, instead of building new ones. In Tanzania, for example, an established healthcare program (UNICEF's Integrated Management of Childhood Illness Program) now dispenses antimalarial drugs—and instruction on how to use them. The UNICEF program was already operating, so it was simple and cheap to add the malaria component. Reported instances of severe malaria and anemia in infants have declined, apparently as a result. In Zambia, the government is planning to use health and prenatal clinics as the network for a coupon system that subsidizes bednets for the poor. Qualifying patients would pick up coupons at the clinics and redeem them at stores for the nets.

4. Assume that sound policy will involve action on many fronts. Malaria is not just a health problem—it's a social problem, an economic problem, an environmental problem, an agricultural problem, an urban planning problem. Health officials alone cannot possibly just make it go away. When the disease flares up, there is a strong and understandable temptation to strap on the spray equipment and douse the mosquitoes. But if this approach actually worked, we wouldn't be in this situation today. Arguably the biggest opportunity for progress against the disease lies, not in our capacity for chemical innovation, but in our capacity for *organizational innovation*—in our ability to build an awareness of the threat across a broad range of policy activities. For example, when government officials are considering loans to irrigation projects, they should be asking: has the potential for malaria been addressed? When foreign donors are designing antipoverty programs, they should be asking: do people need bednets? Routine inquiries of this sort could go a vast distance to reducing the disease.

Where is the DDT in all of this? There isn't any, and that's the point. We now have half a century of evidence that routine use of DDT simply will not prevail against the mosquitoes. Most countries have already absorbed this lesson, and banned the chemical or relegated it to emergency only status. Now the RBM campaign and associated efforts are showing that the frequency and intensity of those emergencies can be reduced through systematic attention to the chronic aspects of the disease. There is less and less justification for DDT, and the futility of using it as a matter of routine is becoming increasingly apparent: in order to control a disease, why should we poison our soils, our waters, and ourselves?

**Donald R. Roberts** **NO**

# Statement before the U.S. Senate Committee on Environment & Public Works, Hearing on the Role of Science in Environmental Policy-Making

Thank you, Chairman Inhofe, and distinguished members of the Committee on Environment and Public Works, for the opportunity to present my views on the misuse of science in public policy. My testimony focuses on misrepresentations of science during decades of environmental campaigning against DDT.

Before discussing how and why DDT science has been misrepresented, you first must understand why this misrepresentation has not helped, but rather harmed, millions of people every year all over the world. Specifically you need to understand why the misrepresentation of DDT science has been and continues to be deadly. By way of explanation, I will tell you something of my experience.

I conducted malaria research in the Amazon Basin in the 1970s. My Brazilian colleague—who is now the Secretary of Health for Amazonas State—and I worked out of Manaus, the capitol of Amazonas State. From Manaus we traveled two days to a study site where we had sufficient numbers of cases for epidemiological studies. There were no cases in Manaus, or anywhere near Manaus. For years before my time there and for years thereafter, there were essentially no cases of malaria in Manaus. However, in the late 1980s, environmentalists and international guidelines forced Brazilians to reduce and then stop spraying small amounts of DDT inside houses for malaria control. As a result, in 2002 and 2003 there were over 100,000 malaria cases in Manaus alone.

Brazil does not stand as the single example of this phenomenon. A similar pattern of declining use of DDT and reemerging malaria occurs in other countries as well, Peru for example. Similar resurgences of malaria have occurred in rural communities, villages, towns, cities, and countries around the world. As illustrated by the return of malaria in Russia, South Korea, urban areas of the Amazon Basin, and increasing frequencies of outbreaks in the United States, our malaria problems are growing worse. Today there are 1 to 2 million malaria deaths each year and hundreds of millions of cases. The

U.S. Senate Committee on Environment & Public Works Hearing on the Tole of Science in Environmental Policy-Making, September 28, 2005.

poorest of the world's people are at greatest risk. Of these, children and pregnant women are the ones most likely to die.

We have long known about DDT's effectiveness in curbing insect-borne disease. Othmar Zeidler, a German chemistry student, first synthesized DDT in 1874. Over sixty years later in Switzerland, Paul Müller discovered the insecticidal property of DDT. Allied forces used DDT during WWII, and the new insecticide gained fame in 1943 by successfully stopping an epidemic of typhus in Naples, an unprecedented achievement. By the end of the war, British, Italian, and American scientists had also demonstrated the effectiveness of DDT in controlling malaria-carrying mosquitoes. DDT's proven efficacy against insect-borne diseases, diseases that had long reigned unchecked throughout the world, won Müller the Nobel Prize for Medicine in 1948. After WWII, the United States conducted a National Malaria Eradication Program, commencing operations on July 1, 1947. The spraying of DDT on internal walls of rural homes in malaria endemic counties was a key component of the program. By the end of 1949, the program had sprayed over 4,650,000 houses. This spraying broke the cycle of malaria transmission, and in 1949 the United States was declared free of malaria as a significant public health problem. Other countries had already adopted DDT to eradicate or control malaria, because wherever malaria control programs sprayed DDT on house walls, the malaria rates dropped precipitously. The effectiveness of DDT stimulated some countries to create, for the first time, a national malaria control program. Countries with pre-existing programs expanded them to accommodate the spraying of houses in rural areas with DDT. Those program expansions highlight what DDT offered then, and still offers now, to the malaria endemic countries. As a 1945 U.S. Public Health Service manual explained about the control of malaria: "Drainage and larviciding are the methods of choice in towns of 2,500 or more people. But malaria is a rural disease. Heretofore there has been no economically feasible method of carrying malaria control to the individual tenant farmer or sharecropper. Now, for the first time, a method is available—the application of DDT residual spray to walls and ceilings of homes." Health workers in the United States were not the only ones to recognize the particular value of DDT. The head of malaria control in Brazil characterized the changes that DDT offered in the following statement: "Until 1945–1946, preventive methods employed against malaria in Brazil, as in the rest of the world, were generally directed against the aquatic phases of the vectors (draining, larvicides, destruction of bromeliads, etc. . . .). These methods, however, were only applied in the principal cities of each state and the only measure available for rural populations exposed to malaria was free distribution of specific drugs."

DDT was a new, effective, and exciting weapon in the battle against malaria. It was cheap, easy to apply, long-lasting once sprayed on house walls, and safe for humans. Wherever and whenever malaria control programs sprayed it on house walls, they achieved rapid and large reductions in malaria rates. Just as there was a rush to quickly make use of DDT to control disease, there was also a rush to judge how DDT actually functioned to control malaria. That rush to judgment turned out to be a disaster. At the heart of the debate—to the extent there was a debate—was a broadly accepted model that established a mathematical framework for using DDT to kill mosquitoes and eradicate malaria. Instead of studying real data to see how DDT actually worked in controlling malaria, some scientists settled upon what they thought was a logical conclusion: DDT worked solely by killing mosquitoes. This conclusion was based on their belief in the model. Scientists who showed that DDT did not function by killing mosquitoes were ignored. Broad acceptance of the mathematical model led to strong convictions about DDT's toxic actions. Since they were convinced that DDT worked only by killing mosquitoes, malaria control specialists became very alarmed when a mosquito was reported to be resistant to DDT's toxic actions. As a result of concern about DDT resistance, officials decided to make rapid use of DDT before problems of resistance could eliminate their option to use DDT to eradicate malaria. This decision led to creation of the global malaria eradication program. The active years of the global malaria eradication program were from 1959 to 1969. Before, during, and after the many years of this program, malaria workers and researchers carried out their responsibilities to conduct studies and report their research. Through those studies, they commonly found that DDT was functioning in ways other than by killing mosquitoes. In essence, they found that DDT was functioning through mechanisms of repellency and irritancy. Eventually, as people forgot early observations of DDT's repellent actions, some erroneously interpreted new findings of repellent actions as the mosquitoes' adaptation to avoid DDT toxicity, even coining a term, "behavioral resistance," to explain what they saw. This new term accommodated their view that toxicity was DDT's primary mode of action and categorized behavioral responses of mosquitoes as mere adaptations to toxic affects. However this interpretation depended upon a highly selective use of scientific data. The truth is that toxicity is not DDT's primary mode of action when sprayed on house walls. Throughout the history of DDT use in malaria control programs there has always been clear and persuasive data that DDT functioned primarily as a spatial repellent. Today we know that there is no insecticide recommended for malaria control that rivals, much less equals, DDT's spatial repellent actions, or that is as long-acting, as cheap, as easy to apply, as safe for human exposure, or as efficacious in the control of malaria as DDT. . . . The 30 years of data from control programs of the Americas plotted . . . illustrate just how effective DDT is in malaria control. The period 1960s through 1979 displays a pattern of malaria controlled through house spraying. In 1979 the World Health Organization (WHO) changed its strategy for malaria control, switching emphasis from spraying houses to case detection and treatment. In other words, the WHO changed emphasis from malaria prevention to malaria treatment. Countries complied with WHO guidelines and started to dismantle their spray programs over the next several years. . . .

I find it amazing that many who oppose the use of DDT describe its earlier use as a failure. Our own citizens who suffered under the burden of malaria, especially in the rural south, would hardly describe it thus.

Malaria was a serious problem in the United States and for some localities, such as Dunklin County, Missouri, it was a very serious problem indeed. For four counties in Missouri, the average malaria mortality from 1910 to 1914 was 168.8 per 100,000 population. For Dunklin County, it was 296.7 per 100,000, a rate almost equal to malaria deaths in Venezuela and actually greater than the mortality rate for Freetown, Sierra Leone. Other localities in other states were equally as malarious. Growing wealth and improved living conditions were gradually reducing malaria rates, but cases resurged during WWII. The advent of DDT, however, quickly eradicated malaria from the United States.

DDT routed malaria from many other countries as well. The Europeans who were freed of malaria would hardly describe its use as a failure. After DDT was introduced to malaria control in Sri Lanka (then Ceylon), the number of malaria cases fell from 2.8 million in 1946 to just 110 in 1961. Similar spectacular decreases in malaria cases and deaths were seen in all the regions that began to use DDT. The newly formed Republic of China (Taiwan) adopted DDT use in malaria control shortly after World War II. In 1945 there were over 1 million cases of malaria on the island. By 1969 there were only 9 cases and shortly thereafter the disease was eradicated from the island and remains so to this day. Some countries were less fortunate. South Korea used DDT to eradicate malaria, but without house spray programs, malaria has returned across the demilitarized zone with North Korea. As DDT was eliminated and control programs reduced, malaria has returned to other countries such as Russia and Argentina. Small outbreaks of malaria are even beginning to appear more frequently in the United States.

These observations have been offered in testimony to document first that there were fundamental misunderstandings about how DDT functioned to exert control over malaria. Second, that regardless of systematic misunderstandings on the part of those who had influence over malaria control strategies and policies, there was an enduring understanding that DDT was the most cost-effective compound yet discovered for protecting poor rural populations from insect-borne diseases like malaria, dengue, yellow fever, and leishmaniasis. I want to emphasize that misunderstanding the mode of DDT action did not lead to the wholesale abandonment of DDT. It took an entirely new dimension in the misuse of science to bring us to the current humanitarian disaster represented by DDT elimination.

The misuse of science to which I refer has found fullest expression in the collection of movements within the environmental movement that seek to stop production and use of specific man-made chemicals. Operatives within these movements employ particular strategies to achieve their objectives. By characterizing and understanding the strategies these operatives use, we can identify their impact in the scientific literature or in the popular press.

The first strategy is to develop and then distribute as widely as possible a broad list of claims of chemical harm. This is a sound strategy because individual scientists can seldom rebut the scientific foundations of multiple and diverse claims. Scientists generally develop expertise in a single, narrow field and are disinclined to engage issues beyond their area of expertise. Even if an authoritative rebuttal of one claim occurs, the other claims still progress. A broad list of claims also allows operatives to tailor platforms for constituencies, advancing one set of claims with one constituency and a different combination for another. Clever though this technique is, a list of multiple claims of harm is hardly sufficient to achieve the objective of a ban. The second strategy then is to mount an argument that the chemical is not needed and propose that alternative chemicals or methods can be used instead. The third strategy is to predict that grave harm will occur if the chemical continues to be used.

The success of Rachel Carson's *Silent Spring* serves as a model for this tricky triad. In *Silent Spring*, Rachel Carson used all three strategies on her primary target, DDT. She described a very large list of potential adverse effects of insecticides, DDT in particular. She argued that insecticides were not really needed and that the use of insecticides produces insects that are insecticide resistant, which only exacerbates the insect control problems. She predicted scary scenarios of severe harm with continued use of DDT and other insecticides. Many have written rebuttals to Rachel Carson and others who have, without scientific justification, broadcast long lists of potential harms of insecticides. . . .

[T]ime and science have discredited most of Carson's claims. Rachel Carson's descriptions of inappropriate uses of insecticides that harmed wildlife are more plausible. However, harm from an inappropriate use does not meet the requirements of anti-pesticide activists. They can hardly lobby for eliminating a chemical because someone used it wrongly. No, success requires that even the proper use of an insecticide will cause a large and systematic adverse effect. However, the proper uses of DDT yield no large and systematic adverse effects. Absent such adverse actions, the activists must then rely on claims about insidious effects, particularly insidious effects that scientists will find difficult to prove one way or the other and that activists can use to predict a future catastrophe.

Rachel Carson relied heavily on possible insidious chemical actions to alarm and frighten the public. Many of those who joined her campaign to ban DDT and other insecticides made extensive use of claims of insidious effects. These claims were amplified by the popular press and became part of the public perception about modern uses of chemicals. For example, four well-publicized claims about DDT were:

1. DDT will cause the obliteration of higher trophic levels. If not obliterated, populations will undergo reproductive failure. Authors of this claim speculated that, even if the use of DDT were stopped, systematic and ongoing obliterations would still occur.
2. DDT causes the death of algae. This report led to speculations that use of DDT could result in global depletion of oxygen.
3. DDT pushed the Bermuda petrel to the verge of extinction and that full extinction might happen by 1978.
4. DDT was a cause of premature births in California sea lions.

*Science* magazine, the most prestigious science journal in the United States, published these and other phantasmagorical allegations and/or predictions of DDT harm. Nonetheless, history has shown that each and every one of these claims and predictions were false.

1. The obliteration of higher trophic levels did not occur; no species became extinct; and levels of DDT in all living organisms declined

precipitously after DDT was de-listed for use in agriculture. How could the prediction have been so wrong? Perhaps it was so wrong because the paper touting this view used a predictive model based on an assumption of no DDT degradation. This was a startling assertion even at the time as *Science* and other journals had previously published papers that showed DDT was ubiquitously degraded in the environment and in living creatures. It was even more startling that *Science* published a paper that flew so comprehensively in the face of previous data and analysis.

2. DDT's action against algae reportedly occurred at concentrations of 500 parts per billion. But DDT cannot reach concentrations in water higher than about 1.2 parts per billion, the saturation point of DDT in water.
3. Data on the Bermuda petrel did not show a cause-and-effect relationship between low numbers of birds and DDT concentrations. DDT had no affect on population numbers, for populations increased before DDT was de-listed for use in agriculture and after DDT was de-listed as well.
4. Data gathered in subsequent years showed that "despite relatively high concentrations [of DDT], no evidence that population growth or the health of individual California sea lions have been compromised. The population has increased throughout the century, including the period when DDT was being manufactured, used, and its wastes discharged off southern California."

If time and science have refuted all these catastrophic predictions, why do many scientists and the public not know these predictions were false? In part, we do not know the predictions were false because the refutations of such claims rarely appear in the literature.

When scientists hear the kinds of claims described above, they initiate research to confirm or refute the claims. After Charles Wurster published his claim that DDT kills algae and impacts photosynthesis, I initiated research on planktonic algae to quantify DDT's effects. From 1968–1969, I spent a year of honest and demanding research effort to discover that not enough DDT would even go into solution for a measurable adverse effect on planktonic algae. In essence, I conducted a confirmatory study that failed to confirm an expected result. I had negative data, and journals rarely accept negative data for publication. My year was practically wasted. Without a doubt, hundreds of other scientists around the world have conducted similar studies and obtained negative results, and they too were unable to publish their experimental findings. Much in the environmental science literature during the last 20–30 years indicates that an enormous research effort went into proving specific insidious effects of DDT and other insecticides. Sadly, the true magnitude of such efforts will never be known because while the positive results of research find their way into the scientific literature, the negative results rarely do. Research on insidious actions that produce negative results all too often ends up only in laboratory and field notebooks and is forgotten. For this reason, I place considerable weight on a published confirmatory study that fails to confirm an expected result.

The use of the tricky triad continues. A . . . recent paper . . . published in *The Lancet* illustrates the triad's modern application. Two scientists at the National Institute of Environmental Health Sciences, Walter Rogan and Aimin Chen, wrote this paper, entitled "Health risks and benefits of bis(4-chlorophenyl)-1,1, 1-trichloroethane (DDT)." It is interesting to see how this single paper spins all three strategies that gained prominence in Rachel Carson's *Silent Spring*.

The journal *Emerging Infectious Diseases* had already published a slim version of this paper, which international colleagues and I promptly rebutted. The authors then filled in some parts, added to the claims of harm, and republished the paper in the British journal, *The Lancet*. To get the paper accepted by editors, the authors described studies that support (positive results) as well as studies that do not support (negative results) each claim. Complying with strategy number 1 of the triad, Rogan and Chen produce a long list of possible harms, including the charge that DDT causes cancer in nonhuman primates. The literature reference for Rogan and Chen's claim that DDT causes cancer in nonhuman primates was a paper by Takayama et al. Takayama and coauthors actually concluded from their research on the carcinogenic effect of DDT in nonhuman primates that "the two cases involving malignant tumors of different types are inconclusive with respect to a carcinogenic effect of DDT in nonhuman primates." Clearly, the people who made the link of DDT with cancer were not the scientists who actually conducted the research.

The authors enacted strategy number two of the triad by conducting a superficial review of the role of DDT in malaria control with the goal of discrediting DDT's value in modern malaria control programs. The authors admitted that DDT had been very effective in the past, but then argued that malaria control programs no longer needed it and should use alternative methods of control. Their use of the second strategy reveals, in my opinion, the greatest danger of granting authority to anti-pesticide activists and their writings. As *The Lancet* paper reveals, the NIEHS scientists assert great authority over the topic of DDT, yet they assume no responsibility for the harm that might result from their erroneous conclusions. After many malaria control specialists have expressed the necessity for DDT in malaria control, it is possible for Rogan and Chen to conclude that DDT is not necessary in malaria control only if they have no sense of responsibility for levels of disease and death that will occur if DDT is not used.

Rogan and Chen also employ the third strategy of environmentalism. Their list of potential harms caused by DDT includes toxic effects, neurobehavior effects, cancers, decrements in various facets of reproductive health, decrements in infant and child development, and immunology and DNA damage. After providing balanced coverage of diverse claims of harm, the authors had no option but to conclude they could not prove that DDT caused harm. However, they then promptly negated this honest conclusion by asserting that if DDT is used for malaria control, then great harm might occur. So, in an amazing turn, they conclude they cannot prove DDT causes harm, but still predict severe harm if it is used.

Rogan and Chen end their paper with a call for more research. One could conclude that the intent of the whole paper is merely to lobby for research to better define DDT harm, and what's the harm in that? Surely increasing knowledge is a fine goal. However, if you look at the specific issue of the relative need for research, you will see that the harm of this technique is great. Millions of children and pregnant women die from malaria every year, and the disease sickens hundreds of millions more. This is an indisputable fact: impoverished people engage in real life and death struggles every day with malaria. This also is a fact: not one death or illness can be attributed to an environmental exposure to DDT. Yet, a National Library of Medicine literature search on DDT reveals over 1,300 published papers from the year 2000 to the present, almost all in the environmental literature and many on potential adverse effects of DDT. A search on malaria and DDT reveals only 159 papers. DDT is a spatial repellent and hardly an insecticide at all, but a search on DDT and repellents will reveal only 7 papers. Is this not an egregiously disproportionate research emphasis on non-sources of harm compared to the enormous harm of malaria? Does not this inequity contribute to the continued suffering of those who struggle with malaria? Is it possibly even more than an inequity? Is it not an active wrong?

Public health officials and scientists should not be silent about enormous investments into the research of theoretical risks while millions die of preventable diseases. We should seriously consider our motivations in apportioning research money as we do. For consider this: the U.S. used DDT to eradicate malaria. After malaria disappeared as an endemic disease in the United States, we became richer. We built better and more enclosed houses. We screened our windows and doors. We air conditioned our homes. We also developed an immense arsenal of mosquito control tools and chemicals. Today, when we have a risk of mosquito borne disease, we can bring this arsenal to bear and quickly eliminate risks. And, as illustrated by aerial spray missions in the aftermath of hurricane Katrina, we can afford to do so. Yet, our modern and very expensive chemicals are not what protect us from introductions of the old diseases. Our arsenal responds to the threat; it does not prevent the appearance of old diseases in our midst. What protects us is our enclosed, screened, air-conditioned housing, the physical representation of our wealth. Our wealth is the factor that stops dengue at the border with Mexico, not our arsenal of new chemicals. Stopping mosquitoes from entering and biting us inside our homes is critical in the prevention of malaria and many other insect-borne diseases. This is what DDT does for poor people in poor countries. It stops large proportions of mosquitoes from entering houses. It is, in fact, a form of chemical screening, and until these people can afford physical screening or it is provided for them, this is the only kind of screening they have.

DDT is a protective tool that has been taken away from countries around the world, mostly due to governments acceding to the whims of the anti-pesticide wing of environmentalism, but it is not only the anti-pesticide wing that lobbies against DDT. The activists have a sympathetic lobbying ally in the pesticide industry. As evidence of insecticide industry working to stop countries from using DDT, I am attaching an email message dated 23rd September and authored by a Bayer official. . . . The Bayer official states

> "[I speak] Not only as the responsible manager for the vector control business in Bayer, being the market leader in vector control and pointing out by that we know what we are talking about and have decades of experiences in the evolution of this very particular market. [but] Also as one of the private sector representatives in the RBM Partnership Board and being confronted with that discussion about DDT in the various WHO, RBM et al circles. So you can take it as a view from the field, from the operational commercial level—but our companies [sic] point of view. I know that all of my colleagues from other primary manufacturers and internationally operating companies are sharing my view."

The official goes on to say that

> "DDT use is for us a commercial threat (which is clear, but it is not that dramatical because of limited use), it is mainly a public image threat."

However the most damning part of this message was the statement that

> "we fully support EU to ban imports of agricultural products coming from countries using DDT"

[There is] . . . clear evidence of international and developed country pressures to stop poor countries from using DDT to control malaria. This message also shows the complicity of the insecticide industry in those internationally orchestrated efforts.

Pressures to eliminate spray programs, and DDT in particular, are wrong. I say this not based on some projection of what might theoretically happen in the future according to some model, or some projection of theoretical harms, I say this based firmly on what has already occurred. The track record of the anti-pesticide lobby is well documented, the pressures on developing countries to abandon their spray programs are well documented, and the struggles of developing countries to maintain their programs or restart their uses of DDT for malaria control are well documented. The tragic results of pressures against the use of DDT, in terms of increasing disease and death, are quantified and well documented. How long will scientists, public health officials, the voting public, and the politicians who lead us continue policies, regulations and funding that have led us to the current state of a global humanitarian disaster? How long will support continue for policies and programs that favor phantoms over facts?

# POSTSCRIPT

## Should DDT Be Banned Worldwide?

Professor Roberts is not alone in his disapproval of the efforts to halt the use of DDT. Alexander Gourevitch, "Better Living Through Chemistry," *Washington Monthly* (March 2003), comes close to accusing environmentalists of condemning DDT on the basis of politics or ideology rather than of science. Angela Logomasini comes even closer in "Chemical Warfare: Ideological Environmentalism's Quixotic Campaign Against Synthetic Chemicals," in Ronald Bailey, ed., *Global Warming and Other Eco-Myths: How the Environmental Movement Uses False Science to Scare Us to Death* (Prima, 2002). Her admission that public health demands have softened some environmentalists' resistance to the use of DDT points to a basic truth about environmental debates: over and over again, they come down to what we should do first: Should we meet human needs regardless of whether or not species die and air and water are contaminated? Or should we protect species, air, water, and other aspects of the environment even if some human needs must go unmet? What if this means endangering the lives of children? In the debate over DDT, the human needs are clear, for insect-borne diseases have killed and continue to kill a great many people. Yet the environmental needs are also clear. The question is one of choosing priorities and balancing risks. See John Danley, "Balancing Risks: Mosquitoes, Malaria, Morality, and DDT," *Business and Society Review* (Spring 2002). It is worth noting that John Beard, "DDT and Human Health," *Science of the Total Environment* (February 2006), finds the evidence for the ill effects of DDT more convincing and says that it is still too early to say it does not contribute to human disease.

The serious nature of the malaria threat is well covered in Michael Finkel, "Bedlam in the Blood: Malaria," *National Geographic* (July 2007). Mosquitoes can be controlled in various ways: Swamps can be drained (which carries its own environmental price), and other breeding opportunities can be eliminated. Fish can be introduced to eat mosquito larvae. And mosquito nets can be used to keep the insects away from people. But these (and other) alternatives do not mean that there does not remain a place for chemical pesticides. In "Pesticides and Public Health: Integrated Methods of Mosquito Management," *Emerging Infectious Diseases* (January–February 2001), Robert I. Rose, an arthropod biotechnologist with the Animal and Plant Health Inspection Service of the U.S. Department of Agriculture, says, "Pesticides have a role in public health as part of sustainable integrated mosquito management. Other components of such management include surveillance, source reduction or prevention, biological control, repellents, traps, and pesticide-resistance management." "The most effective programs today rely on a range of tools," says McGinn in "Combating Malaria," *State of the World 2003* (W. W. Norton, 2003). Still, some countries see

DDT as essential. Today some countries see DDT as essential. See Tina Rosenberg, "What the World Needs Now Is DDT," *New York Times Magazine* (April 11, 2004). However when the World Health Organization (WHO) endorsed the use of DDT to fight malaria, there was immediate outcry; see Allan Schapira, "DDT: A Polluted Debate in Malaria Control," *Lancet* (December 16, 2006). So far, however, the outcry seems fruitless; see Apoorva Mandavilli, "DDT Is Back," *Discover* (January 2007).

It has proven difficult to find effective, affordable drugs against malaria; see Ann M. Thayer, "Fighting Malaria," *Chemical and Engineering News* (October 24, 2005), and Claire Panosian Dunavan, "Tackling Malaria," *Scientific American* (December 2005). A great deal of effort has gone into developing vaccines against malaria, but the parasite has demonstrated a persistent talent for evading all attempts to arm the immune system against it. See Z. H. Reed, M. Friede, and M. P. Kieny, "Malaria Vaccine Development: Progress and Challenges," *Current Molecular Medicine* (March 2006). A newer approach is to develop genetically engineered (transgenic) mosquitoes that either cannot support the malaria parasite or cannot infect humans with it; see George K. Christophides, "Transgenic Mosquitoes and Malarial Transmission," *Cellular Microbiology* (March 2005), and M. J. Friedrich, "Malaria Researchers Target Mosquitoes," *JAMA: Journal of the American Medical Association* (May 23, 2007). On the other hand, Nicholas J. White, "Malaria—Time to Act," *New England Journal of Medicine* (November 9, 2006), argues that rather than wait for a perfect solution, we should recognize that present tools—including DDT—are effective enough now that there is no excuse to avoid using them.

It is worth stressing that malaria is only one of several mosquito-borne diseases that pose threats to public health. Two others are yellow fever and dengue. A recent arrival to the United States is West Nile virus, which mosquitoes can transfer from birds to humans. However, West Nile virus is far less fatal than malaria, yellow fever, or dengue, and a vaccine is in development. See Dwight G. Smith, "A New Disease in the New World," *The World & I* (February 2002) and Michelle Mueller, "The Buzz on West Nile Virus," *Current Health 2* (April/May 2002). But "West Nile Virus Still a Threat," says *Clinical Infectious Diseases* (April 15, 2006).

It is also worth stressing that global warming means climate changes that may increase the geographic range of disease-carrying mosquitoes. Many climate researchers are concerned that malaria, yellow fever, and other now mostly tropical and subtropical diseases may return to temperate-zone nations and even spread into areas where they have never been known. See Atul A. Khasnis and Mary D. Nettleman, "Global Warming and Infectious Disease," *Archives of Medical Research* (November 2005). Sarah DeWeerdt, "Climate Change, Coming Home," *World Watch* (May/June 2007), notes that "by 2020 human-triggered climate change could kill 300,000 people worldwide every year," in large part because of the spread of diseases such as malaria.

# Glossary

## A

**abiotic** Nonliving.

**abundance** The number or amount of something.

**acid precipitation** Acidic rain, snow, or dry particles deposited from the air due to increased acids released by anthropogenic or natural resources.

**acids** Substances that release hydrogen ions (protons) in water.

**active learner** Someone who understands and remembers best by doing things physically.

**active solar system** A mechanical system that actively collects, concentrates, and stores solar energy.

**acute effects** Sudden, severe effects.

**acute poverty** Insufficient income or access to resources needed to provide the basic necessities for life such as food, shelter, sanitation, clean water, medical care, and education.

**adaptation** The acquisition of traits that allow a species to survive and thrive in its environment.

**adaptive management** A management plan designed from the outset to "learn by doing," and to actively test hypotheses and adjust treatments as new information becomes available.

**administrative courts** Courts that hear enforcement cases for agencies or consider appeals to agency rules.

**administrative law** Executive orders, administrative rules and regulations, and enforcement decisions by administrative agencies and special administrative courts.

**aerobic** Living or occurring only in the presence of oxygen.

**aerosols** Minute particles or liquid droplets suspended in the air.

**aesthetic degradation** Changes in environmental quality that offend our aesthetic senses.

**agroecology** Practices, such as low-input farming, community-supported agriculture, and agroecology that use the principles of ecology to show how we can live more harmoniously with the land while still producing the food and fiber we need.

**albedo** A description of a surface's reflective properties.

**allergens** Substances that activate the immune system.

**allopatric speciation** Species that arise from a common ancestor due to geographic barriers that cause reproductive isolation.

**ambient air** The air immediately around us.

**amino acid** An organic compound containing an amino group and a carboxyl group; amino acids are the units or building blocks that make peptide and protein molecules.

**amorphous silicon collectors** Photovoltaic cells that collect solar energy and convert it to electricity using noncrystalline (randomly arranged) thin films of silicon.

**anaerobic respiration** The incomplete intracellular breakdown of sugar or other organic compounds in the absence of oxygen that releases some energy and produces organic acids and/or alcohol.

**analytical thinking** Asks, how can I break this problem down into its constituent parts?

**anemia** Low levels of hemoglobin due to iron deficiency or lack of red blood cells.

**annual** A plant that lives for a single growing season.

**anthropocentric** The belief that humans hold a special place in nature; being centered primarily on humans and human affairs.

**antigens** Chemical compounds to which antibodies bind.

**appropriate technology** Technology that can be made at an affordable price by ordinary people using local materials to do useful work in ways that do the least possible harm to both human society and the environment.

**aquaculture** Growing aquatic species in net pens or tanks.

**aquifers** Porous, water-bearing layers of sand, gravel, and rock below the earth's surface; reservoirs for groundwater.

**arbitration** A formal process of dispute resolution in which there are stringent rules of evidence, cross-examination of witnesses, and a legally binding decision made by the arbitrator that all parties must obey.

**arithmetic scale** One that uses ordinary numbers as units in a linear sequence.

**artesian well** The result of a pressurized aquifer intersecting the surface or being penetrated by a pipe or conduit, from which water gushes without being pumped; also called a spring.

**asthma** A distressing disease characterized by shortness of breath, wheezing, and bronchial muscle spasms.

**atmospheric deposition** Sedimentation of solids, liquids, or gaseous materials from the air.

**atom** The smallest unit of matter that has the characteristics of an element; consists of three main types of subatomic particles: protons, neutrons, and electrons.

**atomic number** The characteristic number of protons per atom of an element. Used as an identifying attribute.

**autotroph** An organism that synthesizes food molecules from inorganic molecules by using an external energy source, such as light energy.

## B

**barrier islands** Low, narrow, sandy islands that form offshore from a coastline.

**bases** Substances that bond readily with hydrogen ions.

**BAT** *See* best available, economically achievable technology.

**Batesian mimicry** Evolution by one species to resemble the coloration, body shape, or behavior of another species that is protected from predators by a venomous stinger, bad taste, or some other defensive adaptation.

**benthic** The bottom of a sea or lake.

**best available, economically achievable technology (BAT)** The best pollution control available.

**best practicable control technology (BPT)** The best technology for pollution control available at reasonable cost and operable under normal conditions.

**beta particles** High-energy electrons released by radioactive decay.

**bill** A piece of legislation introduced in Congress and intended to become law.

**binomials** Two part names (genus and species, usually in Latin) invented by Carl Linneaus to show taxonomic relationships.

**bioaccumulation** The selective absorption and concentration of molecules by cells.

**biocentric preservation** A philosophy that emphasizes the fundamental right of living organisms to exist and to pursue their own goods.

**biocentrism** The belief that all creatures have rights and values; being centered on nature rather than humans.

**biochemical oxygen demand** A standard test of water pollution measured by the amount of dissolved oxygen consumed by aquatic organisms over a given period.

**biocide** A broad-spectrum poison that kills a wide range of organisms.

**biodegradable plastics** Plastics that can be decomposed by microorganisms.

**biodiversity** The genetic, species, and ecological diversity of the organisms in a given area.

**biodiversity hot spots** Areas with exceptionally high numbers of endemic species.

**biofuel** Fuels such as ethanol, methanol, or vegetable oils from crops.

**biogeochemical cycles** Movement of matter within or between ecosystems; caused by living organisms, geological forces, or chemical reactions. The cycling of nitrogen, carbon, sulfur, oxygen, phosphorus, and water are examples.

**biogeographical area** An entire self-contained natural ecosystem and its associated land, water, air, and wildlife resources.

**biological community** The populations of plants, animals, and microorganisms living and interacting in a certain area at a given time.

**biological controls** Use of natural predators, pathogens, or competitors to regulate pest populations.
**biological or biotic factors** Organisms and products of organisms that are part of the environment and potentially affect the life of other organisms.
**biological oxygen demand (BOD)** A standard test for measuring the amount of dissolved oxygen utilized by aquatic microorganisms.
**biological pests** Organisms that reduce the availability, quality, or value of resources useful to humans.
**biological resources** The earth's organisms.
**biomagnification** Increase in concentration of certain stable chemicals (for example, heavy metals or fat-soluble pesticides) in successively higher trophic levels of a food chain or web.
**biomass** The total mass or weight of all the living organisms in a given population or area.
**biomass fuel** Organic material produced by plants, animals, or microorganisms that can be burned directly as a heat source or converted into a gaseous or liquid fuel.
**biomass pyramid** A metaphor or diagram that explains the relationship between the amounts of biomass at different trophic levels.
**biome** A broad, regional type of ecosystem characterized by distinctive climate and soil conditions and a distinctive kind of biological community adapted to those conditions.
**bioremediation** Use of biological organisms to remove or detoxify pollutants from a contaminated area.
**biosphere** The zone of air, land, and water at the surface of the earth that is occupied by organisms.
**biosphere reserves** World heritage sites identified by the IUCN as worthy for national park or wildlife refuge status because of high biological diversity or unique ecological features.
**biota** All organisms in a given area.
**biotic** Pertaining to life; environmental factors created by living organisms.
**biotic potential** The maximum reproductive rate of an organism, given unlimited resources and ideal environmental conditions. Compare with environmental resistance.
**birth control** Any method used to reduce births, including abstinence, delayed marriage, contraception; devices or medication that prevent implantation of fertilized zygotes, and induced abortions.
**black lung disease** Inflammation and fibrosis caused by accumulation of coal dust in the lungs or airways. *See* respiratory fibrotic agents.
**blind experiments** Those in which those carrying out the experiment don't know until after data has been gathered and analyzed which was the experimental treatment and which was the control.
**blue revolution** New techniques of fish farming that may contribute as much to human nutrition as miracle cereal grains but also may create social and environmental problems.
**body burden** The sum total of all persistent toxins in our body that we accumulate from our air, water, diet, and surroundings.
**bog** An area of waterlogged soil that tends to be peaty; fed mainly by precipitation; low productivity; some bogs are acidic.
**boom-and-bust cycles** Population cycles characterized by repeated overshoot of the carrying capacity of the environment followed by population crashes.
**boreal forest** A broad band of mixed coniferous and deciduous trees that stretches across northern North America (and also Europe and Asia); its northernmost edge, the taiga, intergrades with the arctic tundra.
**BPT** *See* best practical control technology.
**breeder reactor** A nuclear reactor that produces fuel by bombarding isotopes of uranium and thorium with high-energy neutrons that convert inert atoms to fissionable ones.
**bronchitis** A persistent inflammation of bronchi and bronchioles (large and small airways in the lungs).
**brownfield development** Building on abandoned or reclaimed polluted industrial sites.
**brownfields** Abandoned or underused urban areas in which redevelopment is blocked by liability or financing issues related to toxic contamination.
**buffalo commons** A large open area proposed for the Great Plains in which wildlife and native people could live as they once did without interference by industrialized society.

# C

**cancer** Invasive, out-of-control cell growth that results in malignant tumors.
**cap-and-trade** An approach to controlling pollution by mandating upper limits (the cap), on how much each country, sector, or specific industry is allowed to emit. Companies that can cut pollution by more than they're required can sell the credit to other companies that have more difficulty meeting their mandated levels.
**capital** Any form of wealth, resources, or knowledge available for use in the production of more wealth.
**captive breeding** Raising plants or animals in zoos or other controlled conditions to produce stock for subsequent release into the wild.
**carbamates** Urethanes such as carbaryl, aldicarb, etc. that are used as pesticides.
**carbohydrate** An organic compound consisting of a ring or chain of carbon atoms with hydrogen and oxygen attached; examples are sugars, starches, cellulose, and glycogen.
**carbon cycle** The circulation and reutilization of carbon atoms, especially via the processes of photosynthesis and respiration.
**carbon neutral** A system or process that doesn't release more carbon to the atmosphere than it consumes.
**carbon management** Storing $CO_2$ or using it in ways that prevent its release into the air.
**carbon monoxide (CO)** Colorless, odorless, nonirritating but highly toxic gas produced by incomplete combustion of fuel, incineration of biomass or solid waste, or partially anaerobic decomposition of organic material.
**carbon sequestration** Storing carbon (usually in the form of $CO_2$) in geological formations or at the bottom of the ocean.
**carbon sink** Places of carbon accumulation, such as in large forests (organic compounds) or ocean sediments (calcium carbonate); carbon is thus removed from the carbon cycle for moderately long to very long periods of time.
**carbon source** Originating point of carbon that reenters the carbon cycle; cellular respiration and combustion.
**carcinogens** Substances that cause cancer.
**carnivores** Organisms that mainly prey upon animals.
**carrying capacity** The maximum number of individuals of any species that can be supported by a particular ecosystem on a long-term basis.
**case law** Precedents from both civil and criminal court cases.
**cash crops** Crops that are sold rather than consumed or bartered.
**catastrophic systems** Dynamic systems that jump abruptly from one seemingly steady state to another without any intermediate stages.
**cell** Minute biological compartments within which the processes of life are carried out.
**cellular respiration** The process in which a cell breaks down sugar or other organic compounds to release energy used for cellular work; may be anaerobic or aerobic, depending on the availability of oxygen.
**cellulosic** Material composed primarily of cellulose.
**chain reaction** A self-sustaining reaction in which the fission of nuclei produces subatomic particles that cause the fission of other nuclei.
**chaotic systems** Systems that exhibit variability, which may not be necessarily random, yet whose complex patterns are not discernible over a normal human time scale.
**chaparral** Thick, dense, thorny evergreen scrub found in Mediterranean climates.
**chemical bond** The force that holds atoms together in molecules and compounds.
**chemical energy** Potential energy stored in chemical bonds of molecules.
**chemosynthesis** The process in which inorganic chemicals, such as hydrogen sulfide (HS) or hydrogen gas ($H^2$), serve as an energy source for synthesis of organic molecules.
**chlorinated hydrocarbons** Hydrocarbon molecules to which chlorine atoms are attached.
**chlorofluorocarbons (CFCs)** Chemical compounds with a carbon skeleton and one or more attached chlorine and fluorine atoms. Commonly used as refrigerants, solvents, fire retardants, and blowing agents.
**chloroplasts** Chlorophyll-containing organelles in eukaryotic organisms; sites of photosynthesis.
**chronic effects** Long-lasting results of exposure to a toxin; can be a permanent change caused by a single, acute exposure or a continuous, low-level exposure.
**chronic food shortages** Long-term undernutrition and malnutrition; usually caused by people's lack of money to buy food or lack of opportunity to grow it themselves.
**chronic obstructive lung disease** Irreversible damage to the linings of the lungs caused by irritants.
**chronically undernourished** Those people whose diet doesn't provide the 2,200 kcal per day, on average, considered necessary for a healthy productive life.
**citizen science** Projects in which trained volunteers work with scientific researchers to answer real-world questions.
**city** A differentiated community with a sufficient population and resource base to allow residents to specialize in arts, crafts, services, and professional occupations.
**civil law** A body of laws regulating relations between individuals or between individuals and corporations concerning property rights, personal dignity and freedom, and personal injury.
**classical economics** Modern, western economic theories of the effects of resource scarcity, monetary policy, and competition on supply and demand of goods and services in the marketplace. This is the basis for the capitalist market system.
**clear-cut** Cutting every tree in a given area, regardless of species or size; an appropriate harvest

method for some species; can be destructive if not carefully controlled.

**climate** A description of the long-term pattern of weather in a particular area.

**climax community** A relatively stable, long-lasting community reached in a successional series; usually determined by climate and soil type.

**closed canopy** A forest where tree crowns spread over 20 percent of the ground; has the potential for commercial timber harvests.

**cloud forests** High mountain forests where temperatures are uniformly cool and fog or mist keeps vegetation wet all the time.

**coal gasification** The heating and partial combustion of coal to release volatile gases, such as methane and carbon monoxide; after pollutants are washed out, these gases become efficient, clean-burning fuel.

**coal-to-liquid (CTL) technology** Turning coal into liquid fuel.

**coal washing** Coal technology that involves crushing coal and washing out soluble sulfur compounds with water or other solvents.

**Coastal Zone Management Act** Legislation of 1972 that gave federal money to 30 seacoast and Great Lakes states for development and restoration projects.

**co-composting** Microbial decomposition of organic materials in solid waste into useful soil additives and fertilizer; often, extra organic material in the form of sewer sludge, animal manure, leaves, and grass clippings are added to solid waste to speed the process and make the product more useful.

**coevolution** The process in which species exert selective pressure on each other and gradually evolve new features or behaviors as a result of those pressures.

**cogeneration** The simultaneous production of electricity and steam or hot water in the same plant.

**cold front** A moving boundary of cooler air displacing warmer air.

**coliform bacteria** Bacteria that live in the intestines (including the colon) of humans and other animals; used as a measure of the presence of feces in water or soil.

**commensalism** A symbiotic relationship in which one member is benefited and the second is neither harmed nor benefited.

**common law** The body of court decisions that constitute a working definition of individual rights and responsibilities where no formal statutes define these issues.

**communal resource management systems** Resources managed by a community for long-term sustainability.

**community-supported agriculture (CSA)** A program in which you make an annual contribution to a local farm in return for weekly deliveries of a "share" of whatever the farm produces.

**competitive exclusion** A theory that no two populations of different species will occupy the same niche and compete for exactly the same resources in the same habitat for very long.

**complexity (ecological)** The number of species at each trophic level and the number of trophic levels in a community.

**composting** The biological degradation of organic material under aerobic (oxygen-rich) conditions to produce compost, a nutrient-rich soil amendment and conditioner.

**compound** A molecule made up of two or more kinds of atoms held together by chemical bonds.

**concentrated animal feeding operations** Facilities in which large numbers of animals spend most or all of their life in confinement.

**concept map** A two-dimensional representation of the relationship between key ideas. A flow chart or graph of ideas.

**conclusion** A statement that follows logically from a set of premises.

**condensation** The aggregation of water molecules from vapor to liquid or solid when the saturation concentration is exceeded.

**condensation nuclei** Tiny particles that float in the air and facilitate the condensation process.

**confidence limits** A statistical measure of the quality of data that tells you how close the sample's average probably is to the average for the entire population of that species.

**conifers** Needle-bearing trees that produce seeds in cones.

**conservation development** Consideration of landscape history, human culture, topography, and ecological values in subdivision design. Using cluster housing, zoning, covenants, and other design features, at least half of a subdivision can be preserved as open space, farmland, or natural areas.

**conservation medicine** A medical field that attempts to understand how environmental changes threaten our own health as well as that of the natural communities on which we depend for ecological services.

**conservation of matter** In any chemical reaction, matter changes form; it is neither created nor destroyed.

**conspicuous consumption** A term coined by economist and social critic Thorstein Veblen to describe buying things we don't want or need to impress others.

**consumer** An organism that obtains energy and nutrients by feeding on other organisms or their remains. *See also* heterotroph.

**consumption** The fraction of withdrawn water that is lost in transmission or that is evaporated, absorbed, chemically transformed, or otherwise made unavailable for other purposes as a result of human use.

**contour plowing** Plowing along hill contours; reduces erosion.

**control rods** Neutron-absorbing material inserted into spaces between fuel assemblies in nuclear reactors to regulate fission reaction.

**controlled studies** Those in which comparisons are made between experimental and control populations that are identical (as far as possible) in every factor except the one variable being studied.

**convection currents** Rising or sinking air currents that stir the atmosphere and transport heat from one area to another. Convection currents also occur in water; *see* spring overturn.

**conventional pollutants** The seven major pollutants (sulfur dioxide, carbon monoxide, particulates, hydrocarbons, nitrogen oxides, photochemical oxidants, and lead) identified and regulated by the U.S. Clean Air Act.

**cool deserts** Deserts such as the American Great Basin characterized by cold winters and sagebrush.

**coral bleaching** Whitening of corals caused by expulsion of symbiotic algae—often resulting from high water temperatures, pollution, or disease.

**coral reefs** Prominent oceanic features composed of hard, limy skeletons produced by coral animals; usually formed along edges of shallow, submerged ocean banks or along shelves in warm, shallow, tropical seas.

**core** The dense, intensely hot mass of molten metal, mostly iron and nickel, thousands of kilometers in diameter at the earth's center.

**core habitat** Essential habitat for a species.

**core region** The primary industrial region of a country; usually located around the capital or largest port; has both the greatest population density and the greatest economic activity of the country.

**Coriolis effect** The influence of friction and drag on air layers near the earth; deflects air currents to the direction of the earth's rotation.

**cornucopian fallacy** The belief that nature is limitless in its abundance and that perpetual growth is not only possible but essential.

**corridor** A strip of natural habitat that connects two adjacent nature preserves to allow migration of organisms from one place to another.

**cost-benefit analysis (CBA)** An evaluation of large-scale public projects by comparing the costs and benefits that accrue from them.

**cover crops** Plants, such as rye, alfalfa, or clover, that can be planted immediately after harvest to hold and protect the soil.

**creative thinking** Asks, how could I do this differently?

**credit** An amount of pollution a company is allowed to sell when they reduce emissions below their allowed cap. *See* cap-and-trade.

**criminal law** A body of court decisions based on federal and state statutes concerning wrongs against persons or society.

**criteria pollutants** *See* conventional pollutants.

**critical factor** The single environmental factor closest to a tolerance limit for a given species at a given time. *See* limiting factors.

**critical thinking** An ability to evaluate information and opinions in a systematic, purposeful, efficient manner.

**croplands** Lands used to grow crops.

**crude birth rate** The number of births in a year divided by the midyear population.

**crude death rate** The number of deaths per thousand persons in a given year; also called crude mortality rate.

**crust** The cool, lightweight, outermost layer of the earth's surface that floats on the soft, pliable underlying layers; similar to the "skin" on a bowl of warm pudding.

**cultural creatives** People who are socially conscious, involve in improving communities and willing to translate values into action.

**cultural eutrophication** An increase in biological productivity and ecosystem succession caused by human activities.

# D

**debt-for-nature swap** Forgiveness of international debt in exchange for nature protection in developing countries.

**deciduous** Trees and shrubs that shed their leaves at the end of the growing season.

**decline spiral** A catastrophic deterioration of a species, community, or whole ecosystem; accelerates as functions are disrupted or lost in a downward cascade.

**decomposers** Fungi and bacteria that break complex organic material into smaller molecules.

**deductive reasoning** Deriving testable predictions about specific cases from general principles.

**deep ecology** A philosophy that calls for a profound shift in our attitudes and behavior toward nature.
**deforestation** Forest removal.
**degradation (of water resource)** Deterioration in water quality due to contamination or pollution; makes water unsuitable for other desirable purposes.
**Delaney Clause** A controversial amendment to the Federal Food, Drug, and Cosmetic Act, added in 1958, prohibiting the addition of any known cancer-causing agent to processed foods, drugs, or cosmetics.
**delta** Fan-shaped sediment deposit found at the mouth of a river.
**demand** The amount of a product that consumers are willing and able to buy at various possible prices, assuming they are free to express their preferences.
**demanufacturing** Disassembly of products so components can be reused or recycled.
**demographic bottleneck** A population founded when just a few members of a species survive a catastrophic event or colonize new habitat geographically isolated from other members of the same species.
**demographic transition** A pattern of falling death rates and birthrates in response to improved living conditions; could be reversed in deteriorating conditions.
**demography** Vital statistics about people: births, marriages, deaths, etc.; the statistical study of human populations relating to growth rate, age structure, geographic distribution, etc., and their effects on social, economic, and environmental conditions.
**denitrifying bacteria** Free-living soil bacteria that converts nitrates to gaseous nitrogen and nitrous oxide.
**density-dependent** Factors affecting population growth that change as population size changes.
**dependency ratio** The number of nonworking members compared to working members for a given population.
**dependent (response) variable** A variable that is affected by the condition being altered in a manipulative experiment.
**desalinization (or desalination)** Removal of salt from water by distillation, freezing, or ultrafiltration.
**desert** A type of biome characterized by low moisture levels and infrequent and unpredictable precipitation. Daily and seasonal temperatures fluctuate widely.
**desertification** Conversion of productive lands to desert.
**detritivore** Organisms that consume organic litter, debris, and dung.
**dew point** The temperature at which condensation occurs for a given concentration of water vapor in the air.
**dieback** A sudden population decline; also called a population crash.
**diminishing returns** A condition in which unrestrained population growth causes the standard of living to decrease to a subsistence level where poverty, misery, vice, and starvation makes life permanently drab and miserable. This dreary prophecy has led economics to be called "the dismal science."
**disability-adjusted life years (DALY)** A measure of premature deaths and losses due to illnesses and disabilities in a population.
**discharge** The amount of water that passes a fixed point in a given amount of time; usually expressed as liters or cubic feet of water per second.
**disclimax community** *See* equilibrium community.
**discount rates** The difference between present value and future value of a resource. Generally equivalent to an interest rate.
**disease** A deleterious change in the body's condition in response to destabilizing factors, such as nutrition, chemicals, or biological agents.
**dissolved oxygen (DO) content** Amount of oxygen dissolved in a given volume of water at a given temperature and atmospheric pressure; usually expressed in parts per million (ppm).
**disturbance-adapted species** Species that depend on disturbances to succeed.
**disturbances** Periodic, destructive events such as fire or floods; changes in an ecosystem that affect (positively or negatively) the organisms living there.
**diversity (species diversity, biological diversity)** The number of species present in a community (species richness), as well as the relative abundance of each species.
**DNA (deoxyribonucleic acid)** A giant molecule composed of millions or billions of nucleotides (sugars and bases called purines and pyramidines held together by phosphate bonds) that form a double helix and store genetic information in all living cells.
**dominant plants** Those plant species in a community that provide a food base for most of the community; they usually take up the most space and have the largest biomass.
**double-blind design** One in which neither the experimenter nor the subjects know until after data has been gathered and analyzed which was the experimental treatment and which was the control.
**downbursts** Sudden, very strong, downdrafts of cold air associated with an advancing storm front.
**drip irrigation** Uses pipe or tubing perforated with very small holes to deliver water one drop at a time directly to the soil around each plant.
**dry alkali injection** Spraying dry sodium bicarbonate into flue gas to absorb and neutralize acidic sulfur compounds.

# E

**Earth Charter** A set of principles for sustainable development, environmental protection, and social justice developed by a council appointed by the United Nations.
**earthquakes** Sudden, violent movement of the earth's crust.
**ecocentric (ecologically centered)** A philosophy that claims moral values and rights for both organisms and ecological systems and processes.
**ecofeminism** A pluralistic, nonhierarchical, relationship-oriented philosophy that suggests how humans could reconceive themselves and their relationships to nature in nondominating ways as an alternative to patriarchal systems of domination.
**ecojustice** Justice in the social order and integrity in the natural order.
**ecological development** A gradual process of environmental modification by organisms.
**ecological diseases** Emergent diseases (new or rarely seen diseases) that cause devastating epidemics among wildlife and domestic animals.
**ecological economics** A relatively new field that brings the insights of ecology to economic analysis.
**ecological equivalents** Different species that occupy similar ecological niches in similar ecosystems in different parts of the world.
**ecological footprint** A measure that computes the demands placed on nature by individuals and nations.
**ecological niche** The functional role and position of a species (population) within a community or ecosystem, including what resources it uses, how and when it uses the resources, and how it interacts with other populations.
**ecological succession** The process by which organisms occupy a site and gradually change environmental conditions so that other species can replace the original inhabitants.
**ecology** The scientific study of relationships between organisms and their environment. It is concerned with the life histories, distribution, and behavior of individual species as well as the structure and function of natural systems at the level of populations, communities, and ecosystems.
**economic development** A rise in real income *per person;* usually associated with new technology that increases productivity or resources.
**economic growth** An increase in the total wealth of a nation; if population grows faster than the economy, there may be real economic growth, but the share per person may decline.
**economic thresholds** In pest management, the point at which the cost of pest damage exceeds the costs of pest control.
**ecosystem** A specific biological community and its physical environment interacting in an exchange of matter and energy.
**ecosystem management** An integration of ecological, economic, and social goals in a unified systems approach to resource management.
**ecosystem restoration** To reinstate an entire community of organisms to as near its natural condition as possible.
**ecotone** A boundary between two types of ecological communities.
**ecotourism** A combination of adventure travel, cultural exploration, and nature appreciation in wild settings.
**edge effects** A change in species composition, physical conditions, or other ecological factors at the boundary between two ecosystems.
**effluent sewerage** A low-cost alternative sewage treatment for cities in poor countries that combines some features of septic systems and centralized municipal treatment systems.
**electron** A negatively charged subatomic particle that orbits around the nucleus of an atom.
**electronic waste** *See* e-waste.
**electrostatic precipitators** The most common particulate controls in power plants; fly ash particles pick up an electrostatic surface charge as they pass between large electrodes in the effluent stream, causing particles to migrate to the oppositely charged plate.
**element** A molecule composed of one kind of atom; cannot be broken into simpler units by chemical reactions.
**El Niño** A climatic change marked by shifting of a large warm water pool from the western Pacific Ocean towards the east. Wind direction and precipitation patterns are changed over much of the Pacific and perhaps around the world.
**emergent diseases** A new disease or one that has been absent for at least 20 years.
**emergent properties** Characteristics of whole, functioning systems that are quantitatively or qualitatively greater than the sum of the systems' parts.

**emigration** The movement of members from a population.

**emission standards** Regulations for restricting the amounts of air pollutants that can be released from specific point sources.

**emissions trading** Programs in which companies that have cut pollution by more than they're required to can sell "credits" to other companies that still exceed allowed levels.

**endangered species** A species considered to be in imminent danger of extinction.

**endemism** A state in which species are restricted to a single region.

**endocrine disrupters** Chemicals that disrupt normal hormone functions.

**energy** The capacity to do work (that is, to change the physical state or motion of an object).

**energy crops** Crops that can be used to make ethanol or diesel fuel.

**energy efficiency** A measure of energy produced compared to energy consumed.

**energy pyramid** A representation of the loss of useful energy at each step in a food chain.

**energy recovery** Incineration of solid waste to produce useful energy.

**entropy** Disorder in a system.

**environment** The circumstances or conditions that surround an organism or group of organisms as well as the complex of social or cultural conditions that affect an individual or community.

**environmental economics** *See* ecological economics.

**environmental ethics** A search for moral values and ethical principles in human relations with the natural world.

**environmental governance** Rules and regulations that govern our impacts on the environment and natural resources.

**environmental health** The science of external factors that cause disease, including elements of the natural, social, cultural and technological worlds in which we live.

**environmental hormones** Chemical pollutants that substitute for, or interfere with, naturally occurring hormones in our bodies; these chemicals may trigger reproductive failure, developmental abnormalities, or tumor promotion.

**environmental impact statement (EIS)** An analysis, required by provisions in the National Environmental Policy Act of 1970, of the effects of any major program a federal agency plans to undertake.

**environmental indicators** Organisms or physical factors that serve as a gauge for environmental changes. More specifically, organisms with these characteristics are called bioindicators.

**Environmental Performance Index (EPI)** A measure that evaluates national sustainability and progress toward achievement of the United Nations Millennium Development Goals.

**environmentalism** Active participation in attempts to solve environmental pollution and resource problems.

**environmental justice** A recognition that access to a clean, healthy environment is a fundamental right of all human beings.

**environmental law** The special body of official rules, decisions, and actions concerning environmental quality, natural resources, and ecological sustainability.

**environmental literacy** Fluency in the principles of ecology that gives us a working knowledge of the basic grammar and underlying syntax of environmental wisdom.

**environmental policy** The official rules or regulations concerning the environment adopted, implemented, and enforced by some governmental agency.

**environmental racism** Decisions that restrict certain people or groups of people to polluted or degraded environments on the basis of race.

**environmental resistance** All the limiting factors that tend to reduce population growth rates and set the maximum allowable population size or carrying capacity of an ecosystem.

**environmental resources** Anything an organism needs that can be taken from the environment.

**environmental science** The systematic, scientific study of our environment as well as our role in it.

**enzymes** Molecules, usually proteins or nucleic acids, that act as catalysts in biochemical reactions.

**epidemiology** The study of the distribution and causes of disease and injuries in human populations.

**epiphyte** A plant that grows on a substrate other than the soil, such as the surface of another organism.

**equilibrium community** Also called a **disclimax community;** a community subject to periodic disruptions, usually by fire, that prevent it from reaching a climax stage.

**estuary** A bay or drowned valley where a river empties into the sea.

**ethics** A branch of philosophy concerned with right and wrong.

**eukaryotic cell** A cell containing a membrane-bounded nucleus and membrane-bounded organelles.

**eutrophic** Rivers and lakes rich in organisms and organic material (*eu* = truly; *trophic* = nutritious).

**evaporation** The process in which a liquid is changed to vapor (gas phase).

**evolution** A theory that explains how random changes in genetic material and competition for scarce resources cause species to change gradually.

**e-waste** Discarded electronic equipment such as computers, cell phones, television sets, etc.

**exhaustible resources** Generally considered the earth's geologic endowment: minerals, nonmineral resources, fossil fuels, and other materials present in fixed amounts in the environment.

**existence value** The importance we place on just knowing that a particular species or a specific organism exists.

**exotic organisms** Alien species introduced by human agency into biological communities where they would not naturally occur.

**exponential growth** Growth at a constant rate of increase per unit of time; can be expressed as a constant fraction or exponent. *See* geometric growth.

**external costs** Expenses, monetary or otherwise, borne by someone other than the individuals or groups who use a resource.

**extinction** The irrevocable elimination of species; can be a normal process of the natural world as species out-compete or kill off others or as environmental conditions change.

**extirpate** To destroy totally; extinction caused by direct human action, such as hunting, trapping, etc.

**extreme poverty** Living on less than $1 (U.S.) per day.

# F

**family planning** Controlling reproduction; planning the timing of birth and having as many babies as are wanted and can be supported.

**famines** Acute food shortages characterized by large-scale loss of life, social disruption, and economic chaos.

**fauna** All of the animals present in a given region.

**fecundity** The physical ability to reproduce.

**fen** An area of waterlogged soil that tends to be peaty; fed mainly by upwelling water; low productivity.

**feral** A domestic animal that has taken up a wild existence.

**fermentation** (alcoholic) A type of anaerobic respiration that yields carbon dioxide and alcohol.

**fertility** Measurement of actual number of offspring produced through sexual reproduction; usually described in terms of number of offspring of females, since paternity can be difficult to determine.

**fetal alcohol syndrome** A tragic set of permanent physical and mental and behavioral birth defects that result when mothers drink alcohol during pregnancy.

**fibrosis** The general name for accumulation of scar tissue in the lung.

**filters** A porous mesh of cotton cloth, spun glass fibers, or asbestos-cellulose that allows air or liquid to pass through but holds back solid particles.

**fire-climax community** An equilibrium community maintained by periodic fires; examples include grasslands, chaparral shrubland, and some pine forests.

**first law of thermodynamics** States that energy is *conserved;* that is, it is neither created nor destroyed under normal conditions.

**flex-fuel vehicles** Vehicles that can burn variable mixtures of gasoline and ethanol.

**floodplains** Low lands along riverbanks, lakes, and coastlines subjected to periodic inundation.

**flora** All of the plants present in a given region.

**flue-gas scrubbing** Treating combustion exhaust gases with chemical agents to remove pollutants. Spraying crushed limestone and water into the exhaust gas stream to remove sulfur is a common scrubbing technique.

**food aid** Financial assistance intended to boost less-developed countries' standards of living.

**food chain** A linked feeding series; in an ecosystem, the sequence of organisms through which energy and materials are transferred, in the form of food, from one trophic level to another.

**food security** The ability of individuals to obtain sufficient food on a day-to-day basis.

**food surpluses** Excess food supplies.

**food web** A complex, interlocking series of individual food chains in an ecosystem.

**forest management** Scientific planning and administration of forest resources for sustainable harvest, multiple use, regeneration, and maintenance of a healthy biological community.

**fossil fuels** Petroleum, natural gas, and coal created by geological forces from organic wastes and dead bodies of formerly living biological organisms.

**founder effect** The effect on a population founded when just a few members of a species survive a catastrophic event or colonize new habitat geographically isolated from other members of the same species.

**freezing condensation** A process that occurs in the clouds when ice crystals trap water vapor. As the ice crystals become larger and heavier, they begin to fall as rain or snow.

**fresh water** Water other than seawater; covers only about 2 percent of earth's surface, including streams, rivers, lakes, ponds, and water associated with several kinds of wetlands.

**freshwater ecosystems** Ecosystems in which the fresh (nonsalty) water of streams, rivers, ponds, or lakes plays a defining role.

**front** The boundary between two air masses of different temperature and density.

**fuel assembly** A bundle of hollow metal rods containing uranium oxide pellets; used to fuel a nuclear reactor.

**fuel cells** Mechanical devices that use hydrogen or hydrogen-containing fuel such as methane to produce an electric current. Fuel cells are clean, quiet, and highly efficient sources of electricity.

**fuel-switching** A change from one fuel to another.

**fuelwood** Branches, twigs, logs, wood chips, and other wood products harvested for use as fuel.

**fugitive emissions** Substances that enter the air without going through a smokestack, such as dust from soil erosion, strip mining, rock crushing, construction, and building demolition.

**fumigants** Toxic gases such as methyl bromine that are used to kill pests.

**fungi** One of the five kingdom classifications; consists of nonphotosynthetic, eukaryotic organisms with cell walls, filamentous bodies, and absorptive nutrition.

**fungicide** A chemical that kills fungi.

# G

**Gaia hypothesis** A theory that the living organisms of the biosphere form a single, complex interacting system that creates and maintains a habitable Earth; named after Gaia, the Greek "Earth mother" goddess.

**gamma rays** Very short wavelength forms of the electromagnetic spectrum.

**gap analysis** A biogeographical technique of mapping biological diversity and endemic species to find gaps between protected areas that leave endangered habitats vulnerable to disruption.

**garden city** A new town with special emphasis on landscaping and rural ambience.

**gasohol** A mixture of gasoline and ethanol.

**gene** A unit of heredity; a segment of DNA nucleus of the cell that contains information for the synthesis of a specific protein, such as an enzyme.

**gene banks** Storage for seed varieties for future breeding experiments.

**general fertility rate** Crude birthrate multiplied by the percentage of reproductive age women.

**genetic assimilation** The disappearance of a species as its genes are diluted through crossbreeding with a closely related species.

**genetic drift** The gradual changes in gene frequencies in a population due to random events.

**genetic engineering** Laboratory manipulation of genetic material using molecular biology techniques to create desired characteristics in organisms.

**genetically modified organisms (GMOs)** Organisms whose genetic code has been altered by artificial means such as interspecies gene transfer.

**genuine progress index (GPI)** An alternative to GNP or GDP for economic accounting that measures real progress in quality of life and sustainability.

**geographic information systems (GIS)** Spatial data, such as boundaries or road networks and computer software to display and analyze those data.

**geographic isolation** *See* allopatric speciation.

**geometric growth** Growth that follows a geometric pattern of increase, such as 2, 4, 8, 16, etc. *See* exponential growth.

**geothermal energy** Energy drawn from the internal heat of the earth, either through geysers, fumaroles, hot springs, or other natural geothermal features, or through deep wells that pump heated groundwater.

**germ plasm** Genetic material that may be preserved for future agricultural, commercial, and ecological values (plant seeds or parts or animal eggs, sperm, and embryos).

**global environmentalism** A concern for, and action to help solve, global environmental problems.

**globalization** The revolution in communications, transportation, finances and commerce that has brought about increasing inter-dependence of national economies.

**grasslands** A biome dominated by grasses and associated herbaceous plants.

**greenfield development** Housing projects built on previously undeveloped farmlands or forests on the outskirts of large cities.

**greenhouse effect** Gases in the atmosphere are transparent to visible light but absorb infrared (heat) waves that are reradiated from the earth's surface.

**green plans** Integrated national environmental plans for reducing pollution and resource consumption while achieving sustainable development and environmental restoration.

**green political parties** Political organizations based on environmental protection, participatory democracy, grassroots organization, and sustainable development.

**green pricing** Setting prices to encourage conservation or renewable energy. Plans that invite customers to pay a premium for energy from renewable sources.

**green revolution** Dramatically increased agricultural production brought about by "miracle" strains of grain; usually requires high inputs of water, plant nutrients, and pesticides.

**gross domestic product (GDP)** The total economic activity within national boundaries.

**gross national product (GNP)** The sum total of all goods and services produced in a national economy. Gross domestic product (GDP) is used to distinguish economic activity within a country from that of off-shore corporations.

**groundwater** Water held in gravel deposits or porous rock below the earth's surface; does not include water or crystallization held by chemical bonds in rocks or moisture in upper soil layers.

**gully erosion** Removal of layers of soil, creating channels or ravines too large to be removed by normal tillage operations.

# H

**habitat** The place or set of environmental conditions in which a particular organism lives.

**habitat conservation plans** Agreements under which property owners are allowed to harvest resources or develop land as long as habitat is conserved or replaced in ways that benefit resident endangered or threatened species in the long run. Some incidental "taking" or loss of endangered species is generally allowed in such plans.

**Hadley cells** Circulation patterns of atmospheric convection currents as they sink and rise in several intermediate bands.

**hazardous** Describes chemicals that are dangerous, including flammables, explosives, irritants, sensitizers, acids, and caustics; may be relatively harmless in diluted concentrations.

**hazardous air pollutants (HAPs)** Especially dangerous air pollutants including carcinogens, neurotoxins, mutagens, teratogens, endocrine system disrupters and other highly toxic compounds.

**hazardous waste** Any discarded material containing substances known to be toxic, mutagenic, carcinogenic, or teratogenic to humans or other life-forms; ignitable, corrosive, explosive, or highly reactive alone or with other materials.

**health** A state of physical and emotional well-being; the absence of disease or ailment.

**heap-leach extraction** A technique for separating gold from extremely low-grade ores. Crushed ore is piled in huge heaps and sprayed with a dilute alkaline-cyanide solution, which percolates through the pile to extract the gold, which is separated from the effluent in a processing plant. This process has a high potential for water pollution.

**heat** A form of energy transferred from one body to another because of a difference in temperatures.

**heat capacity** The amount of heat energy that must be added or subtracted to change the temperature of a body; water has a high heat capacity.

**heat of vaporization** The amount of heat energy required to convert water from a liquid to a gas.

**herbicide** A chemical that kills plants.

**herbivore** An organism that eats only plants.

**heterotroph** An organism that is incapable of synthesizing its own food and, therefore, must feed upon organic compounds produced by other organisms.

**high-level waste repository** A place where intensely radioactive wastes can be buried and remain unexposed to groundwater and earthquakes for tens of thousands of years.

**high-quality energy** Intense, concentrated, and high-temperature energy that is considered high-quality because of its usefulness in carrying out work.

**HIPPO** Habitat destruction, Invasive species, Pollution, Population (human), and Overharvesting, the leading causes of extinction.

**holistic science** The study of entire, integrated systems rather than isolated parts. Often takes a descriptive or interpretive approach.

**homeostasis** Maintaining a dynamic, steady state in a living system through opposing, compensating adjustments.

**Homestead Act** Legislation passed in 1862 allowing any citizen or applicant for citizenship over 21 years old and head of a family to acquire 160 acres of public land by living on it and cultivating it for five years.

**host organism** An organism that provides lodging for a parasite.

**hot desert** Deserts of the American Southwest and Mexico; characterized by extreme summer heat and cacti.

**human development index (HDI)** A measure of quality of life using life expectancy, child survival, adult literacy, childhood education, gender equity and access to clan water and sanitation as well as income.

**human ecology** The study of the interactions of humans with the environment.

**human resources** Human wisdom, experience, skill, labor, and enterprise.

**humus** Sticky, brown, insoluble residue from the bodies of dead plants and animals; gives soil its structure, coating mineral particles and holding them together; serves as a major source of plant nutrients.

**hurricanes** Large cyclonic oceanic storms with heavy rain and winds exceeding 119 km/hr (74 mph).

**hybrid gasoline-electric engines** A small gasoline engine generates electricity that is stored in batteries and powers electric motors that drive vehicle wheels.

**hybrid gasoline-electric vehicles** Automobiles that run on electric power and a small gasoline or diesel engine.

**hydrologic cycle** The natural process by which water is purified and made fresh through evaporation and precipitation. This cycle provides all the freshwater available for biological life.

**hypothesis** A provisional explanation that can be tested scientifically.

# I

**igneous rocks** Crystalline minerals solidified from molten magma from deep in the earth's interior; basalt, rhyolite, andesite, lava, and granite are examples.

**inbreeding depression** In a small population, an accumulation of harmful genetic traits (through random mutations and natural selection) that lowers viability and reproductive success of enough individuals to affect the whole population.

**independent variable** A factor not affected by the condition being altered in a manipulative experiment.

**indicator species** Those whose critical tolerance limits can be used to judge environmental conditions.

**inductive reasoning** Inferring general principles from specific examples.

**industrial revolution** Advances in science and technology that have given us power to understand and change our world.

**industrial timber** Trees used for lumber, plywood, veneer, particleboard, chipboard, and paper; also called roundwood.

**inertial confinement** A nuclear fusion process in which a small pellet of nuclear fuel is bombarded with extremely high-intensity laser light.

**infiltration** The process of water percolation into the soil and pores and hollows of permeable rocks.

**informal economy** Small-scale family businesses in temporary locations outside the control of normal regulatory agencies.

**inherent value** Ethical values or rights that exist as an intrinsic or essential characteristic of a particular thing or class of things simply by the fact of their existence.

**inholdings** Private lands within public parks, forests, or wildlife refuges.

**inorganic pesticides** Inorganic chemicals such as metals, acids, or bases used as pesticides.

**insecticide** A chemical that kills insects.

**insolation** Incoming solar radiation.

**instrumental value** Value or worth of objects that satisfy the needs and wants of moral agents. Objects that can be used as a means to some desirable end.

**intangible resources** Factors such as open space, beauty, serenity, wisdom, diversity, and satisfaction that cannot be grasped or contained. Ironically, these resources can be both infinite and exhaustible.

**integrated gasification combined cycle (IGCC)** A power plant that heats fuel (usually coal, but could be biomass or other sources) to high temperatures and pressures in the presence of 96 percent oxygen. Hydrogen is separated from hydrocarbons and separated from $CO_2$ and other contaminants. The hydrogen is burned in a gas turbine and surplus heat drives a steam turbine, both of which generate electricity.

**integrated pest management (IPM)** An ecologically based pest-control strategy that relies on natural mortality factors, such as natural enemies, weather, cultural control methods, and carefully applied doses of pesticides.

**Intergovernmental Panel on Climate Change (IPCC)** An international organization formed to assess global climate change and its impacts. The IPCC is concerned with social, economic, and environmental impacts of climate change, and it was established by the United Nations Environment Program and the World Meteorological Organization.

**internal costs** The expenses, monetary or otherwise, borne by those who use a resource.

**interplanting** The system of planting two or more crops, either mixed together or in alternating rows, in the same field; protects the soil and makes more efficient use of the land.

**interpretive science** Explanation based on observation and description of entire objects or systems rather than isolated parts.

**interspecific competition** In a community, competition for resources between members of *different* species.

**intraspecific competition** In a community, competition for resources among members of the *same* species.

**invasive species** Organisms that thrive in new territory where they are free of predators, diseases or resource limitations that may have controlled their population in their native habitat.

**ionizing radiation** High-energy electromagnetic radiation or energetic subatomic particles released by nuclear decay.

**ionosphere** The lower part of the thermosphere.

**ions** Electrically charged atoms that have gained or lost electrons.

**I-PAT formula** A formula that says our environmental impacts (I) are the product of our population size (P) times affluence (A) and the technology (T) used to produce the goods and services we consume.

**island biogeography** The study of rates of colonization and extinction of species on islands or other isolated areas based on size, shape, and distance from other inhabited regions.

**isotopes** Forms of a single element that differ in atomic mass due to a different number of neutrons in the nucleus.

# J

**J curve** A growth curve that depicts exponential growth; called a J curve because of its shape.

**jet streams** Powerful winds or currents of air that circulate in shifting flows; similar to oceanic currents in extent and effect on climate.

**joule** A unit of energy. One joule is the energy expended in 1 second by a current of 1 amp flowing through a resistance of 1 ohm.

# K

***k*-selected species** Species that reproduce more slowly, occupy higher trophic levels, have fewer offspring, longer life-spans, and greater intrinsic control of population growth than *r*-selected species.

**keystone species** A species whose impacts on its community or ecosystem are much larger and more influential than would be expected from mere abundance.

**kinetic energy** Energy contained in moving objects such as a rock rolling down a hill, the wind blowing through the trees, or water flowing over a dam.

**known resources** Those that have been located but are not completely mapped but, nevertheless, are likely to become economical in the foreseeable future.

**kwashiorkor** A widespread human protein deficiency disease resulting from a starchy diet low in protein and essential amino acids.

**Kyoto Protocol** An international agreement to reduce greenhouse gas emissions.

# L

**La Niña** The part of a large-scale oscillation in the Pacific (and, perhaps, other oceans) in which trade winds hold warm surface waters in the western part of the basin and cause upwelling of cold, nutrient-rich, deep water in the eastern part of the ocean.

**landfills** Land disposal sites for solid waste; operators compact refuse and cover it with a layer of dirt to minimize rodent and insect infestation, wind-blown debris, and leaching by rain.

**land reform** Democratic redistribution of landownership to recognize the rights of those who actually work the land to a fair share of the products of their labor.

**landscape ecology** The study of the reciprocal effects of spatial pattern on ecological processes. A study of the ways in which landscape history shapes the features of the land and the organisms that inhabit it as well as our reaction to, and interpretation of, the land.

**landslide** The sudden fall of rock and earth from a hill or cliff. Often triggered by an earthquake or heavy rain.

**latent heat** Stored energy in a form that is not sensible (detectable by ordinary senses).

**LD50** A chemical dose lethal to 50 percent of a test population.

**less-developed countries (LDC)** Nonindustrialized nations characterized by low per capita income, high birthrates and death rates, high population growth rates, and low levels of technological development.

**life-cycle analysis** Evaluation of material and energy inputs and outputs at each stage of manufacture, use, and disposal of a product.

**life expectancy** The average age that a newborn infant can expect to attain in a particular time and place.

**life span** The longest period of life reached by a type of organism.

**limiting factors** Chemical or physical factors that limit the existence, growth, abundance, or distribution of an organism.

**lipid** A nonpolar organic compound that is insoluble in water but soluble in solvents, such as alcohol and ether; includes fats, oils, steroids, phospholipids, and carotenoids.

**liquid metal fast breeder** A nuclear power plant that converts uranium 238 to plutonium 239; thus, it creates more nuclear fuel than it consumes. Because of the extreme heat and density of its core, the breeder uses liquid sodium as a coolant.

**lobbying** Using personal contacts, public pressure, or political action to persuade legislators to vote in a particular manner.
**locavore** Someone who eats locally grown, seasonal food.
**logarithmic scale** One that uses logarithms as units in a sequence that progresses by a factor of 10. That is, each subsequent increment on the scale is 10 times the one that precedes it.
**logical learner** Someone who understands and remembers best by thinking through a topic and finding logical reasons for statements.
**logical thinking** Asks, can the rules of logic help understand this?
**logistic growth** Growth rates regulated by internal and external factors that establish an equilibrium with environmental resources.
**longevity** The length or duration of life; compare to survivorship.
**low-head hydropower** Small-scale hydro technology that can extract energy from small headwater dams; causes much less ecological damage.
**low-input high-diversity biofuels** Mixed polycultures of perennial native species that don't require minimal amounts of cultivation, fertilizer, irrigation, or pesticides when grown as energy crops.
**low-quality energy** Diffuse, dispersed energy at a low temperature that is difficult to gather and use for productive purposes.
**LULUs** *L*ocally *U*nwanted *L*and *U*ses such as toxic waste dumps, incinerators, smelters, airports, freeways, and other sources of environmental, economic, or social degradation.

# M

**magma** Molten rock from deep in the earth's interior; called lava when it spews from volcanic vents.
**magnetic confinement** A technique for enclosing a nuclear fusion reaction in a powerful magnetic field inside a vacuum chamber.
**malignant tumor** A mass of cancerous cells that have left their site of origin, migrated through the body, invaded normal tissues, and are growing out of control.
**malnourishment** A nutritional imbalance caused by lack of specific dietary components or inability to absorb or utilize essential nutrients.
**Man and Biosphere (MAB) program** A design for nature preserves that divides protected areas into zones with different purposes. A highly protected core is surrounded by a buffer zone and peripheral regions in which multiple-use resource harvesting is permitted.
**mangroves** Trees from a number of genera that live in salt water.
**manipulative experiment** One in which some conditions are deliberately altered while others are held constant to study cause-and-effect relationships.
**mantle** A hot, pliable layer of rock that surrounds the earth's core and underlies the cool, outer crust.
**marasmus** A widespread human protein deficiency disease caused by a diet low in calories and protein or imbalanced in essential amino acids.
**marginal costs and benefits** The costs and benefits of producing one additional unit of a good or service.
**marine** Living in or pertaining to the sea.
**market equilibrium** The dynamic balance between supply and demand under a given set of conditions in a "free" market (one with no monopolies or government interventions).
**market forces** Depending on capitalist market systems to achieve national goals.
**marsh** Wetland without trees; in North America, this type of land is characterized by cattails and rushes.
**mass burn** Incineration of unsorted solid waste.
**mass wasting** Mass movement of geologic materials downhill caused by rockslides, avalanches, or simple slumping.
**matter** Anything that takes up space and has mass.
**mean** An average.
**mediation** An informal dispute resolution process in which parties are encouraged to discuss issues openly but in which all decisions are reached by consensus and any participant can withdraw at any time.
**Mediterranean climate areas** Specialized landscapes with warm, dry summers; cool, wet winters; many unique plant and animal adaptations; and many levels of endemism.
**megacity** *See* megalopolis.
**megalopolis** Also known as a megacity or supercity; megalopolis indicates an urban area with more than 10 million inhabitants.
**megawatt (MW)** Unit of electrical power equal to 1,000 kilowatts or 1 million watts.
**mesosphere** The atmospheric layer above the stratosphere and below the thermosphere; the middle layer; temperatures are usually very low.
**metabolism** All the energy and matter exchanges that occur within a living cell or organism; collectively, the life processes.
**metamorphic rock** Igneous and sedimentary rocks modified by heat, pressure, and chemical reactions.
**metapopulation** A collection of populations that have regular or intermittent gene flow between geographically separate units.
**methane hydrate** Small bubbles or individual molecules of methane (natural gas) trapped in a crystalline matrix of frozen water.
**micorrhizal symbiosis** An association between the roots of most plant species and certain fungi. The plant provides organic compounds to the fungus, while the fungus provides water and nutrients to the plant.
**microbial agents** Or biological controls, are beneficial microbes (bacteria, fungi) that can be used to suppress or control pests.
**micro-hydro generators** Small power generators that can be used in low-level rivers to provide economical power for four to six homes, freeing them from dependence on large utilities and foreign energy supplies.
**mid-ocean ridges** Mountain ranges on the ocean floor created where molten magma is forced up through cracks in the planet's crust.
**Milankovitch cycles** Periodic variations in tilt, eccentricity, and wobble in the earth's orbit; Milutin Milankovitch suggested that it is responsible for cyclic weather changes.
**millennium assessment** A set of ambitious environmental and human development goals established by the United Nations in 2000.
**milpa agriculture** An ancient farming system in which small patches of tropical forests are cleared and perennial polyculture agriculture practiced and is then followed by many years of fallow to restore the soil; also called **swidden agriculture.**
**mineral** A naturally occurring, inorganic, crystalline solid with definite chemical composition and characteristic physical properties.
**minimum viable population size** The number of individuals needed for long-term survival of rare and endangered species.
**mitigation** Repairing or rehabilitating a damaged ecosystem or compensating for damage by providing a substitute or replacement area.
**mixed perennial polyculture** Growing a mixture of different perennial crop species (where the same plant persists for more than one year) together in the same plot.
**models** Simple representations of more complex systems.
**molecule** A combination of two or more atoms.
**monitored, retrievable storage** Holding wastes in underground mines or secure surface facilities such as dry casks where they can be watched and repackaged, if necessary.
**monoculture agroforestry** Intensive planting of a single species; an efficient wood production approach, but one that encourages pests and disease infestations and conflicts with wildlife habitat or recreation uses.
**monsoon** A seasonal reversal of wind patterns caused by the different heating and cooling rates of the oceans and continents.
**montane coniferous forests** Coniferous forests of the mountains consisting of belts of different forest communities along an altitudinal gradient.
**moral agents** Beings capable of making distinctions between right and wrong and acting accordingly. Those whom we hold responsible for their actions.
**moral extensionism** Expansion of our understanding of inherent value or rights to persons, organisms, or things that might not be considered worthy of value or rights under some ethical philosophies.
**moral subjects** Beings that are not capable of distinguishing between right or wrong or that are not able to act on moral principles and yet are capable of being wronged by others.
**morals** A set of ethical principles that guide our actions and relationships.
**morbidity** Illness or disease.
**more-developed countries (MDC)** Industrialized nations characterized by high per capita incomes, low birth and death rates, low population growth rates, and high levels of industrialization and urbanization.
**mortality** Death rate in a population; the probability of dying.
**Müellerian mimicry** Evolution of two species, both of which are unpalatable and, have poisonous stingers or some other defense mechanism, to resemble each other.
**mulch** Protective ground cover, including both natural products and synthetic materials that protect the soil, save water, and prevent weed growth.
**multiple use** Many uses that occur simultaneously; used in forest management; limited to mutually compatible uses.
**mutagens** Agents, such as chemicals or radiation, that damage or alter genetic material (DNA) in cells.
**mutation** A change, either spontaneous or by external factors, in the genetic material of a cell; mutations in the gametes (sex cells) can be inherited by future generations of organisms.
**mutualism** A symbiotic relationship between individuals of two different species in which both species benefit from the association.

# N

**NAAQS** *N*ational *A*mbient *A*ir *Q*uality *S*tandard; federal standards specifying the maximum allowable levels (averaged over specific time periods) for regulated pollutants in ambient (outdoor) air.

**natality** The production of new individuals by birth, hatching, germination, or cloning.

**natural experiment** A study of events that have already happened.

**natural history** The study of where and how organisms carry out their life cycles.

**natural increase** Crude death rate subtracted from crude birthrate.

**natural organic pesticides** "Botanicals" or organic compounds naturally occurring in plants, animals or microbes that serve as pesticides.

**natural resources** Goods and services supplied by the environment.

**natural selection** The mechanism for evolutionary change in which environmental pressures cause certain genetic combinations in a population to become more abundant; genetic combinations best adapted for present environmental conditions tend to become predominant.

**negative feedback loop** A situation in which a factor or condition causes changes that reduce that factor or condition.

**neoclassical economics** A branch of economics that attempts to apply the principles of modern science to economic analysis in a mathematically rigorous, noncontextual, abstract, predictive manner.

**neo-Luddites** People who reject technology as the cause of environmental degradation and social disruption. Named after the followers of Ned Ludd who tried to turn back the Industrial Revolution in England.

**neo-Malthusian** A belief that the world is characterized by scarcity and competition in which too many people fight for too few resources. Named for Thomas Malthus, who predicted a dismal cycle of misery, vice, and starvation as a result of human overpopulation.

**NEPA** NEPA, the National Environmental Policy Act, is the cornerstone of U.S. environmental policy. It authorizes the Council on Environmental Quality, directs federal agencies to take environmental consequences into account when making decisions, and requires an environmental impact statement for every major federal project likely to have adverse environmental effects.

**net energy yield** Total useful energy produced during the lifetime of an entire energy system minus the energy used, lost, or wasted in making useful energy available.

**neurotoxins** Toxic substances, such as lead or mercury, that specifically poison nerve cells.

**neutron** A subatomic particle, found in the nucleus of the atom, that has no electromagnetic charge.

**new towns** Experimental urban environments that seek to combine the best features of the rural village and the modern city.

**nihilists** Those who believe the world has no meaning or purpose other than a dark, cruel, unceasing struggle for power and existence.

**NIMBY** *N*ot *I*n *M*y *B*ack*Y*ard: the rallying cry of those opposed to LULUs.

**nitrate-forming bacteria** Bacteria that convert nitrites into compounds that can be used by green plants to build proteins.

**nitrite-forming bacteria** Bacteria that combine ammonia with oxygen to form nitrites.

**nitrogen cycle** The circulation and reutilization of nitrogen in both inorganic and organic phases.

**nitrogen-fixing bacteria** Bacteria that convert nitrogen from the atmosphere or soil solution into ammonia that can then be converted to plant nutrients by nitrite- and nitrate-forming bacteria.

**nitrogen oxides** Highly reactive gases formed when nitrogen in fuel or combustion air is heated to over 650°C (1,200°F) in the presence of oxygen or when bacteria in soil or water oxidize nitrogen-containing compounds.

**noncriteria pollutants** *See* unconventional air pollutants.

**nongovernmental organizations (NGOs)** A term referring collectively to pressure and research groups, advisory agencies, political parties, professional societies, and other groups concerned about environmental quality, resource use, and many other issues.

**nonpoint sources** Scattered, diffuse sources of pollutants, such as runoff from farm fields, golf courses, construction sites, etc.

**nonrenewable resources** Minerals, fossil fuels, and other materials present in essentially fixed amounts (within human time scales) in our environment.

**nuclear fission** The radioactive decay process in which isotopes split apart to create two smaller atoms.

**nuclear fusion** A process in which two smaller atomic nuclei fuse into one larger nucleus and release energy; the source of power in a hydrogen bomb.

**nucleic acids** Large organic molecules made of nucleotides that function in the transmission of hereditary traits, in protein synthesis, and in control of cellular activities.

**nucleus** The center of the atom; occupied by protons and neutrons. In cells, the organelle that contains the chromosomes (DNA).

**nuées ardentes** Deadly, denser-than-air mixtures of hot gases and ash ejected from volcanoes.

**numbers pyramid** A diagram showing the relative population sizes at each trophic level in an ecosystem; usually corresponds to the biomass pyramid.

# O

**obese** Generally considered to be a body mass greater than 30 kg/m$^2$, or roughly 30 pounds above normal for an average person.

**ocean shorelines** Rocky coasts and sandy beaches along the oceans; support rich, stratified communities.

**ocean thermal electric conversion (OTEC)** Energy derived from temperature differentials between warm ocean surface waters and cold deep waters. This differential can be used to drive turbines attached to electric generators.

**oceanic islands** Islands in the ocean; formed by breaking away from a continental landmass, volcanic action, coral formation, or a combination of sources; support distinctive communities.

**offset allowances** A controversial component of air quality regulations that allows a polluter to avoid installation of control equipment on one source with an "offsetting" pollution reduction at another source.

**oil shale** A fine-grained sedimentary rock rich in solid organic material called kerogen. When heated, the kerogen liquefies to produce a fluid petroleum fuel.

**old-growth forests** Forests free from disturbance for long enough (generally 150 to 200 years) to have mature trees, physical conditions, species diversity, and other characteristics of equilibrium ecosystems.

**oligotrophic** Condition of rivers and lakes that have clear water and low biological productivity (*oligo* = little; *trophic* = nutrition); are usually clear, cold, infertile headwater lakes and streams.

**omnivore** An organism that eats both plants and animals.

**open access system** A commonly held resource for which there are no management rules.

**open canopy** A forest where tree crowns cover less than 20 percent of the ground; also called woodland.

**open range** Unfenced, natural grazing lands; includes woodland as well as grassland.

**open system** A system that exchanges energy and matter with its environment.

**optimum** The most favorable condition in regard to an environmental factor.

**orbital** The space or path in which an electron orbits the nucleus of an atom.

**organic compounds** Complex molecules organized around skeletons of carbon atoms arranged in rings or chains; includes biomolecules, molecules synthesized by living organisms.

**organophosphates** Organic molecules to which phosphate group(s) are attached.

**overburden** Overlying layers of noncommercial sediments that must be removed to reach a mineral or coal deposit.

**overgrazing** Allowing livestock to eat so much forage that it damages the ecological health of the habitat.

**overharvesting** Harvesting so much of a resource that it threatens its existence.

**overnutrition** Receiving too many calories.

**overshoot** The extent to which a population exceeds the carrying capacity of its environment.

**oxygen cycle** The circulation and reutilization of oxygen in the biosphere.

**oxygen sag** Oxygen decline downstream from a pollution source that introduces materials with high biological oxygen demands.

**ozone** A highly reactive molecule containing three oxygen atoms; a dangerous pollutant in ambient air. In the stratosphere, however, ozone forms an ultraviolet absorbing shield that protects us from mutagenic radiation.

# P

**Pacific Decadal Oscillation (PDO)** A large pool of warm water that moves north and south in the Pacific Ocean every 30 years or so and has large effects on North America's climate.

**parabolic mirrors** Curved mirrors that focus light from a large area onto a single, central point, thereby concentrating solar energy and producing high temperatures.

**paradigm** A model that provides a framework for interpreting observations.

**parasite** An organism that lives in or on another organism, deriving nourishment at the expense of its host, usually without killing it.

**parsimony** If two explanations appear equally plausible, choose the simpler one.

**particulate material** Atmospheric aerosols, such as dust, ash, soot, lint, smoke, pollen, spores, algal cells, and other suspended materials; originally applied only to solid particles but now extended to droplets of liquid.

**parts per billion (ppb)** Number of parts of a chemical found in 1 billion parts of a particular gas, liquid, or solid mixture.

**parts per million (ppm)** Number of parts of a chemical found in 1 million parts of a particular gas, liquid, or solid mixture.

**parts per trillion (ppt)** Number of parts of a chemical found in 1 trillion (1012) parts of a particular gas, liquid, or solid mixture.

**passive heat absorption** The use of natural materials or absorptive structures without moving parts to gather and hold heat; the simplest and oldest use of solar energy.

**pastoralists** People who make a living by herding domestic livestock.

**pasture** Grazing lands suitable for domestic livestock.

**patchiness** Within a larger ecosystem, the presence of smaller areas that differ in some physical conditions and thus support somewhat different communities; a diversity-promoting phenomenon.

**pathogen** An organism that produces disease in a host organism, disease being an alteration of one or more metabolic functions in response to the presence of the organism.

**peat** Deposits of moist, acidic, semidecayed organic matter.

**pelagic** Zones in the vertical water column of a water body.

**pellagra** Lassitude, torpor, dermatitis, diarrhea, dementia, and death brought about by a diet deficient in tryptophan and niacin.

**peptides** Two or more amino acids linked by a peptide bond.

**perennial species** Plants that grow for more than two years.

**permafrost** A permanently frozen layer of soil that underlies the arctic tundra.

**permanent retrievable storage** Placing waste storage containers in a secure building, salt mine, or bedrock cavern where they can be inspected periodically and retrieved, if necessary.

**persistent organic pollutants (POPs)** Chemical compounds that persist in the environment and retain biological activity for long times.

**pest** Any organism that reduces the availability, quality, or value of a useful resource.

**pesticide** Any chemical that kills, controls, drives away, or modifies the behavior of a pest.

**pesticide treadmill** A need for constantly increasing doses or new pesticides to prevent pest resurgence.

**pest resurgence** Rebound of pest populations due to acquired resistance to chemicals and nonspecific destruction of natural predators and competitors by broadscale pesticides.

**pH** A value that indicates the acidity or alkalinity of a solution on a scale of 0 to 14, based on the proportion of $H^+$ ions present.

**phosphorus cycle** The movement of phosphorus atoms from rocks through the biosphere and hydrosphere and back to rocks.

**photochemical oxidants** Products of secondary atmospheric reactions. *See* smog.

**photodegradable plastics** Plastics that break down when exposed to sunlight or to a specific wavelength of light.

**photosynthesis** The biochemical process by which green plants and some bacteria capture light energy and use it to produce chemical bonds. Carbon dioxide and water are consumed while oxygen and simple sugars are produced.

**photosynthetic efficiency** The percentage of available light captured by plants and used to make useful products.

**photovoltaic cell** An energy-conversion device that captures solar energy and directly converts it to electrical current.

**physical or abiotic factors** Nonliving factors, such as temperature, light, water, minerals, and climate, that influence an organism.

**phytoplankton** Microscopic, free-floating, autotrophic organisms that function as producers in aquatic ecosystems.

**pioneer species** In primary succession on a terrestrial site, the plants, lichens, and microbes that first colonize the site.

**plankton** Primarily microscopic organisms that occupy the upper water layers in both freshwater and marine ecosystems.

**plasma** A hot, electrically neutral gas of ions and free electrons.

**plug-in hybrids** Vehicles with hybrid gasoline-electric engines adapted with a larger battery array (enough to propel the vehicle for 50 km or so on the batteries alone) and a plug-in to recharge the batteries from a standard electric outlet.

**poachers** Those who hunt wildlife illegally.

**point sources** Specific locations of highly concentrated pollution discharge, such as factories, power plants, sewage treatment plants, underground coal mines, and oil wells.

**policy** A societal plan or statement of intentions intended to accomplish some social good.

**policy cycle** The process by which problems are identified and acted upon in the public arena.

**political economy** The branch of economics concerned with modes of production, distribution of benefits, social institutions, and class relationships.

**pollution** To make foul, unclean, dirty; any physical, chemical, or biological change that adversely affects the health, survival, or activities of living organisms or that alters the environment in undesirable ways.

**pollution charges** Fees assessed per unit of pollution based on the "polluter pays" principle.

**polycentric complex** Cities with several urban cores surrounding a once dominant central core.

**population** A group of individuals of the same species occupying a given area.

**population crash** A sudden population decline caused by predation, waste accumulation, or resource depletion; also called a dieback.

**population explosion** Growth of a population at exponential rates to a size that exceeds environmental carrying capacity; usually followed by a population crash.

**population momentum** A potential for increased population growth as young members reach reproductive age.

**positive feedback loop** A situation in which a factor or condition causes changes that further enhance that factor or condition.

**postmaterialist values** A philosophy that emphasizes quality of life over acquisition of material goods.

**post-modernism** A philosophy that rejects the optimism and universal claims of modern positivism.

**potential energy** Stored energy that is latent but available for use. A rock poised at the top of a hill or water stored behind a dam are examples of potential energy.

**power** The rate of energy delivery; measured in horsepower or watts.

**precautionary principle** The decision to leave a margin of safety for unexpected developments.

**precedent** An act or decision that can be used as an example in dealing with subsequent similar situations.

**precycling** Making environmentally sound decisions at the store and reducing waste before we buy.

**predation** The act of feeding by a predator.

**predator** An organism that feeds directly on other organisms in order to survive; live-feeders, such as herbivores and carnivores.

**predator-mediated competition** A situation in which predation reduces prey populations and gives an advantage to competitors that might not otherwise be successful.

**premises** Introductory statements that set up or define a problem. Those things taken as given.

**prevention of significant deterioration** A clause of the Clean Air Act that prevents degradation of existing clean air; opposed by industry as an unnecessary barrier to development.

**price elasticity** A situation in which supply and demand of a commodity respond to price.

**primary pollutants** Chemicals released directly into the air in a harmful form.

**primary productivity** Synthesis of organic materials (biomass) by green plants using the energy captured in photosynthesis.

**primary standards** Regulations of the 1970 Clean Air Act; intended to protect human health.

**primary succession** An ecological succession that begins in an area where no biotic community previously existed.

**primary treatment** A process that removes solids from sewage before it is discharged or treated further.

**principle of competitive exclusion** A result of natural selection whereby two similar species in a community occupy different ecological niches, thereby reducing competition for food.

**producer** An organism that synthesizes food molecules from inorganic compounds by using an external energy source; most producers are photosynthetic.

**production frontier** The maximum output of two competing commodities at different levels of production.

**productivity** The synthesis of new organic material. That done by green plants using solar energy is called primary productivity.

**prokaryotic** Cells that do not have a membrane-bounded nucleus or membrane-bounded organelles.

**promoters** Agents that are not carcinogenic but that assist in the progression and spread of tumors; sometimes called cocarcinogens.

**pronatalist pressures** Influences that encourage people to have children.

**proteins** Chains of amino acids linked by peptide bonds.

**proton** A positively charged subatomic particle found in the nucleus of an atom.

**proven reserves** *See* proven resources.

**proven resources** Those that have been thoroughly mapped and are economical to recover at current prices with available technology.

**public trust** A doctrine obligating the government to maintain public lands in a natural state as guardians of the public interest.

**pull factors** (in urbanization) Conditions that draw people from the country into the city.
**push factors** (in urbanization) Conditions that force people out of the country and into the city.

# R

***r*-selected species** Species that tend to have rapid reproduction and high offspring mortality. They frequently overshoot carrying capacity of their environment and display boom and bust cycles. They lack intrinsic population controls and tend to occupy lower trophic levels in food webs than *k*-selected species.
**radiative evolution** Divergence from a common ancestor into two or more new species.
**radioactive** An unstable isotope that decays spontaneously and releases subatomic particles or units of energy.
**radioactive decay** A change in the nuclei of radioactive isotopes that spontaneously emit high-energy electromagnetic radiation and/or subatomic particles while gradually changing into another isotope or different element.
**radionucleides** Isotopes that exhibit radioactive decay.
**rainforest** A forest with high humidity, constant temperature, and abundant rainfall (generally over 380 cm [150 in] per year); can be tropical or temperate.
**rain shadow** Dry area on the downwind side of a mountain.
**rangeland** Grasslands and open woodlands suitable for livestock grazing.
**rational choice** Public decision making based on reason, logic, and science-based management.
**recharge zone** Area where water infiltrates into an aquifer.
**reclamation** Chemical, biological, or physical cleanup and reconstruction of severely contaminated or degraded sites to return them to something like their original topography and vegetation.
**recoverable resources** Those accessible with current technology but not economical under current conditions.
**re-creation** Construction of an entirely new biological community to replace one that has been destroyed on that or another site.
**recycling** Reprocessing of discarded materials into new, useful products; not the same as reuse of materials for their original purpose, but the terms are often used interchangeably.
**red tide** A population explosion or bloom of minute, single-celled marine organisms called dinoflagellates. Billions of these cells can accumulate in protected bays where the toxins they contain can poison other marine life.
**reduced tillage systems** Systems, such as minimum till, conserve-till, and no-till, that preserve soil, save energy and water, and increase crop yields.
**reflective thinking** Asks, what does this all mean?
**reformer** A device that strips hydrogen from fuels such as natural gas, methanol, ammonia, gasoline, or vegetable oil so they can be used in a fuel cell.
**refuse-derived fuel** Processing of solid waste to remove metal, glass, and other unburnable materials; organic residue is shredded, formed into pellets, and dried to make fuel for power plants.
**regenerative farming** Farming techniques and land stewardship that restore the health and productivity of the soil by rotating crops, planting ground cover, protecting the surface with crop residue, and reducing synthetic chemical inputs and mechanical compaction.
**regolith** The parent material between bedrock and subsoil (sometimes called the C horizon) that is made of weathered rock fragments and very little organic material and from which soil is formed.
**regulations** Rules established by administrative agencies; regulations can be more important than statutory law in the day-to-day management of resources.
**rehabilitate land** A utilitarian program to make an area useful to humans.
**rehabilitation** To rebuild elements of structure or function in an ecological system without necessarily achieving complete restoration to its original condition.
**relative humidity** At any given temperature, a comparison of the actual water content of the air with the amount of water that could be held at saturation.
**relativists** Those who believe moral principles are always dependent on the particular situation.
**remediation** Cleaning up chemical contaminants from a polluted area.
**renewable resources** Resources normally replaced or replenished by natural processes; resources not depleted by moderate use; examples include solar energy, biological resources such as forests and fisheries, biological organisms, and some biogeochemical cycles.
**renewable water supplies** Annual freshwater surface runoff plus annual infiltration into underground freshwater aquifers that are accessible for human use.
**replication** Repeating studies or tests to verify reliability.
**reproducibility** Making an observation or obtaining a particular result more than once.
**reproductive isolation** Barriers (geographical, behavioral, or biological) that prevent gene flow between members of a species.
**residence time** The length of time a component, such as an individual water molecule, spends in a particular compartment or location before it moves on through a particular process or cycle.
**resilience** The ability of a community or ecosystem to recover from disturbances.
**resistance** (inertia) The ability of a community to resist being changed by potentially disruptive events.
**resource** In economic terms, anything with potential use in creating wealth or giving satisfaction.
**resource partitioning** In a biological community, various populations sharing environmental resources through specialization, thereby reducing direct competition. *See also* ecological niche.
**resource scarcity** A shortage or deficit in some resource.
**restoration** To bring something back to a former condition. Ecological restoration involves active manipulation of nature to re-create conditions that existed before human disturbance.
**restoration ecology** Seeks to repair or reconstruct ecosystems damaged by human actions.
**riders** Amendments attached to bills in conference committee, often completely unrelated to the bill to which they are added.
**rill erosion** The removing of thin layers of soil as little rivulets of running water gather and cut small channels in the soil.
**risk** Probability that something undesirable will happen as a consequence of exposure to a hazard.
**risk assessment** Evaluation of the short-term and long-term risks associated with a particular activity or hazard; usually compared to benefits in a cost-benefit analysis.
**RNA** Ribonucleic acid; nucleic acid used for transcription and translation of the genetic code found on DNA molecules.
**rock** A solid, cohesive, aggregate of one or more crystalline minerals.
**rock cycle** The process whereby rocks are broken down by chemical and physical forces; sediments are moved by wind, water, and gravity, sedimented and reformed into rock, and then crushed, folded, melted, and recrystallized into new forms.
**rotational grazing** Confining animals to a small area for a short time (often only a day or two) before shifting them to a new location.
**ruminant animals** Cud-chewing animals, such as cattle, sheep, goats, and buffalo, with multichambered stomachs in which cellulose is digested with the aid of bacteria.
**runoff** The excess of precipitation over evaporation; the main source of surface water and, in broad terms, the water available for human use.
**run-of-the-river flow** Ordinary river flow not accelerated by dams, flumes, etc. Some small, modern, high-efficiency turbines can generate useful power with run-of-the-river flow or with a current of only a few kilometers per hour.
**rural area** An area in which most residents depend on agriculture or the harvesting of natural resources for their livelihood.

# S

**S curve** A curve that depicts logistic growth; called an S curve because of its shape.
**salinity** Amount of dissolved salts (especially sodium chloride) in a given volume of water.
**salinization** A process in which mineral salts accumulate in the soil, killing plants; occurs when soils in dry climates are irrigated profusely.
**salt marsh** Shallow wetlands along coastlines that are flooded regularly or occasionally with seawater.
**saltwater intrusion** Movement of saltwater into freshwater aquifers in coastal areas where groundwater is withdrawn faster than it is replenished.
**salvage logging** Harvesting timber killed by fire, disease, or windthrow.
**sample** To analyze a small but representative portion of a population to estimate the characteristics of the entire class.
**sanitary landfills** A landfill in which garbage and municipal waste is buried every day under enough soil or fill to eliminate odors, vermin, and litter.
**saturation point** The maximum concentration of water vapor the air can hold at a given temperature.
**savannas** Open grasslands with sparse tree cover.
**scavenger** An organism that feeds on the dead bodies of other organisms.
**science** A process for producing knowledge methodically and logically.
**scientific consensus** A general agreement among informed scholars.
**scientific method** A systematic, precise, objective study of a problem. Generally this requires observation, hypothesis development and testing, data gathering, and interpretation.

**scientific theory** An explanation supported by many tests and accepted by a general consensus of scientists.

**secondary pollutants** Chemicals modified to a hazardous form after entering the air or that are formed by chemical reactions as components of the air mix and interact.

**secondary recovery technique** Pumping pressurized gas, steam, or chemical-containing water into a well to squeeze more oil from a reservoir.

**secondary standards** Regulations of the 1972 Clean Air Act intended to protect materials, crops, visibility, climate, and personal comfort.

**secondary succession** Succession on a site where an existing community has been disrupted.

**secondary treatment** Bacterial decomposition of suspended particulates and dissolved organic compounds that remain after primary sewage treatment.

**second law of thermodynamics** States that, with each successive energy transfer or transformation in a system, less energy is available to do work.

**secure landfill** A solid waste disposal site lined and capped with an impermeable barrier to prevent leakage or leaching. Drain tiles, sampling wells, and vent systems provide monitoring and pollution control.

**sedimentary rock** Deposited material that remains in place long enough or is covered with enough material to compact into stone; examples include shale, sandstone, breccia, and conglomerates.

**sedimentation** The deposition of organic materials or minerals by chemical, physical, or biological processes.

**selection pressures** Factors in the environment that favor successful reproduction of individuals possessing heritable traits and that reduce viability and fertility of those individuals not possessing those traits.

**selective cutting** Harvesting only mature trees of certain species and size; usually more expensive than clear-cutting, but it is less disruptive for wildlife and often better for forest regeneration.

**seriously undernourished** Those who receive less than 80 percent of their minimum daily caloric requirements.

**shantytowns** Settlements created when people move onto undeveloped lands and build their own shelter with cheap or discarded materials; some are simply illegal subdivisions where a landowner rents land without city approval; others are land invasions.

**sheet erosion** Peeling off thin layers of soil from the land surface; accomplished primarily by wind and water.

**shelterwood harvesting** A harvest method in which mature trees are removed in a series of linear cuts leaving enough trees in unharvested strips to protect from wind storms or other disruptive factors.

**sick building syndrome** Headaches, allergies, chronic fatigue and other symptoms caused by poorly vented indoor air contaminated by pathogens or toxins.

**significant numbers** Meaningful numbers whose accuracy can be verified.

**sinkholes** A large surface crater caused by the collapse of an underground channel or cavern; often triggered by groundwater withdrawal.

**sludge** Semisolid mixture of organic and inorganic materials that settles out of wastewater at a sewage treatment plant.

**slums** Legal but inadequate multifamily tenements or rooming houses; some are custom built for rent to poor people, others are converted from some other use.

**smart growth** Efficient use of land resources and existing urban infrastructure.

**smelting** Heating ores to extract metals.

**smog** The term used to describe the combination of smoke and fog in the stagnant air of London; now often applied to photochemical pollution products or urban air pollution of any kind.

**social justice** Equitable access to resources and the benefits derived from them; a system that recognizes inalienable rights and adheres to what is fair, honest, and moral.

**soil** A complex mixture of weathered mineral materials from rocks, partially decomposed organic molecules, and a host of living organisms.

**soil horizons** Horizontal layers that reveal a soil's history, characteristics, and usefulness.

**soil profile** All the vertical layers or horizons that make up a soil in a particular place.

**southern pine forest** United States coniferous forest ecosystem characterized by a warm, moist climate.

**speciation** The generation of new species.

**species** A population of morphologically similar organisms that can reproduce sexually among themselves but that cannot produce fertile offspring when mated with other organisms.

**species diversity** The number and relative abundance of species present in a community.

**species recovery plan** A plan for restoration of an endangered species through protection, habitat management, captive breeding, disease control, or other techniques that increase populations and encourage survival.

**sprawl** Unlimited outward extension of city boundaries that lowers population density, consumes open space, generates freeway congestion, and causes decay in central cities.

**spring overturn** Springtime lake phenomenon that occurs when the surface ice melts and the surface water temperature warms to its greatest density at 4°C and then sinks, creating a convection current that displaces nutrient-rich bottom waters.

**squatter towns** Shantytowns that occupy land without owner's permission; some are highly organized movements in defiance of authorities; others grow gradually.

**stability** In ecological terms, a dynamic equilibrium among the physical and biological factors in an ecosystem or a community; relative homeostasis.

**stable runoff** The fraction of water available year-round; usually more important than total runoff when determining human uses.

**Standard Metropolitan Statistical Area (SMSA)** An urbanized region with at least 100,000 inhabitants with strong economic and social ties to a central city of at least 50,000 people.

**standing** The right to take part in legal proceedings.

**statistics** Numbers that let you evaluate and compare things.

**statute law** Formal documents or decrees enacted by the legislative branch of government.

**statutory law** Rules passed by a state or national legislature.

**steady-state economy** Characterized by low birth and death rates, use of renewable energy sources, recycling of materials, and emphasis on durability, efficiency, and stability.

**stewardship** A philosophy that holds that humans have a unique responsibility to manage, care for, and improve nature.

**strategic lawsuits against public participation (SLAPP)** Lawsuits that have no merit but are brought merely to intimidate and harass private citizens who act in the public interest.

**strategic metals and minerals** Materials a country cannot produce itself but that it uses for essential materials or processes.

**stratosphere** The zone in the atmosphere extending from the tropopause to about 50 km (30 mi) above the earth's surface; temperatures are stable or rise slightly with altitude; has very little water vapor but is rich in ozone.

**stratospheric ozone** The ozone ($O_3$) occurring in the stratosphere 10 to 50 km above the earth's surface.

**stress-related diseases** Diseases caused or accentuated by social stresses such as crowding.

**strip cutting** Harvesting trees in strips narrow enough to minimize edge effects and to allow natural regeneration of the forest.

**strip farming** Planting different kinds of crops in alternating strips along land contours; when one crop is harvested, the other crop remains to protect the soil and prevent water from running straight down a hill.

**strip mining** Removing surface layers over coal seams using giant, earth-moving equipment; creates a huge open-pit from which coal is scooped by enormous surface-operated machines and transported by trucks; an alternative to deep mines.

**structure** (in ecological terms) Patterns of organization, both spatial and functional, in a community.

**Student Environmental Action Coalition (SEAC)** A student-based environmental organization that is both an umbrella organization and a grassroots network to facilitate environmental action and education on college campuses.

**subduction** The process by which one tectonic plate is pushed down below another as plates crash into each other.

**sublimation** The process by which water can move between solid and gaseous states without ever becoming liquid.

**subsidence** A settling of the ground surface caused by the collapse of porous formations that result from withdrawal of large amounts of groundwater, oil, or other underground materials.

**subsoil** A layer of soil beneath the topsoil that has lower organic content and higher concentrations of fine mineral particles; often contains soluble compounds and clay particles carried down by percolating water.

**sulfur cycle** The chemical and physical reactions by which sulfur moves into or out of storage and through the environment.

**sulfur dioxide** A colorless, corrosive gas directly damaging to both plants and animals.

**Superfund** A fund established by Congress to pay for containment, cleanup, or remediation of abandoned toxic waste sites. The fund is financed by fees paid by toxic waste generators and by cost- recovery from cleanup projects.

**supply** The quantity of that product being offered for sale at various prices, other things being equal.

**surface mining** Some minerals are also mined from surface pits. *See* strip mining.

**surface tension** A condition in which the water surface meets the air and acts like an elastic skin.

**survivorship** The percentage of a population reaching a given age or the proportion of the maximum life span of the species reached by any individual.

**sustainability** Living within the bounds of nature based on renewable resources used in ways that don't deplete nonrenewable resources, harm essential ecological services, or limit the ability of future generations to meet their own needs.

**sustainable agriculture** An ecologically sound, economically viable, socially just, and humane agricultural system. Stewardship, soil conservation, and integrated pest management are essential for sustainability.

**sustainable development** A real increase in well-being and standard of life for the average person that can be maintained over the long-term without degrading the environment or compromising the ability of future generations to meet their own needs.

**sustained yield** Utilization of a renewable resource at a rate that does not impair or damage its ability to be fully renewed on a long-term basis.

**swamp** Wetland with trees, such as the extensive swamp forests of the southern United States.

**swidden agriculture** *See* milpa agriculture.

**symbiosis** The intimate living together of members of two different species; includes **mutualism, commensalism,** and, in some classifications, **parasitism.**

**sympatric speciation** Species that arise from a common ancestor due to biological or behavioral barriers that cause reproductive isolation even though the organisms live in the same place.

**synergism** An interaction in which one substance exacerbates the effects of another. The sum of the interaction is greater than the parts.

**synergistic effects** When an injury caused by exposure to two environmental factors together is greater than the sum of exposure to each factor individually.

**systemic** A condition or process that affects the whole body; many metabolic poisons are systemic.

**systems** Networks of interactions among many interdependent factors.

# T

**taiga** The northernmost edge of the boreal forest, including species-poor woodland and peat deposits; intergrading with the arctic tundra.

**tailings** Mining waste left after mechanical or chemical separation of minerals from crushed ore.

**taking** Unconstitutional confiscation of private property.

**tar sands** Sand deposits containing petroleum or tar.

**technological optimists** Those who believe that technology and human enterprise will find cures for all our problems. Also called **Promethean environmentalism.**

**tectonic plates** Huge blocks of the earth's crust that slide around slowly, pulling apart to open new ocean basins or crashing ponderously into each other to create new, larger landmasses.

**temperate rainforest** The cool, dense, rainy forest of the northern Pacific coast; enshrouded in fog much of the time; dominated by large conifers.

**temperature** A measure of the speed of motion of a typical atom or molecule in a substance.

**temperature inversions** A stable layer of warm air overlays cooler air, trapping pollutants near ground level.

**teratogens** Chemicals or other factors that specifically cause abnormalities during embryonic growth and development.

**terracing** Shaping the land to create level shelves of earth to hold water and soil; requires extensive hand labor or expensive machinery, but it enables farmers to farm very steep hillsides.

**territoriality** An intense form of intraspecific competition in which organisms define an area surrounding their home site or nesting site and defend it, primarily against other members of their own species.

**tertiary treatment** The removal of inorganic minerals and plant nutrients after primary and secondary treatment of sewage.

**thermal plume** A plume of hot water discharged into a stream or lake by a heat source, such as a power plant.

**thermocline** In water, a distinctive temperature transition zone that separates an upper layer that is mixed by the wind (the epilimnion) and a colder, deep layer that is not mixed (the hypolimnion).

**thermodynamics** A branch of physics that deals with transfers and conversions of energy.

**thermodynamics, first law** Energy can be transformed and transferred, but cannot be destroyed or created.

**thermodynamics, second law** With each successive energy transfer or transformation, less energy is available to do work.

**thermosphere** The highest atmospheric zone; a region of hot, dilute gases above the mesosphere extending out to about 1,600 km (1,000 mi) from the earth's surface.

**Third World** Less-developed countries that are not capitalistic and industrialized (First World) or centrally-planned socialist economies (Second World); not intended to be derogatory.

**threatened species** While still abundant in parts of its territorial range, this species has declined significantly in total numbers and may be on the verge of extinction in certain regions or localities.

**tidal station** A dam built across a narrow bay or estuary traps tide water flowing both in and out of the bay. Water flowing through the dam spins turbines attached to electric generators.

**tide pool** Depressions in a rocky shoreline that are flooded at high tide but cut off from the ocean at low tide.

**timberline** In mountains, the highest-altitude edge of forest that marks the beginning of the treeless alpine tundra.

**tolerance limits** *See* limiting factors.

**topsoil** The first true layer of soil; layer in which organic material is mixed with mineral particles; thickness ranges from a meter or more under virgin prairie to zero in some deserts.

**tornado** A violent storm characterized by strong swirling winds and updrafts; tornadoes form when a strong cold front pushes under a warm, moist air mass over the land.

**tort law** Court cases that seek compensation for damages.

**total fertility rate** The number of children born to an average woman in a population during her entire reproductive life.

**total growth rate** The net rate of population growth resulting from births, deaths, immigration, and emigration.

**total maximum daily loads (TMDL)** The amount of particular pollutant that a water body can receive from both point and nonpoint sources and still meet water quality standards.

**toxic colonialism** Shipping toxic wastes to a weaker or poorer nation.

**Toxic Release Inventory** A program created by the Superfund Amendments and Reauthorization Act of 1984 that requires manufacturing facilities and waste handling and disposal sites to report annually on releases of more than 300 toxic materials.

**toxins** Poisonous chemicals that react with specific cellular components to kill cells or to alter growth or development in undesirable ways; often harmful, even in dilute concentrations.

**tradable permits** Pollution quotas or variances that can be bought or sold.

**tragedy of the commons** An inexorable process of degradation of communal resources due to selfish self-interest of "free riders" who use or destroy more than their fair share of common property. *See* open access system.

**transitional zone** A zone in which populations from two or more adjacent communities meet and overlap.

**transpiration** The evaporation of water from plant surfaces, especially through stomates.

**triple bottom line** Corporate accounting that reports social and environmental costs and benefits as well as merely economic ones.

**trophic level** Step in the movement of energy through an ecosystem; an organism's feeding status in an ecosystem.

**tropical rainforests** Forests in which rainfall is abundant—more than 200 cm (80 in) per year—and temperatures are warm to hot year-round.

**tropical seasonal forest** Semievergreen or partly deciduous forests tending toward open woodlands and grassy savannas dotted with scattered, drought-resistant tree species; distinct wet and dry seasons, hot year-round.

**tropopause** The boundary between the troposphere and the stratosphere.

**troposphere** The layer of air nearest to the earth's surface; both temperature and pressure usually decrease with increasing altitude.

**tsunami** Giant seismic sea swells that move rapidly from the center of an earthquake; they can be 10 to 20 meters high when they reach shorelines hundreds or even thousands of kilometers from their source.

**tundra** Treeless arctic or alpine biome characterized by cold, harsh winters, a short growing season, and potential for frost any month of the year; vegetation includes low-growing perennial plants, mosses, and lichens.

# U

**unconventional air pollutants** Toxic or hazardous substances, such as asbestos, benzene, beryllium, mercury, polychlorinated biphenyls, and vinyl chloride, not listed in the original Clean Air Act because they were not released in large quantities; also called noncriteria pollutants.

**unconventional oil** Resources such as shale oil and tar sands that can be liquefied and used like oil.

**undernourished** Those who receive less than 90 percent of the minimum dietary intake over a long-term time period; they lack energy for an active, productive life and are more susceptible to infectious diseases.

**undiscovered resources** Speculative or inferred resources or those that we haven't even thought about.

**universalists** Those who believe that some fundamental ethical principles are universal and unchanging. In this vision, these principles are valid regardless of the context or situation.

**upwelling** Convection currents within a body of water that carry nutrients from bottom sediments toward the surface.

**urban agglomerations** An aggregation of many cities into a large metropolitan area.

**urban area** An area in which a majority of the people are not directly dependent on natural resource-based occupations.

**urbanization** An increasing concentration of the population in cities and a transformation of land use to an urban pattern of organization.

**utilitarian conservation** A philosophy that resources should be used for the greatest good for the greatest number for the longest time.

**utilitarianism** *See* utilitarian conservation.

**utilitarians** Those who hold that an action is right that produces the greatest good for the greatest number of people.

# V

**values** An estimation of the worth of things; a set of ethical beliefs and preferences that determine our sense of right and wrong.

**verbal learner** Someone who understands and remembers best by listening to the spoken word.

**vertical stratification** The vertical distribution of specific subcommunities within a community.

**vertical zonation** Terrestrial vegetation zones determined by altitude.

**village** A collection of rural households linked by culture, custom, and association with the land.

**visible light** A portion of the electromagnetic spectrum that includes the wavelengths used for photosynthesis.

**visual learner** Someone who understands and remembers best by reading, or looking at pictures and diagrams.

**vitamins** Organic molecules essential for life that we cannot make for ourselves; we must get them from our diet; they act as enzyme cofactors.

**volatile organic compounds (VOCs)** Organic chemicals that evaporate readily and exist as gases in the air.

**volcanoes** Vents in the earth's surface through which gases, ash, or molten lava are ejected. Also a mountain formed by this ejecta.

**voluntary simplicity** Deliberately choosing to live at a lower level of consumption as a matter of personal and environmental health.

**vulnerable species** Naturally rare organisms or species whose numbers have been so reduced by human activities that they are susceptible to actions that could push them into threatened or endangered status.

# W

**warm front** A long, wedge-shaped boundary caused when a warmer advancing air mass slides over neighboring cooler air parcels.

**waste stream** The steady flow of varied wastes, from domestic garbage and yard wastes to industrial, commercial, and construction refuse.

**water cycle** The recycling and reutilization of water on earth, including atmospheric, surface, and underground phases and biological and nonbiological components.

**water droplet coalescence** A mechanism of condensation that occurs in clouds too warm for ice crystal formation.

**water scarcity** Annual available freshwater supplies less than 1,000 $m^3$ per person.

**water stress** A situation when residents of a country don't have enough accessible, high-quality water to meet their everyday needs.

**water table** The top layer of the zone of saturation; undulates according to the surface topography and subsurface structure.

**waterlogging** Water saturation of soil that fills all air spaces and causes plant roots to die from lack of oxygen; a result of overirrigation.

**watershed** The land surface and groundwater aquifers drained by a particular river system.

**weather** Description of the physical conditions of the atmosphere (moisture, temperature, pressure, and wind).

**weathering** Changes in rocks brought about by exposure to air, water, changing temperatures, and reactive chemical agents.

**wetland mitigation** Replacing a wetland damaged by development (roads, buildings, etc.) with a new or refurbished wetland.

**wetlands** Ecosystems of several types in which rooted vegetation is surrounded by standing water during part of the year. *See also* swamp, marsh, bog, fen.

**wicked problems** Problems with no simple right or wrong answer where there is no single, generally agreed-on definition of or solution for the particular issue.

**wilderness** An area of undeveloped land affected primarily by the forces of nature; an area where humans are visitors who do not remain.

**Wilderness Act** Legislation of 1964 recognizing that leaving land in its natural state may be the highest and best use of some areas.

**wildlife** Plants, animals, and microbes that live independently of humans; plants, animals, and microbes that are not domesticated.

**wildlife refuges** Areas set aside to shelter, feed, and protect wildlife; due to political and economic pressures, refuges often allow hunting, trapping, mineral exploitation, and other activities that threaten wildlife.

**windbreak** Rows of trees or shrubs planted to block wind flow, reduce soil erosion, and protect sensitive crops from high winds.

**wind farms** Large numbers of windmills concentrated in a single area; usually owned by a utility or large-scale energy producer.

**wise use groups** A coalition of ranchers, loggers, miners, industrialists, hunters, off-road vehicle users, land developers, and others who call for unrestricted access to natural resources and public lands.

**withdrawal** A description of the total amount of water taken from a lake, river, or aquifer.

**woodland** A forest where tree crowns cover less than 20 percent of the ground; also called open canopy.

**work** The application of force through a distance; requires energy input.

**world conservation strategy** A proposal for maintaining essential ecological processes, preserving genetic diversity, and ensuring that utilization of species and ecosystems is sustainable.

**World Trade Organization (WTO)** An association of 135 nations that meet to regulate international trade.

**worldviews** Sets of basic beliefs, images, and understandings that shape how we see the world around us.

# X

**X ray** Very short wavelength in the electromagnetic spectrum; can penetrate soft tissue; although it is useful in medical diagnosis, it also damages tissue and causes mutations.

# Y

**yellowcake** The concentrate of 70 to 90 percent uranium oxide extracted from crushed ore.

# Z

**zero population growth (ZPG)** The number of births at which people are just replacing themselves; also called the replacement level of fertility.

**zone of aeration** Upper soil layers that hold both air and water.

**zone of leaching** The layer of soil just beneath the topsoil where water percolates, removing soluble nutrients that accumulate in the subsoil.

**zone of saturation** Lower soil layers where all spaces are filled with water.

# Credits

## PHOTOGRAPHS

### Chapter L

Opener: p. 1: David L. Hansen, University of Minnesota Agricultural Experiment Station; p. 2: © The McGraw-Hill Companies, Inc./Barry Barker, photographer; p. 4: © Vol. 198/Corbis; p. 6: © PhotoDisc.

### Chapter 1

Opener: © China Photos/Getty Images; 1.1: Courtesy of Dr. Delbert Swanson; 1.3a: Courtesy of the Bancroft Library. University of California, Berkeley; 1.3b: Courtesy of Grey Towers National Historic Landmark; 1.3c: © Bettmann/Corbis; 1.3d: © AP/Wide World Photos; 1.4: © William P. Cunningham; 1.5a: © AP/Wide World Photos; 1.5b: © Tom Turney/The Brower Fund/Earth Island Institute; 1.5c: With permission of the University Archives, Columbia University in the City of New York and Columbia Library; 1.5d: © AP/Wide World Photos; 1.6: © Earth Imaging/Stone/Getty; 1.7: © Vol. 6/Corbis Royalty Free; 1.8: © Corbis Royalty Free; 1.10: © Vol. 244/Corbis Royalty Free; 1.12: © Norbert Schiller/The Image Works; 1.14: © Justin Guariglia/Corbis; 1.17: © The McGraw-Hill Companies, Inc./Barry Barker, photographer; 1.18: © Stone/Getty Images; 1.19: © The McGraw-Hill Companies, Inc./Barry Barker, photographer.

### Chapter 2

Opener: © 2005 Gary Braasch/World View of Global Warming; 2.2 & 2.3: David L. Hansen, University of Minnesota Agricultural Experiment Station; 2.4: © William P. Cunningham; 2.7: © The McGraw-Hill Companies, Inc./John A. Karachewski, photographer; 2.9: © Mary Ann Cunningham; 2.11: © Carr Clifton; 2.12: © William P. Cunningham; 2.13: © Sam Kittner; 2.14: © William P. Cunningham.

### Chapter 3

Opener: © John Warden/Stone/Getty Images; p. 55: © G.I. Bernard/Animals Animals Earth Scenes; 3.8: © William P. Cunningham; 3.9: NOAA; p. 68: The SeaWIFS Project, NASA/Goddard Space Flight Center and ORBIMAGE; 3.22: © David Dennis/Tom Stack & Associates.

### Chapter 4

Opener: © Galen Rowell/Corbis; 4.1: Portrait of Charles Darwin, 1840 by George Richmond (1809–96). Down House, Downe, Kent, UK/Bridgeman Art Library; 4.2: © Vol. 6/Corbis Royalty Free; 4.3: © William P. Cunningham; 4.5: © Digital Vision Ltd./Royalty Free; 4.6 & 4.13: © William P. Cunningham; 4.14: © Ray Coleman/Photo Researchers; 4.15: © D. P. Wilson/Photo Researchers; 4.16: © Michael Fogdon/Animals Animals Earth Scenes; 4.17a & b: © Edward S. Ross; 4.18: © Tom Finkle; 4.19a: © William P. Cunningham; 4.19b: PhotoDisc Royalty Free; 4.19c: © William P. Cunningham; 4.20 © Vol. 6 PhotoDisc/Getty Images; 4.22a: © Corbis Royalty Free; 4.22b: © Eric and David Hosking/Corbis; 4.22c: © Fred Bavindam/Peter Arnold; p. 91: © L. David Mech; 4.24: © Vol. 262/Corbis Royalty Free; 4.27 & 4.28: © William P. Cunningham; 4.29: © Gerals Lacz/Peter Arnold.

### Chapter 5

Opener 5: © William P. Cunningham; 5.1: © Vol. DV13/Getty Images; 5.5: © Vol. 60/PhotoDisc; 5.6: © Vol. 6/Corbis; 5.7: © William P. Cunningham; 5.8: © Mary Ann Cunningham; 5.9: © William P. Cunningham; 5.10: © Vol. 90/Corbis; 5.11: © William P. Cunningham; 5.12: © Vol. 74/PhotoDisc; 5.13: Courtesy of SeaWiFS/NASA; 5.15: NOAA; 5.16a: © Vol. 89/PhotoDisc; 5.16b: © Robert Garvey/Corbis; 5.16c: © Andrew Martinez/Photo Researchers; 5.16d: © Pat O'Hara/Corbis; 5.17: USGS; 5.18: © Stephen Rose; 5.21a, b: © Vol. 16/PhotoDisc; 5.21c: © William P. Cunningham.

### Chapter 6

Opener: © Tom Stewart/Corbis; 6.2: Courtesy of National Museum of Natural History, Smithsonian Institution; p. 121: © Vol. 72/Corbis; 6.7a: © Digital Vision/Getty Images Royalty Free; 6.7b: © Stockbyte Royalty Free; 6.7c: © PhotoDisc; 6.7d: © Corbis Royalty Free; 6.9: © Art Wolfe/Stone/Getty Images; 6.13: © The McGraw-Hill Companies, Inc./Barry Barker, photographer.

### Chapter 7

Opener: © The McGraw-Hill Companies, Inc./Barry Barker, photographer; 7.1: © Apichart Weerawong/AP/Wide World Photos; 7.2: Courtesy of John Cunningham; 7.9: © Alain Le Grasmeur/Corbis; 7.13a: © William P. Cunningham; 7.13b: © Vol. 17/PhotoDisc/Getty Images.

### Chapter 8

Opener 8: © The Carter Center/Emily Staub; 8.1: Courtesy of Donald R. Hopkins; 8.4: © William P. Cunningham; 8.5a: © Vol. 40/Corbis; 8.5b: © Image Source/Getty Royalty Free; 8.5c: Courtesy of Stanley Erlandsen, University of Minnesota; 8.8: © Vol. 72/Corbis; 8.9: © Vol. 53/Corbis; 8.19: © The McGraw-Hill Companies, Inc./Sharon Farmer, photographer.

### Chapter 9

Opener 9: © John Maier/The New York Times; 9.5: © Norbert Schiller/The Image Works; 9.6: © Lester Bergman/Corbis; 9.7: © Scott Daniel Peterson; 9.10 & 9.11a,c: © William P. Cunningham; 9.12: Courtesy U.S. Department of Agriculture Conservation Services; 9.13 & 9.18: © FAO Photo/R. Faidutti; 9.20a: Photo by Lynn Betts, courtesy of USDA Natural Resources Conservation Service; 9.20b: Photo by Jeff Vanuga, courtesy of USDA Natural Resources Conservation Center; 9.20c: © Corbis Royalty Free; 9.21: © William P. Cunningham; 9.22: © Vol. 13/Corbis; 9.23-9.25, p. 200: © William P. Cunningham; 9.29: Photo by Lynn Betts, courtesy of USDA Natural Resources Conservation Service; 9.30: © William P. Cunningham; 9.31: Courtesy of David Hansen, College of Agriculture, University of Minnesota; 9.32: © Tom Sweeney/Minneapolis Star Tribune.

### Chapter 10

Opener: Courtesy of Scott Bauer, U.S. Department of Agriculture; 10.1: © William P. Cunningham; 10.2: © Bettmann/Corbis; 10.5: © William P. Cunningham; 10.6: © Corbis Royalty Free; 10.7: © Stephen Dalton/Photo Researchers; 10.8: © SPL/Photo Researchers; 10.9: © Joe Munroe/Photo Researchers; p. 214: © Howard K. Suzuki/Aquatic Life Structures; 10.11: © Norm Thomas/Photo Researchers; 10.12: Photo by Tim McCabe, courtesy of USDA Natural Resources Conservation Service; 10.15: © William P. Cunningham; 10.16: © FAO Photo/W. Ciesla; 10.18: © W.C. Wood/Tanimura and Antle, Inc.; p. 221: Courtesy of Bill Wilcke; 10.21 & 10.22: © William P. Cunningham.

### Chapter 11

Opener: Courtesy of David Tilman; 11.1: Courtesy of David Tilman; 11.2: © Vol. 112/PhotoDisc; 11.4: © Vol. 244/PhotoDisc; 11.6–11.8 & 11.11: © William P. Cunningham; 11.12: © Corbis Royalty Free; 11.13: Courtesy of Bell Museum, University of MN. Photo taken by Mary Ann Cunningham; 11.15a: Courtesy of Dr. Mitchell Eaton; 11.15b: © William P. Cunningham; 11.15c: © Lynn Funkhouser/Peter Arnold; 11.16: © Vol. 6/Photo Disc; 11.17: © Tom McHugh/Photo Researchers; p. 244 (top): © Digital Vision Royalty Free; p. 244 (bottom): © Mary Ann Cunningham; 11.20 & 11.21: © William P. Cunningham; 11.22: Courtesy of Dr. Ronald Tilson, Minnesota Zoological Garden.

### Chapter 12

Opener: © Ron Thiele; 12.5: © William P. Cunningham; 12.6: © Digital Vision/Getty Images; 12.8a-c: Courtesy United Nations Environment Programme; 12.9: © Steven P. Lynch; 12.10: © Greg Vaughn/Tom Stack & Associates; 12.11: © Gary Braasch/Stone/Getty Images; 12.12: © William P. Cunningham; 12.13: U.S. Forest Service; 12.14: © William P. Cunningham; 12.16: © Mary Ann Cunningham; 12.17: Courtesy David L. Hansen, University of Minnesota, Agriculture Experiment Station; 12.18 & 12.20: © William P. Cunningham; 12.22: Courtesy American Heritage Center, University of Wyoming; 12.23: © William P. Cunningham; 12.24: © Richard Hamilton Smith/Corbis; 12.25: © Corbis/Royalty Free; 12.26: © William P. Cunningham; 12.30: Courtesy of R.O. Bierregaard.

### Chapter 13

Opener: © Mary Ann Cunningham; 13.3, 13.4, 13.5b,c & 13.6: © William P. Cunningham; 13.7: © Mark Edwards/Peter Arnold; 13.8: © Vol. 5/PhotoDisc/Getty Images; 13.9, 13.10 & 13.12: © William P. Cunningham; 13.14a: Courtesy of Wisconsin-Madison Archives; 13.14b: Courtesy University of Wisconsin Arboretum; 13.15: Courtesy Minnesota Department of Natural Resources; 13:16 & 13.17 © William P. Cunningham; 13.19: © Tom Bean/Corbis; 13.20: Courtesy U.S. Army Corps of Engineers; 13.21: © Chris Harris; 13.22: Courtesy U.S. Army Corps of Engineers; 13.23: © Jeff Greenberg/Photo Researchers; p. 293: © Mary Ann Cunningham; 13.25: Courtesy of Sout Florida Water Management District; 13.27: Courtesy NOAA; 13.28: Photo by Don Poggensee, USDA Natural Resources Conservation Service; 13.29: © William P. Cunningham; 13.30: Courtesy Federal Interagency Stream Corridor Restoration Handbook, NRCS, USDA; 13.22a,b, 13.32c & 13.34: © William P. Cunningham.

### Chapter 14

Opener: © William P. Cunningham; 14.7: © Corbis Royalty Free; 14.9: Courtesy of David McGeary; 14.10: © PhotoDisc Royalty Free; p. 309: © Bryan F. Peterson; 14.12: © James P. Blair/National Geographic Image Collection; 14.13: © AP Photo/Bob Bird; 14.15: © Joseph Nettis/Photo Researchers; 14.16a: © Vol. 34/PhotoDisc/Getty; 14.16b,c: © Corbis Royalty Free; 14.17: Charles Mueller, U.S. Geological Survey; p. 317: Department of Energy; 14.20: Chris G. Newhall/U.S. Geological Survey; 14.21: © Los Angeles Times, photo by Al Seib.

### Chapter 15

Opener & 15.1: © William P. Cunningham; 15.8: © Vic Engelbert/ Photo Researchers; 15.10a: Images produced by Hal Pierce, Laboratory for Atmosphere, NASA, Goddard Space Flight Center/NOAA; 15.10b: © Marc Serota/Corbis; 15.10c: © William P. Cunningham; 15.11: Courtesy Candace Kohl, University of California, San Diego; 15.17: © Vol. 244/Corbis; 15.19a, b: Photographer Lisa McKeon, courtesy of Glacier National Park Archives; 15.20: © AP/Wide World Photos; p. 340: © William P. Cunningham; 15.23: © Stonehaven CCS Canada, Inc.; 15.24: © Vol. 23/PhotoDisc/Getty.

## Chapter 16

Opener: © Corbis Royalty Free; 16.2a,b: © William P. Cunningham; 16.3: U.S. Geological Survey; 16.4 & 16.7: © William P. Cunningham; 16.9: Image courtesy of Norman Kuring, SeaWIFS Project; 16.14: © Guang Hui Xie/The Image Bank/Getty Images; 16.18: NASA; 16.20: © McGraw-Hill Companies, Inc., John Thoeming, photographer; 16.22–16.24: © William P. Cunningham; 16.25: © Vol. 44/PhotoDisc; 16.27: © William P. Cunningham; 16.28: © John Cunningham/Visuals Unlimited; 16.30: © Corbis Royalty Free; 16.32: © vario images GmbH & Co.KG/ Alamy; 16.33: © William P. Cunningham.

## Chapter 17

Opener, 17.2 & 17.3: © William P. Cunningham; 17.7: NASA; 17.9: © William P. Cunningham; 17.12: © Ray Ellis/Photo Researchers; 17.15a, c, & d: EROS Data Center, USGS; 17.15b: NASA; 17.16a: Photo by Jeff Vanuga, USDA Natural Resources Conservation Service; 17.16b: Photo by Tim McCabe, USDA Natural Resources Conservation Service; 17.16c: Photo by Lynn Betts, USDA Natural Resources Conservation Service; 17.18: Courtesy NREL/PIX; 17.20: Courtesy of USDA, NRCS, photo by Lynn Betts; 17.21: © William P. Cunningham; 17.22: Courtesy of Tim McCabe, Soil Conservation Services, USDA; 17.23: © William P. Cunningham; 17.24: © Josh Baker/Colorado Whitewater Photography; 17.25: © John Gerlach/Visuals Unlimited; 17.28 & 17.29: © William P. Cunningham.

## Chapter 18

Opener: © Peter Menzel Photography; 18.2: © Simon Fraser/ SPL/Photo Researchers; 18.3: © Joe McDonald/Animals Animals Earth Scenes; 18.4: © Dr. Parvinder Sethi; 18.6: © Roger A. Clark/Photo Researchers; 18.8: © William P. Cunningham; 18.10: © Rob and Melissa Simpson/Valan Photos; 18.12: © Lawrence Lowry/Photo Researchers; 18.15: © William P. Cunningham; 18.16: © Les Stone/ Sygma/Corbis; 18.19: © Steve Austin/Papilo/Corbis; 18.20: Courtesy of Joe Lucas/Marine Entanglement Research Program/National Marine Fisheries Service NOAA; 18.21: © McGraw-Hill Companies, Inc., Bob Coyle, photographer; 18.25: Courtesy FAO/9717/I. de Borhegyi; 18.27: Courtesy of National Renewable Energy Laboratory/NREL/PIX.

## Chapter 19

Opener: © George Steinmetz/Corbis; 19.2: © Vol. 160/ Corbis; 19.4: © Owen Franken/Stock Boston; 19.8: © Vol. 39/PhotoDisc; p. 403: © William P. Cunningham; 19.13: Courtesy of Office of Response and Restoration, NOAA; p. 434: © Mary Ann Cunningham; 19.14 © William P. Cunningham; 19.15 © Regis Martin/The New York Times; p. 437: © Peter Aengst/Wilderness Society/Lighthawk; 19.17: © Corbis Royalty Free; 19.24: Courtesy Office of Civilian Radioactive Waste Management, DOE; 19.26: Courtesy of Department of Energy.

## Chapter 20

Opener: © Mary Ann Cunningham; 20.1: Courtesy of Tom Finkle; 20.3: Courtesy of National Renewable Energy Laboratory/NREL/PIX; 20.4: Courtesy of Carolyn Roberts/ A House of Straw; 20.6: Courtesy Tom Finkle; 20.7: Courtesy of Capstone Micro Turbines; 20.9: © William P. Cunningham; 20.10: Courtesy of National Renewable Energy Laboratory/ NREL/PIX; 20.11: © McGraw-Hill Companies, Inc., Doug Sherman, photographer; 20.12: © William P. Cunningham; 20.14: Courtesy of National Renewable Energy Laboratory/ NREL/PIX; 20.15: Courtesy Warren Gretz, National Renewable Energy Laboratory, DOE; 20.17: Courtesy of Long Island Power Authority; 20.18 & 20.19: © William P. Cunningham; 20.20: Courtesy Warren Gretz, National Renewable Energy Laboratory, DOE; 20.22: Courtesy of Dr. Douglas Pratt; 20.23 & 20.24: William P. Cunningham; 20.25: © The McGraw-Hill Companies, Inc./Barry Barker, photographer; 20.26: Courtesy of Burkhardt Turbines; 20.27: National Renewable Energy Lab; 20.29: © William P. Cunningham; 20.31: © Ocean Power Delivery Ltd.

## Chapter 21

Opener: © Ray Pforther/Peter Arnold; 21.4: © William P. Cunningham; 21.5: © Vol. 31/PhotoDisc; 21.6: © Basel Action Network 2006; 21.10: © William P. Cunningham; p. 482: © Barbara Gauntt; 21.14: David L. Hansen, University of Minnesota Agricultural Experiment Station; 21.15: Courtesy of Urban Ore, Inc., Berkeley, California; 21.17: © Michael Greenlar/The Image Works; 21.22: © Pier van Lier.

## Chapter 22

Opener: © Reuters/Corbis; 22.1: © William P. Cunningham; 22.2 & 22.4: © Corbis Royalty Free; 22.5, 22.8 & 22.10: © William P. Cunningham; 22.11: Courtesy Dr. Helga Leitner; 22.12: © Fred McConnaughey/Photo Researchers; 22.13: © Regents of the University of Minnesota. All Rights Reserved. Used with permission of the Metropolitan Design Center; 22.14a, b: U.S. Geological Survey; 22.15, 22.17, 22.18 & 22.19: © William P. Cunningham; 22.20: © Roofscapes, Inc. Used by permission; all rights reserved; 22.21a-c: © Mary Ann Cunningham; p. 513: © Bill Dunster; 22.23: © William P. Cunningham.

## Chapter 23

Opener, 23.1–23.6 & 23.11: © William P. Cunningham; 23.18: © McGraw-Hill Companies, Inc./Barry Barker, photographer; 23.20: © William P. Cunningham; 23.21: © Mary Ann Cunningham; 23.24: © William P. Cunningham; p. 534: Courtesy Ray Anderson, Interface, Inc.; 23.25: © Mark Luthringer.

## Chapter 24

Opener: © William P. Cunningham; 24.1: U.S. Fish & Wildlife Service; 24.4: © Vol. 25/PhotoDisc; 24.5: © Esther Henderson/ Photo Researchers; 24.6: © Bob Daemmrich/The Image Works; p. 546: © Vol. 16/Corbis; 26.7: © William P. Cunningham; 24.8: © Vol. 25/PhotoDisc; 24.11 & 24.12: © William P. Cunningham; 24.14: Courtesy of Tom Finkle; 24.17: © John Riley/Stone/Getty Images; 24.18: © Phil Degginger/ Animals Animals Earth Scenes; 24.19: © David L. Brown/ Tom Stack & Associates.

## Chapter 25

Opener: © Robert Sobol; 25.2–25.4, p. 570 & 25.7: © William P. Cunningham; 25.9: David L. Hansen, University of Minnesota Agricultural Experiment Station; 25.10–25.13: © William P. Cunningham.

# LINE ART/TEXT CREDITS

## Chapter L

L Table 1.2, p. 4: Source: Dr. Melvin Northrup, Grand Valley State University; L Table 1.4, p. 8: Source: Courtesy of Karen Warren, Philosophy Department, Macaelester College, St. Paul, MN.

## Chapter 1

Table 1.1, p. 25: Source: UNDP Human Development Index 2006; Figure 1.9, p. 21: Source: NASA, 2002; Figure 1.11, p. 24: Source: World Bank, 2000; Figure 1.16, p. 28: Graph from World Energy Assessment, UNDP 2000, Figure 3.10, p. 95; Figure 1.20, p. 30: Source: Norman Myers, Conservation International and Cultural Survival Inc., 2002; TA1.1, p. 23: From LIVING PLANET, 2004, p. 10. Copyright © 2004 World Wide Fund for Nature (Formerly World Wildlife Fund). All rights reserved; TA1.2, p. 23: From LIVING PLANET, 2004, p. 10. Copyright © 2004 World Wide Fund for Nature (Formerly World Wildlife Fund). All rights reserved.

## Chapter 2

TA2.1, p. 38: Source: Data from U.S. Environmental Protection Agency.

## Chapter 3

Figure 3.1, p. 52: From "Salmon and alder as nitrogen sources to riparian forests in a boreal Alaskan watershed" by Robert Naiman and James Helfield in OECOLOGIA (2002) 133:573–582, Figure 3, p. 576. Copyright © 2002. Reprinted by permission of Springer Science and Business Media and the authors, Robert Naiman and James Helfield; Figure 3.10, p. 60: Source: Principles of Ecology: Matter, Energy and Life; Figure 3.16, p. 64: Source: Howard T. Odum, "Trophic Structure and Productivity of Silver Springs, Florida," in Ecological Monographs, 27:55–112, 1957, Ecological Society of America; Figure 3.20, p. 67: Source: Intergovernmental Panel for Climate Change (IPCC), 2002; Figure 3.21, p. 69: Source: Galloway and Cowling, 2002.

## Chapter 4

Figure 4.8, p. 79: Source: Original observation by R.H. MacArthur (1958); Figure 4.25, p. 92: Verner, Jared. WILDLIFE 2000. © 1986. Reprinted by permission of The University of Wisconsin Press; TA4.3, p. 97: Source: Gause, Georgyi Frantsevitch 1934. THE STRUGGLE FOR EXISTENCE; TA4.4, p. 97: Source: Gause, Georgyi Frantsevitch 1934. THE STRUGGLE FOR EXISTENCE.

## Chapter 5

Table 5.1, p. 112: Source: Hahhan, Lee, et al., "Human Disturbance and Natural Habitat: A Biome Level Analysis of a Global Data et," in Biodiversity and Conservation, 1995, Vol. 4: 128–155; Figure 5.2, p. 100: From COMMUNITIES AND ECOSYSTEMS, 2nd Edition by Robert C. Whittaker. Copyright © 1975. Reprinted by permission of Pearson Education, Inc., Upper Saddle River, NJ; Figure 5.3, p. 101: Source: WWF Ecoregions; Figure 5.22, p. 112: Source: United Nations Environment Programme, Global Environmental Outlook, 1997.

## Chapter 6

Figure 6.1, p. 117: Source: World Resources Institute, 2000; Figure 6.5, p. 122: Source: National Vital Statistics Reports 51(3): 2002; Figure 6.8, p. 125: Source: D.A. MacLulich, Fluctuations In the Numbers of the Varying Hare (Lepus americus), Toronto: University of Toronto Press, 1937, reprinted 1974; Figure 6.10, p. 126: Source: Baesd on MacArthur and Wilson, The Theory of Island Biogeography, 1967, Princeton University Press; Figure 6.11, p. 127: Source: H.L. Jones and J. Diamond, "Short-term-base Studies of Turnover in Breeding Bird Populations on the California Coast Island," in Condor, Vol. 78: 526–549, 1976.

## Chapter 7

Table 7.1, p. 134: Source: United Nations Population Division; Table 7.2, p. 137: Source: Population Reference Bureau 2006; Table 7.3, p. 138: Source: Population Reference Bureau 2006; Table 7.4, p. 141: Source: Population Reference Bureau 2006; Table 7.5, p. 149: Source: U.S. Food and Drug Administration, BIRTH CONTROL GUIDE, 2003 Revision; Figure 7.3, p. 134: Source: Figure redrawn with permission from Population Bulletin, Vol. 18, no. 18, 1985. Population Reference Bureau; TA7.3, p. 153: Source: U.S. Census Bureau; TA7.4, p. 153: Source: U.S. Census Bureau; Figure 7.5, p. 136: Source: UN Population Division, 2005; Figure 7.6, p. 137: Source: UN Population Division, 2006; Figure 7.7, p. 138: Source: World Bank, 2000; Figure 7.8, p. 139: Source: UN, ESA/WP, 180, World Population Prospects, 2002 Revision, February 2003, p. 17; Figure 7.10, p. 142: Source: The World Bank, World Development Indicators 1997, the World Bank, Washington DC, 1997; Figure 7.11, p. 142: Source: U.S. Census Bureau, 2006; Figure 7.14, p. 144: Sources: Population Reference Bureau and U.S. Bureau of the Census; Figure 7.15, p. 146: Source: http://www.uwmc.uwc.edu/geography/Demotrans/demtran.htm. Reprinted by permission of Keith Montgomery; Figure 7.17, p. 148: Source: Worldwatch Institute 2003; Figure 7.18, p. 149: Source: UN Population Division, 2004; Figure 7.19, p. 150: Source: Population Reference Bureau, 2003; Figure 7.20, p. 150: Data Source: U.N. Population Division.

## Chapter 8

Table 8.1, p. 157: Source: World Health Organization, 2002; Table 8.2, p. 162: Source: U.S. EPA, 2003; Table 8.3, p. 165: Source: U.S. Department of Health and Human Services, 1995; Table 8.4, p. 174: Source: U.S. National Safety Council, 2003; Table 8.5, p. 175: Source: Environmental Protection Agency; Figure 8.2, p. 156: Source: WHO, 2002; Figure 8.3, p. 156: Data from U.N. Population Division 2006; Figure 8.6, p. 159: Source: Data from U.S. Centers for Disease Control and Prevention; Figure 8.7, p. 159: Source: CDC and USGS; Figure 8.10, p. 161: "How Microbes Acquire Antibiotic Resistance" from TIME, September 12, 1994. Copyright © 1994 TIME Inc.; TA8.2, p. 167; Pickle, J.L., et al ENV HEALTH PERSPECTIVES 1998, 106:745–50. Fata Residential Injuries among U.S. Children, 1985 to 1997 from www.centerforhealthyhousing.org; TA8.3, p. 167: Pickle, J.L., et al ENV HEALTH PERSPECTIVES 1998, 106:745–50. Fata Residential Injuries among U.S. Children, 1985 to 1997 from www.centerforhealthyhousing.org; Figure 8.20, p. 174: Source: D.E. Patton, "USEPA's Framework for Ecological Risk Assessment" in Human Ecological Risk Assessment,

Vol. 1, No. 4; Figure 8.21, p. 175: Reprinted with permission of the Star-Tribune, Minneapolis-St. Paul; TA8.4, p. 177: Source: Data from Slovic, Paul. 1987. Perception of Risk, SCIENCE, 236(4799):286–290.

## Chapter 9

Table 9.1, p. 184: Source: Food and Agriculture Organization (FAO), 2006; Table 9.2, p. 202: Source: Based on 14 years' data From Missouri Experiment Station, Columbia, MO; Figure 9.2, p. 181: FIVIMS Hunger Map from Food and Agriculture Organization of the United Nations website; Figure 9.3, p. 181: Source: Food and Agriculture Organization (FAO), 2002; Figure 9.4, p. 181: Source: Food and Agriculture Organization (FAO), 2002; Figure 9.8, p. 183: Source: Worldwatch Institute 2001; Figure 9.9, p. 184: (a) Data from Willett and Stampfer, 2002, (b) USDA, 2005; Figure 9.11, p. 185: (c) Data from USDA, 2005; Figure 9.15, p. 189: Source: Soil Conservation Service; Figure 9.19, p. 192: Source: Data from USDA; Figure 9.28, p. 198: Source: Data from Information Systems for Biotechnology, Virginia Tech University; TA9.3, p. 206: Source: World Wide Fund for Nature (WWF), 2000.

## Chapter 10

Chapter 10, What can you do?, p. 218: Source: Citizen's Guide to Pest Control and Pesticide Safety, EPA 730-K-95-001; Table 10.1, p. 223: Source: Environmental Working Group, 2002; Figure 10.3, p. 210: Source: U.S. EPA 2000; Figure 10.4, p. 210: Source: U.S. EPA 2002; Figure 10.10, p. 213: Source: Worldwatch Institute, 2003; Figure 10.13, p. 217: Source: From E.A. Guillette, et al., "An Anthropological Approach to the Evaluation of Preschool Children Exposed to Pesticides in Mexico" in Environmental Health Perspective, 106(6):347–353, 1998. U.S. Dept. of Health and Human Services; Figure 10.19, p. 220: Source: Tolba, et al., World Environment, 1972–1992, p. 307, Chapman & Hall, 1992 United Nations Environment Programme; Figure 10.20, p. 222: Source: National Academy of Sciences; TA10.4, p. 227: Data source: C.J. Olson Market Research Inc. for Minnesota Department of Agriculture. QUANTITATIVE RESEARCH REGARDING PEST MANAGEMENT IN MINNESOTA K-12 SCHOOLS, 1999.

## Chapter 11

Figure 11.5, p. 232: Source: Conservation International; Table 11.1, p. 231: Source: IUCN 2006; Table 11.3, p. 235: Source: W.W. Gibbs, 2001.

## Chapter 12

Table 12.1, p. 264: Source: Data from USFS, 2002; Table 12.2, p. 268: Source: Data from World Conservation Union, 1990; Figure 12.2, p. 254: Source: Food and Agriculture Organization (FAO), 2002; Figure 12.3, p. 255: Source: UN Food and Agricultural Organization (FAO); Figure 12.4, p. 255: Source: United Nations Food and Agriculture Organization; Figure 12.7, p. 257: Source: Data from Food and Agriculture Organization of the United Nations, 2002; TA12.1, p. 259: Source: Wildlife Conservation Society; TA12.2, p. 259: Source: Wildlife Conservation Society; TA12.3, p. 259: Source: Wildlife Conservation Society; Figure 12.15, p. 266: Source: World Resource Institute, 2004; Figure 12.19, p. 267: Source: UN World Commission on Protected Areas; Figure 12.21, p. 269: Data from J. Hoekstra, et al, 2005.

## Chapter 13

Figure 13.1, p. 278: Source: USGS; Figure 13.2, p. 279: Data Source: Walker and Moral, 2003; Figure 13.11, p. 286: Data from S. Packard and J. Balaban, 1994; Figure 13.18, p. 290: Source: http://www.peaceparks.org.

## Chapter 14

Table 14.3, p. 313: Source: Blacksmith Institute 2006; Table 14.4, p. 314: Source: E.T. Hayes, Implications of Materials Processing, 1997; Table 14.5, p. 315: Source: The Disaster Center 2005; Table 14.6, p. 316: Source: B. Gutenberg in Earth by F. Pres and R. Siever, 1978, W.H. Freeman & Company; Figure 14.3, p. 305: Sources: U.S. Department of the Interior, U.S. Geological Survey; Figure 14.14, p. 313: Source: George Laycock, Audubon Magazine, Vol. 91(7), July 1989, p. 77; Figure 14.19, p. 318: Source: National Oceanic and Atmospheric Administration, from "U.S. Announces Plans for Improved Tsunami Warning and Detection System."

## Chapter 15

Figure 15.2, p. 324: Source: Courtesy of Dr. William Culver, St. Petersburg Junior College; Figure 15.12, p. 332: Source: Data from United Nations Environment Programme; Figure 15.15, p. 334: Source: Data from NASA Earth Observatory, 2002; TA15.3, p. 346: Data from Peter Reich; TA15.3, p. 346: Climate Change 2007: The Physical Science Basis Summary for Policymakers. Reprinted by permission of Intergovenmental Panel on Climate Change; Figure 15.21, p. 339: Climate Change 2007: The Physical Science Basis Summary for Policymakers. Reprinted by permission of Intergovenmental Panel on Climate Change; Figure 15.22, p. 342: Data from Socolow, et al, ENVIRONMENT, Vol. 46, No. 10, p. 11, 2004.

## Chapter 16

Table 16.1, p. 351: Source: UNEP, 1999; Figure 16.1, p. 348: Source: EPA, 1998; Figure 16.5, p. 351: Source: UNEP, 1999; Figure 16.6, p. 352: Source: UNEP, 1999; Figure 16.8, p. 352: Source: UNEP, 1999; Figure 16.11, p. 355: Source: UNEP, 1999; Figure 16.12, p. 355: Source: Data from B.J. Finlayson-Pitts and J.N. Pitts, Atmospheric Chemistry, 1986, John Wiley & Sons, Inc.; Figure 16.13, p. 356: Source: Environmental Defense Fund based on EPA data, 2003; Figure 16.26, p. 365: Source: National Atmospheric Deposition Program/National Trends Network, 2000. http://nadp.sws.uiuc.edu; Figure 16.31, p. 369: Source: Environmental Protection Agency, 2002.

## Chapter 17

Table 17.1, p. 377: Data from U.S. Geological Survey; Table 17.3, p. 380: Source: World Resources Institute; Figure 17.8, p. 379: Source: U.S. Geological Survey; Figure 17.13, p. 382: Source: Data from U.S. Department of Interior; Figure 17.14, p. 382: Source: UNEP, 2002; Figure 17.17, p. 384: Source: EPA, 2004; Figure 17.19, p. 386: Climate Change 2007: The Physical Science Basis Summary for Policymakers. Reprinted by permission of Intergovenmental Panel on Climate Change; Figure 17.30, p. 394: Source: World Bank estimates based on research paper by Dennis Anderson and William Cavendish, "Efficiency and Substitution in Pollution Abatement: Simulation Studies in Three Sectors;" TA17.3, p. 396: Freshwater Stress and Scarcity in Africa by 2025, VITAL WATER GRAPHICS, reproduced with kind permission from the United Nations Environment Programme (www.unep.org/vitalwater/management.htm).

## Chapter 18

Figure 18.5, p. 401: Source: UNESCO, 2002; Figure 18.14, p. 409: Source: USDA and Natural Resources Conservation Service; Figure 18.18, p. 412: Source: EPA Safe Drinking Water Information System 2001; Figure 18.24, p. 417: © 2001 Sidney Harris, ScienceCartoonsPlus.com. Reprinted by permission; Figure 18.26, p. 418: Source: World Bank estimates based on research paper by Dennis Anderson and William Cavendish, "Efficiency and Substitution in Pollution Abatement: Simulation Studies in Three Sectors;" Figure 18.28, p. 421: Source: B. Cohen and E. Bondoc, "Weed Killers by the Glass," Environmental Working Group 1995.

## Chapter 19

Figure 19.3, p. 426: Source: Data from British Petroleum, 2006; Figure 19.11, p. 432: Source: Jean Laherrere, www.hubbertpeak.org; Figure 19.12, p. 433: Source: British Petroleum, 2006; Figure 19.16, p. 436: Source: British Petroleum, 2006.

## Chapter 20

Table 20.1, p. 451: Source: U.S. Department of Energy; Table 20.2, p. 460: Source: Alan C. Lloyd, 1999; Table 20.3, p. 468: Source: Lester R. Brown, 1991; Figure 20.2, p. 450: Source: U.S. Department of Energy; Figure 20.5, p. 451: Source: U.S. Department of Energy; Figure 20.8, p. 454: Source: National Weather Bureau, U.S. Department of Commerce; p. 465: Tilman et al, 2006. *Science* 314:1596; Figure 20.28, p. 469: Source: Data from U.S. Department of Energy; Figure 20.32, p. 472: Source: World Energy Council, 2002.

## Chapter 21

What can you do? Page 487: Source: Minnesota Pollution Control Agency; Table 21.1, p. 492: Source: EPA, 2005; Figure 21.1, p. 475: Data Source: U.S. EPA; Figure 21.2, p. 476: Source: Data from Environmental Protection Agency, 2006; Figure 21.3, p. 476: Source: Data from U.S. Environmental Protection Agency Office of Solid Waste Management, 2006; Figure 21.8, p. 480: Source: Eurostat, UNEP, 2003; TA21.3, p. 482: Source: M. Lavelle and M. Coyle, The National Law Journal, Vol. 15:52–56, No. 3, September 21, 1992; Figure 21.11, p. 483: Source; Environmental Protection Agency, 2003; Figure 21.13, p. 483: Source: Environmental Protection Agency, 2003; Figure 21.18, p. 487: Source: Data from the U.S. EPA, 2002; Figure 21.20, p. 489: Source: Environmental Protection Agency, 2001; Figure 21.21, p. 489: Source: Environmental Protection Agency.

## Chapter 22

Table 22.1, p. 498: Source: United Nations Population Division, 2004; Table 22.2, p. 500: Source: T. Chandler, Three Thousand Years of Urban Growth, 1974, Academic Press and World Gazetter, 2003; Table 22.3, p. 506: Source: Excerpt from speech by Anthony Downs at the CTS Transportation Research Conference, as appeared on Website by Planners Web, Burlington, VT 2001; Table 22.4, p. 509: Source: Vision 2000, Chattanooga, TN; Figure 22.3, p. 498: Source: United Nations Population Division, World Urbanization Prospects, 2004; Figure 22.6, p. 499: Source: U.S. Census Bureau; Figure 22.9, p. 499: Source: EarthTrends, 2006.

## Chapter 23

Table 23.2, p. 520: Source: R.S. de Groot, Investing in Natural Capital, 1994; Table 23.3, p. 529: Source: Adapted from R. Costanza et al, "The Value of the World's Ecosystem Services and Natural Capital," Nature, Vol. 387 (1997); Figure 23.9, p. 523: Source: T. Prugh, National Capital and Human Economic Survival 1995, ISEE; Figure 23.10, p. 523: Source: Herman Daly in A.M. Jansson, et al, Investing in Natural Capital, ISEE; Figure 23.17, p. 528: Source: Redefining Progress, 2006; Figure 23.22, p. 532: Source: United Nations, 2007; Figure 23.26, p. 537: Source: E.S. Goodstein, Economic Policy Institute, Washington, DC; TA23.3, p. 539: Source: United Nations Development Programme, 2006.

## Chapter 24

Table 24.1, p. 547: Source: N. Vig and M. Kraft, Environmental Policy in the 1990s, 3rd Congressional Quarterly Press; Table 24.3, p. 557: Kai N. Lee, Compass and Gyroscope, 1993, Coyright © 1993 Island Press, Reprinted by permission of Alexander Hoyt & Associates; Table 24.4, p. 558: Source: L. Gunderson, C. Holling and S. Light, Barriers and Bridges to the Renewal of Ecosystems and Institutions, 1995, Columbia University Press; Figure 24.2, p. 542: Source: Modified from Ray Grizzle, Bioscience, Vol. 44(4), April 1994; Figure 24.10, p. 552: Source: U.S. General Accounting Office; Figure 24.13, p. 555: Source: United Nations Environment Programme from Global Environment Outlook 2000; Figure 24.16, p. 558: Source: Buzz Holling, Barriers & Bridges, edited by Gunderson, Holing and Light, 1995. Columbia University Press. Reprinted by permission of the authors; Data Analysis Figure 1, Graphic from PILOT 2006 EPI, Figure 12, p. 78. Reprinted by permission of Environmental Performance Index, Yale University.

## Chapter 25

Table 25.1, p. 568: Source: A Greenprint for Minnesota, Minnesota Office of Environmental Education, 1993; Table 25.2, p. 568: From Robert Merideth, The Environmentalist's Bookshelf: A Guide to the Best Books, 1993 by G.K. Hall, an imprint of Macmillan, Inc. Reprinted by permission; What can you do? p. 572: Based on material by Karen Oberhauser, Bell Museum Imprint, University of Minnesota, 1992. Used by permission. Table 25.3, p. 577: Based on material from "Grassroots Organizing for Everyone" by Claire Greensfelder and Mike Roselle from Call to Action, 1990 Sierra Book Club Books; TA25.3, p. 570: Reprinted by permission of Audubon; Figure 25.5, p. 571: © 1990 Bruce von Alten; Figure 25.6, p. 573: Source: After Society of Environmental Toxicology and Chemistry, 1991; Figure 25.14, p. 580: Source: UN, 2002; Figure 25.15, p. 580: Modified from Raymond Grizzle and Christopher Barrett, personal communication.

# Subject Index